MULTIDIMENSIONAL STOCHASTIC PROCESSES AS ROUGH PATHS

Rough path analysis provides a fresh perspective on Itô's important theory of stochastic differential equations. Key theorems of modern stochastic analysis (existence and limit theorems for stochastic flows, Freidlin–Wentzell theory, the Stroock–Varadhan support description) can be obtained with dramatic simplifications. Classical approximation results and their limitations (Wong–Zakai, McShane's counterexample) receive "obvious" rough path explanations. Evidence is building that rough paths will play an important role in the future analysis of stochastic partial differential equations, and the authors include some first results in this direction. They also emphasize interactions with other parts of mathematics, including Caratheodory geometry, Dirichlet forms and Malliavin calculus.

Based on successful courses at the graduate level, this up-to-date introduction presents the theory of rough paths and its applications to stochastic analysis. Examples, explanations and exercises make the book accessible to graduate students and researchers from a variety of fields.

T0305909

CAMBRIDGE STUDIES IN ADVANCED MATHEMATICS

Editorial Board:

B. Bollobás, W. Fulton, A. Katok, F. Kirwan, P. Sarnak, B. Simon, B. Totaro

All the titles listed below can be obtained from good booksellers or from Cambridge University Press. For a complete series listing visit: http://www.cambridge.org/series/sSeries.asp?code=CSAM

Already published

Multidimensional Stochastic Processes as Rough Paths

Theory and Applications

PETER K. FRIZ

NICOLAS B. VICTOIR

CAMBRIDGE
UNIVERSITY PRESS

CAMBRIDGE
UNIVERSITY PRESS

University Printing House, Cambridge CB2 8BS, United Kingdom

One Liberty Plaza, 20th Floor, New York, NY 10006, USA

477 Williamstown Road, Port Melbourne, VIC 3207, Australia

314-321, 3rd Floor, Plot 3, Splendor Forum, Jasola District Centre, New Delhi - 110025, India

103 Penang Road, #05-06/07, Visioncrest Commercial, Singapore 238467

Cambridge University Press is part of the University of Cambridge.

It furthers the University's mission by disseminating knowledge in the pursuit of education, learning and research at the highest international levels of excellence.

www.cambridge.org
Information on this title: www.cambridge.org/9780521876070

© P. K. Friz and N. B. Victoir 2010

First published 2010

A catalogue record for this publication is available from the British Library

Library of Congress Cataloging in Publication data
Friz, Peter K., 1974–
Multidimensional stochastic processes as rough paths : theory and applications / Peter K. Friz, Nicolas B. Victoir.
p. cm. – (Cambridge studies in advanced mathematics ; 120)
Includes bibliographical references and index.
ISBN 978-0-521-87607-0 (hardback)
1. Stochastic difference equations. 2. Stochastic processes. 3. Random measures.
I. Victoir, Nicolas B. II. Title. III. Series.
QA274.23.F746 2010
519.2 – dc22 2009048605

ISBN 978-0-521-87607-0 Hardback

To Wendy and Laura

Contents

II Abstract theory of rough paths

Appendices

Preface

This book is split into four parts. **Part I** is concerned with basic material about certain ordinary differential equations, paths of Hölder and variation regularity, and the rudiments of Riemann–Stieltjes and Young integration. Nothing here will be new to specialists, but the material seems rather spread out in the literature and we hope it will prove useful to have it collected in one place.

Part II is about the deterministic core of *rough path theory*, à la T. J. Lyons, but actually inspired by the direct approach of A. M. Davie. Although the theory can be formulated in a Banach setting, we have chosen to remain in a finite-dimensional setting; our motivation for this decision comes from the fact that the bulk of classic texts on Brownian motion and stochastic analysis take place in a similar setting, and these are the grounds on which we sought applications.

In essence, with rough paths one attempts to take out probability from the theory of stochastic differential equations – to the extent possible. Probability still matters, but the problems are shifted from the analysis of the actual SDEs to the analysis of elementary stochastic integrals, known as Lévy's stochastic area. In **Part III** we start with a detailed discussion of how multidimensional Brownian motion can be turned into a (random) rough path; followed by a similar study for (continuous) semi-martingales and large classes of multidimensional Gaussian – and Markovian – processes.

In **Part IV** we apply the theory of rough differential equations (RDEs), path-by-path, with the (rough) sample paths constructed in Part III. In the setting of Brownian motion or semi-martingales, the resulting (random) RDE solutions are identified as solutions to classical stochastic differential equations. We then give a selection of applications to stochastic analysis in which rough path techniques have proved useful.

The prerequisites for Parts I and II are essentially a good command of undergraduate analysis. Some knowledge of ordinary differential equations (existence, uniqueness results) and basic geometry (vector fields, geodesics) would be helpful, although everything we need is discussed. In Part III, we assume a general background in measure theoretic probability theory and the basics of stochastic processes, such as Brownian motion. Stochastic area (for Brownian motion) is introduced via stochastic integration, with alternatives described in the text. In the respective chapters on semi-martingales,

Gaussian and Markovian processes, the reader is assumed to have the appropriate background, most of which we have tried to collect in the appendices. Part IV deals with applications to stochastic analysis, stochastic (partial) differential equations in particular. For a full appreciation of the results herein, the reader should be familiar with the relevant background; textbook references are thus given whenever possible at the end of chapters. Exercises are included throughout the text, often with complete (or sketched) solutions.

It is our pleasure to thank our mentors, colleagues and friends. This book would not exist without the teachings of our PhD advisors, S. R. S. Varadhan and T. J. Lyons; both remained available for discussions at various stages throughout the writing process. Once we approached completion, a courageous few offered to do some detailed reading: C. Bayer, M. Caruana, T. Cass, A. Deya, M. Huesmann, H. Oberhauser, J. Teichmann and S. Tindel. Many others offered their time and support in various forms: G. Ben Arous, C. Borrell, F. Baudoin, R. Carmona, D. Chafaï, T. Coulhon, L. Coutin, M. Davis, B. Davies, A. Davie, D. Elworthy, M. Gubinelli, M. Hairer, B. Hambly, A. Iserles, I. Karatzas, A. Lejay, D. Lépingle, P. Malliavin, P. Markowitch, J. Norris, Z. Qian, J. Ramírez, J. Robinson, C. Rogers, M. Sanz-Sole and D. Stroock. This is also a welcome opportunity to thank C. Obtresal, C. Schmeiser, R. Schnabl and W. Wertz for their early teachings. The first author expresses his deep gratitude to the Department of Pure Mathematics and Mathematical Statistics, Cambridge and King's College, Cambridge, where work on this book was carried out under ideal circumstances; he would also like to thank the Radon Institute and his current affiliations, TU and WIAS Berlin, where this book was finalized. Partial support from the Leverhulme Trust and EPSRC grant EP/E048609/1 is gratefully acknowledged. The second author would like to thank the Mathematical Institute, Oxford and Magdalen College, Oxford, where work on the early drafts of this book was undertaken.

Finally, it is our great joy to thank our loving families.

Peter K. Friz (Cambridge, Berlin)
Nicolas B. Victoir (Hong Kong)
June 2009

Introduction

One of the remarkable properties of Brownian motion is that we can use it to construct (stochastic) integrals of the type

$$\int \ldots dB.$$

The reason this is remarkable is that almost every Brownian sample path $(B_t(\omega) : t \in [0, T])$ has infinite variation and there is no help from the classical Stieltjes integration theory. Instead, Itô's theory of stochastic integration relies crucially on the fact that B is a martingale and stochastic integrals themselves are constructed as martingales. If one recalls the elementary interpretation of martingales as fair games one sees that Itô integration is some sort of martingale transform in which the integrand has the meaning of a gambling strategy. Clearly then, the integrand must not anticipate the random movements of the driving Brownian motion and one is led to the class of so-called previsible processes which can be integrated against Brownian motion. When such integration is possible, it allows for a theory of stochastic differential equations (SDEs) of the form[1]

$$dY = \sum_{i=1}^{d} V_i(Y)\, dB^i + V_0(Y)\, dt, \quad Y(0) = y_0. \tag{$*$}$$

Without going into too much detail, it is hard to overstate the importance of Itô's theory: it has a profound impact on modern mathematics, both pure and applied, not to speak of applications in fields such as physics, engineering, biology and finance.

It is natural to ask whether the meaning of $(*)$ can be extended to processes other than Brownian motion. For instance, there is motivation from mathematical finance to generalize the driving process to general (semi-)martingales and luckily Itô's approach can be carried out naturally in this context.

We can also ask for a Gaussian generalization, for instance by considering a differential equation of the form $(*)$ in which the driving signal may be taken from a reasonably general class of Gaussian processes. Such equations have been proposed, often in the setting of fractional Brownian motion of Hurst parameter $H > 1/2$,[2] as toy models to study the ergodic behaviour

---

[1] Here $B = (B^1, \ldots, B^d)$ is a d-dimensional Brownian motion.

[2] Hurst parameter $H = 1/2$ corresponds to Brownian motion. For $H > 1/2$, one has enough sample path regularity to use Young integration.

of non-Markovian systems or to provide new examples of arbitrage-free markets under transactions costs.

Or we can ask for a Markovian generalization. Indeed, it is not hard to think of motivating physical examples (such as heat flow in rough media) in which Brownian motion B may be replaced by a Markov process X^a with uniformly elliptic generator in divergence form, say $\frac{1}{2} \sum_{i,j} \partial_i \left(a^{ij} \partial_j \cdot \right)$, without any regularity assumptions on the symmetric matrix $\left(a^{ij} \right)$.

The Gaussian and Markovian examples have in common that the sample path behaviour can be arbitrarily close to Brownian motion (e.g. by taking $H = 1/2 \pm \varepsilon$ resp. a uniformly ε-close to the identity matrix I). And yet, Itô's theory has a complete breakdown!

It has emerged over recent years, starting with the pioneering work of T. Lyons [116], that differential equations driven by such non-semi-martingales can be solved in the rough path sense. Moreover, the so-obtained solutions are *not* abstract nonsense but have firm probabilistic justification. For instance, if the driving signal converges to Brownian motion (in some reasonable sense which covers $\varepsilon \to 0$ in the aforementioned examples) the corresponding rough path solutions converge to the classical Stratonovich solution of $(*)$, as one would hope.

While this alone seems to allow for flexible and robust stochastic modelling, it is not all about dealing with new types of driving signals. Even in the classical case of Brownian motion, we get some remarkable insights. Namely, the (Stratonovich) solution to $(*)$ can be represented as a deterministic and continuous image of Brownian motion and Lévy's stochastic area

$$A_t^{jk}(\omega) = \frac{1}{2} \left(\int_0^t B^j \, dB^k - \int_0^t B^k \, dB^j \right)$$

alone. In fact, there is a "nice" deterministic map, the *Itô–Lyons map*,

$$(y_0; \mathbf{x}) \mapsto \pi(0, y_0; \mathbf{x})$$

which yields, upon setting $\mathbf{x} = \left(B^i, A^{j,k} : i, j, k \in \{1, \dots, d\} \right)$ a very pleasing version of the solution of $(*)$. Indeed, subject to sufficient regularity of the coefficients, we see that $(*)$ can be solved simultaneously for all starting points y_0, and even all coefficients! Clearly then, one can allow the starting point and coefficients to be random (even dependent on the entire future of the Brownian driving signals) without problems; in stark contrast to Itô's theory which struggles with the integration of non-previsible integrands. Also, construction of stochastic flows becomes a trivial corollary of purely deterministic regularity properties of the Itô–Lyons map.

This brings us to the (deterministic) main result of the theory: continuity of the Itô–Lyons map

$$\mathbf{x} \mapsto \pi(0, y_0; \mathbf{x})$$

in "rough path" topology. When applied in a standard SDE context, it quickly gives an entire catalogue of limit theorems. It also allows us to reduce (highly non-trivial) results, such as the Stroock–Varadhan support theorem or the Freidlin–Wentzell estimates, to relatively simple statements about Brownian motion and Lévy's area. Moreover, and at no extra price, all these results come at the level of stochastic flows. The Itô–Lyons map is also seen to be regular in certain perturbations of **x** which include (but are not restricted to) the usual Cameron–Martin space, and so there is a natural interplay with Malliavin calculus. At last, there is increasing evidence that rough path techniques will play an important role in the theory of stochastic *partial* differential equations and we have included some first results in this direction.

All that said, let us emphasize that the rough path approach to (stochastic) differential equations is *not* set out to replace Itô's point of view. Rather, it complements Itô's theory in precisely those areas where the former runs into difficulties.

We hope that the topics discussed in this book will prove useful to anyone who seeks new tools for robust and flexible stochastic modelling.

The story in a nutshell

1   From ordinary to rough differential equations

Rough path analysis can be viewed as a collection of smart estimates for differential equations of type

$$dy = V(y)\,dx \iff \dot{y} = \sum_{i=1}^{d} V_i(y)\,\dot{x}^i.$$

Although a Banach formulation of the theory is possible, we shall remain in finite dimensions here. For the sake of simplicity, let us assume that the *driving signal* $x \in C^\infty\left([0,T],\mathbb{R}^d\right)$ and that the coefficients $V_1,\ldots,V_d \in C^{\infty,b}\left(\mathbb{R}^e,\mathbb{R}^e\right)$, that is bounded with bounded derivatives of all orders. We are dealing with a simple time-inhomogenous ordinary differential equation (ODE) and there is no question about existence and uniqueness of an $\mathbb{R}^e$-valued solution from every starting point $y_0 \in \mathbb{R}^e$. The usual first-order Euler approximation, from a fixed time-s starting point y_s, is obviously

$$y_t - y_s \approx V_i(y_s)\int_s^t dx^i.$$

(We now adopt the summation convention over repeated up–down indices.) A simple Taylor expansion leads to the following step-2 Euler approximation,

$$y_t - y_s \approx \underbrace{V_i(y_s)\int_s^t dx^i + V_i^k \partial_k V_j(y_s)\int_s^t\int_s^r dx^i dx^j}_{=\,\mathcal{E}(y_s,\mathbf{x}_{s,t})}$$

with

$$\mathbf{x}_{s,t} = \left(\int_s^t dx,\ \int_s^t\int_s^r dx \otimes dx\right) \in \mathbb{R}^d \oplus \mathbb{R}^{d\times d}. \tag{1}$$

Let us now make the following Hölder-type assumption: there exists c_1 and $\alpha \in (0,1]$ such that, for all $s < t$ in $[0,T]$ and all $i,j \in \{1,\ldots,d\}$,

$$(H_\alpha): \left|\int_s^t dx^i\right| \vee \left|\int_s^t\int_s^r dx^i dx^j\right|^{1/2} \leq c_1\,|t-s|^\alpha. \tag{2}$$

Note that $\int_s^t\int_s^r dx^i dx^j$ is readily estimated by $\ell^2\,|t-s|^2$, where $\ell = |\dot{x}|_{\infty;[0,T]}$ is the Lipschitz norm of the driving signal, and so (H_α) holds, somewhat trivially for now, with $c_1 = \ell$ and $\alpha = 1$. [We shall see later that (H_α) also holds for d-dimensional Brownian motion for any $\alpha < 1/2$ and a random variable $c_1(\omega) < \infty$ a.s. provided the double integral is understood in the sense of stochastic integration. Nonetheless, let us keep x deterministic and smooth for now.]

It is natural to ask exactly how good these approximations are. The answer is given by *Davie's lemma* which says that, assuming (H_α) for some $\alpha \in (1/3, 1/2]$, one has the "step-2 Euler estimate"

$$|y_t - y_s - \mathcal{E}(y_s, \mathbf{x}_{s,t})| \le c_2 |t - s|^\theta$$

where $\theta = 3\alpha > 1$. The catch here is *uniformity*: $c_2 = c_2(c_1)$ depends on x only through the Hölder bound c_1 but not on its Lipschitz norm. Since it is easy to see that (H_α) implies

$$\mathcal{E}(y_s, \mathbf{x}_{s,t}) \le c_3 |t - s|^\alpha, \quad c_3 = c_3(c_1),$$

the triangle inequality leads to

$$|y_t - y_s| \le c_4 |t - s|^\alpha, \quad c_4 = c_4(c_1). \tag{3}$$

As often in analysis, uniform bounds allow for passage to the limit. We therefore take $x_n \in C^\infty([0,T], \mathbb{R}^d)$ with *uniform bounds*

$$\sup_n \left| \int_s^t dx_n^i \right| \vee \left| \int_s^t \int_s^r dx_n^i dx_n^j \right|^{1/2} \le c_1 |t - s|^\alpha$$

such that, uniformly in $t \in [0,T]$,

$$\left(\int_0^t dx_n^i, \int_0^t \int_0^r dx_n^i dx_n^j \right) \to \mathbf{x}_t \equiv \left(\mathbf{x}_t^{(1)}, \mathbf{x}_t^{(2)} \right) \in \mathbb{R}^d \oplus \mathbb{R}^{d \times d}.$$

The limiting object $\mathbf{x}$ is a path with values in $\mathbb{R}^d \oplus \mathbb{R}^{d \times d}$ and the class of $\left(\mathbb{R}^d \oplus \mathbb{R}^{d \times d}\right)$-valued paths obtained in this way is precisely what we call the α-Hölder rough paths.[1]

Two important remarks are in order.

(i) The condition $\alpha \in (1/3, 1/2]$ in Davie's estimate is intimately tied to the fact that the condition (H_α) involves the first *two* iterated integrals.

(ii) The space $\mathbb{R}^d \oplus \mathbb{R}^{d \times d}$ is not quite the correct state space for $\mathbf{x}$. Indeed, the calculus product rule $d(x^i x^j) = x^i dx^j + x^j dx^i$ implies that[2]

$$\mathrm{Sym}\left(\int_0^t \int_0^r dx \otimes dx \right) = \frac{1}{2} \left(\int_0^t dx \right) \otimes \left(\int_0^t dx \right).$$

---

[1] To be completely honest, we call this a *weak geometric α-Hölder rough path*.
[2] $\mathrm{Sym}(A) := \frac{1}{2}(A + A^T)$, $\mathrm{Anti}(A) := \frac{1}{2}(A - A^T)$ for $A \in \mathbb{R}^{d \times d}$.

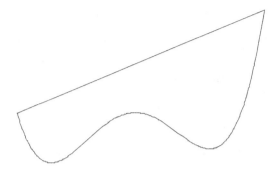

Figure 1. We plot $s \mapsto (x_s^i, x_s^j)$ and the chord which connects (x_0^i, x_0^j), on the lower left side, say, with (x_t^i, x_t^j) on the right side. The (signed) enclosed area (here positive) is precisely $\mathrm{Anti}(\mathbf{x}_t^{(2)})^{i,j}$.

This remains valid in the limit so that $\mathbf{x}(t)$ must take values in

$$\left\{ \mathbf{x} = \left(\mathbf{x}^{(1)}, \mathbf{x}^{(2)} \right) \in \mathbb{R}^d \oplus \mathbb{R}^{d \times d} : \mathrm{Sym}\left(\mathbf{x}^{(2)} \right) = \frac{1}{2} \mathbf{x}^{(1)} \otimes \mathbf{x}^{(1)} \right\}.$$

We can get rid of this algebraic redundancy by switching from $\mathbf{x}$ to[3]

$$\left(\mathbf{x}^{(1)}, \mathrm{Anti}(\mathbf{x}^{(2)}) \right) \in \mathbb{R}^d \oplus so\,(d).$$

At least for a smooth path $x\,(\cdot)$, this has an appealing geometric interpretation. Let (x^i, x^j) denote the projection to two distinct coordinates (i, j); basic multivariable calculus then tells us that

$$\mathrm{Anti}(\mathbf{x}_t^{(2)})^{i,j} = \frac{1}{2} \left(\int_0^t \left(x_s^i - x_0^i \right) dx_s^j - \int_0^t \left(x_s^j - x_0^j \right) dx_s^i \right)$$

is the *area* (with multiplicity and orientation taken into account) between the curve $\{(x_s^i, x_s^j) : s \in [0, t]\}$ and the chord from (x_t^i, x_t^j) to (x_0^i, x_0^j). See Figure 1.

Example 1 Consider $d = 2$ and $x_n(t) = \left(\frac{1}{n} \cos\left(2n^2 t \right), \frac{1}{n} \sin\left(2n^2 t \right) \right) \in \mathbb{R}^2$. Then (H_α) holds with $\alpha = 1/2$, as may be seen by considering separately the cases where $1/n$ is less resp. greater than $(t - s)^{1/2}$. Moreover, the limiting rough path is

$$\mathbf{x}_t \equiv \left(\begin{pmatrix} 0 \\ 0 \end{pmatrix}, \begin{pmatrix} 0 & t \\ -t & 0 \end{pmatrix} \right), \tag{4}$$

since we run around the origin essentially $n^2 t/\pi$ times, sweeping out area π/n^2 at each round. $\square$

---

[3] As will be discussed in Chapter 7, this is precisely switching from the step-2 free nilpotent Lie group (with d generators) to its Lie algebra.

We are now ready for the passage to the limit on the level of ODEs. To this end, consider $(y^n) \subset C([0,T], \mathbb{R}^e)$, obtained by solving, for each n, the ODE

$$dy^n = V(y^n) dx^n, \quad y^n(0) = y_0.$$

By Davie's lemma the sequence (y_n) has a uniform α-Hölder bound c_4 and by Arzela–Ascoli we see that (y_n) has at least one limit point in $C([0,T], \mathbb{R}^e)$. Each such limit point is called a solution to the *rough differential equation* (RDE) which we write as

$$dy = V(y)d\mathbf{x}, \; y(0) = y_0. \tag{5}$$

The present arguments apply immediately for $V \in C^{2,b}$, that is bounded with two bounded derivatives, and more precisely for $V \in \mathrm{Lip}^{\gamma-1}, \gamma > 1/\alpha$, in the sense of Stein.[4] As in classical ODE theory, one additional degree of regularity (e.g. $V \in \mathrm{Lip}^\gamma, \gamma > 1/\alpha$) then gives uniqueness[5] and we will

write

$$y = \pi_{(V)}(0, y_0; \mathbf{x})$$

for such a unique RDE solution. At last, it should not be surprising from our construction that the RDE solution map (a.k.a. *Itô–Lyons map*)

$$\mathbf{x} \mapsto \pi_{(V)}(0, y_0; \mathbf{x})$$

is continuous in $\mathbf{x}$ (e.g. under uniform convergence with uniform Hölder bounds).

Example 2 Assume $\mathbf{x}_t = \left(\int_0^t dx^i, \int_0^t \int_0^r dx^j dx^k \right)_{i,j,k \in \{1,\dots,d\}}$ with smooth x. Then

$$y = \pi_{(V)}(0, y_0; \mathbf{x})$$

is the classical ODE solution to $dy = V(y) dx, \; y(0) = y_0$.   □

Example 3 Assume $\mathbf{x}$ is given by (4) and $V = (V_1, V_2)$. Then

$$y = \pi_{(V)}(0, y_0; \mathbf{x})$$

can be identified as the classical ODE solution to

$$dy = [V_1, V_2](y) dt$$

where $[V_1, V_2] = V_1^i \partial_i V_2 - V_2^i \partial_i V_1$ is the Lie bracket of V_1 and V_2.   □

---

[4] Writing $\gamma = \lfloor \gamma \rfloor + \{\gamma\}$ with integer $\lfloor \gamma \rfloor$ and $\{\gamma\} \in (0,1]$ this means that V is bounded and has up to $\lfloor \gamma \rfloor$ bounded derivatives, the last of which is Hölder with exponent $\{\gamma\}$.

[5] With more effort, uniqueness can be shown under $\mathrm{Lip}^{1/\alpha}$-regularity.

Example 4 Assume $B = \left(B^1, \ldots, B^d\right)$ is a d-dimensional Brownian motion. Define enhanced Brownian motion by

$$\mathbf{B}_t = \left(\int_0^t dB^i, \int_0^t B^j \circ dB^k \right)_{i,j,k \in \{1,\ldots,d\}}$$

(where $\circ$ indicates stochastic integration in the Stratonovich sense). We shall see that $\mathbf{B}$ is an α-Hölder rough path for $\alpha \in (1/3, 1/2)$ and identify

$$Y_t (\omega) := \pi_{(V)} (0, y_0; \mathbf{B})$$

as a solution to the Stratonovich stochastic differential equation[6]

$$dY = \sum_{i=1}^d V_i (Y) \circ dB^i.$$

□

2   Carnot–Caratheodory geometry

We now try to gain a better understanding of the results discussed in the last section. To this end, it helps to understand the more general case of Hölder-type regularity with exponent $\alpha = 1/p \in (0,1]$. As indicated in remark (i), this will require consideration of more iterated integrals and we need suitable notation: given $x \in C^\infty \left([0,T], \mathbb{R}^d\right)$ we generalize (1) to[7]

$$\mathbf{x}_t := S_N (x)_{0,t} := \left(1, \int_0^t dx, \int_{\Delta_{[0,t]}^2} dx \otimes dx, \ldots, \int_{\Delta_{[0,t]}^N} dx \otimes \cdots \otimes dx \right),$$
(6)

called the *step-N signature* of x over the interval $[0,t]$, with values in

$$T^N \left(\mathbb{R}^d\right) := \mathbb{R} \oplus \mathbb{R}^d \oplus \left(\mathbb{R}^d\right)^{\otimes 2} \oplus \cdots \oplus \left(\mathbb{R}^d\right)^{\otimes N}.$$

Observe that we added a zeroth scalar component in our definition of $\mathbf{x}_t$ which is always set to 1. This is pure convention but has some algebraic advantages. To go further, we note that $T^N \left(\mathbb{R}^d\right)$ has the structure of a (truncated) tensor-algebra with tensor-multiplication $\otimes$. (Elements with scalar component equal to 1 are always invertible with respect to $\otimes$.) Computations are simply carried out by considering the standard basis (e_i) of $\mathbb{R}^d$ as non-commutative indeterminants; for instance,

$$\left(a^i e_i\right) \otimes \left(b^j e_j\right) = a^i b^j \left(e_i \otimes e_j\right) \neq a^i b^j \left(e_j \otimes e_i\right).$$

---

[6] A drift term $V_0 (y) \, dt$ can be trivially included by considering the time-space process (t, B).

[7] $\Delta_{[0,t]}^k$ denotes the k-dimensional simplex over $[0,t]$.

The reason we are interested in this sort of algebra is that the trivial

$$x_{s,t} \equiv (-x_s) + x_t = \int_s^t dx =: x_{s,t}$$

generalizes to

$$\mathbf{x}_{s,t} \equiv \mathbf{x}_s^{-1} \otimes \mathbf{x}_t = \left(1, \int_s^t dx, \int_{\Delta^2_{[s,t]}} dx \otimes dx, \dots, \int_{\Delta^N_{[s,t]}} dx \otimes \cdots \otimes dx\right).$$

As a consequence, we have *Chen's relation* $\mathbf{x}_{s,u} = \mathbf{x}_{s,t} \otimes \mathbf{x}_{t,u}$, which tells us precisely how to "patch together" iterated integrals over adjacent intervals $[s,t]$ and $[t,u]$.

Let us now take on remark (ii) of the previous section. One can see that the step-N lift of a smooth path x, as given in (6), takes values in the free step-N nilpotent (Lie) group with d generators, realized as restriction of $T^N\left(\mathbb{R}^d\right)$ to

$$G^N\left(\mathbb{R}^d\right) = \exp\left(\mathbb{R}^d \oplus \left[\mathbb{R}^d, \mathbb{R}^d\right] \oplus \left[\mathbb{R}^d, \left[\mathbb{R}^d, \mathbb{R}^d\right] \oplus \dots\right]\right) \equiv \exp\left(\mathfrak{g}^N\left(\mathbb{R}^d\right)\right)$$

where $\mathfrak{g}^N\left(\mathbb{R}^d\right)$ is the free step-N nilpotent Lie algebra and exp is defined by the usual power-series based on $\otimes$.

Example 5 $[N = 2]$ Note that $\left[\mathbb{R}^d, \mathbb{R}^d\right] = so\left(d\right)$. Then

$$\exp\left(\mathbb{R}^d \oplus \left[\mathbb{R}^d, \mathbb{R}^d\right]\right)$$
$$= \left\{\left(1, v, \frac{1}{2}v \otimes v + A\right) : v \in \mathbb{R}^d, A \in so\left(d\right)\right\}$$

which is precisely the algebraic relation we pointed out in remark (ii) of the previous section. □

If the discussion above tells us that $T^N\left(\mathbb{R}^d\right)$ is too big a state space for lifted smooth paths, *Chow's theorem* tells us that $G^N\left(\mathbb{R}^d\right)$ is the correct state space. It asserts that for all $g \in G^N\left(\mathbb{R}^d\right)$ there exists $\gamma : [0,1] \to \mathbb{R}^d$, which may be taken to be piecewise linear such that $S_N\left(\gamma\right)_{0,1} = g$. One can then define the *Carnot–Caratheodory norm*

$$\|g\| = \inf\left\{\text{length}\left(\gamma|_{[0,1]}\right) : S_N\left(\gamma\right)_{0,1} = g\right\},$$

where the infimum is achieved for some Lipschitz continuous path $\gamma^* : [0,1] \to \mathbb{R}^d$, some sort of *geodesic path* associated with g. The *Carnot–Caratheodory distance* is then simply defined by $d\left(g,h\right) := \left\|g^{-1} \otimes h\right\|$. A Carnot–Caratheodory unit ball is plotted in Figure 2.

Example 6 Take $g = \left(\begin{pmatrix} 0 \\ 0 \end{pmatrix}, \begin{pmatrix} 0 & a \\ -a & 0 \end{pmatrix}\right) \in G^2\left(\mathbb{R}^2\right)$. Then γ^* is the shortest path which returns to its starting point and sweeps out area a. From basic isoperimetry, γ^* must be a circle and $\|g\| = 2\sqrt{\pi}a^{1/2}$. See Figure 3. □

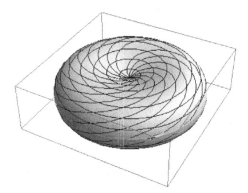

Figure 2. After identifying $G^2\left(\mathbb{R}^d\right)$ with the 3-dimensional Heisenberg group, i.e. $\left(\begin{pmatrix} x \\ y \end{pmatrix}, \begin{pmatrix} 0 & a \\ -a & 0 \end{pmatrix}\right) \equiv (x, y, a)$, we plot the (apple-shaped) unit-ball with respect to the Carnot–Caratheodory distance. It contains (and is contained in) a Euclidean ball.

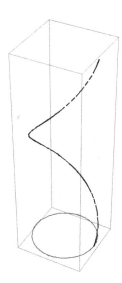

Figure 3. We plot the circle γ^*. The z-axis represents the wiped-out area and runs from 0 to a.

In practice, we rarely need to compute precisely the CC norm of an element $g = \left(1, g^1, \ldots, g^N\right) \in G^N\left(\mathbb{R}^d\right)$. Instead we rely on the so-called equivalence of homogenous norms, which asserts that

$$\exists \kappa > 0 : \frac{1}{\kappa}\, |||g||| \leq \left\|\left(1, g^1, \ldots, g^N\right)\right\| \leq \kappa\, |||g|||$$

where

$$|||g||| := \max_{i=1,\ldots,N} \left|g^i\right|^{1/i}_{\left(\mathbb{R}^d\right)^{\otimes i}}.$$

Here, both "norms" $\|\cdot\|$ and $|||\cdot|||$ are homogenous with respect to *dilation* on $G^N\left(\mathbb{R}^d\right)$,

$$\delta_\lambda : \left(1, g^1, \ldots, g^N\right) \mapsto \left(1, \lambda^1 g^1, \ldots, \lambda^N g^N\right), \quad \lambda \in \mathbb{R}.$$

It is time to make the link to our previous discussion. Recall condition (H_α) from equation (2), which expressed a Hölder-type assumption of the form

$$\int_s^t dx \vee \left|\int_s^t \int_s^r dx \otimes dx\right|^{1/2} \leq c_1 \left|t - s\right|^\alpha.$$

But this says exactly that, for all $0 \leq s < t \leq T$, the corresponding "group" increment $\mathbf{x}_{s,t} = S_2\left(x\right)_{s,t} \in G^2\left(\mathbb{R}^d\right)$ satisfies

$$\left\|\mathbf{x}_{s,t}\right\| = d\left(\mathbf{x}_s, \mathbf{x}_t\right) \lesssim c_1 \left|t - s\right|^\alpha,$$

where d is the Carnot–Caratheodory metric on $G^2\left(\mathbb{R}^d\right)$, which is equivalent to

$$\|\mathbf{x}\|_{\alpha\text{-Höl};[0,T]} \equiv \sup_{s,t\in[0,T]} \frac{d\left(\mathbf{x}_s, \mathbf{x}_t\right)}{\left|t - s\right|^\alpha} \lesssim c_1.$$

This regularity persists under passage to the limit and hence any (weak, geometric) α-Hölder rough path is a genuine α-Hölder path with values in $G^2\left(\mathbb{R}^d\right)$. Conversely, given an abstract α-Hölder path in $G^2\left(\mathbb{R}^d\right)$ equipped with Carnot–Caratheodory distance, we can construct a path x^n by concatenating geodesic paths associated with the increments $\{\mathbf{x}_{t_i,t_{i+1}} : i = 0, \ldots, 2^n\}$, $(t_i) = (i2^{-n}T)$; the resulting sequence (x^n) then satisfies condition (H_α) uniformly and converges uniformly, together with its iterated integrals, to the path $\mathbf{x}\left(\cdot\right)$ with which we started.

Nothing of all this is restricted to $\alpha \in (1/3, 1/2] \longleftrightarrow N = 2$: **for any $\alpha = 1/p \in (0, 1]$ a weak, geometric $1/p$-Hölder rough path x is precisely a $1/p$-Hölder path in the metric space $\left(G^{[p]}\left(\mathbb{R}^d\right), d\right)$ where d denotes the Carnot–Caratheodory distance.** Davie's lemma extends to the step-$[p]$ setting and we are led to a theory of (rough path) differential equations, formally written as

$$dy = V\left(y\right) d\mathbf{x},$$

where $\mathbf{x}$ is a (weak, geometric) $1/p$-Hölder rough path. For $V \in \mathrm{Lip}^{\gamma-1}$ one has existence and $V \in \mathrm{Lip}^{\gamma}$ with $\gamma > p$ uniqueness.[8] Once in possession of a unique solution $y = \pi_{(V)}(0, y_0; \mathbf{x})$ one may ask for regularity of the *Itô–Lyons map*

$$\mathbf{x} \mapsto y.$$

In fact, one can construct the RDE solution as a (weak, geometric) $1/p$-Hölder rough path in its own right, say $\mathbf{y} = \boldsymbol{\pi}_{(V)}(0, y_0; \mathbf{x})$ with values in $G^{[p]}(\mathbb{R}^e)$ and ask for regularity of the *full* Itô–Lyons map

$$(y_0, V, \mathbf{x}) \mapsto \mathbf{y}.$$

It turns out that this solution map is Lipschitz continuous on bounded sets, provided we measure the distance between two driving signals $\mathbf{x}, \tilde{\mathbf{x}}$ with a (non-homogenous[9]) $1/p$-Hölder distance given by

$$\rho_{1/p\text{-Höl}}(\mathbf{x}, \tilde{\mathbf{x}}) := \max_{i=1,\dots,[p]} \sup_{s,t \in [0,T]} \frac{\left| \mathbf{x}_{s,t}^i - \tilde{\mathbf{x}}_{s,t}^i \right|}{|t-s|^{i/p}}.$$

For most applications it is enough to have (uniform) continuity (on bounded sets), in which case one can work with the (homogenous[10]) $1/p$-Hölder distance given by

$$d_{1/p\text{-Höl}}(\mathbf{x}, \tilde{\mathbf{x}}) := \sup_{s,t \in [0,T]} \frac{d(\mathbf{x}_{s,t}, \tilde{\mathbf{x}}_{s,t})}{|t-s|^{i/p}}.$$

The latter often makes computations more transparent and can become indispensible in a probabilistic context (e.g. when studying "exponentially good" approximations in a large deviation context).

But no matter which distance is more practical in a given context, both induce the same "$1/p$-Hölder rough path" topology on the rough path space $C^{1/p\text{-Höl}}([0,T], G^{[p]}(\mathbb{R}^d))$.

---

[8] With more effort, uniqueness can be shown under Lip^p-regularity.

[9] ...with respect to dilation since, in general,

$$\rho_{1/p\text{-Höl}}(\delta_\lambda \mathbf{x}, \delta_\lambda \tilde{\mathbf{x}}) \neq |\lambda| \rho_{1/p\text{-Höl}}(\mathbf{x}, \tilde{\mathbf{x}}).$$

[10] ...again with respect to dilation,

$$d_{1/p\text{-Höl}}(\delta_\lambda \mathbf{x}, \delta_\lambda \tilde{\mathbf{x}}) = |\lambda| d_{1/p\text{-Höl}}(\mathbf{x}, \tilde{\mathbf{x}}).$$

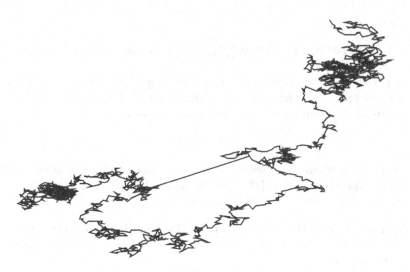

Figure 4. A typical 2-dimensional Brownian sample path. The (signed) area between the straight cord and the sample path corresponds to a typical Lévy area increment.

3 Brownian motion and stochastic analysis

Let B be a d-dimensional Brownian motion. Almost every realization of *enhanced Brownian motion* (EBM)

$$t \mapsto \mathbf{B}_t\left(\omega\right) = \left(1, B_t, \int_0^t B_s \otimes \circ dB_s\right) = \exp\left(B_t + A_{0,t}\right)$$

with $so\left(d\right)$-valued Lévy area $A_{s,t}\left(\omega\right) = \frac{1}{2}\int_s^t\left(B_{s,r} \otimes dB_r - dB_r \otimes B_{s,r}\right)$ is a (weak) geometric rough path, namely

$$\mathbf{B}.\left(\omega\right) \in C^{\alpha\text{-Höl}}\left(\left[0, T\right], \left(G^2\left(\mathbb{R}^d\right), d\right)\right), \quad \alpha \in (1/3, 1/2).$$

In Figure 4 we plot a Brownian path with an associated Lévy area increment. Granted the usual α-Hölder regularity of Brownian motion, this statement is equivalent to the question

$$\text{"Is it true that for } \alpha < 1/2: \quad \sup_{s,t \in [0,1]} \frac{|A_{s,t}|}{|t-s|^{2\alpha}} < \infty \text{ a.s. ?"}$$

The reader is encouraged to think about this before reading on!

Perhaps the most elegant way to establish this "rough path regularity" of Lévy area relies on scaling properties of enhanced Brownian motion. Namely,

$$\mathbf{B}_{s,t} \overset{\mathcal{D}}{=} \mathbf{B}_{0,t-s} \overset{\mathcal{D}}{=} \delta_{(t-s)^{1/2}} \mathbf{B}_{0,1},$$

so that

$$\mathbb{E}\left(d\left(\mathbf{B}_s, \mathbf{B}_t\right)^{2q}\right) = \mathbb{E}\left(\|\mathbf{B}_{s,t}\|^{2q}\right) \leq (\text{const}) \times |t - s|^q$$

for any $q < \infty$. Kolmogorov's criterion applies without any trouble and so $\mathbf{B}$ is indeed a.s. α-Hölder, $\alpha < 1/2$, with respect to d. QED.

Let us also mention a convergence result: we have

$$d_{\alpha\text{-Höl};[0,T]}\left(\mathbf{B}, S_2\left(B^n\right)\right) \to 0$$

in probability where B^n denotes a piecewise linear approximation to B based on dissections $D^n = \{t_i^n : i\}$ with the mesh of D^n tending to 0.

We then have two important conclusions:

(i) Thanks to α-Hölder regularity of $\mathbf{B}$, the (random) RDE $dY = V(Y)\, d\mathbf{B}$ can be solved for a.e. fixed ω and yields a continuous stochastic process

$$Y.\left(\omega\right) = \pi_{(V)}\left(0, y_0; \mathbf{B}\left(\omega\right)\right). \tag{7}$$

(ii) By continuity of the Itô–Lyons map with respect to the rough path metric $d_{\alpha\text{-Höl};[0,T]}$ it follows that

$$\pi_{(V)}\left(0, y_0; B^n\right) \to \pi_{(V)}\left(0, y_0; \mathbf{B}\left(\omega\right)\right)$$

with respect to α-Hölder topology and in probability. Clearly,

$$y^n \equiv \pi_{(V)}\left(0, y_0; B^n\right)$$

is a solution to the (random) ODE

$$dy^n = V\left(y^n\right) dB^n, \ y^n\left(0\right) = y_0$$

and the classical *Wong–Zakai* theorem[11] allows us to identify (7) as the classical Stratonovich solution to

$$dY = V\left(Y\right) \circ dB = \sum_{i=1}^{d} V_i\left(Y\right) \circ dB^i.$$

But why is all this useful? The following list should give some idea ...

- $\pi_{(V)}\left(0, y_0; \mathbf{B}\left(\omega\right)\right)$ is simultaneously defined for *all* starting points y_0 and coefficient vector fields V of suitable regularity. In particular, the construction of stochastic flows is a triviality and this itself can be the starting point for the robust treatment of certain stochastic partial differential equations.

---

[11] For example, the books of Ikeda–Watanabe [88] or Stroock [160].

- Every approximation in rough path topology implies a limit theorem (even on the level of flows). This includes classical piecewise linear approximations and non-standard variations à la McShane, Sussmann. It also includes a variety of weak limit theorems such as a Donsker-type invariance principle.

- Various stochastic Taylor expansions (à la Azencott, Platen, ...) can be obtained via deterministic rough path estimates.

- Support descriptions à la Stroock–Varadhan and large deviation estimates à la Freidlin–Wentzell are reduced to the respective (relatively simple) statements about **B** in the rough path topology.

- The *Young integral* allows us to perturb **B** simultaneously in all $C^{q\text{-var}}\left([0,1], \mathbb{R}^d\right)$-directions with $q < 2$. Since

$$\text{Cameron–Martin} \subset C^{1\text{-var}} \subset C^{q\text{-var}}$$

this implies in particular path space regularity of the SDE solution beyond Malliavin and there is a natural interplay with Malliavin calculus.

- Starting points and vector fields can be fully anticipating.

- At last, for the bulk of these results we can replace Brownian motion at little extra cost by martingales, Gaussian processes or Markov processes provided we can construct a suitable stochastic area and establish the correct rough path regularity!

Part I

Basics

1

Continuous paths of bounded variation

We discuss continuous paths, defined on a fixed time horizon, with values in a metric space E. The emphasis is on paths with nice regularity properties and in particular on continuous paths of bounded variation.[1] We then specialize to the case when $E = \mathbb{R}^d$. Finally, we discuss simple Sobolev-type regularity of paths.

1.1 Continuous paths on metric spaces

We start by defining the supremum or infinity distance.

Definition 1.1 Let (E, d) be a metric space and $[0, T] \subset \mathbb{R}$. Then $C([0, T], E)$ denotes the set of all continuous paths $x : [0, T] \to E$. The supremum or infinity distance of $x, y \in C([0, T], E)$ is defined by

$$d_{\infty;[0,T]}(x, y) := \sup_{t \in [0,T]} d(x_t, y_t).$$

For a single path $x \in C([0, T], E)$, we set

$$|x|_{0;[0,T]} := \sup_{u,v \in [0,T]} d(x_u, x_v),$$

and, given a fixed element $o \in E$, identified with the constant path $\equiv o$,

$$|x|_{\infty;[0,T]} := d_{\infty;[0,T]}(o, x) = \sup_{u \in [0,T]} d(o, x_u). \qquad \square$$

If no confusion is possible we shall omit $[0, T]$ and simply write $d_\infty, |\cdot|_0$ and $|\cdot|_\infty$. If E has a group structure such as $(\mathbb{R}^d, +)$ the neutral element is the usual choice for o. In the present generality, however, the definition of $|\cdot|_\infty$ depends on the choice of o.

Notation 1.2 Of course, $[0, T]$ can be replaced by any other interval $[s, t]$ in which case one considers $x : [s, t] \to E$. All notations adapt by replacing $[0, T]$ by $[s, t]$. Let us also agree that $C_o([s, t], E)$ denotes those paths in $C([s, t], E)$ which start at o, i.e.

$$C_o([s, t], E) = \{x \in C([s, t], E) : x(s) = o\}. \qquad \square$$

---

[1] Also known as rectifiable paths.

Many familiar properties of real-valued functions carry over. For instance, any continuous mapping from $[0,T]$ into E is uniformly continuous.[2] It is also fairly easy to see that $C\left([0,T],E\right)$ is a metric space under d_∞ (the induced topology will be called the uniform or supremum topology). Also, if (E,d) is complete then $(C\left([0,T],E\right),d_\infty)$ is complete.

Definition 1.3 A set $H \subset C\left([0,T],E\right)$ is said to be equicontinuous if, for all $\varepsilon > 0$ there exists δ such that $|t-s| < \delta$ implies $d\left(x_s,x_t\right) < \varepsilon$ for all $x \in H$. It is said to be bounded if $\sup_{x\in H}|x|_\infty < \infty$. $\square$

Theorem 1.4 (Arzela–Ascoli) *Let (E,d) be a complete metric space in which bounded sets have compact closure. Then a set $H \subset C\left([0,T],E\right)$ has compact closure if and only if H is bounded and equicontinuous.*

As a consequence, a bounded, equicontinuous sequence in $C\left([0,T],E\right)$ has a convergent subsequence and, conversely, any convergent sequence in $C\left([0,T],E\right)$ is bounded and equicontinuous.

Proof. Let us recall that a subset of a complete metric space has compact closure if and only if it is totally bounded, i.e. for all $\varepsilon > 0$, it can be covered by finitely many ε-balls.

"$\Longleftarrow$": We show that the assumption "H bounded and equicontinuous" implies total boundedness. We fix $\varepsilon > 0$, and then $\delta > 0$ such that for every $f \in H$,

$$|t-s| < \delta \implies d\left(f_s,f_t\right) < \varepsilon/4. \tag{1.1}$$

Cover $[0,T]$ with a finite number of neighbourhoods $\left(t_i - \frac{\delta}{2}, t_i + \frac{\delta}{2}\right)$, $i = 1,\ldots,m$, and define $H_{t_i} = \{f_{t_i}, f \in H\}$; as $H_{t_i} \subset E$ is bounded, its closure is compact, and so is its union $\bigcup_{1 \le i \le m} \overline{H_{t_i}}$; let $c_1,\ldots,c_n \in \bigcup_{1 \le i \le m} \overline{H_{t_i}}$ be such that $\bigcup_{1 \le i \le m} \overline{H_{t_i}}$ is covered by the union of the $\varepsilon/4$-balls centred around some c_j.

Then, consider Φ, the set of functions from $\{1,\ldots,m\}$ into $\{1,\ldots,n\}$. For each $\varphi \in \Phi$, denote by $L_{\varphi,\varepsilon}$ the set of all functions $f \in C\left([0,T],E\right)$ such that $\max_i d\left(f_{t_i}, c_{\varphi(i)}\right) \le \frac{\varepsilon}{4}$. Observe that from the definition of c_j it follows that H is covered by the union of the $\left(L_{\varphi,\varepsilon}\right)_{\varphi \in \Phi}$.

To end the proof, we need only show that the diameter of each $L_{\varphi,\varepsilon}$ is $\le \varepsilon$.

---

[2] For example, Dieudonné [43], (3.6.15).

If f, g are both in $L_{\varphi,\varepsilon}$, then

$$
\begin{aligned}
d_\infty(f,g) &= \sup_{t \in [0,T]} d(f_t, g_t) \\[2mm]
&\leq \max_{1 \leq i \leq m} \left(d(f_{t_i}, g_{t_i}) + \sup_{s \in (t_i - \frac{\delta}{2}, t_i + \frac{\delta}{2})} d(f_s, f_{t_i}) + d(g_s, g_{t_i}) \right) \\[2mm]
&\leq \max_{1 \leq i \leq m} d(f_{t_i}, g_{t_i}) + \frac{\varepsilon}{2} \text{ from (1.1)} \\[2mm]
&\leq \max_i d(f_{t_i}, c_{\varphi(i)}) + \max_i d(g_{t_i}, c_{\varphi(i)}) + \frac{\varepsilon}{2} \\[2mm]
&\leq \varepsilon \text{ by definition of } L_{\varphi,\varepsilon}.
\end{aligned}
$$

"$\Longrightarrow$": Since compact sets are bounded, only equicontinuity needs proof. By assumption H has compact closure and therefore is totally bounded. Fix $\varepsilon > 0$ and pick $h^1, \ldots, h^n$ such that $H \subset \bigcup_{1 \leq i \leq n} B(h^i, \varepsilon/3)$ where $B(h, \varepsilon)$ denotes the open ε-ball centred at h. By continuity of each $h^i(\cdot)$, there exists $\delta = \delta(\varepsilon)$ such that

$$
|t - s| < \delta \implies \max_{i=1,\ldots,n} d(h^i_s, h^i_t) < \varepsilon/3.
$$

But then, for every $h \in H$, $d(h_s, h_t) \leq \varepsilon/3 + \max_{i=1,\ldots,n} d(h^i_s, h^i_t) + \varepsilon/3 \leq \varepsilon$ provided $|t - s| < \delta$ and so H is equicontinuous.

The consequences for sequences are straightforward and left to the reader. ∎

1.2 Continuous paths of bounded variation on metric spaces

1.2.1 *Bounded variation paths and controls*

Let us write $\mathcal{D}([s,t])$ for the set of all dissections of some interval $[s,t] \subset \mathbb{R}$, thus a typical element in $\mathcal{D}([s,t])$ is written

$$
D = \{ s = t_0 < t_1 < \cdots < t_n = t \}
$$

and consists of $\#D = n$ adjacent intervals $[t_{i-1}, t_i]$. The mesh of D is defined as $|D| := \max_{i=1,\ldots,n} |t_i - t_{i-1}|$ and we shall denote by $\mathcal{D}_\delta([s,t])$ the set of all dissections of $[s,t]$ with mesh less than or equal to δ.

Definition 1.5 Let (E, d) be a metric space and $x : [0, T] \to E$. For $0 \leq s \leq t \leq T$, the 1-variation of x on $[s, t]$ is defined as[3]

$$
|x|_{1\text{-var};[s,t]} = \sup_{(t_i) \in \mathcal{D}([s,t])} \sum_i d(x_{t_i}, x_{t_{i+1}}).
$$

---

[3]Let us agree that $|x|_{1\text{-var};[s,s]} = 0$ for $0 \leq s \leq T$.

If $|x|_{\text{1-var};[s,t]} < \infty$, we say that x is of bounded variation or of finite 1-variation on $[s,t]$. The space of continuous paths of finite 1-variation on $[0,T]$ is denoted by $C^{\text{1-var}}([0,T],E)$, its subset of paths started at $o \in E$ is denoted by $C_o^{\text{1-var}}([0,T],E)$.   □

In the discussion of 1-variation regularity (and later p-variation regularity for $p \geq 1$), the notion of *control* or *control function*, defined on the simplex

$$\Delta := \Delta_T = \{(s,t) : 0 \leq s \leq t \leq T\}$$

turns out to be extremely useful.

Definition 1.6 A map $\omega : \Delta_T \to [0,\infty)$ is called superadditive if for all $s \leq t \leq u$ in $[0,T]$,

$$\omega(s,t) + \omega(t,u) \leq \omega(s,u).$$

If, in addition, ω is continuous and zero on the diagonal, i.e. $\omega(s,s) = 0$ for $0 \leq s \leq T$ we call ω a control or, more precisely, a control function on $[0,T]$.   □

Definition 1.7 We say that the 1-variation of a map $x : [0,T] \to E$ is dominated by the control ω, or controlled by ω, if there exists a constant $C < \infty$ such that for all $s < t$ in $[0,T]$,

$$d(x_s, x_t) \leq C\omega(s,t).$$ □

Simple examples of controls are given by $(s,t) \mapsto |t-s|^\theta$ for $\theta \geq 1$ or the integral of a non-negative function in $L^1([0,T])$ over the interval $[s,t]$. Trivially, a positive linear combination of controls yields another control. If ω is a control and $x : [0,T] \to E$ a map controlled by ω then x is continuous.

Exercise 1.8 *Let $\phi \in C([0,\infty),[0,\infty))$ be increasing, convex with $\phi(0) = 0$. Assuming that ω is a control, show that $\phi \circ \omega : (s,t) \mapsto \phi(\omega(s,t))$ is also a control.*

Solution. Fix $0 < a < b$ and observe that by convexity

$$\frac{\phi(a+b) - \phi(b)}{a} \geq \frac{\phi(a) - \phi(0)}{a}$$

so that $\phi(a+b) \geq \phi(a) + \phi(b)$. Interchanging a,b if needed, this holds for all $a,b \geq 0$ and we conclude that

$$\phi[\omega(s,t)] + \phi[\omega(t,u)] \leq \phi[\omega(s,t) + \omega(t,u)] \leq \phi[\omega(s,u)].$$ □

Exercise 1.9 *Assume $\omega, \tilde{\omega}$ are controls.*
(i) Show that $\omega.\tilde{\omega}$ is a control.
(ii) Show that $\max(\omega, \tilde{\omega})$ need not be a control.
(iii) Given $\alpha, \beta > 0$ with $\alpha + \beta \geq 1$, show that $\omega^\alpha.\tilde{\omega}^\beta$ is a control.

Solution. (iii) By Exercise 1.8, it is enough to consider the case $\alpha + \beta = 1$. But this follows from Hölder's inequality,

$$\forall a, \tilde{a}, b, \tilde{b} \geq 0 : a\tilde{a} + b\tilde{b} \leq \left(a^{\frac{1}{\alpha}} + b^{\frac{1}{\alpha}}\right)^{\alpha} \left(\tilde{a}^{\frac{1}{\beta}} + \tilde{b}^{\frac{1}{\beta}}\right)^{\beta}.$$ $\square$

Exercise 1.10 *Let ω be a control on $[0,T]$ and consider $s < u$ in $[0,T]$. Show that there exists $t \in [s,u]$ such that*

$$\max\{\omega(s,t), \omega(t,u)\} \leq \omega(s,u)/2.$$

Solution. By continuity and monotonicity of controls, there exists t such that $\omega(s,t) = \omega(t,u)$. By super-additivity,

$$2\omega(s,t) = 2\omega(t,u) = \omega(s,t) + \omega(t,u) \leq \omega(s,u)$$

and the proof is finished. $\square$

Proposition 1.11 *Consider $x : [0,T] \to E$ and $\omega = \omega(s,t)$ super-additive, with $s < t$ in $[0,T]$. If $d(x_s, x_t) \leq \omega(s,t)$ for all $s < t$ in $[0,T]$, then $|x|_{1\text{-var};[s,t]} \leq \omega(s,t)$.*

Proof. Let $D = (t_i)$ be a dissection of $[s,t]$. Then, by assumption,

$$\sum_{i=0}^{\#D-1} d\left(x_{t_i, t_{i+1}}\right) \leq \sum_{i=0}^{\#D-1} \omega(t_i, t_{i+1})$$
$$\leq \omega(s,t) \text{ by super-additivity of } \omega.$$

Taking the supremum over all such dissections finishes the proof.   ∎

Proposition 1.12 *Let $x \in C^{1\text{-var}}([0,T], E)$. Then*

$$(s,t) \mapsto \omega_x(s,t) := |x|_{1\text{-var};[s,t]}$$

defines a control on $[0,T]$ such that for all $0 \leq s < t \leq T$

$$d(x_s, x_t) \leq |x|_{1\text{-var};[s,t]}.$$ (1.2)

This control is additive: for all $0 \leq s \leq t \leq u \leq T$,

$$|x|_{1\text{-var};[s,u]} = |x|_{1\text{-var};[s,t]} + |x|_{1\text{-var};[t,u]}.$$

In particular,

$$t \in [0,T] \mapsto \ell(t) := |x|_{1\text{-var};[0,t]} \in \mathbb{R}$$

is continuous, increasing and hence of finite 1-variation.

Proof. Trivially, $|x|_{\text{1-var};[s,s]} = 0$ for all $s \in [0,T]$. To see super-additivity it suffices to take dissections D_1, D_2 of $[s,t]$ and $[t,u]$ respectively; noting that the union of D_1 and D_2 is a dissection of $[s,u]$ we have

$$\sum_{t_i \in D_1} d\left(x_{t_i}, x_{t_{i+1}}\right) + \sum_{t_j \in D_2} d\left(x_{t_j}, x_{x_{j+1}}\right) \le |x|_{\text{1-var};[s,u]}$$

and $\omega_x(s,t) + \omega_x(t,u) \le \omega_x(s,u)$ follows from taking the supremum over all dissections D_1 and D_2. For additivity of ω_x we establish the reverse inequality. Let $D = (v_i)$ be a dissection of $[s,u]$ so that $t \in [v_j, v_{j+1}]$ for some j. We then have

$$\sum_{i=0}^{\#D-1} d\left(x_{v_i,v_{i+1}}\right) = \sum_{i=0}^{j-1} d\left(x_{v_i}, x_{v_{i+1}}\right) + \underbrace{d\left(x_{v_j}, x_{v_{j+1}}\right)}_{\le d\left(x_{v_j}, x_t\right) + d\left(x_t, x_{v_{j+1}}\right)}$$

$$+ \sum_{i=j+1}^{\#D-1} d\left(x_{v_i}, x_{v_{i+1}}\right).$$

But, as

$$\sum_{i=0}^{j-1} d\left(x_{v_i}, x_{v_{i+1}}\right) + d\left(x_{v_j}, x_t\right) \le |x|_{\text{1-var};[s,t]},$$

$$d\left(x_t, x_{v_{j+1}}\right) + \sum_{i=j+1}^{\#D-1} d\left(x_{v_i}, x_{v_{i+1}}\right) \le |x|_{\text{1-var};[t,u]},$$

we have

$$\sum_{i=0}^{\#D-1} d\left(x_{v_i,v_{i+1}}\right) \le |x|_{\text{1-var};[s,t]} + |x|_{\text{1-var};[t,u]}.$$

Taking the supremum over all dissections shows additivity of ω_x. It only remains to prove its continuity.

To this end, fix $s < t$ in $[0,T]$. From monotonocity of ω_x, we see that the limits

$$|x|_{\text{1-var};[s^+,t^-]} := \lim_{h_1,h_2 \searrow 0} |x|_{\text{1-var};[s+h_1,t-h_2]}, |x|_{\text{1-var};[s^-,t^+]}$$

$$:= \lim_{h_1,h_2 \searrow 0} |x|_{\text{1-var};[s-h_1,t+h_2]}$$

exist and that

$$|x|_{\text{1-var};[s^+,t^-]} \le |x|_{\text{1-var};[s,t]} \le |x|_{\text{1-var};[s^-,t^+]}. \tag{1.3}$$

We aim to show the inequalities in (1.3) are actually equalities. To establish "continuity from inside", i.e. $|x|_{\text{1-var};[s^+,t^-]} = |x|_{\text{1-var};[s,t]}$, we define

$\omega\left(s,t\right)=\left|x\right|_{1\text{-var};\left[s^{+},t^{-}\right]}$, and pick $s<t<u$ in $[0,T]$, and h_1,h_2,h_3,h_4 four (small) positive numbers. If (a_i) is a dissection of $[s+h_1,t-h_2]$, and (b_j) a dissection of $[t+h_3,u-h_4]$, then by definition of the 1-variation of x,

$$\sum_i d\left(x_{a_i},x_{a_{i+1}}\right)+\sum_j d\left(x_{b_j},x_{b_{j+1}}\right)\le\left|x\right|_{1\text{-var};\left[s+h_1,u-h_4\right]}.$$

Taking the supremum over all possible dissections (a_i) and (b_i), we obtain

$$\left|x\right|_{1\text{-var};\left[s+h_1,t-h_2\right]}+\left|x\right|_{1\text{-var};\left[t+h_3,u-h_4\right]}\le\left|x\right|_{1\text{-var};\left[s+h_1,u-h_4\right]}.$$

Letting h_1,h_2,h_3,h_4 go to 0 and using the continuity of x we obtain that $(s,t)\mapsto\left|x\right|_{1\text{-var};\left[s^{+},t^{-}\right]}$ is super-additive. We also easily see that for all $s,t\in[0,T]$, $d\left(x_s,x_t\right)\le\left|x\right|_{1\text{-var};\left[s^{+},t^{-}\right]}$. Hence, using Proposition 1.11, we obtain

$$\left|x\right|_{1\text{-var};\left[s^{+},t^{-}\right]}\ge\left|x\right|_{1\text{-var};\left[s,t\right]},$$

and hence we have proved that $\left|x\right|_{1\text{-var};\left[s^{+},t^{-}\right]}=\left|x\right|_{1\text{-var};\left[s,t\right]}$ for all $s<t$ in $[0,T]$. The remaining part of the proof is "continuity from outside", i.e. $\left|x\right|_{1\text{-var};\left[s^{-},t^{+}\right]}=\left|x\right|_{1\text{-var};\left[s,t\right]}$. Using additivity of $\left|x\right|_{1\text{-var};\left[\cdot,\cdot\right]}$ it is easy to see that

$$\begin{aligned}\left|x\right|_{1\text{-var};\left[s^{-},t^{+}\right]}&=&\left|x\right|_{1\text{-var};\left[0,T\right]}-\left|x\right|_{1\text{-var};\left[0,s^{-}\right]}-\left|x\right|_{1\text{-var};\left[t^{+},T\right]}\\&=&\left|x\right|_{1\text{-var};\left[0,T\right]}-\left|x\right|_{1\text{-var};\left[0,s\right]}-\left|x\right|_{1\text{-var};\left[t,T\right]}=\left|x\right|_{1\text{-var};\left[s,t\right]}\end{aligned}$$

and this finishes the proof. ∎

Exercise 1.13 *Assume ω is a control on $[0,T]$. Assume $f\in C\left(\Delta,[0,\infty)\right)$ where $\Delta=\{(s,t):0\le s\le t\le T\}$, non-decreasing in the sense that $[s,t]\subset[u,v]$ implies $f\left(s,t\right)\le f\left(u,v\right)$. Show that*

$$(s,t)\mapsto f\left(s,t\right)\omega\left(s,t\right)$$

is a control. As application, given $x\in C^{1\text{-var}}\left([0,T],E\right)$ and $y\in C\left([0,T],E\right)$, show that

$$(s,t)\mapsto\left|y\right|_{\infty;\left[s,t\right]}\left|x\right|_{1\text{-var};\left[s,t\right]}$$

is a control where $\left|\cdot\right|_{\infty;\left[s,t\right]}$ is defined with respect to some fixed $o\in E$.

Proposition 1.14 *Let $x\in C\left([0,T],E\right)$. Then for all $\delta>0$ and $0\le s\le t\le T$,*

$$\left|x\right|_{1\text{-var};\left[s,t\right]}=\sup_{(t_i)\in\mathcal{D}_{\delta}\left([s,t]\right)}\sum_i d\left(x_{t_i},x_{t_{i+1}}\right)\in[0,\infty].$$

Proof. Clearly,

$$\omega_{x,\delta}\left(s,t\right):=\sup_{(t_i)\in\mathcal{D}_{\delta}\left([s,t]\right)}\sum_i d\left(x_{t_i},x_{t_{i+1}}\right)\le\omega_x\left(s,t\right)=\left|x\right|_{1\text{-var};\left[s,t\right]}.$$

Super-additivity of $\omega_{x,\delta}$ follows from the same argument as for ω_x. Take any $D = (u_i) \in \mathcal{D}_\delta([s,t])$ so that $s = u_0 < u_1 < \cdots < u_n = t$ with $u_{i+1} - u_i < \delta$. It follows that

$$
\begin{aligned}
d(x_s, x_t) &\leq d(x_s, x_{u_1}) + \cdots + d(x_{u_{n-1}}, x_t) \\
&\leq \omega_{x,\delta}(s,t).
\end{aligned}
$$

From Proposition 1.11, we conclude that $|x|_{\text{1-var};[s,t]} \leq \omega_{x,\delta}(s,t)$, which concludes the proof.   ■

We now observe lower semi-continuity of the function $x \mapsto |x|_{\text{1-var}}$ in the following sense.

Lemma 1.15 *Assume (x^n) is a sequence of paths from $[0,T] \rightarrow E$ of finite 1-variation. Assume $x^n \rightarrow x$ pointwise on $[0,T]$. Then, for all $s < t$ in $[0,T]$,*

$$|x|_{\text{1-var};[0,T]} \leq \liminf_{n \to \infty} |x^n|_{\text{1-var};[0,T]} .$$

Proof. Let $D = \{0 = t_0 < t_1 < \cdots < t_K = T\}$ be a dissection of $[0,T]$. By assumption, $x^n \rightarrow x$ pointwise and so

$$
\begin{aligned}
\sum_{i=0}^{K-1} d(x_{t_i}, x_{t_{i+1}}) &= \liminf_{n \to \infty} \sum_i d\left(x_{t_i}^n, x_{t_{i+1}}^n\right) \\
&\leq \liminf_{n \to \infty} |x^n|_{\text{1-var};[0,T]} .
\end{aligned}
$$

Taking the supremum over all the dissections of $[s,t]$ finishes the 1-variation estimate.   ■

In general, the inequality in Lemma 1.15 can be strict. The reader is invited to construct an example in the following exercise.

Exercise 1.16 *Construct $(x^n) \in C^{\text{1-var}}([0,1],\mathbb{R})$ such that $|x^n|_{\infty;[0,1]} \leq 1/n$ but so that $|x^n|_{\text{1-var}} = 1$ for all n. Conclude that the inequality in Lemma 1.15 can be strict.*

1.2.2 Absolute continuity

Definition 1.17 *Let (E,d) be a metric space. A path $x : [0,T] \rightarrow E$ is absolutely continuous if for all $\varepsilon > 0$, there exists $\delta > 0$, such that for all $s_1 < t_1 \leq s_2 < t_2 \leq \cdots < s_n < t_n$ in $[0,T]$ with $\sum_i |t_i - s_i| < \delta$, we have $\sum_i d(x_{s_i}, x_{t_i}) < \varepsilon$.*   □

Proposition 1.18 *Any absolutely continuous path is a continuous path of bounded variation.*

Proof. If $x : [0,T] \rightarrow E$ is absolutely continuous it is obviously continuous. Furthermore, by definition there exists $\delta > 0$, such that for all $s_1 < t_1 \leq s_2 < t_2 \leq \cdots < s_n < t_n \in [0,T]$ with $\sum_i |t_i - s_i| \leq \delta$, we have

$\sum_i d\left(x_{s_i}, x_{t_i}\right) \leq 1$. Pick $D = (t_i)_{1 \leq i \leq n}$ a dissection of $[0,T]$. Then, define $j_0 = 1$ and $j_k = \max\left\{i, t_i - t_{j_{k-1}} \leq \delta\right\}$, and observe that $j_{[T/\delta]+1} = j_k$ for all $k \geq [T/\delta] + 1$.

$$\sum_{i=1}^{n-1} d\left(x_{t_i}, x_{t_{i+1}}\right) \leq \sum_{k=0}^{[T/\delta]+1} \sum_{i=j_k}^{j_{k+1}-1} d\left(x_{t_i}, x_{t_{i+1}}\right).$$

By definition of the j_ks, $\sum_{i=j_k}^{j_{k+1}-1}|t_{i+1} - t_i| = |t_{j_{k+1}} - t_{j_k}| \leq \delta$, hence $\sum_{i=j_k}^{j_{k+1}-1} d\left(x_{t_i}, x_{t_{i+1}}\right) \leq 1$, which implies that $\sum_{i=1}^{n-1} d\left(x_{t_i}, x_{t_{i+1}}\right) \leq [T/\delta] + 1$. Taking the supremum over all dissections finishes the proof. ∎

In general, the converse of the above is not true, as seen in the following.

Example 1.19 (Cantor function) Each $x \in [0,1]$ has a base-3 decimal expansion $x = \sum_{j \geq 1} a_j 3^{-j}$ where $a_j \in \{0,1,2\}$. This expansion is unique unless x is of the form $p3^{-k}$ for some $p, k \in \mathbb{N}$ (we may assume p is not divisible by 3) and in this case x has two expansions: one with $a_j = 0$ for $j > k$ and one with $a_j = 2$ for $j > k$. One of them has $a_k = 1$, the other will have $a_k \in \{0,2\}$. If we agree always to use the latter, we see that

$$a_1 = 1 \text{ iff } x \in (1/3, 2/3)$$
$$a_1 \neq 1, a_2 = 1 \text{ iff } x \in (1/9, 2/9) \cup (7/9, 8/9)$$

and so forth. The Cantor set $\mathcal{C}$ is then defined as the set of all $x \in [0,1]$ that have a base-3 expansion $x = \sum a_j 3^{-j}$ with $a_j \neq 1$ for all j. Thus $\mathcal{C}$ is obtained from $K_{0,1} = [0,1]$ by removing the open middle third, leaving us with the union of $K_{1,1} = [0,1/3]$, $K_{1,2} = [2/3,1]$; followed by removing all open middle thirds, leaving us with the union of

$$K_{2,1} = [0,1/9], K_{2,1} = [2/9,3/9], K_{2,1} = [6/9,7/9], K_{2,1} = [8/9,1]$$

and so forth, so that in the end $\mathcal{C} = \cap_{n=1}^{\infty} \cup_{i=1}^{2^n} K_{n,i}$. Let us now define the Cantor function f on $\mathcal{C}$ by

$$f(x) = \sum_{j \geq 1} \left(\frac{a_j}{2}\right) 2^{-j}, \quad x \in \mathcal{C}.$$

This series is the base-2 expansion of a number in $[0,1]$ and since any number in $[0,1]$ can be obtained this way we see that $f(\mathcal{C}) = [0,1]$. One readily sees that if $x, y \in \mathcal{C}$ and $x < y$, then $f(x) < f(y)$ unless x and y are the endpoints of one of the open intervals removed from $[0,1]$ to obtain $\mathcal{C}$. In this case, $f(x) = p2^{-k}$ for some $p, k \in \mathbb{N}$ and $f(x) = f(y)$, given by the two base-2 expansions of this number. We can therefore extend f to a map from $[0,1]$ to itself by declaring it to be constant on the intervals missing from $\mathcal{C}$. This extended f is still increasing, and since its range is all of $[0,1]$ it cannot have any jump discontinuities, hence it is continuous.

Being increasing, f is obviously of bounded variation on $[0,1]$. We now show that f is not absolutely continuous. Given any $\delta > 0$ we can take s_i, t_i as the boundary points of the intervals $(K_{n,i})_{i=1,\dots,2^n}$ with n chosen large enough so that $\sum_{i=1}^{2^n} (t_i - s_i) < \delta$. Then, since f is constant on $[t_i, s_{i+1}]$ for $i = 1,\dots, 2^n - 1$, we have

$$\sum_{i=1}^{2^n} |f(t_i) - f(s_i)| = f(1) - f(0) = 1.$$

□

1.2.3 Lipschitz or 1-Hölder continuity

Definition 1.20 Let (E, d) be a metric space. A path $x : [0, T] \to E$ is Lipschitz or 1-Hölder continuous[4] if

$$|x|_{1\text{-Höl};[0,T]} := \sup_{s,t \in [0,T]} \frac{d(x_s, x_t)}{|t - s|} < \infty.$$

The space of all such paths is denoted by $C^{1\text{-Höl}}([0,T], E)$, the subset of paths started at $o \in E$ is denoted by $C_o^{1\text{-Höl}}([0,T], E)$. □

We observe that every Lipschitz path is absolutely continuous. In particular, it is of bounded variation and we note

$$|x|_{1\text{-var};[s,t]} \le |x|_{1\text{-Höl};[s,t]} \times |t - s|.$$

Furthermore, $x \in C^{1\text{-var}}([0,T], E)$ is 1-Hölder if and only if it is controlled by $(s,t) \mapsto |t - s|$. It is easy to construct examples which are of bounded variation but not Lipschitz (e.g. $t \mapsto t^{1/2}$). On the other hand, every continuous bounded variation path is a continuous time-change (or reparametrization) of a Lipschitz path.

Proposition 1.21 *A path $x \in C([0,T], E)$ is of finite 1-variation if and only if there exists a continuous non-decreasing function ϕ from $[0,T]$ onto $[0,1]$ and a path $y \in C^{1\text{-Höl}}([0,1], E)$ such that $x = y \circ \phi$.*

Proof. We may assume $|x|_{1\text{-var};[0,T]} \ne 0$ (otherwise, $x|_{[0,T]}$ is constant and there is nothing to show). By Propostion 1.12,

$$\phi(t) = \frac{|x|_{1\text{-var};[0,t]}}{|x|_{1\text{-var};[0,T]}}$$

defines a continuous increasing function from $[0,T]$ onto $[0,1]$. Then, there exists a function y such that $(y \circ \phi)(t) = x(t)$, as $\phi(t_1) = \phi(t_2) \Longrightarrow$

---

[4] ...in view of the later definition of Hölder continuity and in order to avoid redundant notation ...

$x(t_1) = x(t_2)$. Now,

$$\sup_{0 \le u < v \le 1} \frac{d(y(u), y(v))}{|u-v|} = \sup_{0 \le u < v \le T} \frac{d(y(\phi(u)), x(y(v)))}{|\phi(u) - \phi(v)|}$$

$$\le \; |x|_{1\text{-var};[0,T]} \frac{|x|_{1\text{-var};[u,v]}}{\left| |x|_{1\text{-var};[0,u]} - |x|_{1\text{-var};[0,v]} \right|}$$

$$= \; |x|_{1\text{-var};[0,T]} \; .$$

This shows that y is in $C^{1\text{-Höl}}([0,1], E)$. The converse direction is an obvious consequence of the invariance of variation norms under reparametrization. ∎

Remark 1.22 The 1-variation (i.e. length) of a path is obviously invariant under reparametrization and so it is clear that

$$|y|_{1\text{-var};[0,1]} = |x|_{1\text{-var};[0,T]} \; .$$

On the other hand, for the particular parametrization $\phi(\cdot)$ used in the previous proof (essentially the arc-length parametrization) we saw that $|y|_{1\text{-Höl};[0,1]} \le |x|_{1\text{-var};[0,T]}$. With the trivial $|y|_{1\text{-var};[0,1]} \le |y|_{1\text{-Höl};[0,1]}$ we then see that

$$|y|_{1\text{-Höl};[0,1]} = |x|_{1\text{-var};[0,T]} \; .$$ □

Lemma 1.23 *Assume* (x^n) *is a sequence of paths from* $[0,T] \to E$ *of finite 1-variation. Assume* $x^n \to x$ *pointwise on* $[0,T]$. *Then, for all* $s < t$ *in* $[0,T]$,

$$|x|_{1\text{-Höl};\,[s,t]} \le \liminf_{n \to \infty} |x^n|_{1\text{-Höl};[s,t]} \; .$$

Proof. The Hölder statement is a genuine corollary of Lemma 1.15: it suffices to note that for any $u, v \in [s,t]$,

$$d(x_u, x_v) \;\le\; |x|_{1\text{-var};[u,v]}$$
$$\le\; \liminf_{n \to \infty} |x^n|_{1\text{-var};[u,v]} \; .$$
$$\le\; |v - u| \liminf_{n \to \infty} |x^n|_{1\text{-Höl};[s,t]} \; .$$

∎

1.3 Continuous paths of bounded variation on $\mathbb{R}^d$

Unless otherwise stated, $\mathbb{R}^d$ shall be equipped with Euclidean structure. In particular, if $a \in \mathbb{R}^d$ has coordinates $(a^1, \ldots, a^d)$ its norm is given by

$$|a| = \sqrt{|a^1|^2 + \cdots + |a^d|^2} \; .$$

Given a map $x : [0,T] \to \mathbb{R}^d$ the group structure of $(\mathbb{R}^d,+)$ allows us to speak of the *increments* of $x(\cdot)$ and we write[5]

$$x_{s,t} := x_t - x_s.$$

1.3.1   Continuously differentiable paths

We define inductively the set $C^k\left([0,T],\mathbb{R}^d\right)$ of k-times continuously differentiable paths by first defining $C^0\left([0,T],\mathbb{R}^d\right)$ to be $C\left([0,T],\mathbb{R}^d\right)$, and then $C^{k+1}\left([0,T],\mathbb{R}^d\right)$ to be the set of paths with a derivative in $C^k\left([0,T],\mathbb{R}^d\right)$. Finally, we define the set of smooth paths $C^\infty\left([0,T],\mathbb{R}^d\right)$ to be the intersection of all $C^k\left([0,T],\mathbb{R}^d\right)$, for $k \geq 0$.

For continuously differentiable paths, the computation of 1-variation is a simple matter.

Proposition 1.24 *Let $x \in C^1\left([0,T],\mathbb{R}^d\right)$. Then*

$$t \in [0,T] \mapsto \ell(t) := |x|_{1\text{-var};[0,t]} \in \mathbb{R}$$

is continuously differentiable and $\dot{\ell}(t) = |\dot{x}(t)|$ for $t \in (0,T)$. In particular,

$$|x|_{1\text{-var};[s,t]} = \int_s^t |\dot{x}_u|\,du$$

for all $s < t$ in $[0,T]$.

Proof. We first note that $|x_t - x_s| \leq \int_s^t |\dot{x}_u|\,du$; using Proposition 1.11, we obtain that

$$\ell(t) - \ell(s) = |x|_{1\text{-var};[s,t]} \leq \int_s^t |\dot{x}_u|\,du.$$

Equality in the above estimate will follow immediately from $\dot{\ell}(t) = |\dot{x}(t)|$ and this is what we now show. Take $t \in [0,T)$ and h small enough (so that $t + h \leq T$). Clearly,

$$\frac{|x_{t,t+h}|}{h} \leq \frac{|\ell(t+h) - \ell(t)|}{h} \leq \frac{1}{h}\int_t^{t+h} |\dot{x}_u|\,du$$

and upon sending $h \downarrow 0$ we see that ℓ is differentiable at t from the right with derivative equal to $|\dot{x}_t|$. The same argument applies "from the left" and so ℓ is indeed differentiable with derivative $|\dot{x}|$. By assumption on x, this derivative is continuous and the proof is finished. ∎

---

[5] Later on, we shall replace $\mathbb{R}^d$ by a Lie group $(G,\cdot)$ and increments will be defined as $(x_s)^{-1} \cdot x_t$.

1.3.2 Bounded variation

The results of Section 1.2 applied to $\mathbb{R}^d$ equipped with Euclidean distance allow us in particular to consider the space $C^{1\text{-var}}\left([0,T],\mathbb{R}^d\right)$.

Theorem 1.25 $C^{1\text{-var}}\left([0,T],\mathbb{R}^d\right)$ *is Banach with norm* $x \mapsto |x(0)| + |x|_{1\text{-var};[0,T]}$. *The closed subspace of paths in* $C^{1\text{-var}}\left([0,T],\mathbb{R}^d\right)$ *started at* 0, *denoted by* $C_0^{1\text{-var}}\left([0,T],\mathbb{R}^d\right)$, *is also Banach under* $x \mapsto |x|_{1\text{-var};[0,T]}$. *These Banach spaces are not separable.*

Proof. It is easy to see that $C^{1\text{-var}}\left([0,T],\mathbb{R}^d\right), C_0^{1\text{-var}}\left([0,T],\mathbb{R}^d\right)$ are normed linear spaces under the given norms. We thus focus on completeness. Noting that

$$\sup_{t\in[0,T]} |x(t)| \leq |x(0)| + |x|_{1\text{-var};[0,T]} \,,$$

a Cauchy sequence (x^n) with respect to $x \mapsto |x(0)| + |x|_{1\text{-var};[0,T]}$ is also Cauchy in uniform topology and thus (uniformly) convergent to some continuous path $x(\cdot)$. By Lemma 1.15 it is clear that x has finite 1-variation and it only remains to see that $x^n \to x$ in 1-variation norm. To this end, let $D = \{0 = t_0 < \cdots < t_K = t\}$ be an arbitrary dissection of $[0,T]$. For every $\varepsilon > 0$ there exists $N = N(\varepsilon)$ large enough so that for all $n, m \geq N(\varepsilon)$

$$\sup_D \sum_{i=0}^{K-1} d\left(x_{t_i,t_{i+1}}^n, x_{t_i,t_{i+1}}^m\right) < \varepsilon/2.$$

On the other hand, we can fix D and find m large enough so that

$$\sum_{i=0}^{K-1} d\left(x_{t_i,t_{i+1}}^m, x_{t_i,t_{i+1}}\right) < \varepsilon/2$$

which implies that for $n \geq N(\varepsilon)$ large enough

$$\sum_{i=0}^{K-1} d\left(x_{t_i,t_{i+1}}^n, x_{t_i,t_{i+1}}\right) \leq \varepsilon,$$

uniformly over all D. But this precisely says that $x^n \to x$ in 1-variation. Non-separability follows from the example below. ∎

Example 1.26 (non-separability) We give an example of an uncountable family of functions (f_α) in $C^{1\text{-var}}\left([0,1],\mathbb{R}\right)$ for which $|f_\alpha - f_{\alpha'}|_{1\text{-var}} \geq 1$ if $\alpha \neq \alpha'$. To this end, take $\alpha = (\alpha_n)_{n\geq 1}$ to be a $\{0,1\}$-sequence, and write $[0,1)$ as the union of the disjoint interval I_n, $n \geq 1$, where $I_n \equiv \left[1 - \frac{1}{2^{n-1}}, 1 - \frac{1}{2^n}\right)$. If $\alpha_n = 0$ then define f_α to be zero on I_n. Otherwise, define f_α on I_n by

$$f_\alpha(t_n + s) = \frac{1}{2n}\left|\sin\left(n\pi\frac{s}{2^{-n}}\right)\right|$$

so that, using Proposition 1.24, $|f_\alpha|_{1\text{-var};I_n} = 1$. By construction $f_\alpha(t_n) = 0$ for all n and hence f_α is continuous on $[0,1)$. (Left) continuity at 1 is also clear: thanks to the decay factor $1/n$ we see that $f_\alpha(t) \to 0$ as $t \nearrow 1$. $\quad\square$

A simple approximation of a path x on $\mathbb{R}^d$ is given by its piecewise linear approximation.[6]

Definition 1.27 Let $x : [0,T] \to \mathbb{R}^d$, and $D = (t_i)_i$ a dissection of $[0,T]$. We define the piecewise linear approximation to x by

$$x_t^D = x_{t_i} + \frac{t - t_i}{t_{i+1} - t_i} x_{t_i, t_{i+1}} \text{ if } t_i \le t \le t_{i+1}. \qquad\square$$

Proposition 1.28 *Let $x \in C^{1\text{-var}}\left([0,T], \mathbb{R}^d\right)$. Then, for any dissection D of $[0,T]$ and any $s < t$ in $[0,T]$,*

$$\left|x^D\right|_{1\text{-var};[s,t]} \le |x|_{1\text{-var};[s,t]} . \tag{1.4}$$

If (D_n) is an arbitrary sequence of dissections with mesh $|D_n| \to 0$, then x^{D_n} converges uniformly to x. (We can write this more concisely as $x^D \to x$ uniformly on $[0,T]$ as $|D| \to 0$.)

Proof. The estimate (1.4) boils down to the fact that the shortest way to connect two points in $\mathbb{R}^d$ is via a straight line. The convergence result requires the remark that $x(\cdot)$ is *uniformly* continuous on $[0,T]$. The easy details are left to the reader. $\quad\blacksquare$

The question arises if (or when) $x^D \to x$ in 1-variation as $|D| \to 0$. Since piecewise linear approximations are absolutely continuous, the following result tells us that there is no hope unless x is absolutely continuous. (We shall see later that $x^D \to x$ in 1-variation as $|D| \to 0$ indeed holds true provided x is absolutely continuous.)

Proposition 1.29 *The set of absolutely continuous functions from $[0,T] \to \mathbb{R}^d$ is closed in 1-variation and a Banach space under 1-variation norm.*

Proof. We prove that if x^n is absolutely continuous and converges to x in 1-variation norm, then x is absolutely continuous. Fix $\varepsilon > 0$, and $n \in \mathbb{N}$ such that

$$|x - x^n|_{1\text{-var}} + |x_0 - x_0^n| < \frac{\varepsilon}{2}.$$

Then, as x^n is absolutely continuous, there exists $\delta > 0$, such that for all $s_1 < t_1 \le s_2 < t_2 \le \cdots < s_n < t_n$ in $[0,T]$ with $\sum_i |t_i - s_i| < \delta$, we have $\sum_i \left|x_{s_i,t_i}^n\right| < \frac{\varepsilon}{2}$. This implies that

$$\sum_i |x_{s_i,t_i}| \le \sum_i \left|x_{s_i,t_i}^n\right| + \sup_{D=(t_i) \text{ of } [0,T]} \sum_i \left|x_{s_i,t_i} - x_{s_i,t_i}^n\right|$$

$$\le \sum_i \left|x_{s_i,t_i}^n\right| + |x - x^n|_{1\text{-var}} \le \varepsilon$$

and the proof is finished. $\quad\blacksquare$

---

[6]A powerful generalization of this will be discussed in Section 5.2.

Exercise 1.30 *By Proposition 1.29 it is clear that piecewise linear approximations cannot converge (in 1-variation) to the Cantor function* $f :$ $[0,1] \to [0,1]$ *given in Example 1.19. (By Proposition 1.29 any 1-variation limit point is absolutely continuous; but the Cantor function is not absolutely continuous as was seen in Exercise 1.19). Verify this by an explicit computation. More precisely, set* $D_n = \{j3^{-n}; j = 0,\ldots,3^n\}$ *and show that*

$$\left|f - f^{D_n}\right|_{1\text{-var};[0,1]} = |f - I|_{1\text{-var};[0,1]}$$

where $I(x) = x$ *and conclude that* $f^D \not\to f$ *in 1-variation as* $|D| \to 0$.

Solution. f is self-similar, in the sense that for all $n \geq 1, k \in \{0,\ldots,3^n\}$,

$$f\left(k3^{-n} + 3^n x\right) - f\left(k3^{-n}\right) = 2^n f(x).$$

Using self-similarity, we see that, if I denotes the identity function on $[0,1]$,

$$\left|f - f^{D_n}\right|_{1\text{-var};\left[\frac{j}{3^n},\frac{j+1}{3^n}\right]} = 2^n |f - I|_{1\text{-var};[0,1]}.$$

Hence,

$$\begin{aligned}\left|f - f^{D_n}\right|_{1\text{-var};[0,1]} &= \sum_{j=0}^{2^n-1}\left|f - f^{D_n}\right|_{1\text{-var};\left[\frac{j}{3^n},\frac{j+1}{3^n}\right]}\\&= |f - I|_{1\text{-var};[0,1]} > 0.\end{aligned}$$

$\square$

1.3.3 Closure of smooth paths in variation norm

Let us define $C^{0,1\text{-var}}\left([0,T],\mathbb{R}^d\right)$ as the closure of smooth paths from $[0,T] \to \mathbb{R}^d$ in 1-variation norm. Obviously, $C^{0,1\text{-var}}$ is a closed, linear subspace of $C^{1\text{-var}}\left([0,T],\mathbb{R}^d\right)$ and thus a Banach space. Restricting to paths with $x(0) = 0$ yields a further subspace (also Banach) denoted by $C_0^{0,1\text{-var}}\left([0,T],\mathbb{R}^d\right)$. By Proposition 1.29 any element of $C^{0,1\text{-var}}$ must be absolutely continuous (a.c.) and so

$$C^{0,1\text{-var}}\left([0,T],\mathbb{R}^d\right) \subset \left\{x : [0,T] \to \mathbb{R}^d \text{ a.c.}\right\} \subsetneq C^{1\text{-var}}\left([0,T],\mathbb{R}^d\right).$$

We shall show that the first inclusion is in fact an equality.

Proposition 1.31 *The map* $y \mapsto \int_0^{\cdot} y_t dt$ *is a Banach space isomorph from*

$$L^1\left([0,T],\mathbb{R}^d\right) \to C_0^{0,1\text{-var}}\left([0,T],\mathbb{R}^d\right).$$

As a consequence, $x \in C^{0,1\text{-var}}\left([0,T],\mathbb{R}^d\right)$ *if and only if there exists a (uniquely determined)* $\dot{x} \in L^1\left([0,T],\mathbb{R}^d\right)$, *see Remark 1.33, such that*

$$x \equiv x_0 + \int_0^\cdot \dot{x}_t dt$$

and in this case the Banach isometry $|x|_{1\text{-var}} = |\dot{x}|_{L^1}$ *holds.*

Proof. Without loss of generality we consider paths started at $x_0 = 0$. For any smooth $y \in C^\infty\left([0,T],\mathbb{R}^d\right)$ we have $x = \int_0^\cdot y_t dt \in C^\infty\left([0,T],\mathbb{R}^d\right)$ and so, by Proposition 1.24,

$$|x|_{1\text{-var}} = |y|_{L^1}.$$

Obviously, this allows us to extend the map

$$\iota : y \in C^\infty\left([0,T],\mathbb{R}^d\right) \mapsto x = \int_0^\cdot y_t dt \in C^\infty\left([0,T],\mathbb{R}^d\right)$$

to the respective closures. From the very definition of the space $C_0^{0,1\text{-var}}$ and by density of smooth paths in L^1 it follows that i extends to (a Banach space isomorphism)

$$\hat{\iota} : L^1\left([0,T],\mathbb{R}^d\right) \to C_0^{0,1\text{-var}}\left([0,T],\mathbb{R}^d\right).$$

To see that $\hat{\iota}$ still has the simple representation as an indefinite integral, let $y \in L^1\left([0,T],\mathbb{R}^d\right)$, take smooth approximations y^n in L^1 and pass to the limit in

$$\iota\left(y^n\right)_t = \int_0^t y_s^n ds = \int_0^T y_s^n 1_{[0,t]}\left(s\right) ds,$$

using the simple fact that $y^n \to y$ in L^1 implies $y^n 1_{[0,t]} \to y 1_{[0,t]}$ in L^1 for every fixed $t \in [0,T]$. At last, given $x \in \hat{\iota}(L^1\left([0,T],\mathbb{R}^d\right)$ we write $\dot{x}$ rather than y for the uniquely determined $\hat{\iota}^{-1}\left(x\right) \in L^1\left([0,T],\mathbb{R}^d\right)$. ∎

The next proposition requires some background in basic measure theory (Lebesgue–Stieltjes measures, Radon–Nikodym theorem, ...).[7]

Proposition 1.32 *Let* $x : [0,T] \to \mathbb{R}^d$ *be absolutely continuous. Then it can be written in the form* $x_0 + \int_0^\cdot \dot{x}_t dt$ *with* $\dot{x} \in L^1\left([0,T],\mathbb{R}^d\right)$. *As a consequence,*

$$C^{0,1\text{-var}}\left([0,T],\mathbb{R}^d\right) = \left\{x : [0,T] \to \mathbb{R}^d \text{ absolutely continuous}\right\}.$$

Proof. It suffices to consider $d = 1$. The function x determines a signed Borel measure on $\mathbb{R}$ via

$$\mu\left((-\infty,t]\right) = x_{0,t} \equiv x_t - x_0 \text{ for } t \in [0,T]$$

---

[7]See Folland and Stein's book [55] for instance.

and putting zero mass on $\mathbb{R}\backslash[0,T]$. The assumption of absolute continuity of x implies that μ is absolutely continuous (in the sense of measures) with respect to Lebesgue measure λ. By the Radon–Nikodym theorem, there exists an integrable density function $y = d\mu/d\lambda$, an integrable function from $[0,T]$ to $\mathbb{R}$, uniquely defined up to Lebesgue null sets, such that

$$x_t = \mu\left((0,t]\right) = \int_0^t y_s\,ds.$$

Hence, using Proposition 1.31, $x \in C^{0,1\text{-var}}\left([0,T],\mathbb{R}^d\right)$. The converse inclusion follows directly from Proposition 1.29. ∎

Remark 1.33 Our notation for $\dot{x}$ for the unique L^1-function with the property

$$x_t = x_0 + \int_0^t \dot{x}_t\,dt$$

for absolutely continuous x is consistent with the **fundamental theorem of calculus for Lebesgue integrals** (e.g. [55], p. 106). It states that a real-valued function x on $[0,T]$ is absolutely continuous if and only if its derivative

$$\lim_{h\to 0}\frac{x_{t+h} - x_t}{h}$$

exists for almost every $t \in [0,T]$ and gives an L^1-function whose indefinite integral is $x_t - x_0$. We have not shown (and will not use) the fact that $\dot{x}$ is the almost-sure limit of the above difference quotient. □

Corollary 1.34 *Let* $x \in C^{1\text{-var}}\left([0,T],\mathbb{R}^d\right)$. *Then piecewise linear approximations converge in 1-variation,*

$$\left|x - x^D\right|_{1\text{-var};[0,T]} \to 0 \ \text{as} \ |D| \to 0$$

if and only if $x \in C^{0,1\text{-var}}\left([0,T],\mathbb{R}^d\right)$.

Proof. "$\implies$" : Any 1-variation limit of piecewise linear approximation is absolutely continuous and hence in $C^{0,1\text{-var}}$.

"$\impliedby$" : Fix $\varepsilon > 0$, and $x \in C^{0,1\text{-var}}\left([0,T],\mathbb{R}^d\right)$. From the very definition of this space there exists a smooth path y such that

$$|x - y|_{1\text{-var};[0,T]} \le \frac{\varepsilon}{3}.$$

We claim that for all dissections D with small enough mesh (depending on y and ε),

$$|y - y^D|_{1\text{-var};[0,T]} < \frac{\varepsilon}{3}.$$

Indeed, this follows from Proposition 1.24 and the computation

$$
\begin{aligned}
\left| \dot{y} - \dot{y}^D \right|_{L^1[0,1]}
&= \sum_i \int_{t_i}^{t_{i+1}} \left| \dot{y}_s - \frac{y_{t_i,t_{i+1}}}{t_{i+1} - t_i} \right| ds \text{ for } D = \{t_i\} \subset [0,T] \\
&= \sum_i \int_{t_i}^{t_{i+1}} \left| \dot{y}(s) - \dot{y}(\xi_i) \right| ds \text{ with } \xi_i \in (t_i, t_{i+1}) \\
&\leq \left| \ddot{y} \right|_\infty \sum_i \int_{t_i}^{t_{i+1}} |s - \xi_i|\, ds \leq \left| \ddot{y} \right|_\infty |D|\, T.
\end{aligned}
$$

By the triangle inequality and the contraction property of $(\cdot)^D$ as linear map from $C^{1\text{-var}}\left([0,T],\mathbb{R}^d\right)$ into itself, see (1.4), we have

$$
\begin{aligned}
\left| x - x^D \right|_{1\text{-var};[0,T]}
&\leq |x - y|_{1\text{-var};[0,T]} + \left| y - y^D \right|_{1\text{-var};[0,T]} \\
&\quad + \left| x^D - y^D \right|_{1\text{-var};[0,T]} \\
&\leq 2\,|x - y|_{1\text{-var};[0,T]} + \left| y - y^D \right|_{1\text{-var};[0,T]} \\
&\leq \varepsilon
\end{aligned}
$$

and this finishes the proof.   ∎

Corollary 1.35 *The space $C^{0,1\text{-var}}\left([0,T],\mathbb{R}^d\right)$ is a separable Banach space (and hence Polish).*

Proof. Let D_n be the dyadic dissection $\{Tk/2^n : i = k, \ldots, 2^n\}$ and define Ω_n to be the set of paths from $[0,T]$ to $\mathbb{R}^d$, linear on the dyadic intervals of D_n with values at dyadic times in $\mathbb{Q}^d$. Then, $\Omega := \bigcup_n \Omega_n$ is a countable set. If $x \in C^{0,1\text{-var}}\left([0,T],\mathbb{R}^d\right)$ and $\varepsilon > 0$, there exists n such that $\left| x - x^{D_n} \right|_{1\text{-var}} < \varepsilon/2$. It is then easy to find $y \in \Omega_n$ such that $\left| x^{D_n} - y \right|_{1\text{-var}} < \varepsilon/2$, which proves that Ω is dense in $C^{0,1\text{-var}}\left([0,T],\mathbb{R}^d\right)$. This shows that $C^{0,1\text{-var}}\left([0,T],\mathbb{R}^d\right)$ is separable.   ∎

1.3.4 Lipschitz continuity

We now turn to $C^{1\text{-Höl}}\left([0,T],\mathbb{R}^d\right)$, the set of Lipschitz or 1-Hölder paths. It includes, for instance, $C^1\left([0,T],\mathbb{R}^d\right)$ and elementary examples (e.g. $t \mapsto |t|$) show that this inclusion is strict.

Proposition 1.36 *$C^{1\text{-Höl}}\left([0,T],\mathbb{R}^d\right)$ is Banach with norm $x \mapsto |x(0)| + |x|_{1\text{-Höl};[0,T]}$. The closed subspace of paths in $C^{1\text{-Höl}}\left([0,T],\mathbb{R}^d\right)$ started at 0, is also Banach under $x \mapsto |x|_{1\text{-Höl};[0,T]}$. These Banach spaces are not separable.*

Proof. Non-separability follows from Example 1.26 together with $|x|_{1\text{-var};[0,T]} \leq |x|_{1\text{-Höl};[0,T]}$ or using the (well-known) non-separability of

$L^\infty\left([0,T],\mathbb{R}^d\right)$ in conjunction with Proposition 1.37 below. All other parts of the proof are straightforward and left to the reader. ∎

Proposition 1.37 *The map* $y \mapsto \int_0^\cdot y_t dt$ *is a Banach space isomorph from*

$$L^\infty\left([0,T],\mathbb{R}^d\right) \to C_0^{1\text{-Höl}}\left([0,T],\mathbb{R}^d\right).$$

As a consequence, $x \in C^{1\text{-Höl}}\left([0,T],\mathbb{R}^d\right)$ *if and only if there exists a (uniquely determined)* $\dot{x} \in L^\infty\left([0,T],\mathbb{R}^d\right)$ *such that*

$$x \equiv x_0 + \int_0^\cdot \dot{x}_t dt$$

and in this case the Banach isometry $|x|_{1\text{-Höl}} = |\dot{x}|_{L^\infty}$ *holds.*

Proof. Similar to Proposition 1.31 and left to the reader. ∎

From general principles, any continuous path of finite 1-variation can be reparametrized to a 1-Hölder path. In the present context of $\mathbb{R}^d$-valued paths this can be done so that the reparametrized path has constant speed. We have

Proposition 1.38 *Let* $x \in C^{1\text{-var}}\left([0,T],\mathbb{R}^d\right)$, *not constant. Define* $y\left(\cdot\right)$ *by* $y \circ \phi = x$ *where*

$$\phi(t) = |x|_{1\text{-var};[0,t]} \big/ |x|_{1\text{-var};[0,T]}.$$

Then $y \in C^{1\text{-Höl}}\left([0,1],\mathbb{R}^d\right)$ *has constant speed. More precisely,* y *is the indefinite integral of some* $\dot{y} \in L^\infty\left([0,1],\mathbb{R}^d\right)$ *and*

$$|\dot{y}\left(t\right)| \equiv |x|_{1\text{-var};[0,T]} = |y|_{1\text{-Höl};[0,1]}$$

for a.e. $t \in [0,1]$.

Proof. By the precise argument of the proof of Proposition 1.21, y is well-defined and in $C^{1\text{-Höl}}\left([0,1],\mathbb{R}^d\right)$. From the very definition of y and invariance of 1-variation under reparametrization we have

$$|y|_{1\text{-var};[0,\phi(t)]} = |x|_{1\text{-var};[0,t]} = c\phi(t)$$

where $c = |x|_{1\text{-var};[0,T]}$. On the other hand, by Propositions 1.37 and 1.31, y is the indefinite integral of some $\dot{y} \in L^\infty\left([0,1],\mathbb{R}^d\right)$ and

$$|y|_{1\text{-var};[0,\phi(t)]} = \int_0^{\phi(t)} |\dot{y}\left(s\right)| \, ds.$$

It follows that $|\dot{y}| \equiv c$ almost surely. At last, the equality $c = |y|_{1\text{-Höl};[0,1]}$ was noted in Remark 1.22. ∎

Remark 1.39 More generally, the proof shows that x can be reparame-trized to $y \in C^{\text{1-Höl}}\left([0,c],\mathbb{R}^d\right)$ with unit speed, i.e. $|\dot{y}| \equiv 1$ almost surely.   □

The reader will notice that the continuous embedding

$$C^{\text{1-Höl}}\left([0,T],\mathbb{R}^d\right) \hookrightarrow C^{\text{1-var}}\left([0,T],\mathbb{R}^d\right)$$

is a consequence of the trivial estimate

$$\left|x\right|_{\text{1-var};[0,T]} \leq \left|x\right|_{\text{1-Höl};[0,T]} T.$$

As in the previous section it is natural to consider $C^{0,\text{1-Höl}}\left([0,T],\mathbb{R}^d\right)$, defined as the closure of smooth paths in $C^{\text{1-Höl}}\left([0,T],\mathbb{R}^d\right)$. The resulting closure is a space we have already encountered.

Proposition 1.40 *The closure of smooth paths in $C^{\text{1-Höl}}\left([0,T],\mathbb{R}^d\right)$ equals $C^1\left([0,T],\mathbb{R}^d\right)$.*

Proof. Let us first observe that the norm $x \mapsto |x_0| + \sup_{t\in[0,T]}|\dot{x}_t|$ on $C^1\left([0,T],\mathbb{R}^d\right)$ makes $C^1\left([0,T],\mathbb{R}^d\right)$ a Banach space. To avoid trivialities (norms vs semi-norms), let us assume that all paths are null at 0. Using

$$C^1\left([0,T],\mathbb{R}^d\right) \cong \oplus_{i=1}^d C^1\left([0,T],\mathbb{R}\right)$$

and similar for $C^{\text{1-Höl}}$ it suffices to consider $d = 1$. Given a smooth path $x : [0,T] \to \mathbb{R}$ with $x(0) = 0$ we first show that

$$|x|_{\text{1-Höl}} \equiv \sup_{s,t\in[0,T]} \frac{|x(t) - x(s)|}{|t-s|} \quad \text{equals} \quad \sup_{t\in[0,T]} |\dot{x}_t|.$$

Indeed, from $|x(t+h) - x(h)| \leq |x|_{\text{1-Höl}} h$ we see that $|\dot{x}_t| \leq |x|_{\text{1-Höl}}$ for all $t \in [0,T]$ while the converse estimate follows from the intermediate value theorem,

$$\frac{|x(t) - x(s)|}{t-s} = |\dot{x}(\xi)| \quad \text{for } \xi \in (s,t)$$

$$\leq |\dot{x}|_{\infty;[0,T]}.$$

Any sequence (x^n) of smooth paths which converges (in 1-Hölder norm) to some path x is also Cauchy in 1-Hölder. By the previous argument, it is also Cauchy in $C^1\left([0,T],\mathbb{R}\right)$ and so converges to some $\tilde{x} \in C^1\left([0,T],\mathbb{R}\right)$. Since both 1-Hölder and C^1-norm imply pointwise convergence we must have $x = \tilde{x} \in C^1\left([0,T],\mathbb{R}\right)$ and the proof is finished.   ■

1.4 Sobolev spaces of continuous paths of bounded variation

1.4.1 Paths of Sobolev regularity on $\mathbb{R}^d$

We saw in Proposition 1.31 that a path x is in $C^{0,1\text{-var}}\left([0,T],\mathbb{R}^d\right)$ if and only if

$$x_0 + \int_0^{\cdot} \dot{x}_t dt$$

with $\dot{x} \in L^1\left([0,T],\mathbb{R}^d\right)$ and in this case $|x|_{1\text{-var}} = |\dot{x}|_{L^1}$. We then saw, Proposition 1.37, that a path x is Lipschitz, in symbols $x \in C^{1\text{-Höl}}\left([0,T],\mathbb{R}^d\right)$, if and only if

$$x_0 + \int_0^{\cdot} \dot{x}_t dt$$

with $\dot{x} \in L^{\infty}\left([0,T],\mathbb{R}^d\right)$ and in this case $|x|_{1\text{-Höl}} = |\dot{x}|_{L^\infty}$. This suggests considering the following path spaces.

Definition 1.41 For $p \in [1,\infty]$, we define $W^{1,p}\left([0,T],\mathbb{R}^d\right)$ to be the space of $\mathbb{R}^d$-valued functions on $[0,T]$ of the form

$$x\left(\cdot\right) = x_0 + \int_0^{\cdot} y dt \qquad (1.5)$$

with $y \in L^p\left([0,T],\mathbb{R}^d\right)$. Writing $\dot{x}$ instead of y we further define

$$|x|_{W^{1,p};[0,T]} := |\dot{x}|_{L^p;[0,T]} = \left(\int_0^T |\dot{x}|^p du\right)^{1/p}.$$

The set of such paths with $x_0 = o \in \mathbb{R}^d$ is denoted by $W_o^{1,p}\left([0,T],\mathbb{R}^d\right)$. As always, $[0,T]$ may be replaced by any other interval $[s,t] \subset \mathbb{R}$. □

It is clear from the definition that $W^{1,1} = C^{0,1\text{-var}}$ and hence (Proposition 1.32) precisely the set of absolutely continuous paths, while $W^{1,\infty}$ is precisely the set of Lipschitz or 1-Hölder paths. It is also clear from the usual inclusions of L^p-spaces that $W^{1,\infty} \subset W^{1,p} \subset W^{1,1}$. In particular, any path in $W^{1,p}$ is absolutely continuous (and then of course of bounded variation).

Proposition 1.42 *The space* $W^{1,p}\left([0,T],\mathbb{R}^d\right)$ *is a Banach space under the norm*

$$x \mapsto |x_0| + |x|_{W^{1,p};[0,T]}.$$

The closed subspace of paths in $W^{1,p}\left([0,T],\mathbb{R}^d\right)$ *started at* 0, *is also Banach under* $x \mapsto |x|_{W^{1,p};[0,T]}$. *These Banach spaces are separable if and only if* $p \in [1,\infty)$.

Proof. Since $L^p \subset L^1$, we can use Proposition 1.31 to see that the map $x \mapsto \dot{x}$ is well-defined, as is its norm $x \mapsto |x_0| + |\dot{x}|_{L^p[0,T]}$. The closed subspace of paths in $W^{1,p}\left([0,T],\mathbb{R}^d\right)$ started at 0 is isomorphic (as normed space) to $L^p\left([0,T],\mathbb{R}^d\right)$ and hence Banach. The separability statement now follows from well-known facts about L^p-spaces. ∎

Exercise 1.43 *Let $p \in [1,\infty]$ and recall that we equipped $W^{1,p}\left([0,T],\mathbb{R}^d\right)$ with Banach norm*

$$|x\left(0\right)| + |\dot{x}|_{L^p;[0,T]}.$$

Show that an equivalent norm is given by $|x|_{L^q;[0,T]} + |\dot{x}|_{L^p;[0,T]}$, for all $q \in [1,\infty]$.

Solution. L^p-control of $\dot{x}$ gives a modulus for x and in particular $|x_{0,t}| \le |x|_{W^{1,p};[0,T]} \, t^{1-1/p}$ where $1/p = 0$ for $p = \infty$. Using $|x_t| \le |x_0| + |x_{0,t}|$ one controls the supremum of x over $t \in [0,T]$ and then any L^q-norm. □

Every path in $W^{1,p} \subset W^{1,1}$ is continuous and of finite 1-variation. For $p = \infty$, such paths are Lipschitz or 1-Hölder continuous; more precisely

$$|x_{s,t}| \le |x|_{W^{1,\infty};[s,t]} \, |t-s|.$$

Observe that the right-hand side is a control so that $|x_{s,t}|$ in the above estimate can be replaced by $|x|_{\text{1-var};[s,t]}$. In the following theorem we see that a similar statement holds true for all $p > 1$.

Theorem 1.44 *Let $p \in (1,\infty)$. Given $x \in W^{1,p}\left([0,T],\mathbb{R}^d\right)$,*

$$\omega\left(s,t\right) = |x|_{W^{1,p};[s,t]}\left(t-s\right)^{1-1/p}$$

defines a control function on $[0,T]$ and we have $|x|_{\text{1-var};[s,t]} \le \omega\left(s,t\right)$ for all $s < t$ in $[0,T]$. In particular, we have the continuous embedding $W^{1,p}\left([0,T],\mathbb{R}^d\right) \hookrightarrow C^{\text{1-var}}\left([0,T],\mathbb{R}^d\right)$.

Proof. Without loss of generality $x_0 = 0$. By Proposition 1.31, x is the indefinite integral of some $\dot{x} \in L^1$. Define $\alpha = 1 - 1/p$. Using Hölder's inequality with conjugate exponents p and $1/\alpha$

$$\begin{aligned}
|x_{s,t}| &\le \int_s^t |\dot{x}_r| \, dr \le (t-s)^\alpha \left(\int_s^t |\dot{x}_r|^p \, dr\right)^{1/p} \\
&= |x|_{W^{1,p};[s,t]}\left(t-s\right)^\alpha \\
&= \omega\left(s,t\right).
\end{aligned}$$

We show that ω is a control. Continuity of ω is obvious from the fact that $|x|_{W^{1,p};[s,t]}^p$ is the integral of an integrable function, namely $|\dot{x}|^p$, over $[s,t]$. Only super-additivity, $\omega\left(s,t\right) + \omega\left(t,u\right) \le \omega\left(s,u\right)$ with $s \le t \le u$, remains to be shown. From Hölder's inequality with conjugate exponents p and $p/(p-1) = 1/\alpha$ we obtain

$$|x|_{W^{1,p};[s,t]} (t-s)^\alpha + |x|_{W^{1,p};[t,u]} (u-t)^\alpha$$

$$\leq \left(|x|^p_{W^{1,p};[s,t]} + |x|^p_{W^{1,p};[t,u]} \right)^{1/p} \left[(t-s)^{\alpha \frac{p}{p-1}} + (u-t)^{\alpha \frac{p}{p-1}} \right]^{(p-1)/p}$$

$$= |x|_{W^{1,p};[s,u]} (u-t)^\alpha .$$

By Proposition 1.11, we conclude that $|x|_{1\text{-var};[s,t]} \leq \omega(s,t)$. In particular,

$$|x|_{1\text{-var};[0,T]} \leq \omega(0,T) = |x|_{W^{1,p};[0,T]} T^{1-1/p}$$

which gives the continuous embedding. ∎

Proposition 1.45 *Let* $p \in (1,\infty)$. *A function* $x : [0,T] \to \mathbb{R}^d$ *is in* $W^{1,p}([0,T],\mathbb{R}^d)$ *if and only if* $M_p(x) < \infty$ *where*

$$M_p(x) \quad := \quad \sup_{(t_i) \in \mathcal{D}([0,T])} \sum_i \frac{|x_{t_i,t_{i+1}}|^p}{|t_{i+1} - t_i|^{p-1}}$$

$$= \quad \lim_{\delta \to 0} \sup_{(t_i) \in \mathcal{D}_\delta([0,T])} \sum_i \frac{|x_{t_i,t_{i+1}}|^p}{|t_{i+1} - t_i|^{p-1}}$$

and in this case

$$|x|^p_{W^{1,p};[0,T]} = M_p(x) .$$

Proof. Without loss of generality $x_0 = 0$ and we assume $x \in W^{1,p}$ is the indefinite integral of some $\dot{x} \in L^p$. Then, Hölder's inequality gives

$$|x_{t_i,t_{i+1}}| \leq |t_{i+1} - t_i|^{1/p'} \left(\int_{t_i}^{t_{i+1}} |\dot{x}_u|^p \, du \right)^{1/p}$$

where $1/p' + 1/p = 1$. It immediately follows that

$$M_p(x) \leq \int_0^T |\dot{x}|^p \, du = |x|^p_{W^{1,p};[0,T]} . \qquad (1.6)$$

Conversely, suppose that $M_p(x) < \infty$; given $s_1 < t_1 \leq s_2 < t_2 \leq \cdots < s_n < t_n$ in $[0,T]$, Hölder's inequality yields

$$\sum_{i=1}^n |x_{t_i} - x_{s_i}| = \sum_{i=1}^n \frac{|x_{t_i} - x_{s_i}|}{|t_{i+1} - t_i|^{1/p'}} |t_{i+1} - t_i|^{1/p'}$$

$$\leq (M_p(x))^{1/p} \left(\sum_i |t_{i+1} - t_i| \right)^{1/p'}$$

which shows that x is absolutely continuous, hence precisely in $C_0^{0,1\text{-var}}$, and (Proposition 1.32) the indefinite integral of some $\dot{x} \in L^1[0,T]$. We show that $\dot{x} \in L^p[0,T]$, with $\int_0^T |\dot{x}_u|^p\, du$ bounded by $M_p(x)$. Let $D_n = \left\{ \frac{i}{n}T : i = 0, \ldots, n \right\}$. By Corollary 1.34, $x^{D_n} \to x$ in 1-variation norm, and therefore we have the convergence

$$\dot{x}^{D_n} = \frac{n}{T} \sum_{i=1}^{n} x_{\frac{(i-1)T}{n}, \frac{iT}{n}} 1_{\left[\frac{(i-1)T}{n}, \frac{iT}{n} \right)} \;\to\; \dot{x} \in L^1[0,T].$$

By passing to a subsequence $\left(\tilde{D}_k \right) = (D_{n_k})$ we can achieve that $\dot{x}_t^{\tilde{D}_k} \to_{k\to\infty} \dot{x}_t$ for almost every $t \in [0,T]$ with respect to Lebesgue measure. By Fatou's lemma we then see that

$$
\begin{aligned}
\int_0^T |\dot{x}_u|^p\, du &\leq \liminf_{k\to\infty} \int_0^T \left| \dot{x}_t^{\tilde{D}_k} \right|^p du \\
&= \liminf_{k\to\infty} \sum_{i:t_i \in \tilde{D}_k} \frac{\left| x_{t_i, t_{i+1}} \right|^p}{|t_{i+1} - t_i|^{p-1}} \\
&\leq \lim_{\delta \to 0} \sup_{|D| \leq \delta} \sum_{i:t_i \in D} \frac{\left| x_{t_i, t_{i+1}} \right|^p}{|t_{i+1} - t_i|^{p-1}} =: \tilde{M}_p(x).
\end{aligned}
$$

Recalling (1.6) we get $M_p(x) \leq |x|_{W^{1,p};[0,T]}^p \leq \tilde{M}_p(x)$ and with the trivial $\tilde{M}_p(x) \leq M_p(x)$ we must have equality throughout. This finishes the proof. ∎

1.4.2 Paths of Sobolev regularity on metric spaces

We already remarked that $W^{1,1}$ (resp. $W^{1,\infty}$) coincides with the set of absolutely continuous (resp. 1-Hölder) paths and this kind of regularity only requires paths with values in an abstract metric space (E, d). Proposition 1.45 suggests how to define $W^{1,p}$-regularity in a metric setting. Although we shall only need $p = 2$ in later chapters (in particular, in our discussions of large deviations), the case $p \in (1, \infty)$ is covered without extra effort and has applications in large deviation-type results for diffusions on fractals (see comments below).

Definition 1.46 For $p \in (1, \infty)$ we define $W^{1,p}([0,T], E)$ as those paths $x : [0,T] \to (E, d)$ for which

$$|x|_{W^{1,p};[0,T]} := \left(\sup_{(t_i) \in \mathcal{D}([0,T])} \sum_i \frac{\left| d\left(x_{t_i}, x_{t_{i+1}} \right) \right|^p}{|t_{i+1} - t_i|^{p-1}} \right)^{1/p} < \infty.$$

The subset of paths started at $o \in E$ is denoted by $W_o^{1,p}([0,T], E)$. As always, $[0,T]$ may be replaced by any other interval $[s,t]$. □

We now give a generalization of Theorem 1.44.

Theorem 1.47 *For any* $x \in W^{1,p}\left([0,T],E\right)$ *we have for all* $s,t \in [0,T]$,

$$d\left(x_s, x_t\right) \le |x|_{1\text{-var};[s,t]} \le |x|_{W^{1,p};[s,t]} \left(t-s\right)^{1-1/p}. \qquad (1.7)$$

In particular, $W^{1,p}\left([0,T],E\right) \subset C^{1\text{-var}}\left([0,T],E\right)$.

Proof. From the very definition of $|x|_{W^{1,p};[0,T]}$ we have

$$d\left(x_s, x_t\right)^p \le |x|^p_{W^{1,p};[s,t]} |t-s|^{p-1}$$

and the estimate on $d\left(x_s, x_t\right)$ follows. We then show, exactly as in the proof of Theorem 1.44, that the map

$$(s,t) \mapsto |x|_{W^{1,p};[s,t]} \left(t-s\right)^{1-1/p} \qquad (1.8)$$

is super-additive and the estimate on $|x|_{1\text{-var};[s,t]}$ follows by Proposition 1.11. ∎

Remark 1.48 To see that (1.8) is actually a control function, one would have to undergo a similar continuity consideration as in Proposition 1.12. □

As in the case of 1-variation (cf. Proposition 1.14) it is enough in the definition of $|x|_{W^{1,p};[0,T]}$ to look at dissections with small mesh.

Proposition 1.49 *For every* $x \in C\left([0,T],E\right)$,

$$|x|^p_{W^{1,p};[0,T]} = \lim_{\delta \to 0} \sup_{(t_i) \in \mathcal{D}_\delta([0,T])} \sum_{i:t_i \in D} \frac{\left|d\left(x_{t_i}, x_{t_{i+1}}\right)\right|^p}{|t_{i+1} - t_i|^{p-1}} \in [0,\infty].$$

Proof. We assume $|x|_{W^{1,p};[0,T]} < \infty$, leaving the case $|x|_{W^{1,p};[0,T]} = \infty$ to the reader. It suffices to show that, for any $s < t < u$ in $[0,T]$,

$$\frac{\left|d\left(x_s, x_u\right)\right|^p}{|u-s|^{p-1}} \le \frac{\left|d\left(x_s, x_t\right)\right|^p}{|t-s|^{p-1}} + \frac{\left|d\left(x_t, x_u\right)\right|^p}{|u-t|^{p-1}} \qquad (1.9)$$

as this will allow us to replace a given dissection D with a refinement $\tilde{D}$ with $\left|\tilde{D}\right| < \delta$. (We used a similar argument in the proof of Proposition 1.14.) To this end, recall the elementary inequality $(\theta a + (1-\theta) b)^p \le \theta a^p + (1-\theta) b^p$ for $a,b > 0$ and $\theta \in (0,1)$. Replacing θa by a and $(1-\theta)b$ by b gives

$$(a+b)^p \le \frac{a^p}{\theta^{p-1}} + \frac{b^p}{(1-\theta)^{p-1}}$$

and this implies (1.9) with $\theta = (t-s)/(u-s)$ and

$$d\left(x_s, x_u\right) \le d\left(x_s, x_t\right) + d\left(x_t, x_u\right) \equiv a+b.$$

∎

Exercise 1.50 *As usual, let $C\left([0,T],E\right)$ be equipped with the uniform topology. Let $p \in (1,\infty)$.*
(i) Show that

$$x \in C\left([0,T],E\right) \mapsto M_p\left(x\right) := |x|^p_{W^{1,p};[0,T]} \in [0,\infty]$$

is lower semi-continuous.
(ii) Assume that E has the Heine–Borel property, i.e. bounded sets have compact closure. Show that the level sets

$$\{x \in C_o\left([0,T],E\right) : M_p\left(x\right) \le \Lambda\} \ \text{with} \ \Lambda \in [0,\infty) \ \text{and} \ o \in E$$

are compact. (Hint: Arzela–Ascoli.)

Solution. (i) Assume $x^n \to x$ uniformly (or even pointwise) on $[0,T]$ and fix a dissection $D \subset [0,T]$. Then

$$\sum_{i:t_i \in D} \frac{\left|d\left(x_{t_i},x_{t_{i+1}}\right)\right|^p}{|t_{i+1}-t_i|^{p-1}} = \lim_{n\to\infty} \inf \sum_{i:t_i \in D} \frac{\left|d\left(x^n_{t_i},x^n_{t_{i+1}}\right)\right|^p}{|t_{i+1}-t_i|^{p-1}} \le \lim_{n\to\infty} \inf M_p\left(x^n\right)$$

and taking the sup over all dissections finishes the proof.
(ii) By (i) it is clear that level sets are closed. Thanks to Theorem 1.47 we know that $M_p\left(x\right) \le \Lambda$ implies

$$d\left(x_s,x_t\right) \le \Lambda^{1/p}\left(t-s\right)^{1-1/p}$$

with equicontinuity and boundedness of $\{x \in C_o\left([0,T],E\right) : M_p\left(x\right) \le \Lambda\}$. Conclude with Arzela–Ascoli.   □

1.5   Comments

Continuous paths of finite variation, also known as *rectifiable paths*, arise in many areas of analysis and geometry. Ultimately the focus of this book is on non-rectifiable paths and so we avoid the notion of rectifiability altogether. Topics such as absolute continuity of real-valued functions on $\mathbb{R}$, the fundamental theorem of calculus for Lebesgue integrals or the Radon–Nikodym theorem are found in many textbooks on real analysis such as Rudin [149], Folland [55] or Driver [45].

The interplay between variation, Hölder and $W^{1,p}$-spaces was studied in Musielak and Semadeni [132]; in particular, the authors of [132] attribute Proposition 1.45 to Riesz. A nice martingale proof of this can be found in Revuz and Yor [143]. The extension of $W^{1,p}$-regularity to paths in metric spaces is not for the sake of generality but arises, for instance, in the context of sample path large deviation(-type) estimates for symmetric diffusions; see Bass and Kumagai [6] and more specifically our later discussion of large deviation for Markov processes lifted to rough paths, Section 16.7.

2

Riemann–Stieltjes integration

In this chapter we give a brief exposition of the Riemann–Stieltjes integral and its basic properties.

2.1 Basic Riemann–Stieltjes integration

We will use the notation $L\left(\mathbb{R}^d, \mathbb{R}^e\right)$ for the space of linear maps from $\mathbb{R}^d$ into $\mathbb{R}^e$. We will always equip this space with its operator norm, that is if $f \in L\left(\mathbb{R}^d, \mathbb{R}^e\right)$, then

$$|f| = \sup_{\substack{x \in \mathbb{R}^d \\ |x|_{\mathbb{R}^d} = 1}} |fx|_{\mathbb{R}^e}.$$

Definition 2.1 Let x and y be two functions from $[0, T]$ into $\mathbb{R}^d$ and $L\left(\mathbb{R}^d, \mathbb{R}^e\right)$. Let $D_n = (t_i^n : i)$ be a sequence of dissections of $[0, T]$ with $|D_n| \to 0$, and ξ_i^n some points in $\left[t_i^n, t_{i+1}^n\right]$. Assume $\sum_{i=0}^{\#D_n - 1} y\left(\xi_i^n\right) x_{t_i^n, t_{i+1}^n}$ converges when n tends to ∞ to a limit I independent of the choice of ξ_i^n and the sequence (D_n). Then we say that the Riemann–Stieltjes integral of y against x (on $[0, T]$) exists and write

$$\int_0^T y dx := \int_0^T y_u dx_u := I.$$

We call y the integrand and x the integrator. Of course, $[0, T]$ may be replaced by any other interval $[s, t]$. $\square$

Proposition 2.2 Let $x \in C^{1\text{-var}}\left([0, T], \mathbb{R}^d\right)$ and $y : [0, T] \to L\left(\mathbb{R}^d, \mathbb{R}^e\right)$ piecewise continuous.[1] Then the Riemann–Stieltjes integral $\int_0^T y dx$ exists, is linear in y and x, and we have the estimate

$$\left| \int_0^T y dx \right| \leq |y|_{\infty;[0,T]} |x|_{1\text{-var};[0,T]}.$$

Moreover,[2]

$$\int_0^t y_u dx_u - \int_0^s y_u dx_u = \int_s^t y_u dx_u \text{ for all } 0 \leq s < t \leq T. \qquad (2.1)$$

---

[1] This will cover all our applications.

[2] All integrals in (2.1) are understood in the sense of Definition 2.1 with $[0, T]$ replaced by the intervals $[0, t], [0, s], [s, t]$ respectively.

Proof. Let us say that a real-valued function y is dx-integrable (on the fixed time interval $[0, T]$) if the Riemann–Stieltjes integral $\int_0^T y\, dx$ exists.

Step 1: Step-functions, i.e. functions of the form

$$g(t) = a_0 1_{[0,t_1]} + \sum_{i=1}^{n-1} a_i 1_{(t_i, t_{i+1}]} (t),$$

with $0 < t_1 < \cdots < t_n = T$ and $a_i \in L\left(\mathbb{R}^d, \mathbb{R}^e\right)$, are dx-integrable and

$$\int_0^T g\, dx = \sum_{i=0}^{n-1} a_i x_{t_i, t_{i+1}}.$$

Step 2: The set of dx-integrable functions is a linear space, i.e. if g and h are dx-integrable, then so is $\alpha g + \beta h$, with $\alpha, \beta \in \mathbb{R}$ which readily implies that

$$\int_0^T (\alpha g + \beta h)\, dx = \alpha \int_0^T g\, dx + \beta \int_0^T h\, dx.$$

Step 3: If y is dx-integrable then

$$\left| \int_0^T y\, dx \right| \le |y|_{\infty;[0,T]} |x|_{1\text{-var};[0,T]}.$$

Step 4: The space of dx-integrable functions is closed in supremum topology on $[0, T]$. Indeed, assume (y^n) is a sequence of dx-integrable functions such that

$$|y - y^n|_{\infty;[0,T]} \to 0 \text{ as } n \to \infty.$$

By steps 2 and 3,

$$\left| \int_0^T y^n\, dx - \int_0^T y^m\, dx \right| \le |y_n - y_m|_{\infty;[0,T]} |x|_{1\text{-var};[0,T]}$$

and so $I_n = \int_0^T y^n\, dx$ defines a Cauchy sequence whose limit we denote by I. Let $D_n = (t_i^n)_i$ be a sequence of dissections of $[0, T]$ with mesh $|D_n| \to 0$ and ξ_i^n an arbitrary point in $\left[t_i^n, t_{i+1}^n\right]$ for all i, n. Then,

$$\left| I - \sum_{i=0}^{\#D_m - 1} y(\xi_i^m) x_{t_i, t_{i+1}} \right| \le |I - I_n| + \left| \sum_{i=0}^{\#D_m - 1} (y(\xi_i^m) - y^n(\xi_i^m)) x_{t_i, t_{i+1}} \right|$$

$$+ \left| I_n - \sum_{i=0}^{\#D_m - 1} y^n(\xi_i^m) x_{t_i, t_{i+1}} \right|$$

$$\le |I - I_n| + |y - y^n|_{\infty;[0,T]} |x|_{1\text{-var};[0,T]}$$

$$+ \left| I_n - \sum_{i=0}^{\#D_m - 1} y^n(\xi_i^m) x_{t_i, t_{i+1}} \right|. \qquad (2.2)$$

Fixing $\varepsilon > 0$ we can pick n large enough so that

$$|I - I_n| + |y - y^n|_{\infty;[0,T]} |x|_{1\text{-var};[0,T]} < \varepsilon/2.$$

Then, since y^n is dx-integrable, there exists $M > 0$ such that $m > M$ implies that (2.2) $< \varepsilon/2$ and hence

$$\left| I - \sum_{i=0}^{\#D_m - 1} y(\xi_i^m) x_{t_i, t_{i+1}} \right| < \varepsilon.$$

But this shows precisely that the Riemann–Stieltjes integral $\int_0^T y \, dx$ exists and so y is dx-integrable.

Step 5: Any $y \in C\left([0,T], L\left(\mathbb{R}^d, \mathbb{R}^e\right)\right)$ is dx-integrable. Indeed, take $\xi_i^n \in \left(t_i^n, t_{i+1}^n\right)$ where $D_n = (t_i^n)_i$ is as in the previous step and set

$$y^n(t) := y(\xi_0^n) 1_{[0,t_1]}(t) + \sum_{i=1}^{(\#D_n)-1} y(\xi_i^n) 1_{(t_i, t_{i+1}]}(t).$$

It then suffices to observe that $\lim_{n\to\infty} |y - y^n|_{\infty;[0,T]} = 0$ because y is uniformly continuous on $[0,T]$ and we conclude with step 4. If y is only piecewise continuous (i.e. bounded with finitely many points of discontinuity) it suffices to choose D_n such that it contains all points of discontinuity.

Step 6: Given $x, \tilde{x} \in C^{1\text{-var}}\left([0,T], \mathbb{R}^d\right)$, the last step shows that any $y \in C\left([0,T], L\left(\mathbb{R}^d, \mathbb{R}^e\right)\right)$ is $d(\alpha x + \beta \tilde{x})$-integrable, for any $\alpha, \beta \in \mathbb{R}$. This easily implies linearity of

$$x \in C^{1\text{-var}}\left([0,T], \mathbb{R}^d\right) \mapsto \int_0^T y \, dx$$

and in conjunction with step 2 we obtain bilinearity of $(x, y) \mapsto \int_0^T y \, dx$.

Step 7: Fix s, t with $0 \le s < t \le T$. If y is piecewise continuous then so is $1_{[0,t]} y_u$ and

$$\int_0^T 1_{[0,t]}(u) y_u \, dx = \int_0^t y_u \, dx.$$

Relation (2.1) then follows from $1_{[0,t]} y_u = 1_{[0,s]}(u) y_u + 1_{(s,t]}(u) y_u$. The details are left to the reader. ∎

Exercise 2.3 *Assume $x \in C^1\left([0,T], \mathbb{R}^d\right)$ and $y \in C\left([0,T], L\left(\mathbb{R}^d, \mathbb{R}^e\right)\right)$. Show that*

$$\int_0^T y_u \, dx_u = \int_0^T y_u \dot{x}_u \, du.$$

We then have the classical integration-by-parts formula. It can be obtained by a simple passage to the limit in an elementary partial summation formula for finite sums. The details are left to the reader.

Proposition 2.4 (integration by parts) *Let $x \in C^{\text{1-var}}\left([0,T],\mathbb{R}^d\right)$ and $y \in C^{\text{1-var}}\left([0,T],L\left(\mathbb{R}^d,\mathbb{R}^e\right)\right)$. Then*

$$\int_0^T y_u \, dx_u + \int_0^T (dy_u) \, dx_u = y_T x_T - y_0 x_0.$$

Exercise 2.5 *Take $(x,y) \in C^{\text{1-var}}\left([0,T_1],\mathbb{R}^d\right) \times C\left([0,T_1],L\left(\mathbb{R}^d,\mathbb{R}^e\right)\right)$ and assume ϕ a continuous non-decreasing function ϕ from $[0,T_2]$ onto $[0,T_1]$. Show that*

$$\int_0^t y_{\phi(\cdot)} \, d\left(x \circ \phi\right) = \int_0^{\phi(t)} y \, dx \text{ for all } t \in [0,T_2].$$

Exercise 2.6 *Let $x \in C^{\text{1-var}}\left([0,T],\mathbb{R}^d\right)$, ϕ a C^{∞} function from $\mathbb{R}$ into $\mathbb{R}^+$, compactly supported on $[-1,1]$ with $\int_{-\infty}^{\infty} \phi(u) \, du = 1$. Define $\Phi_t = \int_{-\infty}^t \phi(u) \, du$ and extend x to a continuous function from $\mathbb{R}$ into $\mathbb{R}^d$ by setting $x \equiv x_0$ on $(-\infty,0)$ and $x \equiv x_T$ on $[T,\infty)$. Define for all $\varepsilon > 0$ the mollifier approximation to x by*

$$x^{\varepsilon} : t \in [0,T] \mapsto x_0 + \int_{\mathbb{R}} \Phi_{(t-s)/\varepsilon} dx_s.$$

Show that
(i) for all $\varepsilon > 0$, x^{ε} is infinitely differentiable;
(ii) for all $\varepsilon > 0$, $|x^{\varepsilon}|_{\text{1-var};[0,T]} \leq |x|_{\text{1-var};[0,T]}$, and also $|x^{\varepsilon}|_{\text{1-Höl};[0,T]} \leq |x|_{\text{1-Höl};[0,T]}$;
(iii) x^{ε} converges to x in supremum topology when ε tends to 0.

Solution. (i) One can easily see that, for $n \geq 1$, the nth derivative of x^{ε} is $t \to \int_{\mathbb{R}} \varepsilon^{-d} \Phi^{(n)}\left(\frac{t-s}{\varepsilon}\right) dx_s$, where $\Phi^{(n)}$ is the nth derivative of Φ. (ii) the 1-variation of x^{ε} is given by

$$\begin{aligned}
|x^{\varepsilon}|_{\text{1-var}} &= \int_0^T \left| \int_{\mathbb{R}} \frac{1}{\varepsilon} \Phi^{(1)}\left(\frac{t-s}{\varepsilon}\right) dx_s \right| dt \\
&\leq \int_{\mathbb{R}} \left(\int_{\mathbb{R}} \frac{1}{\varepsilon} \left| \Phi^{(1)}\left(\frac{t-s}{\varepsilon}\right) \right| \cdot |dx_s| \right) dt \\
&\leq \int_{\mathbb{R}} \left(\int_{\mathbb{R}} \frac{1}{\varepsilon} \Phi^{(1)}\left(\frac{t-s}{\varepsilon}\right) dt \right) \cdot |dx_s| \\
&\leq \int_{\mathbb{R}} |dx_s| = |x|_{\text{1-var};[0,T]}.
\end{aligned}$$

The 1-Hölder bound follows from integration by parts,

$$x_t^{\varepsilon} = \int_{\mathbb{R}} x_s \frac{1}{\varepsilon} \phi\left(\frac{t-s}{\varepsilon}\right) ds = \int_{-1}^1 x_{t+\varepsilon s} \phi(-s) \, ds. \qquad (2.3)$$

(iii) As x is continuous (and hence uniformly continuous),

$$\lim_{\varepsilon \to 0} \sup_{s,t \in [0,1] \times [0,T]} |x_{t+\varepsilon s} - x_t| = 0,$$

and (2.3) implies that $\lim_{\varepsilon \to 0} \sup_{t \in [0,T]} |x_t^\varepsilon - x_t| = 0$.   □

2.2 Continuity properties

Proposition 2.2 obviously implies that $(x,y) \mapsto \int y dx$, viewed as a map from

$$C^{1\text{-var}}\left([0,T],\mathbb{R}^d\right) \times C\left([0,T], L\left(\mathbb{R}^d, \mathbb{R}^e\right)\right) \to C^{1\text{-var}}\left([0,T], \mathbb{R}^e\right),$$

is a bounded, bilinear map and hence continuous (and even Fréchet smooth) in the respective norms.[3] In particular,

$$|y^n - y|_{\infty;[0,T]} \to 0, |x^n - x|_{1\text{-var};[0,T]} \to 0$$

implies that

$$\int_0^\cdot y^n dx^n \to \int_0^\cdot y dx$$

in 1-variation. However, this is not the last word on continuity. For instance, the seemingly harmless assumption that all x^n are piecewise smooth would already force us to restrict attention to x absolutely continuous (cf. Proposition 1.29). We thus formulate continuity statements that are applicable under the weaker assumption of uniform convergence with uniform 1-variation bounds.

Proposition 2.7 *Let $y^n, y : [0,T] \to L\left(\mathbb{R}^d, \mathbb{R}^e\right)$ be continuous functions and assume $y^n \to y$ uniformly. Assume $x^n, x \in C^{1\text{-var}}\left([0,T],\mathbb{R}^d\right)$ and $x^n \to x$ uniformly with*

$$\sup_n |x^n|_{1\text{-var};[0,T]} < \infty.$$

Then

$$\int_0^t y^n dx^n \to \int_0^t y dx \text{ uniformly for } t \in [0,T].$$

---

[3] Observe that what we call $|\cdot|_{1\text{-var}}$ is a only a semi-norm on $C^{1\text{-var}}\left([0,T],\mathbb{R}^d\right)$ but a genuine norm on $C_0^{1\text{-var}}\left([0,T],\mathbb{R}^d\right)$ and we can obviously assume $x(0) = 0$ as only dx is of interest. Alternatively, define an equivalence relation on $C^{1\text{-var}}\left([0,T],\mathbb{R}^d\right)$ by setting $x \sim y$ iff $t \mapsto x_t - y_t$ is constant; the resulting quotient space, say $\widetilde{C^{1\text{-var}}}$, is Banach under $|\cdot|_{1\text{-var}}$ and viewing $(f,x) \mapsto \int f dx$ as a map $C \times \widetilde{C^{1\text{-var}}} \to C^{1\text{-var}}$.

Proof. Set $c = \sup_n |x^n|_{1\text{-var};[0,T]}$. Then

$$\left|\int_0^t y^n\,dx^n - \int_0^t y\,dx\right| \leq \left|\int_0^t (y^n - y)\,dx^n\right| + \left|\int_0^t y\,dx^n - \int_0^t y\,dx\right|$$

$$\leq c\,|y^n - y|_{\infty;[0,T]} + \left|\int_0^t y\,dx^n - \int_0^t y\,dx\right|$$

and so it is enough to show

$$\left|\int_0^t y\,dx^n - \int_0^t y\,dx\right| \to 0 \tag{2.4}$$

uniformly in $t \in [0,T]$ as $n \to \infty$. Fix $\varepsilon > 0$ and pick $m = m(\varepsilon)$ such that

$$\frac{\sup_n |x^n|_{1\text{-var};[0,T]}}{m} < \varepsilon/2.$$

Then, from uniform continuity of y on $[0,T]$, we can find a dissection $D = (t_i)$ such that the step function

$$y^D(t) := \sum_{i=1}^{(\#D)} y(t_{i-1})\,1_{[t_{i-1},t_i)}(t)$$

satisifies $|y - y^D|_\infty \leq 1/m$. Observe that (using Lemma 1.15) $|x|_{1\text{-var};[0,T]}\,/\,m < \varepsilon/2$. We estimate the left-hand side of (2.4) by adding/subtracting the integrals $\int y^D\,dx^n$ and $\int y^D\,dx$. This leaves us with three terms, of which the first two are dealt with by

$$\sup_n \sup_{t\in[0,T]} \left|\int_0^t (y^D - y)\,dx^n\right| + \sup_{t\in[0,T]} \left|\int_0^t (y^D - y)\,dx\right| \leq \varepsilon.$$

On the other hand, y^D is constant over the (finitely many) intervals $[t_i, t_{i+1})$. Fix $t \in [0,T]$ and let $t_D \in D$ be the largest point in D for which $t_D \leq t$. Then

$$\int_0^t y^D\,d(x - x^n) = \left(\sum_i y_{t_{i-1}}\left(x_{t_{i-1},t_i} - x^n_{t_{i-1},t_i}\right)\right) + y_{t_D}\left(x_{t_D,t} - x^n_{t_D,t}\right)$$

where the sum $\sum_i$ runs over all integers $i \geq 1$ for which $t_{i-1} \leq t$. It follows that

$$\sup_{t\in[0,T]} \left|\int_0^t y^D\,d(x - x^n)\right| \leq (\#D) \times 2\,|x - x^n|_{\infty;[0,T]} \tag{2.5}$$

where $\#D$ denotes the number of points in D, dependent on m and hence on ε. It follows that

$$\left| \int_0^t y \, dx^n - \int_0^t y \, dx \right| \leq \varepsilon + (\#D) \times 2 \left| x - x^n \right|_{\infty;[0,T]}.$$

Using $\left| x - x^n \right|_{\infty;[0,T]} \to 0$ as $n \to \infty$ it follows that

$$\limsup_{n \to \infty} \left| \int_0^t y \, dx^n - \int_0^t y \, dx \right| \leq \varepsilon$$

and we conclude by sending $\varepsilon \downarrow 0$. ∎

Another useful property of Riemann–Stieltjes integration is uniform continuity on bounded sets.

Proposition 2.8 *Let* $y, y' \in C\left([0,T], L\left(\mathbb{R}^d, \mathbb{R}^e\right)\right)$ *and* $x, x' \in C^{1\text{-var}}$ $\left([0,T], \mathbb{R}^d\right)$. *Then,*

$$\left| \int_0^{\cdot} y \, dx - \int_0^{\cdot} y' \, dx' \right|_{1\text{-var};[0,T]}$$
$$\leq \quad \left| x \right|_{1\text{-var};[0,T]} \cdot \left| y - y' \right|_{\infty;[0,T]} + \left| y' \right|_{\infty;[0,T]} \cdot \left| x - x' \right|_{1\text{-var};[0,T]}.$$

In particular, the map $(x,y) \in C^{1\text{-var}}\left([0,T], \mathbb{R}^d\right) \times C\left([0,T], L\left(\mathbb{R}^d, \mathbb{R}^e\right)\right) \mapsto \int_0^{\cdot} y \, dx \in C^{1\text{-var}}\left([0,T], \mathbb{R}^e\right)$ *is locally Lipschitz.*

Proof. It suffices to insert and subtract $\int_0^{\cdot} y' \, dx$, followed by the triangle inequality. ∎

In applications, integrands frequently come in the form $\varphi(x_t) \in L\left(\mathbb{R}^d, \mathbb{R}^e\right)$ for $\varphi : \mathbb{R}^d \to L\left(\mathbb{R}^d, \mathbb{R}^e\right)$ or $V(y_t)$ for an $\mathbb{R}^e$-valued path y and $V : \mathbb{R}^e \to L\left(\mathbb{R}^d, \mathbb{R}^e\right)$. With focus on the latter, we state the following *uniform continuity* property; the simple proof is left to the reader.

Corollary 2.9 *Let* $x, x' \in C^{1\text{-var}}\left([0,T], \mathbb{R}^d\right)$, $y, y' \in C\left([0,T], \mathbb{R}^e\right)$ *and* $V : \mathbb{R}^e \to L\left(\mathbb{R}^d, \mathbb{R}^e\right)$ *continuous. Assume*

$$\left| x \right|_{1\text{-var};[0,T]}, \left| x' \right|_{1\text{-var};[0,T]}, \left| y \right|_{\infty;[0,T]}, \left| y' \right|_{\infty;[0,T]} < R$$

and let $\varepsilon > 0$. *Then there exists* $\delta = \delta(\varepsilon, R, V)$ *so that*

$$\left| x - x' \right|_{1\text{-var};[0,T]} + \left| y - y' \right|_{\infty;[0,T]} < \delta$$

implies

$$\left| \int_0^\cdot V(y)\,dx - \int_0^\cdot V(y')\,dx' \right|_{1\text{-var};[0,T]} < \varepsilon.$$

2.3   Comments

Riemann–Stieltjes integration is discussed in many elementary analysis texts, for example Rudin [148] or Protter and Morrey [139].

3

Ordinary differential equations

We develop the basic theory of ordinary differential equations of the form

$$\frac{dy}{dt} = \sum_{i=1}^{d} V_i(y) \frac{dx^i}{dt}$$

on a fixed time horizon $[0, T]$. Here, x and y are paths with values in $\mathbb{R}^d$, $\mathbb{R}^e$ respectively and we have coefficients $V_i : \mathbb{R}^e \to \mathbb{R}^e$, often viewed as "driving" vector fields on $\mathbb{R}^e$. When the driving signal x is continuously differentiable we are dealing with an example of a (time-inhomogenous) ordinary differential equation. We give a direct existence proof, via Euler approximations, that applies to continuous, finite-variation driving signals and continuous vector fields; uniqueness holds for Lipschitz continuous vector fields.

3.1 Preliminaries

Given a collection of (continuous) vector fields $V = (V_1, \ldots, V_d)$ on $\mathbb{R}^e$ and continuous, finite-variation paths x, y with values in $\mathbb{R}^d$, $\mathbb{R}^e$ we set

$$\int_0^t V(y) \, dx := \sum_{i=1}^{d} \int_0^t V_i(y) \, dx^i.$$

From the point of view of vector-valued Riemann–Stieltjes integration, this amounts precisely to viewing V as a map

$$y \in \mathbb{R}^e \mapsto \{a = (a^1, \ldots, a^d) \mapsto \sum_{i=1}^{d} V_i(y) a^i\} \in L(\mathbb{R}^d, \mathbb{R}^e),$$

where $L(\mathbb{R}^d, \mathbb{R}^e)$ is equipped with the operator norm, so that

$$|V(y)| := |V(y)|_{op} := \sup_{a \in \mathbb{R}^d : |a| = 1} \left| \sum_{i=1}^{d} V_i(y) a^i \right|. \tag{3.1}$$

Definition 3.1 A collection of vector fields $V = (V_1, \ldots, V_d)$ on $\mathbb{R}^e$, viewed as $V : \mathbb{R}^e \to L(\mathbb{R}^d, \mathbb{R}^e)$, is called bounded if

$$|V|_\infty := \sup_{y \in \mathbb{R}^e} |V(y)| < \infty.$$

For any $U \subset \mathbb{R}^e$ we define the 1-Lipschitz norm (in the sense of E. M. Stein) by

$$|V|_{\text{Lip}^1(U)} := \max \left\{ \sup_{y,z \in U : y \neq z} \frac{|V(y) - V(z)|}{|y - z|}, \sup_{y \in U} |V(y)| \right\}.$$

We say that $V \in \text{Lip}^1(\mathbb{R}^e)$ if $|V|_{\text{Lip}^1} \equiv |V|_{\text{Lip}^1(\mathbb{R}^e)} < \infty$ and locally 1-Lipschitz if $|V|_{\text{Lip}^1(U)} < \infty$ for all bounded subsets $U \subset \mathbb{R}^e$. □

(The concept of Lip^1 regularity will later be generalized to Lip^γ in the sense of E. M. Stein.) Observe that 1-Lipschitz paths are Lipschitz continuous paths that are bounded. We now state a classical analysis lemma.

Lemma 3.2 (Gronwall's lemma) *Let* $x \in C^{1\text{-var}}\left([0,T],\mathbb{R}^d\right)$, *and* $\phi :$ $[0,T] \to \mathbb{R}^+$ *a bounded measurable function. Assume that for all* $t \in [0,T]$

$$\phi(t) \leq K + L \int_0^t \phi_s |dx_s|, \tag{3.2}$$

for some $K, L \geq 0$. *Then, for all* $t \in [0,T]$

$$\phi_t \leq K \exp\left(L|x|_{1\text{-var};[0,t]}\right).$$

If $t \mapsto K_t$ *is a non-negative, non-decreasing function,* K *may be replaced by* K_t.

Proof. After n iterated uses of (3.2)

$$\phi(t) \leq K + KL \int_0^t |dx_s| + \cdots + KL^n \int_0^t \int_0^{t_1} \cdots \int_0^{t_{n-1}} |dx_{t_n}| \cdots |dx_{t_1}|$$
$$+ L^{n+1} \int_0^t \int_0^{t_1} \cdots \int_0^{t_{n-1}} \int_0^{t_n} \phi(t_{n+1}) |dx_{t_{n+1}}| \cdots |dx_{t_1}|.$$

Since x is continuous,

$$\int_0^t \int_0^{t_1} \cdots \int_0^{t_{n-1}} |dx_{t_n}| \cdots |dx_{t_1}| = \frac{|x|_{1\text{-var};[0,t]}^n}{n!}.$$

Then,

$$\phi(t) \leq K \exp\left(L|x|_{1\text{-var};[0,t]}\right) + |\phi|_{\infty;[0,T]} \frac{\left[L|x|_{1\text{-var};[0,t]}\right]^{n+1}}{(n+1)!}$$

and sending $n \to \infty$ gives the required estimate. The last statement, replacing K by some non-decreasing K_t, comes from the obvious remark that the previous estimate can be applied on the interval $[0,t]$ with $K = K_t$. ∎

3.2 Existence

Let us first define what we mean by solution of a (controlled, ordinary) differential equation:

Definition 3.3 Given a collection of continuous vector fields $V = (V_1, \ldots, V_d)$ on $\mathbb{R}^e$, a driving signal $x \in C^{1\text{-var}}\left([0,T], \mathbb{R}^d\right)$ and an initial condition $y_0 \in \mathbb{R}^e$, we write $\pi_{(V)}(0, y_0; x)$ for the set of all solutions to the ODE[1]

$$dy_t = \sum_{i=1}^{d} V_i(y_t)\, dx_t^i \equiv V(y_t)\, dx_t \qquad (3.3)$$

for $t \in [0, T]$ started at y_0. The above ODE is understood as a Riemann–Stieltjes integral equation, i.e.

$$y_{0,t} := y_t - y_0 = \int_0^t V(y_s)\, dx_s.$$

In case of uniqueness $y = \pi_{(V)}(0, y_0; x)$ denotes <u>the</u> solution. If necessary, $\pi(0, y_0; x)$ is only considered up to some explosion time. Similarly, $\pi_{(V)}(s, y_s; x)$ stands for solutions of (3.3) started at time s from a point $y_s \in \mathbb{R}^e$. $\qquad \square$

We shall frequently describe $\pi_{(V)}(0, y_0; x)$ as an "ODE solution, driven by x along the vector fields V and started from y_0". Existence of a solution holds under minimal regularity conditions on the vector fields.

Theorem 3.4 *(existence) Assume that*
(i) $V = (V_1, \ldots, V_d)$ is a collection of continuous, bounded vector fields on $\mathbb{R}^e$;
(ii) $y_0 \in \mathbb{R}^e$ is an initial condition;
(iii) x is a path in $C^{1\text{-var}}\left([0,T], \mathbb{R}^d\right)$.
Then there exists a (not necessarily unique) solution to the ODE (3.3). Moreover, for all $0 \le s < t \le T$

$$\left|\pi_{(V)}(0, y_0; x)\right|_{1\text{-var};[s,t]} \le |V|_\infty\, |x|_{1\text{-var};[s,t]}. \qquad (3.4)$$

Remark 3.5 *In case of non-uniqueness, we abuse notation in the above estimate in the sense that $\pi_{(V)}(0, y_0; x)$ stands for an arbitrary solution to (3.3) started at y_0.*

Proof. Let $D = (t_i)_i$ be a dissection of the interval $[0, T]$, and define the *Euler approximation* $y^{(D)} : [0, T] \to \mathbb{R}^e$ by

$$\begin{cases} y_0^{(D)} = y_0, \\ y_t^{(D)} = y_{t_i}^{(D)} + V\left(y_{t_i}^{(D)}\right) x_{t_i, t} \text{ for } t \in [t_i, t_{i+1}]. \end{cases}$$

---

[1] The ODE (3.3) is time-inhomogenous unless x_t is proportional to t.

Then, it is easy to see that for all $0 \leq s < t \leq T$

$$\left| y_{s,t}^{(D)} \right| \leq \left| y^{(D)} \right|_{\text{1-var};[s,t]} \leq \left| V \right|_\infty \left| x \right|_{\text{1-var};[s,t]} . \tag{3.5}$$

In particular,

$$\left| y^{(D)} \right|_{\infty;[0,T]} \leq |y_0| + |V|_\infty \left| x \right|_{\text{1-var};[0,T]} .$$

Moreover, if r_D denotes the greatest real number in D less than r then

$$y_{0,t}^{(D)} - \int_0^t V\left(y_r^{(D)} \right) dx_r = \int_0^t \left[V\left(y_{r_D}^{(D)} \right) - V\left(y_r^{(D)} \right) \right] dx_r . \tag{3.6}$$

Now let (D_n) be a sequence of dissections, with mesh $|D_n| \to 0$ as n tends to ∞. Clearly $\left\{ y^{(D_n)} \right\}$ is equicontinuous and bounded. From Arzela–Ascoli's theorem we see that $\left\{ y^{(D_n)} \right\}$ has at least one limit point y. After relabelling our sequence, we can assume that $y^{(D_n)}$ converges to y uniformly on $[0,T]$. Fix $r \in [0,T]$. From (3.5), $\lim_{n \to \infty} \left| y_r^{(D_n)} - y_{r_{D_n}}^{(D_n)} \right| = 0$. On the other hand, $y_r^{(D_n)} \to y_r$ hence $y_{r_{D_n}}^{(D_n)} \to y_r$ and by continuity of V,

$$\lim_{n \to \infty} \left| V\left(y_r^{(D_n)} \right) - V\left(y_{r_{D_n}}^{(D_n)} \right) \right| = 0. \tag{3.7}$$

By dominated convergence,[2] we can pass to the limit in (3.6) to see that

$$y_{0,t} - \int_0^t V(y_r) \, dx_r = 0.$$

Finally, for $s,t \in [0,T]$, for any solution $y \in \pi_{(V)}(0, y_0; x)$,

$$\begin{aligned}
|y_{s,t}| &= \left| \int_s^t V(y_u) \, dx_u \right| \\
&\leq \int_s^t |V|_\infty \cdot |dx_u| \\
&= |V|_\infty \left| x \right|_{\text{1-var};[s,t]} .
\end{aligned}$$

The right-hand side being a control, we obtain inequality (3.4). ∎

If we only assume continuity of the vector fields (without imposing growth conditions) existence holds up to an explosion time:

Theorem 3.6 *Assume that*
(i) $V = (V_1, \ldots, V_d)$ is a collection of continuous vector fields on $\mathbb{R}^e$;

---

[2] Which requires us to know that Riemann–Stieltjes integrals with continuous integrands coincide with Lebesgue–Stieltjes integrals.

(ii) $y_0 \in \mathbb{R}^e$ is an initial condition;

(iii) x is a path in $C^{1\text{-var}}\left([0,T],\mathbb{R}^d\right)$.

Then either there exists a (global) solution $y : [0,T] \to \mathbb{R}^e$ to ODE (3.3) started at y_0 or there exists $\tau \in [0,T]$ and a (local) solution $y : [0,\tau) \to \mathbb{R}^e$ such that y is a solution on $[0,t]$ for any $t \in (0,\tau)$ and

$$\lim_{t \nearrow \tau} |y(t)| = +\infty.$$

Proof. Without loss of generality take $y_0 = 0$. Replace V by compactly supported vector fields V^n which coincide with V on the ball $\{y : |y| \le n\}$. From the preceding existence theorem, there exists a (not necessarily unique) ODE solution $y^1 := \pi_{(V^1)}(0, y_0; x)$ which we consider only up to time

$$\tau_1 = \inf\left\{t \in [0,T] : \left|y_t^1\right| \ge 1\right\} \wedge T > 0.$$

If $\tau_1 = T$ then $y = y^1$ is a solution on $[0,T]$ and we are done. Set $\tau_0 = 0$. We now define τ_n, y_n inductively and assume $\tau_n \in [0,T], y^n \in C\left([\tau_{n-1}, \tau_n], \mathbb{R}^e\right)$ have been defined. We then define

$$y^{n+1} := \pi_{(V^{n+1})}\left(\tau_n, y_{\tau_n}^n; x\right)$$

as an (again, not necessarily unique) ODE solution started from $y^{n+1}(\tau_n) = y_{\tau_n}^n$ driven by x along the vector fields V^{n+1} up to time

$$\tau_{n+1} = \inf\left\{t \in [\tau_n, T] : \left|y_t^{n+1}\right| \ge n+1\right\} \wedge T > 0.$$

If at any step in this induction, $\tau_n = T$, then $(0,T] = \cup_{k=1}^n (\tau_{k-1}, \tau_k]$ and $y(t) = y_t^k$ for $t \in (\tau_{k-1}, \tau_k]$ defines a solution on $[0,T]$ and we find ourselves in case (i) of the statement of the theorem. Otherwise, we obtain an increasing sequence (τ_n) with $\tau = \lim_n \tau_n \to \infty \in (0,T]$. Any interval $(0,t] \subset (0,\tau)$ can be covered by intervals $(\tau_{k-1}, \tau_k]$ and a solution on $(0,t]$ is constructed as above by setting $y(t) = y_t^k$ for $t \in (\tau_{k-1}, \tau_k]$. Moreover, be definition of τ_n we see that $|y(\tau_n)| = n \to \infty$ as $n \to \infty$ and the proof is finished. ∎

Theorem 3.7 *Assume that*

(i) $V = (V_1, \ldots, V_d)$ is a collection of continuous vector fields on $\mathbb{R}^e$ of linear growth, i.e.

$$\exists A \ge 0 : |V_i(y)| \le A(1 + |y|) \text{ for all } y \in \mathbb{R}^e;$$

(ii) $y_0 \in \mathbb{R}^e$ is an initial condition;

(iii) x is a path in $C^{1\text{-var}}\left([0,T],\mathbb{R}^d\right)$, and $\ell \ge A|x|_{1\text{-var};[0,T]}$.

Then explosion cannot happen. Moreover, any solution y to (3.3) satisfies the estimates

$$|y|_{\infty;[0,T]} \le (|y_0| + \ell)\exp(\ell), \tag{3.8}$$

and, for all $0 \le s < t \le T$,

$$|y|_{1\text{-var};[s,t]} \le (1 + |y_0|) \exp(2\ell) A \int_s^t |dx_u|.$$

Proof. For $s \in [0, \min(\tau(y), T))$,

$$
\begin{aligned}
|y_s| &\le |y_0| + \left| \int_0^s V(y_u) \, dx_u \right| \\
&\le |y_0| + A \int_0^s |dx_u| + A \int_0^s |y_u| \cdot |dx_u|.
\end{aligned}
$$

Hence, by Gronwall's inequality, for all $s \in [0, \min(\tau(y), T))$,

$$|y_s| \le \left(|y_0| + A \int_0^s |dx_u| \right) \exp\left(A \int_0^s |dx_u| \right). \tag{3.9}$$

This implies in particular that explosion cannot happen in finite time, and that

$$|y|_{\infty;[0,s]} \le \left(|y_0| + A \int_0^s |dx_u| \right) \exp\left(A \int_0^s |dx_u| \right).$$

Let us now take $s, t \in [0, T]$. Clearly, for all $u, v \in [s, t]$,

$$y_{u,v} = \int_u^v V(y_r) \, dx_r = \int_u^v V(y_u + y_{u,r}) \, dx_r$$

so that

$$|y_{u,v}| \le A(1 + |y_u|) \int_u^v |dx_u| + A \int_u^v |y_{u,r}| \, |dx_r|.$$

By Gronwall's inequality, we obtain

$$|y_{u,v}| \le A(1 + |y_u|) \int_u^v |dx_u| \exp\left(A \int_u^v |dx_r| \right).$$

Now, using inequality (3.9) and $\int_0^u |dx_r| + \int_u^v |dx_r| = \int_0^v |dx_r|$,

$$
\begin{aligned}
(1 + |y_u|) \exp\left(A \int_u^v |dx_r| \right) &\le \left(|y_0| + 1 + A \int_0^u |dx_r| \right) \exp\left(A \int_0^v |dx_r| \right) \\
&\le |y_0| \exp\left(A \int_0^t |dx_r| \right) + \exp\left(2A \int_0^t |dx_r| \right) \\
&\le (1 + |y_0|) \exp\left(2A \int_0^t |dx_r| \right),
\end{aligned}
$$

which gives

$$|y_{u,v}| \le A(1 + |y_0|) \exp\left(2A \int_0^t |dx_r| \right) \int_u^v |dx_u|,$$

and hence

$$|y|_{\text{1-var};[s,t]} \leq A\left(1 + |y_0|\right) \exp\left(2A \int_0^t |dx_r|\right) \int_s^t |dx_u|.$$

∎

3.3 Uniqueness

We now show uniqueness for ODEs driven along Lipschitz vector fields by establishing Lipschitz continuity of the flow.

Theorem 3.8 *Assume that*
(i) $V = (V_1, \ldots, V_d)$ *is a collection of Lipschitz continuous vector fields on* $\mathbb{R}^e$ *such that, for some* $v \geq 0$,

$$v \geq \sup_{y,z \in \mathbb{R}^d} \frac{|V(y) - V(z)|}{|y - z|};$$

(ii) $x \in C^{\text{1-var}}\left([0,T], \mathbb{R}^d\right)$ *with, for some* $\ell \geq 0$,

$$v\,|x|_{\text{1-var};[0,T]} \leq \ell.$$

Then, for every initial condition there exists a unique ODE solution to $dy = V(y)\,dx$ *on* $[0,T]$. *Moreover, the associated flow is Lipschitz continuous in the following sense that, for any initial conditions* $y_0^1, y_0^2 \in \mathbb{R}^e$,

$$\left|\pi_{(V)}\left(0, y_0^1; x\right) - \pi_{(V)}\left(0, y_0^2; x\right)\right|_{\infty;[0,T]} \leq \left|y_0^1 - y_0^2\right| \exp\left(\ell\right). \tag{3.10}$$

Moreover, for all $s < t$ *in* $[0,T]$ *we have*

$$\left|\pi_{(V)}\left(0, y_0^1; x\right) - \pi_{(V)}\left(0, y_0^2; x\right)\right|_{\text{1-var};[s,t]} \leq \left|y_0^1 - y_0^2\right| \exp\left(2\ell\right).v\,|x|_{\text{1-var};[s,t]}.$$

Proof. Lipschitz continuous vector fields are of linear growth and existence of solutions on $[0,T]$ is guaranteed by Theorem 3.7. Let us write $y^i \in \pi_{(V)}\left(0, y_0^i; x\right), i = 1, 2$ for an arbitrary solution started from y_0^1, y_0^2 respectively and set $\bar{y} = y^1 - y^2$. Then

$$\bar{y}_t = \bar{y}_0 + \int_0^t \left(V\left(y_s^1\right) - V\left(y_s^2\right)\right) dx_s,$$

and hence

$$|\bar{y}_t| \leq |\bar{y}_0| + v \int_0^t |\bar{y}_s| \cdot |dx_s|.$$

Gronwall's inequality then leads to the first stated estimate. Moreover, taking $y_0^1 = y_0^2$ shows that $y^1 \equiv y^2$ and there is indeed a unique solution. For the second estimate, we have

$$
\begin{aligned}
|\bar{y}_{s,t}| &= \left| \int_s^t \left(V\left(y_r^1\right) - V\left(y_r^2\right) \right) dx_s \right| \\
&\leq v \int_s^t |\bar{y}_r| \, |dx_r| \\
&\leq |\bar{y}_s| \, . v \int_s^t |dx_r| + v \int_s^t |\bar{y}_{s,r}| \, . \, |dx_r| \, .
\end{aligned}
$$

Applying Gronwall gives

$$
|\bar{y}_{s,t}| \leq |\bar{y}_s| \, . v \int_s^t |dx_r| \, . \exp\left(\ell\right) .
$$

Using the estimate (3.10), we obtain

$$
|\bar{y}_{s,t}| \leq |\bar{y}_0| \, . v \int_s^t |dx_r| \, . \exp\left(2\ell\right) .
$$

Noting that the right-hand side is a control, we obtain our estimate. ∎
Since uniqueness is a local property we immediately have

Corollary 3.9 *Given* $x \in C^{\text{1-var}}\left([0,T], \mathbb{R}^d\right)$*, there is a unique solution to* $dy = V\left(y\right) dx$ *started at* y_0 *along locally 1-Lipschitz vector fields* $V = (V_1, \ldots, V_d)$ *up to its possible explosion time. If explosion can be ruled out (e.g. under an additional linear growth condition, cf. Theorem 3.7) then there exists a unique solution on* $[0,T]$*.*

3.4 A few consequences of uniqueness

We first show that time-change commutes with solving differential equations.

Proposition 3.10 *Let* $x \in C^{\text{1-var}}\left([0,T_1], \mathbb{R}^d\right)$ *and* $V = (V_1, \ldots, V_d)$ *a collection of locally Lipschitz continuous vector fields on* $\mathbb{R}^e$ *of linear growth. Assume* ϕ *is a continuous non-decreasing function from* $[0,T_2]$ *onto* $[0,T_1]$ *so that*

$$
x \circ \phi \in C^{\text{1-var}}\left([0,T_2], \mathbb{R}^d\right) .
$$

Then

$$
\pi_{(V)}\left(0, y_0; x\right)_{\phi(.)} \equiv \pi_{(V)}\left(0, y_0; x \circ \phi\right) \quad on \quad [0,T_2] .
$$

Proof. Let $y = \pi_{(V)}(0, y_0; x)$ denote the (unique) ODE solution. For all $t \in [0, T_2]$

$$y_{\phi(t)} = y_0 + \sum_{i=1}^{d} \int_0^{\phi(t)} V_i(y_r) \, dx_r^i.$$

By a change of variable $r = \phi(s)$ for Riemann–Stieltjes integrals, we obtain

$$y_{\phi(t)} = y_0 + \sum_{i=1}^{d} \int_0^t V_i(y_{\phi(s)}) \, dx_{\phi(s)}^i$$

which says precisely that $t \mapsto y_{\phi(t)}$ is an ODE solution driven $x \circ \phi$ along vector fields $V_1, \ldots, V_d$ started at y_0. By uniqueness, we therefore have

$$\pi_{(V)}(0, y_0; x)_{\phi(\cdot)} = \pi_{(V)}(0, y_0; x \circ \phi).$$

∎

Definition 3.11 (concatenation, time-reversal) (i) Given $x \in C\left([0, T], \mathbb{R}^d\right)$ and $\tilde{x} \in C\left([T, U], \mathbb{R}^d\right)$ we define the concatenation $x \sqcup \tilde{x}$ as a path in $C\left([0, U], \mathbb{R}^d\right)$ defined by[3]

$$(x \sqcup \tilde{x})(t) = x_t \text{ if } t \in [0, T]$$
$$(x \sqcup \tilde{x})(t) = (x_S - \tilde{x}_S) + \tilde{x}_t \text{ if } t \in [T, U].$$

(ii) Next, the time-inverse of a path $x \in C\left([0, T], \mathbb{R}^d\right)$ is defined as the path x run backwards on $[0, T]$, i.e.

$$\overleftarrow{x}^T : t \in [0, T] \to x_{T-t} \in \mathbb{R}^d.$$

When $[0, T]$ is fixed and no confusion is possible, we simply write $\overleftarrow{x}$ for the time-inverse of x. □

As a simple consequence of uniqueness we have the following two propositions.

Proposition 3.12 *Let $x \in C^{\text{1-var}}\left([0, S], \mathbb{R}^d\right), \tilde{x} \in C^{\text{1-var}}\left([S, T], \mathbb{R}^d\right)$ and $V = (V_1, \ldots, V_d)$ a collection of locally Lipschitz continuous vector fields on $\mathbb{R}^e$ of linear growth. Then*

$$\pi_{(V)}(0, y_0; x) \equiv \pi_{(V)}(0, y_0; x \sqcup \tilde{x}) \text{ on } [0, S]$$

and

$$\pi_{(V)}\left(S, \pi_{(V)}(0, y_0; x)_S; \tilde{x}\right) \equiv \pi_{(V)}(0, y_0; x \sqcup \tilde{x}) \text{ on } [S, T].$$

---

[3] Of course, $x, \tilde{x}$ need not be defined on adjacent intervals but a simple reparametrization will bring things back to the above definition. Formally speaking, concatenation is an operation on paths modulo their parametrization.

Proof. Obvious.   ∎

Proposition 3.13 *Let $x \in C^{1\text{-var}}\left([0,T], \mathbb{R}^d\right)$ and $V = (V_1, \ldots, V_d)$ a collection of locally Lipschitz continuous vector fields on $\mathbb{R}^e$ of linear growth so that there is a (unique) ODE solution $y = \pi_{(V)}(0, y_0; x)$. Then for all $0 \leq t \leq T$,*

$$\pi_{(V)}\left(0, y_T; \overleftarrow{x}^T\right)_{T-t} = y_t.$$

Proof. Same as for Proposition 3.10, just use $\phi(t) = T - t$.   ∎

We record a simple corollary. (As a preview to an application discussed later on: when applied to the left-invariant vector fields $U_1, \ldots, U_d$ on the step-N nilpotent group it implies that the *signature* of $x \sqcup \overleftarrow{x}^T$ over $[0, T]$ is trivial.)

Corollary 3.14 *Let $x \in C^{1\text{-var}}\left([0,T], \mathbb{R}^d\right)$ and $V = (V_1, \ldots, V_d)$ a collection of locally Lipschitz continuous vector fields on $\mathbb{R}^e$ of linear growth so that there is a (unique) ODE solution $y = \pi_{(V)}(0, y_0; x)$. Reparametrize $\overleftarrow{x}$ as a path on $[T, 2T]$, i.e. $\overleftarrow{x}(t) = x_{2T-t}$. Then,*

$$\pi_{(V)}\left(0, y_0; x \sqcup \overleftarrow{x}^T\right)_{2T} = y_0.$$

3.5   Continuity of the solution map

We now investigate continuity properties of the solution map, i.e. the map $(y_0, x) \mapsto y$, the ODE solution to

$$dy = V(y)\, dx = \sum_{i=1}^{d} V_i(y)\, dx^i$$

started at time 0 at $y_0 \in \mathbb{R}^e$. In fact, it will not complicate things to consider the map $(y_0, V, x) \mapsto y$.

3.5.1   Limit theorem for 1-variation signals

Let us recall that our notion of 1-Lipschitz regularity includes the assumption of boundedness, cf. Definition 3.1. We start with our first continuity statement for solutions of ordinary differential equations.

Theorem 3.15 *We consider*
(i) $V^1 = \left(V_1^1, \ldots, V_d^1\right)$ and $V^2 = \left(V_1^2, \ldots, V_d^2\right)$ are two collections of Lip^1 vector fields on $\mathbb{R}^e$ with, for some $\upsilon \geq 0$,

$$\max_{i=1,2} \left|V^i\right|_{\mathrm{Lip}^1} \leq \upsilon;$$

(ii) $y_0^1, y_0^2 \in \mathbb{R}^e$ *are initial conditions;*

(iii) x^1 *and* x^2 *are two paths in* $C^{1\text{-var}}\left([0,T],\mathbb{R}^d\right)$ *with, for some* $\ell \geq 0$,

$$\max_{i=1,2} \left|x^i\right|_{1\text{-var};[0,T]} \leq \ell.$$

Then, if $y^i = \pi_{(V^i)}\left(0, y_0^i; x^i\right)$ *for* $i = 1, 2$, *we have*

$$\left|y^1 - y^2\right|_{\infty;[0,T]} \leq \left(\left|y_0^1 - y_0^2\right| + \upsilon\left|x^1 - x^2\right|_{0;[0,T]} + \left|V^1 - V^2\right|_\infty \ell\right) \exp\left(2\upsilon\ell\right).$$

Proof. Without loss of generality, $x_0^1 = x_0^2 = 0$ so that $\frac{1}{2}\left|x^1 - x^2\right|_{0;[0,T]} \leq \left|x^1 - x^2\right|_{\infty;[0,T]} \leq \left|x^1 - x^2\right|_{0,[0,T]}$. First note that for $i = 1, 2$ we have $\left|y^i\right|_{1\text{-var};[0,T]} \leq \upsilon\ell$. Now, write for $t \in [0,T]$,

$$
\begin{aligned}
\left|y_t^1 - y_t^2\right| &\leq \left|y_0^1 - y_0^2\right| + \left|\int_0^t \left[V^1\left(y_r^1\right) - V^1\left(y_r^2\right)\right] dx_r^1\right| \\
&\quad + \left|\int_0^t V^1\left(y_r^2\right) d\left(x_r^1 - x_r^2\right)\right| \\
&\quad + \left|\int_0^t \left[V^2\left(y_r^2\right) - V^1\left(y_r^2\right)\right] dx_r^2\right| \\
&\leq \left|y_0^1 - y_0^2\right| + \upsilon \int_0^t \left|y_r^1 - y_r^2\right| \cdot \left|dx_r^1\right| \qquad (3.11) \\
&\quad + \left|\int_0^t V^1\left(y_r^2\right) d\left(x_r^1 - x_r^2\right)\right| + \left|V^1 - V^2\right|_\infty \ell.
\end{aligned}
$$

We deduce from the integration-by-parts formula

$$\int_0^t V^1\left(y_r^2\right) d\left(x_r^1 - x_r^2\right) = \int_0^t \left(x_r^1 - x_r^2\right) dV^1\left(y_r^2\right) + V^1\left(y_t^2\right) \cdot \left(x_t^1 - x_t^2\right)$$

the bound

$$
\begin{aligned}
\left|\int_0^t V^1\left(y_r^2\right) d\left(x_r^1 - x_r^2\right)\right| &\leq \left|x^1 - x^2\right|_{\infty;[0,t]} \left|V^1\left(y_\cdot^2\right)\right|_{1\text{-var};[0,t]} \\
&\quad + \left|V^1\left(y_t^2\right)\right| \cdot \left|x^1 - x^2\right|_{\infty;[0,t]} \\
&\leq \upsilon\left|x^1 - x^2\right|_{\infty;[0,t]} \left(1 + \left|y^2\right|_{1\text{-var};[0,t]}\right) \\
&\leq \upsilon\left|x^1 - x^2\right|_{\infty;[0,t]} \left(1 + \upsilon\ell\right).
\end{aligned}
$$

This last inequality and inequality (3.11) give, for $t \in [0,T]$,

$$
\begin{aligned}
\left|y_t^1 - y_t^2\right| &\leq \left|y_0^1 - y_0^2\right| + \upsilon\left(1 + \upsilon\ell\right)\left|x^1 - x^2\right|_{\infty;[0,t]} \\
&\quad + \upsilon \int_0^t \left|y_r^1 - y_r^2\right| \cdot \left|dx_r^1\right| + \left|V^1 - V^2\right|_\infty \ell
\end{aligned}
$$

which implies, using Gronwall's lemma, that

$$\begin{aligned}
\left| y_t^1 - y_t^2 \right| &\leq \left(\left| y_0^1 - y_0^2 \right| + v \left| x^1 - x^2 \right|_{\infty;[0,t]} \right) (1 + v\ell) \\
&\quad + \left| V^1 - V^2 \right|_\infty \ell \right) \exp \left(|V|_{\mathrm{Lip}^1} \int_0^t \left| dx_r^1 \right| \right) \\
&\leq \left(\left| y_0^1 - y_0^2 \right| + v \left| x^1 - x^2 \right|_{\infty;[0,t]} + \left| V^1 - V^2 \right|_\infty \ell \right) \exp \left(2v\ell \right).
\end{aligned}$$

∎

From the above theorem we see in particular that the solution map $(y_0, x) \mapsto \pi_{(V)}(0, y_0; x)$ is uniformly continuous in the sense of "uniform convergence with uniform 1-variation bounds". By a localization argument we now weaken the boundedness assumption inherent to Lip^1 regularity:

Corollary 3.16 *We consider*
(i) $V^1 = \left(V_1^1, \ldots, V_d^1 \right)$ and $V^2 = \left(V_1^2, \ldots, V_d^2 \right)$ are two collections of locally Lip^1 vector fields on $\mathbb{R}^e$, with linear growth;
(ii) $y_0^1, y_0^2 \in \mathbb{R}^e$ are initial conditions, with $\left| y_0^i \right| \leq R$ for some $R \geq 0$;
(iii) x^1 and x^2 are two paths in $C^{1\text{-var}}([0,T], \mathbb{R}^d)$, with $\max_{i=1,2} \left| x^i \right|_{1\text{-var};[0,T]} \leq \ell$ for some $\ell \geq 0$.
Then, if $y^i = \pi_{(V^i)}\left(0, y_0^i; x^i \right)$ for $i = 1, 2$, there exist constants C, M depending only on R, ℓ and the vector fields, such that for all $s, t \in [0, T]$

$$\left| y^1 - y^2 \right|_{\infty;[0,T]} \leq C \left(\left| y_0^1 - y_0^2 \right| + \left| x^1 - x^2 \right|_{0;[0,T]} + \left| V^1 - V^2 \right|_{\infty;B(0,M)} \right).$$

Proof. We saw in Corollary 3.9 that under locally Lipschitz and linear growth assumptions on the vector fields, there is indeed a unique, non-exploding solution. In fact, thanks to the explicit estimate (3.8) there exists $M = M(\ell, R)$ so that $\max_{i=1,2} \left| y^i \right|_{\infty;[0,T]} \leq M$. We now modify the vector fields V^i outside a ball of radius R such as to make them Lip^1-vector fields, say $\tilde{V}^i$, and note that

$$y^i = \pi_{(V^i)}\left(0, y_0^i; x^i \right) = \pi_{(\tilde{V}^i)}\left(0, y_0^i; x^i \right).$$

This allows us to use Theorem 3.15 to finish the proof. ∎

Exercise 3.17 *(change-of-variable formula) Assume f is $C^1(\mathbb{R}^e)$ and $y = \pi_{(V)}(0, y_0; x)$ the unique solution to*

$$dy = V(y)\, dx, \quad y(0) = y_0 \in \mathbb{R}^e$$

along locally 1-Lipschitz vector fields $V = (V_1, \ldots, V_d)$ on $\mathbb{R}^e$, with linear growth, and $x \in C^{1\text{-var}}([0,T], \mathbb{R}^d)$. Show that

$$f(y_T) - f(y_0) = \int_0^T (Vf)(y_s)\, dx_s$$

where $Vf = (V_1 f, \ldots, V_d f)$ and each V_i is identified with a first-order differential operator

$$V_i f = \sum_{k=1}^{e} V_i^k \partial_k f.$$

Solution. For $x \in C^1 \left([0,T], \mathbb{R}^d\right)$, this is just the fundamental theorem of calculus. For $x \in C^{1\text{-var}} \left([0,T], \mathbb{R}^d\right)$ we approximate (uniformly, with uniform 1-variation bounds) and use the limit theorem. One can also appeal to the direct change-of-variable formulae for Riemann–Stieltjes integrals ... $\qquad\square$

3.5.2 Continuity under 1-variation distance

Given a collection V of Lip^1-vector fields we first show that

$$(y_0, x) \in \mathbb{R}^e \times C^{1\text{-var}} \left([0,T], \mathbb{R}^d\right) \mapsto \pi_{(V)} (0, y_0; x) \in C^{1\text{-var}} \left([0,T], \mathbb{R}^e\right)$$

is Lipschitz continuous on bounded sets. Again, it will not complicate things to include V in the following continuity result.

Theorem 3.18 *We consider*
(i) $V^1 = \left(V_1^1, \ldots, V_d^1\right)$ *and* $V^2 = \left(V_1^2, \ldots, V_d^2\right)$ *are two collections of* Lip^1-*vector fields on* $\mathbb{R}^e$, *such that, for some* $\upsilon \geq 0$,

$$\max_{i=1,2} \left|V^i\right|_{\text{Lip}^1} \leq \upsilon;$$

(ii) $y_0^1, y_0^2 \in \mathbb{R}^e$ *are viewed as two time-0 initial conditions;*
(iii) x^1 *and* x^2 *are two paths in* $C^{1\text{-var}} \left([0,T], \mathbb{R}^d\right)$ *such that, for some* $\ell \geq 0$,

$$\max_{i=1,2} \left|x^i\right|_{1\text{-var};[0,T]} \leq \ell.$$

Then, if $y^i = \pi_{(V^i)} \left(0, y_0^i; x^i\right)$ *for* $i = 1, 2$, *we have*

$$\left|y^1 - y^2\right|_{1\text{-var};[0,T]} \leq 2\left(\left|y_0^1 - y_0^2\right| \upsilon \ell + \upsilon \left|x^1 - x^2\right|_{1\text{-var};[0,T]} + \left|V^1 - V^2\right|_\infty \ell\right)e^{3\upsilon\ell}.$$
$$(3.12)$$

Proof. Take $s < t$ in $[0, T]$ and observe that

$$
\begin{aligned}
\left| y_{s,t}^1 - y_{s,t}^2 \right| &= \left| \int_s^t V^1 \left(y_r^1 \right) dx_r^1 - \int_s^t V^2 \left(y_r^2 \right) dx_r^2 \right| \\
&\leq \left| \int_s^t \left(V^1 \left(y_r^1 \right) - V^1 \left(y_r^2 \right) \right) dx_r^2 \right| + \left| \int_s^t V^1 \left(y_r^1 \right) d \left(x_r^1 - x_r^2 \right) \right| \\
&\quad + \left| \int_s^t \left(V^1 \left(y_r^2 \right) - V^2 \left(y_r^2 \right) \right) dx_r^2 \right| \\
&\leq \left(v \left| y^1 - y^2 \right|_{\infty;[0,T]} + \left| V^1 - V^2 \right|_\infty \right) \left| x^2 \right|_{\text{1-var};[s,t]} \\
&\quad + v \left| x^1 - x^2 \right|_{\text{1-var};[s,t]}.
\end{aligned}
$$

As the right-hand side is a control, it follows that

$$
\begin{aligned}
\left| y^1 - y^2 \right|_{\text{1-var};[s,t]} &\leq \left(v \left| y^1 - y^2 \right|_{\infty;[0,T]} + \left| V^1 - V^2 \right|_\infty \right) \left| x^2 \right|_{\text{1-var};[s,t]} \\
&\quad + v \left| x^1 - x^2 \right|_{\text{1-var};[s,t]}.
\end{aligned}
$$

Using Theorem 3.15, and replacing s, t by $0, T$, we then obain (3.12), as claimed. ∎

Remark 3.19 The interval $[0, T]$ in the above theorem is of course arbitrary. In particular, this means that we also have for all $s, t \in [0, T]$,

$$
\left| y^1 - y^2 \right|_{\text{1-var};[s,t]} \leq 2 \left(\left| y_s^1 - y_s^2 \right| v\ell + v \left| x^1 - x^2 \right|_{\text{1-var};[s,t]} + \left| V^1 - V^2 \right|_\infty \ell \right) e^{3v\ell}, \tag{3.13}
$$

where ℓ is a bound on $\max \left\{ \left| x^2 \right|_{\text{1-var};[s,t]}, \left| x^1 \right|_{\text{1-var};[s,t]} \right\}$. □

As before, we can relax the assumption on the vector fields and still keep a uniform Lipschitz bound on bounded sets.

Corollary 3.20 *We consider*
(i) $V^1 = \left(V_1^1, \ldots, V_d^1 \right)$ *and* $V^2 = \left(V_1^2, \ldots, V_d^2 \right)$ *are two collections of locally* Lip^1 *vector fields on* $\mathbb{R}^e$, *with linear growth;*
(ii) $y_0^1, y_0^2 \in \mathbb{R}^e$ *are initial conditions, with* $\left| y_0^i \right| \leq R$ *for some* $R \geq 0$;
(iii) x^1 *and* x^2 *are two paths in* $C^{\text{1-var}} \left([0, T], \mathbb{R}^d \right)$, *with* $\max_{i=1,2} \left| x^i \right|_{\text{1-var};[0,T]} \leq \ell$ *for some* $\ell \geq 0$.
Then, if $y^i = \pi_{(V^i)} \left(0, y_0^i; x^i \right)$ *for* $i = 1, 2$, *there exist constants* C, M *depending only on* R, ℓ *and the vector fields, such that*

$$
\left| y^1 - y^2 \right|_{\text{1-var};[0,T]} \leq C \left(\left| y_0^1 - y_0^2 \right| + \left| x^1 - x^2 \right|_{0;[0,T]} + \left| V^1 - V^2 \right|_{\infty;B(0,M)} \right).
$$

The 1-variation estimates imply 1-Hölder estimates:

Exercise 3.21 *Under the same assumptions as the one in Corollary 3.20, and assuming* x^1 *and* x^2 *to be 1-Hölder, prove the existence of constants*

C, M *depending on* $\max_i |y_0^i|$ *and* $\max_{i=1,2} |x^i|_{1\text{-Höl};[0,T]}$ *and the vector fields, such that*

$$|y^1 - y^2|_{1\text{-Höl};[0,T]} \leq C \left(|y_0^1 - y_0^2| + |x^1 - x^2|_{1\text{-Höl};[0,T]} + |V^1 - V^2|_{\infty;B(0,M)} \right).$$

Solution. We will use $|V|_{\mathrm{Lip}^1} |x|_{1\text{-var};[s,t]} \leq |V|_{\mathrm{Lip}^1} |x|_{1\text{-Höl};[s,t]} |t - s|$. We may take the vector fields to be 1-Lipschitz, as the result then follows by a localization argument. Define $\ell = \max_i |x^i|_{1\text{-Höl};[0,T]}$. From (3.13) we obtain

$$\frac{|y_{s,t}^1 - y_{s,t}^2|}{t - s} \leq 2 \left(|y_s^1 - y_s^2| \cdot v\ell + v |x^1 - x^2|_{1\text{-Höl};[0,T]} + |V^1 - V^2|_{\infty} \ell \right) e^{3v\ell T}.$$

Replacing $|y_s^1 - y_s^2|$ on the right-hand side by $|y^1 - y^2|_{\infty}$, followed by taking the supremum over all $s < t$ in $[0, T]$, leads to an estimate of the form

$$|y^1 - y^2|_{1\text{-Höl};[0,T]} \leq 2 \left(|y^1 - y^2|_{\infty} v\ell + \cdots \right) \exp\left(3v\ell T\right).$$

We then conclude with Theorem 3.15. □

3.6 Comments

There are many books on ODE theory, such as the authorative Hartman [84]; for a concise treatment, see the relevant chapters of Driver [45]. The class of ODEs studied here, where the time-inhomogeneity factorizes in the form of a multidimensional driving signal, is particularly important in (non-linear) control theory, see the relevant contributions in Agrachev [1] for instance. Continuity in the starting point (the "flow") is well known, and its further regularity is discussed in Chapter 4. Continuity in the driving signal is harder to find in the literature but also well known, see Lyons and coworkers [120, 123] and the references cited therein.

4

ODEs: smoothness

We remain in the ODE setting of the previous chapter; that is, we consider differential equations of the form

$$dy = V(y)\, dx, \quad y(0) = y_0,$$

where $x = x(t)$ is a $\mathbb{R}^d$-valued continuous path of bounded variation. In the present chapter we investigate various smoothness properties of the solution, in particular as a function of y_0 and x.

4.1   Smoothness of the solution map

We saw in the last chapter (cf. Theorem 3.18) that Lip^1-regularity of the vector fields leads to (local Lipschitz) continuity of the solution map $\pi_{(V)}(0, y_0; x)$ as a function of the initial condition y_0, the driving signal x and the vector fields $V = (V_1, \dots, V_d)$. Under the slightly stronger regularity assumption of C^1-boundedness we now show that $\pi_{(V)}(0, y_0; x)$ is differentiable in y_0 and x. (For simplicity, we do not discuss differentiability in V.) In fact, we shall see that C^k-boundedness allows for k derivatives of $\pi_{(V)}(0, y_0; x)$ in y_0 and x. As earlier, in the following definition $V = (V_1, \dots, V_d)$ is regarded as a map from $\mathbb{R}^e$ to $L(\mathbb{R}^d, \mathbb{R}^e)$, equipped with operator norm.

Definition 4.1 We say that $V : \mathbb{R}^e \to L(\mathbb{R}^d, \mathbb{R}^e)$ is C^k-bounded if (i) it is k-times Fréchet differentiable and (ii) $(V, DV, \dots, D^k V)$ is a bounded function on $\mathbb{R}^e$. We then set

$$|V|_{\mathrm{C}^k} := \max_{i=0,\dots,k} \left| D^i V \right|_\infty.$$

If only (i) holds, we write $V \in \mathrm{C}^k_{\mathrm{loc}}$. □

4.1.1   Directional derivatives

Lemma 4.2 *Let $V = (V_1, \dots, V_d)$ be a collection of continuously differentiable vector fields, that is $V \in \mathrm{C}^1(\mathbb{R}^e, L(\mathbb{R}^d, \mathbb{R}^e))$. Then, for all $\varepsilon > 0$ and for all bounded sets $U \subset \mathbb{R}^e$, there exists δ such that for all $b, a \in U$,*

$$|b - a| < \delta \implies |V(b) - V(a) - DV(a) \cdot (b - a)| \le \varepsilon |b - a|.$$

Proof. By the fundamental theorem of calculus and the chain rule,

$$|V(b) - V(a) - DV(a) \cdot (b - a)|$$

$$= \left| \left(\int_0^1 [DV(a + t(b - a)) - DV(a)] \, dt \right) \cdot (b - a) \right|$$

$$\leq |b - a| \int_0^1 |DV(a + t(b - a)) - DV(a)| \, dt.$$

We conclude using the fact that DV is uniformly continuous on bounded sets in $\mathbb{R}^e$. ∎

Condition 4.3 (non-explosion) We say that a collection of vector fields $V = (V_1, \ldots, V_d)$ on $\mathbb{R}^e$ satisfies the non-explosion condition if for all $R > 0$, there exists $M > 0$ such that if $(y_0, x) \in \mathbb{R}^e \times C^{1\text{-var}}([0, T], \mathbb{R}^d)$ with $|x|_{1\text{-var}} + |y_0| \leq R$, then

$$\left| \pi_{(V)}(0, y_0; x) \right|_{\infty; [0, T]} < M. \qquad \square$$

Following our usual convention, we agree that, in the case of non-uniqueness, $\pi_{(V)}(0, y_0; x)$ stands for any ODE solutions driven by x along vector fields V started at y_0. For example, a collection of continuous vector fields of linear growth satisfies the non-explosion condition.

Theorem 4.4 *(directional derivatives in starting point and driving signal)* *We fix a collection of C^1_{loc}-vector fields on $\mathbb{R}^e$, $V = (V_1, \ldots, V_d)$, satisfying the non-explosion condition. Then,*
(i) the map[1]

$$(y_0, x) \in \mathbb{R}^e \times C^{1\text{-var}}([0, T], \mathbb{R}^d) \mapsto y \equiv \pi(0, y_0; x) \in C^{1\text{-var}}([0, T], \mathbb{R}^e)$$

has directional derivatives[2]

$$D_{(v,h)}\pi_{(V)}(0, y_0; x) := \left\{ \frac{d}{d\varepsilon} \pi_{(V)}(0, y_0 + \varepsilon v; x + \varepsilon h) \right\}_{\varepsilon = 0} \in C^{1\text{-var}}([0, T], \mathbb{R}^e)$$

in all directions $(v, h) \in \mathbb{R}^e \times C^{1\text{-var}}([0, T], \mathbb{R}^d)$;
(ii) define the bounded variations paths

$$t \mapsto M_t := M_t^{y_0, x} := \sum_{i=1}^d \int_0^t DV_i(y_r) \, dx_r^i \in M_e(\mathbb{R}) \qquad (4.1)$$

(where $M_e(\mathbb{R})$ denotes real $(e \times e)$-matrices) and also

$$t \mapsto H_t := H_t^{y_0, x; h} := \sum_{i=1}^d \int_0^t V_i(y_r) \, dh_r^i \in \mathbb{R}^e \qquad (4.2)$$

---

[1] Since V remains fixed we write π instead of $\pi_{(V)}$.
[2] The derivate exists as a (strong) limit in the Banach space $C^{1\text{-var}}([0, T], \mathbb{R}^e)$.

then $z = D_{(v,h)} \pi_{(V)} (0, y_0; x)$ is the (unique) solution of the linear ODE

$$\begin{cases} dz_t = dM_t^{y_0, x} \cdot z_t + dH_t^{y_0, x; h}, \\ z_0 = v. \end{cases} \tag{4.3}$$

Remark 4.5 Observe that $(y, z) = \left(\pi_{(V)} (0, y_0; x), D_{(v,h)} \pi_{(V)} (0, y_0; x) \right)$ solves the ODE driven by (x, h) given by formal differentiation, namely $dy = V(y) \, dx$, $dz = (DV(y) \, dx) \cdot z + V(y) \, dh$ or, in more detail,

$$\begin{cases} dy_t = \sum_{i=1}^{d} V_i(y_t) \, dx_t^i \\ dz_t = \sum_{i=1}^{d} (DV_i(y_t) \cdot z_t) \, dx_t^i + \sum_{i=1}^{d} V_i(y_t) \, dh_t^i \end{cases}$$

started at $(y_0, z_0) = (y_0, v)$. With $V \in C_{\mathrm{loc}}^1$, DV is continuous and so the vector fields of the ODE for (y, z) are continuous but in general not C_{loc}^1. Nonetheless, it has a unique solution (thanks to the specific structure: first solve for y, then M, H, then z) which satisfies the non-explosion condition. Indeed, this is a straightforward application of the estimates for ODE solutions and Riemann–Stieltjes integrals: estimate y in terms of $(y_0, x.)$, then M, H in terms of $(x., h., y.)$ and finally z in terms of $(v, M., H.)$. □

Proof. We first notice that by a localization argument, we can assume that V is compactly supported. With $(y_0, x), (v, h) \in \mathbb{X} \equiv \mathbb{R}^e \times C^{1\text{-var}} \left([0, T], \mathbb{R}^d \right)$ fixed write

$$y_t^\varepsilon = \pi_{(V)} (0, y_0 + \varepsilon v; x + \varepsilon h)_t, \quad y \equiv y^0$$

and also $z^\varepsilon = (y^\varepsilon - y) / \varepsilon$ for $\varepsilon > 0$. Define $z \in C^{1\text{-var}} \left([0, T], \mathbb{R}^e \right)$ as the (unique) ODE solution to (4.3).

Step 1: We first establish that

$$\lim_{\varepsilon \to 0} z^\varepsilon = z \text{ in } \mathbb{Y}^\infty \tag{4.4}$$

with $\mathbb{Y}^\infty := C \left([0, T], \mathbb{R}^d \right)$, a Banach space under the ∞-norm. From the respective ODEs for y, y^ε and z,

$$\begin{aligned} z_t^\varepsilon - z_t &= \sum_{i=1}^{d} \int_0^t \left[\frac{1}{\varepsilon} (V_i(y_s^\varepsilon) - V_i(y_s)) - DV_i(y_s) \cdot z_s \right] dx_s^i \\ &\quad + \sum_{i=1}^{d} \int_0^t (V_i(y_s^\varepsilon) - V_i(y_s)) \, dh_s^i \\ &= \sum_{i=1}^{d} (\Delta_1^i(0, t) + \Delta_2^i(0, t) + \Delta_3^i(0, t)) \end{aligned}$$

with

$$\Delta_1^i(s,t) = \int_s^t DV_i(y_r) \cdot (z_r^\varepsilon - z_r)\, dx_r^i,$$

$$\Delta_2^i(s,t) = \int_s^t \frac{1}{\varepsilon}\left[V_i(y_r^\varepsilon) - V_i(y_r) - DV_i(y_r)\cdot(y_r^\varepsilon - y_r)\right] dx_r^i,$$

$$\Delta_3^i(s,t) = \int_s^t (V_i(y_r^\varepsilon) - V_i(y_r))\, dh_r^i.$$

First observe that Theorems 3.4 and 3.18 apply (as $V \in C_c^1 \subset \mathrm{Lip}^1$) and we have

$$R := \sup_{\substack{t\in[0,T] \\ \varepsilon\in[0,1]}} |y_t^\varepsilon| < \infty,$$

$$|z^\varepsilon|_{1\text{-var};[0,T]} \le c_1\left(|v| + |h|_{1\text{-var};[0,T]}\right) =: c_2.$$

Fix $\eta > 0$. From $R < \infty$ and Lemma 4.2, we see that there exists $\delta > 0$ such that

$$|y_r^\varepsilon - y_r| < \delta \text{ implies } \frac{1}{\varepsilon}|V_i(y_r^\varepsilon) - V_i(y_r) - DV_i(y_r)\cdot(y_r^\varepsilon - y_r)| \le \eta\,|z_r^\varepsilon|.$$

Using $|y_r^\varepsilon - y_r| \le \varepsilon c_2$, this means that there exists $\varepsilon_0 > 0$ such that $\varepsilon < \varepsilon_0$ implies

$$\sup_{r\in[0,T]} \frac{1}{\varepsilon}|V_i(y_r^\varepsilon) - V_i(y_r) - DV_i(y_r)\cdot(y_r^\varepsilon - y_r)| \le \eta\,|z_r^\varepsilon| \le c_2\eta.$$

In particular, we obtain that

$$\sum_{i=1}^d |\Delta_2^i(s,t)| \le c_3\eta\,|x|_{1\text{-var};[s,t]}.$$

Bounding $|\Delta_3^i(s,t)|$ is even easier; indeed,

$$\sum_{i=1}^d |\Delta_3^i(s,t)| \le \sum_{i=1}^d |V_i|_{\mathrm{Lip}^1}\sup_{r\in[s,t]}|y_r^\varepsilon - y_r|.|h|_{1\text{-var};[s,t]}$$

$$\le c_4\varepsilon\,|h|_{1\text{-var};[s,t]}.$$

Finally, as the vector fields are Lipschitz, we have

$$\sum_{i=1}^d |\Delta_1^i(s,t)| = c_5\int_s^t |z_r^\varepsilon - z_r|.|dx_r|. \tag{4.5}$$

Putting things together, we obtain that for $\varepsilon < \varepsilon_0$,

$$|z_t^\varepsilon - z_t| \le c_5 \int_0^t |z_s^\varepsilon - z_s| \cdot |dx_s| + \left(c_4 \varepsilon \, |h|_{1\text{-var};[0,t]} + c_3 \eta \, |x|_{1\text{-var};[0,t]} \right).$$

By Gronwall's lemma, we obtain that

$$\sup_{t \in [0,T]} |z_t^\varepsilon - z_t| \le \left(c_4 \varepsilon \, |h|_{1\text{-var};[0,T]} + c_3 \eta \, |x|_{1\text{-var};[0,T]} \right) \exp \left(c_5 \, |x|_{1\text{-var};[0,T]} \right),$$

(4.6)

so that $\overline{\lim}_{\varepsilon \to 0} |z^\varepsilon - z|_{\infty;[0,T]} \le c_6 \, (\eta + \varepsilon)$ and since $\eta > 0$ was arbitrary it follows that $\lim_{\varepsilon \to 0} |z^\varepsilon - z|_{\infty;[0,T]} = 0$.

Step 2: Define $\hat{z}^\varepsilon$ to be the solution of

$$\begin{cases} d\hat{z}_t^\varepsilon = dM_t^{y_0 + \varepsilon, x + \varepsilon h} \cdot \hat{z}_t^\varepsilon + dH_t^{y_0 + \varepsilon, x + \varepsilon h; h}, \\ \hat{z}_0^\varepsilon = v. \end{cases}$$

As there was nothing special about $\varepsilon = 0$ in the first step, we have actually just showed that

$$\varepsilon \in [0,1] \mapsto \pi_{(V)}(0, y_0 + \varepsilon v; x + \varepsilon h) \in \mathbb{Y}^\infty := C([0,T], \mathbb{R}^e)$$

is differentiable with derivative $\hat{z}^\varepsilon$. Now

$$\varepsilon \mapsto \left(M_t^{y_0 + \varepsilon, x + \varepsilon h}, H_t^{y_0 + \varepsilon, x + \varepsilon h; h} \right) \mapsto \hat{z}^\varepsilon \in \mathbb{Y}^1 = C^{1\text{-var}}([0,T], \mathbb{R}^e)$$

is continuous (from the continuity properties of the solution map and Riemann–Stieltjes integration respectively). Therefore, from Proposition B.1 in Appendix B,

$$\varepsilon \in [0,1] \mapsto \pi_{(V)}(0, y_0 + \varepsilon v; x + \varepsilon h) \in \mathbb{Y}^1$$

is differentiable; that is, the limit when $\varepsilon \to 0$ of

$$\varepsilon^{-1} \left(\pi_{(V)}(0, y_0 + \varepsilon v; x + \varepsilon h) - \pi_{(V)}(0, y_0 + v; x + h) \right)$$

exists in $\mathbb{Y}^1$. The proof is now finished. ∎

Proposition 4.6 *(higher-order directional derivatives) Let $k \in \{1, 2, \dots\}$. Assume $V = (V_1, \dots, V_d)$ is a collection of C_{loc}^k-vector fields on $\mathbb{R}^e$ satisfying the non-explosion condition. Then*

$$(y_0, x) \mapsto \pi_{(V)}(0, y_0; x)$$

has (up to) kth-order directional derivatives in the sense that, for all $(v_i,$ $h_i)_{1 \le i \le k} \in \left(\mathbb{R}^e \times C^{1\text{-var}}([0,T], \mathbb{R}^d) \right)^{\times k}$,

$$D_{(v_i, h_i)_{1 \le i \le k}}^k \pi_{(V)}(0, y_0; x) := \left\{ \frac{\partial^k}{\partial \varepsilon_1 \dots \partial \varepsilon_k} \pi \left(0, y_0 + \sum_{j=1}^k \varepsilon_j v_j, x + \sum_{j=1}^k \varepsilon_j h_j \right) \right\}_{\varepsilon = 0}$$

exists as a strong limit in the Banach space $C^{1\text{-var}}\left(\left[0,T\right],\mathbb{R}^{e}\right)$. Furthermore, the directional derivatives satisfy the control ODEs obtained by formal differentiation.

Proof. This follows by simple induction: for $j \geq 1$, a solution of an ODE driven by C_{loc}^{j}-vector fields satisfying the non-explosion condition admits a derivative in any arbitrary direction in its starting point and driving signal, and the derivative in such directions, together with the driving signal, satisfies an ODE driven along C_{loc}^{j-1} vector fields that satisfies the non-explosion condition. ∎

4.1.2 Fréchet differentiability

We now show that the solution map to $dy = V\left(y\right)dx$ is continuously Fréchet differentiable in the starting point and driving signal.

Theorem 4.7 *Let $V = \left(V_{1},\ldots,V_{d}\right)$ be a collection of C_{loc}^{1}-vector fields on $\mathbb{R}^{e}$ satisfying the non-explosion condition. Then the map*

$$\left(y_{0},x\right) \in \mathbb{R}^{e} \times C^{1\text{-var}}\left(\left[0,T\right],\mathbb{R}^{e}\right) \mapsto y \equiv \pi_{(V)}\left(0,y_{0};x\right) \in C^{1\text{-var}}\left(\left[0,T\right],\mathbb{R}^{e}\right)$$

is C^{1} in the Fréchet sense.

Proof. From Corollary B.5, we only need to show that the map $\left(y_{0},x\right),\left(v,h\right)$ $\to D_{(v,h)}\pi_{(V)}\left(0,y_{0};x\right)$ from $\left(\mathbb{R}^{e} \times C^{1\text{-var}}\left(\left[0,T\right],\mathbb{R}^{d}\right)\right)^{\times 2}$ into $C^{1\text{-var}}\left(\left[0,T\right],\mathbb{R}^{e}\right)$ is uniformly continuous on bounded sets. This follows from the uniform continuity on bounded sets of the maps

$$\left(y_{0},x\right),\left(v,h\right) \xrightarrow{\phi_{1}} \left(\begin{array}{c}\left(y_{0},x\right),\left(v,h\right) \\ \pi_{(V)}\left(0,y_{0};x\right)\end{array}\right) \xrightarrow{\phi_{2}} \left(\begin{array}{c}v \\ M^{y_{0},x} \\ H^{y_{0},x}\end{array}\right) \xrightarrow{\phi_{3}} D_{(v,h)}\pi_{(V)}\left(0,y_{0};x\right);$$

ϕ_{1} and ϕ_{3} because of Theorem 3.20, and ϕ_{2} because of Corollary 2.8. ∎

We now discuss C^{k}-Fréchet differentiability of the map $\left(y_{0},x\right) \mapsto \pi_{(V)}\left(0,y_{0};x\right)$.

Proposition 4.8 *(higher-order Fréchet) Let $k \geq 1$, and $V = \left(V_{1},\ldots,V_{d}\right)$ a collection of C_{loc}^{k}-vector fields on $\mathbb{R}^{e}$ satisfying the non-explosion condition. Then the map*

$$\left(y_{0},x\right) \in \mathbb{R}^{e} \times C^{1\text{-var}}\left(\left[0,T\right],\mathbb{R}^{e}\right) \mapsto y \equiv \pi_{(V)}\left(0,y_{0};x\right) \in C^{1\text{-var}}\left(\left[0,T\right],\mathbb{R}^{e}\right)$$

is C^{k} in the Fréchet sense.

Proof. The map $\left(y_{0},x\right),\left(v_{i},h_{i}\right)_{1 \leq i \leq k} \mapsto D_{(v_{i},h_{i})_{1 \leq i \leq k}}^{k}\pi_{(V)}\left(0,y_{0};x\right)$ is uniformly continuous on bounded sets because of the uniform continuity on bounded sets of the solution map and the integral. This is enough to conclude the proof using Corollary B.11 in Appendix B. ∎

It can be convenient in applications to view $\pi_{(V)}(0, \cdot; x)$ as a *flow of C^k-diffeomorphisms,* that is, an element in the space of all $\phi : [0, T] \times \mathbb{R}^e \to \mathbb{R}^e : (t, y) \mapsto \phi_t(y)$ such that

$$\begin{cases} \forall t \in [0, T] : \phi_t \text{ is a } C^k\text{-diffeomorphism of } \mathbb{R}^e, \\ \forall \alpha : |\alpha| \leq k : \partial_\alpha \phi_t(y), \partial_\alpha \phi_t^{-1}(y) \text{ are continuous in } (t, y). \end{cases}$$

Corollary 4.9 *Under the assumptions of Proposition 4.8, the map $(t, y_0) \mapsto \pi_{(V)}(0, y_0; x)$ is a flow of C^k-diffeomorphisms.*

Proof. It is clear from Proposition 4.8 that $y_0 \in \mathbb{R}^e \mapsto \pi_{(V)}(0, y_0; x)_t$ is in $C^k(\mathbb{R}^e, \mathbb{R}^e)$. Moreover, it follows from Proposition 3.13 that

$$\pi_{(V)}(0, \cdot; x)_t^{-1} = \pi_{(V)}(0, \cdot; \overleftarrow{x})_t$$

where $\overleftarrow{x}(\cdot) = x(t - \cdot) \in C^{1\text{-var}}([0, t], \mathbb{R}^d)$; we see that $\pi_{(V)}(0, \cdot; x)_t$ is a bijection whose inverse is also in $C^k(\mathbb{R}^e, \mathbb{R}^e)$ and conclude that each $\pi_{(V)}(0, \cdot; \mathbf{x})_t$ is indeed a C^k-diffeomorphism of $\mathbb{R}^e$. At last, each ∂_α-derivative of $\pi_{(V)}(0, \cdot; x)_t$ resp. $\pi_{(V)}(0, \cdot; x)_t^{-1}$ can be represented as a (non-explosive) ODE solution which plainly implies joint continuity in t and y_0. ∎

Exercise 4.10 *Prove Proposition 4.8 with $C^{1\text{-var}}$ replaced throughout by (i) $C^{1\text{-Höl}}$ and (ii) $W^{1,2}$.*

We finish this section with a representation formula for directional derivatives.

Proposition 4.11 (Duhamel's principle) *Consider $(y_0, x) \in \mathbb{R}^e \times C^{1\text{-var}}\left([0, T], \mathbb{R}^d\right)$, a collection of C^1_{loc}-vector fields on $\mathbb{R}^e$, $V = (V_1, \ldots, V_d)$ satisfying the non-explosion condition and write $y \equiv \pi_{(V)}(0, y_0; x) \in C^{1\text{-var}}([0, T], \mathbb{R}^e)$ for the unique ODE solution. Define*

$$M_t = \sum_{i=1}^d \int_0^t DV_i(y_r)\, dx_r^i \in M_e(\mathbb{R})$$

and $J.$ as the $M_e(\mathbb{R})$-valued (unique) solution to the linear ODE,

$$dJ_t = dM_t \cdot J_t, \quad J_0 = I \tag{4.7}$$

(where $\cdot$ denotes matrix multiplication and I the identity matrix). More generally, given $0 \leq s \leq t \leq T$ write $J_{t \leftarrow s}$ for the solution of this ODE started at I at time s. Then $J_{t \leftarrow s}$ is the Jacobian of $\pi_{(V)}(s, \cdot; x)_t : \mathbb{R}^e \to \mathbb{R}^e$ at y_s and we may write $J_{t \leftarrow s} =: J_{t \leftarrow s}^{y_s, x}$ to indicate this. Moreover, the following representation formula holds:

$$D_{(v,h)}\pi_{(V)}(0, y_0; x)_t = D_{(v,h)}y_t = J_{t \leftarrow 0}^{y_0, x} \cdot v + \sum_{i=1}^d \int_0^t J_{t \leftarrow s}^{y_s, x} \cdot V_i(y_s)\, dh_s^i. \tag{4.8}$$

Proof. By Theorem 4.4, for $0 \leq t \leq T$, the flow map $y_0 \mapsto \pi_{(V)}(0, y_0; x)_t$ from $\mathbb{R}^e \to \mathbb{R}^e$ admits partial derivatives in all directions. These are easily seen to be continuous (much more will be shown soon) and so $\pi_{(V)}(0, \cdot; x)_t \in C^1(\mathbb{R}^e, \mathbb{R}^e)$. Its differential (the "Jacobian") at some point y_0, viewed as an $\mathbb{R}^{e \times e}$-matrix, is of the form $\tilde{J}_t = (z_1 | \ldots | z_e)$ where $z_i = z_i(t) = D_{(\mathfrak{b}_i, 0)} \pi_{(V)}(0, y_0; x)_t$ and $(\mathfrak{b}_i)$ denotes the canonical basis of $\mathbb{R}^e$. From Theorem 4.4, $z_i = z_i(t)$ is the solution of a linear ODE of the form $dz_i(t) = dM_t \cdot z_i(t)$ with $z_i(0) = \mathfrak{b}_i$. Equivalently, $\tilde{J}_t$ is the solution of (4.7) started at I at time 0 and by ODE uniqueness, $\tilde{J}_t = J_t$.

The matrix J_t remains invertible for all $t \in [0, T]$. Indeed, its inverse is constructed explictly as the (unique) ODE solution to $dK_t = -K_t \cdot dM_t$ with $K_0 = I$. To see this, we just observe that $d(K_t J_t) = -K_t dM_t J_t + K_t dM_t J_t = 0$.

Of course, there is nothing special about time 0 and the same reasoning shows that, for $0 \leq s \leq t \leq T$, the flow map $\pi_{(V)}(s, \cdot; x)_t$ is in $C^1(\mathbb{R}^e, \mathbb{R}^e)$ with Jacobian given by (the invertible matrix) $J_{t \leftarrow s}$. The chain rule in conjunction with

$$\pi_{(V)}(s, y_s; x)_u = \pi_{(V)}(t, \pi(s, y_s, x)_t; x)_u, \quad 0 \leq s \leq t \leq u \leq T$$

implies[3]

$$J_{u \leftarrow s} = J_{u \leftarrow t} \cdot J_{t \leftarrow s}, \quad 0 \leq s \leq t \leq u \leq T$$

and by defining $J_{s \leftarrow t} := (J_{t \leftarrow s})^{-1}$, $0 \leq s \leq t$, this remains valid for all $s, t, u \in [0, T]$. The validity of (4.8) is nothing more than a variation-of-constants ODE argument (also known as Duhamel's principle) which represents the solution to the inhomogenous equation

$$dz_t = dM_t \cdot z_t + dH_t, \quad z_0 = v,$$

which is precisely $D_{(v,h)} \pi_{(V)}(0, y_0; x)$, in terms of the solution of the homogenous equation, i.e. the ODE satisfied by the Jacobian. More precisely, it suffices to observe that

$$J_{0 \leftarrow t} z_t - v = \int_0^t d(J_{0 \leftarrow s} z_s) = \int_0^t J_{0 \leftarrow s} dH_s = \sum_{i=1}^d \int_0^t J_{0 \leftarrow s} V_i(y_s) \, dh_s^i.$$

Using $(J_{0 \leftarrow t}^x)^{-1} \cdot J_{0 \leftarrow s}^x = J_{t \leftarrow s}^x$ the representation formula (4.8) now follows from simple algebra. ∎

Remark 4.12 The underlying geometry helps to "read" these equations. The flow $\pi_{(V)}(s, \cdot; x)_t$ maps $y_s \mapsto y_t$ (where y is the solution of the ODE

---

[3] The notation $J_{t \leftarrow s}$ (rather than $J_{s \rightarrow t}$) has the advantage of suggesting the right order of matrix multiplication.

driven by x) and its matrix-valued Jacobian should be viewed as a linear map between the respective tangent spaces, i.e.

$$J_{t \leftarrow s} = J_{t \leftarrow s}^{y_s, x} \in L\left(\mathcal{T}_{y_s} \mathbb{R}^e, \mathcal{T}_{y_t} \mathbb{R}^e\right).$$

From the very nature of vector fields, $V_i\left(y_s\right) = V_i|_{y_s} \in \mathcal{T}_{y_s} \mathbb{R}^e$ and $J_{t \leftarrow s} \cdot V_i\left(y_s\right) \in \mathcal{T}_{y_t} \mathbb{R}^e$. In particular, we should think of (4.8) as equality between elements in $\mathcal{T}_{y_t} \mathbb{R}^e$ rather than just $\mathbb{R}^e$.   □

4.2   Comments

Although (or maybe because) the results are unsurprising we are unaware of good references to the smoothness topics discussed here. In a more general Young context, related smoothness properties have been discussed by Li and Lyons [109]. Differential equations driven by $W^{1,2}$-paths (a special case of Exercise 4.10) was Bismut's starting point in [15]; the resulting Hilbert structure of the input signal is convenient in discussing non-degeneracy properties of the solution map. Differential equations driven by $W^{1,2}$-paths also arise naturally in support and large-deviation statements for stochastic differential equations, which we shall encounter in Part IV.

5

Variation and Hölder spaces

We return to the abstract setting of Section 1.1, where we introduced $C\left([0,T],E\right)$, the space continuous paths defined on $[0,T]$ with values in a metric space (E,d), followed by a detailed discussion of continuous paths of finite 1-variation ("bounded variation"). The purpose of the present chapter is to carry out a similar discussion for p-variation and $1/p$-Hölder regularity, $p \in [1,\infty)$. In the later applications to rough paths, E will be a Lie group whose dimension depends on $[p]$, the integer part of p.

5.1 Hölder and p-variation paths on metric spaces

5.1.1 Definition and first properties

We start by defining α-Hölder and p-variation distances.

Definition 5.1 Let (E,d) be a metric space. A path $x : [0,T] \to E$ is said to be
(i) Hölder continuous with exponent $\alpha \geq 0$, or simply α-Hölder, if

$$|x|_{\alpha\text{-Höl};[0,T]} := \sup_{0 \leq s < t \leq T} \frac{d\left(x_s, x_t\right)}{|t-s|^\alpha} < \infty; \tag{5.1}$$

(ii) of finite p-variation for some $p > 0$ if

$$|x|_{p\text{-var};[0,T]} := \left(\sup_{(t_i) \in \mathcal{D}([0,T])} \sum_i d\left(x_{t_i}, x_{t_{i+1}}\right)^p \right)^{1/p} < \infty. \tag{5.2}$$

We will use the notations $C^{\alpha\text{-Höl}}([0,T],E)$ for the set of α-Hölder paths x and $C^{p\text{-var}}([0,T],E)$ for the set of continuous paths $x : [0,T] \to E$ of finite p-variation. $\qquad\square$

It is obvious from these definitions that a path $x : [0,T] \to E$ is constant, i.e. $x_t \equiv o$ for some $o \in E$, if and only if $|x|_{\alpha\text{-Höl};[0,T]} = 0$ and if and only if $|x|_{p\text{-var};[0,T]} = 0$. (In particular, if $E = \mathbb{R}^d$ our quantities (5.1), (5.2) are only *semi-norms*.)

Observe that $C^{0\text{-Höl}}([0,T],E)$ is nothing but the set of continuous paths from $[0,T]$ into E and $|x|_{0\text{-Höl};[0,T]} = |x|_{0;[0,T]}$, where the latter was defined in Section 1.1. Any $\alpha > 0$ can be written as $\alpha = 1/p$ and it is obvious that any $(1/p)$-Hölder path is a continuous path of finite p-variation. Although a path of finite p-variation need not be continuous (e.g. a step-function), our focus is on continuous paths. The following simple proposition then explains why our main interest lies in

$$\alpha \in [0,1] \text{ and } p \geq 1.$$

Proposition 5.2 *Assume* $x : [0,T] \to E$ *is* α-*Hölder continuous, with* $\alpha \in (1,\infty)$, *or continuous of finite p-variation with* $p \in (0,1)$. *Then* x *is constant, i.e.* $x\left(\cdot\right) \equiv x_0$.

Proof. Since α-Hölder paths have finite p-variation with $p = 1/\alpha$ it suffices to consider the case when x is continuous of finite p-variation with $p < 1$. Consider a dissection $D = (t_i) \in \mathcal{D}\left([0,T]\right)$ with mesh $|D|$. Then

$$
\begin{aligned}
d\left(x_0, x_T\right) &\leq \sum_i d\left(x_{t_i}, x_{t_{i+1}}\right) \\
&\leq \max_i d\left(x_{t_i}, x_{t_{i+1}}\right)^{1-p} M^p
\end{aligned}
$$

where $M = |x|_{p\text{-var};[0,T]} < \infty$. Using uniform continuity of x on $[0,T]$, we can make $\max_i d\left(x_{t_i}, x_{t_{i+1}}\right)$ arbitrarily small by taking a dissection with small enough mesh $|D| = \max_i |t_{i+1} - t_i|$. ∎

The case $p = 1$ resp. $\alpha = 1$ was already discussed in detail in Section 1.2 and heavily used in our discussion of ODEs driven by continuous paths of bounded variation. We now begin a systematic study of p-variation, generalizing much of the familiar $p = 1$ case.

Proposition 5.3 *Let* $x \in C\left([0,T],E\right)$. *Then, if* $1 \leq p \leq p' < \infty$,

$$|x|_{p'\text{-var};[0,T]} \leq |x|_{p\text{-var};[0,T]} \cdot$$

In particular, $C^{p\text{-var}}([0,T],E) \subset C^{p'\text{-var}}\left([0,T],E\right)$.

Proof. This follows from the elementary inequality

$$\left(\sum |a_i|^{p'}\right)^{1/p'} \leq \left(\sum |a_i|^p\right)^{1/p} \cdot$$

∎

Exercise 5.4 *Formulate and prove the Hölder version of Proposition 5.3.*

Proposition 5.5 *(interpolation) Let* $x \in C\left([0,T],E\right)$. *(i) For* $1 \leq p < p' < \infty$, *we have*

$$|x|_{p'\text{-var};[0,T]} \leq \left(|x|_{p\text{-var};[0,T]}\right)^{p/p'} \left(|x|_{0;[0,T]}\right)^{1-p/p'} \cdot$$

(ii) For $1 \geq \alpha > \alpha' \geq 0$, *we have*

$$|x|_{\alpha'\text{-Höl};[0,T]} \leq \left(|x|_{\alpha\text{-Höl};[0,T]}\right)^{\alpha'/\alpha} \left(|x|_{0;[0,T]}\right)^{1-\alpha'/\alpha} \cdot$$

Proof. (i) Observe

$$\sum_i d\left(x_{t_i}, x_{t_{i+1}}\right)^{p'} = \sum_i \left(d\left(x_{t_i}, x_{t_{i+1}}\right)^p d\left(x_{t_i}, x_{t_{i+1}}\right)^{p'-p}\right)$$

$$\leq |x|_0^{p'-p} \sum_i d\left(x_{t_i}, x_{t_{i+1}}\right)^p,$$

then pass to the respective suprema over all dissections (t_i), and raise to the power $1/p'$.

(ii) Follows from

$$\frac{d\left(x_s, x_t\right)}{|t-s|^{\alpha'}} \leq \left(\frac{d\left(x_s, x_t\right)}{|t-s|^{\alpha}}\right)^{\alpha'/\alpha} d\left(x_s, x_t\right)^{1-\alpha'/\alpha} \leq \left(\frac{d\left(x_s, x_t\right)}{|t-s|^{\alpha}}\right)^{\alpha'/\alpha} |x|_0^{1-\alpha'/\alpha}$$

and passing to the respective suprema. ∎

Proposition 5.6 *Let $p \geq 1$ and $x \in C\left([0,T], E\right)$.*
(i) $x \in C^{p\text{-var}}\left([0,T], E\right)$ is equivalent to

$$\overline{\lim}_{\delta \to 0} \sup_{(t_i) \in \mathcal{D}_\delta([0,T])} \sum_i d\left(x_{t_i}, x_{t_{i+1}}\right)^p < \infty. \tag{5.3}$$

(ii) If $1 \leq q < p < \infty$ and $x \in C^{q\text{-var}}\left([0,T], E\right)$ then

$$\overline{\lim}_{\delta \to 0} \sup_{(t_i) \in \mathcal{D}_\delta([0,T])} \sum_i d\left(x_{t_i}, x_{t_{i+1}}\right)^p = 0. \tag{5.4}$$

Remark 5.7 The forthcoming Proposition 5.9 implies that one can replace $d\left(x_{t_i}, x_{t_{i+1}}\right)^p$ by $|x|^p_{p\text{-var};[t_i,t_{i+1}]}$ in both (5.3) and (5.4). □

Proof. (i) If x is of finite p-variation then, trivially, (5.3) holds. Conversely, let us write $\varphi(x) = x^p$; it follows from (5.3) that we can find $\delta > 0$ small enough and $c < \infty$ so that

$$\sum_{t_i \in D} \varphi\left[d\left(x_{t_i}, x_{t_{i+1}}\right)\right] < c$$

for any dissection $D = (t_i)$ of $[0,T]$ with $|D| < \delta$. Then, for an arbitrary dissection D of $[0,T]$, the number of intervals of length at least δ cannot be more than $T\delta^{-1}$ and each of these contributes at most $\varphi\left(|x|_{0;[0,T]}\right)$ where, by continuity of x,

$$|x|_{0;[0,T]} \equiv \sup_{s,t \in [0,T]} d\left(x_s, x_t\right) < \infty.$$

Hence, for any dissection D of $[0,T]$,

$$\sum_{t_i \in D} \varphi\left[d\left(x_{t_i}, x_{t_{i+1}}\right)\right] < c + T\delta^{-1}\varphi\left(|x|_{0;[0,T]}\right) < \infty$$

which implies that $x \in C^{p\text{-var}}([0,T], E)$.
(ii) Introduce the modulus of continuity,

$$\operatorname{osc}(x,\delta) = \sup\{d(x_s, x_t) : s,t \in [0,T], |t-s| \le \delta\}.$$

By *uniform* continuity of $x : [0,T] \to E$ we have $\operatorname{osc}(x,\delta) \to 0$ as $\delta \searrow 0$. The estimate

$$\sum_{t_i \in D} d(x_{t_i}, x_{t_{i+1}})^p \le \left(\sum_{t_i \in D} d(x_{t_i}, x_{t_{i+1}})^q\right) \operatorname{osc}(x, |D|)^{p-q}$$

then implies

$$\sup_{(t_i) \in \mathcal{D}_\delta([0,T])} \sum_i d(x_{t_i}, x_{t_{i+1}})^p \le \left(|x|_{q\text{-var};[0,T]}^q\right) \operatorname{osc}(x,\delta)^{p-q}$$

which converges to 0 with $\delta \searrow 0$ as required. ∎

As in the discussion of 1-variation regularity, the notion of *control* or *control function* is extremely useful. Let us recall that a control (on $[0,T]$) is a continuous map ω of $s,t \in [0,T], s \le t$, into the non-negative reals, 0 on the diagonal, and super-additive, i.e for all $s \le t \le u \in [0,T]$,

$$\omega(s,t) + \omega(t,u) \le \omega(s,u).$$

The perhaps most important example of a control is given by $|x|_{p\text{-var};[s,t]}^p$ for $x \in C^{p\text{-var}}([0,T], E)$. This is the content of the following proposition:

Proposition 5.8 *Let (E,d) be a metric space, $p \ge 1$ and $x : [0,T] \to E$ be a continuous path of finite p-variation. Then*

$$\omega_{x,p}(s,t) := |x|_{p\text{-var};[s,t]}^p$$

defines a control.

Proof. We dealt with the case $p = 1$ in Proposition 1.12 and thus can focus on $p > 1$.
Step 1: The same argument which gave super-additivity in the case $p = 1$ gives super-additivity of $\omega_{x,p}$ in the present setting. The proof of continuity of $\omega_{x,p}$ splits up into showing (i) "continuity from inside"

$$\omega_{x,p}(s+,t-) \equiv \lim_{h_1,h_2 \searrow 0} \omega_{x,p}(s+h_1, t-h_2) = \omega_{x,p}(s,t),$$

which follows from the same argument as in the case $p = 1$, and (ii) "continuity from outside"

$$\omega_{x,p}(s-,t+) \equiv \lim_{h_1,h_2 \searrow 0} \omega_{x,p}(s-h_1, t+h_2) = \omega_{x,p}(s,t),$$

for all $s < t$. Remark that $\omega_{x,p}(s,t+), \omega_{x,p}(s-,t)$, etc. are defined in the obvious way and that all limits here exist by monotonicty of $\omega_{x,p}$. In fact, this reduces the proof of (ii) to showing

$$\omega_{x,p}(s-,t+) \leq \omega_{x,p}(s,t)$$

and this requires a careful analysis which is not covered by our previous "$p = 1$" discussion.[1] As a further reduction, it is enough to establish "one-sided continuity from outside", i.e.

$$\omega_{x,p}(s,t) \geq \omega_{x,p}(s,t+) \text{ and } \omega_{x,p}(s,t) \geq \omega_{x,p}(s-,t) . \tag{5.5}$$

We only discuss $\omega_{x,p}(s,t) \geq \omega_{x,p}(s,t+)$, the other inequality following from the same argument, and show how to deduce it from continuity of $\omega_{x,p}$ at the diagonal, i.e.

$$\omega_{x,p}(t,t+) = 0. \tag{5.6}$$

(The proof of (5.6) is left to step 2 below.) Fixing $s < t$ and $h, \varepsilon > 0$ we consider $D = (s = t_0 < t_1 < \cdots < t_{n-1} < t_n = t + h)$ such that

$$\sum_{i=0}^{n-1} d\left(x_{t_i}, x_{t_{i+1}}\right)^p > \omega_{x,p}(s, t+h) - \varepsilon;$$

splitting $D = D_1 \cup D_2$ so that all points in $[s,t]$ are contained in D_1 (clearly, D_1 is a dissection of $[s,t]$) yields

$$\sum_{t_i \in D_1} d\left(x_{t_i}, x_{t_{i+1}}\right)^p + \omega_{x,p}(t, t+h) > \omega_{x,p}(s, t+h) - \varepsilon$$

and after sending h to 0, using $\omega_{x,p}(t,t+) = 0$,

$$\omega_{x,p}(s,t) \geq \sum_{i:t_i \in D_1} d\left(x_{t_i}, x_{t_{i+1}}\right)^p > \omega_{x,p}(s,t+) - \varepsilon$$

and upon sending ε to 0 we see that it is indeed enough to prove right-continuity of $\omega_{x,p}$ at the diagonal.

Step 2: To see (5.6) we seek a contradiction to

$$\lim_{h \searrow 0} \omega_{x,p}(t, t+h) =: \delta > 0.$$

Observe that the limit exists by monotonicity. Keeping t fixed throughout, thanks to continuity of x, we can find h_1 such that for all $h \in [0, h_1]$,

$$d\left(x_t, x_{t+h}\right)^p < \delta/8. \tag{5.7}$$

---

[1] In the case $p = 1$ we used additivity of $\omega_{x,1}$ to obtain continuity from outside. In general, when $p > 1$ a control $\omega_{x,p}$ is not additive.

Fix $h_0 \in [0, h_1]$ and a dissection $(t = \tau_0 < \tau_1 < \cdots < \tau_{k-1} < \tau_k = t + h_0)$ of $[t, t + h_0]$ such that

$$\sum_{i=0}^{k-1} d\left(x_{\tau_i}, x_{\tau_{i+1}}\right)^p > 7\delta/8,$$

which is possible since $\omega_{x,p}(t, t + h_0) \geq \omega_{x,p}(t, t+) = \delta$. Using (5.7), we have

$$\sum_{i=1}^{k-1} d\left(x_{\tau_i}, x_{\tau_{i+1}}\right)^p > 7\delta/8 - \delta/8 = 3\delta/4.$$

Doing the same with τ_1 in place of $t+h_0$ yields $(t = \sigma_0 < \sigma_1 < \cdots < \sigma_{l-1} < \sigma_l = \tau_1)$, a dissection of $[t, \tau_1]$, such that

$$\sum_{j=1}^{l-1} d\left(x_{\sigma_j}, x_{\sigma_{j+1}}\right)^p > 3\delta/4.$$

Combining the previous two sums, over non-overlapping intervals of the form $[\sigma_j, \sigma_{j+1}], [\tau_i, \tau_{i+1}] \subset [t, t + h_0]$, yields

$$\omega_{x,p}(t, t + h_0) \geq 3\delta/4 + 3\delta/4 = 3\delta/2,$$

which implies $\omega(t, t+) \geq 3\delta/2$, which contradicts $\lim_{h \searrow 0} \omega_{x,p}(t, t+h)$. That concludes the proof. ∎

Proposition 5.9 *Let (E, d) be a metric space, $p \geq 1$ and $x : [0, T] \to E$ be a continuous path of finite p-variation and $\delta > 0$. Then*
(i)

$$\omega_{x,\delta,p}(s, t) := \sup_{(t_i) \in \mathcal{D}_\delta([s,t])} \sum_i d\left(x_{t_i}, x_{t_{i+1}}\right)^p \leq |x|_{p\text{-var};[s,t]}^p$$

defines a control.
(ii) We have

$$\sup_{(t_i) \in \mathcal{D}_\delta([0,T])} \sum_i d\left(x_{t_i}, x_{t_{i+1}}\right)^p = \sup_{(t_i) \in \mathcal{D}_\delta([0,T])} \sum_i |x|_{p\text{-var};[t_i, t_{i+1}]}^p$$

as well as

$$\sup_{|t-s|<\delta} \frac{d(x_s, x_t)}{|t - s|^{1/p}} = \sup_{|t-s|<\delta} |x|_{1/p\text{-Höl};[s,t]}.$$

Proof. (i) The proof follows along the same lines as the proof of Proposition 5.8.
(ii) In both cases, the $\leq$ part is obvious. Using the fact that $\omega_{x,\delta,p}$ is a control, we obtain

$$\sup_{(t_i) \in \mathcal{D}_\delta([0,T])} \sum_i d\left(x_{t_i}, x_{t_{i+1}}\right)^p \leq \sup_{(t_i) \in \mathcal{D}_\delta([0,T])} \sum_i |x|_{p\text{-var};[t_i, t_{i+1}]}^p$$

$$\leq \sup_{(t_i) \in \mathcal{D}_\delta([0,T])} \sum_i \omega_{x,\delta,p}(t_i, t_{i+1})$$

$$\leq \omega_{x,\delta,p}([0, T])$$

and, by the very definition of $\omega_{x,\delta,p}([0,T])$, equality must hold through-out. The $1/p$-Hölder statement is also simple to prove and left to the reader. ∎

The following proposition is extremely important. We shall use part (i) below (without further notice) throughout the book; part (ii) says that a modulus of continuity on small intervals gives quantitative control over large intervals.

Proposition 5.10 *Let (E, d) be a metric space, ω a control on $[0, T]$, $p \geq 1, C > 0$, and $x : [0, T] \to E$ a continuous path.*
(i) The pointwise estimate

$$d\left(x_s, x_t\right) \leq C\,\omega\left(s,t\right)^{1/p} \quad \text{for all } s < t \text{ in } [0, T]$$

implies the p-variation estimate

$$|x|_{p\text{-var};[s,t]} \leq C\,\omega\left(s,t\right)^{1/p} \quad \text{for all } s < t \text{ in } [0, T].$$

(We say that x is of finite p-variation controlled by ω.)
(ii) Under the weaker assumption

$$d\left(x_s, x_t\right) \leq C\,\omega\left(s,t\right)^{1/p} \quad \text{for all } s < t \text{ in } [0, T] \text{ such that } \omega\left(s,t\right) \leq 1$$

we have

$$|x|_{p\text{-var};[s,t]} \leq 2C\left(\omega\left(s,t\right)^{1/p} \vee \omega\left(s,t\right)\right) \quad \text{for all } s < t \text{ in } [0, T].$$

Proof. (Remark that only the super-additivity of ω is used in the proof.) Ad (i). By assumption, $d\left(x_s, x_t\right)^p \leq C^p\,\omega\left(s,t\right)$. Then for any dissection $D = \{t_i\}$ of $[s, t]$, super-additivity implies

$$\sum_i d\left(x_{t_i}, x_{t_{i+1}}\right)^p \leq C^p \sum_i \omega\left(t_i, t_{i+1}\right) \leq C^p \omega\left(s,t\right).$$

Taking the supremum over all such dissections finishes the proof of the first part.
(ii) Defining $\phi_p\left(x\right) = x \vee x^p$ we see (cf. Exercise 1.8) that $(s, t) \mapsto \phi_p\left(\omega\left(s,t\right)\right)$ is a control. In view of part (i) we only need to prove

$$d\left(x_s, x_t\right) \leq 2C\phi_p\left(\omega\left(s,t\right)\right)^{1/p}.$$

If s, t are such that $\omega\left(s,t\right) \leq 1$, there is nothing to prove, so we fix s, t such that $\omega(s,t) > 1$. Define $t_0 = s$, and $t_{i+1} = \inf\{u > t_i, \omega(t_i, u) = 1\} \wedge t$. From super-additivity of ω it follows that $t_N = t$ for $N \geq \omega(s, t)$. We conclude with

$$d\left(x_s, x_t\right) \leq \sum_{0 \leq i < \omega(s,t)} d\left(x_{t_i}, x_{t_{i+1}}\right)$$

$$\leq \sum_{0 \leq i < \omega(s,t)} C\omega\left(t_i, t_{i+1}\right)^{1/p}$$

$$\leq C\left(1 + \omega\left(s, t\right)\right)$$

$$\leq 2C\omega\left(s, t\right).$$

∎

Exercise 5.11 *Let* $x \in C^{p\text{-var}}\left([0,T], E\right), p \geq 1,$ *with associated control function*

$$\omega_{x,p}\left(s, t\right) = |x|_{p\text{-var};[s,t]}^p.$$

Show that, for any $s < t < u$ *in* $[0, T]$,

$$\omega_{x,p}\left(s, t\right) + \omega_{x,p}\left(t, u\right) \leq \omega_{x,p}\left(s, u\right) \leq 2^{p-1}\left[\omega_{x,p}\left(s, t\right) + \omega_{x,p}\left(t, u\right)\right].$$

Solution. The first inequality is immediate. For the second, if $s < s' < t < t' < u$ we have

$$d\left(x_{s'}, x_{t'}\right) \leq d\left(x_{s'}, x_t\right) + d\left(x_t, x_{t'}\right)$$

and since $(a+b)^p \leq 2^{p-1}\left(a^p + b^p\right)$ for $a, b \geq 0$ the conclusion follows. □

The very same argument used in Lemma 1.15 shows lower semi-continuity of $x \mapsto |x|_{p\text{-var}}$ in the following sense.

Lemma 5.12 *Let* (x^n) *be a sequence of paths from* $[0,T] \to E$ *of finite p-variation. Assume* $x^n \to x$ *pointwise on* $[0,T]$. *Then, for all* $s < t$ *in* $[0,T]$,

$$|x|_{p\text{-var};[s,t]} \leq \liminf_{n \to \infty} |x^n|_{p\text{-var};[s,t]}.$$

In particular,

$$|x|_{1/p\text{-Höl};[s,t]} \leq \liminf_{n \to \infty} |x^n|_{1/p\text{-Höl};[s,t]}.$$

In a similar spirit, the following lemma says that $|x|_{p\text{-var}}$ is a right-continuous function of p.

Lemma 5.13 *Let* $x: [0,T] \to E$ *be a continuous path of finite p-variation. Then, for all* $s < t$ *in* $[0,T]$, *the map* $p' \in [p, \infty) \mapsto |x|_{p'\text{-var};[s,t]}$ *is non-increasing and*

$$\lim_{p' \searrow p} |x|_{p'\text{-var};[s,t]} = |x|_{p\text{-var};[s,t]}. \tag{5.8}$$

Proof. The non-increasing statement was proved in Proposition 5.3. In particular, it implies that

$$\omega\left(s,t\right)^{1/p} := \lim_{p' \searrow p} |x|_{p'\text{-var};[s,t]}$$

exists and satisfies $\omega\left(s,t\right)^{1/p} \le |x|_{p\text{-var};[s,t]}$; we only need to show the converse inequality. Clearly, $d\left(x_s,x_t\right) \le |x|_{p'\text{-var};[s,t]}$ and sending $p' \searrow p$ we have

$$d\left(x_s,x_t\right) \le \omega\left(s,t\right)^{1/p} \tag{5.9}$$

for all $s < t$ in $[0,T]$. Let us show that ω is super-additive. First observe that

$$\omega\left(s,t\right) = \left(\lim_{p' \searrow p} |x|_{p'\text{-var};[s,t]}\right)^p = \lim_{p' \to p} |x|^{p'}_{p'\text{-var};[s,t]}.$$

Then, for $s \le t \le u$ and using super-additivity of $(s,t) \mapsto |x|^{p'}_{p'\text{-var};[s,t]}$,

$$
\begin{aligned}
\omega\left(s,t\right) + \omega\left(t,u\right) &= \lim_{p' \to p} \left(|x|^{p'}_{p'\text{-var};[s,t]} + |x|^{p'}_{p'\text{-var};[t,u]}\right) \\
&\le \lim_{p' \to p} |x|^{p'}_{p'\text{-var};[s,u]} = \omega\left(s,u\right).
\end{aligned}
$$

But (5.9) and super-additivity of ω imply $|x|_{p\text{-var};[s,t]} \le \omega\left(s,t\right)^{1/p}$. We conclude that $\omega\left(s,t\right)^{1/p} = |x|_{p\text{-var};[s,t]}$ as required. ∎

5.1.2 On some path-spaces contained in $C^{p\text{-var}}\left([0,T],E\right)$

Observe that x is a path of finite p-variation controlled by $(s,t) \mapsto |t-s|$ if and only if x is $1/p$-Hölder. Hence, $1/p$-Hölder paths are of finite p-variation. Conversely, we now show that every finite p-variation path is the time-change of a $1/p$-Hölder path.

Proposition 5.14 *Let (E,d) be a metric space, and let $x : [0,T] \to E$ be a continuous path. Then x is of finite p-variation if and only if there exists a continuous increasing function h from $[0,T]$ onto $[0,1]$ and a $1/p$-Hölder path g such that $x = g \circ h$.*

Proof. Let x be of finite p-variation, non-zero. Then,

$$h(t) = \frac{\omega_{x,p}\left(0,t\right)}{\omega_{x,p}\left(0,T\right)}$$

defines a continuous (from Proposition 5.8) increasing function from $[0,T]$ onto $[0,1]$. Then, there exists a function g such that $g \circ h\left(t\right) = x\left(t\right)$, as

$h(t_1) = h(t_2) \implies x(t_1) = x(t_2)$. Now,

$$
\sup_{u,v \in [0,1]} \frac{|g(u) - g(v)|}{|u-v|^{1/p}} = \sup_{u,v \in [0,T]} \frac{|g(h(u)) - g(h(v))|}{|h(u) - h(v)|^{1/p}}
$$

$$
\leq \omega_{x,p}(0,T)^{1/p} \frac{\omega_{x,p}(u,v)^{1/p}}{|\omega_{x,p}(0,u) - \omega_{x,p}(0,v)|^{1/p}}.
$$

From the sub-additivity of $\omega_{x,p}$, $|\omega_{x,p}(0,u) - \omega_{x,p}(0,v)| \geq |\omega_{x,p}(u,v)|$, so that

$$
\sup_{u,v} \frac{|g(u) - g(v)|}{|u-v|^{1/p}} \leq \omega_{x,p}(0,T)^{1/p},
$$

i.e. g is $1/p$-Hölder. ∎

Exercise 5.15 (absolute continuity of order p) *We say that $x : [0,T]$ $\to E$ is "absolutely continuous of order p" if, for all $\varepsilon > 0$, there exists $\delta > 0$, such that for all $s_1 < t_1 \leq s_2 < t_2 \leq \cdots < s_n < t_n$ in $[0,T]$ with $\sum_i |t_i - s_i|^p < \delta$, we have*

$$
\sum_i d(x_{s_i}, x_{t_i})^p < \varepsilon. \tag{5.10}
$$

(i) Assume $p \geq 1$. Show that in the definition of absolute continuity of order p one can replace (5.10) by

$$
\sum_i |x|_{p\text{-var};[s_i,t_i]}^p < \varepsilon.
$$

(ii) Assume $p > 1$, and show that x is absolutely continuous of order p if and only if

$$
\overline{\lim}_{\delta \to 0} \sup_{D \in \mathcal{D}_\delta([0,T])} \sum_i d(x_{t_i}, x_{t_{i+1}})^p = 0. \tag{5.11}
$$

Solution. (i) Consider $s_1 < t_1 \leq s_2 < t_2 \leq \cdots < s_n < t_n$ in $[0,T]$ with $\sum_i |t_i - s_i|^p < \delta$. Let $(u_j^i) \in \mathcal{D}([s_i,t_i])$ be a dissection of $[s_i,t_i]$ and observe that

$$
\left(\sum_j |u_{j+1}^i - u_j^i|^p \right)^{1/p} \leq \sum_j |u_{j+1}^i - u_j^i| = t_i - s_i.
$$

It follows that $\sum_i \sum_j |u_{j+1}^i - u_j^i|^p \leq \sum_i |t_i - s_i|^p < \delta$ and so

$$
\sum_i \sum_j d\left(x_{u_{j+1}^i}, x_{u_j^i}\right)^p < \varepsilon
$$

and we conclude by taking the supremum over all possible dissections of $[s_i, t_i]$, $i = 1, \ldots, n$.

(ii) " $\Leftarrow$ " : Condition (5.11) implies that

$$\forall \varepsilon > 0 : \exists \tilde{\delta} : \quad \sup_{D \in \mathcal{D}_{\tilde{\delta}}([0,T])} \sum_i d\left(x_{u_i}, x_{u_{i+1}}\right)^p < \varepsilon,$$

for any dissection D with $|D| \leq \tilde{\delta}$. Fix $\varepsilon > 0$ and take $s_1 < t_1 \leq s_2 < t_2 \leq \cdots < s_n < t_n$ in $[0,T]$ such that

$$\sum_i |t_i - s_i|^p < \delta := \tilde{\delta}^p$$

which plainly implies $\max_i |t_i - s_i| \leq \tilde{\delta}$. Take $D = (u_i)$ to be a refinement of $\{0 \leq s_1 < t_1 \leq \cdots < s_n < t_n \leq T\}$ with mesh $|D| \leq \tilde{\delta}$, without adding any (unnecessary) points in the intervals $[s_i, t_i]$. It then follows that

$$\sum_i d\left(x_{s_i}, x_{t_i}\right)^p \leq \sum_{i : u_i \in D} d\left(x_{u_i}, x_{u_{i+1}}\right)^p < \varepsilon,$$

which shows that x is absolutely continuous of order p.
" $\Longrightarrow$ " : Fix x an absolutely continuous path of order p, and $\varepsilon > 0$. We may write an arbitrary dissection $D = (s_i)$ of $[0,T]$ in the form

$$D = \{0 \leq s_1 < t_1 = s_2 < \cdots < t_{n-1} = s_n < t_n \leq T\}$$

and furthermore assume $|D|$ is small enough so that $\sum_i |t_i - s_i|^p \leq T |D|^{p-1} < \delta$, where δ is chosen so that this implies, using the assumption of absolute continuity of order p of the path x,

$$\sum_i d\left(x_{s_i}, x_{s_{i+1}}\right)^p < \varepsilon.$$

This estimate is uniform over all dissections D with $|D| < (\delta/T)^{1/(p-1)} = \tilde{\delta}$. It follows that $\overline{\lim}_{\delta \to 0} \sup_{D \in \mathcal{D}_\delta([0,T])} \sum_i d\left(x_{t_i}, x_{t_{i+1}}\right)^p = 0$. $\quad\square$

Example 5.16 (Besov spaces) In Section 1.4 we introduced the (Sobolev) path spaces $W^{1,q}\left([0,T], E\right)$ which provided examples of finite 1-variation paths with precise Hölder modulus $|t - s|^{1-1/q}$. We now introduce the fractional Sobolev – or Besov – spaces $W^{\delta,q}\left([0,T], E\right)$ with $\delta < 1$ whose elements are paths having finite p-variation with $p = 1/\delta > 1$ and precise Hölder modulus $|t - s|^{\delta - 1/q}$. More precisely, we make the following definition. Given $q \in [1, \infty)$ and $\delta \in (1/q, 1)$, the space $W^{\delta,q}\left([0,T], E\right)$ is the set of all $x \in C\left([0,T], E\right)$ for which

$$|x|_{W^{\delta,q};[0,T]} := \left(\int_0^T \int_0^T \left(\frac{d\left(x_u, x_v\right)}{|v - u|^{\delta + 1/q}} \right)^q du\, dv \right)^{1/q} < \infty.$$

Following Section A.2 in Appendix A, the Garsia–Rodemich–Rumsey estimate leads quickly to a Besov–Hölder resp. variation "embedding", by which we mean

$$|x|_{\delta-1/q\text{-Höl};[0,T]} \leq (\text{const})\, |x|_{W^{\delta,q};[0,T]},$$

$$|x|_{(1/\delta)\text{-var};[0,T]} \leq (\text{const})\, |x|_{W^{\delta,q};[0,T]} \qquad\qquad \square$$

5.2   Approximations in geodesic spaces

For a continuous path $x : [0,T] \to \mathbb{R}^d$, and a dissection $D = (t_i)_i$ of $[0,T]$, we constructed the piecewise linear approximation x^D by defining $x^D_{t_i} = x_{t_i}$ and connecting by straight lines in between. Straight lines in $\mathbb{R}^d$ are geodesics in the sense of the following definition.

Definition 5.17 In a metric space (E,d) a geodesic (or geodesic path) joining two points $a, b \in E$ is a continuous path $\Upsilon^{a,b} : [0,1] \to E$ such that $\Upsilon^{a,b}(0) = a$, $\Upsilon^{a,b}(1) = b$ and

$$d\left(\Upsilon^{a,b}_s, \Upsilon^{a,b}_t\right) = |t - s|\, d(a,b) \qquad\qquad (5.12)$$

for all $s < t$ in $[0,1]$. If any two points in E are joined by a (not necessarily unique) geodesic, we call E a geodesic space. $\qquad \square$

Equation (5.12) expresses that there are no shortcuts between any two points on the geodesic path. Even if E is complete and connected, it need not be a geodesic space; for example, the unit circle $S^1 \subset \mathbb{R}^2$ *with metric induced from* $\mathbb{R}^2$ is not geodesic. However, S^1 is a geodesic space under arclength distance. Readers with some background in Riemannian geometry will recall the Hopf–Rinow theorem;[2] it says precisely that a complete connected Riemannian manifold is a geodesic space. The main example of a geodesic space to have in mind for our purposes is the free step-N nilpotent group equipped with Carnot–Caratheodory metric, to be discussed in detail later on.

Geodesic spaces have exactly the structure that allows us to generalize the idea of piecewise linear approximations.

To simplify, when considering a geodesic space E and two points $a, b \in E$, we will define $\Upsilon^{a,b}$ to be an arbitrary geodesic between a and b.

Definition 5.18 (piecewise geodesic approximation) Let x be a continuous path from $[0,T]$ into some geodesic space (E,d). Given a dissection $D = \{t_0 = 0 < t_1 < \cdots < t_n = T\}$ of $[0,T]$ we define x^D as the concatenation of geodesics connecting x_{t_i} and $x_{t_{i+1}}$ for $i = 1, \ldots, n-1$. More precisely, set

$$x^D_t = x_t \text{ for all } t \in D$$

---

[2] For example, Bishop and Crittenden [14, p. 154].

and for $t \in (t_i, t_{i+1})$,

$$x_t^D = \Upsilon^{x_{t_i}, x_{t_{i+1}}} \left(\frac{t - t_i}{t_{i+1} - t_i} \right).$$

□

Lemma 5.19 *Let E be a geodesic space and $x \in C([0,T], E)$. Then, x^D converges to x uniformly on $[0,T]$. That is,*

$$\sup_{t \in [0,T]} d\left(x_t^D, x_t\right) \to 0 \text{ as } |D| \to 0.$$

Proof. Fix two consecutive points $t_i < t_{i+1}$ in D and note that it is enough to show that $d\left(x_t^D, x_t\right) \to 0$ uniformly for $t \in [t_i, t_{i+1}]$. To see this, fix $\varepsilon > 0$ and pick $\delta = \delta(\varepsilon)$ so that

$$\mathrm{osc}\,(x; \delta) \equiv \sup_{\substack{s < t \text{ in } [0,T]: \\ t - s < \delta}} d(x_s, x_t) < \varepsilon/2$$

(which is possible since x is continuous on the compact $[0,T]$ and hence uniformly continuous). Then, for $t \in [t_i, t_{i+1}]$ and provided that $|D| < \delta$, we have

$$
\begin{aligned}
d\left(x_t^D, x_t\right) &\leq d\left(x_t^D, x_{t_i}\right) + d(x_{t_i}, x_t) \\
&= \left| \frac{t - t_i}{t_{i+1} - t_i} \right| d\left(x_{t_i}, x_{t_{i+1}}\right) + d(x_{t_i}, x_t) \\
&\leq 2\mathrm{osc}\,(x; \delta) < \varepsilon
\end{aligned}
$$

which already finishes the proof. ∎

Proposition 5.20 *Let E be a geodesic space and $x \in C^{p\text{-var}}([0,T], E)$, $p \geq 1$ and $D = \{0 = t_0 < t_1 < \cdots < t_n = T\}$ a dissection of $[0,T]$. Then,*

$$\left|x^D\right|_{p\text{-var};[0,T]} \leq 3^{1-1/p} |x|_{p\text{-var};[0,T]}. \tag{5.13}$$

If x is $1/p$-Hölder,

$$\left|x^D\right|_{1/p\text{-Höl};[0,T]} \leq 3^{1-1/p} |x|_{1/p\text{-Höl};[0,T]}. \tag{5.14}$$

Remark 5.21 D *induces a dissection of any interval $[t_i, t_j]$ with endpoints $t_i, t_j \in D$. It follows that $[0,T]$ in (5.13) may be replaced by any interval $[t_i, t_j]$ with $t_i, t_j \in D$.* □

Proof. *p-variation estimate:* To prove (5.13) we use the control $\omega(s,t) = |x|_{p\text{-var};[s,t]}^p$ and then define ω_D first on the intervals of D by

$$\omega_D(s,t) = \left(\frac{t - s}{t_{i+1} - t_i} \right)^p \omega(t_i, t_{i+1}) \text{ for } t_i \leq s \leq t \leq t_{i+1}, \ 1 \leq i < \#D$$

and then for arbitary $s < t$, say $1 \leq i < j < \#D$ and $t_i \leq s \leq t_{i+1} \leq t_j \leq t \leq t_{j+1}$, by

$$\omega_D(s,t) = \omega_D(s,t_{i+1}) + \omega(t_{i+1},t_j) + \omega_D(t_j,t). \qquad (5.15)$$

Clearly, if $t_i \leq s \leq t \leq t_{i+1}$,

$$
\begin{aligned}
d\left(x_s^D, x_t^D\right) &= d\left(\Upsilon^{x_{t_i}, x_{t_{i+1}}}\left(\frac{s - t_i}{t_{i+1} - t_i}\right), \Upsilon^{x_{t_i}, x_{t_{i+1}}}\left(\frac{t - t_i}{t_{i+1} - t_i}\right)\right) \\
&= \left|\frac{t - s}{t_{i+1} - t_i}\right| d\left(x_{t_i}, x_{t_{i+1}}\right) \quad \text{using (5.12)} \\
&\leq \omega_D(s,t)^{1/p}.
\end{aligned}
$$

On the other hand, if $t_i \leq s \leq t_{i+1} \leq t_j \leq t \leq t_{j+1}$,

$$
\begin{aligned}
d\left(x_s^D, x_t^D\right) &\leq d\left(x_s^D, x_{t_{i+1}}^D\right) + d\left(x_{t_{i+1}}, x_{t_j}\right) + d(x_{t_j}^D, x_t) \\
&\leq \omega_D(s,t_{i+1})^{1/p} + \omega(t_{i+1},t_j)^{1/p} + \omega_D(t_j,t)^{1/p} \\
&\leq 3^{1-1/p}\left(\omega_D(s,t_{i+1}) + \omega(t_{i+1},t_j) + \omega_D(t_j,t)\right)^{1/p} \\
&= 3^{1-1/p}\omega_D(s,t)^{1/p}
\end{aligned}
$$

using (5.15) in the last line. Hence, for all $s < t$ in $[0,T]$,

$$d\left(x_s^D, x_t^D\right)^p \leq 3^{p-1}\omega_D(s,t). \qquad (5.16)$$

It now suffices to show that ω_D is a control (only super-additivity is non-trivial) to obtain the desired conclusion, namely

$$
\begin{aligned}
\left|x^D\right|_{p\text{-var};[0,T]} &\leq 3^{1-1/p}\omega_D(0,T)^{1/p} \\
&= 3^{1-1/p}\left|x\right|_{p\text{-var};[0,T]}.
\end{aligned}
$$

To see super-additivity, $\omega_D(s,t) + \omega_D(t,u) \leq \omega_D(s,u)$ for $s \leq t \leq u$ in $[0,T]$ we first consider the case when s,t,u are contained in one interval, say $t_i \leq s < t < u \leq t_{i+1}$. Then

$$
\begin{aligned}
\omega_D(s,t) + \omega_D(t,u) &= \left(\frac{t - s}{t_{i+1} - t_i}\right)^p \omega(t_i,t_{i+1}) + \left(\frac{u - t}{t_{i+1} - t_i}\right)^p \omega(t_i,t_{i+1}) \\
&\leq \left(\frac{u - s}{t_{i+1} - t_i}\right)^p \omega(t_i,t_{i+1}) = \omega_D(s,u).
\end{aligned}
$$

Consider the case that s,t are contained in one interval, say $t_i \leq s < t \leq t_{i+1} \leq t_k \leq u \leq t_{k+1}$. Then

$$\omega_D(s,t) + \omega_D(t,u) = \underbrace{\omega_D(s,t) + \omega_D(t,t_{i+1})}_{\leq \omega_D(s,t_{i+1})} + \omega(t_{i+1},t_k) + \omega_D(t_k,u)$$

(using the first case!) and conclude the defining equality $\omega_D\,(s, u) = \omega_D\,(s, t_{i+1}) + \omega\,(t_{i+1}, t_k) + \omega_D(t_k, u)$. The case that t, u are contained in one interval is similar. At last, if s, t, u are in three different intervals, say

$$t_i \leq s \leq t_{i+1} \leq t_j \leq t \leq t_{j+1} \leq t_k \leq u \leq t_{k+1},$$

then $\omega_D(s, t) + \omega_D(t, u)$ equals

$$\omega_D(s, t_{i+1}) + \omega(t_{i+1}, t_j) + \underbrace{\underbrace{\omega_D(t_j, t) + \omega_D(t, t_{j+1})}_{\leq \omega_D\,(t_j, t_{j+1}) = \omega(t_j, t_{j+1})} + \omega(t_{j+1}, t_k)}_{\leq \omega(t_{i+1}, t_k)} + \omega_D(t_k, u)$$

and we conclude again with the defining equality for $\omega_D\,(s, u)$. This covers all cases and we have established that ω_D is a control.

1/p-Hölder estimate: If x is actually $1/p$-Hölder then

$$\omega\,(s, t)^{1/p} = |x|_{p\text{-var};[s,t]} \leq |x|_{1/p\text{-Höl};[0,T]}\,|t - s|^{1/p}$$

and so for $t_i \leq s \leq t \leq t_{i+1}$

$$\omega_D(s, t) = \left(\frac{t - s}{t_{i+1} - t_i}\right)^p \omega\,(t_i, t_{i+1}) \leq |x|^p_{1/p\text{-Höl};[0,T]}\,|t - s|\,.$$

For general $s < t$, say $t_i \leq s \leq t_{i+1} \leq t_j \leq t \leq t_{j+1}$, we have

$$\begin{aligned}
\omega_D(s, t) &= \omega_D(s, t_{i+1}) + \omega(t_{i+1}, t_j) + \omega_D(t_j, t) \\
&\leq |x|^p_{1/p\text{-Höl};[0,T]}\,(|t_{i+1} - s| - |t_j - t_{i+1}| - |t - t_j|) \\
&= |x|^p_{1/p\text{-Höl};[0,T]}\,|t - s|
\end{aligned}$$

The claimed estimate (5.14) now follows immediately from

$$d\left(x_s^D, x_t^D\right) \leq 3^{1-1/p} \omega_D^{1/p}(s, t) \leq 3^{1-1/p}\,|x|_{1/p\text{-Höl};[0,T]}\,|t - s|^{1/p}\,.$$

∎

Remark 5.22 The above proof actually shows that

$$\left|x^D\right|^p_{p\text{-var};[0,T]} \leq 3^{p-1} \sup_{(t_i) \in \mathcal{D}_{|D|}([0,T])} \sum_i |x|^p_{p\text{-var};[t_i, t_{i+1}]}\,, \tag{5.17}$$

and a slight extension shows that for $|D| < \delta$ one has the estimate

$$\sup_{(t_j) \in \mathcal{D}_\delta([0,T])} \sum_i \left|x^D\right|^p_{p\text{-var};[t_j, t_{j+1}]} \leq 3^{p-1} \sup_{(t_i) \in \mathcal{D}_\delta([0,T])} \sum_i |x|^p_{p\text{-var};[t_i, t_{i+1}]}\,. \tag{5.18}$$

□

Combining Lemma 5.19 and Proposition 5.20 gives immediately the following important approximation result.

Theorem 5.23 *Let E be a geodesic space and $x \in C^{p\text{-var}}([0,T],E)$, $p \geq 1$. Let (D_n) be a sequence of dissection of $[0,T]$ such that its mesh $|D_n|$ converges to 0. Then, x^{D_n} converges to x "uniformly with uniform p-variation bounds". That is,*

$$\sup_{t \in [0,T]} d\left(x_t^{D_n}, x_t\right) \to 0$$

and

$$\sup_n \left|x^{D_n}\right|_{p\text{-var};[0,T]} \leq 3^{1-1/p} \left|x\right|_{p\text{-var};[0,T]}.$$

If x is $1/p$-Hölder then

$$\sup_n \left|x^{D_n}\right|_{1/p\text{-Höl};[0,T]} \leq 3^{1-1/p} \left|x\right|_{1/p\text{-Höl};[0,T]}.$$

Exercise 5.24 *Let E be a geodesic space, $p \in (1,\infty)$ and $x \in W^{1,p}([0,T], E)$ as defined in Section 1.4.2. Show that*

$$\left|x^D\right|_{W^{1,p};[0,T]}^p = \sum_{i:t_i \in D} \frac{\left|d\left(x_{t_i}, x_{t_{i+1}}\right)\right|^p}{\left|t_{i+1} - t_i\right|^{p-1}}.$$

5.3 Hölder and p-variation paths on $\mathbb{R}^d$

5.3.1 Hölder and p-variation Banach spaces

We now turn to $\mathbb{R}^d$ (equipped with Euclidean distance) as our most familiar example of a metric (and geodesic) space.

Theorem 5.25 *(i) $C^{p\text{-var}}\left([0,T],\mathbb{R}^d\right)$ is Banach with norm $x \mapsto |x(0)| + |x|_{p\text{-var};[0,T]}$. The closed subspace of paths in $C^{p\text{-var}}\left([0,T],\mathbb{R}^d\right)$ started at 0 is also Banach under $x \mapsto |x|_{p\text{-var};[0,T]}$.*
(ii) $C^{1/p\text{-Höl}}\left([0,T],\mathbb{R}^d\right)$ is Banach with norm $x \mapsto |x(0)| + |x|_{1/p\text{-Höl};[0,T]}$. The closed subspace of paths in $C^{1/p\text{-Höl}}\left([0,T],\mathbb{R}^d\right)$ started at 0 is also Banach under $x \mapsto |x|_{1/p\text{-Höl};[0,T]}$.
These Banach spaces are not separable.

Proof. The case $p = 1$ was dealt with in Section 1.3. Leaving straightforward details to the reader, let us say that completeness in the case $p > 1$ is proved as in the case $p = 1$; non-separability follows from the following example. ∎

Example 5.26 We construct an uncountable family of functions so that the distance of any two $f \neq f'$ remains bounded below by a fixed positive real. An uncountable subset of $C\left([0,1],\mathbb{R}\right)$ is given by

$$f_\varepsilon(t) = \sum_{k \geq 1} \varepsilon_k 2^{-k/p} \sin\left(2^k \pi t\right), \quad t \in [0,1],$$

where ε is a ± 1 sequence, that is, $\varepsilon_k \in \{-1,1\}$ for all k. We show (i) that

$$f_\varepsilon \in C^{1/p\text{-Höl}}\left([0,1],\mathbb{R}^d\right) \subset C^{p\text{-var}}\left([0,1],\mathbb{R}^d\right)$$

and (ii) if $\varepsilon \neq \varepsilon'$ then

$$2 < |f_\varepsilon - f_{\varepsilon'}|_{p\text{-var};[0,1]} \leq |f_\varepsilon - f_{\varepsilon'}|_{1/p\text{-Höl};[0,1]}.$$

$\square$

Proof. Ad (i). For $0 \leq s < t \leq 1$ we have

$$|f_\varepsilon(t) - f_\varepsilon(s)| \leq \sum_{1 \leq k \leq |\log_{(2)}(t-s)|} \varepsilon_k 2^{-k/p}\left(\sin\left(2^k \pi t\right) - \sin\left(2^k \pi s\right)\right)$$

$$+ \sum_{k > |\log_{(2)}(t-s)|} \varepsilon_k 2^{-k/p}\left(\sin\left(2^k \pi t\right) - \sin\left(2^k \pi s\right)\right)$$

where $\log_{(2)}$ is the logarithm with base 2. Using $|\varepsilon|_{l^\infty} \leq 1$, we obtain $\left|\sin\left(2^k \pi t\right) - \sin\left(2^k \pi s\right)\right| \leq 2^k \pi |t - s|$ for the first sum and $|\sin(\cdots)| \leq 1$ for the second, and hence

$$|f_\varepsilon(t) - f_\varepsilon(s)| \leq \pi |t-s| \sum_{1 \leq k \leq |\log_{(2)}(t-s)|} 2^{-k/p} 2^k + \sum_{k > |\log_{(2)}(t-s)|} 2 \cdot 2^{-k/p}$$

$$\leq c_1 |t-s|^{1/p}$$

for some constant $c_1 = c_1(p)$, independent of s, t and ε. This proves (i). Ad (ii). Assume $\varepsilon \neq \varepsilon'$ and let $j \geq 1$ be the first index for which $\varepsilon_j \neq \varepsilon'_j$, i.e.

$$\varepsilon_1 = \varepsilon'_1, \ldots, \varepsilon_{j-1} = \varepsilon'_{j-1} \text{ but } \varepsilon_j \neq \varepsilon'_j.$$

Consider then a dissection D of $[0,1]$ given by $t_i = i 2^{-j-1} : i = 0, \ldots, 2^{j+1}$. From

$$\left|\sin\left(2^j \pi t_{i+1}\right) - \sin\left(2^j \pi t_i\right)\right|^p = 1$$

it follows readily that $\left|\sin\left(2^j \pi \cdot\right)\right|_{p\text{-var};[0,1]} \geq 2^{j/p}$. Moreover,

$$\left|(f_\varepsilon - f_{\varepsilon'})(t_{i+1}) - (f_\varepsilon - f_{\varepsilon'})(t_i)\right| = \left|\varepsilon_j - \varepsilon'_j\right| 2^{-j/p} \left|\sin\left(2^j \pi t_{i+1}\right) - \sin\left(2^j \pi t_i\right)\right|$$

$$= 2 \cdot 2^{-j/p}.$$

This shows that $|f_\varepsilon - f_{\varepsilon'}|_{p\text{-var};[0,1]} \geq 2$. ∎

5.3.2 Compactness

Lemma 5.27 *Consider* $(x_n) \subset C\left([0,T],\mathbb{R}^d\right)$ *and assume* $x_n \to x \in$ $C\left([0,T],\mathbb{R}^d\right)$ *uniformly.*
(i) Assume $\sup_n |x_n|_{p\text{-var};[0,T]} < \infty$. *Then* $x_n \to x$ *in* p'-*variation for any* $p' > p$.
(ii) Assume $\sup_n |x_n|_{\alpha\text{-Höl};[0,T]} < \infty$. *Then* $x_n \to x$ *in* α'-*Hölder norm for any* $\alpha' < \alpha$.

Proof. By Lemma 5.12 we see that x is of finite p-variation. It then suffices to apply the interpolation result (Proposition 5.5) to the difference $x - x_n$. ∎

Proposition 5.28 *(compactness) Consider* $(x_n) \subset C\left([0,T],\mathbb{R}^d\right)$.
(i) Assume (x_n) *is equicontinuous, bounded and* $\sup_n |x_n|_{p\text{-var};[0,T]} < \infty$.
Then x_n *converges (in* $p' > p$ *variation, along a subsequence) to some* $x \in C^{p\text{-var}}\left([0,T],\mathbb{R}^d\right)$.
(ii) Assume (x_n) *is bounded and* $\sup_n |x_n|_{\alpha\text{-Höl};[0,T]} < \infty$. *Then* x_n *converges (in* $\alpha' < \alpha$ *Hölder topology, along a subsequence) to some* $x \in C^{\alpha\text{-Höl}}\left([0,T],\mathbb{R}^d\right)$.

Proof. An obvious consequence of Arzela–Ascoli and the previous lemma. ∎

The following corollary will be useful, for example, in the proof of the forthcoming Theorem 6.8.

Corollary 5.29 *(i) Assume* $(x^n), x$ *are in* $C^{p\text{-var}}\left([0,T],\mathbb{R}^d\right)$ *such that* $\sup_n |x_n|_{p\text{-var};[0,T]} < \infty$ *and* $x^n \to x$ *uniformly, then for* $p' > p$,

$$\sup_{(s,t)\in\Delta_T} \left| |x^n|_{p'\text{-var};[s,t]} - |x|_{p'\text{-var};[s,t]} \right| \to 0 \text{ as } n \to \infty$$

where $\Delta_T = \{(s,t) : 0 \le s \le t \le T\}$. *Furthermore,* $\left\{ |x^n|_{p'\text{-var};[\cdot,\cdot]} : n \in \mathbb{N} \right\}$ *is equicontinuous in the sense that for every* $\varepsilon > 0$ *there exists* δ *such that* $|t - s| < \delta$ *implies*

$$\sup_n |x^n|_{p'\text{-var};[s,t]} < \varepsilon. \tag{5.19}$$

(ii) If $(x^n), x$ *are in* $C^{\alpha\text{-Höl}}\left([0,T],E\right)$ *such that* $\sup_n |x^n|_{\alpha\text{-Höl};[0,T]} < \infty$ *and* $x^n \to x$ *uniformly, then for all* $s < t$ *in* $[0,T]$, *then for* $\alpha' < \alpha$,

$$\sup_{(s,t)\in\Delta_T} \left| |x^n|_{\alpha'\text{-Höl};[s,t]} - |x|_{\alpha'\text{-Höl};[s,t]} \right| \to 0 \text{ as } n \to \infty$$

and $\left\{ |x^n|_{\alpha'\text{-Höl};[\cdot,\cdot]} : n \in \mathbb{N} \right\}$ *is equicontinuous, similar to part (i).*

Proof. (i) Proposition 5.5, applied to $x^n - x$, actually shows that

$$\lim_{n \to \infty} \sup_{s,t} |x^n - x|_{p'\text{-var};[s,t]} = 0.$$

Hence, $|x^n|_{p'\text{-var};[\cdot,\cdot]}$ converges uniformly on Δ_T and, by Arzela–Ascoli's theorem, is equicontinuous. That is, for any $\varepsilon > 0$ there exists δ such that $|(s,t) - (s',t')| < \delta$ implies

$$\sup_n \left| |x^n|_{p'\text{-var};[s,t]} - |x^n|_{p'\text{-var};[s',t']} \right| < \varepsilon.$$

In particular, this applies to $(s,t) \in \Delta_T$ with $|t - s| < \delta$ and $s' := t' := s$, and using $|x^n|_{p'\text{-var};[s',s']} = 0$ we see that

$$\sup_n |x^n|_{p'\text{-var};[s,t]} < \varepsilon$$

which concludes the proof of (i). The proof of (ii) follows similar lines. ∎

5.3.3 Closure of smooth paths in variation norm

For $p \geq 1$ we define $C^{0,p\text{-var}}\left([0,T],\mathbb{R}^d\right)$ resp. $C^{0,1/p\text{-Höl}}\left([0,T],\mathbb{R}^d\right)$ as the closure of smooth paths from $[0,T] \to \mathbb{R}^d$ in p-variation resp. $1/p$-Hölder norm. In symbols,

$$C^{0,p\text{-var}}\left([0,T],\mathbb{R}^d\right) : = \overline{C^\infty\left([0,T],\mathbb{R}^d\right)}^{p\text{-var}},$$

$$C^{0,1/p\text{-Höl}}\left([0,T],\mathbb{R}^d\right) : = \overline{C^\infty\left([0,T],\mathbb{R}^d\right)}^{1/p\text{-Höl}}.$$

Obviously, these are closed, linear subspaces of $C^{p\text{-var}}\left([0,T],\mathbb{R}^d\right)$ resp. $C^{1/p\text{-Höl}}\left([0,T],\mathbb{R}^d\right)$ and thus Banach spaces and so is the restriction to paths with $x(0) = 0$, denoted by $C_o^{0,p\text{-var}}\left([0,T],\mathbb{R}^d\right)$ resp. $C_o^{0,1/p\text{-Höl}}\left([0,T],\mathbb{R}^d\right)$. The case $p = 1$ was already discussed earlier in Section 1.3 where, among other things, we identified $C^{0,1\text{-var}}$ as absolutely continuous paths and $C^{0,1\text{-Höl}}\left([0,T],\mathbb{R}^d\right)$ as $C^1\left([0,T],\mathbb{R}^d\right)$. For $p > 1$ we have

Lemma 5.30 *Let* $p > 1$.
(i) Let Ω *be a set in* $C^{1\text{-var}}\left([0,T],\mathbb{R}^d\right)$ *such that* $C^{0,1\text{-var}}\left([0,T],\mathbb{R}^d\right) \subset \overline{\Omega}^{1\text{-var}}$. *Then,*

$$\overline{\Omega}^{p\text{-var}} = C^{0,p\text{-var}}\left([0,T],\mathbb{R}^d\right).$$

(ii) Let Ω *be a set in* $C^{1\text{-Höl}}\left([0,T],\mathbb{R}^d\right)$ *such that* $C^1\left([0,T],\mathbb{R}^d\right) \subset \overline{\Omega}^{1\text{-Höl}}$. *Then,*

$$\overline{\Omega}^{1/p\text{-Höl}} = C^{0,1/p\text{-Höl}}\left([0,T],\mathbb{R}^d\right).$$

Proof. (i) First, $C^{0,p\text{-var}} \subset \overline{\Omega}^{p\text{-var}}$ follows immediately from

$$C^{\infty} \subset C^{0,1\text{-var}} \subset \overline{\Omega}^{1\text{-var}} \subset \overline{\Omega}^{p\text{-var}}.$$

The converse inclusion follows readily from $C^{1\text{-var}} \subset \overline{C^{\infty}}^{p\text{-var}}$; indeed,

$$\Omega \subset C^{1\text{-var}} \subset \overline{C^{\infty}}^{p\text{-var}} \implies \overline{\Omega}^{p\text{-var}} \subset \overline{C^{\infty}}^{p\text{-var}}.$$

To see $C^{1\text{-var}} \subset \overline{C^{\infty}}^{p\text{-var}}$, recall from Exercise 2.6 that any $x \in C^{1\text{-var}}$ can be approximated by $x_n \in C^{\infty}$ in uniform norm with uniform 1-variation bounds, i.e.

$$|x - x_n|_{\infty;[0,T]} \to 0, \quad \sup_n |x_n|_{1\text{-var};[0,T]} < \infty;$$

then interpolation (Proposition 5.5 applied to $x - x_n$) gives $x_n \to x$ in p-variation, which is what we had to prove.
(ii) Similar and left to the reader. ∎

Theorem 5.31 (Wiener's characterization) *Let* $x \in C^{p\text{-var}}([0,T], \mathbb{R}^d)$, *with* $p > 1$. *The following statements are equivalent.*
(i.1) $x \in C^{0,p\text{-var}}([0,T], \mathbb{R}^d)$.
(i.2a) $\lim_{\delta \to 0} \sup_{D=(t_i), |D|<\delta} \sum_i |x|^p_{p\text{-var};[t_i, t_{i+1}]} = 0$.
(i.2b) $\lim_{\delta \to 0} \sup_{D=(t_i), |D|<\delta} \sum_i d\left(x_{t_i}, x_{t_{i+1}}\right)^p = 0$.
(i.3) $\lim_{|D| \to 0} d_{p\text{-var}}\left(x^D, x\right) = 0$.
Secondly, let $x \in C^{1/p\text{-Höl}}\left([0,T], \mathbb{R}^d\right)$, *with* $p > 1$. *The following statements are equivalent.*
(ii.1) $x \in C^{0,1/p\text{-Höl}}\left([0,T], \mathbb{R}^d\right)$.
(ii.2a) $\lim_{\delta \to 0} \sup_{|t-s|<\delta} |x|_{1/p\text{-Höl};[s,t]} = 0$.
(ii.2b) $\lim_{\delta \to 0} \sup_{|t-s|<\delta} d(x_s, x_t)/|t-s|^{1/p} = 0$.
(ii.3) $\lim_{|D| \to 0} d_{1/p\text{-Höl}}\left(x^D, x\right) = 0$.

Remark 5.32 From purely metric considerations, we have seen in Exercise 5.15 that (i.2b) is equivalent to "absolute continuity of order p". Remark also that the case $p = 1$ requires special care: by Corollary 1.34 (i.1) ⇔ (i.3) holds true. On the other hand, Proposition 1.14 tells us that in the case $p = 1$ condition (i.2) is tantamount to saying x is constant; in particular, conditions (i.1), (i.3) do not imply (i.2). Similar comments apply in the Hölder case. □

Proof. We only prove the p-variation statements, as the $1/p$-Hölder ones follow the same logic.
From Lemma 5.30, the $d_{p\text{-var}}$-closure of $C^{1\text{-var}}([0,T], \mathbb{R}^d)$ is $C^{0,p\text{-var}}([0,T], \mathbb{R}^d)$, which implies $(i.3) \Rightarrow (i.1)$. The reverse proof of $(i.1) \Rightarrow (i.3)$ follows the same lines as the proof in the case $p = 1$, i.e. the proof of Corollary 1.34.

We already proved in Proposition 5.9 that $(i.2a) \Leftrightarrow (i.2b)$ and now turn to $(i.1) \Rightarrow (i.2b)$.

Let us fix $\varepsilon > 0$, and a smooth path y such that $d_{p\text{-var}}(x,y)^p \leq \varepsilon 2^{-p}$. For a dissection D, we obtain from the triangle inequality

$$\sum_{t_i \in D} d\left(x_{t_i}, x_{t_{i+1}}\right)^p \leq 2^{p-1} \sum_{t_i \in D} d\left(y_{t_i}, y_{t_{i+1}}\right)^p + 2^{p-1} d_{p\text{-var};[0,T]}(y,x)^p.$$

Since y is smooth, there exists $\delta > 0$ (that depends on n) such that for all dissections $|D| < \delta$ implies that

$$\sum_{t_i \in D} d\left(y_{t_i}, y_{t_{i+1}}\right)^p < \varepsilon 2^{-p}.$$

Hence, we obtain that for all dissections D with $|D| < \delta$,

$$\sum_{t_i \in D} d\left(x_{t_i}, x_{t_{i+1}}\right)^p \leq \varepsilon.$$

We finish by proving $(i.2a) \Rightarrow (i.3)$. First, if x and y are two paths, and δ is some fixed positive real, observe that for all subdivisions $D = (t_i)$,

$$\sum_i d\left(x_{t_i,t_{i+1}}, y_{t_i,t_{i+1}}\right)^p$$

$$\leq \sum_{i, |t_{i+1}-t_i| \leq \delta} d\left(x_{t_i,t_{i+1}}, y_{t_i,t_{i+1}}\right)^p + \sum_{i, |t_{i+1}-t_i| > \delta} d\left(x_{t_i,t_{i+1}}, y_{t_i,t_{i+1}}\right)^p$$

$$\leq 2^{p-1} \left(\sum_{i, |t_{i+1}-t_i| \leq \delta} |x|^p_{p\text{-var};[t_i,t_{i+1}]} + \sum_{i, |t_{i+1}-t_i| \leq \delta} |y|^p_{p\text{-var};[t_i,t_{i+1}]} \right)$$
$$+ \frac{T}{\delta} d_0(x,y).$$

Taking the supremum over all dissections, we obtain

$$d_{p\text{-var}}(x,y)^p \leq 2^{p-1} \sup_{(t_i) \in \mathcal{D}_\delta([0,T])} \sum_i |x|^p_{p\text{-var};[t_i,t_{i+1}]}$$

$$+ 2^{p-1} \sup_{(t_i) \in \mathcal{D}_\delta([0,T])} \sum_i |y|^p_{p\text{-var};[t_i,t_{i+1}]}$$

$$+ \frac{T}{\delta} d_0(x,y).$$

Taking a bounded variation path x and its piecewise linear approximation x^D for some dissection D with $|D| < \delta$, we obtain, using inequality (5.18),

$$d_{p\text{-var}}\left(x, x^D\right)^p \leq c_p \sup_{(t_i) \in \mathcal{D}_\delta([0,T])} \sum_i |x|^p_{p\text{-var};[t_i,t_{i+1}]}$$

$$+ \frac{T}{\delta} d_0\left(x, x^D\right).$$

First fix $\delta > 0$ such that $c_p \sup_{(t_i) \in \mathcal{D}_\delta([0,T])} \sum_i |x|^p_{p\text{-var};[t_i,t_{i+1}]} < \varepsilon/2$. Then as x^D converges to x in uniform topology when $|D| \to 0$, there exists $\delta_2 < \delta$ such that for all dissections D with $|D| < \delta_2$,

$$\frac{T}{\delta} d_0\left(x, x^D\right) < \varepsilon/2.$$

Hence, for all dissections D with $|D| < \delta_2$, we have $d_{p\text{-var}}\left(x, x^D\right)^p < \varepsilon$ and the proof is finished. ∎

Corollary 5.33 *For $p > 1$, we have the following set inclusions:*

$$\bigcup_{1 \leq q < p} C^{q\text{-var}}\left([0,T], \mathbb{R}^d\right) \ \subset \ C^{0,p\text{-var}}\left([0,T], \mathbb{R}^d\right)$$

$$\subset \ C^{p\text{-var}}\left([0,T], \mathbb{R}^d\right) \subset \bigcap_{q > p} C^{q\text{-var}}\left([0,T], \mathbb{R}^d\right).$$

Proof. Recalling basic inclusions between p- and q-variation spaces (Proposition 5.3), only the inclusion

$$\bigcup_{q < p} C^{q\text{-var}}\left([0,T], \mathbb{R}\right) \subset C^{0,p\text{-var}}\left([0,T], \mathbb{R}\right)$$

requires an argument. Thanks to Proposition 5.6,

$$x \in \bigcup_{1 \leq q < p} C^{q\text{-var}}\left([0,T], \mathbb{R}\right) \implies \lim_{\delta \to 0} \sup_{D=(t_i), |D| < \delta} \sum_i d\left(x_{t_i}, x_{t_{i+1}}\right)^p = 0$$

and we conclude using Theorem 5.31. ∎

Example 5.34 An example of a function in $C^{1/2\text{-Höl}}\left([0,1], \mathbb{R}\right)$ but not in $C^{0,1/2\text{-Höl}}\left([0,1], \mathbb{R}\right)$ is given by $t \mapsto t^{1/2}$, as follows immediately from Wiener's characterization, Theorem 5.31.

Exercise 5.35 *(i) Define $g(x) = \sum_{i=1}^\infty c^{-i/p} \sin\left(c^i x\right)$. If c is a sufficiently large positive integer, show that $g \in C^{p\text{-var}}\left([0,1], \mathbb{R}\right)$ but $g \notin C^{0,p\text{-var}}\left([0,1], \mathbb{R}\right)$.*
(ii) Define $h(x) = x^{1/p} \cos^2\left(\pi/x\right) / \log x$ for $x > 0$, $h(0) = 0$. Show that $h \in C^{0,p\text{-var}}\left([0,1], \mathbb{R}\right)$ and

$$h \notin \cup_{q < p} C^{q\text{-var}}\left([0,1], \mathbb{R}\right).$$

Proposition 5.36 *Let $p \geq 1$. The spaces $C^{0,p\text{-var}}\left([0,T], \mathbb{R}^d\right), C^{0,1/p\text{-Höl}}\left([0, T], \mathbb{R}^d\right)$ are separable Banach spaces (and hence Polish).*

Proof. From Proposition 1.35, there is a countable space Ω that is dense in $C^{0,1\text{-var}}\left([0,T], \mathbb{R}^d\right)$. We conclude using Lemma 5.30. ∎

5.4 Generalized variation

5.4.1 Definition and basic properties

The concept of variation (and then p-variation) allows for an obvious generalization:

Definition 5.37 Let (E, d) be a metric space, $\varphi \in C\left([0, \infty), [0, \infty)\right)$, 0 at 0, strictly increasing and onto. A path $x : [0, T] \to E$ is said to be of finite φ-variation on the interval $[s, t]$ if

$$|x|_{\varphi\text{-var};[0,T]} := \inf \left\{ M > 0, \sup_{D \in \mathcal{D}([0,T])} \sum_{t_i \in D} \varphi \left[\frac{d\left(x_{t_i}, x_{t_{i+1}}\right)}{M} \right] \leq 1 \right\} < \infty.$$

We will use the notation $C^{\varphi\text{-var}}([0, T], E)$ for the set of continuous paths $x : [0, T] \to E$ of finite φ-variation. The set of paths pinned at time zero to some fixed element $o \in E$ is denoted by $C_o^{\varphi\text{-var}}([0, T], E)$. $\square$

For $\varphi(x) = x^p$, the definition of $|x|_{\varphi\text{-var};[0,T]}$ coincides with $|x|_{p\text{-var};[0,T]}$. If (E, d) is a normed space and φ is (globally) convex, $|\cdot|_{\varphi\text{-var};[0,T]}$ is a seminorm. Several variation functions of interest are not convex (including the class $\psi_{p,q}$ to be introduced in the forthcoming Definition 5.45, which will be convenient for our later applications).

A first interest in φ-variation comes from the fact that (sharp) sample path properties for stochastic processes are often available in this form. For example, a classical result of Taylor (cf. the forthcoming Theorem 13.15) states that Brownian motion has a.s. finite $\psi_{2,1}$-variation on any compact interval $[0, T]$ and this is optimal (cf. Theorem 13.69). A wide class of (enhanced) Gaussian processes have a.s. finite $\psi_{p,p/2}$-variation, while (enhanced) Markov processes (with uniformly elliptic generator in divergence form) have the "Brownian" $\psi_{2,1}$-variation regularity. The other reason for our interest in φ-variation is that it is intimately related to the uniqueness of solutions to rough differential equations under minimal regularity assumptions; as will be discussed in Section 10.5.

Lemma 5.38 Let $x \in C^{\varphi\text{-var}}([0, T], E)$. Then, for all $M \geq |x|_{\varphi\text{-var};[0,T]}$ we have

$$\sup_{D \in \mathcal{D}([0,T])} \sum_{t_i \in D} \varphi \left[\frac{d\left(x_{t_i}, x_{t_{i+1}}\right)}{M} \right] \leq 1. \tag{5.20}$$

Proof. Only the case $M = |x|_{\varphi\text{-var};[0,T]}$ requires a proof. By definition, there exists a sequence $M_n \downarrow M$ such that (5.20) holds with M replaced by M_n. In particular, for a fixed dissection D,

$$\sum_{t_i \in D} \varphi \left[\frac{d\left(x_{t_i}, x_{t_{i+1}}\right)}{M_n} \right] \leq 1, \text{ and hence } \sum_{t_i \in D} \varphi \left[\frac{d\left(x_{t_i}, x_{t_{i+1}}\right)}{M} \right] \leq 1$$

by continuity of φ. Taking the supremum over all $D \in \mathcal{D}([0,T])$ finishes the proof. ∎

Just as in the common case of p-variation, controls are a very useful concept.

Proposition 5.39 *Let (E,d) be a metric space, $\varphi \in C([0,\infty),[0,\infty))$, 0 at 0, strictly increasing and onto. Then the following are equivalent.*
(i) $x \in C^{\varphi\text{-var}}([0,T],E)$ with $|x|_{\varphi\text{-var};[0,T]} \le M$ for some $M \ge 0$.
(ii) There exists a control ω with $\omega(0,T) \le 1$ such that for all $s < t$ in $[0,T]$,

$$d(x_s,x_t) \le M\varphi^{-1}(\omega(s,t)).$$

Proof. Define

$$\omega_{x,\varphi}(s,t) := \sup_{D \in \mathcal{D}([s,t])} \sum_{t_i \in D} \varphi\left[\frac{d(x_{t_i},x_{t_{i+1}})}{M}\right].$$

Working as in the proof of Proposition 5.8, we see that $\omega_{x,\varphi}$ is a control with $\omega_{x,\varphi}(0,T) \le 1$. We then, by definition of $\omega_{x,\varphi}$, have that $d(x_s,x_t) \le M\varphi^{-1}(\omega_{x,\varphi}(s,t))$.

Conversely, assume that for all $s < t$ in $[0,T]$,

$$d(x_s,x_t) \le M\varphi^{-1}(\omega(s,t))$$

for some $M > 0$ and control ω with $\omega(0,T) \le 1$. Then, for a dissection D, we have

$$
\begin{aligned}
\sum_{t_i \in D} \varphi\left[\frac{d(x_{t_i},x_{t_{i+1}})}{M}\right] &\le \sum_{t_i \in D} \varphi \circ \varphi^{-1}[\omega(t_i,t_{i+1})] \\
&\le \omega(0,T) \le 1,
\end{aligned}
$$

and hence $|x|_{\varphi\text{-var};[0,T]} \le M$. ∎

For simplicity, we will only look at φ-variation of paths for functions φ satisfying the following condition.

Condition 5.40 [Δ_c] *Assume $\varphi \in C([0,\infty),[0,\infty))$, 0 at 0, strictly increasing and onto. We say φ satisfies condition Δ_c if for all $c > 0$, there exists $\Delta_c \ge 0$ such that $\forall x \in [0,\infty) : \varphi(cx) \le \Delta_c\varphi(x)$ and $\lim_{c\to 0}\Delta_c = 0$.* □

The condition Δ_c leads to the following convenient equivalences.

Proposition 5.41 *Let (E,d) be a metric space, and let $x : [0,T] \to E$ be a continuous path. Assume the variation function φ satisfies condition (Δ_c). Then the following conditions are equivalent.*
(i) The path x is of finite φ-variation.
(ii) There exists $M > 0$ such that

$$\sup_{D \in \mathcal{D}([0,T])} \sum_{t_i \in D} \varphi\left[\frac{d(x_{t_i},x_{t_{i+1}})}{M}\right] < \infty.$$

(iii) for all $K > 0$,

$$\sup_{D \in \mathcal{D}([0,T])} \sum_{t_i \in D} \varphi \left[\frac{d\left(x_{t_i}, x_{t_{i+1}}\right)}{K} \right] < \infty.$$

Proof. Trivially, (i) $\Longrightarrow$ (ii) and (iii) $\Longrightarrow$ (ii). We show that (ii) implies (i) and (iii). For any $K > 0$, using condition Δ_c, we have

$$\sup_{D \in \mathcal{D}([0,T])} \sum_{t_i \in D} \varphi \left[\frac{d\left(x_{t_i}, x_{t_{i+1}}\right)}{K} \right] \leq \Delta_{M/K} \sup_{D \in \mathcal{D}([0,T])} \sum_{t_i \in D} \varphi \left[\frac{d\left(x_{t_i}, x_{t_{i+1}}\right)}{M} \right] < \infty$$

which proves (ii) $\Longrightarrow$ (iii). Similarly,

$$\sup_{D \in \mathcal{D}([0,T])} \sum_{t_i \in D} \varphi \left[\frac{d\left(x_{t_i}, x_{t_{i+1}}\right)}{K} \right] \leq \Delta_{M/K} \sup_{D \in \mathcal{D}([0,T])} \sum_{t_i \in D} \varphi \left[\frac{d\left(x_{t_i}, x_{t_{i+1}}\right)}{M} \right]$$

$$\leq 1 \text{ for } K \text{ large enough (as } \Delta_{M/K} \to_{K \to \infty} 0)$$

and so (ii) $\Longrightarrow$ (i). The proof is finished. $\blacksquare$

Recall $C^{p\text{-var}}\left([0,T], E\right) \subset C^{\tilde{p}\text{-var}}\left([0,T], E\right)$, $\tilde{p} \geq p$. This generalizes to

Lemma 5.42 *Assume $\varphi, \tilde{\varphi}$ satisfies condition (Δ_c) and $\tilde{\varphi} = O\left(\varphi\right)$ at $0+$. Then*

$$C^{\varphi\text{-var}}\left([0,T], E\right) \subset C^{\tilde{\varphi}\text{-var}}\left([0,T], E\right).$$

Proof. Let $x \in C^{\varphi\text{-var}}\left([0,T], E\right)$, then for all $K > 0$, and in particular $K := |x|_{0;[0,T]}$, we have

$$\sup_{D \in \mathcal{D}([0,T])} \sum_{t_i \in D} \varphi \left[\frac{d\left(x_{t_i}, x_{t_{i+1}}\right)}{K} \right] < \infty.$$

For all i, $d\left(x_{t_i}, x_{t_{i+1}}\right)/K \leq 1$, and by assumption, on $[0, 1]$, there exists a finite constant c such that $\tilde{\varphi}\left(u\right) \leq c\varphi\left(u\right)$ for all $u \in [0, 1]$. Hence

$$\sup_{D \in \mathcal{D}([0,T])} \sum_{t_i \in D} \tilde{\varphi} \left[\frac{d\left(x_{t_i}, x_{t_{i+1}}\right)}{K} \right] \leq c \sup_{D \in \mathcal{D}([0,T])} \sum_{t_i \in D} \varphi \left[\frac{d\left(x_{t_i}, x_{t_{i+1}}\right)}{K} \right] < \infty$$

and so $x \in C^{\tilde{\varphi}\text{-var}}\left([0,T], E\right)$, using Proposition 5.41. $\blacksquare$

We now make some quantitative relations between φ-variation and p-variation. As will be seen in Corollary 5.44 below, p-variation estimates often imply φ-variation estimates with no extra work.

Theorem 5.43 *Fix $p \geq 1$, and assume that φ satisfies condition Δ_c. Assume also that $\varphi^{-1}\left(\cdot\right)^p$ is convex on $[0, \delta]$ for some $\delta \in (0, 1]$. Let*

$x \in C^{\varphi\text{-var}}([0,T], E)$. *Then, the control[3]*

$$\omega(s,t) := \sup_{D \in \mathcal{D}([s,t])} \sum_{t_i \in D} \varphi\left[\frac{d\left(x_{t_i}, x_{t_{i+1}}\right)}{|x|_{\varphi\text{-var};[0,T]}}\right]$$

satisifies $\omega(0,T) \leq 1$. *Moreover, for* $C = C(\varphi, p)$ *and all* $s < t$ *in* $[0,T]$,

$$|x|_{p\text{-var};[s,t]} \leq C |x|_{\varphi\text{-var};[0,T]} \, \varphi^{-1}\left(\omega(s,t)\right).$$

Proof. It suffices to consider the case of non-constant $x(\cdot)$ so that $|x|_{\varphi\text{-var};[0,T]} > 0$. By Lemma 5.38, $\omega(0,T) \leq 1$. Then, as φ satisfies condition Δ_c, there exists $\kappa > 0$ such that $\Delta_\kappa \leq \delta$, and we have

$$\varphi\left[\kappa \frac{d(x_s, x_t)}{|x|_{\varphi\text{-var};[0,T]}}\right] \leq \Delta_\kappa \varphi\left[\frac{d(x_s, x_t)}{|x|_{\varphi\text{-var};[0,T]}}\right] \leq \delta\omega(s,t)$$

from which $d(x_s, x_t)^p \leq \kappa^{-p} |x|_{\varphi\text{-var};[0,T]}^p \, \varphi^{-1}\left(\delta\omega(s,t)\right)^p$. Note that $\delta\omega(s,t) \in [0,\delta]$ for any $s < t$ in $[0,T]$ and so, from convexity of $\varphi^{-1}(\cdot)^p$ on $[0,\delta]$,

$$(s,t) \mapsto \varphi^{-1}\left(\delta\omega(s,t)\right)^p$$

is a control. It then follows from basic super-additivity properties of controls that

$$|x|_{p\text{-var};[s,t]}^p \leq \frac{|x|_{\varphi\text{-var};[0,T]}^p}{\kappa^p} \varphi^{-1}\left(\delta\omega(s,t)\right)^p \leq \frac{|x|_{\varphi\text{-var};[0,T]}^p}{\kappa^p} \varphi^{-1}\left(\omega(s,t)\right)^p$$

where we used $\delta \leq 1$ in the final step, and the proof is finished. ∎

We now consider a second path y with values in some metric space $(\tilde{E}, \tilde{d})$, whose p-variation is dominated by the p-variation of x. (This situation will be typical for solutions of (rough) differential equations.)

Corollary 5.44 *Fix* $p \geq 1$, *and assume that* φ *satisfies condition* Δ_c. *Assume also that* $\varphi^{-1}(\cdot)^p$ *is convex on* $[0,\delta]$ *for some* $\delta \in (0,1]$. *Let* $x \in C([0,T], E)$, $y \in C([0,T], \tilde{E})$ *be such that for all* $s < t$ *in* $[0,T]$,

$$|y|_{p\text{-var};[s,t]} \leq K |x|_{p\text{-var};[s,t]} . \tag{5.21}$$

Then, for some constant $C = C(p, \varphi)$ *and all* $s < t$ *in* $[0,T]$,

$$|y|_{\varphi\text{-var};[s,t]} \leq CK |x|_{\varphi\text{-var};[s,t]} .$$

Proof. From Theorem 5.43, we have

$$\begin{aligned} d(y_s, y_t) &\leq |y|_{p\text{-var};[s,t]} \leq K |x|_{p\text{-var};[s,t]} \\ &\leq CK |x|_{\varphi\text{-var};[0,T]} \, \varphi^{-1}\left(\omega(s,t)\right). \end{aligned}$$

---

[3]With convention $0/0 = 0$.

Hence, if $D = (t_i)$ is a dissection of $[0, T]$, we have

$$\sum_{t_i \in D} \varphi \left[\frac{d \left(y_{t_i}, y_{t_{i+1}} \right)}{CK \left| x \right|_{\varphi\text{-var};[0,T]}} \right] \leq \sum_{t_i \in D} \omega \left(t_i, t_{i+1} \right) \leq \omega \left(0, T \right) \leq 1$$

which implies that $|y|_{\varphi\text{-var};[0,T]} \leq CK \left| x \right|_{\varphi\text{-var};[0,T]}$. There is nothing special about the interval $[0, T]$ and by a simple reparametrization argument we see that $|y|_{\varphi\text{-var};[s,t]} \leq CK \left| x \right|_{\varphi\text{-var};[s,t]}$. ∎

5.4.2 Some explicit estimates for $\psi_{p,q}$

We now apply all these abstract considerations to the following class of variation function.

Definition 5.45 For any $(p, q) \in \mathbb{R}^+ \times \mathbb{R}$ set $\psi_{p,q}(0) = 0$ and

$$\psi_{p,q}(t) := \left\{ \begin{array}{l} \frac{t^p}{(\ln \ln 1/t)^q} \text{ for } t \in (0, e^{-e}) \\ t^p \text{ for } t \geq e^{-e} \end{array} \right.$$

or, equivalently, $\psi_{p,q}(t) = t^p / (\ln^* \ln^* 1/t)^q$ where $\ln^* = \max(1, \ln)$. □

Exercise 5.46 *Show that, for any* $(p, q) \in \mathbb{R}^+ \times \mathbb{R}$, *the function* $\psi_{p,q}(\cdot)$ *satisfies condition* Δ_c.

Solution. A possible choice is $\Delta_c = 4/\psi_{p,q}(1/c)$. The details are left to the reader. □

Exercise 5.47 *For* $(p_1, q_1), (p_2, q_2)$ *in* $\mathbb{R}^+ \times \mathbb{R}$, *we say that* $(p_1, q_1) \leq (p_2, q_2)$ *if* $p_1 \leq p_2$ *or if* $p_1 = p_2$ *and* $q_1 \leq q_2$. *Show that for* $(p_1, q_1) \leq (p_2, q_2)$,
$$C^{\psi_{p_1,q_1}\text{-var}} \left([0, T], E \right) \subset C^{\psi_{p_2,q_2}\text{-var}} \left([0, T], E \right).$$

Solution. In all cases, we have

$$\limsup_{t \to 0+} \frac{\psi_{p_2,q_2}(t)}{\psi_{p_1,q_1}(t)} = \limsup_{t \to 0+} t^{p_2 - p_1} \left(\frac{1}{\ln \ln 1/t} \right)^{q_2 - q_1}$$

and this limit is bounded as $t \to 0+$ if $(p_1, q_1) \leq (p_2, q_2)$. The result follows from Lemma 5.42. □

The following estimates on the inverse of $\psi_{p,q}^{-1}$ will be useful to us later on.

Lemma 5.48 *There exists* $C = C(p, q)$ *such that for all* $t \in [0, \infty)$,

$$\frac{1}{C} \psi_{1/p, -q/p}(t) \leq \psi_{p,q}^{-1}(t) \leq C \psi_{1/p, -q/p}(t).$$

Proof. For t large enough $\psi_{p,q}(t) = t^p$, $\psi_{1/p,-q/p} = t^{1/p}$ and there is nothing to show. For t small it suffices to observe that $\psi_{1/p,-q/p}$ is the asymptotic inverse of $\psi_{p,q}$ at $0+$.   ∎

Finally, the following proposition will allow us to use Theorem 5.43 with the functions $\psi_{p,q}$.

Proposition 5.49 *For any $p' > p > 0$ and $q \in \mathbb{R}$, the function $\left(\psi_{p,q}^{-1}(\cdot)\right)^{p'}$ is locally convex in a positive neighbourhood of 0.*

Proof. Obvious from an explicit computation of the second derivative of $\psi_{p,q}^{-1}(\cdot)^{p'}$ near $0+$.   ∎

5.5   Higher-dimensional variation

5.5.1   *Definition and basic properties*

We now discuss p-variation regularity of a function

$$f \quad : \quad [0,T]^2 \to \left(\mathbb{R}^d, |\cdot|\right)$$

$$\begin{pmatrix} s \\ u \end{pmatrix} \quad \mapsto \quad f\begin{pmatrix} s \\ u \end{pmatrix}.$$

The generalization to $[0,T]^n$ with $n > 2$ follows the same arguments but will not be relevant to us. Given a rectangle $R = [s,t] \times [u,v] \subset [0,T]^2$ we write

$$f(R) := f\begin{pmatrix} s,t \\ u,v \end{pmatrix} := f\begin{pmatrix} s \\ u \end{pmatrix} + f\begin{pmatrix} t \\ v \end{pmatrix} - f\begin{pmatrix} s \\ v \end{pmatrix} - f\begin{pmatrix} t \\ u \end{pmatrix}. \quad (5.22)$$

If $d = 1$ and f is smooth, this is precisely $\int_R \frac{\partial^2 f}{\partial a \partial b}(a,b)\, da\, db$. Also, if $f(s,t) = g_s \otimes h_t$, then

$$f\begin{pmatrix} s,t \\ u,v \end{pmatrix} = g_{s,t} \otimes h_{u,v}.$$

We will also use the following notations, consistent with our 1-dimensional increment notation.

$$f\begin{pmatrix} t \\ u,v \end{pmatrix} := f\begin{pmatrix} t \\ v \end{pmatrix} - f\begin{pmatrix} t \\ u \end{pmatrix},$$

$$f\begin{pmatrix} s,t \\ v \end{pmatrix} := f\begin{pmatrix} t \\ v \end{pmatrix} - f\begin{pmatrix} s \\ v \end{pmatrix}.$$

We will also frequently use the notation $|R|$ for the area of R.

Definition 5.50 Let $f : [0,T]^2 \to (\mathbb{R}^d, |\cdot|)$ and $p \in [1,\infty)$. We say that f has finite p-variation if $|f|_{p\text{-var};[0,T]^2} < \infty$, where

$$|f|_{p\text{-var};[s,t]\times[u,v]} = \sup_{\substack{(t_i)\in\mathcal{D}([s,t]) \\ (t_i')\in\mathcal{D}([u,v])}} \left(\sum_{i,j} \left| f\left(\begin{array}{c} t_i, t_{i+1} \\ t_j', t_{j+1}' \end{array} \right) \right|^p \right)^{1/p}$$

and write $f \in C^{p\text{-var}}\left([0,T], \mathbb{R}^d\right)$. $\qquad\square$

In the 1-dimensional (1D) case, i.e. functions defined on $[0,T]$, the notion of *control* is fundamental. In the 2-dimensional (2D) case, controls are defined on $\Delta_T \times \Delta_T$ where we recall that

$$\Delta := \Delta_T = \{(s,t) : 0 \le s \le t \le T\}.$$

We think of elements in $\Delta_T \times \Delta_T$ as *rectangles* contained in the square $[0,T]^2$ and write $[s,t]\times[u,v]$ rather than $((s,t),(u,v))$ for a generic element.

Definition 5.51 Let $\Delta_T = \{(s,t) : 0 \le s \le t \le T\}$. A *2D control* (more precisely, *2D control function on* $[0,T]^2$) is a continuous map $\omega : \Delta_T \times \Delta_T \to [0,\infty)$ which is super-additive in the sense that for all rectangles R_1, R_2, and R with $R_1 \cup R_2 \subset R$ and $R_1 \cap R_2 = \varnothing$,

$$\omega(R_1) + \omega(R_2) \le \omega(R)$$

and such that for all rectangles of zero area,

$$\omega(R) = 0.$$

A 2D control ω is said to be *Hölder-dominated* if there exists a constant C such that for all $s < t$ in $[0,T]$,

$$\omega\left([s,t]^2\right) \le C|t - s|.$$

$\qquad\square$

The proof of the following lemma is a straightforward adaption of the 1D case treated in Section 5.1 and left to the reader.

Lemma 5.52 *Let* $f \in C\left([0,T]^2, \mathbb{R}^d\right)$. *Then*
(i) If f is of finite p-variation for some $p \ge 1$,

$$R \text{ rectangles in } [0,T]^2 \mapsto |f|^p_{p\text{-var};R}$$

is a 2D control.
(i) f is of finite p-variation on $[0,T]^2$ if and only if there exists a 2D control ω such that for all rectangles $R \subset [0,T]^2$,

$$|f(R)|^p \le \omega(R)$$

and we say that "ω controls the p-variation of f".

Remark 5.53 If $f : [0,T]^2 \to \mathbb{R}^d$ is symmetric (i.e. $f(s,u) = f(u,s)$ for all s,u) and of finite p-variation then $[s,t] \times [u,v] \mapsto |f|^p_{p\text{-var};[s,t] \times [u,v]}$ is symmetric. In fact, one can always work with symmetric controls, it suffices to replace a given ω with $[s,t] \times [u,v] \mapsto \omega([s,t] \times [u,v]) + \omega([u,v] \times [s,t])$. □

Lemma 5.54 *A function* $f \in C\left([0,T]^2, \mathbb{R}^d\right)$ *is of finite p-variation if and only if*

$$\sup_{(t_j) \in \mathcal{D}([0,T])} \sum_{i,j} \left| f\left(\begin{array}{c} t_i, t_{i+1} \\ t_j, t_{j+1} \end{array} \right) \right|^p < \infty.$$

Moreover, the p-variation of f is controlled by 3^{p-1} times

$$\omega([s,t] \times [u,v]) := \sup_{(t_j) \in \mathcal{D}([0,T])} \sum_{\substack{i,j \\ [t_i,t_{i+1}] \subset [s,t] \\ [t_j,t_{j+1}] \subset [u,v]}} \left| f\left(\begin{array}{c} t_i, t_{i+1} \\ t_j, t_{j+1} \end{array} \right) \right|^p.$$

Proof. Assuming that $\omega\left([0,T]^2\right)$ is finite, it is easy to check that ω is a $2D$ control. Then, for any given $[s,t]$ and $[u,v]$ which do not intersect or such that $[s,t] = [u,v]$,

$$\left| f\left(\begin{array}{c} s,t \\ u,v \end{array} \right) \right|^p \leq \omega([s,t] \times [u,v]).$$

Take now $s \leq u \leq t \leq v$, then,

$$f\left(\begin{array}{c} s,t \\ u,v \end{array} \right) = f\left(\begin{array}{c} s,u \\ u,v \end{array} \right) + f\left(\begin{array}{c} u,t \\ u,v \end{array} \right)$$

$$= f\left(\begin{array}{c} s,u \\ u,v \end{array} \right) + f\left(\begin{array}{c} u,t \\ u,t \end{array} \right) + f\left(\begin{array}{c} s,u \\ t,v \end{array} \right).$$

Hence,

$$\left| f\left(\begin{array}{c} s,t \\ u,v \end{array} \right) \right|^p \leq 3^{p-1}\left(\omega([s,u] \times [u,v]) + \omega\left([u,t]^2\right) + \omega([s,u] \times [t,v]) \right)$$

$$\leq 3^{p-1} \omega([s,t] \times [u,v]).$$

The other cases are dealt with similarly, and we find at the end that for all $s \leq t, u \leq v$,

$$\left| f\left(\begin{array}{c} s,t \\ u,v \end{array} \right) \right|^p \leq 3^{p-1}\left(\omega[s,t] \times [u,v] \right).$$

This concludes the proof.   ∎

Example 5.55 Given two functions $g, h \in C^{p\text{-var}}\left([0,T], \mathbb{R}^d\right)$ we can define

$$(g \otimes h)\left(\begin{array}{c} s \\ t \end{array} \right) := g(s) \otimes h(t) \in \mathbb{R}^d \otimes \mathbb{R}^d$$

and $g \otimes h$ has finite $2D$ p-variation.[4] More precisely,

$$\left| (g \otimes h) \begin{pmatrix} s,t \\ u,v \end{pmatrix} \right|^p \leq |g|^p_{p\text{-var};[s,t]} \, |h|^p_{p\text{-var};[u,v]} =: \omega\left([s,t] \times [u,v]\right)$$

and since ω is indeed a $2D$ control function (as a product of two $1D$ control functions!) we see that

$$|g \otimes h|_{p\text{-var};[s,t] \times [u,v]} \leq |g|_{p\text{-var};[s,t]} \, |h|_{p\text{-var};[u,v]} \, .$$

$\square$

Exercise 5.56 *Given $f \in C^{p\text{-var}}\left([0,T]^2, \mathbb{R}^d\right)$, for any fixed $[s,t] \times [u,v] \in [0,T]^2$, prove that the (1-dimensional) p-variation of $r \in [s,t] \to f\begin{pmatrix} r \\ u,v \end{pmatrix}$ is bounded by $|f|_{p\text{-var};[s,t] \times [u,v]}$. Similarly, prove that the (1-dimensional) p-variation of $r \in [u,v] \to f\begin{pmatrix} s,t \\ r \end{pmatrix}$ is bounded by $|f|_{p\text{-var};[s,t] \times [u,v]}$.*

Remark 5.57 If ω is a $2D$ control function, then

$$(s,t) \mapsto \omega\left([s,t]^2\right)$$

is a $1D$ control function, i.e. $\omega\left([s,t]^2\right) + \omega\left([t,u]^2\right) \leq \omega\left([s,u]^2\right)$, and $(s,t) \to \omega\left([s,t]^2\right)$ is continuous and zero on the diagonal. $\square$

Remark 5.58 A function $f \in C\left([0,T]^2, \mathbb{R}^d\right)$ of finite p-variation can also be considered as a path $t \mapsto f(t, \cdot)$ with values in the space $C^{p\text{-var}}([0,T], \mathbb{R}^d)$ with p-variation (semi-)norm. It is instructive to observe that $t \mapsto f(t, \cdot)$ has finite p-variation if and only if f has finite $2D$ p-variation. $\square$

5.5.2 Approximations to 2D functions

Piecewise linear-type approximations

Recall from Section 5.3 that a continuous path of finite p-variation can be approximated by smooth and/or piecewise linear paths in the sense of "uniform convergence with uniform p-variation bounds", but not, in general, in p-variation norm. The same is true in the 2D case and the approximations defined below are the natural 2D analogue of piecewise linear approximations.

---

[4] $X \otimes X$ is equipped with a compatible tensor norm.

Definition 5.59 $(D, \tilde{D}$ piecewise linear-type approximation of a 2D function) Assume $f \in C\left([0,T]^2, E\right)$ where E is a normed space. Let

$$D = (\tau_i),\, \tilde{D} = (\tilde{\tau}_j) \in \mathcal{D}\,[0,T]\,.$$

A function $f^{D,\tilde{D}} \in C\left([0,T]^2, E\right)$ with the property that

$$f^{D,\tilde{D}}\begin{pmatrix} s \\ t \end{pmatrix} = f\begin{pmatrix} s \\ t \end{pmatrix} \qquad \text{for all } (s,t) \in D \times \tilde{D}$$

is uniquely defined by requiring that

$$
\begin{aligned}
f^{(D,\tilde{D})}(\cdot,0) &= f^D(\cdot,0)\,, \\
f^{(D,\tilde{D})}(0,\cdot) &= f^{\tilde{D}}(0,\cdot)\,,
\end{aligned}
$$

and, for $(s,t) \times (u,v) \subset (\tau_i, \tau_{i+1}) \times (\tilde{\tau}_j, \tilde{\tau}_{j+1})$,

$$f^{(D,\tilde{D})}\begin{pmatrix} s,t \\ u,v \end{pmatrix} = \frac{t-s}{\tau_{i+1} - \tau_i} \times \frac{v-u}{\tilde{\tau}_{j+1} - \tilde{\tau}_j} f\begin{pmatrix} \tau_i, \tau_{i+1} \\ \tilde{\tau}_j, \tilde{\tau}_{j+1} \end{pmatrix}.$$

$\square$

Proposition 5.60 *Let* $f \in C^{\rho\text{-var}}\left([0,T]^2, E\right)$ *and* $D, \tilde{D} \in \mathcal{D}\,[0,T]$. *Then, for all* $s < t$ *in* D *and* $u < v$ *in* $\tilde{D}$ *we have*

$$\left|f^{D,\tilde{D}}\right|^{\rho}_{\rho\text{-var};[s,t]\times[u,v]} \leq 9^{\rho-1}\,|f|^{\rho}_{\rho\text{-var};[s,t]\times[u,v]}\,. \tag{5.23}$$

Moreover, $f^{D,\tilde{D}} \to f$ *uniformly as* $|D|, \left|\tilde{D}\right| \to 0.$

Remark 5.61 It need not be true that $f^{D,\tilde{D}} \to f$ as $|D|, \left|\tilde{D}\right| \to 0$ in ρ-variation. However, by interpolation this holds true when ρ is replaced by $\rho' > \rho$. $\square$

Proof. Without loss of generality, $[s,t] \times [u,v] = [0,1]^2$. Given $D = (\tau_i), \tilde{D} = (\tilde{\tau}_j) \in \mathcal{D}\,[0,1]$ we now define $\omega_{D,\tilde{D}}$ on $[0,1]^2$ as follows: for small rectangles $[s,t] \times [u,v] \subset I \times J \equiv [\tau_i, \tau_{i+1}] \times [\tilde{\tau}_j, \tilde{\tau}_{j+1}]$ we set

$$\omega_D\left([s,t] \times [u,v]\right) := \frac{(t-s)(v-u)}{|I \times J|}\,|f|^{\rho}_{\rho\text{-var};I \times J} \qquad (\text{with } s,t \in I;\, u,v \in J);$$

then, for vertical "strips" of the form $[s,t] \times (J_1 \cup \cdots \cup J_n)$ with $s,t \in I \equiv [\tau_i, \tau_{i+1}]$ and $J_l = [\tilde{\tau}_{j+l-1}, \tilde{\tau}_{j+l}]$,

$$\omega_D\left([s,t] \times (J_1 \cup \cdots \cup J_n)\right) := \frac{(t-s)}{|I|}\,|f|^{\rho}_{\rho\text{-var};I \times (J_1 \cup \cdots \cup J_n)}$$

for $s,t \in I;\, u,v \in J$.

We use a similar definition for horizontal strips; at last, for a (possibly) large rectangle with endpoints in D we set

$$\omega_D\left((I_1 \cup \cdots \cup I_m) \times (J_1 \cup \cdots \cup J_n)\right) := |f|^\rho_{\rho\text{-var};(I_1 \cup \cdots \cup I_m) \times (J_1 \cup \cdots \cup J_n)}.$$

Now, an arbitrary rectangle $A = [a, b] \times [c, d] \subset [0, 1]^2$ decomposes uniquely into (at most) 9 rectangles $A_1, \ldots, A_9$ of the above type (4 small rectangles in the corners, 2 vertical and 2 horizontal strips and 1 rectangle with endpoints in D) and we define $\omega_{D,\tilde{D}}(A) = \sum_{i=1}^{9} \omega_{D,\tilde{D}}(A_i)$. We leave it to the reader to check that $\omega_{D,\tilde{D}}$ is indeed a 2D control function on $[0, 1]^2$. On the other hand, it is clear from the definition of $f^{D,\tilde{D}}$ that $\left|f^{D,\tilde{D}}(A_i)\right|^\rho \leq \omega_{D,\tilde{D}}(A_i)$ for $i = 1, \ldots, 9$ and so

$$\left|f^{D,\tilde{D}}(A)\right|^\rho = \left|\sum_{i=1}^{9} f^{D,\tilde{D}}(A_i)\right|^\rho \leq 9^{\rho-1} \sum_{i=1}^{9} \left|f^{D,\tilde{D}}(A_i)\right|^\rho = 9^{\rho-1} \omega_{D,\tilde{D}}(A).$$

The proof of (5.23) is then finished with the remark that $\omega_{D,\tilde{D}}\left([0,1]^2\right) = |R|^\rho_{\rho\text{-var};[0,1]^2}$. At last, uniform convergence of $f^{D,\tilde{D}} \to f$ as $|D|, \left|\tilde{D}\right| \to 0$ is a simple consequence of (uniform) continuity of f on $[0,1]^2$. ∎

Mollifier approximations

We now turn to another class of well-known smooth approximations: mollifier approximations.

Notation 5.62 (continuous extension of 2D functions) Whenever necessary, we shall extend a continuous function f defined on $[0, T]^2$ to a continuous function $f = f(s, t)$ defined on $\mathbb{R}^2$ by setting

$$f(0, 0) \quad \text{for } s, t \; < \; 0, \; f(0, T) \text{ for } s < 0, t > T,$$
$$f(T, T) \quad \text{for } s, t \; > \; T, \; f(T, 0) \text{ for } s > T, t < 0$$

and, for $s \in [0, T]$ resp. $t \in [0, T]$,

$$f(s, t) = \begin{cases} f(s, 0) & \text{if } t < 0 \\ f(s, T) & \text{if } t > T \end{cases} \quad \text{resp. } f(s, t) = \begin{cases} f(0, t) & \text{if } s < 0 \\ f(T, t) & \text{if } s > T \end{cases}.$$

Note that, as a consequence of this definition, we have

$$f(R) = f\left(R \cap [0, T]^2\right)$$

for all rectangles in $\mathbb{R}^2$. □

Definition 5.63 ($\mu, \tilde{\mu}$ mollifier approximation of a 2D function) Assume $f \in C\left([0,T]^2, E\right)$ where E is a normed space. Let $\mu, \tilde{\mu}$ be two compactly supported probability measures on $\mathbb{R}$. We define $f^{\mu,\tilde{\mu}} \in C\left([0,T]^2, E\right)$ by

$$f^{\mu,\tilde{\mu}} \left(\begin{array}{c} s \\ u \end{array} \right) = \int \int f \left(\begin{array}{c} s-a \\ u-b \end{array} \right) d\mu(a) \, d\tilde{\mu}(b),$$

noting that the same relation remains valid for rectangular increments,

$$f^{\mu,\tilde{\mu}} \left(\begin{array}{c} s,t \\ u,v \end{array} \right) = \int \int f \left(\begin{array}{c} s-a, t-a \\ u-b, v-b \end{array} \right) d\mu(a) \, d\tilde{\mu}(b).$$ □

Proposition 5.64 *Let $\mu, \tilde{\mu}$ be two compactly supported probability measures on $\mathbb{R}$ and $f \in C^{\rho\text{-var}}\left([0,T]^2, E\right)$, extended to a continuous function on $\mathbb{R}^2$ (cf. notation above) with ρ-variation controlled by*

$$\omega \left(\begin{array}{c} s,t \\ u,v \end{array} \right) = |f|^\rho_{\rho\text{-var};[s,t]\times[u,v]}.$$

Then $f^{\mu,\tilde{\mu}}$ is also of finite ρ-variation, controlled by the 2D control

$$\omega^{\mu,\tilde{\mu}} \left(\begin{array}{c} s,t \\ u,v \end{array} \right) = \int \int \omega \left(\begin{array}{c} s-a, t-a \\ u-b, v-b \end{array} \right) d\mu(a) \, d\tilde{\mu}(b),$$

and

$$\left|f^{\mu,\tilde{\mu}}\right|^\rho_{\rho\text{-var};[0,T]^2} \le \omega^{\mu,\tilde{\mu}}\left([0,T]^2\right) \le |f|^\rho_{\rho\text{-var};[0,T]^2}. \tag{5.24}$$

Moreover, $f^{\mu_n,\tilde{\mu}_n} \to f$ uniformly on $[0,T]^2$ whenever $\mu_n, \tilde{\mu}_n$ converge weakly to the Dirac measure at zero.[5]

Remark 5.65 There is nothing special about the interval $[0,T]$. However, we <u>cannot</u> deduce from (5.24) that

$$\left|f^{\mu,\tilde{\mu}}\right|^\rho_{\rho\text{-var};[s,t]^2} = |f|^\rho_{\rho\text{-var};[s,t]^2} \quad \text{for all } s < t \text{ in } [0,T].$$

The reason is that $f^{\mu,\tilde{\mu}}$ depends on our extension of f from $[0,T]^2$ to $\mathbb{R}^2$. Thus, if we construct $\hat{f}^{\mu,\tilde{\mu}}$ from $\hat{f} = f|_{[s,t]^2}$, extended to $\mathbb{R}^2$, it will not, in general, coincide with $f^{\mu,\tilde{\mu}}$. □

---

[5] That is, $\int \varphi d\mu_n \to \varphi(0)$ for all continuous, bounded φ.

Proof. Given $(s_i) \in \mathcal{D}[0,T]$, $(t_j) \in \mathcal{D}[0,T]$ we have, using Jensen's inequality,

$$
\begin{aligned}
\left| f^{\mu,\tilde{\mu}} \left(\begin{array}{c} s_i, s_{i+1} \\ t_j, t_{j+1} \end{array} \right) \right|^\rho
&= \left| \int \int f \left(\begin{array}{c} s_i - a, s_{i+1} - a \\ t_j - b, t_{j+1} - b \end{array} \right) d\mu(a) \, d\tilde{\mu}(b) \right|^\rho \\
&\leq \int \int \left| f \left(\begin{array}{c} s_i - a, s_{i+1} - a \\ t_j - b, t_{j+1} - b \end{array} \right) \right|^\rho d\mu(a) \, d\tilde{\mu}(b) \\
&= \int \int \left| f \left(\left(\begin{array}{c} s_i - a, s_{i+1} - a \\ t_j - b, t_{j+1} - b \end{array} \right) \right) \right|^\rho d\mu(a) \, d\tilde{\mu}(b) \\
&\leq \omega^{\mu,\tilde{\mu}} \left([s_i, s_{i+1}] \times [t_j, t_{j+1}] \right),
\end{aligned}
$$

which shows that f has finite ρ-variation controlled by $\omega^{\mu,\tilde{\mu}}$. Moreover, since $f(R) = f\left(R \cap [0,T]^2 \right)$ for all rectangles,

$$
|f|^\rho_{\rho\text{-var};[0-a,T-a] \times [0-b,T-b]} \leq |f|^\rho_{\rho\text{-var};[0,T]^2}
$$

and so $\omega^{\mu,\tilde{\mu}} \left([0,T]^2 \right) \leq |f|^\rho_{\rho\text{-var};[0,T]^2}$, which concludes the proof. $\blacksquare$

5.6 Comments

Proposition 5.8 appears in Lyons and Qian [120]; our (complete) proof partially follows Dudley and Norvaiša [47], p. 93. Continuity properties of the type discussed in Lemma 5.13 appear in Musielak and Semadeni [132]. The notion of paths which are "absolutely continuous of order p", cf. Exercise 5.15, is due to Love [114]. Fractional Sobolev spaces (discussed in Example 5.16) are also known as Besov or Slobodetzki spaces and arise in many areas of analysis. The notion of "geodesic space" and its variations (length space, etc.) is now well understood, for example Gromov [72] or Burago *et al.* [21] and the references cited therein. Exercise 5.35 is taken from Dudley and Norvaiša [48], p. 28. An almost complete list of references for generalized variation, or φ-variation, is found in Dudley and Norvaiša [47]. Comments on higher-dimensional p-variation will be given in Chapter 6.

6

Young integration

We construct $\int_0^{\cdot} y\,dx$, the Young integral of y against x where

$$x \in C^{p\text{-var}}\left([0,T],\mathbb{R}^d\right),\, y \in C^{q\text{-var}}\left([0,T],L\left(\mathbb{R}^d,\mathbb{R}^e\right)\right)$$

with $1/p+1/q > 1$. Although the results here are well known, our approach is novel and extends – without much conceptual effort! – to rough path estimates for ordinary – and then rough – differential equations.

6.1 Young–Lóeve estimates

We start with two elementary analysis lemmas, tailor-made for obtaining the Young–Lóeve estimate in Proposition 6.4 below.

Lemma 6.1 *Let $\xi > 0$ and $\theta > 1$. Consider $\varrho : [0,T] \to \mathbb{R}$ with*

$$\varrho\left(r\right) \leq 2\varrho\left(r/2\right) + \xi r^{\theta} \quad \text{for all } r \in [0,T] \tag{6.1}$$

and such that $\varrho\left(r\right) = o\left(r\right)$ as $r \to 0+$, i.e.

$$\lim_{r \to 0+} \frac{\varrho\left(r\right)}{r} = 0. \tag{6.2}$$

Then, for all $r \in [0,T]$,

$$\varrho\left(r\right) \leq \frac{\xi}{1 - 2^{1-\theta}}\, r^{\theta}.$$

Proof. We define $\phi\left(r\right) = \varrho\left(r\right) / \left(\xi r^{\theta}\right)$ and note that (6.1) implies

$$\phi\left(r\right) \leq 1 + 2^{1-\theta} \phi\left(r/2\right).$$

Iterated use of this inequality shows that for all $n \in \mathbb{N}$,

$$\phi\left(r\right) \leq 1 + \sum_{k=1}^{n-1} \left(2^{1-\theta}\right)^k + \left(2^{1-\theta}\right)^n \phi\left(\frac{r}{2^n}\right). \tag{6.3}$$

We now send $n \to \infty$. The last term on the right-hand side above tends to zero since

$$
\begin{aligned}
\left(2^{1-\theta}\right)^n \phi\left(\frac{r}{2^n}\right) &= \left(2^{1-\theta}\right)^n \frac{\varrho\left(r2^{-n}\right)}{\xi r^{\theta} 2^{-\theta n}} \\
&\leq \frac{r^{1-\theta}}{\xi} \frac{\varrho\left(r2^{-n}\right)}{r2^{-n}} \\
&\to \quad 0 \quad \text{by assumption (6.2)}
\end{aligned}
$$

and the proof is finished with

$$\phi\left(r\right) \leq \sum_{k=0}^{\infty} 2^{-\theta k} 2^{k} = \frac{1}{1 - 2^{1-\theta}}.$$

∎

Lemma 6.2 *Let* $\Gamma : \Delta \equiv \{0 \leq s < t \leq T\} \to \mathbb{R}^{e}$ *and assume*
(i) there exists a control $\hat{\omega}$ such that

$$\lim_{r \to 0} \sup_{(s,t) \in \Delta : \hat{\omega}(s,t) \leq r} \frac{|\Gamma_{s,t}|}{r} = 0; \tag{6.4}$$

(ii) there exists a control ω and $\theta > 1, \xi > 0$ such that

$$|\Gamma_{s,u}| \leq |\Gamma_{s,t}| + |\Gamma_{t,u}| + \xi \omega\left(s, u\right)^{\theta} \tag{6.5}$$

holds for $0 \leq s \leq t \leq u \leq T$. Then, for all $0 \leq s < t \leq T$,

$$|\Gamma_{s,t}| \leq \frac{\xi}{1 - 2^{1-\theta}} \omega\left(s, t\right)^{\theta}.$$

Remark 6.3 It is important to notice that the control $\hat{\omega}$ does not appear in the conclusion. □

Proof. At the cost of replacing Γ by Γ/ξ, we can and will take $\xi = 1$. We assume that $\hat{\omega} \leq \frac{1}{\varepsilon} \omega$ for some $\varepsilon > 0$. If this is not the case, we replace ω by $\omega + \varepsilon \hat{\omega}$ and let ε tend to 0 at the end. Define for all $r \in [0, \omega\left(0, T\right)]$,

$$\varrho\left(r\right) = \sup_{s,t \text{ such that } \omega(s,t) \leq r} |\Gamma_{s,t}|.$$

Consider any fixed pair (s, u) with $0 \leq s < u \leq T$ such that $\omega\left(s, u\right) \leq r$, and pick t such that

$$\omega\left(s, t\right), \omega\left(t, u\right) \leq \omega\left(s, u\right)/2.$$

(This is possible thanks to basic properties of a control function, see Exercise 1.10). By definition of ϱ,

$$|\Gamma_{s,t}| \leq \varrho\left(r/2\right), \quad |\Gamma_{s,u}| \leq \varrho\left(r/2\right)$$

and it follows from the assumption (6.5) that

$$|\Gamma_{s,u}| \leq 2\varrho\left(\frac{r}{2}\right) + r^{\theta}.$$

Taking the supremum over all $s < u$ for which $\omega\left(s, u\right) \leq r$ yields that for $r \in [0, \omega\left(0, T\right)]$,

$$\varrho\left(r\right) \leq 2\varrho\left(\frac{r}{2}\right) + r^{\theta}.$$

On the other hand, assumption (6.4) implies that

$$\lim_{r \to 0} \frac{\varrho(r)}{r} = 0.$$

It then suffices to apply Lemma 6.1 to see that for all $r \in [0, \omega(0,T)]$,

$$\varrho(r) \le \frac{1}{1 - 2^{1-\theta}} r^\theta$$

and this readily translates to the statement that, for all $0 \le s < t \le T$,

$$|\Gamma_{s,t}| \le \frac{1}{1 - 2^{1-\theta}} \omega(s,t)^\theta.$$

∎

Proposition 6.4 (Young–Lóeve estimate) *Assume*

$$x \in C^{\text{1-var}}\left([0,T],\mathbb{R}^d\right), \quad y \in C^{\text{1-var}}\left([0,T], L\left(\mathbb{R}^d, \mathbb{R}^e\right)\right)$$

for $p, q \ge 1$ with $\theta := 1/p + 1/q > 1$. With the definition

$$\Gamma_{s,t} := \int_s^t y_u\, dx_u - y_s x_{s,t} = \int_s^t y_{s,u}\, dx_u$$

we have

$$|\Gamma_{s,t}| \le \frac{1}{1 - 2^{1-\theta}} |x|_{p\text{-var};[s,t]} |y|_{q\text{-var};[s,t]}. \qquad (6.6)$$

Remark 6.5 It is instructive to think of $y_s x_{s,t}$ as a first-order Euler approximation to the Riemann–Stieltjes integral $\int_s^t y\, dx$ so that (6.6) is nothing but a "first-order Euler error" estimate. The point of this estimate is its uniformity: although 1-variation was assumed to have a well-defined Riemann–Stieltjes integral, the final estimate only depends on the respective p, q-variation. □

Proof. From Exercise 1.9,

$$\omega(s,t) := |x|_{p\text{-var};[s,t]}^{1/\theta} |y|_{q\text{-var};[s,t]}^{1/\theta}$$

is a control. For all $s < t$ in $[0,T]$ we define

$$\Gamma_{s,t} = \int_s^t y_u\, dx_u - y_s x_{s,t} = \int_s^t y_{s,u}\, dx_u.$$

Then, for fixed $s < t < u$ in $[0,T]$, we have

$$\begin{aligned}
\Gamma_{s,u} - \Gamma_{s,t} - \Gamma_{t,u} &= \int_s^u y_{s,r}\, dx_r - \int_s^t y_{s,r}\, dx_r - \int_t^u y_{t,r}\, dx_r \\
&= y_{s,t} x_{t,u}
\end{aligned}$$

and hence

$$\begin{aligned}
|\Gamma_{s,u}| &\leq |\Gamma_{s,t}| + |\Gamma_{t,u}| + |x|_{p\text{-var};[s,t]}\, |y|_{q\text{-var};[t,u]} \\
&= |\Gamma_{s,t}| + |\Gamma_{t,u}| + \omega\left(s,u\right)^{\theta}.
\end{aligned}$$

Defining $\tilde{\omega}\left(s,t\right) = |x|_{1\text{-var};[s,t]} + |y|_{1\text{-var};[s,t]}$, elementary Riemann–Stieltjes integral estimates show that

$$|\Gamma_{s,t}| \leq |y_{s,\cdot}|_{\infty;[s,t]}\, |x|_{1\text{-var};[s,t]} \leq \tilde{\omega}\left(s,t\right)^{2}.$$

It only remains to apply Lemma 6.2 and the proof is finished. ∎

In the following section we shall use the Young–Lóeve estimate to define the Young integral for $x \in C^{p\text{-var}}$ and $y \in C^{q\text{-var}}$.

Remark 6.6 We could have assumed $y \in C^{q\text{-var}}$ right away in Proposition 6.4. Indeed, as long as $x \in C^{1\text{-var}}$, $\Gamma_{s,t}$ remains a well-defined Riemann–Stieltjes integral and the only change in the argument is to use

$$\tilde{\omega}\left(s,t\right) := |x|_{1\text{-var};[s,t]} + |y|^{q}_{q\text{-var};[s,t]} \implies |\Gamma_{s,t}| \leq \tilde{\omega}\left(s,t\right)^{1+1/q}$$

in the final lines of the proof. □

6.2   Young integrals

The Young–Lóeve estimate clearly implies that

$$(x,y) \mapsto \int_{0}^{\cdot} y\,dx$$

is bilinear (as a function of smooth $\mathbb{R}^{d}$ resp. $L\left(\mathbb{R}^{d},\mathbb{R}^{e}\right)$-valued paths x,y) and continuous in the respective p- and q-variation norm. The (unique, continuous) extension of this map to

$$x \in C^{0,p\text{-var}}\left([0,T],\mathbb{R}^{d}\right),\ \ y \in C^{0,q\text{-var}}\left([0,T],L\left(\mathbb{R}^{d},\mathbb{R}^{e}\right)\right)$$

is immediate from general principles and by squeezing p,q to $p+\varepsilon, q+\varepsilon$ so that $1/\left(p+\varepsilon\right) + 1/\left(q+\varepsilon\right) > 1$ one covers genuine p-variation and q-variation regularity, i.e. $x \in C^{p\text{-var}}$ and $y \in C^{q\text{-var}}$. That said, we shall proceed in a slightly different way which will motivate our later definition of rough differential equations. To this end, recall that any $x \in C^{p\text{-var}}\left([0,T],\mathbb{R}^{d}\right)$ can be approximated "uniformly with uniform variation bounds" by bounded variation paths x^{n}, i.e.

$$d_{\infty;[0,T]}\left(x^{n},x\right) \to 0 \text{ and } \sup_{n} |x^{n}|_{p\text{-var};[0,T]} < \infty.$$

(For instance, piecewise geodesic=linear approximations will do.)

Definition 6.7 (Young integral) Given $x \in C^{p\text{-var}}\left([0,T],\mathbb{R}^d\right)$ and $y \in C^{q\text{-var}}\left([0,T],L\left(\mathbb{R}^d,\mathbb{R}^e\right)\right)$ we say that $z \in C\left([0,T],\mathbb{R}^e\right)$ is an (indefinite) Young integral of y against x if there exists a sequence $(x^n,y^n) \subset C^{1\text{-var}}\left([0,T],\mathbb{R}^d\right) \times C^{1\text{-var}}\left([0,T],L\left(\mathbb{R}^d,\mathbb{R}^e\right)\right)$ which converges uniformly with uniform variation bounds in the sense

$$|x^n - x|_{\infty;[0,T]} \quad \to \quad 0 \text{ and } \sup_n |x^n|_{p\text{-var};[0,T]} < \infty,$$

$$|y^n - y|_{\infty;[0,T]} \quad \to \quad 0 \text{ and } \sup_n |y^n|_{q\text{-var};[0,T]} < \infty,$$

and

$$\int_0^\cdot y^n \, dx^n \to z \text{ uniformly on } [0,T] \text{ as } n \to \infty.$$

If z is unique we write $\int_0^\cdot y\,dx$ instead of z and set $\int_s^t y\,dx := \int_0^t y\,dx - \int_0^s y\,dx$. □

Theorem 6.8 (Young–Lóeve) *Given* $x \in C^{p\text{-var}}\left([0,T],\mathbb{R}^d\right)$ *and* $y \in C^{q\text{-var}}\left([0,T],L\left(\mathbb{R}^d,\mathbb{R}^e\right)\right)$ *with* $\theta = 1/p + 1/q > 1$, *there exists a unique (indefinite) Young integral of* y *against* x, *denoted by* $\int_0^\cdot y\,dx$ *and the Young–Lóeve estimate*

$$\forall 0 \le s \le t \le T : \left|\int_s^t y\,dx - y_s x_{s,t}\right| \le \frac{1}{1-2^{1-\theta}}\,|x|_{p\text{-var};[s,t]}\,|y|_{q\text{-var};[s,t]}$$

remains valid. Moreover, the indefinite Young integral has finite p-variation and

$$\left|\int_0^\cdot y\,dx\right|_{p\text{-var};[s,t]} \le C\,|x|_{p\text{-var};[s,t]}\left(|y|_{q\text{-var};[s,t]} + |y|_{\infty;[s,t]}\right)$$

$$\le 2C\,|x|_{p\text{-var};[s,t]}\left(|y|_{q\text{-var};[0,T]} + |y_0|\right) \qquad (6.7)$$

where $C = C(p,q)$.

Proof. Let us first argue that any limit point z of $\int_0^\cdot y^n\,dx^n$ (in uniform topology on $[0,T]$) satisfies the Young–Lóeve estimate. For every $\varepsilon > 0$ small enough so that

$$\theta_\varepsilon := 1/(p+\varepsilon) + 1/(q+\varepsilon) > 1$$

the Young–Lóeve estimate of Proposition 6.4 gives

$$\left|\int_s^t y_r^n \, dx_r^n - y_s^n x_{s,t}^n\right| \le \frac{1}{1-2^{1-\theta_\varepsilon}}\,|x^n|_{(p+\varepsilon)\text{-var};[s,t]}\,|y^n|_{(q+\varepsilon)\text{-var};[s,t]}. \qquad (6.8)$$

By Corollary 5.29, the right-hand side above can be made arbitrarily small, uniformly in n, provided $t-s$ is small enough; this readily leads to equicontinuity of the indefinite Riemann–Stieltjes integrals

$$\left\{\int_0^\cdot y_r^n \, dx_r^n : n \in \mathbb{N}\right\}.$$

Boundedness is clear and so, by Arzela–Ascoli, we have uniform convergence along a subsequence to some $z \in C([0,T],\mathbb{R}^e)$, which proves existence of the Young integral. Using the first part of Corollary 5.29, we let n tend to ∞ in (6.8) and obtain

$$|z_{s,t} - y_s x_{s,t}| \le \frac{1}{1 - 2^{1-\theta_\varepsilon}} |x|_{(p+\varepsilon)\text{-var};[s,t]} |y|_{(q+\varepsilon)\text{-var};[s,t]}.$$

Then, an application of Lemma 5.13 justifies the passage $\varepsilon \searrow 0$ which shows validity of the Young–Lóeve estimate,

$$|z_{s,t} - y_s x_{s,t}| \le \frac{1}{1 - 2^{1-\theta}} |x|_{p\text{-var};[s,t]} |y|_{q\text{-var};[s,t]}. \tag{6.9}$$

To prove uniqueness we use the control $\omega(s,t) := |x|^{1/\theta}_{p\text{-var};[s,t]} |y|^{1/\theta}_{q\text{-var};[s,t]}$ (cf. Exercise 1.9). Assume $z, \tilde{z}$ are two limit points of $\int_0^\cdot y^n dx^n$ so that $z_0 = \tilde{z}_0 = 0$. Fix a dissection (t_i) of $[0,T]$ and observe

$$
\begin{aligned}
|z_T - \tilde{z}_T| &\le \sum_i \left| z_{t_i,t_{i+1}} - x_{t_i} y_{t_i,t_{i+1}} + x_{t_i} y_{t_i,t_{i+1}} - \tilde{z}_{t_i,t_{i+1}} \right| \\
&\le \frac{2}{1 - 2^{1-\theta}} \sum_i \omega(t_i, t_{i+1})^\theta \\
&\le \frac{2}{1 - 2^{1-\theta}} \omega(0,T) \max_i \omega(t_i, t_{i+1})^{\theta-1}.
\end{aligned}
$$

Applying this to a sequence of dissections with mesh ($=\max_i |t_{i+1} - t_i|$) tending to zero we see that $|z_T - \tilde{z}_T|$ can be made arbitrarily small and hence must be zero. This shows $z_T = \tilde{z}_T$ and, as T was arbitrary, $z \equiv \tilde{z}$.

At last, set $c = 1/(1 - 2^{1-\theta})$ and observe that (6.9) implies

$$|z_{s,t}|^p \le 2^{1-1/p} \left(c^p \omega(s,t)^{\theta p} + |y|^p_{\infty;[s,t]} |x|^p_{p\text{-var};[s,t]} \right).$$

We observe that the right-hand side above is super-additive in (s,t) and so

$$\left| \int_0^\cdot y dx \right|^p_{p\text{-var};[s,t]} \le 2^{1-1/p} \left(c^p \omega(s,t)^{\theta p} + |y|^p_{\infty;[s,t]} |x|^p_{p\text{-var};[s,t]} \right)$$

and the proof is easily finished. ∎

Exercise 6.9 *The purpose of this exercise is to show that our Definition 6.7 is consistent with the "usual" definition of Young integrals as limits of Riemann–Stieltjes sums. To this end, let $x \in C^{p\text{-var}}([0,T],\mathbb{R})$, $y \in C^{q\text{-var}}([0,T],\mathbb{R})$ with $\theta := 1/p + 1/q > 1$ and let $D_n = (t_i^n)_i$ be a sequence of dissections of $[0,T]$ with $|D_n| \to 0$, and ξ_i^n some points in $[t_i^n, t_{i+1}^n]$. Show that $\sum_{i=0}^{|D_n|-1} y(\xi_i^n) x_{t_i,t_{i+1}}$ converges when n tends to ∞ to a limit I independent of the choice of ξ_i^n and the sequence (D_n). Identify I as a Young integral $\int_0^T y dx$ in the sense of Definition 6.7.*

Exercise 6.10 *Let* $\varphi = (\varphi_i)_{i=1,\dots,d}$ *be a collection of maps from* $\mathbb{R}^d$ *to* $\mathbb{R}^e$ *and assume* $x \in C^{p\text{-var}}\left([0,T],\mathbb{R}^d\right)$. *Show that*

$$\int_0^{\cdot} \varphi(x)\, dx$$

is a well-defined Young integral provided φ, *viewed as a map* $\mathbb{R}^d \to$ $L\left(\mathbb{R}^d, \mathbb{R}^e\right)$ *is* $(\gamma - 1)$-*Hölder provided*

$$\gamma > p$$

and $\gamma - 1 \in (0,1]$, *to avoid trivialities.(We shall encounter this type of regularity assumption in our forthcoming discussion of rough integrals.)*

Solution. The path $\varphi(x.)$ has finite $q = p/(\gamma - 1)$-variation and we see that the Young integral is well-defined since

$$\frac{1}{q} + \frac{1}{p} = \frac{\gamma}{p} > 1.$$

$\square$

6.3   Continuity properties of Young integrals

Proposition 6.11 *Given* $x \in C^{p\text{-var}}\left([0,T],\mathbb{R}^d\right)$ *and* $y \in C^{q\text{-var}}\left([0,T],\right.$ $\left.L\left(\mathbb{R}^d,\mathbb{R}^e\right)\right)$ *with* $1/p + 1/q > 1$ *the map*

$$(x,y) \mapsto \int_0^{\cdot} y\, dx$$

from $C^{p\text{-var}}\left([0,T],\mathbb{R}^d\right) \times C^{q\text{-var}}\left([0,T],L\left(\mathbb{R}^d,\mathbb{R}^e\right)\right) \to C^{p\text{-var}}\left([0,T],\mathbb{R}^e\right)$ *equipped with the respective* p, q-*variation norms is a bilinear and continuous map. As a consequence, it is Lipschitz continuous on bounded sets and Fréchet smooth.*

Proof. Bilinearity of $(x,y) \mapsto \int y\, dx$ follows from bilinearity of the approximations in the definition and uniqueness; continuity in the sense of bilinear maps is immediate from Young–Lóeve and Fréchet smoothness for bilinear, continuous maps is trivial. $\blacksquare$

The following property deals with continuity with respect to "uniform convergence with uniform bounds".

Proposition 6.12 *Assume given* $x^n, x \in C^{p\text{-var}}\left([0,T],\mathbb{R}^d\right)$ *and* $y^n, y \in C^{q\text{-var}}\left([0,T],L\left(\mathbb{R}^d,\mathbb{R}^e\right)\right)$ *such that*

$$\lim_{n\to\infty} |x^n - x|_{\infty;[0,T]} = 0 \ and \ \sup_n |x^n|_{p\text{-var};[0,T]} < \infty,$$

$$\lim_{n\to\infty} |y^n - y|_{\infty;[0,T]} = 0 \ and \ \sup_n |y^n|_{q\text{-var};[0,T]} < \infty$$

and $1/p + 1/q > 1$. Then

$$\lim_{n \to \infty} \left| \int_0^\cdot y^n \, dx^n - \int_0^\cdot y \, dx \right|_{\infty;[0,T]} = 0 \text{ and } \sup_n \left| \int_0^\cdot y^n \, dx^n \right|_{p\text{-var};[0,T]} < \infty.$$

Proof. Increase p, q by ε small enough so that $1/(p + \varepsilon) + 1/(q + \varepsilon) > 1$. By interpolation, $x^n \to x$ in $(p + \varepsilon)$-variation and similarly $y^n \to y$ in $(q + \varepsilon)$-variation. By the preceding proposition, $\int_0^\cdot y^n \, dx^n$ converges to $\int_0^\cdot y \, dx$ in $(p + \varepsilon)$-variation and hence in ∞-norm. The uniform p-variation bounds on $\int_0^\cdot y^n \, dx^n$ follow immediately from the estimate in the Young–Lóeve theorem. ∎

Exercise 6.13 *Using continuity properties, establish an integration-by-parts formula for Young integrals.*

Exercise 6.14 *Fix $p, q \geq 1$ with $1/p + 1/q > 1$, $R > 0$, and fix $x \in C^{p\text{-var}}\left([0,T], \mathbb{R}^d\right)$. Define $B_R = \left\{ y \in C^{q\text{-var}}\left([0,T], L\left(\mathbb{R}^d, \mathbb{R}^e\right)\right), |y|_{q\text{-var};[0,T]} \leq R \right\}$. Prove that the map*

$$B_R \quad \to \quad C^{p\text{-var}}\left([0,T], \mathbb{R}^e\right)$$

$$y \quad \mapsto \quad \int_0^\cdot y \, dx$$

is Lipschitz using the $d_{q\text{-var}}$-metric. Prove it is uniformly continuous with respect to $d_{q'\text{-var}}$-metric with $q' > q$. Prove also it is uniformly continuous with respect to d_∞-metric.
[Hint: Lipschitz with respect to $d_{q\text{-var}}$-metric is just the Young–Lóeve estimate. For $q' > q$, use the first case plus interpolation.]

6.4 Young–Lóeve–Towghi estimates and 2D Young integrals

Young integrals extend naturally to higher dimensions but only the 2D case will be relevant to us. 2D statements which extend line by line from the 1D statements will not be discussed in detail. In particular, we shall use the Riemann–Stieltjes integral $\int_{[0,T]^2} y \, dx$ of a continuous function $y \in C\left([0,T]^2, L\left(\mathbb{R}^d, \mathbb{R}^e\right)\right)$ with respect to a bounded variation function $x \in C^{1\text{-var}}\left([0,T]^2, \mathbb{R}^d\right)$. We will also show that the estimate

$$\left| \int_R y \, dx \right| \leq |y|_{\infty;R} \, |x|_{1\text{-var};R}$$

is valid for any rectangle $R \subset [0,T]^2$. We first need to extend Lemma 6.2 to its 2-dimensional version.

Lemma 6.15 *Let* $\Gamma : \{0 \leq s < t \leq T\}^2 \to \mathbb{R}^e$ *be such that*[1]
(i) for some control $\hat{\omega}$,

$$\lim_{r \to 0} \sup_{\text{rectangle } R \text{ s.t. } \hat{\omega}(R) \leq r} \frac{|\Gamma(R)|}{r} = 0; \tag{6.10}$$

(ii) for some control ω *and some real* $\theta > 1$, *for all rectangles* R *being the union of two (essentially) disjoint rectangles* R_1, R_2,

$$|\Gamma(R)| \leq |\Gamma(R_1)| + |\Gamma(R_2)| + \xi \omega(R)^\theta. \tag{6.11}$$

Then, for all rectangles $R \in [0, T]^2$,

$$|\Gamma(R)| \leq \frac{\xi}{1 - 2^{1-\theta}} \omega(R)^\theta.$$

Proof. The proof follows exactly the same lines as the 1-dimensional proof.   ∎

We now proceed as in the 1D case and prove uniform Young–Lóeve-type estimates, for integrand and driving signal which are assumed to be of bounded variation.

Proposition 6.16 *Let*

$$x \in C^{\text{1-var}}\left([0,T]^2, \mathbb{R}^d\right), \ y \in C^{\text{1-var}}\left([0,T]^2, L\left(\mathbb{R}^d, \mathbb{R}^e\right)\right).$$

Given $R = [s,t] \times [u,v] \subset [0,T]^2$, *define*

$$\Gamma(R) = \int_R y\left(\begin{matrix} s, \cdot \\ u, \cdot \end{matrix}\right) dx.$$

Then, for all such rectangles $R \subset [0,T]^2$ *and* $p, q \geq 1$ *such that* $\theta := p^{-1} + q^{-1} > 1$,

$$|\Gamma(R)| \leq \left(\frac{1}{1 - 2^{1-\theta}}\right)^2 |y|_{q\text{-var};R} \, |x|_{p\text{-var};R}.$$

Proof. Define the control $\omega(R) = |x|_{p\text{-var};R}^{\frac{1}{\theta}} |y|_{q\text{-var};R}^{\frac{1}{\theta}}$. We consider a generic rectangle $R = [r,t] \times [u,v]$ cut into two non-intersecting rectangles, say $R_1 = [r,s] \times [u,v]$ and $R_2 = [s,t] \times [u,v]$. Observe that from the definition

---

[1] We identify once again $\{0 \leq s < t \leq T\}^2$ with rectangles contained in $[0,T]^2$.

of $\Gamma(R)$, we see

$$
\begin{aligned}
\Gamma(R) &= \int_R y\begin{pmatrix} r, \cdot \\ u, \cdot \end{pmatrix} dx \\
&= \int_{R_1} y\begin{pmatrix} r, \cdot \\ u, \cdot \end{pmatrix} dx + \int_{R_2} y\begin{pmatrix} r, \cdot \\ u, \cdot \end{pmatrix} dx \\
&= \Gamma(R_1) + \Gamma(R_2) + \int_{R_2} \left\{ y\begin{pmatrix} r, \cdot \\ u, \cdot \end{pmatrix} - y\begin{pmatrix} s, \cdot \\ u, \cdot \end{pmatrix} \right\} dx \\
&= \Gamma(R_1) + \Gamma(R_2) + \int_{R_2} y\begin{pmatrix} r, s \\ u, \cdot \end{pmatrix} dx \\
&= \Gamma(R_1) + \Gamma(R_2) + \int_{[u,v]} y\begin{pmatrix} r, s \\ u, \cdot \end{pmatrix} d\left(x\begin{pmatrix} s, t \\ \cdot \end{pmatrix} \right).
\end{aligned}
$$

But from Exercise 5.56, the 1-dimensional q-variation of $y\begin{pmatrix} r, s \\ \cdot \end{pmatrix}$ over $[u,v]$ is bounded by $|y|_{q\text{-var};[r,s],[u,v]}$ and the 1-dimensional p-variation of $x\begin{pmatrix} s, t \\ \cdot \end{pmatrix}$ over $[u,v]$ is bounded by $|x|_{p\text{-var};[s,t],[u,v]}$. Hence, using Young–Lóeve 1 dimensional estimates, we obtain

$$
|\Gamma(R)| \le |\Gamma(R_1)| + |\Gamma(R_2)| + \frac{1}{1 - 2^{1-\theta}} \omega(R)^\theta.
$$

Defining $\tilde{\omega}$ to be a 2D control dominating the 1-variation of x and y, we see that

$$
|\Gamma(R)| \le \tilde{\omega}(R)^2.
$$

It only remains to apply Lemma 6.15.   ∎

With the notation of the above theorem, we see that

$$
\begin{aligned}
\int_R y\,dx &= \int_R y\begin{pmatrix} s, \cdot \\ u, \cdot \end{pmatrix} dx + \int_R y\begin{pmatrix} s \\ u, \cdot \end{pmatrix} dx \\
&\quad + \int_R y\begin{pmatrix} s, \cdot \\ u \end{pmatrix} dx + y\begin{pmatrix} s \\ u \end{pmatrix} x(R).
\end{aligned}
$$

We see that the second and third integrals will be well defined when y and x are of finite q- and p-variation if their 1D projections $y\begin{pmatrix} s \\ \cdot \end{pmatrix}$ and $y\begin{pmatrix} \cdot \\ u \end{pmatrix}$ are of finite (1D) q-variation. This is actually satisfied if $y\begin{pmatrix} 0 \\ \cdot \end{pmatrix}$ and $y\begin{pmatrix} \cdot \\ 0 \end{pmatrix}$ are of finite (1D) q-variation and y is of finite (2D) q-variation. To simplify, we therefore restrict ourselves to paths y such that $y\begin{pmatrix} 0 \\ \cdot \end{pmatrix} = y\begin{pmatrix} \cdot \\ 0 \end{pmatrix} = 0$. In particular, we define $C_{0,0}^{q\text{-var}}\left([0,T]^2, L\left(\mathbb{R}^d, \mathbb{R}^e\right)\right)$ the set of functions y such that $y_{0,\cdot} = y_{\cdot,0} = 0$.

Definition 6.17 (Young integral) Given $x \in C^{p\text{-var}}\left([0,T]^2, \mathbb{R}^d\right), y \in C_{0,0}^{q\text{-var}}\left([0,T]^2, L\left(\mathbb{R}^d, \mathbb{R}^e\right)\right)$ we say that $z \in C\left([0,T]^2, \mathbb{R}^e\right)$ is an (indefinite) Young integral of y against x if there exists a sequence $(x^n, y^n) \subset$

$C^{\text{1-var}}\left([0,T]^2,\mathbb{R}^d\right)\times C^{\text{1-var}}_{0,0}\left([0,T]^2,L\left(\mathbb{R}^d,\mathbb{R}^e\right)\right)$ which converges uniformly with uniform variation bounds in the sense

$$\lim_{n\to\infty} d_{\infty;[0,T]^2}\left(x^n,x\right) = 0 \text{ and } \sup_n |x^n|_{p\text{-var};[0,T]^2} < \infty,$$

$$\lim_{n\to\infty} d_{\infty;[0,T]^2}\left(y^n,y\right) = 0 \text{ and } \sup_n |y^n|_{q\text{-var};[0,T]^2} < \infty,$$

and

$$\lim_{n\to\infty} \int_{[0,s]\times[0,u]} y^n\,dx^n = z \text{ uniformly on } (s,u)\in[0,T]^2 \text{ as } n\to\infty.$$

If z is unique we write $\int_0^{\cdot\cdot} y\,dx$ instead of z.   □

Following the same lines as the 1-dimensional case (which involves generalizing a few analysis and p-variation lemmas from 1D to 2D), we obtain

Theorem 6.18 (Young–Lóeve–Towghi) *Given* $x \in C^{p\text{-var}}\left([0,T]^2,\mathbb{R}^d\right)$, $y \in C^{q\text{-var}}_{0,0}\left([0,T]^2,L\left(\mathbb{R}^d,\mathbb{R}^e\right)\right)$ *with* $\theta = 1/p + 1/q > 1$, *there exists a unique (indefinite) Young integral of* y *against* x, *denoted by* $\int_0^{\cdot\cdot} y\,dx$ *and we have*

$$\left|\int_R y\left(\begin{matrix} s & u \\ \cdot & \cdot \end{matrix}\right)dx\right| \le \left(\frac{1}{1-2^{1-\theta}}\right)^2 |x|_{p\text{-var};R}\,|y|_{q\text{-var};R} \qquad (6.12)$$

for all rectangles $R = [s,t]\times[u,v] \subset [0,T]^2$.

One can also check, just as in the 1D case, that $(x,y)\mapsto \int_0^{\cdot\cdot} y_u\,dx_u$ is a bilinear continuous map from

$$C^{p\text{-var}}\left([0,T]^2,\mathbb{R}^d\right) \times C^{q\text{-var}}_{0,0}\left([0,T]^2,L\left(\mathbb{R}^d,\mathbb{R}^e\right)\right) \to C^{p\text{-var}}\left([0,T]^2,\mathbb{R}^e\right)$$

(and hence Lipschitz on bounded sets and Fréchet smooth).

6.5   Comments

Young integration goes back to Young [177]. The higher-dimensional case was partially discussed in Young [177] and then in Towghi [170] in the form which is relevant to us.

Part II

Abstract theory of rough paths

7

Free nilpotent groups

Motivated by simple higher-order Euler schemes for ODEs we give a systematic and self-contained account of the "algebra of iterated integrals". Tensor algebras play a natural role. However, thanks to algebraic relations between iterated integrals the "correct" state-space will be seen to be a (so-called) free nilpotent Lie group, faithfully represented as a subset of the tensor algebra. It becomes a metric (and even geodesic) space under the so-called Carnot–Caratheodory metric and will later serve as a natural state-space for geometric rough paths.

7.1 Motivation: iterated integrals and higher-order Euler schemes

Let x be an $\mathbb{R}^d$-valued continuous path of bounded variation and define the kth iterated integrals of the path segment $x|_{[s,t]}$ as

$$\mathbf{g}^{k;i_1,\ldots,i_k} := \int_s^t \int_s^{u_k} \cdots \int_s^{u_2} dx_{u_1}^{i_1} \cdots dx_{u_k}^{i_k}. \tag{7.1}$$

The collection of all such iterated integrals,

$$\mathbf{g} = \left(\mathbf{g}^{k;i_1,\ldots,i_k} : 1 \le k \le N; \, i_1,\ldots,i_k \in \{1,\ldots,d\} \right) \tag{7.2}$$

is called the step-N signature of the path segment $x|_{[s,t]}$ and is denoted by $S_N(x)_{s,t}$. Postponing (semi-obvious) algebraic formalities to the next section, let us consider (higher-order) Euler schemes for the ODE

$$dy = V(y) \, dx = \sum_{i=1}^d V_i(y) \, dx^i$$

with $V \in C\left(\mathbb{R}^e, L\left(\mathbb{R}^d, \mathbb{R}^e\right)\right)$ and recall that $\pi_{(V)}(0, y_0; x)$ stands for any (not necessarily unique) solution started at y_0, possibly only defined up to some explosion time. Let I denote the identity function on $\mathbb{R}^e$ and recall the identification of a vector field $W = \left(W^1, \ldots, W^e\right)^T : \mathbb{R}^e \to \mathbb{R}^e$ with the first-order differential operator

$$\sum_{k=1}^e W^k(y) \frac{\partial}{\partial y^k}.$$

Granted sufficient regularity of V, a simple Taylor expansion suggests, at least for $0 < t - s << 1$, a step-N approximation of the form

$$y_t \approx y_s + \sum_{i=1}^{d} V_i(y_s) x_{s,t}^i$$

$$+ \ldots$$

$$+ \sum_{\substack{i_1,\ldots,i_N \\ \in\{1,\ldots,d\}}} V_{i_1} \cdots V_{i_N} I(y_s) \int_s^t \int_s^{u_N} \cdots \int_s^{u_2} dx_{u_1}^{i_1} \ldots dx_{u_N}^{i_N}.$$

Having made plain the importance of iterated integrals in higher-order Euler schemes, let us observe the presence of non-linear constraints between the iterated integrals (7.1). This happens already at the "second" level of iterated integrals.

Example 7.1 Let $x \in C_0^{\text{1-var}}([0,T], \mathbb{R}^d)$ and write $\mathbf{x} = S_2(x)$ for its step-2 lift.
(i) Using integration by parts it readily follows that, for all $i, j \in \{1,\ldots,d\}$

$$\mathbf{x}_{s,t}^{2;i,j} + \mathbf{x}_{s,t}^{2;j,i} = \int_s^t x_{s,r}^i dx_r^j + \int_s^t x_{s,r}^j dx_r^i = x_{s,t}^i x_{s,t}^j. \qquad (7.3)$$

(ii) As a trivial consequence of $\mathbf{x}_{s,t}^1 := x_t - x_s$ we have

$$\mathbf{x}_{s,t}^{1;i} + \mathbf{x}_{t,u}^{1;i} = \mathbf{x}_{s,u}^{1;i}, \quad i = 1,\ldots,d. \qquad (7.4)$$

More interestingly, an elementary computation[1] gives

$$\mathbf{x}_{s,u}^{2;i,j} = \mathbf{x}_{s,t}^{2;i,j} + \mathbf{x}_{t,u}^{2;i,j} + \mathbf{x}_{s,t}^{1;i} \mathbf{x}_{t,u}^{1;j}, \quad i,j = 1,\ldots,d. \qquad (7.5)$$

This (matrix) equation can be expressed in terms of equations of the respective symmetric and anti-symmetric parts. Adding the equation obtained by interchanging i, j we see, using (7.3)

$$x_{s,u}^i x_{s,u}^j = x_{s,t}^i x_{s,t}^j + x_{t,u}^i x_{t,u}^j + x_{s,t}^i x_{t,u}^j + x_{s,t}^j x_{t,u}^i$$

which just (re-)expresses the additivity of vector increments (7.4). On the other hand, subtracting the equation obtained by interchanging i, j, followed by multiplication with $1/2$, yields

$$A_{s,u}^{i,j} = A_{s,t}^{i,j} + A_{t,u}^{i,j} + \frac{1}{2}\left(x_{s,t}^i x_{t,u}^j - x_{s,t}^j x_{t,u}^i\right) \qquad (7.6)$$

---

[1] ...to be compared with the forthcoming Theorem 7.11.

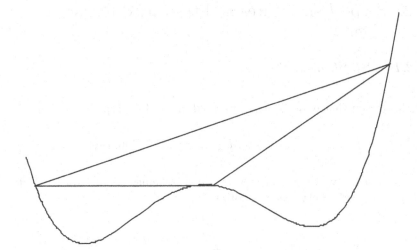

Figure 7.1. We plot (x^i, x^j). The triangle connects the points: (x_s^i, x_s^j) on the lower left side, (x_t^i, x_t^j) in the middle and (x_u^i, x_u^j) on the right side. Note that the respective area increments (signed area between the path and a linear chord) are not additive. In fact, $A_{s,u}^{i,j} = A_{s,t}^{i,j} + A_{t,u}^{i,j} + \Delta_{s,t,u}^{i,j}$, where $\Delta_{s,t,u}^{i,j}$ is the area of the triangle as indicated in the figure.

where

$$A_{s,t}^{i,j} := \frac{1}{2}\left(\mathbf{x}_{s,t}^{2;i,j} - \mathbf{x}_{s,t}^{2;j,i}\right) = \frac{1}{2}\left(\int_s^t x_{s,r}^i \, dx_r^j - \int_s^t x_{s,r}^j \, dx_r^i\right)$$

has an appealing geometric interpretation as seen in Figure 7.1.

Let us draw some first conclusions.

(i) The naive state space $\mathbb{R}^{d+d^2+\cdots+d^N}$ for iterated integrals of the form (7.2) is too big. For instance, there is no need to store the symmetric part of $\mathbf{x}_{s,t}^2$.

(ii) Any analysis which is based on higher-order Euler approximations over finer and finer intervals, e.g. starting with some interval $[s, u]$ then $[s, t], [t, u]$, etc., must acknowledge the non-linear nature of "higher-order increments" as seen in (7.5) and (7.6).

7.2   Step-N signatures and truncated tensor algebras

7.2.1   Definition of S_N

We have seen that expressions of iterated integrals of type

$$\int_{s<u_1<\ldots<u_k<t} dx_{u_1}^{i_1} \ldots dx_{u_k}^{i_k}, \quad x \in C^{\text{1-var}}\left([0,T],\mathbb{R}^d\right)$$

appear naturally when considering Euler schemes. A first-order Euler scheme over the interval $[s,t]$ involves

$$\left(\int_{s<u<t} dx_u^i\right)_{i=1,\ldots,d} \;\leftrightarrow\; \sum_{i=1}^{d}\left(\int_{s<u<t} dx_u^i\right) e_i \in \mathbb{R}^d$$

where $(e_i)_{i=1,\ldots,d}$ denotes the canonical basis of $\mathbb{R}^d$. For a second-order scheme one needs additionally

$$\left(\int_{s<u<v<t} dx_u^i dx_v^j,\right)_{i,j=1,\ldots,d}$$

$$\leftrightarrow\; \sum_{i,j=1}^{d}\left(\int_{s<u<v<t} dx_u^i dx_v^j,\right)(e_i \otimes e_j) \in \mathbb{R}^d \otimes \mathbb{R}^d$$

where $\mathbb{R}^d \otimes \mathbb{R}^d$ with basis $(e_i \otimes e_j)_{i,j=1,\ldots,d}$ can be viewed as the set of real-valued $d \times d$ matrices with canonical basis.[2] The (obvious) generalization to $k \geq 2$ reads

$$\left(\int_{s<u_1<\ldots<u_k<t} dx_{u_1}^{i_1} \ldots dx_{u_k}^{i_k}\right)_{i_1,\ldots,i_k}$$

$$\leftrightarrow\; \sum_{i_1,\ldots,i_d}\left(\int_{s<u_1<\ldots<u_k<t} dx_{u_1}^{i_1} \ldots dx_{u_k}^{i_k}\right)(e_{i_1} \otimes \cdots \otimes e_{i_k}) \quad (7.7)$$

where $(e_{i_1} \otimes \cdots \otimes e_{i_k})$, $i_1,\ldots,i_k \in \{1,\ldots,d\}$, is the canonical basis of $\left(\mathbb{R}^d\right)^{\otimes k}$, the space of k-tensors over $\mathbb{R}^d$. Life is easier without many indices and we shall write the right-hand side of (7.7) simply as

$$\int_{s<u_1<\ldots<u_k<t} dx_{u_1} \otimes \cdots \otimes dx_{u_k} \in \left(\mathbb{R}^d\right)^{\otimes k}.$$

---

[2] $e_i \otimes e_j$ corresponds to the matrix with entry 1 in the ith line, jth column and 0 everywhere else.

We note that, as vector spaces $\left(\mathbb{R}^d\right)^{\otimes k} \cong \mathbb{R}^{d^k}$, and it is a convenient convention to set

$$\left(\mathbb{R}^d\right)^{\otimes 0} := \mathbb{R}.$$

To any $\mathbb{R}^d$-valued path γ of finite length defined on some interval $[s,t]$, we may associate the collection of its iterated integrals. We have

Definition 7.2 The step-N signature of $\gamma \in C^{1\text{-var}}\left([s,t],\mathbb{R}^d\right)$ is given by

$$S_N(\gamma)_{s,t} \equiv \left(1, \int_{s<u<t} dx_u, \ldots, \int_{s<u_1<\ldots<u_k<t} dx_{u_1} \otimes \cdots \otimes dx_{u_k}\right)$$

$$\in \oplus_{k=0}^N \left(\mathbb{R}^d\right)^{\otimes k}.$$

The path $u \mapsto S_N(\gamma)_{s,u}$ is called the (step-N) lift of x. □

Remark 7.3 This notation is further justified by Chen's theorem below: the step-N lift of some $\gamma \in C^{1\text{-var}}\left([0,T],\mathbb{R}^d\right)$, $t \mapsto S_N(\gamma)_t$ takes values in a group so that $S_N(\gamma)_{s,t}$ is the natural increment of this path, i.e. the product of $(S_N(\gamma)_s)^{-1}$ with $S_N(\gamma)_t$. □

Given two vectors $a,b \in \mathbb{R}^d$ with coordinates $\left(a^i\right)_{i=1,\ldots,d}$ and $\left(b^i\right)_{i=1,\ldots,d}$ one can construct the matrix $\left(a^i b^j\right)_{i,j=1,\ldots,d}$ and hence the 2-tensor

$$a \otimes b := \sum_{i,j=1}^d \left(a^i b^j\right) e_i \otimes e_j \in \mathbb{R}^d \otimes \mathbb{R}^d.$$

(In fact, this is the linear extension of the map $\otimes : \mathbb{R}^d \times \mathbb{R}^d \to \mathbb{R}^d \otimes \mathbb{R}^d$ which maps the pair (e_i, e_j) to the (i,j)th basis element of $\mathbb{R}^d \otimes \mathbb{R}^d$, for which we already used the suggestive notation $e_i \otimes e_j$.) More generally, given

$$a = \sum_{i_1,\ldots,i_k} a^{i_1,\ldots,i_k} e_{i_1} \otimes \cdots \otimes e_{i_k} \in \left(\mathbb{R}^d\right)^{\otimes k} \tag{7.8}$$

then, with similar notation $b \in \left(\mathbb{R}^d\right)^{\otimes l}$, we agree that $a \otimes b$ is defined by

$$a \otimes b = \sum a^{i_1,\ldots,i_k} b^{j_1,\ldots,j_l} e_{i_1} \otimes \cdots \otimes e_{i_k} \otimes e_{j_1} \otimes \ldots e_{j_l} \tag{7.9}$$

$$\in \left(\mathbb{R}^d\right)^{\otimes k} \otimes \left(\mathbb{R}^d\right)^{\otimes l} \cong \left(\mathbb{R}^d\right)^{\otimes(k+l)}.$$

We now define

$$T^N\left(\mathbb{R}^d\right) := \oplus_{k=0}^N \left(\mathbb{R}^d\right)^{\otimes k}, \tag{7.10}$$

and write $\pi_k : T^N\left(\mathbb{R}^d\right) \to \left(\mathbb{R}^d\right)^{\otimes k}$ for the projection to the kth tensor level. We shall also use the projection

$$\pi_{0,k} : T^N\left(\mathbb{R}^d\right) \to T^k\left(\mathbb{R}^d\right), \text{ for } k \leq N, \tag{7.11}$$

which maps $g = \left(g^0, \ldots, g^N\right) \in T^N\left(\mathbb{R}^d\right)$ into $\left(g^0, \ldots, g^k\right) \in T^N\left(\mathbb{R}^d\right)$. Here, $g^k = \pi_k\left(g\right) \in \left(\mathbb{R}^d\right)^{\otimes k}$ is sometimes refered to as the kth level of g. Given $g, h \in T^N\left(\mathbb{R}^d\right)$, one extends (7.9) to $T^N\left(\mathbb{R}^d\right)$ by setting

$$g \otimes h = \sum_{\substack{i+j \leq N \\ i,j \geq 0}} g^i \otimes h^j \Leftrightarrow \forall k \in \{0, \ldots, N\} : \pi_k\left(g \otimes h\right) = \sum_{i=0}^{k} g^{k-i} \otimes h^i.$$

The vector space $T^N\left(\mathbb{R}^d\right)$ becomes an (associative) algebra under $\otimes$. More precisely, we have

Proposition 7.4 *The space* $\left(T^N\left(\mathbb{R}^d\right), +, .; \otimes\right)$ *is an associative algebra with neutral element*

$$1 := (1, 0, \ldots, 0) = 1 + 0 + \cdots + 0 \in T^N\left(\mathbb{R}^d\right).$$

(The unit element for $+$ is $0 = (0, 0, \ldots, 0)$, of course.) We will call $T^N\left(\mathbb{R}^d\right)$ *the truncated tensor algebra of level N.*

Proof. Straightforward and left to the reader. ∎

Remark 7.5 We shall see below that the set of all $g \in T^N\left(\mathbb{R}^d\right)$ with $\pi_0\left(g\right) = 1$ forms a (Lie) group under $\otimes$ with unit element 1. When $N = 1$ this group is isomorphic to $\left(\mathbb{R}^d, +\right)$ with the usual unit element 0 and this persuades us to set

$$o := (1, 0, \ldots, 0).$$ □

Similar to the algebra of square matrices, the algebra product is not commutative (unless $N = 1$ or $d = 1$), indeed

$$\begin{pmatrix} 1 \\ 0 \end{pmatrix} \otimes \begin{pmatrix} 1 \\ 1 \end{pmatrix} = \begin{pmatrix} 1 & 1 \\ 0 & 0 \end{pmatrix} \neq \begin{pmatrix} 1 \\ 1 \end{pmatrix} \otimes \begin{pmatrix} 1 \\ 0 \end{pmatrix} = \begin{pmatrix} 1 & 1 \\ 1 & 0 \end{pmatrix}.$$

Let us now define a norm on $T^N\left(\mathbb{R}^d\right) := \oplus_{k=0}^{N} \left(\mathbb{R}^d\right)^{\otimes k}$. To this end, we equip each tensor level $\left(\mathbb{R}^d\right)^{\otimes k}$ with Euclidean structure, which amounts to declaring the canonical basis $\{e_{i_1} \otimes \cdots \otimes e_{i_k} : i_1, \ldots, i_k \in \{1, \ldots, d\}\}$ to be orthonormal so that for any $a \in \left(\mathbb{R}^d\right)^{\otimes k}$ of form (7.8)

$$|a|_{\left(\mathbb{R}^d\right)^{\otimes k}} = \sqrt{\sum_{i_1, \ldots, i_k} \left|a^{i_1, \ldots, i_k}\right|^2}$$

and when no confusion is possible we shall simply write $|a|$. Let us also observe that for $0 \leq i \leq k \leq N$,

$$(a, b) \in \left(\mathbb{R}^d\right)^{\otimes i} \times \left(\mathbb{R}^d\right)^{\otimes k-i}, \quad |a \otimes b|_{\left(\mathbb{R}^d\right)^{\otimes k}} = |a|_{\left(\mathbb{R}^d\right)^{\otimes i}} |b|_{\left(\mathbb{R}^d\right)^{\otimes (k-i)}}$$

which is a *compatibility* relation between the tensor norms on the respective tensor levels. Then for any $g = \sum_{k=0}^{N} \pi_k(g) \in T^N(\mathbb{R}^d)$ we set

$$|g|_{T^N(\mathbb{R}^d)} := \max_{k=0,\dots,N} |\pi_k(g)|$$

which makes $T^N(\mathbb{R}^d)$ a Banach space (of finite dimension $1 + d + d^2 + \cdots + d^N$); again we shall write $|g|$ if no confusion is possible. We remark that there are other choices of norms on $T^N(\mathbb{R}^d)$, of course all equivalent, but this one will turn out to be convenient later on (cf. definition of $\rho_{p-\omega}$ later).

Exercise 7.6 *Consider the (infinite) tensor algebra* $T^\infty(\mathbb{R}^d) := \oplus_{k=0}^{\infty} (\mathbb{R}^d)^{\otimes k}$. *Show that* $T^N(\mathbb{R}^d)$ *is the algebra obtained by factorization by the ideal* $\{g \in T^\infty(\mathbb{R}^d) : \pi_i(g) = 0 \text{ for } 0 \leq i \leq N\}$.

Exercise 7.7 *Identify* $T^\infty(\mathbb{R}^d) \equiv \mathbb{R}\langle e_1, \dots, e_d \rangle$ *with the algebra of polynomials in d non-commutative indeterminants $e_1, \dots, e_d$. Conclude that $T^N(\mathbb{R}^d)$ can be viewed as an algebra of polynomials in d non-commutative indeterminants for which $e_{i_1} \dots e_{i_n} \equiv 0$ whenever $n \geq N$.*

7.2.2 Basic properties of S_N

Given $x : [0, T] \to \mathbb{R}^d$, continuous of bounded variation, and a fixed $s \in [0, T)$ the path $S_N(x)_{s,\cdot}$ takes values in $T^N(\mathbb{R}^d) \cong \mathbb{R}^{1+d+\cdots+d^N}$, as vector spaces. Almost by definition, the path $S_N(x)_{s,t}$ then satisfies an ODE on $T^N(\mathbb{R}^d)$ driven by x.

Proposition 7.8 *Let* $x : [0, T] \to \mathbb{R}^d$ *be a continuous path of bounded variation. Then, for fixed $s \in [0, T)$,*

$$\begin{cases} dS_N(x)_{s,t} = S_N(x)_{s,t} \otimes dx_t, \\ S_N(x)_{s,s} = 1. \end{cases}$$

Remark 7.9 *If we write* $x(\cdot) = \sum_{i=1}^{d} x^i(\cdot) e_i$ *and define the (linear!) vector fields $U_i : T^N(\mathbb{R}^d) \to T^N(\mathbb{R}^d)$ by $g \mapsto g \otimes e_i$, $i \in \{1, \dots, d\}$ this ODE can be rewritten in the more familiar form*

$$dS_N(x)_{s,t} = \sum_{i=1}^{d} U_i \left(S_N(x)_{s,\cdot} \right) dx_t^i.$$

$\square$

Proof. Let us look at level k, $k \geq 1$, of $S_N(x)_{s,t}$,

$$\int_{s < r_1 < \ldots < r_k < t} dx_{r_1} \otimes \cdots \otimes dx_{r_k}$$

$$= \int_{r_k = s}^{t} \left(\int_{s < r_1 < \ldots < r_{k-1} < r_k} dx_{r_1} \otimes \cdots \otimes dx_{r_{k-1}} \right) \otimes dx_{r_k}$$

$$= \int_{r = s}^{t} \pi_{k-1} \left(S_N(x)_{s,r} \right) \otimes dx_r.$$

Hence, we see that

$$S_N(x)_{s,t} = 1 + \int_s^t S_N(x)_{s,r} \otimes dx_r.$$

                                                                                        ■

    Proposition 7.8 tells us that $S_N(x)$ satisfies an ODE of the type discussed in Part I of this book. A number of interesting properties of signatures are then direct consequences of the corresponding ODE statements. We first describe what happens under reparametrization.

Proposition 7.10 *Let* $x : [0, T] \to \mathbb{R}^d$ *be a continuous path of bounded variation,* $\phi : [0, T] \to [T_1, T_2]$ *a non-decreasing surjection, and write* $x_t^\phi := x_{\phi(t)}$ *for the reparametrization of* x *under* ϕ. *Then, for all* $s, t \in [0, T]$,

$$S_N(x)_{\phi(s), \phi(t)} = S_N\left(x^\phi\right)_{s,t}.$$

Proof. A consequence of Propositions 3.10 and 7.8.   ■

    Simple as it is, Proposition 7.10 has an appealing interpretation. If $[x]$ denotes the equivalence class of some $x \in C^{1\text{-var}}\left([0, T], \mathbb{R}^d\right)$ obtained by all possible reparametrizations, then the signature-map

$$x \in C^{1\text{-var}}\left([0, T], \mathbb{R}^d\right) \mapsto S_N(x)_{0,T}$$

is really a function of $[x]$. We now discuss the signature of the concatenation of two paths. Recall that, given $\gamma \in C^{1\text{-var}}\left([0, T], \mathbb{R}^d\right), \eta \in C^{1\text{-var}}\left([T, 2T], \mathbb{R}^d\right)$, we set

$$\gamma \sqcup \eta \equiv \begin{cases} \gamma(\cdot) \text{ on } [0, T] \\ \eta(\cdot) - \eta(0) + \gamma(T) \text{ on } [T, 2T] \end{cases}$$

so that $\gamma \sqcup \eta \in C^{1\text{-var}}\left([0, 2T], \mathbb{R}^d\right)$. If η was defined on $[0, T]$, rather than $[T, 2T]$, one may prefer to reparametrize such that $\gamma \sqcup \eta$ too is defined on $[0, T]$. Whatever parametrization one chooses, thanks to the previous proposition the signature of the entire path $\gamma \sqcup \eta$ is intrinsically defined. The following theorem says the signature of $\gamma \sqcup \eta$ is precisely the tensor product of the respective signatures of γ and η.

Theorem 7.11 (Chen) *Given $\gamma \in C^{1\text{-var}}\left([0,T],\mathbb{R}^d\right), \eta \in C^{1\text{-var}}\left([T,2T],\mathbb{R}^d\right)$,*

$$S_N\left(\gamma \sqcup \eta\right)_{0,2T} = S_N\left(\gamma\right)_{0,T} \otimes S_N\left(\eta\right)_{T,2T}.$$

Equivalently, given $x \in C^{1\text{-var}}\left([0,T],\mathbb{R}^d\right)$ and $0 \leq s < t < u \leq T$ we have

$$S_N\left(x\right)_{s,u} = S_N\left(x\right)_{s,t} \otimes S_N\left(x\right)_{t,u}.$$

Proof. By induction on N. For $N = 0$, it just reads $1 = 1.1$. Assume it is true for N and all $s < t < u \in [0,T]$, and let us prove it is true for $N + 1$. First observe, in $T^{(N+1)}\left(\mathbb{R}^d\right)$,

$$S_{N+1}\left(x\right)_{s,u} = 1 + \int_s^u S_{N+1}\left(x\right)_{s,r} \otimes dx_r = 1 + \int_s^u S_N\left(x\right)_{s,r} \otimes dx_r,$$

where the second equality follows from truncation beyond level $(N+1)$. For similar reasons,

$$S_N\left(x\right)_{s,t} \otimes \int_t^u S_N\left(x\right)_{t,r} \otimes dx_r = S_{N+1}\left(x\right)_{s,t} \otimes \int_t^u S_N\left(x\right)_{t,r} \otimes dx_r.$$

Hence, using the induction hypothesis to split up $S_N\left(x\right)_{s,r}$ when $s < t < r < u$,

$$
\begin{aligned}
S_{N+1}\left(x\right)_{s,u} &= 1 + \int_s^t S_N\left(x\right)_{s,r} \otimes dx_r + \int_t^u S_N\left(x\right)_{s,t} \otimes S_N\left(x\right)_{t,r} \otimes dx_r \\
&= S_{N+1}\left(x\right)_{s,t} + S_{N+1}\left(x\right)_{s,t} \otimes \left(\int_t^u S_N\left(x\right)_{t,r} \otimes dx_r\right) \\
&= S_{N+1}\left(x\right)_{s,t} \otimes \left(1 + \left(S_{N+1}\left(x\right)_{t,u} - 1\right)\right) \\
&= S_{N+1}\left(x\right)_{s,t} \otimes S_{N+1}\left(x\right)_{t,u}.
\end{aligned}
$$

∎

We now show that the inverse (with respect to $\otimes$) of the signature of a path is precisely the signature of that path with time reversed.

Proposition 7.12 *Let $x \in C^{1\text{-var}}\left([0,T],\mathbb{R}^d\right)$. Then, if $\overleftarrow{x}$ denotes the path $x : t \in [0,T] \mapsto x_{T-t} \in \mathbb{R}^d$,*

$$S_N\left(x\right)_{0,T} \otimes S_N\left(\overleftarrow{x}\right)_{0,T} = S_N\left(\overleftarrow{x}\right)_{0,T} \otimes S_N\left(x\right)_{0,T} = 1.$$

Proof. Using the fact that $t \mapsto S_N\left(x\right)_{0,t}$ is the solution to an ODE driven by x (Proposition 7.8) this follows immediately from the corresponding results on ODEs with time-reversed driving signal (Proposition 3.13). ∎

Definition 7.13 For $\lambda \in \mathbb{R}$, we define the dilation map

$$\delta_\lambda : T^N\left(\mathbb{R}^d\right) \to T^N\left(\mathbb{R}^d\right)$$

such that $\pi_k\left(\delta_\lambda\left(g\right)\right) = \lambda^k \pi_k\left(g\right)$. □

Exercise 7.14 *Check that if λ is a real, $x : [0,1] \to \mathbb{R}^d$ is a continuous path of bounded variation, λx the path x scaled by λ, then*

$$S_N \left(\lambda x \right)_{s,t} = \delta_\lambda S_N \left(x \right)_{s,t}.$$

Proposition 7.15 *Let $(x_n) \subset C^{1\text{-var}} \left([0,1], \mathbb{R}^d \right)$ with $\sup_n |x_n|_{1\text{-var};[0,1]} < \infty$, uniformly convergent to some $x \in C^{1\text{-var}} \left([0,1], \mathbb{R}^d \right)$. Then, $S_N \left(x_n \right)_{0,\cdot}$ converges uniformly to $S_N \left(x \right)_{0,\cdot}$. In particular,*

$$\lim_{n \to \infty} S_N \left(x_n \right)_{0,1} = S_N \left(x \right)_{0,1}.$$

Proof. Using the fact that $t \mapsto S_N \left(x \right)_{0,t}$ is the solution to an ODE driven by x (Proposition 7.8) this follows immediately from the ODE results on continuity of the solution map (cf. Corollary 3.16). ∎

7.3 Lie algebra $\mathfrak{t}^N \left(\mathbb{R}^d \right)$ and Lie group $1 + \mathfrak{t}^N \left(\mathbb{R}^d \right)$

The space $\left(T^N \left(\mathbb{R}^d \right), +, ., \otimes \right)$ is an associative algebra. We introduce two simple subspaces (linear and affine-linear, respectively) which will guide us towards the (crucial) free nilpotent Lie algebra and group and their Lie algebra. Let us set

$$\mathfrak{t}^N \left(\mathbb{R}^d \right) \equiv \left\{ g \in T^N \left(\mathbb{R}^d \right) : \pi_0 \left(g \right) = 0 \right\}$$

so that

$$1 + \mathfrak{t}^N \left(\mathbb{R}^d \right) = \left\{ g \in T^N \left(\mathbb{R}^d \right) : \pi_0 \left(g \right) = 1 \right\}.$$

7.3.1 The group $1 + \mathfrak{t}^N \left(\mathbb{R}^d \right)$

We first show that elements in $1 + \mathfrak{t}^N \left(\mathbb{R}^d \right)$ are invertible with respect to the tensor product $\otimes$.

Lemma 7.16 *Any $g = 1 + a \in 1 + \mathfrak{t}^N \left(\mathbb{R}^d \right)$ has an inverse with respect to $\otimes$ given by*

$$g^{-1} = (1 + a)^{-1} = \sum_{k=0}^{N} (-1)^k a^{\otimes k},$$

that is, $g \otimes g^{-1} = g^{-1} \otimes g = 1$.

Proof. We have

$$
\begin{aligned}
(1 + a) \sum_{k=0}^{N} (-1)^k a^k &= \sum_{k=1}^{N+1} (-1)^{k+1} a^k + \sum_{k=0}^{N} (-1)^k a^k \\
&= a^{N+1} + 1.
\end{aligned}
$$

Now because we set to zero any elements in $\left(\mathbb{R}^d\right)^{\otimes n}$ for $n > N$, we see that $a^{N+1} = 0$. ∎

It is also obvious that if g and h are in $1 + \mathfrak{t}^N\left(\mathbb{R}^d\right)$, then $g \otimes h \in 1 + \mathfrak{t}^N\left(\mathbb{R}^d\right)$. Finally, $1 + \mathfrak{t}^N\left(\mathbb{R}^d\right)$ is an affine-linear subspace of $T^N\left(\mathbb{R}^d\right)$ hence a smooth manifold, trivially diffeomorphic to $\mathfrak{t}^N\left(\mathbb{R}^d\right) \cong \mathbb{R}^{d+\cdots+d^N}$. Noting that the group operations $\otimes$, $^{-1}$ are smooth maps (in fact, polynomial when written out in coordinates) we have

Proposition 7.17 *The space* $1 + \mathfrak{t}^N\left(\mathbb{R}^d\right)$ *is a Lie group*[3] *with respect to tensor multiplication* $\otimes$.

Let us remark that the *manifold topology* of $1 + \mathfrak{t}^N\left(\mathbb{R}^d\right)$ is, of course, induced by the metric

$$\rho\left(g, h\right) := |g - h|_{T^N\left(\mathbb{R}^d\right)} = |g - h| = \max_{i=1,\ldots,N} |\pi_i\left(g - h\right)|, \qquad (7.12)$$

which arises from the norm on $T^N\left(\mathbb{R}^d\right) \supset 1 + \mathfrak{t}^N\left(\mathbb{R}^d\right)$. We end this subsection with a simple observation.

Proposition 7.18 *Assume* $g_n, g \subset 1 + \mathfrak{t}^N\left(\mathbb{R}^d\right)$. *Then,* $\lim_{n\to\infty} |g_n - g| = 0$ *if and only if* $\lim_{n\to\infty} \left|g_n^{-1} \otimes g - 1\right| = 0$.

Proof. All group operations in the Lie group $1 + \mathfrak{t}^N\left(\mathbb{R}^d\right)$ are continuous in the manifold topology of $1 + \mathfrak{t}^N\left(\mathbb{R}^d\right)$. In particular,

$$g_n \to g \Leftrightarrow g_n^{-1} \to g^{-1} \Leftrightarrow g_n^{-1} \otimes g \to g^{-1} \otimes g = 1$$

and we conclude with the remark that, as $n \to \infty$,

$$g_n \to g \Leftrightarrow |g_n - g| \to 0 \text{ and } g_n^{-1} \otimes g \to 1 \Leftrightarrow \left|g_n^{-1} \otimes g - 1\right| \to 0.$$

∎

7.3.2 The Lie algebra $\mathfrak{t}^N\left(\mathbb{R}^d\right)$ and the exponential map

The vector space $\left(\mathfrak{t}^N\left(\mathbb{R}^d\right), +, .\right)$ becomes itself an algebra under $\otimes$. As in every algebra, the *commutator*, in our case

$$(g, h) \mapsto [g, h] := g \otimes h - h \otimes g \in \mathfrak{t}^N\left(\mathbb{R}^d\right)$$

for $g, h \in \mathfrak{t}^N\left(\mathbb{R}^d\right)$, defines a bilinear map which is easily seen to be *anticommutative*, i.e.

$$[g, h] = -[h, g], \text{ for all } g, h \in \mathfrak{t}^N\left(\mathbb{R}^d\right)$$

---

[3]Recall that a *Lie group* is by definition a group which is also a smooth manifold and in which the group operations are smooth maps.

and to satisfy the *Jacobi identity* for all $g, h, k \in \mathfrak{t}^N \left(\mathbb{R}^d \right)$; that is,

$$[g, [h, k]] + [h, [k, g]] + [k, [g, h]] = 0 \quad \text{for all } g, h, k \in \mathfrak{t}^N \left(\mathbb{R}^d \right).$$

Recalling that a vector space $\mathcal{V} = (\mathcal{V}, +, .)$ equipped with a bilinear, anti-commutative map $[\cdot, \cdot] : \mathcal{V} \times \mathcal{V} \to \mathcal{V}$ which satisfies the Jacobi identity is called a *Lie algebra* (the map $[\cdot, \cdot]$ is called the *Lie bracket*), this can be summarized as

Proposition 7.19 $\left(\mathfrak{t}^N \left(\mathbb{R}^d \right), +, ., [\cdot, \cdot] \right)$ *is a Lie algebra.*

We now define the exponential and logarithm maps via their power series:

Definition 7.20 The exponential map is defined by

$$\exp \quad : \quad \mathfrak{t}^N \left(\mathbb{R}^d \right) \to 1 + \mathfrak{t}^N \left(\mathbb{R}^d \right)$$

$$a \quad \mapsto \quad 1 + \sum_{k=1}^{N} \frac{a^{\otimes k}}{k!}$$

while the logarithm map is defined by

$$\log \quad : \quad 1 + \mathfrak{t}^N \left(\mathbb{R}^d \right) \to \mathfrak{t}^N \left(\mathbb{R}^d \right)$$

$$(1 + a) \quad \mapsto \quad \sum_{k=1}^{N} (-1)^{k+1} \frac{a^{\otimes k}}{k}.$$

□

The definitions of exp and log are precisely via their classical power series with (i) usual powers replaced by "tensor powers" and (ii) the infinite sums replaced by $\sum_{k=1}^{N}$, thanks to working within the tensor algebra with truncation beyond level N. A direct calculation shows that

$$\exp \left(\log \left(1 + a \right) \right) = a, \quad \log \left(\exp \left(a \right) \right) = a, \quad \text{for all } a \in \mathfrak{t}^N \left(\mathbb{R}^d \right).$$

We emphasize that $\log \left(\cdot \right)$ is globally defined and there are no convergence issues whatsoever.

Example 7.21 Fix $a \in \mathbb{R}^d \cong \pi_1 \left(\mathfrak{t}^N \left(\mathbb{R}^d \right) \right)$. The step-$N$ signature of $x \left(\cdot \right)$ given by $t \in [0, 1] \mapsto ta$ computes to

$$
\begin{aligned}
S_N \left(x \right)_{0,1} &= 1 + \sum_{k=1}^{N} \int_{0 < r_1 < \ldots < r_k < 1} dx_{r_1} \otimes \cdots \otimes dx_{r_k} \\
&= 1 + \sum_{k=1}^{N} a^{\otimes k} \int_{0 < r_1 < \ldots < r_k < 1} dr_1 \ldots dr_k \\
&= 1 + \sum_{k=1}^{N} \frac{a^{\otimes k}}{k!} = \exp \left(a \right).
\end{aligned}
\tag{7.13}
$$

□

7.3.3   The Campbell–Baker–Hausdorff formula

Recall that exp maps $t^N\left(\mathbb{R}^d\right)$ one-to-one and onto $1 + t^N\left(\mathbb{R}^d\right)$. In general, $e^a e^b \neq e^{a+b}$, but one has

$$\forall a,b \in t^N\left(\mathbb{R}^d\right) : e^a e^b = e^{a+b+\frac{1}{2}[a,b]+\frac{1}{12}[a,[a,b]]+\frac{1}{12}[b,[b,a]]+\ldots} \qquad (7.14)$$

where ... stands for (a linear combination, *with universal coefficients*, of) higher iterated brackets of a and b. Thanks to truncation beyond tensor-level N, all terms involving N or more iterated brackets must be zero. For $N = 2$, this formula is obtained by simple computation; indeed, given $a,b \in t^2\left(\mathbb{R}^d\right)$ we have

$$
\begin{aligned}
\exp\left(a\right) \otimes \exp\left(b\right) &= \left(1 + a + \frac{a^{\otimes 2}}{2}\right) \otimes \left(1 + b + \frac{b^{\otimes 2}}{2}\right) \\
&= 1 + a + b + \frac{1}{2}\left(a \otimes b - b \otimes a\right) + \frac{(a+b)^{\otimes 2}}{2} \\
&= \exp\left(a + b + \frac{1}{2}[a,b]\right).
\end{aligned}
$$

The same computation is possible, if tedious, for $N = 3$ and allows us to recover the next set of bracket terms as seen in (7.14). The general case, however, requires a different argument and we shall give a proof based on ordinary differential equations.

Given a linear operator $A : t^N\left(\mathbb{R}^d\right) \mapsto t^N\left(\mathbb{R}^d\right)$ define A^0 as identity map and set $A^n = A \circ \cdots \circ A$ (n times). An arbitrary real-analytic function $\sum_{n \geq 0} \gamma_n z^n$ gives rise to the operator $\sum_{n \geq 0} \gamma_n A^n$ provided this series converges in operator norm. Fix $a \in t^N\left(\mathbb{R}^d\right)$ and define the linear map

$$t^N\left(\mathbb{R}^d\right) \ni d \mapsto (\operatorname{ad} a)\,d \equiv [a,d] \in t^N\left(\mathbb{R}^d\right).$$

Observe that $(\operatorname{ad} a)^n \equiv 0$ for $n > N$.

Lemma 7.22 *For all $a,d \in t^N\left(\mathbb{R}^d\right)$ we have*

$$\exp\left(a\right) \otimes d \otimes \exp\left(-a\right) = e^{\operatorname{ad} a}\,d$$

where

$$e^{\operatorname{ad} a} \equiv \sum_{n \geq 0} \frac{1}{n!}\left(\operatorname{ad} a\right)^n = \sum_{n=0}^{N} \frac{1}{n!}\left(\operatorname{ad} a\right)^n.$$

As a consequence we have the following operator identity on $t^N\left(\mathbb{R}^d\right)$:

$$e^{\operatorname{ad} c} = e^{\operatorname{ad} a} \circ e^{\operatorname{ad} b}$$

where $c = \log\left(\exp\left(a\right) \otimes \exp\left(b\right)\right)$.

Proof. Define the linear map ρ_t from $\mathfrak{t}^N\left(\mathbb{R}^d\right)$ into itself by

$$\rho_t : d \mapsto \exp\left(t\,a\right) \otimes d \otimes \exp\left(-t\,a\right).$$

Obviously, $t \mapsto \rho_t\left(d\right) \in \mathfrak{t}^N\left(\mathbb{R}^d\right)$ is differentiable and

$$\frac{d}{dt}\rho_t\left(d\right) = \left[a, \rho_t\left(d\right)\right] = \left(\operatorname{ad} a\right) \rho_t\left(d\right).$$

Noting $\rho_0\left(d\right) = d$ the solution is given by $\rho_t\left(d\right) = e^{t\,\operatorname{ad} a}d$ and we conclude by setting $t = 1$. The last statement is immediately verified by considering the image of some $d \in \mathfrak{t}^N\left(\mathbb{R}^d\right)$ under both $e^{\operatorname{ad} c}$ and $e^{\operatorname{ad} a} \circ e^{\operatorname{ad} b}$. ∎

Lemma 7.23 *Assume* $t \mapsto c_t \in \mathfrak{t}^N\left(\mathbb{R}^d\right)$ *is continuously differentiable. Then*

$$\exp\left(c_t\right) \otimes \frac{d}{dt}\exp\left(-c_t\right) = -G\left(\operatorname{ad} c_t\right) \dot{c}_t$$

where $\dot{c}_t = dc_t/dt$ *and*

$$G\left(z\right) = \frac{e^z - 1}{z} = \sum_{n \geq 0}\frac{1}{\left(n + 1\right)!}z^n.$$

Proof. Set

$$b_{s,t} = \exp\left(s\,c_t\right) \otimes \frac{d}{dt}\exp\left(-s\,c_t\right).$$

Then $b_{0,t} = 1 \otimes 0 = 0$. Taking derivatives with respect to s, a short computation shows that

$$\begin{aligned} \frac{d}{ds}b_{s,t} &= \left[c_t, b_{s,t}\right] - \dot{c}_t \\ &= \left(\operatorname{ad} c_t\right) b_{s,t} - \dot{c}_t \end{aligned}$$

where $\dot{c}_t = dc_t/dt$. Keeping t fixed, the solution is given by

$$\begin{aligned} b_{s,t} &= e^{s\,\operatorname{ad} c_t}b_{0,t} - f\left(s, \operatorname{ad} c_t\right)\dot{c}_t \\ &= -f\left(s, \operatorname{ad} c_t\right)\dot{c}_t \end{aligned}$$

where $f\left(s, z\right) = \left(e^{s\,z} - 1\right)/z$, entire in z. Setting $s = 1$ finishes the proof. ∎

Theorem 7.24 (Campbell–Baker–Hausdorff) *Let* $a, b \in \mathfrak{t}^N(\mathbb{R}^d)$. *Then*

$$\log\left[\exp\left(a\right) \otimes \exp\left(b\right)\right] = b + \int_0^1 H\left(e^{t\,\operatorname{ad} a} \circ e^{\operatorname{ad} b}\right) a\,dt \qquad (7.15)$$

where

$$H\left(z\right) = \frac{\ln z}{z - 1} = \sum_{n \geq 0}\frac{\left(-1\right)^n}{n + 1}\left(z - 1\right)^n.$$

In particular, $\log\left(\exp\left(a\right)\otimes\exp\left(b\right)\right)$ *equals a sum of iterated brackets of a and b, with universal[4] coefficients.*

Proof. First observe that the linear operator $H\left(e^{t\,\mathrm{ad}\,a}\circ e^{\mathrm{ad}\,b}\right)$ is in fact given by a finite series. Indeed, $e^{t\,\mathrm{ad}\,a}\circ e^{\mathrm{ad}\,b}$ minus the identity map is a finite sum of N or less iterated applications of $\mathrm{ad}\,a$, $\mathrm{ad}\,b$. Consequently, only the first N terms of the expansion of H are needed. We also observe that both sides of (7.15) are polynomials of degree less than or equal to N in the coordinates $\left\{a^{n;i_1,\ldots,i_n}\right\},\left\{b^{n;i_1,\ldots,i_n}\right\},1\le n\le N$. Therefore, it is sufficient to check (7.15) for a,b in a neighbourhood of 0. For $t\in[0,1]$ define

$$c_t = \log\left(e^{t\,a}\otimes e^{b}\right).$$

Then

$$e^{c_t}\otimes\frac{d}{dt}e^{-c_t} = e^{t\,a}\otimes e^{b}\otimes\frac{d}{dt}\left(e^{-b}\otimes e^{-t\,a}\right) = -a$$

which, combined with Lemma 7.23, shows that

$$a = G\left(\mathrm{ad}\,c_t\right)\dot{c}_t. \tag{7.16}$$

On the other hand, Lemma 7.22 implies the operator identity

$$\mathrm{ad}\,c_t = \ln\left(e^{t\,\mathrm{ad}\,a}\circ e^{\mathrm{ad}\,b}\right)$$

which certainly holds when a,b are small enough. Since $G(\ln z)H(z)=1$, at least near $z=1$, the operator identity

$$G\left(\mathrm{ad}\,c_t\right) = G\left(\ln\left(e^{t\,\mathrm{ad}\,a}\circ e^{\mathrm{ad}\,b}\right)\right) = H\left(e^{t\,\mathrm{ad}\,a}\circ e^{\mathrm{ad}\,b}\right)^{-1}$$

holds for a,b small enough. This allows us to rewrite (7.16) as

$$\dot{c}_t = H\left(e^{t\,\mathrm{ad}\,a}\circ e^{\mathrm{ad}\,b}\right)a$$

and by integration

$$c_1 = b+\int_0^1 H\left(e^{t\,\mathrm{ad}\,a}\circ e^{\mathrm{ad}\,b}\right)a\,dt$$

which is exactly what we wanted to show. ∎

Definition 7.25 Define $\mathfrak{g}^N\left(\mathbb{R}^d\right)\subset\mathfrak{t}^N\left(\mathbb{R}^d\right)$ as the smallest sub-Lie algebra of $\mathfrak{t}^N\left(\mathbb{R}^d\right)$ which contains $\pi_1\left(\mathfrak{t}^N\left(\mathbb{R}^d\right)\right)\cong\mathbb{R}^d$. That is,

$$\mathfrak{g}^N\left(\mathbb{R}^d\right) = \mathbb{R}^d\oplus\left[\mathbb{R}^d,\mathbb{R}^d\right]\oplus\cdots\oplus\underbrace{\left[\mathbb{R}^d,\left[\ldots,\left[\mathbb{R}^d,\mathbb{R}^d\right]\right]\right]}_{(N-1)\text{ brackets}}.$$

We call it the free step-N nilpotent Lie algebra. □

---

[4]The coefficients are given by numerical constants, computable from H. In particular, they do not depend on N or $d=\dim\mathbb{R}^d$.

Remark 7.26 (universal property of free Lie algebras) Let i denote the (linear) inclusion map

$$i : \pi_1 \left(\mathfrak{t}^N \left(\mathbb{R}^d \right) \right) \cong \mathbb{R}^d \to \mathfrak{g}^N \left(\mathbb{R}^d \right).$$

Then $\mathfrak{g} = \mathfrak{g}^N \left(\mathbb{R}^d \right)$ has the property that for any (step-N nilpotent) Lie algeba $\mathfrak{a}$ and any linear map

$$f : \mathbb{R}^d \to \mathfrak{a}$$

there is a Lie algebra homomorphism $\phi : \mathfrak{g} \to \mathfrak{a}$ so that $f = \phi \circ i$. Indeed, we can take a basis of $\mathfrak{g}$ where each element is of the form

$$\begin{aligned} e_\alpha &= \left[e_{\alpha_1}, \left[e_{\alpha_2}, \left[\ldots \left[e_{\alpha_{k-1}}, e_{\alpha_k} \right] \right] \right] \right] \\ &\in \left[\mathbb{R}^d, \left[\ldots, \left[\mathbb{R}^d, \mathbb{R}^d \right] \right] \right] \subset \left(\mathbb{R}^d \right)^{\otimes k}, k \leq N \end{aligned}$$

and one checks that

$$\phi \left(\sum c_\alpha e_\alpha \right) := \sum c_\alpha \left[f \left(e_{\alpha_1} \right), \left[f \left(e_{\alpha_2} \right), \left[\ldots \left[f \left(e_{\alpha_{k-1}} \right), f \left(e_{\alpha_k} \right) \right] \right] \right] \right]$$

has the desired properties. □

Corollary 7.27 Let $a, b \in \mathfrak{g}^N \left(\mathbb{R}^d \right)$. Then $\log \left(e^a \otimes e^b \right) \in \mathfrak{g}^N \left(\mathbb{R}^d \right)$. In other words, $\exp \left(\mathfrak{g}^N \left(\mathbb{R}^d \right) \right)$ is a subgroup of $\left(1 + \mathfrak{t}^N \left(\mathbb{R}^d \right), \otimes, 1 \right)$

Proof. An obvious corollary of the Campbell–Baker–Hausdorff formula. ■

7.4 Chow's theorem

As was seen in Example 7.21, the step-N signature of the path $t \in [0,1] \mapsto vt$, $v \in \mathbb{R}^d$, is precisely $\exp(v) \in 1 + \mathfrak{t}^N \left(\mathbb{R}^d \right)$. A piecewise linear path is just the concatenation of such paths (up to irrelevant reparametrization) and by Chen's theorem its step-N signature is of the form

$$e^{v_1} \otimes \cdots \otimes e^{v_m} \in 1 + \mathfrak{t}^N \left(\mathbb{R}^d \right)$$

with $v_1, \ldots, v_m \in \mathbb{R}^d$. Conversely, any element of this form arises as the step-N signature of a piecewise linear path, e.g. of $x : [0, m] \to \mathbb{R}^d$ with $x_{i-1,i} = v_i$, $i = 1, \ldots, m$ and linear between integer times. (If one prefers, the trivial reparametrization $\tilde{x}(t) = x(tm)$ defines a piecewise linear path on $[0,1]$ with identical signature.)

Theorem 7.28 (Chow) Let $g \in \exp \left(\mathfrak{g}^N \left(\mathbb{R}^d \right) \right)$. Then, there exist $v_1, \ldots,$ $v_m \in \mathbb{R}^d$ such that

$$g = e^{v_1} \otimes \cdots \otimes e^{v_m}.$$

Equivalently, there exists a piecewise linear path $x : [0,1] \to \mathbb{R}^d$ with signature g, by which we mean $g = S_N(x)_{0,1}$.

Proof. It is enough to show this for $\log(g)$ in an open neighbourhood of $0 \in \mathfrak{g}^N(\mathbb{R}^d)$ or, equivalently, for g in an open neighbourhood A of the unit element $1 = \exp(0)$. Indeed, given $g \in \exp(\mathfrak{g}^N(\mathbb{R}^d))$, it is clear that $\delta_\varepsilon g \in A$ for ε small enough and if x denotes the (piecewise linear) path whose signature equals $\delta_\varepsilon g$, the scaled path $x(\cdot)/\varepsilon$, which is still piecewise linear, has signature g.

With $(e_i : i = 1,\dots,d)$ denoting the standard basis in $\mathbb{R}^d$, let us define the $\exp(\mathfrak{g}^N(\mathbb{R}^d))$-valued paths

$$\phi_t^i = \exp(te_i).$$

An easy application of the Campbell–Baker–Hausdorff formula shows that[5]

$$\begin{aligned}\phi_t^{ij} &: = \exp(-te_j) \otimes \exp(-te_i) \otimes \exp(te_j) \otimes \exp(te_i)\\ &= \exp(t^2[e_i,e_j] + o(t^2)).\end{aligned}$$

For a multi-index $I = iJ$, we define by induction

$$\begin{aligned}\phi_t^I &: = \phi_{-t}^J \otimes \exp(-te_i) \otimes \phi_t^J \otimes \exp(te_i)\\ &= \exp\left(t^{|I|}e_I + o\left(t^{|I|}\right)\right)\end{aligned}$$

where $e_I = [e_{i_1},[\dots[e_{i_{k-1}},e_{i_k}]]]$ for $I = (i_1,\dots,i_k) \in \{1,\dots,d\}^k$. We then define $\psi_t^I := \phi_{t^{1/|I|}}^I$ for $t \geq 0$ and for $t < 0$,

$$\psi_t^I := \begin{cases} \phi_{-|t|^{1/|I|}}^I & \text{for } |I| \text{ odd}\\ \phi_{-|t|^{1/|I|}}^J \otimes \exp\left(-|t|^{1/|I|}e_i\right) \otimes \phi_{|t|^{1/|I|}}^J \otimes \exp\left(|t|^{1/|I|}e_i\right) & \text{for } |I| \text{ even}\end{cases}$$

so that

$$\psi_t^I = \exp(te_I + o(t)). \tag{7.17}$$

We now choose a vector space basis $(e_{I_k} : k = 1,\dots,n)$ of $\mathfrak{g}^N(\mathbb{R}^d)$ and define

$$\varphi(t_1,\dots,t_n) = \log\left(\psi_{t_n}^{I_n} \otimes \cdots \otimes \psi_{t_1}^{I_1}\right) \in \mathfrak{g}^N(\mathbb{R}^d)$$

and note that $\varphi(0) = \log(1) = 0$. We also observe that, thanks to (7.17), $\varphi : \mathbb{R}^n \to \mathfrak{g}^N(\mathbb{R}^d) \cong \mathbb{R}^n$ (as vector spaces) is continuously differentiable near 0, with non-degenerate derivative at the origin.[6] This implies that the (one-to-one) image under φ of a small enough neighbourhood of $0 \in \mathbb{R}^n$, say Ω, contains an open neighbourhood of $0 \in \mathfrak{g}^N(\mathbb{R}^d)$. On

---

[5] Observe that the $o(t^2)$ term is an element in $\mathfrak{g}^N(\mathbb{R}^d) \cap \left\{(\mathbb{R}^d)^{\otimes 3} \oplus \cdots \oplus (\mathbb{R}^d)^{\otimes N}\right\}$.

[6] The differential of $\log$ at 1 is the identity map and hence plays no role in the non-degeneracy at 0.

the other hand, unwrapping the very definition of φ shows any element of the form

$$\exp \varphi\left(t_1, \ldots, t_n\right) \in \exp\left(\mathfrak{g}^N\left(\mathbb{R}^d\right)\right), \quad \left(t_1, \ldots, t_n\right) \in \Omega,$$

is the concatenation of piecewise linear path segments in basis directions $(e_i : i = 1, \ldots, d)$ and so the proof is finished. ∎

Corollary 7.29 *Assume* $(g_k) \subset \exp\left(\mathfrak{g}^N\left(\mathbb{R}^d\right)\right)$ *converges to* $1 \in \exp$ $\left(\mathfrak{g}^N\left(\mathbb{R}^d\right)\right)$ *as* $k \to \infty$. *Let* x_k *denote the piecewise linear path with signature* g_k *as constructed in the previous theorem. Then the length of* x_k *converges to* 0 *as* $k \to \infty$.

Proof. The proof of the previous theorem shows that any element $g \in \exp\left(\mathfrak{g}^N\left(\mathbb{R}^d\right)\right)$, close enough to 1, can be written as

$$g = \exp \varphi\left(t_1, \ldots, t_n\right)$$

and the length of the associated (piecewise linear) path $x = x^g$ is bounded by

$$K\left(|t_1|^{1/|I_1|} + \cdots + |t_n|^{1/|I_n|}\right)$$

where K counts the maximal number of concatenations involved in the $\psi_t^{I_j}$'s. Observe that K depends on the space $\mathfrak{g}^N\left(\mathbb{R}^d\right)$ but can be chosen independent of g. Since $g \to 1$ is equivalent to $(t_1, \ldots, t_n) \to 0$, we obtain the desired continuity statement

$$\text{length}\left(x^g\right) \to 0 \text{ as } g \to 1.$$

∎

7.5   Free nilpotent groups

7.5.1   Definition and characterization

Let us consider

(i) the set of all step-N signatures of continuous paths of finite length,

$$G^N\left(\mathbb{R}^d\right) := \left\{S_N\left(x\right)_{0,1} : x \in C^{\text{1-var}}\left([0, 1], \mathbb{R}^d\right)\right\};$$

(ii) the image of the sub-Lie algebra $\mathfrak{g}^N\left(\mathbb{R}^d\right) \subset \mathfrak{t}^N\left(\mathbb{R}^d\right)$, cf. Definition 7.25, under the exponential map

$$\exp\left(\mathfrak{g}^N\left(\mathbb{R}^d\right)\right) \subset 1 + \mathfrak{t}^N\left(\mathbb{R}^d\right);$$

(iii) the subgroup $\langle \exp (\mathbb{R}^d) \rangle$ of $1 + \mathfrak{t}^N (\mathbb{R}^d)$ generated by elements in $\exp (\mathbb{R}^d)$, i.e.

$$\langle \exp (\mathbb{R}^d) \rangle := \left\{ \bigotimes_{i=1}^m \exp (v_i) : m \geq 1, v_1, \ldots, v_m \in \mathbb{R}^d \right\}.$$

Theorem 7.30 (and definition) *We have*

$$G^N (\mathbb{R}^d) = \exp (\mathfrak{g}^N (\mathbb{R}^d)) = \langle \exp (\mathbb{R}^d) \rangle$$

and $G^N (\mathbb{R}^d)$ is a (closed) sub-Lie group of $(1 + \mathfrak{t}^N (\mathbb{R}^d), \otimes)$, called the free nilpotent group of step N over $\mathbb{R}^d$.

Proof. Step 1: We show $\exp (\mathfrak{g}^N (\mathbb{R}^d)) = \langle \exp (\mathbb{R}^d) \rangle$. Obviously, $\exp (\mathbb{R}^d) \subset \exp (\mathfrak{g}^N (\mathbb{R}^d))$ and it follows from the CBH formula that

$$\langle \exp (\mathbb{R}^d) \rangle \subset \exp (\mathfrak{g}^N (\mathbb{R}^d)) .$$

Conversely, Chow's theorem tells us that any element of $\exp (\mathfrak{g}^N (\mathbb{R}^d))$ can be expressed in the form $e^{v_1} \otimes \cdots \otimes e^{v_m}$ with $v_i \in \mathbb{R}^d$, which plainly implies

$$\exp (\mathfrak{g}^N (\mathbb{R}^d)) \subset \langle \exp (\mathbb{R}^d) \rangle .$$

Step 2: We show that $\langle \exp (\mathbb{R}^d) \rangle$ is a closed subset of $1 + \mathfrak{t}^N (\mathbb{R}^d)$. By the previous step, it suffices to check that $\exp (\mathfrak{g}^N (\mathbb{R}^d))$ is closed. But this follows from (obvious!) closedness of $\mathfrak{g}^N (\mathbb{R}^d) \subset \mathfrak{t}^N (\mathbb{R}^d)$, $\exp (\mathfrak{g}^N (\mathbb{R}^d)) = \log^{-1} (\mathfrak{g}^N (\mathbb{R}^d))$ and continuity of log.

Step 3: We show $G^N (\mathbb{R}^d) = \langle \exp (\mathbb{R}^d) \rangle$. The inclusion

$$\langle \exp (\mathbb{R}^d) \rangle \subset G^N (\mathbb{R}^d)$$

is clear from Example 7.21 and Chen's theorem. Indeed, any $e^{v_1} \otimes \cdots \otimes e^{v_m}$ is the step-N signature of $\tilde{x} : [0,1] \to \mathbb{R}^d$ where $\tilde{x}(t) = x(tm)$ and $x : [0, m] \to \mathbb{R}^d$ with $x_{i-1,i} = v_i$, $i = 1, \ldots, m$ and linear between integer times. (This was already pointed out in our discussion of Theorem 7.28.) By step 2, $\overline{\langle \exp (\mathbb{R}^d) \rangle} = \langle \exp (\mathbb{R}^d) \rangle$ and so the other inclusion will follow from

$$G^N (\mathbb{R}^d) \subset \overline{\langle \exp (\mathbb{R}^d) \rangle}.$$

To prove this inclusion, take $g = S_N (x)_{0,1} \in G^N (\mathbb{R}^d)$, for some $x \in C^{1\text{-var}} ([0, 1], \mathbb{R}^d)$. We know that piecewise linear approximations (x^n) satisfy $\sup_n |x^n|_{1\text{-var};[0,1]} < \infty$ and converge to x, uniformly on $[0, 1]$. By Chen's theorem, $g_n = S_N (x_n)_{0,1} \in \langle \exp (\mathbb{R}^d) \rangle$ and from Proposition 7.15, g_n converges to g. This shows that $g \in \overline{\langle \exp (\mathbb{R}^d) \rangle}$, which is what we wanted.

<u>Step 4:</u> We show that the set $G := G^N \left(\mathbb{R}^d \right) = \exp \left(\mathfrak{g}^N \left(\mathbb{R}^d \right) \right) = \left\langle \exp \left(\mathbb{R}^d \right) \right\rangle$ is a closed sub-Lie group of $\left(1 + \mathfrak{t}^N \left(\mathbb{R}^d \right), \otimes \right)$. Topological closedness was already seen in step 2. It is also clear that $G = \left\langle \exp \left(\mathbb{R}^d \right) \right\rangle$ is an abstract subgroup of $\left(1 + \mathfrak{t}^N \left(\mathbb{R}^d \right), \otimes \right)$. To see that we have an actual sub-*Lie* group we have to check that G is a submanifold of $1 + \mathfrak{t}^N \left(\mathbb{R}^d \right)$. It is not hard to deduce this from $d \left(\log \circ \exp \right) = \text{Id}$, so that by the chain rule, $d \exp$ is one-to-one at every point. Alternatively, one can appeal to a standard theorem in Lie group theory[7] which asserts that a closed abstract subgroup is automatically a (closed) Lie subgroup. ∎

Remark 7.31 (manifold topology of $G^N \left(\mathbb{R}^d \right)$) In equation (7.12) we defined the metric ρ on $1 + \mathfrak{t}^N \left(\mathbb{R}^d \right)$, induced from the norm $\left| \cdot \right|_{T^N \left(\mathbb{R}^d \right)}$. By trivial restriction, ρ is also a metric on $G^N \left(\mathbb{R}^d \right)$, given by

$$\rho \left(g, h \right) = \max_{i=1,\dots,N} \left| \pi_i \left(g - h \right) \right|,$$

which induces the (sub)manifold topology on $G^N \left(\mathbb{R}^d \right)$. □

7.5.2 Geodesic existence

Chow's theorem tells us that for all elements $g \in G^N \left(\mathbb{R}^d \right)$, there exists a continuous path x of finite length such that $S_N \left(x \right)_{0,1} = g$. One may ask for the shortest path (and its length) which has the correct signature. For instance, given $a > 0$, we can ask for the shortest path with signature

$$\exp \left(\begin{pmatrix} 0 \\ 0 \end{pmatrix} + \begin{pmatrix} 0 & a \\ -a & 0 \end{pmatrix} \right) \in G^2 \left(\mathbb{R}^2 \right),$$

or, equivalently, the shortest path in $\mathbb{R}^2$ which ends where it starts and wipes out area a. As is well known from basic isoperimetry, the shortest such path is given by a circle (with area a) with easily computed length given by $2\sqrt{\pi a}$. With this motivating example in mind we now state

Theorem 7.32 (geodesic existence) *For every* $g \in G^N \left(\mathbb{R}^d \right)$, *the so-called "Carnot–Caratheodory norm"[8]*

$$\left\| g \right\| := \inf \left\{ \int_0^1 \left| d\gamma \right| : \gamma \in C^{1\text{-var}} \left([0,1], \mathbb{R}^d \right) \ \text{and} \ S_N \left(\gamma \right)_{0,1} = g \right\}$$

is finite and achieved at some minimizing path γ^*, *i.e.*

$$\left\| g \right\| = \int_0^1 \left| d\gamma^* \right| \ \text{and} \ S_N \left(\gamma^* \right)_{0,1} = g.$$

---

[7] See, for example, Warner [175].

[8] As usual, $\mathbb{R}^d$ is equipped with Euclidean structure so that $\int_0^1 \left| d\gamma \right|$ is the length of $\gamma \in C^{1\text{-var}} \left([0,1], \mathbb{R}^d \right)$, based on the Euclidean distance on $\mathbb{R}^d$.

Moreover, this minimizer can (and will) be parametrized to be Lipschitz (i.e. 1-Hölder) continuous and of constant speed, i.e. $|\dot{\gamma}^*(r)| \equiv$ (const) for a.e. $r \in [0,1]$.

Remark 7.33 By invariance of length and signatures under reparametrization, γ^* need not by defined on $[0,1]$ but may be defined for any interval $[s,t]$ with non-empty interior. $\square$

Proof. From Chow's theorem, the inf is taken over a non-empty set so that $\|g\| < \infty$. By definition of inf, there is a sequence (γ^n) with signature g and we can assume (by reparametrization, cf. Proposition 1.38) that each $\gamma^n = \gamma^n(t)$ has a.s. constant speed

$$|\dot{\gamma}^n| \equiv |\gamma^n|_{1\text{-Höl};[0,1]} = c_n$$

where c_n is the length of the path γ^n and $c_n \downarrow \|g\|$. Clearly,

$$\sup_n |\gamma^n|_{1\text{-Höl};[0,1]} = \sup_n c_n < \infty$$

and from Arzela–Ascoli, after relabelling the sequence, γ^n converges uniformly to some (continuous) limit path γ^*. By Lemma 1.23,

$$|\gamma^*|_{1\text{-Höl};[0,1]} \leq \liminf_n |\gamma^n|_{1\text{-Höl};[0,1]} \qquad (7.18)$$

which shows in particular that γ^* itself is 1-Hölder, hence absolutely continuous, so that

$$\int_0^1 |d\gamma^*| = \int_0^1 |\dot{\gamma}_t^*| \, dt.$$

From basic continuity properties of the signature (Proposition 7.15)

$$g \equiv S_N(\gamma^n)_{0,1} \to S_N(\gamma^*)_{0,1}$$

which shows that $S_N(\gamma^*)_{0,1} = g$. It remains to see that

$$\|g\| = \int_0^1 |\dot{\gamma}_t^*| \, dt.$$

First, $\|g\| \leq \int_0^1 |\dot{\gamma}_t^*| \, dt$ is obvious from the definition of $\|g\|$. On the other hand, using (7.18) we have

$$\int_0^1 |\dot{\gamma}_t^*| \, dt = |\gamma^*|_{1\text{-Höl};[0,1]} \leq \liminf_n c_n = \|g\|$$

and the proof is finished. ∎

7.5.3 Homogenous norms

Let us now define the important concept of a homogenous norm on $G^N \left(\mathbb{R}^d \right)$.

Definition 7.34 A homogenous norm is a continuous map $|||.||| : G^N \left(\mathbb{R}^d \right)$ $\rightarrow \mathbb{R}^+$ which satisfies
(i) $|||g||| = 0$ if and only if g equals the unit element $1 \in G^N \left(\mathbb{R}^d \right)$,
(ii) homogeneity with respect to the dilation operator δ_λ,

$$|||\delta_\lambda g||| = |\lambda| \cdot |||g||| \text{ for all } \lambda \in \mathbb{R}.$$

A homogenous norm is said to be symmetric if $|||g||| = |||g^{-1}|||$, and sub-additive if $|||g \otimes h||| \leq |||g||| + |||h|||$. $\square$

Remark 7.35 If $|||.|||$ is a non-symmetric homogenous norm, then $g \rightarrow |||g||| + |||g^{-1}|||$ is a symmetric homogenous norm. Sub-additivity is preserved under such symmetrization. $\square$

Proposition 7.36 *Every symmetric sub-additive homogenous norm* $|||.|||$ *leads to a genuine metric* $G^N \left(\mathbb{R}^d \right)$ *via*

$$(g, h) \mapsto \left|\left|\left| g^{-1} \otimes h \right|\right|\right|.$$

Moreover, this metric is left-invariant.[9]

Proof. We write $d(g, h) = \left|\left|\left| g^{-1} \otimes h \right|\right|\right|$. Property (i) in Definition 7.34 implies $d(g, h) = 0$ iff $g = h$. Sub-additivity of $|||.|||$ implies the triangle inequality for d and symmetry $|||.|||$ implies $d(g, h) = d(h, g)$. At last, left-invariance of d, i.e.

$$d(g \otimes h, g \otimes k) = d(h, k),$$

follows from $(g \otimes h)^{-1} \otimes (g \otimes k) = h^{-1} \otimes k$. $\blacksquare$

Example 7.37 The simplest example of a homogenous norm is the map

$$g \in G^N \left(\mathbb{R}^d \right) \mapsto |||g||| = \max_{i=1,\dots,N} |\pi_i(g)|^{1/i}.$$

In general, it is neither symmetric nor sub-additive. $\square$

Exercise 7.38 *Prove that*
(i) $|||.|||_1 : g \in G^N \left(\mathbb{R}^d \right) \mapsto (N!)^{-1/N} \max_{i=1,\dots,N} (i! |\pi_i(g)|)^{1/i}$ *is a sub-additive homogenous norm.*
(ii) $|||.|||_2 : g \in G^N \left(\mathbb{R}^d \right) \mapsto \max_{i=1,\dots,N} |\pi_i(\log g)|^{1/i}$ *is a symmetric homogenous norm.*

---

[9] ... but in general not right-invariant. A right-invariant metric could be defined by $\left|\left|\left| g \otimes h^{-1} \right|\right|\right|$.

Exercise 7.39 *Compute the minimal length of all paths with signature*

$$\exp\left(\begin{pmatrix} x \\ y \end{pmatrix} + \begin{pmatrix} 0 & a \\ -a & 0 \end{pmatrix}\right).$$

(This is precisely the Carnot–Caratheodory norm of $(x, y, a) \in \mathbb{H}$, the 3-dimensional Heisenberg group.) Check that this gives

$$\sqrt{x^2 + y^2} \text{ when } a = 0,$$
$$2\sqrt{\pi |a|} \text{ when } x = y = 0.$$

(See [157] for instance.)

7.5.4 Carnot–Caratheodory metric

We now check that the Carnot–Caratheodory norm $\|\cdot\|$, which we introduced in Theorem 7.32, defines a homogenous norm in the sense of Definition 7.34.

The geodesic existence result came with a map, the Carnot–Caratheodory norm, from $G^N(\mathbb{R}^d) \to [0, \infty)$. In conjunction with the group structure of $G^N(\mathbb{R}^d)$, it is only a small step to define a genuine metric on $G^N(\mathbb{R}^d)$.

Proposition 7.40 *Let $g, h \in G^N(\mathbb{R}^d)$. We have*
(i) $\|g\| = 0$ if and only if $g = 1$, the unit element in $G^N(\mathbb{R}^d)$;
(ii) homogeneity $\|\delta_\lambda g\| = |\lambda| \, \|g\|$ for all $\lambda \in \mathbb{R}$;
(iii) symmetry $\|g\| = \|g^{-1}\|$;
(iv) sub-additivity $\|g \otimes h\| \le \|g\| + \|h\|$;
(v) continuity: $g \mapsto \|g\|$ is continuous.

Proof. Notation: for $g \in G$ let $\gamma_g^* = \gamma^*$ denote an arbitrary minimizer from the geodesic existence theorem.
(i) If $\|g\| = 0$, γ_g^* has almost everywhere zero derivative, hence $g = S_N\left(\gamma_g^*\right)_{0,1} = 1$. If $g = 1$, it is obvious that $\|g\| = 0$.
(ii) The case $\lambda = 0$ is easy, so we assume $\lambda \ne 0$. The path $\lambda \gamma_g^*$ satisfies $S_N\left(\lambda \gamma_g^*\right)_{0,1} = \delta_\lambda g$. Hence $\|\delta_\lambda g\| \le \text{length}(\lambda \gamma_g^*) = |\lambda| \times \text{length}(\gamma_g^*) = |\lambda| \, \|g\|$. The opposite inequality follows from replacing λ by $1/\lambda$ and g by $\delta_\lambda g$.
(iii) Using the fact that $S_N\left(\overleftarrow{\gamma_g^*}\right)_{0,1} = g^{-1}$ we obtain

$$\left\|g^{-1}\right\| \le \text{length}\left(\overleftarrow{\gamma_g^*}\right) = \text{length}\left(\gamma_g^*\right) = \|g\|.$$

The opposite inequality follows from replacing g by g^{-1}.
(iv) If γ_g^*, γ_h^* denote the resp. geodesics then, from Chen's theorem,

$$g \otimes h = S_N\left(\gamma_{g,h}^*\right)_{0,1}$$

where $\gamma_{g,h}^*$ is the (Lipschitz continuous) concatenation of γ_g^* and γ_h^* with obvious length $\|g\| + \|h\|$. Hence, $\|g \otimes h\|$ must be less than or equal to the length of $\gamma_{g,h}^*$.

(v) Consider a sequence g_n such that $|g_n - g| \to_{n \to \infty} 0$. (Here $|\cdot|$ denotes a norm on the tensor algebra which induces the "original" topology on $G^N(\mathbb{R}^d)$.) By continuity of the group operations $\otimes$ and $(\cdot)^{-1}$, all of which are polynomial in the coordinates, $g_n^{-1} \otimes g \to 1$ is an obvious consequence. From sub-additivity,

$$\left| \|g_n\| - \|g\| \right| \leq \left\| g_n^{-1} \otimes g \right\|$$

and since $\left\| g_n^{-1} \otimes g \right\|$ is dominated by the length of any path with correct signature (namely $g_n^{-1} \otimes g$), it follows from Corollary 7.29 that

$$\left\| g_n^{-1} \otimes g \right\| \to 0.$$

As a consequence, $\left| \|g_n\| - \|g\| \right| \to 0$ which implies continuity. ∎

Definition 7.41 The Carnot–Caratheodory norm on $G^N(\mathbb{R}^d)$ induces (via Proposition 7.36) a genuine (left-invariant, continuous)[10] metric d on $G^N(\mathbb{R}^d)$, called the Carnot–Caratheodory metric. □

The space $\left(G^N(\mathbb{R}^d), d \right)$ is not only a metric but a *geodesic* space (in the sense of Definition 5.17). To this end recall that, given $g \in G^N(\mathbb{R}^d)$, Theorem 7.32 provides us with an associated Lipschitz path $\gamma^* : [0,1] \to \mathbb{R}^d$ of minimal length[11] equal to $\|g\|$ such that $t \in [0,1] \mapsto S_N(\gamma^*)_{0,t} \in G^N(\mathbb{R}^d)$ connects the unit element in $G^N(\mathbb{R}^d)$ with g.

Proposition 7.42 $G^N(\mathbb{R}^d)$ *equipped with Carnot metric d is a geodesic space. Given $g,h \in G^N(\mathbb{R}^d)$, a connecting geodesic is given by*

$$t \in [0,1] \mapsto \Upsilon_t := g \otimes S_N(\gamma^*)_{0,t}$$

where γ^* is the geodesic associated with $g^{-1} \otimes h$.

Proof. Obviously, Υ is continuous and $\Upsilon_0 = g$, $\Upsilon_1 = h$. For any $s < t$ in $[0,1]$,

$$
\begin{aligned}
d\left(\Upsilon_s, \Upsilon_t \right) \ &:= \ \left\| S_N(\gamma^*)_{s,t} \right\| \\
&\leq \ \int_s^t |d\gamma^*| \qquad\qquad (7.19) \\
&= \ (t-s) \int_0^1 |d\gamma^*| \\
&= \ (t-s) \left\| g^{-1} \otimes h \right\| = |t-s|\, d(g,h).
\end{aligned}
$$

---

[10] By Proposition 7.40, part (v).

[11] By reparametrization, the speed $|\dot{\gamma}_t^*|$ may be taken constant for a.e. $t \in [0,1]$.

In fact, the inequality cannot be strict; there would be a strict inequality in

$$\begin{aligned} d(g,h) &\leq d(\Upsilon_0, \Upsilon_s) + d(\Upsilon_s, \Upsilon_t) + d(\Upsilon_t, \Upsilon_1) \\ &\leq (|s| + |t - s| + |1 - t|)\, d(g,h) \\ &= d(g,h) \end{aligned}$$

which is not possible. We conclude that equality holds in (7.19), which shows that Υ is the desired connecting geodesic. ∎

Remark 7.43 [sub-Riemannian structure of $G^N(\mathbb{R}^d)$] The geodesic constructed above satisfies the differential equation

$$d\Upsilon_t = \sum_{i=1}^{d} U_i(\Upsilon_t)\, d\gamma^i$$

where the $U_i(g) = g \otimes e_i, i = 1, \ldots, d$ are easily seen to be left-invariant vector fields on $G^N(\mathbb{R}^d)$. In fact,

$$\mathrm{Lie}\,[U_1, \ldots, U_d]\,|_g = T_g G^N(\mathbb{R}^d)$$

for all $g \in G^N(\mathbb{R}^d)$, where $\mathrm{Lie}[\ldots]$ stands for the generated Lie algebra and $T_g G^N(\mathbb{R}^d)$ denotes the tangent space to $G^N(\mathbb{R}^d)$ at the point g. Chow's theorem can now be understood as the statement that any two points in $G^N(\mathbb{R}^d)$ can be joined by a path that remains tangent to the $\{U_1, \ldots, U_d\}$. A sub-Riemannian metric on $T_g G^N(\mathbb{R}^d)$ is given by declaring the $\{U_1, \ldots, U_d\}$ to be orthonormal. This induces a natural length for any path which remains tangent to $\mathrm{span}\{U_1, \ldots, U_d\}$. This applies in particular to the geodesic Υ of the previous proposition ($\Upsilon_0 = g, \Upsilon_1 = h$) and this natural length is precisely the Carnot–Caratheodory distance $d(g,h)$. □

7.5.5 Equivalence of homogenous norms

Similar to the case of norms on $\mathbb{R}^d$, all homogenous norms on $G^N(\mathbb{R}^d)$ are equivalent. The proof relies crucially on the continuity of homogenous norms (which was part of their definition).

Theorem 7.44 *All homogenous norms on $G^N(\mathbb{R}^d)$ are equivalent. More precisely, if $\|.\|_1$ and $\|.\|_2$ are two homogenous norms, there exists $C \geq 1$ such that for all $g \in G^N(\mathbb{R}^d)$, we have*

$$\frac{1}{C}\|g\|_1 \leq \|g\|_2 \leq C\|g\|_1. \tag{7.20}$$

Proof. It is enough to consider the case when $\|g\|_1$ is given by

$$|||g||| := \max_{i=1,\ldots,N} |\pi_i(g)|^{1/i}.$$

Let $B = \left\{ g \in G^N\left(\mathbb{R}^d\right), |||g||| = 1 \right\}$. Clearly, B is a compact set by continuity and $\|.\|_2$ attains a (positive) minimum m and maximum M, i.e. for all $g \in B$,

$$m \leq \|g\|_2 \leq M.$$

Since (7.20) holds trivially true when $g = 1$, the unit element of $G^N\left(\mathbb{R}^d\right)$, we only need to consider $g \neq 1$. We define $\varepsilon = 1/|||g|||$ so that $|||\delta_\varepsilon g||| = 1$. In particular,

$$m \leq \|\delta_\varepsilon g\|_2 \leq M$$

and by using homogeneity of $\|.\|_2$ we obtain $m \leq \|g\|_2 / |||g||| \leq M$ and (7.20) follows. ∎

Let us recall that the metric ρ on $G^N\left(\mathbb{R}^d\right)$, induced from the norm $|\cdot|_{T^N\left(\mathbb{R}^d\right)}$, is given by

$$\rho(g,h) = |g - h| = \max_{i=1,\ldots,N} |\pi_i(g) - \pi_i(h))|. \qquad (7.21)$$

When $h = 1$ this reduces to

$$|g - 1| = \max_{i=1,\ldots,N} |\pi_i(g)|.$$

Proposition 7.45 *Let $|||.|||$ be a homogenous norm on $G^N\left(\mathbb{R}^d\right)$. Then, there exists a constant $C > 0$ such that for all $g \in G^N\left(\mathbb{R}^d\right)$*

$$\frac{1}{C} \min\left\{ |||g|||, |||g|||^N \right\} \leq |g - 1| \leq C \max\left\{ |||g|||, |||g|||^N \right\}$$

and

$$\frac{1}{C} \min\left\{ |g - 1|, |g - 1|^{1/N} \right\} \leq |||g||| \leq C \max\left\{ |g - 1|, |g - 1|^{1/N} \right\}.$$

Proof. By equivalence of homogenous norms, it suffices to consider the case when

$$|||g||| := \max_{i=1,\ldots,N} |\pi_i(g)|^{1/i}.$$

But then, obviously, $|||g||| \leq \max\left\{ |g - 1|, |g - 1|^{1/N} \right\}$ which implies that

$$\min\left\{ |||g|||, |||g|||^N \right\} \leq |g - 1|.$$

On the other hand

$$|g - 1| = \max_{i=1,\ldots,N} |\pi_i(g)| \leq \max_{i=1,\ldots,N} |||g|||^i = \max\left\{ |||g|||, |||g|||^N \right\}$$

and together these imply all the stated inequalities. ∎

Corollary 7.46 *The topology on $G^N(\mathbb{R}^d)$ induced by Carnot–Caratheodory distance (in fact, by any metric associated with a symmetric sub-additive homogenous norm) coincides with the original[12] topology of $G^N(\mathbb{R}^d)$.*

Proof. Let $|||.|||$ be any symmetric sub-additive homogenous norm on $G^N(\mathbb{R}^d)$ and write $d(g,h) = |||g^{-1} \otimes h|||$ for the associated metric. Given a sequence $(g_n) \subset G^N(\mathbb{R}^d)$ and $g \in G^N(\mathbb{R}^d)$, Proposition 7.45 implies that $|g_n^{-1} \otimes g - 1| \to 0$ if and only if $d(g_n, g) = |||g_n^{-1} \otimes g||| \to 0$. On the other hand, we saw in Proposition 7.18 that $|g_n^{-1} \otimes g - 1| \to 0$ if and only if $|g_n - g| \to 0$. ∎

Remark 7.47 There are more geometric arguments for this. A Riemannian taming argument easily gives that convergence with respect to the CC distance implies convergence in the original topology. For the converse, continuity of the CC norm implies $\|g_n^{-1} \otimes g\| \to 0$ as $g_n^{-1} \otimes g \to 1$, which (by Proposition 7.18) is equivalent to $g_n \to g$ in the original topology. □

We can improve Corollary 7.46 towards a quantitative comparison of the Carnot–Caratheodory distance with the "Euclidean" distance on $G^N(\mathbb{R}^d)$ as given in (7.21). To this end, we need

Lemma 7.48 *Let $g, h \in G^N(\mathbb{R}^d)$, of form $g = 1 + g^1 + \cdots + g^N$, $g^i \in (\mathbb{R}^d)^{\otimes i}$ and similarly for h. The following equations then hold in $(\mathbb{R}^d)^{\otimes k}$ $k = 1, \ldots, N$:*

$$\left(g^{-1} \otimes h\right)^k = \sum_{i=1}^{k} \left(g^{-1}\right)^{k-i} \otimes \left(h^i - g^i\right) \tag{7.22}$$

and

$$h^k - g^k = \sum_{i=1}^{k} g^{k-i} \otimes \left(g^{-1} \otimes h\right)^i. \tag{7.23}$$

Proof. Set $g^0 = h^0 = 1$. By definition of the tensor product in $G^N(\mathbb{R}^d) \subset T^N(\mathbb{R}^d)$, $\left(g^{-1} \otimes h\right)^k = \sum_{i=0}^{k} \left(g^{-1}\right)^{k-i} \otimes h^i$. The result follows by subtracting from the previous expression $0 = \left(g^{-1} \otimes g\right)^k = \sum_{i=0}^{k} \left(g^{-1}\right)^{k-i} \otimes g^i$. The other equality follows from

$$h - g = g \otimes \left(g^{-1} \otimes h - 1\right).$$

∎

Proposition 7.49 *(ball-box estimate) Consider $g, h \in G^N(\mathbb{R}^d)$. There exists a constant $C = C(N) > 0$ such that*

$$d(g,h) \leq C \max\left\{|h - g|, |h - g|^{1/N} \max\left\{1, \|g\|^{1-\frac{1}{N}}\right\}\right\} \tag{7.24}$$

---

[12] Cf. Remark 7.31.

and

$$|h - g| \leq C \max \left\{ d(g, h) \max \left\{ 1, \|g\|^{N-1} \right\}, d(g, h)^N \right\}. \tag{7.25}$$

In particular, recalling from (7.21) that $\rho(g, h) \equiv |h - g|$,

$$\mathrm{Id} : \left(G^N\left(\mathbb{R}^d \right), d \right) \leftrightarrows \left(G^N\left(\mathbb{R}^d \right), \rho \right)$$

is Lipschitz on bounded sets in $\rightarrow$ direction and $1/N$-Hölder on bounded sets in $\leftarrow$ direction.

Proof. Equation (7.22) implies

$$
\begin{aligned}
\left| \left(g^{-1} \otimes h \right)^k \right| &\leq c_1 \sum_{i=1}^{k} \left\| g^{-1} \right\|^{k-i} \cdot \left| h^i - g^i \right| \\
&= c_1 \sum_{i=1}^{k} \|g\|^{k-i} \cdot \left| h^i - g^i \right| \quad \text{by symmetry of } \|\cdot\| \\
&\leq c_2 |h - g| \max \left(1, \|g\|^{k-1} \right).
\end{aligned}
$$

Hence,

$$
\begin{aligned}
\max_k \left| \left(g^{-1} \otimes h \right)^k \right|^{1/k} &\leq c_3 \max_{k=1,\ldots,N} \left[\max \left\{ 1, \|g\|^{1-\frac{1}{k}} \right\} |h - g|^{1/k} \right] \\
&\leq c_4 \max \left\{ |h - g|, |h - g|^{1/N} \max \left\{ 1, \|g\|^{1-\frac{1}{N}} \right\} \right\}.
\end{aligned}
$$

Conversely, from (7.23),

$$\left| h^k - g^k \right| \leq c_5 \sum_{i=1}^{k} \|g\|^{k-i} \left\| g^{-1} \otimes h \right\|^i.$$

Hence,

$$
\begin{aligned}
|h - g| &\leq c_6 \sum_{k=1}^{N} \sum_{i=1}^{k} \|g\|^{k-i} \left\| g^{-1} \otimes h \right\|^i \\
&\leq c_7 \sum_{i=1}^{N} d(g, h)^i \max \left\{ 1, \|g\|^{N-i} \right\} \\
&\leq c_8 \max \left\{ d(g, h) \max \left\{ 1, \|g\|^{N-1} \right\}, d(g, h)^N \right\}.
\end{aligned}
$$

∎

Corollary 7.50 *Let d denote the Carnot–Caratheodory distance on $G^N\left(\mathbb{R}^d \right)$. Then $\left(G^N\left(\mathbb{R}^d \right), d \right)$ is a Polish space in which closed bounded sets are compact.*

Proof. Completeness, separability and compactness of closed, bounded sets are obvious for $G^N \left(\mathbb{R}^d \right)$ under ρ, the metric induced from $\left| \cdot \right|_{T^N \left(\mathbb{R}^d \right)}$. It then suffices to apply the previous proposition. ∎

Exercise 7.51 *(i) Let $x \in \mathbb{R}^d$. Show that $\left\| \exp \left(x \right) \right\| = \left| x \right|$, the Euclidean length of x.*
(ii) Assume $\left\| . \right\|_1$ is a sub-additive, homogenous norm on $G^N \left(\mathbb{R}^d \right)$ such that for all $x \in \mathbb{R}^d$,

$$(*) : \left\| \exp \left(x \right) \right\|_1 = \left| x \right| .$$

Show that $\left\| g \right\|_1 \leq \left\| g \right\|$ for all $g \in G^N \left(\mathbb{R}^d \right)$. This says that the Carnot-Caratheodory norm is the largest sub-additive, homogenous norm which satisfies $(*)$.

Solution. Let $g \in G^N \left(\mathbb{R}^d \right)$, γ a geodesic associated with g, and $\left(\gamma^n \right)$ a sequence of piecewise linear approximations. Then, if $\left(t_i^n \right)$ are the discontinuity points of the derivative of γ^n,

$$
\begin{aligned}
\left\| S_N \left(\gamma^n \right)_{0,1} \right\|_1 &= \left\| \bigotimes_i S_N \left(\gamma^n \right)_{t_i^n, t_{i+1}^n} \right\|_1 \\
&\leq \sum_i \left\| S_N \left(\gamma^n \right)_{t_i^n, t_{i+1}^n} \right\|_1 \\
&= \sum_i \left\| \exp \left(\gamma_{t_i^n, t_{i+1}^n}^n \right) \right\|_1 \\
&= \sum_i \left| \gamma_{t_i^n, t_{i+1}^n}^n \right| \\
&= \int_0^1 \left| d\gamma_u^n \right| .
\end{aligned}
$$

Letting n tend to ∞, we have by continuity of the map $S_N \left(. \right)_{0,1}$, which follows from Theorem 3.15,

$$\left\| S_N \left(\gamma \right)_{0,1} \right\|_1 \leq \int_0^1 \left| d\gamma_u \right| ,$$

which reads $\left\| g \right\|_1 \leq \left\| g \right\|$. □

7.5.6 From linear maps to group homomorphisms

Linear maps from $\mathbb{R}^n$ to $\mathbb{R}^d$ can always be written as $x \to Ax$ where A is a $n \times d$ matrix. With a slight abuse of notation, we will call A the linear map itself. It is obvious that A is a homomorphism from the group $\left(\mathbb{R}^n, + \right)$ into the group $\left(\mathbb{R}^d, + \right)$ (in fact, the linear maps describe the set of all such homomorphisms), i.e. that for all $x, y \in \mathbb{R}^n$, $A \left(x + y \right) = Ax + Ay$. It will be useful to extend A to a homomorphism from $G^N \left(\mathbb{R}^n \right)$ to $G^N \left(\mathbb{R}^d \right)$. To this end, we recall that $t^N (\mathbb{R}^n)$ is generated by $\mathbb{R}^n$ in the sense that a vector space basis of $t^N (\mathbb{R}^n)$ is given by

$$\cup_{m=1}^{N} \left\{ \bigotimes_{i=1}^{m} e_{j_i} : j_i \in \{1, \dots, n\} \right\}$$

where $e_1, \dots, e_n$ is the canonical basis of $\mathbb{R}^n$. We can then (uniquely) extend A to a homomorphism $t^N (\mathbb{R}^n) \to t^N (\mathbb{R}^d)$ by requiring that it is compatible with $\otimes$, i.e.

$$A \left(\bigotimes_{i=1}^{m} e_{j_i} \right) := \bigotimes_{i=1}^{m} (A e_{j_i}) \in (\mathbb{R}^d)^{\otimes m} \subset t^N (\mathbb{R}^d)$$

and then extend A by linearity to all of $t^N (\mathbb{R}^n)$. On the other hand, $t^N (\mathbb{R}^n)$ is a Lie algebra with bracket $[a, b] = a \otimes b - b \otimes a$ and so A is clearly compatible with the bracket, which is to say a Lie algebra homomorphism. From the Campell–Baker–Hausdorff formula,

$$\mathbf{A} (\cdot) := \exp (A (\log (\cdot)))$$

is then a group homomorphism between the Lie groups $1 + t^N (\mathbb{R}^n)$ and $1 + t^N (\mathbb{R}^d)$. Equivalently, one can define directly

$$\mathbf{A} \left(1 + a^1 + \cdots + a^N \right) := 1 + A a^1 + A^{\otimes 2} a^2 + \cdots + A^{\otimes N} a^N,$$

where $a^k = \pi_k (a) \in (\mathbb{R}^n)^{\otimes k}$ and $A^{\otimes k} : (\mathbb{R}^n)^{\otimes k} \to (\mathbb{R}^d)^{\otimes k}$ is defined by linearity from

$$A^{\otimes k} (e_{j_1} \otimes \cdots \otimes e_{j_k}) := A e_{j_1} \otimes \cdots \otimes A e_{j_k} \text{ with } j_i \in \{1, \dots, n\},$$

and check that this defines a group homomorphism. By sheer restriction, this yields the group homomorphism $\mathbf{A}$ between $G^N (\mathbb{R}^n)$ and $G^N (\mathbb{R}^d)$. That said, we will find it convenient in the sequel to have a direct construction of $\mathbf{A}$ based on step-N signatures. We have

Proposition 7.52 *Let A be a linear map from $\mathbb{R}^n$ into $\mathbb{R}^d$. There exists a unique homomorphism from $G^N (\mathbb{R}^n)$ to $G^N (\mathbb{R}^d)$, denoted by $\mathbf{A}$, such that for all $x \in \mathbb{R}^d$,*

$$\mathbf{A} \exp (x) = \exp (Ax).$$

For all $g \in G^N (\mathbb{R}^n)$ we have[13]

$$\|\mathbf{A} g\| \leq |A|_{op} \|g\|$$

and if $g \in G^N (\mathbb{R}^n)$ is written as the step-N signature of some $x \in C^{\text{1-var}}$ $([0, 1], \mathbb{R}^n)$, i.e. $g = S_N (x)_{0,1}$, then

$$\mathbf{A} g = \mathbf{A} S_N (x)_{0,1} = S_N (Ax)_{0,1}.$$

---

[13]$|\cdot|_{op}$ denotes the operator (matrix) norm.

Proof. Let $x, \tilde{x}$ be continuous paths of bounded variation such that $g = S_N(x)_{0,1} = S_N(\tilde{x})_{0,1}$. Then, writing $S_N(Ax)_{0,1}$ and $S_N(A\tilde{x})_{0,1}$ in coordinates shows that they are equal. Hence, it is possible to define

$$\mathbf{A}g = S_N(Ax)_{0,1}.$$

We establish that $\mathbf{A}$ is a homomorphism, which may be done by checking $\mathbf{A}\left(g^{-1} \otimes h\right) = (\mathbf{A}g)^{-1} \otimes \mathbf{A}h$ for arbitrary elements $g, h \in G^N(\mathbb{R}^n)$, which we may assume to be of form

$$g = S_N(x)_{0,1}, \ h = S_N(y)_{0,1}$$

$(x, y$ are continuous paths of bounded variation). We recall that $g^{-1} = S_N(\overleftarrow{x})_{0,1}$, the signature of $\overleftarrow{x} = x(1 - \cdot)$, and define z to be the concatenation of $\overleftarrow{x}$ and y. Then, we have

$$\mathbf{A}\left(g^{-1} \otimes h\right) = S_N(Az)_{0,1}.$$

On the other hand, Az is the concatenation of $\overleftarrow{Ax} = A\overleftarrow{x}$ and Ay and the proof is finished by observing that

$$S_N(Az)_{0,1} = S_N\left(\overleftarrow{Ax}\right)_{0,1} \otimes S_N(Ay)_{0,1} = (\mathbf{A}g)^{-1} \otimes \mathbf{A}h.$$

Finally, we discuss the estimate on $\|\mathbf{A}g\|$. Let $\gamma : [0,1] \to \mathbb{R}^d$ be a geodesic path associated with g, i.e. a path such that $S_N(\gamma)_{0,1} = g$ and $\int_0^1 |d\gamma| = \|g\|$. Then,

$$\|\mathbf{A}g\| = \left\|S_N(A\gamma)_{0,1}\right\| \leq \int_0^1 |d(A\gamma)| \leq |A|_{op} \int_0^1 |d\gamma| = |A|_{op} \|g\|$$

and the proof is finished. ∎

Example 7.53 One simple linear map from $\mathbb{R}^n \oplus \mathbb{R}^n$ into $\mathbb{R}^n$ is the addition map, i.e. $plus(x, y) = x + y$ where $x, y \in \mathbb{R}^n$. It extends to a homomorphism **plus** from $G^N(\mathbb{R}^n \oplus \mathbb{R}^n)$ into $G^N(\mathbb{R}^n)$. □

Example 7.54 *Another simple linear map from $\mathbb{R}^d \oplus \mathbb{R}^{d'}$ onto $\mathbb{R}^d$ is the projection p onto the first d coordinates. It then extends to a homomorphism* **p** *from $G^N\left(\mathbb{R}^d \oplus \mathbb{R}^{d'}\right)$ into $G^N(\mathbb{R}^d)$. For example, if (x, h) is a $\mathbb{R}^d \oplus \mathbb{R}^{d'}$-valued path,*

$$\mathbf{p} \circ S_N(x, h)_{0,1} = S_N(x)_{0,1}.$$ □

Exercise 7.55 *Another simple linear map from $\mathbb{R}^d$ to $\mathbb{R}^d$ is the map $x \to \lambda x$ for a given $\lambda \in \mathbb{R}$. Prove that its homomorphism extension is the restriction of the dilation map δ_λ to $G^N(\mathbb{R}^d)$.*

Exercise 7.56 *Consider for* $\lambda = (\lambda_i)_{1 \leq i \leq d} \in \mathbb{R}^d$, *the map* $\delta_\lambda : 1 + t^N(\mathbb{R}^d) \to 1 + t^N(\mathbb{R}^d)$ *defined by*

$$\delta_\lambda \left(1 + \sum_{k=1}^N \sum_{1 \leq i_1, \ldots, i_k \leq d} x_{i_1, \ldots, i_k} e_{i_1} \otimes \ldots \otimes e_{i_k} \right)$$

$$= 1 + \sum_{k=1}^N \sum_{1 \leq i_1, \ldots, i_k \leq d} x_{i_1, \ldots, i_k} \lambda_{i_1} \ldots \lambda_{i_k} e_{i_1} \otimes \ldots \otimes e_{i_k}.$$

(i) Prove that δ_λ is the extension of the linear map $\sum_{i=1}^d x_i e_i \mapsto \sum_{i=1}^d x_i \lambda_i e_i$.
(ii) Prove that if all the λ_i are equal to some scalar, then the restriction of δ_λ to $G^N(\mathbb{R}^d)$ is the dilation map from Exercise 7.55.

Exercise 7.57 *Show that*

$$\sup_{g \in G^N(\mathbb{R}^n), \|g\| > 0} \|Ag\| / \|g\| = |A|_{op}.$$

Solution. $\leq$ is clear from Proposition 7.52 and equality is achieved at $g = \exp(x)$ where $x \in \mathbb{R}^n$, non-zero, is such that $|Ax| / |x| = |A|_{op}$. $\qquad \square$

Exercise 7.58 *Prove that for all $(\lambda, g) \in \mathbb{R}_+^d \times G^N(\mathbb{R}^d)$, $\|\delta_\lambda(g)\| \leq (\max_{i=1,\ldots,d} \lambda_i) \|g\|$ and $\|g\| \leq (\max_{i=1,\ldots,d} 1/\lambda_i) \|\delta_\lambda(g)\|$.*

7.6 The lift of continuous bounded variation paths on $\mathbb{R}^d$

7.6.1 Quantitative bound on S_N

Recall from Section 7.2.1 that S_N maps a continuous path x of finite 1-variation with values in $\mathbb{R}^d$ to a path $\{t \mapsto S_N(x)_t \equiv S_N(x)_{0,t}\}$ simply by computing all iterated (Riemann–Stieltjes) integrals up to order N. Recall also that $S_N(x)$ was seen to take values in the (free, step-N nilpotent) group $G^N(\mathbb{R}^d) \subset T^N(\mathbb{R}^d)$. We call $S_N(x)$ the *canonical lift* of x to a $G^N(\mathbb{R}^d)$-valued path, since[14]

$$\pi_1\left(S_N(x)_{0,t} \right) = x_{0,t}$$

for all $t \in [0, T]$. As we shall now see, $S_N(x)$ is not only of finite length (i.e. 1-variation) with respect to Carnot–Caratheodory metric on $G^N(\mathbb{R}^d)$ but has the *same* length as the $\mathbb{R}^d$-valued path x.

---

[14] $\pi_1\left(S_N(x)_t \right) = x_t$ for all $t \in [0, T]$ only holds if $x(0) = 0$.

Proposition 7.59 *Let $x \in C^{1\text{-var}}\left([0,T],\mathbb{R}^d\right)$. Then,*

$$\|S_N\left(x\right)\|_{1\text{-var};[0,T]} = |x|_{1\text{-var};[0,T]} \,.$$

Proof. From the very definition of the Carnot–Caratheodory norm, $\left\|S_N\left(x\right)_{s,t}\right\| \geq |x_{s,t}|$ for all $0 \leq s < t \leq T$ and thus

$$d\left(S_N\left(x\right)_s, S_N\left(x\right)_t\right) = \left\|S_N\left(x\right)_{s,t}\right\| \geq |x_{s,t}|\,.$$

Clearly then, $\|S_N\left(x\right)\|_{1\text{-var};[s,t]} \geq |x|_{1\text{-var};[s,t]}$ for all $0 \leq s < t \leq T$. Conversely,

$$\left\|S_N\left(x\right)_{s,t}\right\| \leq \int_s^t |dx| = |x|_{1\text{-var};[s,t]}$$

and since $(s,t) \mapsto |x|_{1\text{-var};[s,t]}$ is a (super-additive) control function, we immediately obtain that for all $0 \leq s < t \leq T$,

$$\|S_N\left(x\right)\|_{1\text{-var};[s,t]} \leq |x|_{1\text{-var};[s,t]}$$

and the proof is finished. ∎

Exercise 7.60 *The purpose of this exercise is to replace $C^{1\text{-var}}$-regularity in Proposition 7.59 by $W^{1,p}$-regularity with $p \in (1,\infty)$. Following Section 1.4.2, the space of all $\mathbf{x} : [0,T] \to G^N\left(\mathbb{R}^d\right)$ with*

$$\|\mathbf{x}\|_{W^{1,p};[0,T]} := \left(\sup_{D \subset [0,T]} \sum_{i:t_i \in D} \frac{\left\|\mathbf{x}_{t_i,t_{i+1}}\right\|^p}{|t_{i+1}-t_i|^{p-1}}\right)^{1/p} < \infty$$

is denoted by $W^{1,p}\left([0,T],G^N\left(\mathbb{R}^d\right)\right)$, a subset of $C^{1\text{-var}}\left([0,T],G^N\left(\mathbb{R}^d\right)\right)$. Let $x \in W^{1,p}\left([0,T],\mathbb{R}^d\right)$ and recall from Section 1.4.1 that its $W^{1,p}$-(semi-) norm is given by

$$|x|_{W^{1,p};[0,T]} = \left(\int_0^T |\dot{x}_t|^p \, dt\right)^{1/p}.$$

Show that

$$\|S_N\left(x\right)\|_{W^{1,p};[0,T]} = |x|_{W^{1,p};[0,T]}\,.$$

(The case of $W^{1,2}\left([0,T],G^2\left(\mathbb{R}^d\right)\right)$ is important as it allows an intrinsic definition of the rate function of enhanced Brownian motion viewed as a symmetric diffusion process on $G^2\left(\mathbb{R}^d\right)$.)

Solution. From the results in Section 1.4.1 we know that x is the indefinite integral of some $\dot{x} \in L^p\left([0,T],\mathbb{R}^d\right)$ and that

$$|x|_{W^{1,p};[0,T]}^p = \sup_{D \subset [0,T]} \sum_{i:t_i \in D} \frac{\left|x_{t_i,t_{i+1}}\right|^p}{|t_{i+1}-t_i|^{p-1}} = \int_0^T |\dot{x}_t|^p \, dt.$$

Next, $\|S_N(x)\|_{W^{1,p};[0,T]} \geq |x|_{W^{1,p};[0,T]}$ follows readily from $\left\|S_N(x)_{s,t}\right\| \geq |x_{s,t}|$. For the converse inequality, we first observe that

$$\left\|S_N(x)_{s,t}\right\| \leq \int_s^t |\dot{x}_u|\, du \leq |x|_{W^{1,p};[s,t]}\, |t-s|^{1-1/p}.$$

Hence

$$\frac{\left\|S_N(x)_{s,t}\right\|^p}{|t-s|^{p-1}} \leq |x|^p_{W^{1,p};[s,t]} \implies \|S_N(x)\|^p_{W^{1,p};[s,t]} \leq |x|^p_{W^{1,p};[s,t]},$$

where we used super-additivity of $(s,t) \mapsto |x|^p_{W^{1,p};[s,t]}$. $\qquad\square$

We aim now to show that continuous $G^N\left(\mathbb{R}^d\right)$-valued paths of bounded variation are in one-to-one correspondence with elements of $C^{1\text{-var}}\left([0,T],\mathbb{R}^d\right)$. We first need a simple lemma; recall that $o \equiv 1$ stands for the unit element in $G^N\left(\mathbb{R}^d\right)$. We note that the projection map $\pi_{0,k}$ (cf. (7.11)) may be restricted to yield a projection map $\pi_{0,k} : G^N\left(\mathbb{R}^d\right) \to G^k\left(\mathbb{R}^d\right)$, whenever $k \leq N$.

Lemma 7.61 *Let* $\mathbf{x},\mathbf{y}$ *be two elements of* $C_o\left([0,T],G^N\left(\mathbb{R}^d\right)\right)$ *such that* $\pi_{0,N-1}\left(\mathbf{x}\right) = \pi_{0,N-1}\left(\mathbf{y}\right)$. *Then, the path* h *defined by*

$$h_t := \log\left(\mathbf{x}_t^{-1} \otimes \mathbf{y}_t\right) \in \mathfrak{g}^N\left(\mathbb{R}^d\right)$$

is such that for some constant C *depending only on* N, *for all* $s,t \in [0,T]$,

$$|h_{s,t}| \leq C\left(\|\mathbf{x}_{s,t}\| + \|\mathbf{y}_{s,t}\|\right)^N.$$

Proof. Note that only the projection of h to $\left(\mathbb{R}^d\right)^{\otimes N}$ is non-zero so that $\exp(h_t)$ commutes with all elements in $G^N\left(\mathbb{R}^d\right)$. In particular,

$$
\begin{aligned}
\mathbf{y}_{s,t} &= \mathbf{y}_s^{-1} \otimes \mathbf{y}_t \\
&= \exp\left(-h_s\right) \otimes \mathbf{x}_s^{-1} \otimes \mathbf{x}_t \otimes \exp\left(h_t\right) \\
&= \mathbf{x}_s^{-1} \otimes \mathbf{x}_t \otimes \exp\left(-h_s\right) \otimes \exp\left(h_t\right) \\
&= \mathbf{x}_{s,t} \otimes \exp\left(h_{s,t}\right).
\end{aligned}
$$

Using the equivalence of homogenous norms, we obtain

$$
\begin{aligned}
|h_{s,t}| &\leq c\left\|\exp\left(h_{s,t}\right)\right\|^N \\
&\leq c\left(\left\|\mathbf{x}_{s,t}^{-1} \otimes \mathbf{y}_{s,t}\right\|^N\right) \\
&\leq c\left(\|\mathbf{x}_{s,t}\| + \|\mathbf{y}_{s,t}\|\right)^N.
\end{aligned}
$$

∎

Theorem 7.62 *Let $N \geq 1$ and $x \in C_o^{\text{1-var}}\left([0,T],\mathbb{R}^d\right)$. Then $\mathbf{x} = S_N(x)$ is the unique "lift" of x in the sense that $\pi_1(\mathbf{x}) = x$ and such that $\mathbf{x} \in C_o^{\text{1-var}}\left([0,T],G^N\left(\mathbb{R}^d\right)\right)$. Moreover,*

$$S_N : C_o^{\text{1-var}}\left([0,T],\mathbb{R}^d\right) \to C_o^{\text{1-var}}\left([0,T],G^N\left(\mathbb{R}^d\right)\right)$$

is a bijection with inverse π_1 and, for all $0 \leq s < t \leq T$,

$$\|\mathbf{x}\|_{\text{1-var};[s,t]} = |x|_{\text{1-var};[s,t]} \, .$$

Proof. It is obvious that $\mathbf{x} = S_N(x)$ has the lifting property, i.e. that $\pi_1 \circ S_N$ is the identity map on $C_o^{\text{1-var}}\left([0,T],\mathbb{R}^d\right)$, and from Proposition 7.59 we see that $\mathbf{x} \in C_o^{\text{1-var}}\left([0,T],G^N\left(\mathbb{R}^d\right)\right)$. To see uniqueness it is enough to show that $S_N \circ \pi_1$ is the identity map on $C_o^{\text{1-var}}\left([0,T],G^N\left(\mathbb{R}^d\right)\right)$. This is trivially true when $N = 1$. By induction, we now assume the statement is true at level $N - 1$, for $N \geq 2$. Define $\mathbf{y} = S_N \circ \pi_1(\mathbf{x})$ and the path h by

$$h_t := \log\left(\mathbf{x}_t^{-1} \otimes \mathbf{y}_t\right) \in \mathfrak{g}^N\left(\mathbb{R}^d\right).$$

From Lemma 7.61, $|h_{s,t}| \leq c\left(\|\mathbf{x}_{s,t}\| + \|\mathbf{y}_{s,t}\|\right)^N \leq c\omega(s,t)^N$ where $\omega(s,t)$ is the super-additive (control) function $\|\mathbf{x}\|_{\text{1-var};[s,t]} + \|\mathbf{y}\|_{\text{1-var};[s,t]}$. In particular, h is of finite $\frac{1}{N}$-variation. As $1/N < 1$ and $h_0 = 0$, this implies that $h \equiv 0$, i.e. that $\mathbf{y} = \mathbf{x}$. ∎

7.6.2 Modulus of continuity for the map S_N

Proposition 7.63 *Let $x^1, x^2 \in C^{\text{1-var}}\left([0,T],\mathbb{R}^d\right)$, and $\ell \geq \max_{i=1,2} |x^i|_{\text{1-var};[s,t]}$. Then, for all $N \geq 1$,*

$$\exists C_N : \forall 0 \leq s < t \leq T : \left|\pi_N\left(S_N\left(x^1\right)_{s,t} - S_N\left(x^2\right)_{s,t}\right)\right|$$
$$\leq C_N \ell^{N-1} \left|x^1 - x^2\right|_{\text{1-var};[s,t]}. \tag{7.26}$$

In particular, if ω is a fixed control and ε is a positive real such that for all $s,t \in [0,T]$,

$$\max_{i=1,2} \left|x^i\right|_{\text{1-var};[s,t]} \leq \omega(s,t) \ \text{and} \ \left|x^1 - x^2\right|_{\text{1-var};[s,t]} \leq \varepsilon\omega(s,t),$$

we have[15]

$$\max_{k=1,\ldots,N} \sup_{0 \leq s < t \leq T} \frac{\left|\pi_k\left(\mathbf{x}_{s,t} - \mathbf{y}_{s,t}\right)\right|}{\omega(s,t)^k} \leq C_N \varepsilon.$$

---

[15] In the terminology of the forthcoming Definition 8.6, this is equivalent to $\rho_{1,\omega}\left(S_N\left(x^1\right), S_N\left(x^2\right)\right) \leq C_N \varepsilon$.

Proof. Obviously, (7.26) holds true with $C_1 = 1$ for $N = 1$. We proceed by induction, assuming (7.26) holds. Then

$$\pi_{N+1}\left(S_{N+1}\left(x^1\right)_{s,t} - S_{N+1}\left(x^2\right)_{s,t}\right) = \int_s^t \pi_N\left(S_N\left(x^1\right)_{s,r} - S_N\left(x^2\right)_{s,r}\right) \otimes dx_r^1$$
$$+ \int_s^t \pi_N\left(S_N\left(x^2\right)_{s,r}\right) \otimes d\left(x_r^1 - x_r^2\right).$$

From the induction hypothesis,

$$\sup_{r \in [s,t]} \left|\pi_N\left(S_N\left(x^1\right)_{s,r} - S_N\left(x^2\right)_{s,r}\right)\right|$$
$$\leq C_N \left(\max_{i=1,2} \left|x^i\right|_{1\text{-var};[s,t]}\right)^{N-1} \left|x^1 - x^2\right|_{1\text{-var};[s,t]},$$

hence the first integral on the right-hand side above is estimated by

$$\left|\int_s^t \pi_N\left(S_N\left(x^1\right)_{s,r} - S_N\left(x^2\right)_{s,r}\right) \otimes dx_r^1\right|$$
$$\leq C_N \left(\max_{i=1,2} \left|x^i\right|_{1\text{-var};[s,t]}\right)^N \left|x^1 - x^2\right|_{1\text{-var};[s,t]}.$$

On the other hand, $\sup_{r \in [s,t]} \left|\pi_N\left(S_N\left(x^2\right)_{s,r}\right)\right| \leq (1/N!) \left|x^2\right|_{1\text{-var};[s,t]}^N$ so that we can estimate the second integral on the right-hand side above:

$$\left|\int_s^t \pi_N\left(S_N\left(x^2\right)_{s,r}\right) \otimes d\left(x_r^1 - x_r^2\right)\right| \leq \frac{1}{N!} \left|x^2\right|_{1\text{-var};[s,t]}^N \left|x^1 - x^2\right|_{1\text{-var};[s,t]}.$$

Combining the two estimates finishes the induction step and thus concludes the proof of the first part of the proposition. The second part is an obvious corollary of the first part. ∎

The previous proposition implies in particular that if x_1 and x_2 are two continuous paths of length bounded by 1, and such that $|x_1 - x_2|_{1\text{-var};[0,1]} \leq \varepsilon$, then for all $N \in \{1, 2, 3, \dots\}$,

$$\max_{k=1,\dots,N} \left|\pi_k\left(g_1 - g_2\right)\right| \leq C_N \varepsilon$$

where $g_i := S_N\left(x_i\right)_{0,1} \in G^N\left(\mathbb{R}^d\right)$, $i = 1, 2$. We now prove in some sense the converse statement; the basic idea of the proof of the following proposition is illustrated in Figure 7.2.

Proposition 7.64 *Let $g_1, g_2 \in G^N\left(\mathbb{R}^d\right)$, with $\|g_1\|, \|g_2\| \leq C_1$ and*

$$\max_{k=1,\dots,N} \left|\pi_k\left(g_1 - g_2\right)\right| \leq \varepsilon, \quad k = 1, \dots, N.$$

Then, there exists $x_i \in C^{1\text{-var}}\left([0,1],\mathbb{R}^d\right)$, such that

$$S_N\left(x_i\right)_{0,1} = g_i, \ i = 1,2,$$

and a constant $C_2 = C_2\left(C_1,N\right)$, such that

$$\max_{i=1,2}|x_i|_{1\text{-var};[0,1]} \leq C_2,$$

$$|x_1 - x_2|_{1\text{-var};[0,1]} \leq \varepsilon C_2.$$

Proof. <u>First case:</u> Assume that $\pi_{1,N-1}\left(g_1\right) = \pi_{1,N-1}\left(g_2\right) = 0$. In such a case, we can write $g_i = \exp\left(\ell_i\right) = 1 + \ell_i$; the hypothesis implies that $|\ell_i| \leq c_1$, and that $\ell_2 = \ell_1 + \varepsilon m$, with $|m| \leq c_2$. We write g_1 and g_2 in the following way :

$$
\begin{aligned}
g_1 &= \exp\left(\ell_1 - m\right) \otimes \exp\left(m\right), \\
g_2 &= \exp\left(\ell_1 - m\right) \otimes \exp\left((1+\varepsilon)\,m\right) \\
&= \exp\left(\ell_1 - m\right) \otimes \delta_{(1+\varepsilon)^{1/N}} \exp\left(m\right).
\end{aligned}
$$

Define $z : [0,1] \to \mathbb{R}^d$ to be a geodesic associated with the group element $\exp\left(\ell_1 - m\right)$ (observe that the length of z is bounded by a constant independent of ε). Define also $y : [0,1] \to \mathbb{R}^d$ to be a geodesic associated with the group element $\exp\left(m\right)$ (the length of y is bounded by a constant independent of ε). Define $x_1 : [0,2] \to \mathbb{R}^d$ to be the concatenation of z and y, and $x_2 : [0,2] \to \mathbb{R}^d$ to be the concatenation of z and $(1+\varepsilon)^{1/N}\,y$. Observe that the 1-variation distance between x_1 and x_2 is equal to $\left[(1+\varepsilon)^{1/N} - 1\right]$ times the length of y, i.e. it is bounded by a constant times $\left((1+\varepsilon)^{1/N} - 1\right) \leq \varepsilon$. Reparametrizing the paths x_1 and x_2 to be from $[0,1]$ into $\mathbb{R}^d$ finishes the first case.

<u>General case:</u> We prove the general case by induction. The case $N=1$ can be solved using the first case (or more simply with straight lines). Assuming that the proposition holds for elements in $G^N\left(\mathbb{R}^d\right)$ we now prove that it also holds for elements in $G^{N+1}\left(\mathbb{R}^d\right)$. To this end, take two arbitrary elements $g_1, g_2 \in G^{N+1}\left(\mathbb{R}^d\right)$ with

$$\max_{k=1,\dots,N+1}\left|\pi_k\left(g_1\right) - \pi_k\left(g_2\right)\right| < \varepsilon.$$

Set $c_3 := \|g_1\| \vee \|g_2\|$ and define $h_i \in G^N\left(\mathbb{R}^d\right)$ by projection of g_i to the first N levels, i.e. so that $\pi_{0,N}\left(g_i\right) = \pi_{0,N}\left(h_i\right)$. Obviously $\|h_i\| \leq c_3$ for $i=1,2$ and

$$\max_{k=1,\dots,N}\left|\pi_k\left(h_1\right) - \pi_k\left(h_2\right)\right| < \varepsilon.$$

By the induction hypothesis, there exist two paths z_1, z_2 (which we may take to be defined on $[0,1]$) of length bounded by c_4, with the length of

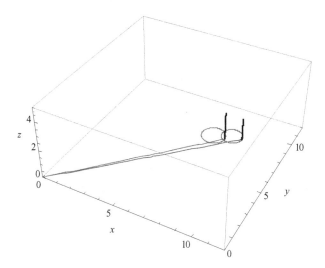

Figure 7.2. We illustrate the basic idea of the proof of Proposition 7.64. Two points of $G^2\left(\mathbb{R}^2\right)$, identified with the 3-dimensional Heisenberg group, are given as $g_1 = (9,9,1.5)$ and $g_2 = (8,8.5,2)$. The corresponding paths x_1, x_2 are the concatenation of straight lines, connecting the origin with $(9,9)$ and $(8,8.5)$ respectively, followed by a circle which wipes out the prescribed area, 1.5 and 2 respectively.

$z_1 - z_2$ bounded by $c_4 \varepsilon$, where $c_4 = c_4\left(N, c_3\right)$, and with the property that $S_N\left(z_i\right)_{0,1} = h_i$, $i = 1,2$. We now define

$$k_i = S_{N+1}\left(z_i\right)_{0,1}^{-1} \otimes g_i \in G^{N+1}\left(\mathbb{R}^d\right); \quad i = 1,2.$$

Clearly, $\|k_i\| \le c_4 + c_3 =: c_5$ and $\pi_{1,N}\left(k_i\right) = 0$. Also, from Proposition 7.63, we have, for all $j \le N+1$,

$$\left|\pi_j\left(S_{N+1}\left(z_1\right)_{0,1}^{-1} - S_{N+1}\left(z_2\right)_{0,1}^{-1}\right)\right| = \left|\pi_j\left(S_{N+1}\left(\overleftarrow{z_1}\right)_{0,1} - S_{N+1}\left(\overleftarrow{z_2}\right)_{0,1}\right)\right|$$
$$\le c_6 \varepsilon.$$

It is easy to see that, for any $a, b \in T^{N+1}\left(\mathbb{R}^d\right)$,

$$\pi_{0,N}\left(a\right) = \pi_{0,N}\left(b\right) \implies a^{-1} \otimes b = 1 + \pi_{N+1}\left(a^{-1}\right) + \pi_{N+1}\left(b\right).$$

Applied in our context this gives $k_i = 1 + \pi_{N+1}\left(S_{N+1}\left(z_i\right)^{-1}\right) + \pi_{N+1}\left(g_i\right)$, and hence, for all $j \le N+1$, we have $\left|\pi_j\left(k_1 - k_2\right)\right| \le \left(1 + c_6\right)\varepsilon$. Therefore, using the first case, there exist two paths y_1, y_2 (defined on $[0,1]$) of length bounded by c_7, with length of $y_1 - y_2$ bounded by $c_7 \varepsilon$, and with the property that $S_{N+1}\left(y_i\right) = k_i$; $i = 1,2$. We conclude the proof by observing that the paths $x_i = z_i \sqcup y_i$, (re)parametrized to $[0,1]$, satisfy $S_{N+1}\left(x_i\right) = g_i$, are of length bounded by c_8, and with length of $x_1 - x_2$ bounded by $c_8 \varepsilon$. The proof is now finished.   ∎

As a first corollary of this powerful lemma, we prove a modulus of continuity for homomorphism on $G^N\left(\mathbb{R}^d\right)$ that extends the linear map on $\mathbb{R}^d$ (see Section 7.5.6).

Proposition 7.65 *Let A be a linear map from $\mathbb{R}^n$ into $\mathbb{R}^d$. Then, for $g, h \in G^N\left(\mathbb{R}^n\right)$ with $\|g\|$ and $\|h\|$ bounded by 1, we have for some constant $C = C\left(N\right)$,*

$$\left|\mathbf{A}g - \mathbf{A}h\right| \leq C\left|A\right|_{op}\left|g - h\right|.$$

Proof. Using Lemma 7.64, we take $g = S_N\left(x\right)_{0,1}$ and $h = S_N\left(y\right)_{0,1}$, where x and y are some bounded variation paths of length bounded by some constant c_1 and such that

$$\left|x - y\right|_{1\text{-var};[0,1]} \leq c_2\left|g - h\right|.$$

Then, using Proposition 7.63,

$$
\begin{aligned}
\left|\mathbf{A}g - \mathbf{A}h\right| &= \left|S_N(Ax)_{0,1} - S_N(Ay)_{0,1}\right| \\
&\leq c_3\left|A\left(x - y\right)\right|_{1-\text{var};[0,1]} \\
&\leq c_3\left|A\right|_{op}\left|\left(x - y\right)\right|_{1-\text{var};[0,1]} \\
&\leq c_4\left|g - h\right|.
\end{aligned}
$$

∎

7.7 Comments

Section 7.2: The *signature* map goes back to the classical work of Chen [28, 29] and it was clear from his work that truncation at step N leads to (a presentation) of the step-N free nilpotent Lie group. This point of view was in particular adopted by Lyons [115, 116]; see Lyons *et al.* [123] for references.

Section 7.3: All this is standard, see Lyons *et al.* [123] for references. The proof of the Campbell–Baker–Hausdorff formula via differential equations is also well-known and appears, for instance, in Strichartz [157]. If one only cares about the qualitative statement Corollary 7.27, an algebraic proof is possible via *Friedrich's criterion*, see Reutenauer [142] for instance.

Section 7.4: In our setting, *Chow's* theorem plays the role of a converse to the Campbell–Baker–Hausdorff formula. See Baudoin [7], Folland and Stein [54] for related points of view. It can be formulated in sub-Riemannian geometry (Montgomery [131] and the references cited therein), see also Varopouols *et al.* [174].

Section 7.5: Again, geodesics are essentially a sub-Riemannian concept but the details are simpler in our setting. Equivalence of homogenous norms, on the other hand, is a group concept and allows for considerable simplifcation when it comes to topological consistency of the Carnot–Caratheodory metric with the original topology. Exercise 7.39 is taken from Strichartz [157].

Section 7.6 contains some "lifting" estimates which correspond, in essence, to the case $p = 1$ in the forthcoming estimates for the Lyons lift (see Section 9.1). Proposition 7.64 appears to be new; the construction of "almost" geodesic paths associated with a pair of "nearby" group elements will be a key ingredient (cf. the proof of the forthcoming Theorem 10.26) in establishing local Lipschitzness of the Itô–Lyons map.

8

Variation and Hölder spaces on free groups

In the general setting of a (continuous) path with values in a metric space, say $x : [0,T] \to E$, we defined its p-variation "norm" over $[0,T]$, in symbols $|x|_{p\text{-var};[0,T]}$. This applies in particular to a $E = G^N\left(\mathbb{R}^d\right)$-valued path $\mathbf{x}\left(\cdot\right)$, where $G^N\left(\mathbb{R}^d\right)$ is the free step-N nilpotent group discussed at length in the previous chapter. As a constant reminder that the (Carnot–Caratheodory) metric d on $G^N\left(\mathbb{R}^d\right)$ was derived from $\|\cdot\|$, the Carnot–Caratheodory norm, we shall then use the notation

$$
\begin{aligned}
\|\mathbf{x}\|_{p\text{-var};[0,T]} &= \sup_{(t_i)\subset[0,T]} \left(\sum_i d\left(\mathbf{x}_{t_i},\mathbf{x}_{t_{i+1}}\right)^p\right)^{1/p} \qquad (8.1)\\
&= \sup_{(t_i)\subset[0,T]} \left(\sum_i \left\|\mathbf{x}_{t_i,t_{i+1}}\right\|^p\right)^{1/p}
\end{aligned}
$$

and, thanks to homogeneity of $\|\cdot\|$ with respect to dilation, speak of a *homogenous p-variation norm*. As a special case of Definition 5.1,

$$
C^{p\text{-var}}\left([0,T],G^N\left(\mathbb{R}^d\right)\right) = \left\{\mathbf{x} \in C\left([0,T],G^N\left(\mathbb{R}^d\right)\right) : \|\mathbf{x}\|_{p\text{-var};[0,T]} < \infty\right\}
$$

and we shall assume $p \geq 1$ unless otherwise stated. When $E = \mathbb{R}^d$, (8.1) is precisely the usual p-variation (semi-)norm and

$$
(x,y) \mapsto |x_0 - y_0| + |x - y|_{p\text{-var};[0,T]}
$$

defines a genuine metric, the p-variation metric, on path space. Recalling that $|x - y|_{p\text{-var};[0,T]}^p$ is of the form

$$
\sup_{(t_i)\subset[0,T]} \sum_i \left|x_{t_i,t_{i+1}} - y_{t_i,t_{i+1}}\right|^p
$$

(with $x_{s,t} \equiv x_t - x_s \in \mathbb{R}^d$), a convenient extension to a $G^N\left(\mathbb{R}^d\right)$-valued path is to replace $|x_{s,t} - y_{s,t}|$ by $d\left(\mathbf{x}_{s,t},\mathbf{y}_{s,t}\right)$, where now

$$
\mathbf{x}_{s,t} \equiv \mathbf{x}_s^{-1} \otimes \mathbf{x}_t \in G^N\left(\mathbb{R}^d\right).
$$

Alternatively, we may replace $\left|x_{t_i,t_{i+1}} - y_{t_i,t_{i+1}}\right|^p$ by $\left|\pi_k\left(\mathbf{x}_{s,t} - \mathbf{y}_{s,t}\right)\right|^{p/k}$ for all $k = 1,\ldots,N$, using the fact that $G^N\left(\mathbb{R}^d\right) \subset T^N\left(\mathbb{R}^d\right)$; recalling that

$\pi_k : T^N \left(\mathbb{R}^d \right) \to \left(\mathbb{R}^d \right)^{\otimes k}$ denotes projection to the kth tensor level. When $N = 1$, the two notations coincide; for $N \geq 2$, they do not. However, both resulting p-variation distances remain "locally uniformly" comparable (and in particular induce the same topology and the same notion of Cauchy sequences) and both are useful. The first allows us to discuss the properties of the space $C^{p\text{-var}} \left([0,T], G^N \left(\mathbb{R}^d \right) \right)$ with often identical arguments as in the $\mathbb{R}^d$ case. The second arises naturally in Lipschitz estimates of the Lyons lift and the Itô–Lyons map, discussed in later chapters. Of course, everything said here applies in a Hölder context. In particular, the *homogenous $1/p$-Hölder norm* is given by

$$\|\mathbf{x}\|_{1/p\text{-Höl};[0,T]} = \sup_{0 \leq s < t \leq T} \frac{d\left(\mathbf{x}_s, \mathbf{x}_t \right)}{|t-s|^{1/p}} = \sup_{0 \leq s < t \leq T} \frac{\|\mathbf{x}_{s,t}\|}{|t-s|^{1/p}} \qquad (8.2)$$

and

$$C^{1/p\text{-Höl}} \left([0,T], G^N \left(\mathbb{R}^d \right) \right) = \left\{ \mathbf{x} \in C \left([0,T], G^N \left(\mathbb{R}^d \right) \right) : \|\mathbf{x}\|_{1/p\text{-Höl};[0,T]} < \infty \right\}.$$

8.1 p-Variation and $1/p$-Hölder topology

8.1.1 *Homogenous p-variation and Hölder distances*

As usual in the discussion of p-variation, we assume $p \geq 1$. We then have

Definition 8.1 (homogenous variation and Hölder distance) Given $\mathbf{x}, \mathbf{y} \in C \left([0,T], G^N \left(\mathbb{R}^d \right) \right)$ we define

$$d_{p\text{-var};[0,T]} \left(\mathbf{x}, \mathbf{y} \right) := \left(\sup_D \sum_{t_i \in D} d\left(\mathbf{x}_{t_i,t_{i+1}}, \mathbf{y}_{t_i,t_{i+1}} \right)^p \right)^{1/p}$$

and

$$d_{1/p\text{-Höl};[0,T]} \left(\mathbf{x}, \mathbf{y} \right) := \sup_{0 \leq s < t \leq T} \frac{d\left(\mathbf{x}_{s,t}, \mathbf{y}_{s,t} \right)}{|t-s|^{1/p}},$$

where it is understood that

$$d_{0;[0,T]} \left(\mathbf{x}, \mathbf{y} \right) := d_{0\text{-Höl};[0,T]} \left(\mathbf{x}, \mathbf{y} \right) := \sup_{0 \leq s < t \leq T} d\left(\mathbf{x}_{s,t}, \mathbf{y}_{s,t} \right).$$

□

With $o = 1$, the unit element in $G^N \left(\mathbb{R}^d \right)$, we note that $d_{p\text{-var};[0,T]}(\mathbf{x}, o) = \|\mathbf{x}\|_{p\text{-var};[0,T]}$, the homogenous p-variation norm of $\mathbf{x}$ and similarly in the Hölder case. It is obvious that

$$(\mathbf{x}, \mathbf{y}) \mapsto d_{p\text{-var};[0,T]}(\mathbf{x}, \mathbf{y}) \text{ resp. } d_{1/p\text{-Höl};[0,T]}(\mathbf{x}, \mathbf{y}) \qquad (8.3)$$

is non-negative, symmetric and one sees (precisely as the case of $\mathbb{R}^d$-valued paths) that the triangle inequality is satisfied. However, when the right hand side above is 0 this only tells us that $\mathbf{x}_t = c \otimes \mathbf{y}_t$ with $c = \mathbf{x}_0^{-1} \otimes \mathbf{y}_0$. If attention is restricted to paths with fixed starting point, which is the case for $C_o^{p\text{-var}}\left([0,T],G^N\left(\mathbb{R}^d\right)\right)$ resp. $C_o^{1/p\text{-Höl}}\left([0,T],G^N\left(\mathbb{R}^d\right)\right)$, then (8.3) defines a genuine metric. Otherwise, it suffices to add the distance of the starting points, in which case

$$(\mathbf{x}, \mathbf{y}) \quad \mapsto \quad d(\mathbf{x}_0, \mathbf{y}_0) + d_{p\text{-var};[0,T]}(\mathbf{x}, \mathbf{y})$$

resp. $\quad d(\mathbf{x}_0, \mathbf{y}_0) + d_{1/p\text{-Höl};[0,T]}(\mathbf{x}, \mathbf{y})$

gives a genuine metric on $C^{p\text{-var}}\left([0,T],G^N\left(\mathbb{R}^d\right)\right)$ resp. $C^{1/p\text{-Höl}}\left([0,T], G^N\left(\mathbb{R}^d\right)\right)$. Let us observe that for any $\lambda \in \mathbb{R}$,

$$d_{p\text{-var};[0,T]}(\delta_\lambda \mathbf{x}, \delta_\lambda \mathbf{y}) = |\lambda|\, d_{p\text{-var};[0,T]}(\mathbf{x}, \mathbf{y}),$$
$$d_{1/p\text{-Höl};[0,T]}(\delta_\lambda \mathbf{x}, \delta_\lambda \mathbf{y}) = |\lambda|\, d_{1/p\text{-Höl};[0,T]}(\mathbf{x}, \mathbf{y}),$$

where δ_λ denotes dilation by λ on $G^N\left(\mathbb{R}^d\right)$, which explains the terminology *homogenous p-variation* (resp. *1/p-Hölder*) distance; we also note that

$$(s,t) \mapsto d_{p\text{-var};[s,t]}(\mathbf{x}, \mathbf{y})^p$$

is a control function.

Definition 8.2 (Homogenous p-ω distance) (i) Given a control function ω on $[0,T]$ and

$$\mathbf{x} \in C\left([0,T],G^N\left(\mathbb{R}^d\right)\right)$$

we define[1] the homogenous p-ω norm

$$\|\mathbf{x}\|_{p\text{-}\omega;[0,T]} := \sup_{0 \le s < t \le T} \frac{d\left(\mathbf{x}_s, \mathbf{x}_t\right)}{\omega\left(s,t\right)^{1/p}} = \sup_{0 \le s < t \le T} \frac{\|\mathbf{x}_{s,t}\|}{\omega\left(s,t\right)^{1/p}} \qquad (8.4)$$

and

$$C^{p\text{-}\omega}\left([0,T],G^N\left(\mathbb{R}^d\right)\right) = \left\{\mathbf{x} \in C\left([0,T],G^N\left(\mathbb{R}^d\right)\right) : \|\mathbf{x}\|_{p\text{-}\omega;[0,T]} < \infty\right\}.$$

(ii) Given $\mathbf{x}, \mathbf{y} \in C^{p\text{-}\omega}\left([0,T],G^N\left(\mathbb{R}^d\right)\right)$ and a control function ω on $[0,T]$, we define

$$d_{p\text{-}\omega;[0,T]}(\mathbf{x}, \mathbf{y}) = \sup_{0 \le s < t \le T} \frac{d\left(\mathbf{x}_{s,t}, \mathbf{y}_{s,t}\right)}{\omega\left(s,t\right)^{1/p}}.$$

$\square$

---

[1] The definition of $\|\cdot\|_{p\text{-}\omega;[0,T]}$ and $C^{p\text{-}\omega}$ would make sense for paths with values in an abstract metric space.

Remark 8.3 Whenever $\omega(s,t) = 0$ for some $s < t$, the right-hand side of (8.4) is infinity unless $\|\mathbf{x}_{s,t}\| = 0$, in which case we use the convention $0/0 = 0$. Equivalently, one can define

$$\|\mathbf{x}\|_{p\text{-}\omega;[0,T]} = \inf\left\{M \geq 0 : \|\mathbf{x}_{s,t}\| \leq M\omega(s,t)^{1/p} \text{ for all } 0 \leq s < t \leq T\right\}.$$

Similar remarks apply to our definition of $d_{p\text{-}\omega;[0,T]}(\mathbf{x},\mathbf{y})$.   □

It is clear from the definition that

$$(\mathbf{x},\mathbf{y}) \mapsto d(\mathbf{x}_0,\mathbf{y}_0) + d_{p\text{-}\omega;[0,T]}(\mathbf{x},\mathbf{y})$$

is a metric on $C^{p\text{-}\omega}\left([0,T],G^N\left(\mathbb{R}^d\right)\right)$ and, as above, one can omit the term $d(\mathbf{x}_0,\mathbf{y}_0)$ if attention is restricted to paths pinned at time 0. Let us also observe, as an elementary consequence of super-additivity of controls,

$$d_{p\text{-var};[0,T]}(\mathbf{x},\mathbf{y}) \leq \omega(0,T)^{1/p}\, d_{p\text{-}\omega;[0,T]}(\mathbf{x},\mathbf{y}); \qquad (8.5)$$

in the special case $\omega(s,t) = t - s$ this reads

$$d_{p\text{-var};[0,T]}(\mathbf{x},\mathbf{y}) \leq T^{1/p}d_{1/p\text{-Höl};[0,T]}(\mathbf{x},\mathbf{y}).$$

Proposition 8.4 *For all* $\mathbf{x}^1,\mathbf{x}^2 \in C^{p\text{-var}}\left([0,T],G^N\left(\mathbb{R}^d\right)\right)$ *there exists a control* ω *with* $\omega(0,T) = 1$ *such that*

$$\begin{aligned} \|\mathbf{x}^i\|_{p\text{-}\omega;[0,T]} &\leq 3^{1/p}\|\mathbf{x}^i\|_{p\text{-var};[0,T]}, \quad i = 1,2; \\ d_{p\text{-}\omega;[0,T]}(\mathbf{x}^1,\mathbf{x}^2) &\leq 3^{1/p}d_{p\text{-var};[0,T]}(\mathbf{x}^1,\mathbf{x}^2). \end{aligned}$$

Proof. Given $\mathbf{x},\mathbf{y} \in C^{p\text{-var}}\left([0,T],G^N\left(\mathbb{R}^d\right)\right)$ define the control

$$\omega_{\mathbf{x},\mathbf{y}}(s,t) := \left(\frac{d_{p\text{-var};[s,t]}(\mathbf{x},\mathbf{y})}{d_{p\text{-var};[0,T]}(\mathbf{x},\mathbf{y})}\right)^p$$

(using the convention $0/0 = 0$ if necessary). Note $\omega_{\mathbf{x},o}(s,t) = \|\mathbf{x}\|^p_{p\text{-var};[s,t]} / \|\mathbf{x}\|^p_{p\text{-var};[0,T]}$ where o denotes the trivial path constant equal to the unit element in $G^N\left(\mathbb{R}^d\right)$. Then

$$\omega(s,t) := \frac{1}{3}\omega_{\mathbf{x}^1,\mathbf{x}^2}(s,t) + \frac{1}{3}\omega_{\mathbf{x}^1,o}(s,t) + \frac{1}{3}\omega_{\mathbf{x}^2,o}(s,t)$$

has the desired properties. For instance,

$$d(\mathbf{x}^1_{s,t},\mathbf{x}^2_{s,t}) \leq d_{p\text{-var};[s,t]}(\mathbf{x}^1,\mathbf{x}^2) \leq d_{p\text{-var};[0,T]}(\mathbf{x}^1,\mathbf{x}^2)\underbrace{\omega_{\mathbf{x}^1,\mathbf{x}^2}(s,t)^{1/p}}_{\leq (3\omega(s,t))^{1/p}}$$

and similar for $\mathbf{x}^1,o$ and $\mathbf{x}^2,o$.   ■

We now show that a map which is uniformly continuous on bounded sets in the metric $d_{p\text{-}\omega}$ for all possible control functions ω, is also uniformly continuous on bounded sets in the metric $d_{p\text{-var}}$. To this end, for paths $\mathbf{x} \in C\left([0,T], G^M\left(\mathbb{R}^d\right)\right)$ we define the (homogenous) balls

$$\begin{aligned} B_{p\text{-}\omega}\left(R\right) &= \left\{\mathbf{x} : d_{p\text{-}\omega;[0,T]}\left(\mathbf{x}, o\right) < R\right\}, \\ B_{p\text{-var}}\left(R\right) &= \left\{\mathbf{x} : d_{p\text{-var};[0,T]}\left(\mathbf{x}, o\right) < R\right\}. \end{aligned}$$

Corollary 8.5 *Consider a map*

$$\phi : C^{p\text{-var}}\left([0,T], G^M\left(\mathbb{R}^d\right)\right) \to C^{p\text{-var}}\left([0,T], G^N\left(\mathbb{R}^e\right)\right).$$

Assume that for any control function ω, any $\varepsilon > 0$ and $R > 0$ there exists $\delta = \delta\left(\varepsilon, R; \omega\right)$ such that

$$\mathbf{x}, \mathbf{y} \in B_{p\text{-}\omega}\left(R\right), \, d_{p,\omega}\left(\mathbf{x}, \mathbf{y}\right) < \delta \implies d_{p\text{-}\omega}\left(\phi\left(\mathbf{x}\right), \phi\left(\mathbf{y}\right)\right) < \varepsilon.$$

Then, for any $\varepsilon > 0$ and $R > 0$ there exists $\eta = \eta\left(\varepsilon, R\right)$ such that

$$\mathbf{x}, \mathbf{y} \in B_{p\text{-var}}\left(R\right), \, d_{p\text{-var}}\left(\mathbf{x}, \mathbf{y}\right) < \eta \implies d_{p\text{-var}}\left(\phi\left(\mathbf{x}\right), \phi\left(\mathbf{y}\right)\right) < \varepsilon.$$

In fact, we can choose $\eta = \delta\left(\varepsilon, CR; \omega\right)/C$ with ω defined as in Proposition 8.4 and $C = 3^{1/p}$. (This shows that if ϕ is (locally) γ-Hölder on bounded sets in the metric $d_{p\text{-}\omega}$, for all possible control functions ω, it is also (locally) γ-Hölder on bounded sets in the metric $d_{p\text{-var}}$.)

Proof. Given $\varepsilon > 0$ and $R > 0$ and $\mathbf{x}, \mathbf{y} \in B_{p\text{-var}}\left(R\right)$ we take the corresponding control ω with the properties as stated in Proposition 8.4. Taking $\eta = \delta\left(\varepsilon, 3^{1/p}R; \omega\right)/3^{1/p}$ then shows that

$$d_{p\text{-var}}\left(\mathbf{x}, \mathbf{y}\right) < \eta \implies d_{p\text{-}\omega}\left(\mathbf{x}, \mathbf{y}\right) < \delta$$

so that $d_{p\text{-}\omega}\left(\phi\left(\mathbf{x}\right), \phi\left(\mathbf{y}\right)\right) < \varepsilon$ and we conclude with (8.5). ∎

8.1.2 *Inhomogenous p-variation and Hölder distances*

Our definition of homogenous (variation, Hölder, ω-modulus) distance was based on measuring the distance of increments $\mathbf{x}_{s,t}, \mathbf{y}_{s,t} \in G^N\left(\mathbb{R}^d\right)$ using the Carnot–Caratheodory distance. Alternatively, recalling that $G^N\left(\mathbb{R}^d\right) \subset T^N\left(\mathbb{R}^d\right)$ we can use the (vector space) norm defined on the latter which leads to the distance of increments given by

$$\left|\mathbf{x}_{s,t} - \mathbf{y}_{s,t}\right|_{T^N\left(\mathbb{R}^d\right)} = \max_{k=1,\dots,N} \left|\pi_k\left(\mathbf{x}_{s,t} - \mathbf{y}_{s,t}\right)\right|.$$

Observe that, for $N > 1$, this distance is *not* homogenous with respect to dilation on $G^N\left(\mathbb{R}^d\right)$.

Definition 8.6 (inhomogenous variation and Hölder distance)
Given $\mathbf{x}, \mathbf{y} \in C\left([0,T], G^N\left(\mathbb{R}^d\right)\right)$ we define
(i) for $k = 1, \ldots, N$,

$$
\rho^{(k)}_{p\text{-var};[0,T]}(\mathbf{x}, \mathbf{y}) = \sup_{(t_i) \subset [0,T]} \left(\sum_i \left| \pi_k \left(\mathbf{x}_{t_i, t_{i+1}} - \mathbf{y}_{t_i, t_{i+1}} \right) \right|^{p/k} \right)^{k/p}
$$

and

$$
\rho_{p\text{-var};[0,T]}(\mathbf{x}, \mathbf{y}) = \max_{k=1,\ldots,N} \rho^{(k)}_{p\text{-var};[0,T]}(\mathbf{x}, \mathbf{y});
$$

(ii) for any control function ω on $[0,T]$, for $k = 1, \ldots, N$,

$$
\rho^{(k)}_{p\text{-}\omega;[0,T]}(\mathbf{x}, \mathbf{y}) = \sup_{0 \le s < t \le T} \frac{\left| \pi_k \left(\mathbf{x}_{s,t} - \mathbf{y}_{s,t} \right) \right|}{\omega(s,t)^{k/p}}
$$

and

$$
\rho_{p\text{-}\omega;[0,T]}(\mathbf{x}, \mathbf{y}) = \max_{k=1,\ldots,N} \rho^{(k)}_{p\text{-}\omega;[0,T]}(\mathbf{x}, \mathbf{y});
$$

(iii) for $k = 1, \ldots, N$,

$$
\rho^{(k)}_{1/p\text{-Höl};[0,T]}(\mathbf{x}, \mathbf{y}) = \sup_{0 \le s < t \le T} \frac{\left| \pi_k \left(\mathbf{x}_{s,t} - \mathbf{y}_{s,t} \right) \right|}{|t - s|^{k/p}}
$$

and

$$
\rho_{1/p\text{-Höl};[0,T]}(\mathbf{x}, \mathbf{y}) = \max_{k=1,\ldots,N} \rho^{(k)}_{1/p\text{-Höl};[0,T]}(\mathbf{x}, \mathbf{y}).
$$

$\square$

Some remarks are in order. By taking $\mathbf{y} \equiv o$ we have a notion of an *inhomogenous* (variation, Hölder, ω-modulus) "norm", but this will play no role in the sequel. Let us also remark that

$$
\|\mathbf{x}\|_{p\text{-var};[0,T]} = d_{p\text{-var};[0,T]}(\mathbf{x}, o) < \infty \text{ iff } \rho_{p\text{-var};[0,T]}(\mathbf{x}, o) < \infty
$$

so that $C^{p\text{-var}}\left([0,T], G^N\left(\mathbb{R}^d\right)\right)$ is precisely the set of paths $\mathbf{x}$ with finite $\rho_{p\text{-var};[0,T]}$ distance between $\mathbf{x}$ and o. The map

$$
(\mathbf{x}, \mathbf{y}) \mapsto \rho_{p\text{-var};[0,T]}(\mathbf{x}, \mathbf{y})
$$

is obviously non-negative, symmetric and one easily sees that the triangle inequality is satisfied. Then, precisely as in the previous section, adding $|\mathbf{x}_0 - \mathbf{y}_0|_{T^N(\mathbb{R}^d)}$ gives rise to a genuine metric on $C^{p\text{-var}}\left([0,T], G^N\left(\mathbb{R}^d\right)\right)$; if attention is restricted to a path with pinned starting point, $\rho_{p\text{-var};[0,T]}$

is already a metric. Of course, all this applies mutatis mutandis in a $1/p$-Hölder resp. p-ω context and $\rho_{1/p\text{-Höl}}$ resp. $\rho_{p\text{-}\omega}$ gives rise to metrics on the $C^{1/p\text{-Höl}}$ resp. $C^{p\text{-}\omega}$ spaces. Super-additivity of controls leads easily to

$$\rho_{p\text{-var};[0,T]}\left(\mathbf{x}^1,\mathbf{x}^2\right) \leq \rho_{p\text{-}\omega;[0,T]}\left(\mathbf{x}^1,\mathbf{x}^2\right) \max\left(\omega\left(0,T\right)^{1/p},\omega\left(0,T\right)^{N/p}\right),$$
(8.6)

which should be compared with (8.5).

Proposition 8.7 *For all* $\mathbf{x}^1,\mathbf{x}^2 \in C^{p\text{-var}}\left([0,T],G^N\left(\mathbb{R}^d\right)\right)$ *there exists a control* ω *with* $\omega\left(0,T\right)=1$ *and a constant* $C=C\left(p,N\right)$ *such that*

$$\rho_{p\text{-}\omega;[0,T]}(\mathbf{x}^i,o) \leq C\rho_{p\text{-var};[0,T]}(\mathbf{x}^i,o),\quad i=1,2;$$
$$\rho_{p\text{-}\omega;[0,T]}(\mathbf{x}^1,\mathbf{x}^2) \leq C\rho_{p\text{-var};[0,T]}(\mathbf{x}^1,\mathbf{x}^2).$$

Proof. Given $\mathbf{x},\mathbf{y} \in C^{p\text{-var}}\left([0,T],G^N\left(\mathbb{R}^d\right)\right)$, we define for convenience, for $k=1,\ldots,N$,

$$\omega_{\mathbf{x},\mathbf{y}}^{(k)}\left(s,t\right) = \rho_{p\text{-var};[s,t]}^{(k)}\left(\mathbf{x},\mathbf{y}\right)^{p/k}.$$

We then define

$$\begin{aligned}
\bar{\omega}^{(k)}\left(s,t\right) &= \frac{1}{3}\omega_{\mathbf{x}^1,\mathbf{x}^2}^{(k)}\left(s,t\right)/\omega_{\mathbf{x}^1,\mathbf{x}^2}^{(k)}\left(0,T\right) \\
&\quad + \frac{1}{3}\omega_{\mathbf{x}^1,o}^{(k)}\left(s,t\right)/\omega_{\mathbf{x}^1,o}^{(k)}\left(0,T\right) \\
&\quad + \frac{1}{3}\omega_{\mathbf{x}^2,o}^{(k)}\left(s,t\right)/\omega_{\mathbf{x}^2,o}^{(k)}\left(0,T\right),
\end{aligned}$$

so that $\bar{\omega}^{(k)}$ is a control with $\bar{\omega}^{(k)}\left(0,T\right)=1$, and finally

$$\omega\left(s,t\right) = \frac{1}{N}\sum_{k=1}^N \bar{\omega}^{(k)}\left(s,t\right).$$

To see that this definition of ω does the job, observe that for all $k=1,\ldots,N$ and all $0\leq s<t\leq T$

$$\begin{aligned}
\left|\pi_k\left(\mathbf{x}_{s,t}^1-\mathbf{x}_{s,t}^2\right)\right|^{p/k} &\leq \omega_{\mathbf{x}^1,\mathbf{x}^2}^{(k)}\left(s,t\right) \\
&\leq 3\bar{\omega}^{(k)}\left(s,t\right)\times\omega_{\mathbf{x}^1,\mathbf{x}^2}^{(k)}\left(0,T\right) \\
&\leq 3N\omega\left(s,t\right)\times\omega_{\mathbf{x}^1,\mathbf{x}^2}^{(k)}\left(0,T\right)
\end{aligned}$$

from which we see that

$$\begin{aligned}
\left|\pi_k\left(\mathbf{x}_{s,t}^1-\mathbf{x}_{s,t}^2\right)\right| &\leq \left(3N\right)^{k/p}\omega\left(s,t\right)^{k/p}\times\left(\omega_{\mathbf{x}^1,\mathbf{x}^2}^{(k)}\left(0,T\right)\right)^{k/p} \\
&\leq \left(3N\right)^{k/p}\omega\left(s,t\right)^{k/p}\times\max_{k=1,\ldots,N}\left[\omega_{\mathbf{x}^1,\mathbf{x}^2}^{(k)}\left(0,T\right)\right]^{k/p} \\
&\leq \left(3N\right)^{N/p}\omega\left(s,t\right)^{k/p}\rho_{p\text{-var};[0,T]}\left(\mathbf{x}^1,\mathbf{x}^2\right)
\end{aligned}$$

which says precisely that

$$\rho_{p\text{-}\omega;[0,T]}\left(\mathbf{x}^1,\mathbf{x}^2\right) \le (3N)^{N/p}\,\rho_{p\text{-var};[0,T]}\left(\mathbf{x}^1,\mathbf{x}^2\right).$$

The same argument applies to $\left(\mathbf{x}^1,o\right)$ and $\left(\mathbf{x}^2,o\right)$ and the proof is finished.   ∎

We now show that a map uniformly continuous on bounded sets in the $\rho_{p\text{-}\omega}$ metric, for all ω, is uniformly continuous on bounded sets in the $\rho_{p\text{-var}}$ metric. To this end, for paths $\mathbf{x} \in C\left([0,T],G^M\left(\mathbb{R}^d\right)\right)$ we define the (inhomogenous) balls

$$\beta_{p\text{-}\omega}\left(R\right) = \left\{\mathbf{x}: \rho_{p\text{-}\omega;[0,T]}\left(\mathbf{x},o\right) < R\right\},$$

$$\beta_{p\text{-var}}\left(R\right) = \left\{\mathbf{x}: \rho_{p\text{-var};[0,T]}\left(\mathbf{x},o\right) < R\right\}.$$

Corollary 8.8 *Consider a map*

$$\phi: C^{p\text{-var}}\left([0,T],G^M\left(\mathbb{R}^d\right)\right) \to C^{p\text{-var}}\left([0,T],G^N\left(\mathbb{R}^e\right)\right)$$

such that for any control function ω, any $\varepsilon > 0$ and $R > 0$ there exists $\delta = \delta\left(\varepsilon,R;\omega\right)$ such that

$$\mathbf{x},\mathbf{y} \in \beta_{p\text{-}\omega}\left(R\right),\ \rho_{p\text{-}\omega}\left(\mathbf{x},\mathbf{y}\right) < \delta \implies \rho_{p\text{-}\omega}\left(\phi\left(\mathbf{x}\right),\phi\left(\mathbf{y}\right)\right) < \varepsilon.$$

Then, for any $\varepsilon > 0$ and $R > 0$ there exists $\eta = \eta\left(\varepsilon,R\right)$ such that

$$\mathbf{x},\mathbf{y} \in \beta_{p\text{-var}}\left(R\right),\ \rho_{p\text{-var}}\left(\mathbf{x},\mathbf{y}\right) < \eta \implies \rho_{p\text{-var}}\left(\phi\left(\mathbf{x}\right),\phi\left(\mathbf{y}\right)\right) < \varepsilon.$$

In fact, we can choose $\eta = \delta\left(\varepsilon,CR;\omega\right)/C$ with ω defined as in Proposition 8.7 and $C = C\left(N,p\right)$. (This shows that if ϕ is (locally) γ-Hölder on bounded sets in the $\rho_{p\text{-}\omega}$ metric, for all ω, it is also (locally) γ-Hölder on bounded sets in the $\rho_{p\text{-var}}$ metric.)

Proof. Obvious from Proposition 8.7.   ∎

8.1.3 Homogenous vs inhomogenous distances

Proposition 8.9 *Let ω be a control function on $[0,T]$. For all paths $\mathbf{x},\mathbf{y}$ in $C^{p\text{-}\omega}\left([0,T],G^N\left(\mathbb{R}^d\right)\right)$*

$$d_{p\text{-}\omega}(\mathbf{x},\mathbf{y}) \le C\max\left\{\rho_{p\text{-}\omega}\left(\mathbf{x},\mathbf{y}\right),\rho_{p\text{-}\omega}\left(\mathbf{x},\mathbf{y}\right)^{1/N}\max\left\{1,\|\mathbf{x}\|_{p\text{-}\omega}^{1-\frac{1}{N}}\right\}\right\} \quad (8.7)$$

and

$$\rho_{p\text{-}\omega}\left(\mathbf{x},\mathbf{y}\right) \le C\max\left\{d_{p\text{-}\omega}(\mathbf{x},\mathbf{y})\max\left\{1,\|\mathbf{x}\|_{p\text{-}\omega}^{N-1}\right\},d_{p\text{-}\omega}(\mathbf{x},\mathbf{y})^N\right\} \quad (8.8)$$

where $C = C\left(N\right)$. The corresponding Hölder estimates are obtained by taking $\omega\left(s,t\right) = t - s$.

Proof. First we see that[2]

$$d_{p\text{-}\omega;[0,T]}(\mathbf{x},\mathbf{y}) = \sup_{0\leq s<t\leq T} d\left(\delta_{\frac{1}{\omega(s,t)^{1/p}}}\mathbf{x}_{s,t}, \delta_{\frac{1}{\omega(s,t)^{1/p}}}\mathbf{y}_{s,t}\right),$$

$$\rho_{p\text{-}\omega;[0,T]}(\mathbf{x},\mathbf{y}) = \sup_{0\leq s<t\leq T} \left|\delta_{\frac{1}{\omega(s,t)^{1/p}}}\mathbf{x}_{s,t} - \delta_{\frac{1}{\omega(s,t)^{1/p}}}\mathbf{y}_{s,t}\right|_{T^N(\mathbb{R}^d)}$$

so that these definitions indeed only differ in how to measure distance in $G^N(\mathbb{R}^d)$. A quantitative comparison of these distances was given in Proposition 7.49, an application of which finishes the proof. ∎

For a concise formulation of the next theorem, let us set

$$\tilde{d}_{p\text{-}\omega;[0,T]}(\mathbf{x},\mathbf{y}) := d_{p\text{-}\omega;[0,T]}(\mathbf{x},\mathbf{y}) + d(\mathbf{x}_0,\mathbf{y}_0)$$

$$\tilde{\rho}_{p\text{-}\omega;[0,T]}(\mathbf{x},\mathbf{y}) := \rho_{p\text{-}\omega;[0,T]}(\mathbf{x},\mathbf{y}) + \rho(\mathbf{x}_0,\mathbf{y}_0)$$

and similarly for $d_{1/p\text{-}\text{Höl}}, d_{p\text{-}\text{var}}$ and $\rho_{1/p\text{-}\text{Höl}}, \rho_{p\text{-}\text{var}}$ where we have already started to omit $[0,T]$ in the notation when no confusion is possible. We have

Theorem 8.10 *Let ω be an arbitrary control function on $[0,T]$. Each identity map*

$$\text{Id}: \left(C^{p\text{-}\omega}\left([0,T],G^N(\mathbb{R}^d)\right), \tilde{d}_{p\text{-}\omega}\right) \leftrightarrows \left(C^{p\text{-}\omega}\left([0,T],G^N(\mathbb{R}^d)\right), \tilde{\rho}_{p\text{-}\omega}\right)$$

$$\text{Id}: \left(C^{\frac{1}{p}\text{-}\text{Höl}}\left([0,T],G^N(\mathbb{R}^d)\right), \tilde{d}_{p\text{-}\text{Höl}}\right) \leftrightarrows \left(C^{\frac{1}{p}\text{-}\text{Höl}}\left([0,T],G^N(\mathbb{R}^d)\right), \tilde{\rho}_{p\text{-}\text{Höl}}\right)$$

$$\text{Id}: \left(C^{p\text{-}\text{var}}\left([0,T],G^N(\mathbb{R}^d)\right), \tilde{d}_{p\text{-}\text{var}}\right) \leftrightarrows \left(C^{p\text{-}\text{var}}\left([0,T],G^N(\mathbb{R}^d)\right), \tilde{\rho}_{p\text{-}\text{var}}\right)$$

is Lipschitz on bounded sets in $\to$ direction and $1/N$-Hölder on bounded sets in $\leftarrow$ direction.

Proof. The relevant estimates between "$d(\mathbf{x}_0,\mathbf{y}_0)$ and $\rho(\mathbf{x}_0,\mathbf{y}_0)$" follow directly from Proposition 7.49 and we focus on the path-space distance without tilde: the case of the identity map from $C^{p\text{-}\omega}\left([0,T],G^N(\mathbb{R}^d)\right)$ into itself (equipped with homogenous resp. inhomogenous distance $d_{p\text{-}\omega}$ resp. $\rho_{p\text{-}\omega}$) is covered directly by Proposition 8.9, and the case of $1/p$-Hölder paths is a special case, namely $\omega(s,t) = t - s$. We thus turn to p-variation.

$\to$ direction: Let $\mathbf{x}^1, \mathbf{x}^2 \in B_{p\text{-}\text{var}}(R)$, the "homogenous" p-variation ball of radius R as defined before Corollary 8.5, and let ω denote the corresponding control constructed in Proposition 8.4. Then, with constants c_1, c_2 which may depend on N, R, p we have

$$\rho_{p\text{-}\text{var}}\left(\mathbf{x}^1, \mathbf{x}^2\right) \leq \rho_{p\text{-}\omega}\left(\mathbf{x}^1, \mathbf{x}^2\right) \text{ by (8.6)}$$

$$\leq c_1 d_{p\text{-}\omega}\left(\mathbf{x}^1, \mathbf{x}^2\right) \text{ by (8.8)}$$

$$\leq c_2 d_{p\text{-}\text{var}}\left(\mathbf{x}^1, \mathbf{x}^2\right) \text{ by the very choice of } \omega.$$

---

[2]If $\omega(s,t) = 0$ and $\mathbf{x} \in C^{p\text{-}\omega}$ then $\mathbf{x}_{s,t} = o$, the unit element in $G^N(\mathbb{R}^d)$; we agree that $\delta_{\frac{1}{0}}o = o$.

The $\leftarrow$ direction follows the same logic, but now we rely on (8.5), (8.7) and ω as constructed in Proposition 8.7.   ■

We finish this section with a simple proposition (it will serve as a technical ingredient in our discussion of RDE smoothness later on).

Proposition 8.11 *Let* $\mathbf{A}$ *denote the canonical lift of* $A \in L\left(\mathbb{R}^n, \mathbb{R}^d\right)$ *to* $\mathrm{Hom}\left(G^N\left(\mathbb{R}^n\right), G^N\left(\mathbb{R}^d\right)\right)$. *Then, for fixed* $\mathbf{x} \in C^{p\text{-var}}\left([0,T], G^N\left(\mathbb{R}^n\right)\right)$ *the map*

$$A \in L\left(\mathbb{R}^n, \mathbb{R}^d\right) \mapsto \mathbf{A}\mathbf{x} := \{t \in [0,T] \mapsto \mathbf{A}\mathbf{x}_t\} \in C^{p\text{-var}}\left([0,T], G^N\left(\mathbb{R}^d\right)\right)$$

is continuous.

Proof. We first prove this for $\mathbf{x} \in C^{p\text{-}\omega}\left([0,T], G^N\left(\mathbb{R}^n\right)\right)$. Recall that $d_{p\text{-}\omega}$ and the inhomogenous distance $\rho_{p\text{-}\omega}$ are locally Hölder equivalent (and in particular induce the same topology). Consider $A, B \in L\left(\mathbb{R}^n, \mathbb{R}^d\right)$ with operator norm bounded by some M. Controlling $\rho_{p\text{-}\omega}\left(\mathbf{A}\mathbf{x}, \mathbf{B}\mathbf{x}\right)$ amounts to controlling $\left|(\mathbf{A}\mathbf{x})^i - (\mathbf{B}\mathbf{x})^i\right|$ for $i = 1, \ldots, N$. Every $(\mathbf{A}\mathbf{x})^i$ can be written out as a contraction of $A \otimes \cdots \otimes A$ (i times) against the i-tensor $\mathbf{x}^i$. It is then easy to see that

$$\left|(\mathbf{A}\mathbf{x})^i - (\mathbf{B}\mathbf{x})^i\right| \leq C_M |A - B| \left|\mathbf{x}^i\right|.$$

This implies that $A \mapsto \mathbf{A}\mathbf{x}$ is even Lipschitz when $C^{p\text{-var}}\left([0,T], G^N\left(\mathbb{R}^d\right)\right)$ is equipped with $\rho_{p\text{-}\omega}$. Switching to $d_{p\text{-}\omega}$ we still have continuity, and hence $d_{p\text{-var}}$ continuity.   ■

8.2   Geodesic approximations

Our interest in $G^N\left(\mathbb{R}^d\right)$-valued paths comes from the fact (cf. Section 7.2.1) that any continuous path x of finite 1-variation with values in $\mathbb{R}^d$ can be lifted to a path $\{t \mapsto S_N\left(x\right)_t \equiv S_N\left(x\right)_{0,t}\}$ with values in $G^N\left(\mathbb{R}^d\right)$ simply by computing all iterated (Riemann–Stieltjes) integrals up to order N.

It is natural to ask some sort of converse: how can an *abstract* path $\mathbf{x} : [0,T] \to G^N\left(\mathbb{R}^d\right)$ be approximated by a sequence $(S_N\left(x^n\right))$? We have

Proposition 8.12 *Let* $\mathbf{x} \in C_o^{p\text{-var}}\left([0,T], G^N\left(\mathbb{R}^d\right)\right)$, $p \geq 1$. *Then there exists* $(x^n) \subset C^{1\text{-Höl}}\left([0,T], \mathbb{R}^d\right)$, *such that*

$$d_{\infty;[0,T]}\left(\mathbf{x}, S_N\left(x^n\right)\right) \to 0 \ as \ n \to \infty$$

and

$$\sup_n \|S_N\left(x^n\right)\|_{p\text{-var};[0,T]} \leq 3^{1-1/p} \|\mathbf{x}\|_{p\text{-var};[0,T]} < \infty.$$

For $\mathbf{x} \in C_o^{1/p\text{-Höl}}\left([0,T],G^N\left(\mathbb{R}^d\right)\right)$ *we have*

$$\sup_n \|S_N\left(x^n\right)\|_{1/p\text{-Höl};[0,T]} \leq 3^{1-1/p} \|\mathbf{x}\|_{1/p\text{-Höl};[0,T]} < \infty.$$

Proof. Given the fact that $G^N\left(\mathbb{R}^d\right)$ is a geodesic space under Carnot–Caratheodory distance, this follows readily from the approximation results in geodesic spaces, Section 5.2. ∎

8.3 Completeness and non-separability

By Theorem 8.10 we can equip the space $C^{p\text{-var}}([0,T],G^N\left(\mathbb{R}^d\right))$ with either homogenous or inhomogenous p-variation distance and not only obtain the same topology but also the same "metric" notions of bounded sets or Cauchy-sequences. The same holds for Hölder paths, of course, and we have

Theorem 8.13 *(i) Let $p \geq 1$. The space $C^{p\text{-var}}([0,T],G^N\left(\mathbb{R}^d\right))$ is a complete, non-separable metric space (with respect to either homogenous or inhomogenous p-variation distance).*
(ii) The space $C^{1/p\text{-Höl}}([0,T],G^N\left(\mathbb{R}^d\right))$ is a complete, non-separable metric space (with respect to either homogenous or inhomogenous $1/p$-Hölder distance).

Proof. (i) It suffices to consider the homogenous p-variation distance, and more precisely

$$\tilde{d}_{p\text{-var}}\left(\mathbf{x},\mathbf{y}\right) \equiv d_{p\text{-var}}\left(\mathbf{x},\mathbf{y}\right) + d\left(\mathbf{x}_0,\mathbf{y}_0\right)$$

for $\mathbf{x},\mathbf{y} \in C^{p\text{-var}}([0,T],G^N\left(\mathbb{R}^d\right))$. The completeness proof follows exactly from the arguments used to establish completeness of $C^{p\text{-var}}([0,T],\mathbb{R}^d)$. Then, if $C^{p\text{-var}}([0,T],G^N\left(\mathbb{R}^d\right))$ were separable for some $N = 1,2,\ldots$ the same would be true for its projection to $N = 1$, but we know that $C^{p\text{-var}}([0,T],\mathbb{R}^d)$ is not separable, cf. Theorem 5.25 it is not.
(ii) Similar and left to the reader. ∎

8.4 The d_0/d_∞ estimate

Lemma 8.14 *Let $g,h \in G^N\left(\mathbb{R}^d\right)$. Then there exists $C = C\left(N,d\right)$ such*

that

$$\|g^{-1} \otimes h \otimes g\| \leq C \max\left\{\|h\|, \|h\|^{1/N} \|g\|^{1-1/N}\right\}.$$

Proof. Viewing g, h as elements in $T^N\left(\mathbb{R}^d\right)$, and writing $(\cdot)^n \equiv \pi_n(\cdot)$ for projection to the nth tensor level, we have for $n = 1, ..., N$

$$\left(g^{-1} \otimes h \otimes g\right)^n = \sum_{\substack{i+j+k=n \\ j>0}} \left(g^{-1}\right)^i \otimes h^j \otimes g^k \in \left(\mathbb{R}^d\right)^{\otimes n}.$$

Since every tensor level $\left(\mathbb{R}^d\right)^{\otimes n}$ is equipped with Euclidean structure, we easily see that

$$\left|\left(g^{-1} \otimes h \otimes g\right)^n\right| \leq \sum_{\substack{i+j+k=n \\ j>0}} \left|\left(g^{-1}\right)^i\right| \left|h^j\right| \left|g^k\right|.$$

Using $\left|g^k\right| \leq c_1 \|g\|^k$ and similar estimates for g^{-1}, h, also recalling $\left\|g^{-1}\right\| = \|g\|$, we find

$$\left|\left(g^{-1} \otimes h \otimes g\right)^n\right| \leq c_2 \sum_{j=1}^{n} \|g\|^{n-j} \|h\|^j$$

which implies

$$
\begin{aligned}
\left|\left(g^{-1} \otimes h \otimes g\right)^n\right|^{1/n} &\leq c_3 \max_{j=1,...,n} \|g\|^{1-j/n} \|h\|^{j/n} \\
&\leq c_3 \sup_{1/N \leq \theta \leq 1} \|g\|^{1-\theta} \|h\|^\theta.
\end{aligned}
$$

By equivalence of homogenous norms,

$$\left\|g^{-1} \otimes h \otimes g\right\| \leq c_4 \sup_{1/N \leq \theta \leq 1} \|g\|^{1-\theta} \|h\|^\theta.$$

The proof is now easily finished. ∎

Proposition 8.15 (d_0/d_∞ **estimate**) *On the path-space $C_0\left([0,1], G^N\left(\mathbb{R}^d\right)\right)$ the distances d_∞ and $d_0 \equiv d_{0\text{-Höl}}$ are locally $1/N$-Hölder equivalent. More precisely, there exists $C = C(N, d)$ such that*

$$d_\infty\left(\mathbf{x}, \mathbf{y}\right) \leq d_0\left(\mathbf{x}, \mathbf{y}\right) \leq C \max \left\{d_\infty\left(\mathbf{x}, \mathbf{y}\right), d_\infty\left(\mathbf{x}, \mathbf{y}\right)^{1/N}\right.$$

$$\left. \left(\|\mathbf{x}\|_\infty + \|\mathbf{y}\|_\infty\right)^{1-1/N}\right\}.$$

Proof. Only the second inequality requires a proof. We write gh instead of $g \otimes h$. For any $s < t$ in $[0,1]$,

$$\mathbf{x}_{s,t}^{-1}\mathbf{y}_{s,t} = \mathbf{x}_{s,t}^{-1}\mathbf{y}_s^{-1}\mathbf{x}_s\mathbf{x}_{s,t}\mathbf{x}_t^{-1}\mathbf{y}_t\mathbf{x}_t^{-1}\mathbf{y}_t\mathbf{y}_t^{-1}\mathbf{x}_t.$$

By sub-additivity,

$$
\begin{aligned}
\left\|\mathbf{x}_{s,t}^{-1}\mathbf{y}_{s,t}\right\| &\leq \left\|\mathbf{x}_{s,t}^{-1}\mathbf{y}_s^{-1}\mathbf{x}_s\mathbf{x}_{s,t}\right\| + \left\|\mathbf{x}_t^{-1}\mathbf{y}_t\mathbf{x}_t^{-1}\mathbf{y}_t\mathbf{y}_t^{-1}\mathbf{x}_t\right\| \\
&= \left\|v^{-1}\mathbf{y}_s^{-1}\mathbf{x}_s v\right\| + \left\|w^{-1}\mathbf{x}_t^{-1}\mathbf{y}_t w\right\|
\end{aligned}
$$

with $v = \mathbf{x}_{s,t}$ and $w = \mathbf{y}_t^{-1}\mathbf{x}_t$. Note that

$$\left\|\mathbf{y}_t^{-1}\mathbf{x}_t\right\| = \left\|\mathbf{x}_t^{-1}\mathbf{y}_t\right\| = d\left(\mathbf{x}_t, \mathbf{y}_t\right)$$

and $\|v\|, \|w\| \leq \|\mathbf{x}\|_\infty + \|\mathbf{y}\|_\infty$. The conclusion now follows from Lemma 8.14. ∎

8.5 Interpolation and compactness

This interpolation result will be used extensively.

Lemma 8.16 *Assume* $p' > p \geq 1$.
(i) Let $\mathbf{x}, \mathbf{y}$ *be two elements of* $C^{p\text{-var}}\left([0,T], G^N\left(\mathbb{R}^d\right)\right)$. *Then,*

$$d_{p'\text{-var}}\left(\mathbf{x}, \mathbf{y}\right) \leq \left(\|\mathbf{x}\|_{p\text{-var}} + \|\mathbf{y}\|_{p\text{-var}}\right)^{p/p'} d_0\left(\mathbf{x}, \mathbf{y}\right)^{1-p/p'}. \qquad (8.9)$$

(ii) Let $\mathbf{x}, \mathbf{y}$ *be two elements of* $C^{p\text{-}\omega}\left([0,T], G^N\left(\mathbb{R}^d\right)\right)$. *Then, for all* $p' > p$,

$$d_{p'\text{-}\omega}\left(\mathbf{x}, \mathbf{y}\right) \leq \left(\|\mathbf{x}\|_{p'\text{-}\omega} + \|\mathbf{y}\|_{p'\text{-}\omega}\right)^{p/p'} d_0\left(\mathbf{x}, \mathbf{y}\right)^{1-p/p'}. \qquad (8.10)$$

Proof. (i) Consider a dissection $(t_i) \subset [0,T]$. Then (8.9) follows from

$$
\begin{aligned}
d\left(\mathbf{x}_{t_i,t_{i+1}}, \mathbf{y}_{t_i,t_{i+1}}\right)^{p'} &\leq d\left(\mathbf{x}_{t_i,t_{i+1}}, \mathbf{y}_{t_i,t_{i+1}}\right)^p d_{0,[0,T]}\left(\mathbf{x},\mathbf{y}\right)^{p'-p} \\
&\leq \left(\left\|\mathbf{x}_{t_i,t_{i+1}}\right\| + \left\|\mathbf{y}_{t_i,t_{i+1}}\right\|\right)^p d_{0,[0,T]}\left(\mathbf{x},\mathbf{y}\right)^{p'-p},
\end{aligned}
$$

followed by summation over i, the elementary $\left(\sum |a_i + b_i|^p\right)^{1/p} \leq \left(\sum |a_i|^p\right)^{1/p} + \left(\sum |b_i|^p\right)^{1/p}$ and taking the supremum over all such dissections.
(ii) Left to the reader. ∎
    As a consequence of the above interpolation result, Lemma 5.12, and the Arzela–Ascoli theorem, we obtain the following compactness result.

Proposition 8.17 *Let* $(\mathbf{x}^n)$ *be a sequence in* $C\left([0,T], G^N\left(\mathbb{R}^d\right)\right)$.
(i) Assume $(\mathbf{x}^n)_n$ *is equicontinuous, bounded and* $\sup_n \|\mathbf{x}^n\|_{p\text{-var};[0,T]} < \infty$.
Then $\mathbf{x}^n$ *converges (in* $p' > p$ *variation, along a subsequence) to some* $\mathbf{x} \in C^{p\text{-var}}\left([0,T], G^N\left(\mathbb{R}^d\right)\right)$.
(ii) Assume $(\mathbf{x}^n)_n$ *is bounded and* $\sup_n \|\mathbf{x}^n\|_{\alpha\text{-Höl};[0,T]} < \infty$. *Then* $\mathbf{x}^n$ *converges (in* $\alpha' < \alpha$ *Hölder topology, along a subsequence) to some* $\mathbf{x} \in C^{\alpha\text{-Höl}}\left([0,T], G^N\left(\mathbb{R}^d\right)\right)$.

    The following corollary will also be useful.

Corollary 8.18 *(i) If* $(\mathbf{x}^n)$, $\mathbf{x}$ *are in* $C^{p\text{-var}}\left([0,T], G^N\left(\mathbb{R}^d\right)\right)$ *such that* $\sup_n |\mathbf{x}^n|_{p\text{-var};[0,T]} < \infty$ *and* $\lim_{n\to\infty} d_{\infty;[0,T]}\left(\mathbf{x}, \mathbf{x}^n\right) = 0$, *then for* $p' > p$,

$$\sup_{(s,t)\in\Delta_T} \left| \|\mathbf{x}^n\|_{p'\text{-var};[s,t]} - \|\mathbf{x}\|_{p'\text{-var};[s,t]} \right| \to 0 \ \text{as } n \to \infty$$

where $\Delta_T = \{(s,t) : 0 \leq s \leq t \leq T\}$. *Furthermore*, $\{\|\mathbf{x}^n\|_{p\text{'-var; }[\cdot,\cdot]} : n \in \mathbb{N}\}$ *is equicontinuous in the sense that for every* $\varepsilon > 0$ *there exists* δ *such that* $|t - s| < \delta$ *implies*

$$\sup_n \|\mathbf{x}^n\|_{p\text{'-var; }[s,t]} < \varepsilon. \qquad (8.11)$$

(ii) If $(\mathbf{x}^n)$, $\mathbf{x}$ *are in* $C^{\alpha\text{-Höl}}\left([0,T], G^N\left(\mathbb{R}^d\right)\right)$ *so that* $\sup_n \|\mathbf{x}^n\|_{\alpha\text{-Höl;}[0,T]} < \infty$ *and* $\lim_{n\to\infty} d_{\infty;[0,T]}(\mathbf{x}, \mathbf{x}^n) = 0$, *then for all* $s < t$ *in* $[0,T]$, *and for* $\alpha' < \alpha$,

$$\sup_{(s,t)\in\Delta_T} \left| \|\mathbf{x}^n\|_{\alpha'\text{-Höl;}[s,t]} - \|\mathbf{x}\|_{\alpha'\text{-Höl;}[s,t]} \right| \to 0 \ \textit{as} \ n \to \infty$$

and $\left\{ \|\mathbf{x}^n\|_{\alpha'\text{-Höl;}[\cdot,\cdot]} : n \in \mathbb{N} \right\}$ *is equicontinuous, similar to part (i).*

Proof. The argument given in the proof of Corollary 5.29 for the case of $\mathbb{R}^d$-valued paths extends line-by-line to the case of $G^N\left(\mathbb{R}^d\right)$-valued paths. $\blacksquare$

8.6 Closure of lifted smooth paths

We now define $C^{0,p\text{-var}}\left([0,T], G^N\left(\mathbb{R}^d\right)\right)$ resp. $C^{0,1/p\text{-Höl}}\left([0,T], G^N\left(\mathbb{R}^d\right)\right)$ as the closure of step-N lifted smooth paths from $[0,T] \to \mathbb{R}^d$ in p-variation resp. $1/p$-Hölder topology. A little care is needed since, by convention, $S_N(x)_0 = 1 \equiv o$, the unit element in $G^N\left(\mathbb{R}^d\right)$.

Definition 8.19 (i) We define $C_o^{0,p\text{-var}}\left([0,T], G^N\left(\mathbb{R}^d\right)\right)$ as the set of continuous paths $\mathbf{x} : [0,T] \to G^N\left(\mathbb{R}^d\right)$ for which there exists a sequence of smooth $\mathbb{R}^d$-valued paths x_n such that

$$d_{p\text{-var}}\left(\mathbf{x}, S_N(x_n)\right) \to_{n\to\infty} 0$$

and $C^{0,p\text{-var}}\left([0,T], G^N\left(\mathbb{R}^d\right)\right)$ as the set of paths $\mathbf{x}$ with

$$\mathbf{x}_{0,\cdot} = \mathbf{x}_0^{-1} \otimes \mathbf{x}. \in C_o^{0,p\text{-var}}\left([0,T], G^N\left(\mathbb{R}^d\right)\right).$$

(ii) Similarly, $C_o^{0,1/p\text{-Höl}}\left([0,T], G^N\left(\mathbb{R}^d\right)\right)$ is the set of paths $\mathbf{x}$ for which there exists a sequence of smooth $\mathbb{R}^d$-valued paths x_n such that

$$d_{1/p\text{-Höl}}\left(\mathbf{x}, S_N(x_n)\right) \to_{n\to\infty} 0$$

and $C^{0,1/p\text{-Höl}}\left([0,T], G^N\left(\mathbb{R}^d\right)\right)$ are those paths $\mathbf{x}$ with $\mathbf{x}_{0,\cdot} \in C_o^{0,1/p\text{-Höl}}\left([0,T], G^N\left(\mathbb{R}^d\right)\right)$. $\square$

Obviously, all these spaces are closed subsets of $C^{p\text{-var}}\left([0,T],G^N\left(\mathbb{R}^d\right)\right)$ resp. $C^{1/p\text{-H\"ol}}\left([0,T],G^N\left(\mathbb{R}^d\right)\right)$ and thus complete. Proposition 7.63 implies a fortiori continuity of S_N, as a map from $C_o^{1\text{-var}}\left([0,T],\mathbb{R}^d\right)$ to $C_o^{1\text{-var}}\left([0,T],G^N\left(\mathbb{R}^d\right)\right)$. Clearly then

$$C^{0,1\text{-var}}\left([0,T],G^N\left(\mathbb{R}^d\right)\right) = S_N\left(C^{0,1\text{-var}}\left([0,T],\mathbb{R}^d\right)\right) \qquad (8.12)$$

and also $C^{0,1\text{-H\"ol}}\left([0,T],G^N\left(\mathbb{R}^d\right)\right) = S_N\left(C^1\left([0,T],\mathbb{R}^d\right)\right)$. The reader will recall from Proposition 1.32 that $C^{0,1\text{-var}}\left([0,T],\mathbb{R}^d\right)$, defined as a 1-variation closure of smooth paths, turned out to be precisely the space of absolutely continuous paths (with respect to Euclidean metric on $\mathbb{R}^d$); in Exercise 8.20 it is seen that the same is true in the case of $G^N\left(\mathbb{R}^d\right)$-valued paths.

Exercise 8.20 *Show that $C^{0,1\text{-var}}\left([0,T],G^N\left(\mathbb{R}^d\right)\right)$ is precisely the space of absolutely continuous paths (with respect to Carnot–Caratheodory metric on $G^N\left(\mathbb{R}^d\right)$).*

Solution. It suffices to consider $\mathbf{x} \in C_o^{0,1\text{-var}}\left([0,T],G^N\left(\mathbb{R}^d\right)\right)$. Any such $\mathbf{x}\left(\cdot\right)$ is of the form $S_N\left(x\right)$, where x is an $\mathbb{R}^d$-valued absolutely continuous path. Hence, for every $\varepsilon > 0$ and $s_1 < t_1 \le s_2 < t_2 \le \cdots < s_n < t_n$ in $[0,T]$ there exists δ so that $\sum_i |t_i - s_i| < \delta$ implies $\sum_i |x_{s_i,t_i}| < \varepsilon$ and in fact (cf. Exercise 5.15)

$$\sum_i |x|_{1\text{-var};[s_i,t_i]} < \varepsilon.$$

But then, thanks to Proposition 7.59

$$\sum_i |d\left(x_{s_i},x_{t_i}\right)| \le \sum_i |\mathbf{x}|_{1\text{-var};[s_i,t_i]} = \sum_i |x|_{1\text{-var};[s_i,t_i]} < \varepsilon.$$

Lemma 8.21 *We fix $p > 1$.*
(i) If $\Omega \subset C^{1\text{-var}}\left([0,T],G^N\left(\mathbb{R}^d\right)\right)$ and if $C^{0,1\text{-var}}\left([0,T],G^N\left(\mathbb{R}^d\right)\right)$ is included in the $d_{1\text{-var}}$-closure of Ω, then the $d_{p\text{-var}}$-closure of Ω is equal to $C^{0,p\text{-var}}\left([0,T],G^N\left(\mathbb{R}^d\right)\right)$.
(ii) If $\Omega \subset C^{1\text{-H\"ol}}\left([0,T],G^N\left(\mathbb{R}^d\right)\right)$ and if $C^{0,1\text{-H\"ol}}\left([0,T],G^N\left(\mathbb{R}^d\right)\right)$ is included in the $d_{1\text{-H\"ol}}$-closure of Ω, then the $d_{1/p\text{-H\"ol}}$-closure of Ω is equal to $C^{0,1/p\text{-H\"ol}}\left([0,T],G^N\left(\mathbb{R}^d\right)\right)$.

Proof. Same proof as the $N = 1$ case (cf. Lemma 5.30). ∎

We can now extend "Wiener's characterization" from the $\mathbb{R}^d$ setting (Theorem 8.22) to the group setting. Recall in particular that $\mathbf{x}^D$ denotes the geodesic approximation to $\mathbf{x}$ based on some dissection $D = (t_i)$ of $[0,T]$. That is, $\mathbf{x}_{t_i}^D = \mathbf{x}_{t_i}$ for all i and $\mathbf{x}^D|_{[t_i,t_{i+1}]}$ is a geodesic connecting $\mathbf{x}_{t_i}$ and $\mathbf{x}_{t_{i+1}}$ as in Proposition 7.42. The proof of the $\mathbb{R}^d$ case then extends without any changes and we have

Theorem 8.22 (Wiener's characterization) *Let* $\mathbf{x} \in C^{p\text{-var}}\left([0,T], G^N\left(\mathbb{R}^d\right)\right)$, *with* $p > 1$. *The following statements are equivalent.*

(i.1) $\mathbf{x} \in C^{0,p\text{-var}}([0,T], G^N\left(\mathbb{R}^d\right))$.

(i.2a) $\lim_{\delta \to 0} \sup_{D=(t_i),|D|<\delta} \sum_i \|\mathbf{x}\|^p_{p\text{-var};[t_i,t_{i+1}]} = 0$.

(i.2b) $\lim_{\delta \to 0} \sup_{D=(t_i),|D|<\delta} \sum_i d\left(\mathbf{x}_{t_i}, \mathbf{x}_{t_{i+1}}\right)^p = 0$.

(i.3) $\lim_{|D|\to 0} d_{p\text{-var}}\left(\mathbf{x}^D, \mathbf{x}\right) = 0$.

Secondly, let $\mathbf{x} \in C^{1/p\text{-Höl}}\left([0,T], G^N\left(\mathbb{R}^d\right)\right)$, *with* $p > 1$. *The following statements are equivalent.*

(ii.1) $\mathbf{x} \in C^{0,1/p\text{-Höl}}\left([0,T],\mathbb{R}^d\right)$.

(ii.2a) $\lim_{\delta \to 0} \sup_{|t-s|<\delta} \|\mathbf{x}\|_{1/p\text{-Höl};[s,t]} = 0$.

(ii.2b) $\lim_{\delta \to 0} \sup_{|t-s|<\delta} d(\mathbf{x}_s, \mathbf{x}_t)/|t-s|^{1/p} = 0$.

(ii.3) $\lim_{|D|\to 0} d_{1/p\text{-Höl}}\left(\mathbf{x}^D, \mathbf{x}\right) = 0$.

Exercise 8.23 *Let* $p \geq 1$. *Show that* $C^{0,p\text{-var}}\left([0,T], G^N\left(\mathbb{R}^d\right)\right)$ *is precisely the space of paths which are absolutely continuous of order* p.

Solution. *The case* $p = 1$ *was dealt with in Exercise 8.20. For* $p > 1$ *we simply combine the result of Exercise 5.15 with Wiener's characterization.*

Corollary 8.24 *For* $p > 1$, *we have the following set inclusions:*

$$\bigcup_{1 \leq q < p} C^{q\text{-var}}\left([0,T], G^N\left(\mathbb{R}^d\right)\right) \quad \subset \quad C^{0,p\text{-var}}\left([0,T], G^N\left(\mathbb{R}^d\right)\right)$$

$$\subset \quad C^{p\text{-var}}\left([0,T], G^N\left(\mathbb{R}^d\right)\right)$$

$$\subset \quad \bigcap_{q>p} C^{q\text{-var}}\left([0,T], G^N\left(\mathbb{R}^d\right)\right).$$

Proof. Similar to Corollary 5.33. ∎

Proposition 8.25 *Let* $p \geq 1$. *The spaces*

$$C^{0,p\text{-var}}\left([0,T], G^N\left(\mathbb{R}^d\right)\right), C^{0,1/p\text{-Höl}}\left([0,T], G^N\left(\mathbb{R}^d\right)\right)$$

are Polish with respect to either homogenous or inhomogenous p-variation resp. 1/p-Hölder distance.

Proof. As remarked in Section 8.3, either choice (homogenous, inhomogenous) of p-variation distance leads to the same topology and notion of Cauchy-sequence. Clearly then, $C^{0,p\text{-var}}$ is complete under either distance. It remains to discuss separability. From Corollary 1.35, there exists a countable space Ω dense in $C^{0,1\text{-var}}\left([0,T],\mathbb{R}^d\right)$; by continuity of S_N, $S_N(\Omega)$ is dense in $C^{0,1\text{-var}}\left([0,T], G^N\left(\mathbb{R}^d\right)\right)$. We conclude using Lemma 5.30. Similar arguments for $1/p$-Hölder spaces are left to the reader. ∎

8.7 Comments

Section 8.1 introduces the basic path space distance for $G^N\left(\mathbb{R}^d\right)$-valued paths (homogenous such as $d_{p\text{-var}}$, inhomogenous such as $\rho_{p\text{-var}}$). As will be discussed in detail in the next chapter, if one chooses $N = [p]$ these distances are "rough path" distances. Noting that both $d_{p\text{-var}}$ and $\rho_{p\text{-var}}$ induce the same topology, both notions are useful. Typical rough path continuity statements are locally Lipschitz continuous in the inhomogenous distance (which is also the distance put forward by Lyons, e.g. Lyons and Qian [120] in the references cited therein). The homogenous distance, on the other hand, comes in handy when establishing large deviation results via exponentially good approximations (as seen in Lemma 13.40 for instance); not to mention its general convenience, which often allows us to write arguments in the same way as for paths on Euclidean space.

The geodesic approximation result of **Section 8.2** appeared in Friz and Victoir [63]; here it is derived as a special case of a general approximation result on geodesic spaces. **Section 8.3** follows Friz and Victoir [63]. The d_0/d_∞ estimate in **Section 8.4** is taken from Friz and Victoir [61]; **Sections 8.5** and **8.6** follow Friz and Victoir [63].

9

Geometric rough path spaces

We have studied $G^N\left(\mathbb{R}^d\right)$-valued paths of finite p-variation for any $N \in \mathbb{N}$ and $p \geq 1$. If one thinks of $G^N\left(\mathbb{R}^d\right)$ as the correct state-space that allows us not only to keep track of the spatial position in $\mathbb{R}^d$ but also to keep track of the accumulated area (and higher indefinite iterated integrals up to order N) then it should not be surprising that N and p stand in some canonical relation. To wit, if $p = 1$, knowledge of the $\mathbb{R}^d$-valued path allows us to compute iterated integrals by Riemann–Stieltjes theory and there is no need to include area and other iterated integrals in the state-space. The same remark applies, more generally, to $p \in [1, 2)$ using iterated Young integration and in this case we should take $N = [p] = 1$. When $p \geq 2$, this is not possible and knowledge of higher indefinite iterated integrals up to order $N = [p]$ must be an a priori information, i.e. assumed to be known.[1]

However, we shall establish that integrals of order greater than $[p]$ are still canonically determined. More precisely, we shall see that for $N \geq [p]$ there exists a canonical bijection[2]

$$S_N : C_o^{p\text{-var}}\left([0,T], G^{[p]}\left(\mathbb{R}^d\right)\right) \to C_o^{p\text{-var}}\left([0,T], G^N\left(\mathbb{R}^d\right)\right)$$

such that for all $\mathbf{x} \in C_o^{p\text{-var}}\left([0,T], G^{[p]}\left(\mathbb{R}^d\right)\right)$ we have

$$\|\mathbf{x}\|_{p\text{-var};[0,T]} \leq \|S_N\left(\mathbf{x}\right)\|_{p\text{-var};[0,T]} \leq C_N \|\mathbf{x}\|_{p\text{-var};[0,T]} .$$

The analogous $1/p$-Hölder estimate also holds, and is a consequence of the p-variation estimate. Indeed, by reparametrization, $[0,T]$ may be replaced by $[s,t]$ so that the Hölder statement follows trivially from $\|\mathbf{x}\|_{p\text{-var};[s,t]} \leq \|\mathbf{x}\|_{1/p\text{-Höl}} |t - s|^{1/p}$.

This gives a first hint on the importance of these so-called *(weak) geometric rough paths* whose regularity (p-variation) is in relation to their state-space $\left(G^{[p]}\left(\mathbb{R}^d\right)\right)$.

---

[1] In a typical probabilistic situation (cf. Part III of this book), $N = 2$ or 3 and the required iterated integrals will be constructed via some *stochastic* integration procedure.

[2] Recall that o in $C_o^{p\text{-var}}$ indicates that all paths start at the unit element of $G^{[p]}$.

9.1   The Lyons-lift map $x \mapsto S_N(x)$

9.1.1   *Quantitative bound on S_N*

We start with two simple technical lemmas.

Lemma 9.1 *Let $g_1, g_2 \in T^{N+1}\left(\mathbb{R}^d\right)$. Then,*

$$g_1 \otimes g_2 - (g_1 + g_2) = \pi_{0,N}\left(g_1\right) \otimes \pi_{0,N}\left(g_2\right) - \pi_{0,N}\left(g_1 + g_2\right).$$

In particular, if $h_1, h_2 \in T^{N+1}\left(\mathbb{R}^d\right)$ are such that $\pi_{0,N}\left(g_1\right) = \pi_{0,N}\left(h_1\right)$ and $\pi_{0,N}\left(g_2\right) = \pi_{0,N}\left(h_2\right)$, we have

$$g_1 \otimes g_2 - (g_1 + g_2) = h_1 \otimes h_2 - (h_1 + h_2).$$

Proof. Simple algebra.   ∎

Lemma 9.2 *Let $x^1, x^2 \in C^{1\text{-var}}\left([s,u],\mathbb{R}^d\right)$, such that for some $N \geq 1$, $S_N\left(x^1\right)_{s,u} = S_N\left(x^2\right)_{s,u}$. Then, if $\ell \geq \max\left\{\int_s^u \left|dx^1\right|, \int_s^u \left|dx^2\right|\right\}$, we have for some constant C depending only on N,*

$$\left|S_{N+1}\left(x^1\right)_{s,u} - S_{N+1}\left(x^2\right)_{s,u}\right| \leq C \ell^{N+1}. \tag{9.1}$$

Proof. By assumption,

$$\left|S_{N+1}\left(x^1\right)_{s,u} - S_{N+1}\left(x^2\right)_{s,u}\right| = \left|\pi_{N+1}\left(S_{N+1}\left(x^1\right)_{s,u} - S_{N+1}\left(x^2\right)_{s,u}\right)\right|$$

and (9.1) follows from

$$\left|\pi_{N+1}\left(S_{N+1}\left(x^i\right)_{s,u}\right)\right| \leq \frac{1}{(N+1)!}\left(\int_s^u \left|dx^i\right|\right)^{N+1}, \quad i = 1, 2.$$

Alternative proof of (9.1) which introduces the useful idea of "reducing two paths to one path": Without loss of generality, assume $(s,u) = (0,1)$ and observe that $S_N\left(x^1\right)_{0,1} = S_N\left(x^2\right)_{0,1}$ implies

$$S_{N+1}\left(x^1\right)_{0,1} - S_{N+1}\left(x^2\right)_{0,1} = S_{N+1}\left(x^1\right)_{0,1} \otimes S_{N+1}\left(x^2\right)_{0,1}^{-1} - 1.$$

Define $x = x^1 \sqcup \overleftarrow{x^2}$, i.e. as the concatenation of $x^1\left(\cdot\right)$ and $x^2\left(1 - \cdot\right)$ and assume x is (re-)parametrized on $[0,1]$. It follows that

$$S_{N+1}\left(x^1\right)_{0,1} - S_{N+1}\left(x^2\right)_{0,1} = S_{N+1}\left(x\right)_{0,1} - 1,$$

with $S_N\left(x\right)_{0,1} = 1$ and $|x|_{1\text{-var};[0,1]} = \left|x^1\right|_{1\text{-var};[0,1]} + \left|x^2\right|_{1\text{-var};[0,1]} \leq 2\ell$. But then

$$\left|S_{N+1}\left(x\right)_{0,1} - 1\right| = \left|\pi_{N+1}\left(S_{N+1}\left(x\right)_{0,1}\right)\right| \quad \text{from } S_N\left(x\right)_{0,1} = 1$$

$$\leq \frac{1}{(N+1)!}(2\ell)^{N+1}$$

and (9.1) follows.   ∎

We are now ready for the crucial quantitative estimate on $\|S_N(x)\|_{p\text{-var}}$ for $N \geq p$.

Proposition 9.3 *Let* $x \in C^{1\text{-var}}\left([0,T],\mathbb{R}^d\right)$. *Then, for all* $N \geq [p]$, *there exists a constant* C *depending only on* N *and* p *(and **not** depending on the 1-variation norm of* x *nor its p-variation) such that for all* $s < t$ *in* $[0,T]$,

$$\|S_N(x)\|_{p\text{-var};[s,t]} \leq C\|S_{[p]}(x)\|_{p\text{-var};[s,t]}. \tag{9.2}$$

The constant C *can be chosen to be right-continuous with respect to* p.

Proof. It is enough to show that for all $N \geq [p]$,

$$\|S_{N+1}(x)\|_{p\text{-var};[s,t]} \leq c_1(p)\|S_N(x)\|_{p\text{-var};[s,t]}, \tag{9.3}$$

where $p \mapsto c_1(p)$ is right continuous. Explicit dependency on p in our constant will be written in this proof.

We define $\mathbf{x} = S_N(x)$, $\mathbf{y} = S_{N+1}(x)$ and $\omega(s,t) = \|\mathbf{x}\|_{p-\text{var};[s,t]}^p$. By Theorem 7.32 we can find, for all $s < t$ in $[0,T]$, a "geodesic" path $x^{s,t}$: $[s,t] \to \mathbb{R}^d$ associated with $\mathbf{x}_{s,t} \in G^N\left(\mathbb{R}^d\right)$ which is the shortest (Lipschitz) path which has step-N signature equal to $\mathbf{x}_{s,t}$. Then, define

$$\Gamma_{s,t} = \mathbf{y}_{s,t} - S_{N+1}\left(x^{s,t}\right)_{s,t}.$$

For $s < t < u$, we also define $x^{s,t,u}$ to be the concatenation of $x^{s,t}$ and $x^{t,u}$, and observe that $S_{N+1}\left(x^{s,t,u}\right)_{s,u} = S_{N+1}\left(x^{s,t}\right)_{s,t} \otimes S_{N+1}\left(x^{t,u}\right)_{t,u}$. Then, as $\mathbf{y}_{s,t} \otimes \mathbf{y}_{t,u} = \mathbf{y}_{s,u}$,

$$\begin{aligned}
\Gamma_{s,u} - (\Gamma_{s,t} + \Gamma_{t,u}) &= \mathbf{y}_{s,t} \otimes \mathbf{y}_{t,u} - (\mathbf{y}_{s,t} + \mathbf{y}_{t,u}) \\
&\quad - S_{N+1}\left(x^{s,t}\right)_{s,t} \otimes S_{N+1}\left(x^{t,u}\right)_{t,u} \\
&\quad + S_{N+1}\left(x^{s,t}\right)_{s,t} + S_{N+1}\left(x^{t,u}\right)_{t,u} \\
&\quad + S_{N+1}\left(x^{s,t}\right)_{s,t} \otimes S_{N+1}\left(x^{t,u}\right)_{t,u} - S_{N+1}\left(x^{s,u}\right)_{s,u}.
\end{aligned}$$

By construction, for all $k \leq N$ and all $s,t \in [0,T]$, $\pi_k(\mathbf{y}_{s,t}) = \pi_k\left(S_{N+1}\left(x^{s,t}\right)_{s,t}\right)$, hence we can apply Lemma 9.1 to see that

$$\begin{aligned}
\mathbf{y}_{s,t} \otimes \mathbf{y}_{t,u} - (\mathbf{y}_{s,t} + \mathbf{y}_{t,u}) &= S_{N+1}\left(x^{s,t}\right)_{s,t} \otimes S_{N+1}\left(x^{t,u}\right)_{t,u} \\
&\quad - \left(S_{N+1}\left(x^{s,t}\right)_{s,t} + S_{N+1}\left(x^{t,u}\right)_{t,u}\right).
\end{aligned}$$

Hence, we are left with

$$\Gamma_{s,u} - (\Gamma_{s,t} + \Gamma_{t,u}) = S_{N+1}\left(x^{s,t,u}\right)_{s,u} - S_{N+1}\left(x^{s,u}\right)_{s,u},$$

which we bound using Lemma 9.2:

$$|\Gamma_{s,u} - (\Gamma_{s,t} + \Gamma_{t,u})| \leq c_1 \max\left\{ \int_s^u |dx_r^{s,t,u}|, \int_s^u |dx_r^{s,u}| \right\}^{N+1}$$

$$\leq c_2 \omega(s,u)^{\frac{N+1}{p}}. \tag{9.4}$$

Secondly, using $\int_s^t |dx^{s,t}| \leq \int_s^t |dx|$, Lemma 9.2 gives

$$|\Gamma_{s,t}| = \left| \mathbf{y}_{s,t} - S_{N+1}\left(x^{s,t}\right)_{s,t} \right| \leq c_3 |x|_{1\text{-var};[s,t]}^{N+1} =: \tilde{\omega}(s,t)^{N+1}.$$

The last two inequalities allow us to apply the same (analysis) Lemma 6.2 (which we have already used for establishing the Young–Lóeve estimate). We get, for all $0 \leq s < t \leq T$,

$$\left| \mathbf{y}_{s,t} - S_{N+1}\left(x^{s,t}\right)_{s,t} \right| \leq c_4 C_{N,p} \, \omega(s,t)^{\frac{N+1}{p}}$$

where $C_{N,p}$ can be taken to be $1/(1 - 2^{1 - \frac{N+1}{p}})$. This implies by the triangle inequality that

$$|\pi_{N+1}(\mathbf{y}_{s,t})| \leq c_4 C_{N,p}\omega(s,t)^{\frac{N+1}{p}} + \left| \pi_{N+1}\left(S_{N+1}\left(x^{s,t}\right)_{s,t} \right) \right|$$

$$\leq c_5(1 + C_{N,p})\,\omega(s,t)^{\frac{N+1}{p}},$$

and hence, using the equivalence of homogenous norms, that for all $s,t \in [0,T]$,

$$\|\mathbf{y}_{s,t}\| \leq c_6 \left(1 + (1 + C_{N,p})^{1/p}\right) \omega(s,t)^{1/p}.$$

This means that

$$\|\mathbf{y}\|_{p\text{-var};[s,t]} \leq c_6 \left(1 + (1 + C_{N,p})^{1/p}\right) \|\mathbf{x}\|_{p\text{-var};[s,t]}$$

for all $0 \leq s < t \leq T$ and the proof is finished. ∎

9.1.2 Definition of the map S_N on $C_o^{p\text{-var}}\left([0,T], G^{[p]}\left(\mathbb{R}^d\right)\right)$

Definition 9.4 Let $N \geq [p] \geq 1$ and $\mathbf{x} \in C_o^{p\text{-var}}\left([0,T], G^{[p]}\left(\mathbb{R}^d\right)\right)$. A path in $C_o^{p\text{-var}}\left([0,T], G^N\left(\mathbb{R}^d\right)\right)$ that projects down onto $\mathbf{x}$ is said to be a p-Lyons lift of $\mathbf{x}$ of order N. When p is fixed, we will simply speak of a Lyons lift. □

We now show that there exists a unique Lyons lift of order N, for all $\mathbf{x} \in C_o^{p\text{-var}}\left([0,T], G^{[p]}\left(\mathbb{R}^d\right)\right)$. In the terminology of the forthcoming Definition 9.15 this says precisely that a *weak geometric p-rough path* admits a unique Lyons lift.

Theorem 9.5 *Let $N \geq [p] \geq 1$ and $\mathbf{x} \in C_o^{p\text{-var}}\left([0,T], G^{[p]}\left(\mathbb{R}^d\right)\right)$. Then there exists a unique Lyons lift of order N of $\mathbf{x}$. By writing $S_N(\mathbf{x})$ for this path, we define the map S_N on $C_o^{p\text{-var}}\left([0,T], G^{[p]}\left(\mathbb{R}^d\right)\right)$. Moreover,*
(i) the map $S_N : C_o^{p\text{-var}}\left([0,T], G^{[p]}\left(\mathbb{R}^d\right)\right) \to C_o^{p\text{-var}}\left([0,T], G^N\left(\mathbb{R}^d\right)\right)$ is a bijection with inverse $\pi_{0,[p]}$;
(ii) we have for some constant $C = C(N,p)$, which may be taken right-continuous in p,

$$\|\mathbf{x}\|_{p\text{-var};[s,t]} \leq \|S_N(\mathbf{x})\|_{p\text{-var};[s,t]} \leq C \|\mathbf{x}\|_{p\text{-var};[s,t]} \qquad (9.5)$$

for all $s < t$ in $[0,T]$.

Remark 9.6 Observe that if $\mathbf{x} \in C^{q\text{-var}}\left([0,T], G^{[q]}\left(\mathbb{R}^d\right)\right)$ for $q \leq p$, $S_N\left(S_{[p]}(\mathbf{x})\right) = S_N(\mathbf{x})$. This justifies using the same notation for all p, and in particular the same notation for $p > 1$ and $p = 1$. For further convenience, we will define for $N \leq [p]$ and for $\mathbf{x} \in C^{p\text{-var}}\left([0,T], G^{[p]}\left(\mathbb{R}^d\right)\right)$ the path $S_N(\mathbf{x})$ which is just the projection of $\mathbf{x}$ onto $G^N\left(\mathbb{R}^d\right)$. In particular, the estimate $\|S_N(\mathbf{x})\|_{p\text{-var};[s,t]} \leq C \|\mathbf{x}\|_{p\text{-var};[s,t]}$ still holds for $N \leq [p]$. □

Proof. <u>First step, existence:</u> Let x_n be a sequence in $C^{1\text{-var}}\left([0,T], \mathbb{R}^d\right)$ such that

$$\sup_n \left\|S_{[p]}(x_n)\right\|_{p\text{-var};[0,T]} < \infty \text{ and } \lim_{n\to\infty} d_\infty\left(S_{[p]}(x_n), \mathbf{x}\right) = 0.$$

By Proposition 9.3, for all $s < t$ in $[0,T]$ and ε small enough (namely, such that $[p+\varepsilon] = [p]$),

$$\left\|S_N(x_n)_{s,t}\right\| \leq c_{p+\varepsilon} \left\|S_{[p]}(x_n)\right\|_{(p+\varepsilon)\text{-var};[s,t]}.$$

By Corollary 8.18 the right-hand side above can be made arbitrarily small, uniformly in n, provided $t - s$ is small enough; this implies readily that $S_N(x_n)$ is equicontinuous. Boundedness is clear and so, by Arzela–Ascoli and switching to a subsequence if necessary, we have the existence of a continuous $G^N\left(\mathbb{R}^d\right)$-valued path $\mathbf{z}$ such that

$$S_N(x^n) \to \mathbf{z} \text{ uniformly on } [0,T] \text{ as } n \to \infty.$$

From the very choice of (x_n) it then follows that the projection of $\mathbf{z}$ to a $G^{[p]}\left(\mathbb{R}^d\right)$-valued path must be equal to $\mathbf{x}$. Then, for all $0 \leq s < t \leq T$,

$$
\begin{aligned}
\|\mathbf{z}_{s,t}\| &= \lim_{n\to\infty} \left\|S_N(x^n)_{s,t}\right\| \\
&\leq \varlimsup_{n\to\infty} \|S_N(x^n)\|_{(p+\varepsilon)\text{-var};[s,t]} \\
&\leq c_{p+\varepsilon} \varlimsup_{n\to\infty} \left\|S_{[p]}(x^n)\right\|_{(p+\varepsilon)\text{-var};[s,t]}
\end{aligned}
$$

where we used (9.2), Proposition 9.3, for the last estimate. On the other hand, from the first part of Corollary 8.18,

$$\lim_{n \to \infty} \left\| S_{[p]}(x^n) \right\|_{(p+\varepsilon)\text{-var};[s,t]} = \|\mathbf{x}\|_{(p+\varepsilon)\text{-var};[s,t]}$$

and hence, for all $0 \le s < t \le T$,

$$\|\mathbf{z}_{s,t}\| \le c_{p+\varepsilon} \|\mathbf{x}\|_{(p+\varepsilon)\text{-var};[s,t]}.$$

Using "right-continuity" of $p \mapsto c_p$ (Proposition 9.3) and also right-continuity of the homogenous p-variation norm with respect to p (Lemma 5.13) we may send $\varepsilon \to 0$ to obtain

$$\|\mathbf{z}_{s,t}\| \le c_p \|\mathbf{x}\|_{p\text{-var};[s,t]}.$$

Super-additivity of $(s,t) \mapsto \|\mathbf{x}\|_{p\text{-var};[s,t]}^p$ implies that

$$\|\mathbf{z}\|_{p\text{-var};[s,t]} \le c_p \|\mathbf{x}\|_{p\text{-var};[s,t]};$$

the converse estimate $\|\mathbf{x}\|_{p\text{-var};[s,t]} \le \|\mathbf{z}\|_{p\text{-var};[s,t]}$ is trivial since we know that $\mathbf{z}$ lifts $\mathbf{x}$.

In particular, we found a (Lyons) lift of $\mathbf{x}$ in $C_o^{p\text{-var}}\left([0,T], G^N(\mathbb{R}^d)\right)$, which satisfies (9.5).

Second step, uniqueness: Given $\mathbf{z}, \tilde{\mathbf{z}} \in C_o^{p\text{-var}}\left([0,T], G^{M+1}(\mathbb{R}^d)\right)$ with $\pi_{0,M}(\mathbf{z}) \equiv \pi_{0,M}(\tilde{\mathbf{z}})$ we show (by induction in M) that $M \ge [p]$ implies $\mathbf{z} \equiv \tilde{\mathbf{z}}$. From Lemma 7.61, $h_t = \log\left(\mathbf{z}_t^{-1} \otimes \tilde{\mathbf{z}}_t\right)$ defines a path in $\mathfrak{g}^{M+1}(\mathbb{R}^d) \cap \left(\mathbb{R}^d\right)^{\otimes(M+1)}$, and for all $s,t \in [0,T]$,

$$|h_{s,t}| \le c_1 \|\mathbf{z}_{s,t}\|^{M+1} + c_1 \|\tilde{\mathbf{z}}_{s,t}\|^{M+1}.$$

We define the control $\omega(s,t) = \|\mathbf{z}\|_{p\text{-var};[s,t]}^p + \|\tilde{\mathbf{z}}\|_{p\text{-var};[s,t]}^p$. The previous inequality now reads

$$|h_{s,t}| \le c_2 \omega(s,t)^{(M+1)/p}.$$

In particular, h is of finite $\frac{p}{M+1}$-variation. As $\frac{p}{M+1} < 1$, we deduce that h is constant equal to $h_0 = 0$, i.e. that $\mathbf{z} = \tilde{\mathbf{z}}$, which is what we wanted to show.

Third step: It remains to see that, as stated in (i), S_N is a bijection with inverse $\pi_{0,[p]}$. Obviously, $\pi_{0,[p]} \circ S_N$ is the identity map on $C_o^{p\text{-var}}\left([0,T], G^{[p]}(\mathbb{R}^d)\right)$. Conversely, given $\mathbf{x} \in C_o^{p\text{-var}}\left([0,T], G^N(\mathbb{R}^d)\right)$ it is clear from (9.5) that $S_N \circ \pi_{0,[p]}(\mathbf{x})$ has finite p-variation. By uniqueness, we see that $S_N \circ \pi_{0,[p]}(\mathbf{x}) = \mathbf{x}$ so that $S_N \circ \pi_{0,[p]}$ acts as identity map on $C_o^{p\text{-var}}\left([0,T], G^N(\mathbb{R}^d)\right)$. This completes the proof. ∎

Exercise 9.7 *Let $N \geq [p] \geq 1$ and $\mathbf{x} \in C_o^{p\text{-var}}\left([0,T], G^{[p]}\left(\mathbb{R}^d\right)\right)$. Prove that there exists $\mathbf{z} \in C_o\left([0,T], G^N\left(\mathbb{R}^d\right)\right)$ such that for all sequences $(x^n) \subset C^{1\text{-var}}\left([0,T], \mathbb{R}^d\right)$ such that*

$$d_{\infty;[0,T]}\left(S_{[p]}\left(x^n\right), \mathbf{x}\right) \to 0 \text{ and } \sup_n \left|S_{[p]}\left(x^n\right)\right|_{p\text{-var};[0,T]} < \infty,$$

we have

$$S_N\left(x^n\right) \to \mathbf{z} \text{ uniformly on } [0,T] \text{ as } n \to \infty.$$

(This exercise shows that we could have defined the Lyons lift as a limit, similar to our definition of the Young integral and the forthcoming definition of the solution to a rough differential equation.)

9.1.3 Modulus of continuity for the map S_N

We shall now establish that the Lyons-lifting map is locally Lipschitz continuous with respect to inhomogenous rough path distances on pathspace. To this end, we need the following two lemmas.

Lemma 9.8 *Let $N \geq [p] \geq 1$ and $\mathbf{x} \in C_o^{p\text{-var}}\left([s,t], G^{[p]}\left(\mathbb{R}^d\right)\right)$, and $x^{s,t} \in C^{1\text{-var}}\left([s,t], \mathbb{R}^d\right)$ with $\int_s^t \left|dx^{s,t}\right| \leq K \|\mathbf{x}\|_{p\text{-var};[s,t]}$ such that $S_N\left(x^{s,t}\right)_{s,t} = S_N\left(\mathbf{x}_{s,t}\right)$. Then, for some constant C depending only on K, N and p,*

$$\left|S_{N+1}(\mathbf{x})_{s,t} - S_{N+1}\left(x^{s,t}\right)_{s,t}\right| \leq c \|\mathbf{x}\|_{p\text{-var};[s,t]}^{N+1}.$$

Proof. This is fairly obvious. As $S_N(\mathbf{x})_{s,t} = S_N\left(x^{s,t}\right)_{s,t}$, we have

$$\left|S_{N+1}(\mathbf{x})_{s,t} - S_{N+1}\left(x^{s,t}\right)_{s,t}\right|$$

$$= \left|\pi_{N+1}\left(S_{N+1}(\mathbf{x})_{s,t}\right) - \pi_{N+1}\left(S_{N+1}\left(x^{s,t}\right)_{s,t}\right)\right|$$

$$\leq \left|\pi_{N+1}\left(S_{N+1}(\mathbf{x})_{s,t}\right)\right| + \left|\pi_{N+1}\left(S_{N+1}\left(x^{s,t}\right)_{s,t}\right)\right|.$$

Using equivalence of homogenous norms and the quantitative estimates on the Lyons lift obtained in Theorem 9.5, we have

$$\left|S_{N+1}(\mathbf{x})_{s,t} - S_{N+1}\left(x^{s,t}\right)_{s,t}\right|$$

$$\leq c_1\left(\left\|S_{N+1}(\mathbf{x})_{s,t}\right\|^{N+1} + \left\|S_{N+1}\left(x^{s,t}\right)_{s,t}\right\|^{N+1}\right)$$

$$\leq c_2\left(\|\mathbf{x}\|_{p\text{-var};[s,t]}^{N+1} + \left\|x^{s,t}\right\|_{1\text{-var};[s,t]}^{N+1}\right)$$

$$\leq c_3 \|\mathbf{x}\|_{p\text{-var};[s,t]}^{N+1}.$$

∎

The following result generalizes Lemma 9.2.

Lemma 9.9 *Let* $x^1, x^2, \tilde{x}^1, \tilde{x}^2 \in C^{1\text{-var}}\left([s,u], \mathbb{R}^d\right)$ *such that*

$$S_N\left(x^1\right)_{s,u} = S_N\left(x^2\right)_{s,u},$$
$$S_N\left(\tilde{x}^1\right)_{s,u} = S_N\left(\tilde{x}^2\right)_{s,u},$$

and assume there exist $\ell \geq 0$, $\varepsilon > 0$ *such that*

$$\max\left\{\int_s^u |dx^1| + \int_s^u |dx^2|, \int_s^u |d\tilde{x}^1| + \int_s^u |d\tilde{x}^2|\right\} \leq \ell,$$
$$\max\left\{\int_s^u |dx_r^1 - d\tilde{x}_r^1|, \int_s^u |dx_r^2 - d\tilde{x}_r^2|\right\} \leq \varepsilon\ell.$$

Then, for some constant C *depending only on* N,

$$\left|\left(S_{N+1}\left(x^1\right)_{s,u} - S_{N+1}\left(x^2\right)_{s,u}\right) - \left(S_{N+1}\left(\tilde{x}^1\right)_{s,u} - S_{N+1}\left(\tilde{x}^2\right)_{s,u}\right)\right| \leq C\varepsilon\ell^{N+1}.$$

Proof. Working as in the proof of Lemma 9.2, we see that we can assume $x^2 = \tilde{x}^2 = 0$ and $(s,u) = (0,1)$. Then, scaling x^1 and $\tilde{x}^1$ by $\frac{1}{\ell}$, we can assume $\ell = 1$. The lemma then follows from Proposition 7.63. ∎

We can now prove local Lipschitzness of the Lyons-lifting map S_N.

Theorem 9.10 *Let* $\mathbf{x}^1, \mathbf{x}^2 \in C^{p\text{-var}}\left([0,T], G^{[p]}\left(\mathbb{R}^d\right)\right)$ *and* ω *a control such that for all* $0 \leq s < t \leq T$ *and* $i = 1,2$,

$$\left\|\mathbf{x}^i\right\|_{p\text{-var};[s,t]}^p \leq \omega(s,t),$$
$$\rho_{p,\omega}\left(\mathbf{x}^1, \mathbf{x}^2\right) \leq \varepsilon.$$

Then, for all $N \geq [p]$, *there exists a constant* C *depending only on* N *and* p *such that for all* $s < t$ *in* $[0,T]$,

$$\rho_{p,\omega}\left(S_N\left(\mathbf{x}^1\right), S_N\left(\mathbf{x}^2\right)\right) \leq C\varepsilon.$$

Proof. It is enough to show that for all $N \geq [p]$, if $\mathbf{x}^1$ and $\mathbf{x}^2$ are two paths in $C^{p\text{-var}}\left([0,T], G^N\left(\mathbb{R}^d\right)\right)$ with $\rho_{p,\omega}\left(\mathbf{x}^1, \mathbf{x}^2\right) \leq \varepsilon$, then, for some constant c_N,

$$\rho_{p,\omega}\left(S_{N+1}\left(\mathbf{x}^1\right), S_{N+1}\left(\mathbf{x}^2\right)\right) \leq c_N\varepsilon.$$

Let $x^{1,s,t}, x^{2,s,t} \in C^{1\text{-var}}\left([s,t], \mathbb{R}^d\right)$ such that

$$S_N\left(x^{i,s,t}\right)_{s,t} = \mathbf{x}_{s,t}^i,$$

and such that

$$\int_s^t d\left|x^{i,s,t}\right| \leq c_1 \omega(s,t)^{1/p},$$
$$\int_s^t d\left|x^{1,s,t} - x^{2,s,t}\right| \leq c_1 \varepsilon\omega(s,t)^{1/p}.$$

This is possible thanks to Proposition 7.64, applied to $\delta_\lambda \mathbf{x}^1_{s,t}$ and $\delta_\lambda \mathbf{x}^2_{s,t}$ with $\lambda = 1/\omega\,(s,t)^{1/p}$. We define similarly $x^{i,s,u}$ and $x^{i,t,u}$, and then $x^{i,s,t,u}$ to be the concatenation of $x^{i,s,t}$ and $x^{i,t,u}$. Observe in particular that

$$\int_s^u d\left|x^{1,s,t,u} - x^{2,s,t,u}\right| \leq 2^{1-1/p} c_1 \varepsilon \omega\,(s,u)^{1/p}.$$

Following the proof of Proposition 9.3, we define for $s < t$,

$$\Gamma^i_{s,t} = S_{N+1}\left(x^{i,s,t}\right)_{s,t} - S_{N+1}\left(\mathbf{x}^i\right)_{s,t}, \; i = 1,2,$$

and $\overline{\Gamma}_{s,t} = \Gamma^1_{s,t} - \Gamma^2_{s,t}$. It is clear from the proof of Proposition 9.3 that

$$\left|\Gamma^i_{s,t}\right| \leq \frac{c_2}{2}\omega\,(s,t)^{\frac{N+1}{p}}, \; i = 1,2,$$

and, by the triangle inequality,

$$\left|\overline{\Gamma}_{s,t}\right| \leq c_2 \omega\,(s,t)^{\frac{N+1}{p}}. \tag{9.6}$$

On the other hand, employing the same logic as in the proof of Proposition 9.3, we see that

$$\Gamma^i_{s,u} - \left(\Gamma^i_{s,t} + \Gamma^i_{t,u}\right) = S_{N+1}\left(x^{i,s,t,u}\right)_{s,u} - S_{N+1}\left(x^{i,s,u}\right)_{s,u}.$$

We can therefore use Lemma 9.9 to see that

$$\left|\overline{\Gamma}_{s,u} - \left(\overline{\Gamma}_{s,t} + \overline{\Gamma}_{t,u}\right)\right| \leq c_3 \varepsilon \omega\,(s,u)^{\frac{N+1}{p}}. \tag{9.7}$$

Inequalities (9.6) and (9.7) allow us to use the (analysis) Lemma 6.2, and we learn that for all $0 \leq s < t \leq T$,

$$\left|\overline{\Gamma}_{s,t}\right| \leq c_4 \varepsilon \omega\,(s,t)^{\frac{N+1}{p}}.$$

From Proposition 7.63,

$$\left|\pi_{N+1}\left(S_{N+1}\left(x^{1,s,t}\right)_{s,t} - S_{N+1}\left(x^{2,s,t}\right)_{s,t}\right)\right| \leq c_5 \varepsilon \omega\,(s,t)^{\frac{N+1}{p}}.$$

This implies by the triangle inequality that

$$\left|\pi_{N+1}\left(S_{N+1}\left(\mathbf{x}^1\right)_{s,t} - S_{N+1}\left(\mathbf{x}^2\right)_{s,t}\right)\right| \leq c_6 \varepsilon \omega\,(s,t)^{\frac{N+1}{p}},$$

i.e. that $\rho_{p,\omega}\left(S_{N+1}\left(\mathbf{x}^1\right), S_{N+1}\left(\mathbf{x}^2\right)\right) \leq \varepsilon$. ∎

From Theorem 8.10 we immediately deduce

Corollary 9.11 *Let $N \geq [p]$.*
(i) The map

$$S_N : C_o^{p\text{-var}}\left([0,T], G^{[p]}\left(\mathbb{R}^d\right)\right) \to C_o^{p\text{-var}}\left([0,T], G^N\left(\mathbb{R}^d\right)\right)$$

is uniformly continuous on bounded sets, using the $d_{p\text{-var}}$-metric.
(ii) The map

$$S_N : C_o^{1/p\text{-Höl}} \left([0,T], G^{[p]}\left(\mathbb{R}^d\right)\right) \to C_o^{1/p\text{-Höl}} \left([0,T], G^N\left(\mathbb{R}^d\right)\right)$$

is uniformly continuous on bounded sets, using the $d_{1/p\text{-Höl}}$-metric.

9.2 Spaces of geometric rough paths

Theorem 9.12 Let $p \geq 1$ and $N \geq [p]$. Let Ω denote either $C_o^{p\text{-var}}$, $C_o^{0,p\text{-var}}$, $C_o^{1/p\text{-Höl}}$ or $C_o^{0,1/p\text{-Höl}}$. Then
(i) the map

$$\mathbf{x} \in \Omega([0,T], G^N\left(\mathbb{R}^d\right)) \to \pi_{0,[p]}(\mathbf{x}) \in \Omega([0,T], G^{[p]}\left(\mathbb{R}^d\right))$$

is a bijection, with inverse the map S_N;
(ii) for $p \geq 2$ and $d \geq 2$, the map

$$\mathbf{x} \in \Omega([0,T], G^N\left(\mathbb{R}^d\right)) \to \pi_{0,[p]-1}(\mathbf{x}) \in \Omega([0,T], G^{[p]-1}\left(\mathbb{R}^d\right))$$

is not a bijection.

Remark 9.13 The proof of part (ii) will show that:
(case (ii a)) $\pi_{0,[p]-1}$ is not an injection, when p is not an integer. It is proven in [122] that it is a surjection when p is not an integer.
(case (ii b1)) $\pi_{0,[p]-1}$ is not an injection when p is an integer, and $\Omega = C^{p\text{-var}}$ or $\Omega = C^{1/p\text{-Höl}}$.
(case (ii b2)) $\pi_{0,[p]-1}$ is not a surjection, when p is an integer, and $\Omega = C^{0,p\text{-var}}$ or $\Omega = C^{0,1/p\text{-Höl}}$; we leave it as an exercise for the reader to prove that it is not an injection in this case.

Proof. (i) The case $\Omega = C^{p\text{-var}}$ follows from Theorem 9.5 and $C^{1/p\text{-Höl}}$ is an obvious corollary from the $C^{p\text{-var}}$ case. The case $\Omega = C^{0,p\text{-var}}$ (resp. $C^{0,1/p\text{-Höl}}$) follows from the case $\Omega = C^{p\text{-var}}$ (resp. $C^{1/p\text{-Höl}}$) by Wiener's characterization (Theorem 8.22).
(ii) (a) We first assume that $\Omega = C^{p\text{-var}}$ or $C^{1/p\text{-Höl}}$. Let h be a non-zero $\mathfrak{g}^{[p]}\left(\mathbb{R}^d\right) \cap \left(\mathbb{R}^d\right)^{\otimes[p]}$-valued path which is $1/q$-Hölder with $q = N/p$. As in the proof of Lemma 7.61, we see that if $\mathbf{x} \in \Omega([0,T], G^{[p]}\left(\mathbb{R}^d\right))$ and $\mathbf{y}$ is defined by

$$\mathbf{y}_t = \mathbf{x}_t \otimes \exp(h_t),$$

then $\mathbf{y} \in \Omega([0,T], G^{[p]}\left(\mathbb{R}^d\right))$ and $\pi_{0,[p-1]}(\mathbf{y}) = \pi_{0,[p-1]}(\mathbf{x})$. This means that $\pi_{0,[p]-1}(\mathbf{y}) = \pi_{0,[p]-1}(\mathbf{x})$, and as $\mathbf{y} \neq \mathbf{x}$, $\pi_{0,[p]-1}$ is not an injection from $\Omega([0,T], G^N\left(\mathbb{R}^d\right))$ into $\Omega([0,T], G^{[p-1]}\left(\mathbb{R}^d\right))$.

(b) We now assume that $\Omega = C^{0,p\text{-var}}$ or $C^{0,1/p\text{-Höl}}$.

(b1) We deal with the case $N > p$ and again take a non-zero

$$h \in C^{0,1/q\text{-Höl}}\left([0,T],\mathfrak{g}^{[p]}\left(\mathbb{R}^d\right) \cap \left(\mathbb{R}^d\right)^{\otimes[p]}\right),$$

with $q = N/p$. Define the path $\mathbf{y}$ as above, $\mathbf{y}_t = \mathbf{x}_t \otimes \exp(h_t)$. We have already seen that $\mathbf{y} \in \Omega([0,T], G^{[p]}(\mathbb{R}^d))$. Using Wiener characterization we actually see that $\mathbf{y} \in \Omega([0,T], G^{[p]}(\mathbb{R}^d))$ (it is at this point that we need $q > 1$, that is $N > p$). Once again, $\pi_{0,[p]-1}(\mathbf{y}) = \pi_{0,[p]-1}(\mathbf{x})$, and as $\mathbf{y} \neq \mathbf{x}$, $\pi_{0,[p]-1}$ is not an injection from $\Omega([0,T], G^N(\mathbb{R}^d))$ into $\Omega([0,T], G^{[p]-1}(\mathbb{R}^d))$.

(b2) It only remains to deal with the case $\Omega = C^{0,p\text{-var}}$ or $C^{0,1/p\text{-Höl}}$, and $N = p$ (which implies that $p \in \{2,3,\dots\}$). We aim to prove in this case that $\pi_{0,[p]-1}$ is not a surjection or, in other words, that there exists a path $\mathbf{x} \in \Omega([0,T], G^{[p]-1}(\mathbb{R}^d))$ which admits no lift to $\Omega([0,T], G^{[p]}(\mathbb{R}^d))$. To this end, assume $T = 1$ for simplicity of notation, and assume we have a path $\mathbf{y} \in \Omega([0,1], G^{[p]}(\mathbb{R}^d))$ that projects down onto $\mathbf{x}$. Let $\omega(s,t) = \|\mathbf{y}\|^p_{p\text{-var};[s,t]}$, which is finite by assumption. By definition of the increment of $\mathbf{y}$, we have

$$\mathbf{y}_{0,1} = \bigotimes_{i=0}^{2^n-1} \mathbf{y}_{\frac{i}{2^n}, \frac{i+1}{2^n}}.$$

Define $\tilde{\mathbf{x}}_{\frac{i}{2^n}, \frac{i+1}{2^n}} \in G^{[p]}(\mathbb{R}^d)$ by $\pi_{0,[p]-1}\left(\log\left(\tilde{\mathbf{x}}_{\frac{i}{2^n}, \frac{i+1}{2^n}}\right)\right) = \log\left(\mathbf{x}_{\frac{i}{2^n}, \frac{i+1}{2^n}}\right)$, and $\pi_{[p]}\left(\log\left(\tilde{\mathbf{x}}_{\frac{i}{2^n}, \frac{i+1}{2^n}}\right)\right) = 0$. Observe that $\tilde{\mathbf{x}}_{\frac{i}{2^n}, \frac{i+1}{2^n}}$ are *not* the increments of a $G^{[p]}(\mathbb{R}^d)$-valued path; we view $\tilde{\mathbf{x}}$ as a map that associates to every dyadic interval of form $\left[\frac{i}{2^n}, \frac{i+1}{2^n}\right]$ an element in $G^{[p]}(\mathbb{R}^d)$. Hence, as we also have $\pi_{0,[p]-1}\left(\mathbf{y}_{\frac{i}{2^n}, \frac{i+1}{2^n}}\right) = \mathbf{x}_{\frac{i}{2^n}, \frac{i+1}{2^n}}$, a short computation gives

$$\mathbf{y}_{0,1} = \bigotimes_{i=0}^{2^n-1} \tilde{\mathbf{x}}_{\frac{i}{2^n}, \frac{i+1}{2^n}} + \sum_{i=0}^{2^n-1} \pi_{[p]}\left(\log\left(\mathbf{y}_{\frac{i}{2^n}, \frac{i+1}{2^n}}\right)\right).$$

Then, by equivalence of homogenous norms, there exists $c > 0$ such that

$$\left|\pi_{[p]}\left(\log\left(\mathbf{y}_{\frac{i}{2^n}, \frac{i+1}{2^n}}\right)\right)\right| \leq c \left\|\mathbf{y}_{\frac{i}{2^n}, \frac{i+1}{2^n}}\right\|^{[p]}$$

$$\leq c\,\omega\left(\frac{i}{2^n}, \frac{i+1}{2^n}\right) \text{ by definition of } \omega.$$

In particular, we obtain that

$$\left|\bigotimes_{i=0}^{2^n-1} \tilde{\mathbf{x}}_{\frac{i}{2^n}, \frac{i+1}{2^n}}\right| \leq \omega(0,1) + |\mathbf{y}_{0,1}| < \infty.$$

Therefore, we proved that a necessary condition for a path $\mathbf{x} \in \Omega([0,T], G^{[p]-1}(\mathbb{R}^d))$ to admit a lift $\mathbf{y} \in \Omega([0,T], G^{[p]}(\mathbb{R}^d))$ is given by

$$\sup_{n \geq 0} \left| \bigotimes_{i=0}^{2^n-1} \tilde{\mathbf{x}}_{\frac{i}{2^n}, \frac{i+1}{2^n}} \right| < \infty.$$

We therefore aim to provide a path $\mathbf{x}$ such that the above expression is infinite. To this end, using $d \geq 2$, we let e_1, e_2 be the first two vectors in the standard basis of $\mathbb{R}^d$. Define $v_1 = e_2$, and $v_{i+1} = [e_1, v_i]$ so that $v_i \in (\mathbb{R}^d)^{\otimes i}$. For two paths, $f \in C_0^{0,1/p\text{-Höl}}([0,T],\mathbb{R})$ and $g \in C_0^{0,\frac{p-1}{p}\text{-Höl}}([0,T],\mathbb{R})$, we define $\mathbf{x} \in \Omega([0,T], G^{[p]-1}(\mathbb{R}^d))$ by

$$\mathbf{x}(t) = e^{f_t e_1 + g_t v_{[p]-1}} = e^{f_t e_1} \otimes e^{g_t v_{[p]-1}} \in G^{[p]-1}(\mathbb{R}^d).$$

We note that $\mathbf{x}$ is indeed in $\Omega = C^{0,p\text{-var}}$, from the definition of f, g and

$$\|\mathbf{x}_{s,t}\| = \left\| e^{f_{s,t} e_1} \otimes e^{g_{s,t} v_{[p]-1}} \right\| \leq (\text{const}) \times \left(|f_{s,t}| + |g_{s,t}|^{\frac{1}{p-1}} \right).$$

Defining $\tilde{\mathbf{x}}_{\frac{i}{2^n}, \frac{i+1}{2^n}} \in G^{[p]}(\mathbb{R}^d)$ as explained above, the Campbell–Baker–Hausdorff formula yields

$$\bigotimes_{i=0}^{2^n-1} \tilde{\mathbf{x}}_{\frac{i}{2^n}, \frac{i+1}{2^n}}$$

$$= \exp\left(f_{0,1} e_1 + g_{0,1} v_{[p]-1} + \sum_{i=0}^{2^n-1} \left(f_{\frac{i}{2^n}} g_{\frac{i}{2^n}, \frac{i+1}{2^n}} - g_{\frac{i}{2^n}} f_{\frac{i}{2^n}, \frac{i+1}{2^n}} \right) v_{[p]} \right).$$

In particular, $\sup_n \left| \bigotimes_{i=0}^{2^n-1} \tilde{\mathbf{x}}_{\frac{i}{2^n}, \frac{i+1}{2^n}} \right| < \infty$ if and only if

$$\sup_{n \geq 0} \left| \sum_{i=0}^{2^n-1} f_{\frac{i}{2^n}} g_{\frac{i}{2^n}, \frac{i+1}{2^n}} - g_{\frac{i}{2^n}} f_{\frac{i}{2^n}, \frac{i+1}{2^n}} \right| < \infty,$$

which itself is equivalent to

$$\sup_{n \geq 0} \left| \sum_{i=0}^{2^n-1} f_{\frac{i}{2^n}} g_{\frac{i}{2^n}, \frac{i+1}{2^n}} \right| < \infty.$$

The following Exercise 9.14 shows that for any $p, q \geq 1$ with $1/p + 1/q = 1$ there exist paths

$$f \in C_0^{0,1/p\text{-Höl}}([0,T],\mathbb{R}), g \in C_0^{0,1/q\text{-Höl}}([0,T],\mathbb{R})$$

such that $\sup_{n \geq 0} \left| \sum_{i=0}^{2^n-1} f_{\frac{i}{2^n}} g_{\frac{i}{2^n}, \frac{i+1}{2^n}} \right| = \infty$. In particular, we now see that there exists a path $\mathbf{x} \in \Omega([0,T], G^{[p]-1}(\mathbb{R}^d))$ which does not admit a lift to $\Omega([0,T], G^{[p]}(\mathbb{R}^d))$. ∎

Exercise 9.14 *Assume that $\phi, \varphi : [0,1] \to [0,1]$ is a continuous increasing bijection such that $\sum_{k=1}^{\infty} \phi^{-1}\left(2^{-k}\right)$ and $\sum_{k=1}^{\infty} \varphi^{-1}\left(2^{-k}\right)$ are convergent series. We then define the functions*

$$x^\phi : t \in [0,1] \to \sum_{k=1}^{\infty} \phi^{-1}\left(2^{-k}\right) \sin\left(2^{k+1}\pi t\right),$$

$$y^\phi : t \in [0,1] \to \sum_{k=1}^{\infty} \phi^{-1}\left(2^{-k}\right) \left(1 - \cos\left(2^{k+1}\pi t\right)\right).$$

(i) Prove that if ϕ is such that $\lim_{x \to 0} x^p / \phi(x) = 0$, then x^ϕ and y^ϕ belong to $C_0^{0, 1/p\text{-Höl}}\left([0,T], \mathbb{R}\right)$.
(ii) Prove that

$$\sum_{i=0}^{2^n-1} x^\phi_{\frac{i}{2^n}} \, y^\varphi_{\frac{i}{2^n}, \frac{i+1}{2^n}} \geq \frac{1}{4} \sum_{j=1}^{n} 2^j \phi^{-1}\left(2^{-j}\right) \varphi^{-1}\left(2^{-j}\right).$$

(iii) Provide an example of functions $f, g \in C_0^{0,1/p\text{-Höl}}\left([0,1], \mathbb{R}\right) \times C_0^{0,1/q\text{-Höl}}$
$\left([0,1], \mathbb{R}\right)$, for $1/p + 1/q = 1$, such that

$$\sup_{n \geq 0} \left| \sum_{i-0}^{2^n-1} f_{\frac{i}{2^n}} \, g_{\frac{i}{2^n}, \frac{i+1}{2^n}} \right| = \infty.$$

Solution. (i) First case. We first assume that $\phi^{-1}(x) \leq K |x|^{1/p}$. Then, for all $s, t \in [0,1]$,

$$\left| x^\phi(t) - x^\phi(s) \right|$$

$$\leq \sum_{k=1}^{\infty} 2^{-k/p} \left| \sin\left(2^{k+1}\pi t\right) - \sin\left(2^{k+1}\pi s\right) \right|$$

$$\leq 2K \left(\sum_{k=1}^{\log_2 \frac{1}{t-s}} \phi^{-1}\left(2^{-k}\right) 2^{k+1}\pi |t-s| + \sum_{k=\log_2 \frac{1}{t-s}}^{+\infty} \phi^{-1}\left(2^{-k}\right) \right)$$

$$\leq 2K \left(\pi |t-s| \sum_{k=1}^{\log_2 \frac{1}{t-s}} 2^{k(1-1/p)} + \sum_{k=\log_2 \frac{1}{t-s}}^{+\infty} 2^{-k/p} \right)$$

$$\leq cK |t-s|^{1/p}.$$

Second case. Now, if $\lim_{x \to 0} \frac{x^p}{\phi(x)} = 0$, $\phi^{-1}(x) = o\left(|x|^{1/p}\right)$. Hence, for all $\varepsilon > 0$, there exists $n \geq 0$ such that $k \geq n$ implies $\phi^{-1}\left(2^{-k}\right) \leq \varepsilon 2^{-k/p}$. We write

$$x^\phi = x^{\phi,0,n} + x^{\phi,n,\infty},$$

where $x^{\phi,i,j} = \sum_{i+1}^{j} \phi^{-1}\left(2^{-k}\right) \sin\left(2^{k+1}\pi t\right)$. Clearly, as $x^{\phi,0,n}$ is smooth, we have

$$\lim_{h \to 0} \sup_{s,t,|t-s| \le h} \frac{\left|x^{\phi,0,n}\left(t\right) - x^{\phi,0,n}\left(s\right)\right|}{|t-s|^{1/p}} = 0.$$

Moreover, working as in the first case, we have

$$\left|x^{\phi}\left(t\right) - x^{\phi}\left(s\right)\right| \le c\varepsilon \left|t-s\right|^{1/p}.$$

Hence, we proved that for all $\varepsilon > 0$,

$$\lim_{h \to 0} \sup_{s,t,|t-s| \le h} \frac{\left|x^{\phi}\left(t\right) - x^{\phi}\left(s\right)\right|}{|t-s|^{1/p}} \le c\varepsilon,$$

which concludes the proof of (i).

(ii) As $\cos\left(2^{k+1}\pi\frac{i+1}{2^n}\right) - \cos\left(2^{k+1}\pi\frac{i}{2^n}\right)$ and $\sin\left(2^{k+1}\pi\frac{i}{2^n}\right)$ are equal to 0, we obtain that

$$\sum_{i=0}^{2^n-1} x^{\phi}_{\frac{i}{2^n}} y^{\varphi}_{\frac{i}{2^n},\frac{i+1}{2^n}} = \sum_{j,k=1}^{n} \phi^{-1}\left(2^{-k}\right)\varphi^{-1}\left(2^{-j}\right).I^{n}_{j,k}$$

where

$$I^{n}_{j,k} = \sum_{i=0}^{2^n-1} \sin\left(2\pi i 2^{k-n}\right) \left[\cos\left(2^{j+1}\pi\frac{i+1}{2^n}\right) - \cos\left(2^{j+1}\pi\frac{i}{2^n}\right)\right].$$

Trigonometric exercises show that $I^{n}_{j,k} = 0$ if $j \ne k$, and that $I^{n}_{j,j} \ge 2^{j-2}$. As ϕ^{-1} and φ^{-1} are positive, we therefore obtain that

$$\sum_{i=0}^{2^n-1} x^{\phi}_{\frac{i}{2^n}} y^{\varphi}_{\frac{i}{2^n},\frac{i+1}{2^n}} \ge \frac{1}{4}\sum_{j=1}^{n} 2^{j}\phi^{-1}\left(2^{-j}\right)\varphi^{-1}\left(2^{-j}\right).$$

(iii) Just take $f,g = x^{\phi}, x^{\varphi}$ with $\phi\left(x\right) = x^p \log \frac{1}{x}$ and $\varphi\left(x\right) = x^q \log \frac{1}{x}$. $\quad\square$

We therefore showed that the sets $C^{p\text{-var}}_o\left([0,T], G^{[p]}\left(\mathbb{R}^d\right)\right)$, $C^{0,p\text{-var}}_o\left([0,T], G^{[p]}\left(\mathbb{R}^d\right)\right)$, $C^{1/p\text{-H\"ol}}_o\left([0,T], G^{[p]}\left(\mathbb{R}^d\right)\right)$ and $C^{0,1/p\text{-H\"ol}}_o\left([0,T], G^{[p]}\left(\mathbb{R}^d\right)\right)$ are quite fundamental! We therefore give their elements names:

Definition 9.15 (i) A weak geometric p-rough path is a continuous path of finite p-variation with values in the free nilpotent group of step $[p]$ over $\mathbb{R}^d$, i.e. an element of $C^{p\text{-var}}\left([0,T], G^{[p]}\left(\mathbb{R}^d\right)\right)$.

(ii) A geometric p-rough path is a continuous path with values in the free nilpotent group of step $[p]$ over $\mathbb{R}^d$ which is in the p-variation closure of the set of bounded variation paths, i.e. an element of $C^{0,p\text{-var}}\left([0,T], G^{[p]}\left(\mathbb{R}^d\right)\right)$.

(iii) A weak geometric $1/p$-Hölder rough path is a $1/p$-Hölder path with values in the free nilpotent group of step $[p]$ over $\mathbb{R}^d$, i.e. an element of

$C^{1/p\text{-Höl}}([0,T], G^{[p]}(\mathbb{R}^d))$.

(iv) A geometric $1/p$-Hölder rough path is a continuous path with values in the free nilpotent group of step $[p]$ over $\mathbb{R}^d$ which is in the $1/p$-Hölder closure of the set of 1-Hölder paths, i.e. an element of $C^{0,1/p\text{-Höl}}([0,T], G^{[p]}(\mathbb{R}^d))$. □

Recall from the interpolation results of the previous chapter that

$$C^{(p+\varepsilon)\text{-var}}([0,T], G^{[p]}(\mathbb{R}^d)) \quad \subset \quad C^{0,p\text{-var}}([0,T], G^{[p]}(\mathbb{R}^d))$$

$$\subset \quad C^{p\text{-var}}([0,T], G^{[p]}(\mathbb{R}^d))$$

and for this reason the difference between weak and genuine geometric p-rough paths is important only when we care about very precise results.

Exercise 9.16 *Identify the $G^2(\mathbb{R}^2)$ with the 3-dimensional Heisenberg group $\mathbb{H} \cong \mathbb{R}^3$. Verify that the "pure-area path"*

$$(0, 0; t)$$

is a weak geometric 2-rough path but not a genuine 2-rough path.

Exercise 9.17 *Assume $g \in \mathfrak{g}^n(\mathbb{R}^d) \cap (\mathbb{R}^d)^{\otimes n}$. Show that $t \mapsto \exp(t g) \subset G^n(\mathbb{R}^d)$ is a weak Hölder geometric n-rough path and compute explicitly*

$$S_N(\exp((\cdot) g)) \subset G^N(\mathbb{R}^d), \qquad n \leq N.$$

Solution. Claim that $S_N(\exp_n(\cdot g))_{0,t} = \exp_N(tg)$ where we write $\exp_k$ for the exp-map in $T^k(\mathbb{R}^d)$ in order to distinguish from $\exp_n$ and $\exp_N$. To see this, note that

$$d(\exp_N(s g), \exp_N(t g)) = \|\exp_N(-sg) \otimes \exp_N(tg)\|$$

and from the Campbell–Baker–Hausdorff formula, this clearly equals $\|\exp_N((t-s) g)\|$, which is bounded by a constant times $(t-s)^{1/n} |g|^{1/n}$. It follows that $\exp_N(\cdot g)$ is $1/n$-Hölder and by uniqueness of the Lyons lift the claim is proved. □

9.3   Invariance under Lipschitz maps

The content of this section is not used directly in the sequel and depends on techniques of the forthcoming Section 10.6 on rough integration. The probabilistic motivation here is the fact that $\Phi \circ M$, the image of a semi-martingale M under a C^2-map Φ, is again a semi-martingale; a manifest consequence of Itô's lemma.

Having defined what we mean by a weak geometric p-rough path, say $\mathbf{x} \in C^{p\text{-var}}([0,T], G^{[p]}(\mathbb{R}^d))$, it is natural to ask whether the (yet to be defined!) image of $\mathbf{x}$ under a sufficiently smooth map $\Phi : \mathbb{R}^d \to \mathbb{R}^e$ is also a weak geometric p-rough path. The following result can then be summarized by saying that the image of a weak geometric p-rough path under a $\mathrm{Lip}_{\mathrm{loc}}^{\gamma}$-map, $\gamma > p$, is indeed another weak geometric p-rough path.

Theorem 9.18 *Assume*
(i) $\mathbf{x} \in C^{p\text{-var}}\left([0,T], G^{[p]}\left(\mathbb{R}^d\right)\right)$;
(ii) $\Phi \in \mathrm{Lip}_{\mathrm{loc}}^{\gamma}\left(\mathbb{R}^d, \mathbb{R}^e\right), \gamma > p$.
Then there exists a unique continuous (in fact, uniformly continuous on bounded sets) map

$$\Phi_* : C^{p\text{-var}}\left([0,T], G^{[p]}\left(\mathbb{R}^d\right)\right) \mapsto C^{p\text{-var}}\left([0,T], G^{[p]}\left(\mathbb{R}^e\right)\right)$$

with the property that, whenever $\mathbf{x} = S_{[p]}(x)$ *for some* $x \in C^{1\text{-var}}\left([0,T], \mathbb{R}^d\right)$ *then*

$$\Phi_* \mathbf{x} = S_{[p]}(\Phi \circ x).$$

Proof. The proof relies on the forthcoming Theorem 10.47 in Section 10.6. Indeed, $\varphi := D\Phi = (\partial_1 \Phi, \dots, \partial_d \Phi) \subset \mathrm{Lip}_{\mathrm{loc}}^{\gamma-1}\left(\mathbb{R}^d, \mathbb{R}^e\right)$ satisfies the assumption of that theorem, so that

$$\Phi_* \mathbf{x} := \int_0^{\cdot} \varphi(x_s) \, d\mathbf{x}_s \quad \text{(rough integral)}$$

is a well-defined geometric p-rough path; more precisely, an element of $C^{p\text{-var}}\left([0,T], G^{[p]}\left(\mathbb{R}^e\right)\right)$ and the rough integral $\Phi_* \mathbf{x}$ is a (uniformly) continuous (on bounded sets) function of the integrator $\mathbf{x}$ in p-variation metric. To see that $\Phi_* \mathbf{x}$ has the claimed "push-forward" behaviour whenever $\mathbf{x} = S_{[p]}(x)$ for some $x \in C^{1\text{-var}}\left([0,T], \mathbb{R}^d\right)$ it suffices to note that, by the fundamental theorem of calculus,

$$\Phi(x)_{0,t} = \int_0^t \varphi(x_s) \, dx_s \quad \text{(classical Riemann–Stieltjes integral)}.$$

By the basic consistency properties of a rough integral $\Phi_* \mathbf{x}$ is the precisely the step-$[p]$ lift of the indefinite Riemann–Stieltjes integral

$$\int_0^{\cdot} \varphi(x_s) \, dx_s$$

and so $\Phi_* \mathbf{x} = S_{[p]}(\Phi \circ x)$ as was claimed. ∎

9.4 Young pairing of weak geometric rough paths

Throughout this section, we fix $p > q \geq 1$ such that $q^{-1} + p^{-1} > 1$. Observe that this implies $q \in [1,2)$.

9.4.1   *Motivation*

Consider a path $x \in C^{p\text{-var}}\left([0,T],\mathbb{R}^d\right)$. It is natural, e.g. in the context of differential equations with drift term, to consider the "space-time" path $t \mapsto (x(t),t)$, plainly an element of $C^{p\text{-var}}\left([0,T],\mathbb{R}^{d+1}\right)$. It can also be important to replace x by $x+h$ where h is a suitable perturbation.[3] Let us now move to a genuine geometric rough path setting and consider $\mathbf{x} \in C^{p\text{-var}}\left([0,T],G^{[p]}\left(\mathbb{R}^d\right)\right)$. Recall the intuition that $\mathbf{x}$ contains the a priori information of up to $[p]$ iterated integrals which are not defined, in general, in Riemann–Stieltjes or Young sense; there is, however, enough regularity built into the definition of such a geometric p-rough path that higher iterated integrals, i.e. beyond level $[p]$, are canonically defined, as was seen in our discussion of the Lyons-lifting map.

Now, even if $x = \pi_1(\mathbf{x})$ has not sufficient regularity to form the integral $\int x \otimes dx$, one surely can form the integral $\int x dt$ or $\int x \otimes dh$ for sufficiently regular h, e.g. when the last integral is a well-defined Young integral as discussed in Section 6.2. In particular, one would hope that, given a sufficiently regular $h : [0,T] \to \mathbb{R}^{d'}$ there is a canonically defined geometric rough path, say $S_{[p]}(\mathbf{x},h)$, with values in $G^{[p]}\left(\mathbb{R}^d \oplus \mathbb{R}^{d'}\right)$ which coincides with $S_{[p]}(z)$ where $z = (x,h) : [0,T] \to \mathbb{R}^d \oplus \mathbb{R}^{d'}$ whenever $\mathbf{x} = S_{[p]}(x)$ for some nice path x. By the same token, given a sufficiently regular $h : [0,T] \to \mathbb{R}^d$ one would hope that there is a canonically defined geometric rough path, say $T_h(\mathbf{x})$, with values in $G^{[p]}\left(\mathbb{R}^d\right)$ which coincides with $S_{[p]}(x+h)$ whenever $\mathbf{x} = S_{[p]}(x)$ for some nice path x. Moreover, such constructions should be robust, for example so that $\mathbf{x} \mapsto T_h(\mathbf{x})$ is continuous in (e.g. p-variation) rough path distance. When $p \in (2,3)$, so that $\mathbf{x}$ takes values in $G^2\left(\mathbb{R}^d\right)$, the reader will have no difficulties in deriving such results by making use of the Young–Lóeve estimates of Section 6.1.

The remainder of this chapter is devoted to handling the general case.

9.4.2   *The space* $C^{(p,q)\text{-var}}\left([0,T],\mathbb{R}^d \oplus \mathbb{R}^{d'}\right)$

We recall from Section 7.5.6 that for fixed $\lambda \in \mathbb{R}$, the dilation map $\delta_\lambda : G^N\left(\mathbb{R}^d\right) \to G^N\left(\mathbb{R}^d\right)$ is the unique group homomorphism which extends scalar mutiplication $a \in \mathbb{R}^d \mapsto \lambda a \in \mathbb{R}^d$. Similarly, for fixed $\gamma_1, \gamma_2 \in \mathbb{R}$, the map $(a,b) \in \mathbb{R}^d \oplus \mathbb{R}^{d'} \mapsto (\gamma_1 a, \gamma_2 b) \in \mathbb{R}^d \oplus \mathbb{R}^{d'}$ lifts to a group homomorphism

$$\delta_{\gamma_1,\gamma_2} : G^N\left(\mathbb{R}^d \oplus \mathbb{R}^{d'}\right) \to G^N\left(\mathbb{R}^d \oplus \mathbb{R}^{d'}\right).$$

(Elements in $G^N\left(\mathbb{R}^d \oplus \mathbb{R}^{d'}\right)$ arise as the step-N signature of a path in $\mathbb{R}^d \oplus \mathbb{R}^{d'}$. Scaling the first d (resp. last d') coordinates of the path by γ_1

---

[3] For example, when adding a Cameron–Martin path to Brownian motion.

(resp. γ_2') gives rise precisely to the $\delta_{\gamma_1,\gamma_2}$-dilation of the original signature.) Observe that almost by definition of $\delta_{\gamma_1,\gamma_2}$, we have for a path $(x, h) \in C^{1\text{-var}}\left([0, 1], \mathbb{R}^d\right) \times C^{1\text{-var}}\left([0, 1], \mathbb{R}^{d'}\right)$,

$$\delta_{\gamma_1,\gamma_2} S_N\left(x \oplus h\right)_{0,1} = S_N\left(\gamma_1 x \oplus \gamma_2 h\right)_{0,1}.$$

Noting that $C^{p\text{-var}}\left([0, T], G^N\left(\mathbb{R}^d\right)\right)$ is the set of continuous $G^N\left(\mathbb{R}^d\right)$-valued paths $\mathbf{x}$ such that for some control function ω (e.g. $(s, t) \mapsto \|\mathbf{x}\|_{p\text{-var};[s,t]}$) one has

$$\sup_{s<t \text{ in } [0,T]} \left\| \delta_{\frac{1}{\omega(s,t)^{1/p}}} \mathbf{x}_{s,t} \right\| < \infty,$$

we are led to the following definition.

Definition 9.19 We say that a continuous path $\mathbf{x} \in C\left([0, T], G^N\left(\mathbb{R}^d \oplus \mathbb{R}^{d'}\right)\right)$ is of finite mixed (p, q)-variation, and write $\mathbf{x} \in C^{(p,q)\text{-var}}\left([0, T], G^N\left(\mathbb{R}^d \oplus \mathbb{R}^{d'}\right)\right)$ if, for some control ω,

$$\|\mathbf{x}\|_{p,q\text{-}\omega;[0,T]} := \sup_{s<t \text{ in } [0,T]} \left\| \delta_{\frac{1}{\omega(s,t)^{1/p}}, \frac{1}{\omega(s,t)^{1/q}}} \left(\mathbf{x}_{s,t}\right) \right\| < \infty.$$

(A convention of type $0/0 = 0$ is in place, to deal with $s < t$ such that $\omega\left(s, t\right) = 0$.) If we can take $\omega\left(s, t\right) = |t - s|$, we say that $\mathbf{x}$ is $\left(\frac{1}{p}, \frac{1}{q}\right)$-Hölder and write $\mathbf{x} \in C^{\left(\frac{1}{p}, \frac{1}{q}\right)\text{-Höl}}\left([0, T], G^N\left(\mathbb{R}^d \oplus \mathbb{R}^{d'}\right)\right)$.

As usual, we write $C_o^{(p,q)\text{-var}}\left([0, T], G^N\left(\mathbb{R}^d \oplus \mathbb{R}^{d'}\right)\right)$, etc. if we only consider paths started at o, the unit element in $G^N\left(\mathbb{R}^d \oplus \mathbb{R}^{d'}\right)$. □

Definition 9.20 *For a pair of controls ω_1, ω_2, we also define*

$$\|\mathbf{x}\|_{p,q\text{-}\omega_1,\omega_2;[0,T]} = \sup_{s<t \text{ in } [0,T]} \left\| \delta_{\frac{1}{\omega_1(s,t)^{1/p}}, \frac{1}{\omega_2(s,t)^{1/q}}} \left(\mathbf{x}_{s,t}\right) \right\|,$$

and, for two paths $\mathbf{x}^1, \mathbf{x}^2 \in C^{(p,q)\text{-var}}\left([0, T], G^N\left(\mathbb{R}^d \oplus \mathbb{R}^{d'}\right)\right)$,

$$\rho_{p,q\text{-}\omega_1,\omega_2;[0,T]}\left(\mathbf{x}^1, \mathbf{x}^2\right) =$$

$$\sup_{s<t \text{ in } [0,T]} \left| \delta_{\frac{1}{\omega_1(s,t)^{1/p}}, \frac{1}{\omega_2(s,t)^{1/q}}} \left(\mathbf{x}_{s,t}^1\right) - \delta_{\frac{1}{\omega_1(s,t)^{1/p}}, \frac{1}{\omega_2(s,t)^{1/q}}} \left(\mathbf{x}_{s,t}^2\right) \right|.$$
□

Exercise 9.21 *Let $\mathbf{x} \in C^{(p,q)\text{-var}}\left([0, T], G^N\left(\mathbb{R}^d \oplus \mathbb{R}^{d'}\right)\right)$ and assume ω_1 and ω_2 are control functions. Show that, for all $s < t$ in $[0, T]$,*

$$\|\mathbf{x}_{s,t}\| \leq c \|\mathbf{x}\|_{p,q\text{-}\omega_1,\omega_2;[0,T]} \left(\omega_1\left(s, t\right)^{1/p} + \omega_2\left(s, t\right)^{1/q}\right).$$

9.4.3 Quantitative bounds on S_N

For $(x, h) \in C^{\text{1-var}}\left([0, T], \mathbb{R}^d \oplus \mathbb{R}^{d'}\right)$, we now aim to show that for every $N \in \{1, 2, \dots\}$,

$$\left\| S_N \left(x \oplus h\right)_{s,t} \right\| \leq (\text{const}) \times \left(\left\| S_{[p]} \left(x\right) \right\|_{p\text{-var};[s,t]} + |h|_{q\text{-var};[s,t]} \right).$$

As in earlier chapters, the constant here depends only on the p-variation of x and the q-variation of h, allowing for a subsequent passage to the limit. The argument is similar to proving the Lyons-lift estimate

$$\left\| S_N \left(x\right)_{s,t} \right\| \leq (\text{const}) \times \left(\left\| S_{[p]} \left(x\right) \right\|_{p\text{-var};[s,t]} \right),$$

although in the latter case $N \in \{1, \dots, [p]\}$ is trivial. To start, we need a lemma replacing the use of geodesics.

Lemma 9.22 *Let $N \geq 1$, and $(x, h) \in C^{\text{1-var}}\left([s, t], \mathbb{R}^d\right) \times C^{\text{1-var}}\left([s, t], \mathbb{R}^{d'}\right)$ such that, for fixed $\alpha, \beta > 0$,*

$$\left\| S_{N+1} \left(\alpha x\right)_{s,t} \right\| + \left\| S_N \left(\alpha x \oplus \beta h\right)_{s,t} \right\| \leq C_1.$$

Then there exists a path $(x^{s,t}, h^{s,t})$ such that
(i)

$$S_N \left(x \oplus h\right)_{s,t} = S_N \left(x^{s,t} \oplus h^{s,t}\right)_{s,t} \qquad (9.8)$$

$$S_{N+1} \left(x\right)_{s,t} = S_{N+1} \left(x^{s,t}\right)_{s,t}; \qquad (9.9)$$

(ii) there exists a constant C_N which depends only on C_1 and N such that

$$\alpha \int_s^t \left| dx_u^{s,t} \right| + \beta \int_s^t \left| dh_u^{s,t} \right| \leq C_N.$$

Proof. Observe that if $S_N \left(x \oplus h\right)_{s,t} = S_N \left(x^{s,t} \oplus h^{s,t}\right)_{s,t}$, then $S_N \left(\alpha x \oplus \beta h\right)_{s,t} = S_N \left(\alpha x^{s,t} \oplus \beta h^{s,t}\right)_{s,t}$ and $S_{N+1} \left(\alpha x\right)_{s,t} = S_{N+1} \left(\alpha x^{s,t}\right)_{s,t}$. Hence, we can assume without loss of generality that $\alpha = \beta = 1$. By definition of the Carnot–Caratheodory homogenous norm, there exist two paths $x^{1,s,t}, h^{1,s,t}$ such that

$$S_N \left(x \oplus h\right)_{s,t} = S_N \left(x^{1,s,t} \oplus h^{1,s,t}\right)_{s,t},$$

with

$$\int_s^t \left| dx_u^{1,s,t} \right| + \int_s^t \left| dh_u^{1,s,t} \right| \leq c_1 \left\| S_N \left(x \oplus h\right)_{s,t} \right\|.$$

Then, define $g = S_{N+1} \left(x^{1,s,t} \right)_{s,t}^{-1} \otimes S_{N+1} \left(x \right)_{s,t}$, and observe that

$$
\begin{aligned}
\|g\| &\leq c_2 \int_s^t \left| dx_u^{1,s,t} \right| + \left\| S_{N+1} \left(x \right)_{s,t} \right\| \\
&\leq c_3 \left(\left\| S_N \left(x \oplus h \right)_{s,t} \right\| + \left\| S_{N+1} \left(x \right)_{s,t} \right\| \right) \\
&\leq c_4.
\end{aligned}
$$

Define a path $x^{2,s,t}$ such that

$$
S_{N+1} \left(x^{2,s,t} \right)_{s,t} = g,
$$

with

$$
\int_s^t \left| dx_u^{2,s,t} \right| = \|g\| \leq c_4.
$$

We have

$$
S_{N+1} \left(x^{1,s,t} \right)_{s,t} \otimes S_{N+1} \left(x^{2,s,t} \right)_{s,t} = S_{N+1} \left(x \right)_{s,t}
$$

and

$$
S_N \left(x^{1,s,t} \oplus h^{1,s,t} \right) \otimes S_N \left(x^{2,s,t} \oplus 0 \right) = S_N \left(x \oplus h \right)_{s,t}.
$$

Therefore, concatenating $x^{1,s,t} \oplus h^{1,s,t}$ and $x^{2,s,t} \oplus 0$ gives us a path that satisfies the required conditions of the lemma. ∎

We then need a slight generalization of Lemma 9.2.

Lemma 9.23 *Let* $\left(x^1 \oplus h^1, x^2 \oplus h^2 \right)$ *be two paths in* $C^{1\text{-var}} \left([s,u], \mathbb{R}^d \oplus \mathbb{R}^{d'} \right)$. *Assume that*
(i) $S_N \left(x^1 \oplus h^1 \right)_{s,u} = S_N \left(x^2 \oplus h^2 \right)_{s,u}$,
(ii) $S_{N+1} \left(x^1 \right)_{s,u} = S_{N+1} \left(x^2 \right)_{s,u}$,
(iii) $\int_s^u \left| dx_r^1 \right| + \int_s^u \left| dx_r^2 \right| \leq \ell_1$ *and* $\int_s^u \left| dh_r^1 \right| + \int_s^u \left| dh_r^2 \right| \leq \ell_2$.
Then, there exists a constant $C = C(N)$ *such that*

$$
\left| S_{N+1} \left(x^1 \oplus h^1 \right)_{s,u} - S_{N+1} \left(x^2 \oplus h^2 \right)_{s,u} \right| \leq C \sum_{k=1}^{N+1} \ell_1^{N+1-k} \ell_2^k.
$$

Proof. Working as in Lemma 9.2, we can assume without loss of generality that $x^2 \oplus h^2 = 0$, and $(s,u) = (0,1)$. Define for convenience $y = x^1 \oplus h^1$. We have, by definition of S_{N+1} and the triangle inequality

$$
\left| S_{N+1} \left(y \right)_{s,u} - 1 \right| \leq \sum_{i_1, \ldots, i_{N+1}} \left| \int_{s < r_1 < \ldots < r_{N+1} < u} dy_{r_1}^{i_1} \ldots dy_{r_{N+1}}^{i_{N+1}} \right|.
$$

Because $S_{N+1}\left(x^1\right)_{s,u} = 1$, we have

$$\left|S_{N+1}\left(y\right)_{s,u} - 1\right| \le$$

$$\sum_{\substack{i_1,\dots,i_{N+1} \\ \{i_1,\dots,i_{N+1}\}\cap\{d+1,\dots,d+d'\}\neq 0}} \left|\int_{s<r_1<\dots<r_{N+1}<u} dy_{r_1}^{i_1} \dots dy_{r_{N+1}}^{i_{N+1}}\right|.$$

By (iii), $\left|\int_{s<r_1<\dots<r_{N+1}<u} dy_{r_1}^{i_1} \dots dy_{r_{N+1}}^{i_{N+1}}\right|$ is bounded by a constant times $\ell_1^{N+1-k}\ell_2^k$, where k is the cardinal of $\{i_1,\dots,i_{N+1}\} \cap \{d+1,\dots,d+d'\}$. That concludes the proof. ∎

We can now generalize lemma 9.3, to give a quantitative estimate on $\|S_N\left(x \oplus h\right)\|_{p\text{-var}}$ for $N \ge 1$.

Lemma 9.24 *Let* (x,h) *be a path in* $C^{1\text{-var}}\left([0,T],\mathbb{R}^d \oplus \mathbb{R}^{d'}\right)$, *and*

$$\omega_1 = \left\|S_{[p]}\left(x\right)\right\|_{p\text{-var};[\cdot,\cdot]}^p, \qquad \omega_2 = |h|_{q\text{-var};[\cdot,\cdot]}^q.$$

For all $N \ge 1$, *there exists a constant* $C = C\left(N,p,q\right)$ *(and **not** depending on the 1-variation norm of* x *or* h*) such that*

$$\|S_N\left(x \oplus h\right)\|_{p,q\text{-}\omega_1,\omega_2;[0,T]} \le C. \qquad (9.10)$$

Proof. Define $(\mathcal{H}_N)$: for all paths $x \oplus h \in C^{1\text{-var}}\left([0,T],\mathbb{R}^d \oplus \mathbb{R}^{d'}\right)$ with $\left\|S_{[p]}\left(x\right)\right\|_{p\text{-var};[0,T]}$ and $|h|_{q\text{-var};[0,T]}$ bounded above by 1, we have for some constant c_N, for all $s < t$ in $[0,T]$,

$$\|S_N\left(x \oplus h\right)\|_{p,q\text{-}\omega_1,\omega_2;[0,T]} \le c_N.$$

First observe that, for $N = 1$, we have for $s < t \in [0,T]$,

$$\left|\frac{x_{s,t}}{\omega_1\left(s,t\right)^{1/p}} \oplus \frac{h_{s,t}}{\omega_2\left(s,t\right)^{1/q}}\right| \le 2,$$

i.e. $(\mathcal{H}_1)$ is satisfied.

Then, notice it is enough to prove that $(\mathcal{H}_N)$ implies for all $x \oplus h \in C^{1\text{-var}}\left([0,T],\mathbb{R}^d \oplus \mathbb{R}^{d'}\right)$ with $\left\|S_{[p]}\left(x\right)\right\|_{p\text{-var};[0,T]}$ and $|h|_{q\text{-var};[0,T]}$ bounded above by 1, we have

$$\left\|S_N\left(x \oplus h\right)_{0,T}\right\| \le c_{N+1}.$$

Indeed, for an arbitrary $x \oplus h \in C^{1\text{-var}}\left([0,T],\mathbb{R}^d \oplus \mathbb{R}^{d'}\right)$, applying the above to $\left(\frac{x}{\left\|S_{[p]}(x)\right\|_{p\text{-var};[0,T]}}, \frac{h}{|h|_{q\text{-var};[0,T]}}\right)$ would imply that

$$\left\|S_N\left(\frac{x}{\left\|S_{[p]}\left(x\right)\right\|_{p\text{-var};[0,T]}} \oplus \frac{h}{|h|_{q\text{-var};[0,T]}}\right)\right\| \le c_N.$$

By time change, the above would also hold for all s, t, when replacing $[0, T]$ by $[s, t]$. This is precisely saying that

$$\|S_N(x \oplus h)\|_{p, q\text{-}\omega_1, \omega_2; [0, T]} \leq C.$$

Let us therefore fix some path $x \oplus h \in C^{1\text{-var}}\left([0, T], \mathbb{R}^d \oplus \mathbb{R}^{d'}\right)$ with $\left\|S_{[p]}(x)\right\|_{p\text{-var}; [0, T]}$ and $|h|_{q\text{-var}; [0, T]}$ less than 1. We define the control ω by

$$\omega(s, t) = \frac{1}{2}\left(\left\|S_{[p]}(x)\right\|_{p\text{-var}; [0, T]}^p + |h|_{q\text{-var}; [0, T]}^q\right),$$

which satisfies $\omega(0, T) \leq 1$. The induction the hypothesis tells us that

$$\left\|S_N\left(\frac{x}{\omega(s, t)^{1/p}} \oplus \frac{h}{\omega(s, t)^{1/q}}\right)\right\|_{s, t} \leq c_1.$$

If $N + 1 \leq [p]$, the hypothesis tells us that $\left\|S_{N+1}\left(\frac{x}{\omega(s, t)^{1/p}}\right)\right\|_{s, t} \leq c_2$. If $N + 1 > [p]$, then Theorem 9.5 tells us that $\left\|S_{N+1}\left(\frac{x}{\omega(s, t)^{1/p}}\right)\right\|_{s, t} \leq c_2$.

Hence, following Lemma 9.22, we can define the path $x^{s, t} \oplus h^{s, t}$, such that

$$S_N(x \oplus h)_{s, t} = S_N\left(x^{s, t} \oplus h^{s, t}\right)_{s, t}$$
$$S_{N+1}(x)_{s, t} = S_{N+1}\left(x^{s, t}\right)_{s, t}$$

and such that $\frac{1}{\omega(s, t)^{1/p}} \int_s^t |dx_u^{s, t}| + \frac{1}{\omega(s, t)^{1/q}} \int_s^t |dh_u^{s, t}|$ is bounded above by a constant c_2. Define similarly the paths $x^{t, u}, h^{t, u}, x^{s, u}, h^{s, u}$, and define $x^{s, t, u}$ (resp. $h^{s, t, u}$) to be the concatenation of $x^{s, t}$ and $x^{t, u}$ (resp. $h^{s, t}$ and $h^{t, u}$). Then, define

$$\Gamma_{s, t} = S_{N+1}(x \oplus h)_{s, t} - S_{N+1}\left(x^{s, t} \oplus h^{s, t}\right)_{s, t}.$$

Working as in Proposition 9.3,

$$|\Gamma_{s, u} - (\Gamma_{s, t} + \Gamma_{t, u})| \leq \left|S_{N+1}\left(x^{s, t, u} \oplus h^{s, t, u}\right)_{s, u} - S_{N+1}\left(x^{s, u} \oplus h^{s, u}\right)_{s, u}\right|,$$

which we bound using Lemma 9.23:

$$|\Gamma_{s, u} - (\Gamma_{s, t} + \Gamma_{t, u})| \leq c_3 \sum_{k=1}^{N+1} \omega(s, u)^{\frac{N+1-k}{p}} \omega(s, u)^{\frac{k}{q}}$$
$$\leq c_4 \omega(s, u)^\theta$$

where $\theta = \min_{1 \leq k \leq N+1} \left\{ \frac{N+1-k}{p} + \frac{k}{q} \right\} > 1$. Using, as in the proof of Propostion 9.3, that x and h are actually of bounded variation, we obtain using the (analysis) Lemma 6.2 that for all s, t,

$$|\Gamma_{s,t}| \leq c_5 \omega (s,t)^\theta .$$

We then deduce from the triangle inequality that

$$\left| \pi_{N+1} \left(S_{N+1} (x \oplus h)_{0,T} \right) \right| \leq c_6,$$

which concludes the proof. ∎

9.4.4 Definition of Young pairing map

We define
(i) $\mathbf{p}_N$ to be the extension of the projection onto the first d coordinates of $\mathbb{R}^d \oplus \mathbb{R}^{d'}$ to a homomorphism from $G^N \left(\mathbb{R}^d \oplus \mathbb{R}^{d'} \right)$ onto $G^N \left(\mathbb{R}^d \right)$;
(ii) $\mathbf{p}'_N$ to be the extension of the projection onto the last d' coordinates of $\mathbb{R}^d \oplus \mathbb{R}^{d'}$ to a homomorphism from $G^N \left(\mathbb{R}^d \oplus \mathbb{R}^{d'} \right)$ onto $G^N \left(\mathbb{R}^{d'} \right)$.

Definition 9.25 Let $(\mathbf{x}, h) \in C_o^{p\text{-var}} \left([0,T], G^{[p]} \left(\mathbb{R}^d \right) \right) \times C_o^{q\text{-var}} \left([0,T], \mathbb{R}^{d'} \right)$. A path $\mathbf{z}$ in $C_o^{p,q\text{-var}} \left([0,T], G^N \left(\mathbb{R}^d \oplus \mathbb{R}^{d'} \right) \right)$ such that $\mathbf{p}_N (\mathbf{z}) = S_N (\mathbf{x})$ and $\mathbf{p}'_N (\mathbf{z}) = S_N (h)$ is said to be a (p,q)-Lyons lift or Young pairing of $(\mathbf{x}, h)$ of order N. We shall see in the following theorem that such a Young pairing is unique, and will denote it by $S_N (\mathbf{x}, h)$ or $S_N (\mathbf{x} \oplus h)$. □

Theorem 9.26 Let $(\mathbf{x}, h) \in C_o^{p\text{-var}} \left([0,T], G^{[p]} (\mathbb{R}^d) \right) \times C_o^{q\text{-var}} \left([0,T], \mathbb{R}^{d'} \right)$, and $N \geq 1$. Then there exists a unique (p,q)-Lyons lift of order N of $(\mathbf{x}, h)$. By writing $S_N (\mathbf{x}, h)$ for this path, we define the map S_N on $C_o^{p\text{-var}} \left([0,T], G^{[p]} (\mathbb{R}^d) \right) \times C_o^{q\text{-var}} \left([0,T], \mathbb{R}^{d'} \right)$. Moreover, if

$$\omega_1 = \left\| S_{[p]} (\mathbf{x}) \right\|_{p\text{-var};[.,.]}, \quad \omega_2 = |h|_{q\text{-var};[.,.]},$$

then, for some constant $C = C(N, p, q)$,

$$\| S_N (\mathbf{x}, h) \|_{p,q\text{-}\omega_1,\omega_2;[s,t]} \leq C.$$

Proof. The existence of such a lift follows the same lines as the homogenous case, but using Lemma 9.24 rather than Lemma 9.3. Uniqueness follows from the following lemma. ∎

Lemma 9.27 Let $\mathbf{y}$ and $\mathbf{z}$ be two elements of $C_o^{p,q\text{-var}} \left([0,T], G^N \left(\mathbb{R}^d \oplus \mathbb{R}^{d'} \right) \right)$ such that $\mathbf{p}_N (\mathbf{z}) = \mathbf{p}_N (\mathbf{y})$ and $\mathbf{p}'_N (\mathbf{z}) = \mathbf{p}'_N (\mathbf{y})$. Then we have $\mathbf{z} = \mathbf{y}$.

Proof. Let (H_N) be the induction hypothesis that the lemma holds true for level-N paths, i.e. as written in the statement. For $N = 1$, there is nothing to prove, hence (H_1) is true. Assume now that (H_{N-1}) is true, and let us prove that (H_N) is true. Define $\omega_1 = \|\mathbf{y}\|_{p\text{-var};[.,.]}^p$, $\omega_2 = \|\mathbf{z}\|_{q\text{-var};[.,.]}^q$.

We fix $e_1, \ldots, e_d$ a basis of $\mathbb{R}^d$ and $e_{d+1}, \ldots, e_{d+d'}$ a basis of $\mathbb{R}^{d'}$, so that $e_1, \ldots, e_{d+d'}$ is a basis of $\mathbb{R}^d \oplus \mathbb{R}^{d'}$.

Define the $\left(\mathbb{R}^d \oplus \mathbb{R}^{d'}\right)^{\otimes N} \cap \mathfrak{g}_N\left(\mathbb{R}^d \oplus \mathbb{R}^{d'}\right)$-valued path f by

$$f(t) = \log\left(\mathbf{z}_t^{-1} \otimes \mathbf{y}_t\right).$$

With $f(s,t) = f(t) \equiv f(s)$ we have $\exp(f(s,t)) = \mathbf{z}_{s,t}^{-1} \otimes \mathbf{y}_{s,t}$; we may also write

$$f(t) = \sum_{1 \leq i_1, \ldots, i_N \leq d+d'} f_{i_1, \ldots, i_N}(t)\, e_1 \otimes \ldots \otimes e_{d+d'}.$$

Hypothesis (H_{N-1}) implies that $f_{i_1, \ldots, i_N} = 0$ if $\{i_1, \ldots, i_N\} \cap \{d+1, \ldots, d+d'\} = \varnothing$ and from Lemma 7.61,

$$\left|f_{i_1, \ldots, i_N}(s,t)\right| \leq c_1 \omega_1(s,t)^{\frac{N-a}{p}} \omega_2(s,t)^{\frac{a}{q}},$$

where a is the cardinal of the set $\{i_1, \ldots, i_N\} \cap \{d+1, \ldots, d+d'\}$. Set $\omega = \omega_1 + \omega_2$ so that

$$\left|f_{i_1, \ldots, i_N}(s,t)\right| \leq c_1 \left[2\omega(s,t)\right]^{\frac{N-a}{p} + \frac{a}{q}}.$$

But for all $i_1, \ldots, i_N$ such that the cardinal of $\{i_1, \ldots, i_N\} \cap \{d+1, \ldots, d+d'\}$ is greater than or equal to 1, $\frac{N-a}{p} + \frac{a}{q} > 1$, which implies that for the respective component $f_{i_1, \ldots, i_N}$ is of finite r-variation for some $r < 1$, i.e. it is equal to $f_{i_1, \ldots, i_N}(0) = 0$. ∎

9.4.5 Modulus of continuity for the map S_N

We go quickly in this section as there are no new ideas required. First, we need to generalize Lemmas 9.22 and 9.23 to handle the "difference of paths".

Lemma 9.28 *Let* $N \geq 1$, *and* $\left(x^i, h^i\right)_{i=1,2}$ *be two paths in* $C^{1\text{-var}}\left([s,t], \mathbb{R}^d\right) \times C^{1\text{-var}}\left([s,t], \mathbb{R}^{d'}\right)$ *such that, for fixed* $\alpha, \beta > 0$,

$$\left\|S_{N+1}(\alpha x)_{s,t}\right\| + \left\|S_N(\alpha x \oplus \beta h)_{s,t}\right\| \leq C.$$

Then there exist paths $\left\{x^{i,s,t}, h^{i,s,t} : i = 1, 2\right\}$ such that
(i) we have

$$S_N \left(x^i \oplus h^i\right)_{s,t} = S_N \left(x^{i,s,t} \oplus h^{i,s,t}\right)_{s,t}$$

$$S_{N+1} \left(x^i\right)_{s,t} = S_{N+1} \left(x^{i,s,t}\right)_{s,t};$$

(ii) there exists a constant C_N depending only on C and N such that

$$\alpha \int_s^t \left|dx_u^{i,s,t}\right| + \beta \int_s^t \left|dh_u^{i,s,t}\right| \leq C_N;$$

(iii) with

$$\varepsilon = \left| S_{N+1} \left(\alpha x^1\right)_{s,t} - S_{N+1}(\alpha x^2)_{s,t} \right| + \left| S_N \left(\alpha x^1 \oplus \beta h^1\right)_{s,t} - S_N \left(\alpha x^2 \oplus \beta h^2\right)_{s,t} \right|,$$

we have

$$\alpha \int_s^t \left|dx_u^{1,s,t} - dx_u^{2,s,t}\right| + \beta \int_s^t \left|dh_u^{1,s,t} - dh_u^{2,s,t}\right| \leq C_N \varepsilon.$$

Proof. As in the proof of Lemma 9.22, we can assume $\alpha = \beta = 1$. Using Proposition 7.64, there exist two paths $\left(x^{i,1,s,t}, h^{i,1,s,t}\right)_{i=1,2}$ such that

$$S_N \left(x^i \oplus h^i\right)_{s,t} = S_N \left(x^{i,1,s,t} \oplus h^{i,1,s,t}\right)_{s,t},$$

with

$$\int_s^t \left|dx_u^{i,1,s,t}\right| + \int_s^t \left|dh_u^{i,1,s,t}\right| \leq c_1,$$

$$\int_s^t \left|dx_u^{1,1,s,t} - dx_u^{2,1,s,t}\right| + \int_s^t \left|dh_u^{1,1,s,t} - dh_u^{2,1,s,t}\right|$$

$$\leq c_1 \left| S_N \left(x^i \oplus h^i\right)_{s,t} - S_N \left(x^i \oplus h^i\right)_{s,t} \right|.$$

Then, define $g^i = S_{N+1} \left(x^{i,1,s,t}\right)_{s,t}^{-1} \otimes S_{N+1} \left(x^i\right)_{s,t}$. Observe that $\|g^i\| \leq c_2$. Then, from Lemma 9.1

$$\left|g^1 - g^2\right| =$$

$$\left| \left(S_{N+1} \left(x^{1,1,s,t}\right)_{s,t} - S_{N+1} \left(x^{2,1,s,t}\right)_{s,t}\right) - \left(S_{N+1} \left(x^1\right)_{s,t} - S_{N+1} \left(x^2\right)_{s,t}\right) \right|$$

$$\leq \left| S_{N+1} \left(x^1\right)_{s,t} - S_{N+1} \left(x^2\right)_{s,t} \right| + \left| S_{N+1} \left(x^{1,1,s,t}\right)_{s,t} - S_{N+1} \left(x^{2,1,s,t}\right)_{s,t} \right|$$

$$\leq c_3 \varepsilon$$

using Proposition 7.63. Using Proposition 7.64 one more time, we define the paths $\left(x^{i,2,s,t}\right)_{i=1,2}$ by

$$S_{N+1} \left(x^{i,2,s,t}\right)_{s,t} = g^i,$$

with

$$\int_s^t \left| dx_u^{i,2,s,t} \right| = \left\| g^i \right\| \le c_2$$

$$\int_s^t \left| dx_u^{1,2,s,t} - dx_u^{2,2,s,t} \right| = \left| g^1 - g^2 \right| \le c_3 \varepsilon.$$

Concatenating the paths $x^{i,1,s,t} \oplus h^{i,1,s,t}$ and $x^{i,2,s,t} \oplus 0$ gives us two paths that satisfy the required conditions of the lemma. ∎

We leave the proof of the next lemma, extending Lemma 9.23, to the reader.

Lemma 9.29 *Let* $\left(x^i, h^i \right)_{i=1,2}, \left(\tilde{x}^i, \tilde{h}^i \right)_{i=1,2}$ *be four pairs in* $C^{1\text{-var}}$ $\left([s,u], \mathbb{R}^d \oplus \mathbb{R}^{d'} \right)$. *Assume that*

(i) $S_N \left(x^1 \oplus h^1 \right)_{s,u} = S_N \left(x^2 \oplus h^2 \right)_{s,u}$ *and* $S_N \left(\tilde{x}^1 \oplus \tilde{h}^1 \right)_{s,u} = S_N \left(\tilde{x}^2 \oplus \tilde{h}^2 \right)_{s,u}$;
(ii) $S_{N+1} \left(x^1 \right)_{s,u} = S_{N+1} \left(x^2 \right)_{s,u}$ *and* $S_{N+1} \left(\tilde{x}^1 \right)_{s,u} = S_{N+1} \left(\tilde{x}^2 \right)_{s,u}$;
(iii)

$$\int_s^u \left| dx_r^1 \right| + \int_s^u \left| dx_r^2 \right| \le \ell_1 \text{ and } \int_s^u \left| dh_r^1 \right| + \int_s^u \left| dh_r^2 \right| \le \ell_2,$$

$$\int_s^u \left| d\tilde{x}_r^1 \right| + \int_s^u \left| d\tilde{x}_r^2 \right| \le \ell_1 \text{ and } \int_s^u \left| d\tilde{h}_r^1 \right| + \int_s^u \left| d\tilde{h}_r^2 \right| \le \ell_2;$$

(iv)

$$\int_s^u \left| dx_r^1 - d\tilde{x}_r^1 \right| + \int_s^u \left| dx_r^2 - d\tilde{x}_r^2 \right| \le \varepsilon \ell_1 \text{ and}$$

$$\int_s^u \left| dh_r^1 - d\tilde{h}_r^1 \right| + \int_s^u \left| dh_r^2 - d\tilde{h}_r^2 \right| \le \varepsilon \ell_2.$$

Then,

$$\left| \left(S_{N+1} \left(x^1 \right)_{s,u} - S_{N+1} \left(x^2 \right)_{s,u} \right) - \left(S_{N+1} \left(x^3 \right)_{s,u} - S_{N+1} \left(x^4 \right)_{s,u} \right) \right|$$

$$\le C\varepsilon \sum_{k=1}^{N+1} \ell_1^{N+1-k} \ell_2^k.$$

Similar to the proof of Theorem 9.10 we are also led to

Theorem 9.30 *Let* $1 \le q \le p$ *so that* $1/p + 1/q > 1$. *Assume* $\left(\mathbf{x}^i, h^i \right)_{i=1,2}$ *are two pairs of elements in* $C_o^{p\text{-var}} \left([0,T], G^{[p]} \left(\mathbb{R}^d \right) \right) \times C_o^{q\text{-var}} \left([0,T], \mathbb{R}^{d'} \right)$, *and* ω *a control such that for all* $s,t \in [0,T]$, *for* $i = 1,2$,

$$\left\| \mathbf{x}^i \right\|_{p\text{-var};[s,t]}^p + \left| h^i \right|_{q\text{-var};[s,t]}^q \le \omega(s,t)$$

$$\rho_{p\text{-}\omega} \left(\mathbf{x}^1, \mathbf{x}^2 \right) + \rho_{q\text{-}\omega} \left(h^1, h^2 \right) \le \varepsilon.$$

Then, for all $N \geq 1$, there exists a constant C depending only on N, p, and q such that

$$\rho_{p,q\text{-}\omega}\left(S_N\left(\mathbf{x}^1, h^1\right), S_N\left(\mathbf{x}^2, h^2\right)\right) \leq C\varepsilon.$$

Corollary 9.31 *Let ω be a control, $1 \leq q \leq q', 1 \leq p \leq p', 1/p + 1/q > 1$. Then, for fixed $R > 0$, the maps[4]*

$$\left(\left\{\|\mathbf{x}\|_{p\text{-}\omega} \leq R\right\}, d_{p'\text{-}\omega}\right) \quad \times \quad \left(\left\{|h|_{q\text{-}\omega} \leq R\right\}, d_{q'\text{-}\omega}\right)$$

$$\rightarrow \quad \left(C^{p\text{-}\omega}\left([0,T], G^{[p]}\left(\mathbb{R}^d \oplus \mathbb{R}^{d'}\right)\right), d_{p'\text{-}\omega}\right)$$

$$(\mathbf{x}, h) \quad \mapsto \quad S_N(\mathbf{x}, h)$$

and

$$\left(\left\{\|\mathbf{x}\|_{p\text{-var}} \leq R\right\}, d_{p'\text{-var}}\right) \quad \times \quad \left(\left\{|h|_{q\text{-var}} \leq R\right\}, d_{q'\text{-var}}\right)$$

$$\rightarrow \quad \left(C^{p\text{-var}}\left([0,T], G^{[p]}\left(\mathbb{R}^d \oplus \mathbb{R}^{d'}\right)\right), d_{p'\text{-var}}\right)$$

$$(\mathbf{x}, h) \quad \mapsto \quad S_N(\mathbf{x}, h)$$

and

$$\left(\left\{\|\mathbf{x}\|_{p\text{-var}} \leq R\right\}, d_\infty\right) \quad \times \quad \left(\left\{|h|_{q\text{-var}} \leq R\right\}, d_\infty\right)$$

$$\rightarrow \quad \left(C^{p\text{-var}}\left([0,T], G^{[p]}\left(\mathbb{R}^d \oplus \mathbb{R}^{d'}\right)\right), d_\infty\right)$$

$$(\mathbf{x}, h) \quad \mapsto \quad S_N(\mathbf{x}, h)$$

are uniformly continuous.

Proof. A consequence of the previous theorem and an interpolation argument. ∎

Remark 9.32 As a typical application, we see that the Young pairing $(\mathbf{x}, h) \mapsto S_N(\mathbf{x}, h)$ is also continuous in the sense of "uniform convergence with uniform bounds". Indeed, take any sequence of paths $(x_n, h_n) \in C^{1\text{-var}}\left([0,T], \mathbb{R}^d\right)$ such that

$$\sup_n \left(\left\|S_{[p]}(x_n)\right\|_{p\text{-var};[0,T]} + |h_n|_{q\text{-var};[0,T]}\right) \quad < \quad \infty$$

$$\lim_{n\to\infty} d_\infty\left(S_{[p]}(x_n), \mathbf{x}\right) + d_\infty\left(S_{[p]}(h_n), \mathbf{h}\right) \quad = \quad 0.$$

Then, by Theorem 9.26 and the last part of the previous corollary above it, it follows that

$$\sup_n \left\|S_{[p]}(x_n \oplus h_n)\right\|_{p\text{-var};[0,T]} \quad < \quad \infty,$$

$$\lim_{n\to\infty} d_\infty\left(S_{[p]}(x_n \oplus h_n), S_{[p]}(\mathbf{x} \oplus h)\right) \quad = \quad 0.$$

---

[4] $d_{q'\text{-}\omega}\left(h^1, h^2\right) = \left|h^1 - h^2\right|_{q'\text{-}\omega}.$

9.4.6 Translation of rough paths

In Section 7.5.6, we defined the map **plus** from $G^N\left(\mathbb{R}^d \oplus \mathbb{R}^d\right)$ to $G^N\left(\mathbb{R}^d\right)$ to be the unique homomorphism such that for all $x, y \in \mathbb{R}^d$, $\mathbf{plus}\left(\exp\left(x \oplus y\right)\right) = \exp\left(x + y\right)$. If $\mathbf{x}$ is a $G^N\left(\mathbb{R}^d \oplus \mathbb{R}^d\right)$-valued path, we can therefore define the $G^N\left(\mathbb{R}^d\right)$-valued path $\mathbf{plus}\left(\mathbf{x}\right) : t \in [0,T] \mapsto \mathbf{plus}\left(\mathbf{x}_t\right) \in G^N\left(\mathbb{R}^d\right)$. When $\mathbf{x}$ is the weak geometric rough path equal to $\mathbf{S}_{[p]}\left(\mathbf{y} \oplus h\right)$, where $\mathbf{y}$ is a weak geometric p-rough path and h a weak geometric q-rough path, $\mathbf{plus}\left(\mathbf{x}\right)$ is then a canonical notion of addition of two paths.

Theorem 9.33 *Let* $(\mathbf{x}, h) \in C^{p\text{-var}}\left([0,T], G^{[p]}\left(\mathbb{R}^d\right)\right) \times C^{q\text{-var}}\left([0,T], \mathbb{R}^d\right)$. *The **translation** of $\mathbf{x}$ by h, denoted $T_h\left(\mathbf{x}\right) \in C^{p\text{-var}}\left([0,T], G^{[p]}\left(\mathbb{R}^d\right)\right)$, is defined by*

$$T_h\left(\mathbf{x}\right)_t = \mathbf{plus}\left(S_{[p]}\left(\mathbf{x} \oplus h\right)_t\right).$$

(i) We have for some constant C_1 depending only on p and q,

$$\left\|T_h\left(\mathbf{x}\right)\right\|_{p\text{-var};[0,T]} \leq C_1\left(\left\|\mathbf{x}\right\|_{p\text{-var};[0,T]} + \left|h\right|_{q\text{-var};[0,T]}\right). \tag{9.11}$$

(ii) Let $\left(\mathbf{x}^i, h^i\right)_{i=1,2} \in C^{p\text{-var}}\left([0,T], G^{[p]}\left(\mathbb{R}^d\right)\right) \times C^{q\text{-var}}\left([0,T], \mathbb{R}^d\right)$, and ω a control. If we have for all $s, t \in [0,T]$,

$$\left\|\mathbf{x}^i\right\|^p_{p\text{-var};[s,t]} + \left|h^i\right|^q_{q\text{-var};[s,t]} \leq \omega\left(s,t\right),$$

then

$$\rho_{p\text{-}\omega}\left(T_{h^1}\left(\mathbf{x}^1\right), T_{h^2}\left(\mathbf{x}^2\right)\right) \leq C_2\left(\rho_{p\text{-}\omega}\left(\mathbf{x}^1, \mathbf{x}^2\right) + \omega\left(0,T\right)^{1/q-1/p}\rho_{q\text{-}\omega}\left(h^1, h^2\right)\right)$$

for some constant C_2 depending only on p and q.

Remark 9.34 If $\mathbf{x}^i = S_{[p]}\left(x^i\right)$ where x^i is of bounded variation and if h^i is also of bounded variation, then $T_{h^i}\left(\mathbf{x}\right) = S_{[p]}\left(h^i + x^i\right)$, i.e. $T_{h^i}\left(\mathbf{x}^i\right)$ is just the canonical lift of the sum of the paths x^i and h^i. □

Proof. We first prove the quantitative bound on $\left\|T_{h^i}\left(\mathbf{x}^i\right)\right\|_{p\text{-var};[0,T]}$

$$
\begin{aligned}
\left\|T_{h^i}\left(\mathbf{x}^i\right)_{s,t}\right\| &= \left\|\mathbf{plus}\left(S_{[p]}\left(\mathbf{x}^i \oplus h^i\right)_{s,t}\right)\right\| \\
&\leq c_1\left\|S_{[p]}\left(\mathbf{x}^i \oplus h^i\right)_{s,t}\right\|.
\end{aligned}
$$

From Exercise 9.21, defining

$$\omega_1 = \left\|S_{[p]}\left(x\right)\right\|^p_{p\text{-var};[.,.]}, \qquad \omega_2 = \left|h\right|^q_{q\text{-var};[.,.]},$$

we have

$$\left\|S_{[p]}\left(\mathbf{x}^i \oplus h^i\right)_{s,t}\right\| \leq c_2\left\|S_{[p]}\left(\mathbf{x}^i \oplus h^i\right)\right\|_{p,q\text{-}\omega_1,\omega_2}\left(\omega_1\left(s,t\right)^{1/p} + \omega_2\left(s,t\right)^{1/q}\right).$$

From Theorem 9.26, $\left\| S_{[p]} \left(\mathbf{x}^i \oplus h^i \right) \right\|_{p,q\text{-}\omega_1,\omega_2}$ is bounded, which proves (9.11).

Then, for $s, t \in [0, T]$, defining $\varepsilon_{s,t} = \delta_{\frac{1}{\omega(s,t)^{1/p}}} T_{h^1} \left(\mathbf{x}^1 \right)_{s,t} - \delta_{\frac{1}{\omega(s,t)^{1/p}}} T_{h^2}$ $\left(\mathbf{x}^2 \right)_{s,t}$ and using in the third line Proposition 7.65, we have

$$|\varepsilon_{s,t}| =$$

$$\left| \delta_{\frac{1}{\omega(s,t)^{1/p}}} \text{plus} \left(S_{[p]} \left(\mathbf{x}^1 \oplus h^1 \right)_{s,t} \right) - \delta_{\frac{1}{\omega(s,t)^{1/p}}} \text{plus} \left(S_{[p]} \left(\mathbf{x}^2 \oplus h^2 \right)_{s,t} \right) \right|$$

$$= \left| \text{plus} \left(\delta_{\frac{1}{\omega(s,t)^{1/p}}} S_{[p]} \left(\mathbf{x}^1 \oplus h^1 \right)_{s,t} \right) - \text{plus} \left(\delta_{\frac{1}{\omega(s,t)^{1/p}}} S_{[p]} \left(\mathbf{x}^2 \oplus h^2 \right)_{s,t} \right) \right|$$

$$\leq c_1 \left| \delta_{\frac{1}{\omega(s,t)^{1/p}}} S_{[p]} \left(\mathbf{x}^1 \oplus h^1 \right)_{s,t} - \delta_{\frac{1}{\omega(s,t)^{1/p}}} S_{[p]} \left(\mathbf{x}^2 \oplus h^2 \right)_{s,t} \right|$$

$$\leq \begin{array}{l} c_1 \left| \delta_{\frac{1}{\omega(s,t)^{1/p}}, \frac{1}{\omega(s,t)^{1/q}}} S_{[p]} \left(\mathbf{x}^1 \oplus \omega(s,t)^{1/q-1/p} h^1 \right)_{s,t} \right. \\ \left. \quad - \delta_{\frac{1}{\omega(s,t)^{1/p}}, \frac{1}{\omega(s,t)^{1/q}}} S_{[p]} \left(\mathbf{x}^2 \oplus \omega(s,t)^{1/q-1/p} h^2 \right)_{s,t} \right| \end{array} .$$

Using Theorem 9.30, we then obtain

$$|\varepsilon_{s,t}| \leq c_2 \left(\rho_{p\text{-}\omega} \left(\mathbf{x}^1, \mathbf{x}^2 \right) + \rho_{q\text{-}\omega} \left(\omega(s,t)^{1/q-1/p} h^1, \omega(s,t)^{1/q-1/p} h^2 \right) \right)$$

$$\leq c_2 \left(\rho_{p\text{-}\omega} \left(\mathbf{x}^1, \mathbf{x}^2 \right) + \omega(s,t)^{1/q-1/p} \rho_{q\text{-}\omega} \left(h^1, h^2 \right) \right).$$

Hence, as $q < p$, taking supremum over all $s, t \in [0, T]$, we have

$$\rho_{p\text{-}\omega} \left(T_{h^1} \left(\mathbf{x}^1 \right), T_{h^2} \left(\mathbf{x}^2 \right) \right) \leq c_2 \left(\rho_{p\text{-}\omega} \left(\mathbf{x}^1, \mathbf{x}^2 \right) + \omega(0,T)^{1/q-1/p} \rho_{q\text{-}\omega} \left(h^1, h^2 \right) \right).$$

∎

As a corollary, interpolation provides us with the following uniform continuity on bounded sets result.

Corollary 9.35 *The rough path translation* $(\mathbf{x}, h) \mapsto T_h(\mathbf{x})$ *as a map from*

$$C_o^{p\text{-var}} \left([0,T], G^{[p]} \left(\mathbb{R}^d \right) \right) \times C_o^{q\text{-var}} \left([0,T], \mathbb{R}^d \right) \to C_o^{p\text{-var}} \left([0,T], G^{[p]} \left(\mathbb{R}^d \right) \right)$$

is uniformly continuous on bounded sets, using the $d_{p\text{-var}}$*-metric. This is also true as a map from*

$$C_o^{1/p\text{-Höl}} \left([0,T], G^{[p]} \left(\mathbb{R}^d \right) \right) \times C_o^{1/p\text{-Höl}} \left([0,T], \mathbb{R}^d \right)$$

$$\to C_o^{1/p\text{-Höl}} \left([0,T], G^{[p]} \left(\mathbb{R}^d \right) \right).$$

Exercise 9.36 *Assume* $T_{-x^n}(\mathbf{x}) \to 0$. *Show that this is, in general, not equivalent to* $S_2(x^n) \to \mathbf{x}$ *and neither implies the other.*

Exercise 9.37 *This exercise will demonstrate (again!) the power of p-variation estimates in the sense that they immediately imply non-trivial estimates in terms of Hölder and Besov norm. Recall from Example 5.16 that for $\delta \in (1/2, 1]$ and $q = 1/\delta$,*

$$|h|_{q\text{-var};[s,t]} \leq (\text{const}) \times |h|_{W^{\delta,2}\text{-var};[s,t]} |t - s|^{\delta - 1/2}.$$

Assume $\mathbf{x} \in C^{\alpha\text{-Höl}}\left([0,T], G^{[1/\alpha]}\left(\mathbb{R}^d\right)\right)$, $h \in W^{\delta,2}\left([0,T], \mathbb{R}^d\right)$ with $\alpha \in (1/4, 1/2)$ and $\delta := \alpha + 1/2$. Show that

$$\|T_h(\mathbf{x})\|_{\alpha\text{-Höl};[s,t]} \leq (\text{const}) \times \left(\|\mathbf{x}\|_{\alpha\text{-Höl};[s,t]} + |h|_{W^{\delta,2};[s,t]}\right).$$

Explain the restriction $\alpha > 1/4$.

Solution. Set $q = 1/\delta = 1/(\alpha + 1/2)$ and $p = 1/\alpha$ so that $h \in C^{q\text{-var}}$ and $\mathbf{x} \in C^{p\text{-var}}$. To apply the above corollary we need

$$1/p + 1/q > 1 \iff \alpha + (\alpha + 1/2) > 1,$$

which explains the restriction $\alpha > 1/4$. The actual estimate then immediately follows from

$$\begin{aligned} T_h(\mathbf{x})_{s,t} &\leq c\|\mathbf{x}\|_{p\text{-var};[s,t]} + c|h|_{q\text{-var};[s,t]} \\ &\leq c\|\mathbf{x}\|_{\alpha\text{-Höl};[s,t]}|t-s|^\alpha + c|h|_{W^{\delta,2};[s,t]}|t-s|^\alpha. \end{aligned}$$

□

9.5 Comments

The main results of this chapter can be found in Lyons [116], see also Lyons and Qian [120] and Lyons *et al.* [123], although some of our proofs are new. Let us note that the estimate for the Lyons lift (Theorem 9.5) can be made independent of N, a consequence of Lyons' "neo-classical inequality", see Lyons [116] and, for a sharpened version, forthcoming work by Hara and Hino. The necessity to distinguish between geometric rough paths and *weak* geometric rough paths was recognized in Friz and Victoir [63]. Exercise 9.37 is taken from Friz and Victoir [64].

10

Rough differential equations

Our construction of Young's integral was based on estimates for classical Riemann–Stieltjes integrals with constants depending only on the p- and q-variation of integrand and integrator respectively, followed by a limit argument. The same approach works for ordinary differential equations: in this chapter we establish estimates for ordinary differential equations with constants only depending on a suitable p-variation bound of the driving signal. A limiting procedure then leads us naturally to "rough differential equations".

10.1 Preliminaries

As was pointed out in Section 7.1, for a fixed starting time s, a natural step-N approximation for the solution of the ODE

$$dy = V(y)\, dx = \sum_{i=1}^{d} V_i(y)\, dx^i, \ y_s \in \mathbb{R}^e,$$

is given by

$$y_t \approx y_s + \mathcal{E}_{(V)}\left(y, S_N(x)_{s,t}\right) \tag{10.1}$$

where

Definition 10.1 (Euler scheme) Let $N \in \mathbb{N}$. Given $(N-1)$ times continuously differentiable vector fields $V = (V_1, \dots, V_d)$ on $\mathbb{R}^e$, $\mathbf{g} \in T^{(N)}(\mathbb{R}^d)$ and $y \in \mathbb{R}^e$ we call

$$\mathcal{E}_{(V)}(y, \mathbf{g}) := \sum_{k=1}^{N} \sum_{\substack{i_1, \dots, i_k \\ \in \{1, \dots, d\}}} V_{i_1} \dots V_{i_k} I(y)\, \mathbf{g}^{k, i_1, \dots, i_k}$$

the (increment of the) step-N Euler scheme. □

When $\mathbf{g} = S_N(x)_{s,t}$, the (step-$N$) signature of a path segment $x|_{[s,t]}$, we call $\mathcal{E}_{(V)}\left(y_s, S_N(x)_{s,t}\right)$ the (increment of the) step-N Euler scheme for $dy = V(y)\, dx$ over the time interval $[s,t]$.

We now prove a simple error estimate for the step-N scheme. To this end, it is convenient to assume Lipschitz regularity of the vector fields in the sense of E. Stein. To prepare for the following definition, given a real $\gamma > 0$, we agree that $\lfloor \gamma \rfloor$ is the largest integer *strictly* smaller than γ so that

$$\gamma = \lfloor \gamma \rfloor + \{\gamma\} \text{ with } \lfloor \gamma \rfloor \in \mathbb{N} \text{ and } \{\gamma\} \in (0,1].$$

Definition 10.2 (Lipschitz map) A map $V : E \to F$ between two normed spaces E, F is called γ-Lipschitz (in the sense of E. Stein), in symbols

$$V \in \text{Lip}^\gamma (E, F) \text{ or simply } V \in \text{Lip}^\gamma (E) \text{ if } E = F,$$

if V is $\lfloor \gamma \rfloor$ times continuously differentiable and such that there exists a constant $0 \le M < \infty$ such that the supremum norm of its kth derivatives, $k = 0, \ldots, \lfloor \gamma \rfloor$, and the $\{\gamma\}$-Hölder norm of its $\lfloor \gamma \rfloor$th derivative are bounded by M. The smallest M satisfying the above conditions is the γ-Lipschitz norm of V, denoted $|V|_{\text{Lip}^\gamma}$. □

(It should be noted that Lip^N maps have $(N-1)$ bounded derivatives, with the $(N-1)$th derivative being Lipschitz, but need not be N times continuously differentiable.) This definition applies in particular to a collection of vector fields $V = (V_1, \ldots, V_d)$ on $\mathbb{R}^e$, which we can view as a map $y \mapsto \{a = (a^1, \ldots, a^d) \mapsto \sum_{i=1}^d V(y) a^i\}$ from $\mathbb{R}^e$ into $L(\mathbb{R}^d, \mathbb{R}^e)$, equipped with operator norm. Saying that $V \in \text{Lip}^\gamma (\mathbb{R}^e, L(\mathbb{R}^d, \mathbb{R}^e))$ is equivalent to $V_1, \ldots, V_d \in \text{Lip}^\gamma (\mathbb{R}^e)$, but it is usually the γ-Lipschitz norm of V which comes up naturally in estimates.

We are now ready to state a first error estimate for the Euler approximation in (10.1).

Proposition 10.3 (Euler ODE estimate) *Let* $\gamma > 1$, $V = (V_i)_{1 \le i \le d}$ *be a collection of vector fields in* $\text{Lip}^{\gamma-1} (\mathbb{R}^e)$ *and* $x \in C^{1\text{-var}} ([s,t], \mathbb{R}^d)$. *Then there exists a constant* $C = C(\gamma)$ *such that*

$$\left| \pi_{(V)} (s, y_s; x)_{s,t} - \mathcal{E}_{(V)} \left(y_s, S_{\lfloor \gamma \rfloor} (x)_{s,t} \right) \right| \le C \left(|V|_{\text{Lip}^{\gamma-1}} \int_s^t |dx_r| \right)^\gamma.$$
$$(10.2)$$

Proof. At the cost of replacing x by $|V|_{\text{Lip}^{\gamma-1}} x$ and V_i by $\frac{1}{|V|_{\text{Lip}^{\gamma-1}}} V_i$, we can and will assume that $|V|_{\text{Lip}^{\gamma-1}} = 1$. We set $n := \lfloor \gamma \rfloor$ and first show that

$$y_{s,t} - \mathcal{E}_{(V)} \left(y_s, S_n (x)_{s,t} \right)$$

$$= \sum_{\substack{i_1, \ldots, i_n \\ \in \{1, \ldots, d\}}} \int_{s < r_1 < \cdots < r_n < t} [V_{i_1} \cdots V_{i_n} I(y_{r_1}) - V_{i_1} \cdots V_{i_n} I(y_s)] \, dx_{r_1}^{i_1} \cdots dx_{r_n}^{i_n}.$$

To this end, consider a smooth function f and note that for any $k \leq n-1$, $V_{i_1} \cdots V_{i_k} f \in C^1$. By iterated use of the change-of-variable formula (cf. Exercise 3.17),

$$f(y_t) = f(y_s) + \sum_{k=1}^{n-1} \sum_{\substack{i_1,\ldots,i_k \\ \in \{1,\ldots,d\}}} \int_{s<r_1<\cdots<r_k<t} V_{i_1} \cdots V_{i_k} f(y_s) \, dx_{r_1}^{i_1} \cdots dx_{r_k}^{i_k}$$

$$+ \sum_{\substack{i_1,\ldots,i_k \\ \in \{1,\ldots,d\}}} \int_{s<r_1<\cdots<r_n<t} V_{i_1} \cdots V_{i_n} f(y_{r_1}) \, dx_{r_1}^{i_1} \cdots dx_{r_n}^{i_n}$$

and the claim follows from specializing to $f = I$, the identity function. Clearly,

$$|y_{s,t}| = \left| \int_s^t V(y) \, dx \right| \leq c_1 \int_s^t |dx_r|.$$

$\text{Lip}^{\gamma-1}$-regularity of the vector fields implies that $V_{i_1} \cdots V_{i_n} I(\cdot)$ is Hölder continuous with exponent $\{\gamma\} \equiv \gamma - n$. Hence, for all $r \in [s,t]$,

$$|V_{i_1} \cdots V_{i_n} I(y_r) - V_{i_1} \cdots V_{i_n} I(y_s)| \leq c_2 \left(\int_s^t |dx_r| \right)^{\{\gamma\}}$$

and after integration, using $\gamma = n + \{\gamma\}$,

$$\left| \int_{s<r_1<\cdots<r_n<t} [V_{i_1} \cdots V_{i_n} I(y_{r_1}) - V_{i_1} \cdots V_{i_n} I(y_s)] \, dx_{r_1}^{i_1} \cdots dx_{r_n}^{i_n} \right|$$

$$\leq c_3 \left(\int_s^t |dx_r| \right)^\gamma.$$

Summation over the indices finishes the estimate. ∎

Remark 10.4 The proof also showed that, keeping the notation $n = \lfloor \gamma \rfloor$,

$$\pi_{(V)}(0, y_0; x)_{0,T} - \mathcal{E}_{(V)}\left(y_0, S_{\lfloor \gamma \rfloor}(x)_{0,T} \right)$$

$$= \sum_{\substack{i_1,\ldots,i_k \\ \in \{1,\ldots,d\}}} \int_{0<r_1<\cdots<r_n<T} [V_{i_1} \cdots V_{i_n} I(y_{r_1}) - V_{i_1} \cdots V_{i_n} I(y_0)] \, dx_{r_1}^{i_1} \cdots dx_{r_n}^{i_n}$$

$$= \sum_{\substack{i_1,\ldots,i_k \\ \in \{1,\ldots,d\}}} \int_{0<\underline{r}<T} [V_{i_1} \cdots V_{i_n} I(y_r) - V_{i_1} \cdots V_{i_n} I(y_0)]$$

$$dx_r^{i_1} \underbrace{\int_{r<\underline{r_2}\cdots<\underline{r_n}<T} dx_{r_2}^{i_2} \cdots dx_{r_n}^{i_n}}_{\equiv d\left(\mathbf{x}_{r,T}^{n;i_1,\ldots,i_n} \right)}$$

(we underlined integration variables here)

$$\equiv \int_{0<r<T} [V^n(y_r) - V^n(y_0)] \, d\left(\mathbf{x}_{r,T}^n \right).$$

□

10.2   Davie's estimate

The main result in this section will require a quantitative understanding of

(A) the difference of ODE solutions started at the same point, with different driving signals (but with common iterated integrals up to a given order);

(B) the difference of ODE solutions started at different points but with identical driving signals.

This is the content of the following two lemmas.

Lemma 10.5 (Lemma A) *Assume that*
(i) $V = (V_i)_{1 \le i \le d}$ is a collection of vector fields in $\mathrm{Lip}^{\gamma-1}(\mathbb{R}^e)$, with $\gamma > 1$;
(ii) $s < u$ are some elements in $[0, T]$;
(iii) $y_s \in \mathbb{R}^e$ (thought of as a "time-s" initial condition);
(iv) x and $\tilde{x}$ are some paths in $C^{1\text{-var}}([s, u], \mathbb{R}^d)$ such that $S_{\lfloor \gamma \rfloor}(x)_{s,u} = S_{\lfloor \gamma \rfloor}(\tilde{x})_{s,u}$;
(v) $\ell \ge 0$ is a bound on $|V|_{\mathrm{Lip}^{\gamma-1}} \left(\int_s^u |dx| + \int_s^u |d\tilde{x}| \right)$.
Then we have, for some constant $C = C(\gamma)$,

$$\left| \pi_{(V)}(s, y_s; x)_{s,u} - \pi_{(V)}(s, y_s; \tilde{x})_{s,u} \right| \le C\ell^{\gamma}.$$

Proof. We do not give the most straightforward proof (which would be to insert the Euler approximation of order $\lfloor \gamma \rfloor$ and use the triangle inequality), but provide a (still simple) proof that will be more instructive later on. By reparametrization of time, we can assume $(s, u) = (0, 1)$. Define the concatenation of $\tilde{x}(1 - \cdot)$ and $x(\cdot)$, in symbols

$$z := \overleftarrow{\tilde{x}} \sqcup x,$$

reparametrized so that $z : [0, 1] \to \mathbb{R}^d$. Then,

$$
\begin{aligned}
\pi_{(V)}(0, y_0; x)_{0,1} - \pi_{(V)}(0, y_0; \tilde{x})_{0,1} &= \pi_{(V)}(0, y_0; x)_1 - \pi_{(V)}(0, y_0; \tilde{x})_1 \\
&= \pi_{(V)}(0, \pi_{(V)}(0, y_0; \tilde{x})_1; z)_1 - \pi_{(V)}(0, y_0; \tilde{x})_1 \\
&= \pi_{(V)}(0, \pi_{(V)}(0, y_0; \tilde{x})_1; z)_{0,1}.
\end{aligned}
$$

By assumption (iv) and Chen's theorem,

$$S_{\lfloor \gamma \rfloor}(z)_{0,1} = S_{\lfloor \gamma \rfloor}(\tilde{x})_{0,1}^{-1} \otimes S_{\lfloor \gamma \rfloor}(x)_{0,1} = 1.$$

Hence, $\mathcal{E}_{(V)}\left(\cdot, S_{\lfloor \gamma \rfloor}(z)_{0,1} \right) \equiv 0$ and the proof is finished with the ODE

Euler estimates from Proposition 10.3,

$$
\begin{aligned}
\left| \pi_{(V)}(0, \cdot; z)_{0,1} \right| &= \left| \pi_{(V)}(0, \cdot; z)_{0,1} - \mathcal{E}_{(V)} \left(\cdot, S_{\lfloor \gamma \rfloor}(z)_{0,1} \right) \right| \\
&\leq c_1 \left[|V|_{\mathrm{Lip}^{\gamma-1}} \int_0^1 |dz| \right]^{\gamma} \\
&= c_1 \left(|V|_{\mathrm{Lip}^{\gamma-1}} \left(\int_0^1 |dx| + \int_0^1 |d\tilde{x}| \right) \right)^{\gamma} \leq c_1 \ell^{\gamma}.
\end{aligned}
$$

∎

Lemma 10.6 (Lemma B) *Assume that*
(i) $V = (V_i)_{1 \leq i \leq d}$ is a collection of vector fields in $\mathrm{Lip}^1(\mathbb{R}^e)$;
(ii) $t < u$ are some elements in $[0, T]$;
(iii) $y_t, \tilde{y}_t \in \mathbb{R}^e$ (thought of as "time-t" initial conditions);
(iv) x is a path in $C^{1\text{-var}}([t, u], \mathbb{R}^d)$;
(v) $\ell \geq 0$ is a bound on $|V|_{\mathrm{Lip}^1} \int_t^u |dx|$.
Then, if $\pi_{(V)}(t, \cdot; x)$ denotes the unique solution to $dy = V(y)\, dx$ from some time-t initial condition, we have

$$
\left| \pi_{(V)}(t, y_t; x)_{t,u} - \pi_{(V)}(t, \tilde{y}_t; x)_{t,u} \right| \leq |y_t - \tilde{y}_t| . \ell \exp(\ell).
$$

In particular, the flow associated with $dy = V(y)\, dx$ is Lipschitz continuous.

Proof. This was seen in Theorem 3.8. ∎

Equipped with these two simple lemmas, and the technical Lemma 10.59 in Appendix 10.8, we are now ready to provide the crucial p-variation estimate of an ODE solution in terms of the p-variation of the driving signal.

Lemma 10.7 (Davie's lemma) *Let $\gamma > p \geq 1$. Assume that*
(i) $V = (V_i)_{1 \leq i \leq d}$ is a collection of vector fields in $\mathrm{Lip}^{\gamma-1}(\mathbb{R}^e)$;
(ii) x is a path in $C^{1\text{-var}}([0, T], \mathbb{R}^d)$, and $\mathbf{x} := S_{[p]}(x)$ is its canonical lift to a $G^{[p]}(\mathbb{R}^d)$-valued path;
(iii) $y_0 \in \mathbb{R}^e$ is an initial condition.
*Then there exists a constant C_1 depending on p, γ (and **not** depending on the 1-variation norm of x) such that for all $s < t$ in $[0, T]$,*

$$
\left| \pi_{(V)}(0, y_0; x) \right|_{p\text{-var};[s,t]} \leq C_1 \left(|V|_{\mathrm{Lip}^{\gamma-1}} \|\mathbf{x}\|_{p\text{-var};[s,t]} \vee |V|_{\mathrm{Lip}^{\gamma-1}}^p \|\mathbf{x}\|_{p\text{-var};[s,t]}^p \right).
\tag{10.3}
$$

Moreover, if $x^{s,t} \in C^{1\text{-var}}([s, t], \mathbb{R}^d)$ is a path such that

$$
S_{\lfloor \gamma \rfloor}\left(x^{s,t}\right)_{s,t} = S_{\lfloor \gamma \rfloor}(x)_{s,t} \quad and \quad \int_s^t \left| dx^{s,t} \right| \leq K \|\mathbf{x}\|_{p\text{-var};[s,t]}
\tag{10.4}
$$

for some $K \geq 1$ then, for any time-s initial condition $y_s \in \mathbb{R}^e$,

$$
\left| \pi_{(V)}(s, y_s; x)_{s,t} - \pi_{(V)}\left(s, y_s; x^{s,t}\right)_{s,t} \right| \leq C_2 \left(K |V|_{\mathrm{Lip}^{\gamma-1}} \|\mathbf{x}\|_{p\text{-var};[s,t]} \right)^{\gamma}
\tag{10.5}
$$

where C_2 depends on p and γ.

Remark 10.8 In case of non-uniqueness we abuse notation in the sense that $\pi_{(V)}(0, y_0; x)$ resp. $\pi_{(V)}(s, y_s; x)$ in the above estimates stands for any choice of ODE solution to $dy = V(y)\,dx$ with the indicated initial conditions at times $0, s$ respectively. □

Remark 10.9 From Proposition 10.3, inequality (10.5) is equivalent to the Euler estimate

$$\left|\pi_{(V)}(s, y_s; x)_{s,t} - \mathcal{E}_{(V)}\left(s, y_s; S_{\lfloor\gamma\rfloor}(x^{s,t})_{s,t}\right)\right| \leq C_3 \left(|V|_{\mathrm{Lip}^{\gamma-1}} \|\mathbf{x}\|_{p\text{-var};[s,t]}\right)^{\gamma}.$$
(10.6)
□

Remark 10.10 A finite variation path (and even Lipschitz continuous) with the properties (10.4) always exists. Indeed, it suffices to take $x^{s,t}$ as geodesic associated with the element $\mathbf{g} = S_{\lfloor\gamma\rfloor}(x)_{s,t} \in G^{\lfloor\gamma\rfloor}(\mathbb{R}^d)$, parametrized on the interval $[s, t]$. The length of this curve is precisely equal to $\|\mathbf{g}\|$, the Carnot–Caratheodory norm of $\mathbf{g} \in G^{\lfloor\gamma\rfloor}(\mathbb{R}^d)$, and so

$$\int_s^t |dx^{s,t}| = \left\|S_{\lfloor\gamma\rfloor}(x)_{s,t}\right\| \leq \left\|S_{\lfloor\gamma\rfloor}(x)\right\|_{p\text{-var};[s,t]}$$

$$\leq C_{(\gamma,p)} \left\|S_{[p]}(x)\right\|_{p\text{-var};[s,t]}$$

where in the last step we used the estimates for the Lyons-lift map, Proposition 9.3, applicable as $\gamma > p$ and so $\lfloor\gamma\rfloor \geq [p]$. Let us also note that

$$\int_s^t |dx^{s,t}| \leq K \int_s^t |dx|$$
(10.7)

since $\int_s^t |dx^{s,t}| \leq K \|\mathbf{x}\|_{p\text{-var};[s,t]} \leq K \|\mathbf{x}\|_{1\text{-var};[s,t]} = K |x|_{1\text{-var};[s,t]}$. □

Proof. The case $p < \gamma < 2$ is discussed in Exercises 10.12, 10.13. We assume here $\gamma \geq 2$, so that $dy = V(y)\,dx$ with time-s initial condition y_s has a unique solution, denoted as usual by $\pi(s, y_s; x)$. We define $\mathbf{x} = S_{[p]}(x)$, and

$$\omega(s, t) = \left(K|V|_{\mathrm{Lip}^{\gamma-1}} \|\mathbf{x}\|_{p\text{-var};[s,t]}\right)^p.$$

Thanks to $\int_s^t |dx^{s,t}| \leq K \|\mathbf{x}\|_{p\text{-var};[s,t]}$ and an elementary ODE estimate (Theorem 3.4), we have

$$\left|\pi_{(V)}\left(s, y_s; x^{s,t}\right)_{s,t}\right| \leq c_1 \omega(s,t)^{\frac{1}{p}}.$$
(10.8)

Then, for all $s < t$ in $[0, T]$, we define

$$\Gamma_{s,t} = y_t - \pi\left(s, y_s; x^{s,t}\right)_t = y_{s,t} - \pi\left(s, y_s; x^{s,t}\right)_{s,t}.$$

Then, for fixed $s < t < u$ in $[0, T]$, we have

$$\Gamma_{s,u} - \Gamma_{s,t} - \Gamma_{t,u} = -\pi \left(s, y_s; x^{s,u}\right)_{s,u} + \pi \left(s, y_s; x^{s,t}\right)_{s,t} + \pi \left(t, y_t; x^{t,u}\right)_{t,u}.$$

Define $x^{s,t,u}$ to be the concatenation of $x^{s,t}$ and $x^{t,u}$ and set, for better readability,

$$
\begin{aligned}
A &:= \pi \left(s, y_s; x^{s,t,u}\right)_{s,u} - \pi \left(s, y_s; x^{s,u}\right)_{s,u} \\
B &:= \pi \left(t, y_t; x^{t,u}\right)_{t,u} - \pi \left(t, \pi \left(s, y_s; x^{s,t}\right)_t; x^{t,u}\right)_{t,u} \\
&= \pi \left(t, y_t; x^{t,u}\right)_{t,u} - \pi \left(t, y_t - \Gamma_{s,t}; x^{t,u}\right)_{t,u}
\end{aligned}
$$

which then allows us to write

$$\Gamma_{s,u} - \Gamma_{s,t} - \Gamma_{t,u} = A + B. \tag{10.9}$$

The term A is estimated by – nomen est omen – *Lemma A*, noting that

$$\int_s^u \left|dx^{s,t,u}\right| = \int_s^t \left|dx^{s,t}\right| + \int_t^u \left|dx^{t,u}\right| \le 2K \left\|\mathbf{x}\right\|_{p\text{-var};[s,u]}.$$

Similarly, *Lemma B* was tailor-made to estimate B and we are led to

$$
\begin{aligned}
\left|\Gamma_{s,u} - \Gamma_{s,t} - \Gamma_{t,u}\right| &\le c_1 \omega \left(s, u\right)^{\gamma/p} + c_2 \left|\pi \left(s, y_s; x^{s,t}\right)_t - y_t\right| \omega \left(t, u\right)^{1/p} \\
&\qquad \exp \left(c_2 \omega \left(t, u\right)^{1/p}\right) \\
&\le c_1 \omega \left(s, u\right)^{\gamma/p} + c_2 \left|\Gamma_{s,t}\right| \omega \left(t, u\right)^{1/p} \exp \left(c_2 \omega \left(t, u\right)^{1/p}\right).
\end{aligned}
$$

The elementary inequality

$$1 + c_2 \omega \left(t, u\right)^{1/p} \exp \left(c_2 \omega \left(t, u\right)^{1/p}\right) \le \exp \left(2 c_2 \omega \left(s, u\right)^{1/p}\right),$$

combined with the triangle inequality, then gives

$$\left|\Gamma_{s,u}\right| \le \left|\Gamma_{s,t}\right| \exp \left(2 c_2 \omega \left(s, u\right)^{1/p}\right) + \left|\Gamma_{t,u}\right| + c_1 \omega \left(s, u\right)^{\frac{\gamma}{p}}. \tag{10.10}$$

On the other hand, using again *Lemma A*,

$$
\begin{aligned}
\left|\Gamma_{s,t}\right| &= \left|y_{s,t} - \pi \left(s, y_s; x^{s,t}\right)_{s,t}\right| \\
&\le c_3 \left[\left|V\right|_{\text{Lip}^{\gamma-1}} \int_s^t \left|dx\right| + \left|V\right|_{\text{Lip}^{\gamma-1}} \int_s^t \left|dx^{s,t}\right|\right]^\gamma \\
&\le c_3 \underbrace{\left[\left|V\right|_{\text{Lip}^{\gamma-1}} \left(K+1\right) \int_s^t \left|dx\right|\right]^\gamma}_{=: \bar{\omega}(s,t)} \qquad \text{thanks to (10.7).}
\end{aligned}
$$

Obviously, $\tilde{\omega}$ is a control function whose finiteness depends crucially on the a priori assumption that x has finite 1-variation and we summarize the previous estimate in writing:

$$|\Gamma_{s,t}| = O\left(\tilde{\omega}\left(s,t\right)^{\gamma}\right) \text{ where } \gamma > 1. \tag{10.11}$$

The two estimates (10.10), (10.11) are precisely what is needed to apply (the elementary analysis) Lemma 10.59 (found in the appendix of this chapter): it follows that, for all $s < t$ in $[0,T]$,

$$|\Gamma_{s,t}| \leq c_4\omega\left(s,t\right)^{\gamma/p} \exp\left(c_4\omega\left(s,t\right)^{1/p}\right)$$

and we emphasize that c_4 does not depend on $\tilde{\omega}$ and in particular not on the 1-variation of x. From (10.8) and the triangle inequality, we therefore have for all $s < t$ in $[0,T]$,

$$|y_{s,t}| \leq c_1\omega\left(s,t\right)^{1/p} + c_4\omega\left(s,t\right)^{\gamma/p} \exp\left(c_4\omega\left(s,t\right)^{1/p}\right)$$

and if attention is restricted to s,t such that $\omega\left(s,t\right) \leq 1$, we obviously have

$$|y_{s,t}| \leq \left(c_1 + c_4 e^{c_4}\right)\omega\left(s,t\right)^{1/p}.$$

But then it follows from Proposition 5.10 that for all $s < t$ in $[0,T]$,

$$|y|_{p\text{-var};[s,t]} \leq c_5 \left(\omega\left(s,t\right)^{1/p} \vee \omega\left(s,t\right)\right).$$

That also leads to

$$
\begin{aligned}
|\Gamma_{s,t}| &= \left|y_{s,t} - \pi\left(s,y_s;x^{s,t}\right)_{s,t}\right| \\
&\leq c_5 \left(\omega\left(s,t\right)^{1/p} \vee \omega\left(s,t\right)\right) + c_1\omega\left(s,t\right)^{1/p}
\end{aligned}
$$

and hence

$$
\begin{aligned}
|\Gamma_{s,t}| &\leq \min\left\{c_4\omega\left(s,t\right)^{\gamma/p} \exp\left(c_4\omega\left(s,t\right)^{1/p}\right), c_5 \left(\omega\left(s,t\right)^{1/p} \vee \omega\left(s,t\right)\right)\right. \\
&\quad \left. + c_1\omega\left(s,t\right)^{1/p}\right\} \\
&\leq c_6\omega\left(s,t\right)^{\gamma/p}.
\end{aligned}
$$

The proof is now finished. ∎

Exercise 10.11 *Prove that we can take the constants C_1 and C_2 in Lemma 10.7 to be continuous in p, for $p \in [1,\gamma)$.*

The following exercise deals with the case $p < \gamma < 2$ in Lemma 10.7.

Exercise 10.12 *(i) for $1 < \gamma < 2$, and $a, b > 0$, prove that*

$$a^{\gamma-1} b \le (\gamma - 1) a b^{\gamma-1} + (2 - \gamma) b^{\gamma}. \tag{10.12}$$

(ii) Under the assumption of Lemma 10.7 for $p < \gamma < 2$, prove that if $\Gamma_{s,t} = y_{s,t} - V(y_s) x_{s,t}$ and $\omega(s,t) = |V|_{\mathrm{Lip}^{\gamma-1}} |x|_{p\text{-var},[s,t]}$, we have for $s < t < u$,

$$|\Gamma_{s,u} - \Gamma_{s,t} - \Gamma_{t,u}| \le c(\gamma - 1) |\Gamma_{s,t}| \, \omega(t,u)^{\frac{\gamma-1}{p}} + c(2 - \gamma) \omega(t,u)^{\gamma/p}.$$

(iii) Prove Lemma 10.7 in the case $\gamma < 2$.

Solution. (i) Write $x = a/b$, and dividing by b^{γ}, we see that (10.12) is equivalent to

$$x^{\gamma-1} \le (\gamma - 1) x + (2 - \gamma) \text{ for } x > 0,$$

which is checked by basic calculus.
(ii) For $s < t < u$,

$$
\begin{aligned}
|\Gamma_{s,u} - \Gamma_{s,t} - \Gamma_{t,u}| &= |[V(y_t) - V(y_s)] x_{t,u}| \\
&\le c |\Gamma_{s,t}|^{\gamma-1} \omega(t,u)^{1/p} \\
&\le c(\gamma - 1) |\Gamma_{s,t}| \omega(t,u)^{\frac{\gamma-1}{p}} + c(2 - \gamma) \omega(t,u)^{\gamma/p} \\
&\quad \text{using (i).}
\end{aligned}
$$

(iii) Exactly the same argument as in the proof of Lemma 10.7.   □

Exercise 10.13 *(i) Under the assumption of Lemma 10.7 for $p < \gamma < 2$, prove using Young estimates that for all s, t*

$$|y|_{p\text{-var};[s,t]} \le c_1 |V|_{\mathrm{Lip}^{\gamma-1}} |x|_{p\text{-var};[s,t]} \left\{ 1 + |y|_{p\text{-var};[s,t]} \right\}.$$

(ii) Using proposition 5.10, prove inequality (10.3) of Lemma 10.7.

Solution. From Young's inequality,

$$
\begin{aligned}
|y_{s,t}| &= \left| \int_s^t V(y) \, dx \right| \\
&\le c_1 |x|_{p\text{-var};[s,t]} \left\{ |V|_\infty + |V|_{\mathrm{Lip}^{\gamma-1}} |y|_{p\text{-var};[s,t]}^{\gamma-1} \right\} \\
&\le c_1 |V|_{\mathrm{Lip}^{\gamma-1}} |x|_{p\text{-var};[s,t]} \left\{ 1 + |y|_{p\text{-var};[s,t]}^{\gamma-1} \right\} \\
&\le c_2 |V|_{\mathrm{Lip}^{\gamma-1}} |x|_{p\text{-var};[s,t]} \left\{ 1 + |y|_{p\text{-var};[s,t]} \right\} \\
&\le c_3 |V|_{\mathrm{Lip}^{\gamma-1}} |x|_{p\text{-var};[s,t]} \left\{ 1 + |y|_{p\text{-var};[s,t]}^p \right\}^{1/p}.
\end{aligned}
$$

The p-variation of x is controlled by $\omega_{x;p} \equiv |x|^p_{p\text{-var};[\cdot,\cdot]}$. Using similar notation for y we see that $|y_{s,t}|^p$ is estimated by a constant times $\omega_{x;p} + \omega_{x;p}\omega_{y;p}$ which is a control (cf. Exercise 1.9) and from the basic super-additivity property of controls,

$$
\begin{aligned}
|y|_{p\text{-var};[s,t]} &\leq c_3 |V|_{\text{Lip}^{\gamma-1}} |x|_{p\text{-var};[s,t]} \left\{ 1 + |y|^p_{p\text{-var};[s,t]} \right\}^{1/p} \\
&\leq c_3 |V|_{\text{Lip}^{\gamma-1}} |x|_{p\text{-var};[s,t]} \left\{ 1 + |y|_{p\text{-var};[s,t]} \right\}.
\end{aligned}
$$

(ii) For s,t such that $2c_2 |V|_{\text{Lip}^{\gamma-1}} |x|_{p\text{-var};[s,t]} < 1$, we obtain

$$
\begin{aligned}
|y|_{p\text{-var};[s,t]} &\leq \frac{c_2 |V|_{\text{Lip}^{\gamma-1}} |x|_{p\text{-var};[s,t]}}{\left(1 - c_2 |V|_{\text{Lip}^{\gamma-1}} |x|_{p\text{-var};[s,t]} \right)} \\
&\leq 2c_2 |V|_{\text{Lip}^{\gamma-1}} |x|_{p\text{-var};[s,t]}.
\end{aligned}
$$

We then obtain estimates on $|y|_{p\text{-var};[s,t]}$ for s,t such that $2c_2 |V|_{\text{Lip}^{\gamma-1}}$ $|x|_{p\text{-var};[s,t]} \geq 1$ using Proposition 5.10. $\qquad\square$

10.3 RDE solutions

Davie's lemma gives us *uniform* estimates for ODE solutions which depend only on the rough path regularity (e.g. p-variation or $1/p$-Hölder) of the canonical lift of a "nice" driving signal $x \in C^{1\text{-var}}\left([0,T],\mathbb{R}^d\right)$. It should therefore come as no surprise that a careful passage to the limit will yield a sensible notion of differential equations driven by a "generalized" driving signal, given as a limit of nice driving signals (in p-variation or $1/p$-Hölder rough path sense ...). This class of generalized driving signals is precisely the class of weak geometric p-rough paths introduced in the previous chapter. Indeed, we saw in Section 8.2 that for any $\mathbf{x} \in C^{p\text{-var}}\left([0,T],G^{[p]}\left(\mathbb{R}^d\right)\right)$, there exist $(x_n) \subset C^{1\text{-var}}\left([0,T],\mathbb{R}^d\right)$ which approximate $\mathbf{x}$ uniformly with uniform p-variation bounds,

$$
\lim_{n\to\infty} d_{0;[0,T]}\left(S_{[p]}(x_n),\mathbf{x}\right) = 0 \text{ and } \sup_n \left\|S_{[p]}(x_n)\right\|_{p\text{-var};[0,T]} < \infty. \quad (10.13)
$$

10.3.1 Passage to the limit with uniform estimates

Our aim is now to make precise the meaning of the *rough differential equation* (RDE)

$$
dy = V(y)\,d\mathbf{x}, \ y(0) = y_0 \in \mathbb{R}^e \quad (10.14)
$$

where $V = (V_i)_{1 \leq i \leq d}$ is a family of sufficiently nice vector fields and $\mathbf{x} : [0,T] \to G^{[p]}\left(\mathbb{R}^d\right)$ is a weak geometric p-rough path. The following is essentially an existence result for such RDEs; since our precise definition is very much motivated by this result, the precise definition of an RDE solution (together with remarks on alternative definitions) is postponed till the next section.

Theorem 10.14 (existence) *Assume that*
(i) $V = (V_i)_{1 \leq i \leq d}$ is a collection of vector fields in $\mathrm{Lip}^{\gamma-1}(\mathbb{R}^e)$, where $\gamma > p$;
(ii) (x_n) is a sequence in $C^{1\text{-var}}([0,T], \mathbb{R}^d)$, and $\mathbf{x}$ is a weak geometric p-rough path such that

$$\lim_{n \to \infty} d_{0;[0,T]}\left(S_{[p]}(x_n), \mathbf{x}\right) \text{ and } \sup_n \left\|S_{[p]}(x_n)\right\|_{p\text{-var};[0,T]} < \infty;$$

(iii) $y_0^n \in \mathbb{R}^e$ is a sequence converging to some y_0.
Then, at least along a subsequence, $\pi_{(V)}(0, y_0^n; x_n)$ converges in uniform topology to some limit, say $y \in C([0,T], \mathbb{R}^d)$. Any such limit point satisfies the following estimates: there exists a constant C depending on p, γ such that for all $s < t$ in $[0,T]$,

$$|y|_{p\text{-var};[s,t]} \leq C\left(|V|_{\mathrm{Lip}^{\gamma-1}} \|\mathbf{x}\|_{p\text{-var};[s,t]} \vee |V|_{\mathrm{Lip}^{\gamma-1}}^p \|\mathbf{x}\|_{p\text{-var};[s,t]}^p\right). \quad (10.15)$$

Moreover, if $x^{s,t} : [s,t] \to \mathbb{R}^d$ is any continuous bounded variation path such that

$$S_{\lfloor \gamma \rfloor}\left(x^{s,t}\right)_{s,t} = S_{\lfloor \gamma \rfloor}(\mathbf{x})_{s,t} \text{ and } \int_s^t \left|dx^{s,t}\right| \leq K \|\mathbf{x}\|_{p\text{-var};[s,t]}$$

for some constant $K \geq 1$ then, again for all $s < t$ in $[0,T]$,

$$\left|y_{s,t} - \pi_{(V)}\left(s, y_s; x^{s,t}\right)_{s,t}\right| \leq C'\left(K |V|_{\mathrm{Lip}^{\gamma-1}} \|\mathbf{x}\|_{p\text{-var};[s,t]}\right)^{\gamma}, \quad (10.16)$$

where C' depends on p, γ.

Proof. Let $\varepsilon > 0$ be small enough such that $p + \varepsilon < \gamma$, and let us define for convenience the function $\phi_p(x) = x \vee x^p$. By Davie's Lemma (i.e. Lemma 10.7) for all $s, t \in [0,T]$, there exists $c_1 = c_{1,p+\varepsilon}$ such that

$$\left|\pi_{(V)}(0, y_0^n; x_n)_{s,t}\right| \leq c_{1,p+\varepsilon} \phi_{p+\epsilon}\left(|V|_{\mathrm{Lip}^{\gamma-1}} \left\|S_{[p]}(x_n)\right\|_{(p+\varepsilon)\text{-var};[s,t]}\right), \quad (10.17)$$

where c_1 is independent of n. From Corollary 5.29, $\left\|S_{[p]}(x_n)\right\|_{(p+\varepsilon)\text{-var};[\cdot,\cdot]}$ is equicontinuous in the sense that for all $\varepsilon > 0$, there exists δ such that for all s, t with $|t - s| < \delta$,

$$\left\|S_{[p]}(x_n)\right\|_{(p+\varepsilon)\text{-var};[s,t]} < \varepsilon.$$

This implies that $\pi_{(V)}(0, y_0^n; x_n)$ is equicontinuous and hence converges (along a subsequence) to a path y. We therefore obtain that

$$|y_{s,t}| \leq c_{1,p+\varepsilon} \phi_{p+\varepsilon} \left(|V|_{\mathrm{Lip}^{\gamma-1}} \lim_{n\to\infty} \left\| S_{[p]}(x_n) \right\|_{(p+\varepsilon)\text{-var};[s,t]} \right)$$

$$\leq c_{1,p+\varepsilon} \phi_{p+\varepsilon} \left(|V|_{\mathrm{Lip}^{\gamma-1}} \|\mathbf{x}\|_{(p+\varepsilon)\text{-var};[s,t]} \right).$$

By Exercise 10.11, $\lim_{\varepsilon\to 0} c_{1,p+\varepsilon} = c_{1,p}$, and by Lemma 5.13, $\lim_{\varepsilon\to 0} \|\mathbf{x}\|_{(p+\varepsilon)\text{-var};[s,t]} = \|\mathbf{x}\|_{p\text{-var};[s,t]}$. Hence, for all $s, t \in [0, T]$,

$$|y_{s,t}| \leq c_{1,p} \phi_p \left(|V|_{\mathrm{Lip}^{\gamma-1}} \|\mathbf{x}\|_{p\text{-var};[s,t]} \right).$$

The right-hand side of the last expression raised to power p defines a control, hence we obtain (10.15).

From Remark 10.9, proving inequality (10.16) is equivalent to the proof of the following inequality:

$$\left| y_{s,t} - \mathcal{E}_{(V)} \left(s, y_s; S_{\lfloor\gamma\rfloor}(x^{s,t})_{s,t} \right) \right| \leq C \left(|V|_{\mathrm{Lip}^{\gamma-1}} \|\mathbf{x}\|_{p\text{-var};[s,t]} \right)^\gamma.$$

By Davie's Lemma (i.e. Lemma 10.7), for all $s, t \in [0, T]$,

$$\left| \pi_{(V)}(0, y_0^n; x_n)_{s,t} - \mathcal{E}_{(V)} \left(s, y_s; S_{\lfloor\gamma\rfloor}(x_n^{s,t})_{s,t} \right) \right|$$

$$\leq c_{1,p+\varepsilon} \left(|V|_{\mathrm{Lip}^{\gamma-1}} \left\| S_{[p]}(x_n) \right\|_{(p+\varepsilon)\text{-var};[s,t]} \right)^\gamma,$$

where c_1 is independent of n. Letting n tend to infinity (along the subsequence that allows us to have convergence of $\pi_{(V)}(0, y_0^n; x_n)$ to y), we obtain

$$\left| y_{s,t} - \mathcal{E}_{(V)} \left(s, y_s; S_{\lfloor\gamma\rfloor}(\mathbf{x})_{s,t} \right) \right| \leq c_{1,p+\varepsilon} \left(|V|_{\mathrm{Lip}^{\gamma-1}} \|\mathbf{x}\|_{(p+\varepsilon)\text{-var};[s,t]} \right)^\gamma.$$

Letting ε converge to 0 finishes the proof. ∎

Let us point out explicitly the error estimate for the Euler scheme, which was established in the final step of the previous proof.

Corollary 10.15 (Euler RDE estimates) *Under the assumptions of Theorem 10.14 (in particular* $\mathbf{x} \in C^{p\text{-var}}\left([0, T], G^{[p]}(\mathbb{R}^d)\right)$, $V \in \mathrm{Lip}^{\gamma-1}(\mathbb{R}^e)$, $\gamma > p$) *we have*

$$\left| y_{s,t} - \mathcal{E}_{(V)} \left(y_s, S_{\lfloor\gamma\rfloor}(\mathbf{x})_{s,t} \right) \right| \leq C \left(|V|_{\mathrm{Lip}^{\gamma-1}} \|\mathbf{x}\|_{p\text{-var};[s,t]} \right)^\gamma$$

where C only depends on p, γ. If $\mathbf{x}$ is $1/p$-Hölder,

$$\left| y_{s,t} - \mathcal{E}_{(V)} \left(y_s, S_{\lfloor\gamma\rfloor}(\mathbf{x})_{s,t} \right) \right| \leq \left(C |V|_{\mathrm{Lip}^{\gamma-1}}^\gamma \|\mathbf{x}\|_{1/p\text{-Höl};[0,T]}^\gamma \right) \times |t - s|^{\gamma/p}.$$

Remark 10.16 Note $\lfloor\gamma\rfloor \geq [p]$ so that $S_{\lfloor\gamma\rfloor}(\mathbf{x})$ is the Lyons lift of $\mathbf{x}$. For γ close enough to p, $S_{\lfloor\gamma\rfloor}(\mathbf{x}) \equiv \mathbf{x}$. □

10.3.2   Definition of RDE solution and existence

Perhaps the simplest way to turn Theorem 10.14 into a sensible definition of what we mean by

$$dy = V(y) \, d\mathbf{x}, \qquad (10.18)$$

started at $y_0 \in \mathbb{R}^e$, is the following:

Definition 10.17 Let $\mathbf{x} \in C^{p\text{-var}}\left([0,T], G^{[p]}\left(\mathbb{R}^d\right)\right)$ be a weak geometric p-rough path. We say that $y \in C\left([0,T], \mathbb{R}^e\right)$ is a solution to the rough differential equation (short: RDE solution) driven by $\mathbf{x}$ along the collection of $\mathbb{R}^e$-vector fields $V = (V_i)_{i=1,\dots,d}$ and started at y_0 if there exists a sequence $(x_n)_n$ in $C^{1\text{-var}}\left([0,T], \mathbb{R}^d\right)$ such that (10.13) holds, and ODE solutions $y_n \in \pi_{(V)}(0, y_0; x_n)$ such that

$$y_n \to y \text{ uniformly on } [0,T] \text{ as } n \to \infty .$$

The (formal) equation (10.18) is referred to as a rough differential equation (short: RDE).   □

This definition generalizes immediately to other time intervals such as $[s,t]$, and we define $\pi_{(V)}(s, y_s; \mathbf{x}) \subset C\left([s,t], \mathbb{R}^e\right)$ to be the set of all solutions to the above RDE starting at y_s at time s driven by $\mathbf{x} \in C^{p\text{-var}}\left([s,t], G^{[p]}\left(\mathbb{R}^d\right)\right)$. In case of uniqueness, $\pi_{(V)}(s, y_s; \mathbf{x})$ is *the* solution of the RDE.

Let us note that Theorem 10.14 is now indeed an existence result for RDE solutions. That said, there are some possible variations on the theme of the RDE definition on which we wish to comment.

Remark 10.18 (RDE, Davie definition) Theorem 10.14 and Corollary 10.15 allow us to pick other "defining" properties of RDE solutions. For instance, A. M. Davie [37] defines y to be an RDE solution if there exists a control function $\tilde{\omega}$ and a function $\theta(\delta) = o(\delta)$ as $\delta \to 0$ such that for all $s < t$ in $[0,T]$,

$$\left| y_{s,t} - \mathcal{E}_{(V)}(y_s, \mathbf{x}_{s,t}) \right| \le \theta(\tilde{\omega}(s,t)). \qquad (10.19)$$

(Note that (10.19) contains an implicit regularity assumption on V so that the Euler scheme $\mathcal{E}$ is well-defined.) Applying Corollary 10.15 (take γ small enough so that $\lfloor \gamma \rfloor \ge [p]$, hence $S_{\lfloor \gamma \rfloor}(\mathbf{x}) \equiv \mathbf{x}$, then $\theta(\delta) = \delta^{\gamma/p}$ and $\tilde{\omega}(s,t) = (\text{const}) \times \|\mathbf{x}\|_{p\text{-var};[s,t]}^p$) shows that any RDE solution in the sense of Definition 10.17 is also a solution in Davie's sense. With either definition, let us note that (10.19) leads immediately to the statement that y satisfies some sort of compensated Riemann–Stieltjes integral equation,

$$y_t - y_0 = \lim_{n \to \infty} \sum_{t_i \in D_n} \mathcal{E}_{(V)}\left(y_{t_i}, \mathbf{x}_{t_i, t_{i+1}}\right),$$

for any sequence of dissection (D_n) of $[0,t]$ with mesh tending to zero.   □

Remark 10.19 (RDE, Lyons definition) As one expects, RDE solutions can also be defined as the solution to a "rough" integral equation and this is Lyons' original approach [116, 120, 123]. To this end, one first needs a notion of rough integration (cf. Section 10.6) which allows, for sufficiently smooth $\varphi = (\varphi_1, \ldots, \varphi_d)$, defined on $\mathbb{R}^d$, the definition of an (indefinite) rough integral

$$\int_0^\cdot \varphi(z) \, d\mathbf{z} \text{ with } z = \pi_1(\mathbf{z})$$

such that, in the case when $\mathbf{z} = S_{[p]}(z)$ for some $z \in C^{1\text{-var}}\left([0,T], \mathbb{R}^d\right)$, it coincides with $S_{[p]}(\xi)$ where ξ is the classical Riemann–Stieltjes integral $\int_0^\cdot \varphi(z) \, dz$.

Note that (10.18) cannot be rewritten as an integral equation of the above form (for y is not part of the integrating signal $\mathbf{x}$). Nonetheless, the "enhanced" differential equation (in which the input signal is carried along to the output)

$$dx = dx$$
$$dy = V(y)\, dx$$

can be written in the desired form

$$z_{0,\cdot} = \int_0^\cdot \varphi(z) \, dz$$

provided we set $z = (x, y)$ and

$$\varphi(z) = \begin{pmatrix} 1 & 0 \\ V(y) & 0 \end{pmatrix}.$$

The above integral equation indeed makes sense as a rough integral equation (with implicit regularity assumption on V so the rough integral is well-defined) replacing z by a genuine geometric p-rough path $\mathbf{z} \in C^{p\text{-var}}\left([0,T], G^{[p]}\left(\mathbb{R}^d \oplus \mathbb{R}^e\right)\right)$ and solutions can be constructed, for instance, by a Picard–iteration [116, 120, 123]. An $\mathbb{R}^e$-valued solution is then recovered by projection

$$\mathbf{z} \mapsto \boldsymbol{\pi}_1(\mathbf{z}) = z = (x, y) \mapsto y$$

and again one can see that an RDE solution in the sense of Definition 10.17 is also a solution in this sense. $\qquad\square$

10.3.3 Local existence

As in ODE theory, if the vector fields have only locally the necessary regularity for existence, we get local solutions.

Exercise 10.20 *We keep the notation of Theorem 10.14. Fix $s < t$ in $[0,T]$ and assume that for some open set Ω, we have for all $u \in [s,t]$, $y_u \in \Omega$. Prove that*

$$|y|_{p\text{-var};[s,t]} \le C \left(|V|_{\mathrm{Lip}^{\gamma-1}(\Omega)} \|\mathbf{x}\|_{p\text{-var};[s,t]} \vee |V|_{\mathrm{Lip}^{\gamma-1}(\Omega)}^p \|\mathbf{x}\|_{p\text{-var};[s,t]}^p \right)$$

where $C = C(p,\gamma)$.

As a consequence of this result, we obtain the following:

Theorem 10.21 (local existence) *Asssume that*
(i) $V = (V_i)_{1 \leq i \leq d}$ *is a collection of vector fields* $\mathrm{Lip}_{loc}^{\gamma-1} (\mathbb{R}^e)$, *with* $\gamma > p$;
(ii) $\mathbf{x} : [0, T] \to G^{[p]} (\mathbb{R}^d)$ *is a weak geometric p-rough path;*
(iii) $y_0 \in \mathbb{R}^e$ *is an initial condition.*
Then either there exists a (global) solution $y : [0, T] \to \mathbb{R}^e$ *to* $dy = V(y) d\mathbf{x}$
with initial condition y_0, *or there exists* $\tau \in [0, T]$ *and a (local) solution*
$y : [0, \tau) \to \mathbb{R}^e$ *such that* y *is a solution on* $[0, t]$ *for any* $t \in (0, \tau)$ *and*

$$\lim_{t \nearrow \tau} |y(t)| = +\infty.$$

Proof. It simplifies the argument to have unique ODE solutions, we thus
assume $\gamma \geq 2$. (Otherwise $p < \gamma < 2$ and some care, similar to the proof
of Theorem 3.6, is needed; we leave this extension to the reader.) Without
loss of generality, $y_0 = 0$. Pick $(x^n) \subset C^{1\text{-var}} ([0, T], \mathbb{R}^d)$ so that

$$S_{[p]} (x_n) \to \mathbf{x} \text{ uniformly}$$

with $\sup_n \left\| S_{[p]} (x_n) \right\|_{p\text{-var};[0,T]} < \infty$. Replace V by compactly supported
$\mathrm{Lip}^{\gamma-1}$-vector fields V^n which coincide with V on the ball $\{y : |y| \leq n\}$.
From the preceding existence theorem,

$$\pi_{(V^{(1)})} (0, y_0; x^n) \to y^{(1)} \in \pi_{(V^1)} (0, y_0; \mathbf{x}),$$

where strictly speaking we have replaced x^n by a subsequence along which
this convergence holds. If

$$\left. |y^{(1)}| \right|_{\infty;[0,T]} \leq 1$$

then we can replace V_1 by V and hence find a global solution, in which case
we are done. Otherwise,

$$\tau_1 = \inf \left\{ t \geq 0 : \left| y_t^{(1)} \right| \geq 1 \right\} \in [0, T]$$

and we switch to another subsequence so that

$$\pi_{(V^{(2)})} (0, y_0; x^n) \to y^{(2)};$$

observing that $y^{(1)} \equiv y^{(2)}$ on $[0, \tau_1]$. Again, if $\left. |y^{(2)}| \right|_{\infty;[0,T]} \leq 2$ we find a
global solution, otherwise we define another $\tau_2 \geq \tau_1$ by

$$\tau_2 = \inf \left\{ t \geq 0 : \left| y^{(2)} \right| \geq 2 \right\} \in [0, T]$$

and so on. Iterating this either yields a global solution or a family of RDE
solutions

$$y^{(n)} \in \pi_{(V^n)} (0, y_0; \mathbf{x}),$$

consistent in the sense that $y^{(n)} \equiv y^{(n+1)}$ on $[0, \tau_n]$. Moreover, by definition
of τ_n we see that $|y(\tau_n)| = n \to \infty$ as $n \to \infty$ and the proof is finished. ∎

10.3.4 Uniqueness and continuity

For ordinary differential equations, we saw that existence is guaranteed for continuous vector fields, while uniqueness requires Lipschitz vector fields; that is, one additional degree of smoothness. In essence, this remains true for RDEs driven by p-rough paths: we saw that RDE solutions exist for $\mathrm{Lip}^{\gamma-1}$-vector fields when $\gamma > p$ and we will now see that uniqueness holds, still assuming $\gamma > p$, for Lip^{γ}-vector fields (or $\mathrm{Lip}^{\gamma}_{loc}$ since uniqueness is a local issue!).

We will show uniqueness by establishing Lipschitz regularity of the RDE flows. Later, we will see that uniqueness also holds for $\gamma = p$ (in this case, we will only prove uniform continuity on bounded sets of the RDE flow rather than Lipschitzness), and that uniqueness still holds when we relax the rough path regularity of the driving signal in a way that gives uniqueness under optimal regularity assumptions for RDEs driven by Brownian motion and Lévy's area.

The reader may find it useful to quickly revise the proof of Davie's lemma, which is similar to arguments in this section. In particular, the following *Lemmas $\overline{A}$ and $\overline{B}$* are essentially straightforward generalizations of what we called *Lemmas A and B* in Section 10.2. We start with the elementary yet useful

Lemma 10.22 *Let $g, \tilde{g} \in \mathrm{Lip}^{\beta}\left(\mathbb{R}^e, L(\mathbb{R}^f, \mathbb{R}^e)\right)$ with $\beta \in [1, 2]$, and $a, b, \tilde{a}, \tilde{b} : [0, 1] \to \mathbb{R}^e$, and $x, \tilde{x} \in C^{1\text{-var}}\left([0, 1], \mathbb{R}^f\right)$. Then, with*

$$\Delta \equiv \int_0^1 \left(g\left(a_r\right) - g\left(b_r\right)\right) dx_r - \int_0^1 \left(\tilde{g}\left(\tilde{a}_r\right) - \tilde{g}\left(\tilde{b}_r\right)\right) d\tilde{x}_r,$$

we have[1]

$$
\begin{aligned}
|\Delta| \ \le \ & |g|_{\mathrm{Lip}^{\beta}} \int_0^1 \left|\left(a_r - b_r\right) - \left(\tilde{a}_r - \tilde{b}_r\right)\right| \cdot |dx_r| \\
& + \left(\left|a - b\right|_{\infty;[0,1]} + \left|\tilde{a} - \tilde{b}\right|_{\infty;[0,1]}\right)^{\beta-1} |x|_{1\text{-var};[0,1]} \\
& \quad \times \left(|g|_{\mathrm{Lip}^{\beta}} \left|b - \tilde{b}\right|_{\infty;[0,1]} + |g - \tilde{g}|_{\mathrm{Lip}^{\beta-1}}\right) \\
& + |\tilde{g}|_{\mathrm{Lip}^{\beta}} \cdot \left|\tilde{a} - \tilde{b}\right|_{\infty;[0,1]} |x - \tilde{x}|_{1\text{-var};[0,1]}.
\end{aligned}
$$

---

[1] In the case $\beta = 1$, $|g - \tilde{g}|_{\mathrm{Lip}^{\beta-1}}$ has to be replaced by $2|g - \tilde{g}|_{\infty}$.

Proof. <u>Step 1:</u> Fix $a, b, \tilde{a}, \tilde{b} \in \mathbb{R}^e$. When $\beta > 1$, $g \in C^1$ and we can write $g(a) - g(\tilde{a}) - \left(g(b) - g\left(\tilde{b}\right)\right)$ as

$$\int_0^1 g'(ta + (1-t)\tilde{a})(a - \tilde{a})\,dt - \int_0^1 g'\left(tb + (1-t)\tilde{b}\right)\left(b - \tilde{b}\right)dt$$

$$= \int_0^1 g'(ta + (1-t)\tilde{a})\left(a - \tilde{a} - \left(b - \tilde{b}\right)\right)dt$$

$$+ \int_0^1 \left[g'(ta + (1-t)\tilde{a}) - g'\left(tb + (1-t)\tilde{b}\right)\right]\left(b - \tilde{b}\right)dt$$

to obtain, using $|g'|_\infty \le |g|_{\mathrm{Lip}^\beta}$ for $\beta > 1$,

$$\left|(g(a) - g(b)) - \left(g(\tilde{a}) - g\left(\tilde{b}\right)\right)\right| \le |g|_{\mathrm{Lip}^\beta}\left|(a - b) - \left(\tilde{a} - \tilde{b}\right)\right|$$

$$+ |g|_{\mathrm{Lip}^\beta}\left(|a - b| + \left|\tilde{a} - \tilde{b}\right|\right)^{\beta - 1}\left|b - \tilde{b}\right|.$$

In fact, this argument remains valid for $\beta = 1$ since $g \in \mathrm{Lip}^1$ implies absolute continuity of $t \mapsto g(ta + (1-t)\tilde{a})$.

<u>Step 2:</u> Obviously $\left|\left(\tilde{g}(\tilde{a}) - \tilde{g}\left(\tilde{b}\right)\right) - \left(g(\tilde{a}) - g\left(\tilde{b}\right)\right)\right| \le |g - \tilde{g}|_{\mathrm{Lip}^{\beta - 1}}\left|\tilde{a} - \tilde{b}\right|^{\beta - 1}$ so that, by the triangle inequality,

$$\left|(g(a) - g(b)) - \left(\tilde{g}(\tilde{a}) - \tilde{g}\left(\tilde{b}\right)\right)\right| \le |g|_{\mathrm{Lip}^\beta}\left|(a - b) - \left(\tilde{a} - \tilde{b}\right)\right|$$

$$+ \left(|a - b| + \left|\tilde{a} - \tilde{b}\right|\right)^{\beta - 1}\left(|g|_{\mathrm{Lip}^\beta}\left|b - \tilde{b}\right| + |g - \tilde{g}|_{\mathrm{Lip}^{\beta - 1}}\right).$$

<u>Step 3:</u> At last, we write $\Delta = \Delta_1 + \Delta_2$ where

$$\Delta_1 = \int_0^1 \left[g(a_r) - g(b_r) - \left(\tilde{g}(\tilde{a}_r) - \tilde{g}\left(\tilde{b}_r\right)\right)\right]dx_r,$$

$$\Delta_2 = \int_0^1 \left[\tilde{g}(\tilde{a}_r) - \tilde{g}\left(\tilde{b}_r\right)\right]d(x_r - \tilde{x}_r).$$

Using the elementary $\left|\int \dots dx_r\right| \le \int |\dots| |dx_r| \le |\dots|_\infty |x|_{1\text{-var}}$ we bound Δ_1 using the step-2 estimate, while $\left|\tilde{g}(\tilde{a}_r) - \tilde{g}\left(\tilde{b}_r\right)\right|_\infty \le |\tilde{g}|_{\mathrm{Lip}^\beta} \cdot \left|\tilde{a} - \tilde{b}\right|_{\infty;[0,1]}$ allows us to bound Δ_2. Together, they imply the claimed estimate. ∎

We now turn to *Lemma $\overline{A}$* and note the assumption of Lip^γ-regularity, $\gamma \ge 1$, in contrast to *Lemma A* which was formulated for $\mathrm{Lip}^{\gamma - 1}$ vector fields, $\gamma > 1$.

Lemma 10.23 (Lemma $\overline{A}$) *Assume that*
(i) $\left(V_i^1\right)_{1 \le i \le d}$ and $\left(V_i^2\right)_{1 \le i \le d}$ are two collections of vector fields in $\mathrm{Lip}^\gamma(\mathbb{R}^e)$ vector fields, with $\gamma \ge 1$;

(ii) s < u are two elements of $[0, T]$;
(iii) $y_s^1, y_s^2 \in \mathbb{R}^e$ (thought of as "time-s" initial conditions);
(iv) $x^1, \tilde{x}^1$ and $x^2, \tilde{x}^2$ are driving signals in $C^{1\text{-var}}\left([s, u], \mathbb{R}^d\right)$ such that

$$S_{[\gamma]}\left(x^1\right)_{s,u} = S_{[\gamma]}\left(\tilde{x}^1\right)_{s,u},$$
$$S_{[\gamma]}\left(x^2\right)_{s,u} = S_{[\gamma]}\left(\tilde{x}^2\right)_{s,u};$$

(v) $\ell \geq 0, \delta > 0$ are such that

$$\max\left\{\int_s^u |dx^1| + \int_s^u |d\tilde{x}^1|, \int_s^u |dx^2| + \int_s^u |d\tilde{x}^2|\right\} \leq \ell,$$
$$\max\left\{\int_s^u |dx^1 - dx^2|, \int_s^u |d\tilde{x}^1 - d\tilde{x}^2|\right\} \leq \delta;$$

(vi) $v \geq 0$ is a bound on $\left|V^1\right|_{\text{Lip}^\gamma}$ and $\left|V^2\right|_{\text{Lip}^\gamma}$.
Then, for some constant C depending only on γ, we have[2]

$$\left|\left(\pi_{(V^1)}\left(s, y_s^1; x^1\right)_{s,u} - \pi_{(V^1)}\left(s, y_s^1; \tilde{x}^1\right)_{s,u}\right) - \left(\pi_{(V^2)}\left(s, y_s^2; x^2\right)_{s,u}\right.\right.$$
$$\left.\left. - \pi_{(V^2)}\left(s, y_s^2; \tilde{x}^2\right)_{s,u}\right)\right|$$
$$\leq C\left(\left|y_s^1 - y_s^2\right| + \frac{1}{v}\left|V^1 - V^2\right|_{\text{Lip}^{\gamma-1}}\right) \cdot (v\ell)^\gamma \exp\left(Cv\ell\right)$$
$$+ C\delta \cdot (v\ell)^{[\gamma]} \exp\left(Cv\ell\right).$$

Proof. First, as in the proof of Lemma A, we take $(s, u) = (0, 1)$ and observe

$$\pi_{(V^i)}(0, y_0^i; x^i)_{0,1} - \pi_{(V^i)}(0, y_0^i; \tilde{x}^i)_{0,1} = \pi_{(V^i)}(0, \pi_{(V^i)}(0, y_0^i \tilde{x}^i)_1; z^i)_{0,1}; \ i=1,2,$$

where $z^1 = \overleftarrow{\tilde{x}^1} \sqcup x^1$ and $z^2 = \overleftarrow{\tilde{x}^2} \sqcup x^2$ are reparametrized in the same way to a path from $[0, 1]$ to $\mathbb{R}^d$. By assumption (iv) and Chen's theorem z^1 has trivial step-$[\gamma]$ signature, i.e.

$$S_{[\gamma]}\left(z^1\right)_{0,1} = S_{[\gamma]}\left(\tilde{x}^1\right)_{0,1}^{-1} \otimes S_{[\gamma]}\left(x^1\right)_{0,1} = 1, \tag{10.20}$$

and similarly for z^2. Next, by assumption (v) we have $\int_0^1 |dz_r^1 - dz_r^2| \leq 2\delta$. Using the Lipschitzness of the flow of ODEs, in the quantitative form of Theorem 3.8, we see that it is enough to prove the above lemma with $\tilde{x}^2 = \tilde{x}^1 = 0$. We can thus assume $z^1 = x^1$ and $z^2 = x^2$. To simplify the notation, define

$$y_u^i = \pi_{(V^i)}\left(0, y_0^i; x^i\right)_u, \ i = 1, 2,$$

---

[2]For $\gamma = 1$, $\left|V^1 - V^2\right|_{\text{Lip}^\gamma - 1}$ is replaced by $2\left|V^1 - V^2\right|_\infty$.

and, for $N := [\gamma]$,

$$\mathbf{x}_{u,1}^{i,N} = \int_{u<r_1<\cdots<r_n<1} dx_{r_1}^i \otimes \cdots \otimes dx_{r_N}^i \in \left(\mathbb{R}^d\right)^{\otimes N}, \quad i = 1, 2.$$

Observe that

$$\max_{i=1,2} \left(\int_0^1 \left| d\mathbf{x}_{u,1}^{i,N} \right| \right) \leq \ell^N.$$

From (10.20) and Remark 10.4 concerning the remainder representation of an Euler approximation,[3] we have

$$y_{0,1}^i = y_{0,1}^i - \mathcal{E}_{(V^i)}\left(y_0, S_N \left(x^i\right)_{0,1}\right) = \int_0^1 \left[V^{i,N}\left(y_u^i\right) - V^{i,N}\left(y_0^i\right)\right] d\mathbf{x}_{u,1}^{i,N}, \quad i=1,2,$$

and so we have

$$\begin{aligned} y_{0,1}^1 - y_{0,1}^2 &= \int_0^1 \left[V^{1,N}\left(y_u^1\right) - V^{1,N}\left(y_0^1\right)\right] d\mathbf{x}_{u,1}^{1,N}, \\ &\quad - \int_0^1 \left[V^{2,N}\left(y_u^2\right) - V^{2,N}\left(y_0^2\right)\right] d\mathbf{x}_{u,1}^{2,N}. \end{aligned}$$

Note that $V^{i,N} \in \text{Lip}^\beta$ where $\beta := \gamma - N + 1 \in [1,2)$ since $N = [\gamma]$. Using Lemma 10.22 to the paths $u \mapsto y_u^1, y_0^1, y_u^2, y_0^2$, we then obtain, using the bound $\max_{i=1,2} \left(\left|V^{i,N}\right|_{\text{Lip}^{\gamma-N+1}}\right) \leq c_1 v^N$,

$$\begin{aligned} \left|y_{0,1}^1 - y_{0,1}^2\right| &\leq c_1 v^N \left|y_{0,.}^1 - y_{0,.}^2\right|_{\infty;[0,1]} \left|\mathbf{x}_{.,1}^{1,N}\right|_{1\text{-var};[0,1]} \\ &\quad + c_1 \left(\left|y_{0,.}^1\right|_{\infty;[0,1]} + \left|y_{0,.}^2\right|_{\infty;[0,1]}\right)^{\gamma-1} \left|\mathbf{x}_{.,1}^{1,N}\right|_{1\text{-var};[0,1]} \\ &\quad \times \left(v^N \left|y_0^1 - y_0^2\right| + \left|V^{1,N} - V^{2,N}\right|_{\text{Lip}^{\gamma-N}}\right) \\ &\quad + c_1 v^N \cdot \left|y_{0,.}^2\right|_{\infty;[0,1]} \left|\mathbf{x}_{.,1}^{1,N} - \mathbf{x}_{.,1}^{2,N}\right|_{1\text{-var};[0,1]}. \end{aligned}$$

We first observe, by Theorem 3.18,

$$\left|y_{0,1}^1 - y_{0,1}^2\right|_{\infty;[0,1]} \leq c_2 \left(\left|y_0^1 - y_0^2\right| v\ell + \delta v + \left|V^1 - V^2\right|_\infty \ell\right) e^{c_2 v\ell}$$

and the ODE estimate of Theorem 3.4 gives

$$\left|y_{0,.}^1\right|_{\infty;[0,1]} + \left|y_{0,.}^2\right|_{\infty;[0,1]} \leq c_3 v\ell.$$

---

[3] If $V = (V_1, \ldots, V_d)$ we think of $V^N \equiv \left(V_{j_1} \ldots V_{j_N}\right)_{j_1,\ldots,j_N \in \{1,\ldots,d\}}$ as an element of the (Euclidean) space $\left(\mathbb{R}^d\right)^{\otimes N}$ which contracts naturally with elements of the form $\mathbf{x}_{u,1}^{i,N} \in \left(\mathbb{R}^d\right)^{\otimes N}$.

Moreover, we easily see that

$$\left|V^{1,N} - V^{2,N}\right|_{\mathrm{Lip}^{\gamma}-N} \leq c_4 v^{N-1} \left|V^1 - V^2\right|_{\mathrm{Lip}^{\gamma}-1}$$

and, from Proposition 7.63, we have $\left|\mathbf{x}_{\cdot,1}^{1,N} - \mathbf{x}_{\cdot,1}^{2,N}\right|_{1\text{-var};[0,1]} \leq c_5 \delta \ell^{N-1}$. We also have $\left|\mathbf{x}_{\cdot,1}^{N}\right|_{1\text{-var};[0,1]} \leq \ell^N$ of course. Putting all these inequalities together gives the desired estimate. ∎

Exercise 10.24 *In the final step of the proof of Lemma $\bar{A}$, detail how all the estimates are put together.*

Solution. Assume $V^1 = V^2$ at first. Using

$$\left|y_{0,\cdot}^1 - y_{0,\cdot}^2\right|_{\infty;[0,1]} \leq c_2 \left(\left|y_0^1 - y_0^2\right| v\ell + \delta v\right) e^{c_2 v\ell}$$

we get

$$\begin{aligned}
\left|y_{0,1}^1 - y_{0,1}^2\right| &\lesssim v^N \left(\left|y_0^1 - y_0^2\right| v\ell + \delta v\right) e^{cv\ell} \ell^N \\
&\quad + (v\ell)^{\gamma-1} \ell^N v^N \left|y_0^1 - y_0^2\right| \\
&\quad + v^N . \left(v\ell . \delta\ell^{N-1}\right) \\
&\equiv \left|y_0^1 - y_0^2\right| \Delta_1 + \delta\Delta_2
\end{aligned}$$

with

$$\begin{aligned}
\Delta_1 &= v^N v\ell e^{cv\ell} \ell^N + (v\ell)^{\gamma-1} \ell^N v^N \\
&\leq (\text{const}) \times \left[(v\ell)^{N+1} + (v\ell)^{N+\gamma-1}\right] e^{cv\ell}.
\end{aligned}$$

From $N = [\gamma]$ it is clear that $\min(N+1, N+\gamma-1) \geq \gamma$ and so

$$\Delta_1 \leq (\text{const}) \times (v\ell)^{\gamma} e^{cv\ell}.$$

Similarly,

$$\Delta_2 = v^{N+1} \ell^N e^{cv\ell} + v^{N+1} \ell^N \lesssim v(v\ell)^N e^{cv\ell}.$$

When $V^1 \neq V^2$ we have

$$\left|y_{0,1}^1 - y_{0,1}^2\right| \leq \left|y_0^1 - y_0^2\right| \Delta_1 + \delta\Delta_2 + \left|V^1 - V^2\right|_{\mathrm{Lip}^{\gamma}-1} \Delta_3$$

with

$$\begin{aligned}
\Delta_3 &= (v\ell)^{\gamma-1} \ell^N \leq (v\ell)^{\gamma-1} \ell^N e^{cv\ell} \\
&\lesssim (v\ell)^{\gamma-1} \ell^1 e^{cv\ell} = \frac{1}{v}(v\ell)^{\gamma} e^{cv\ell}.
\end{aligned}$$

□

Lemma 10.25 (Lemma $\overline{B}$) *Assume that*
(i) $(V_i^1)_{1 \leq i \leq d}$ *and* $(V_i^2)_{1 \leq i \leq d}$ *are two collections of vector fields in* $\mathrm{Lip}^\gamma (\mathbb{R}^e)$
with $\gamma \geq 1$;
(ii) $t < u$ *are some elements of* $[0, T]$;
(iii) $y_t^1, y_t^2, \tilde{y}_t^1, \tilde{y}_t^2 \in \mathbb{R}^e$ *(thought of as "time-t" initial conditions);*
(iv) x^1, x^2 *are two driving signals in* $C^{1\text{-var}} ([t, u], \mathbb{R}^d)$;
(v) $\ell \geq 0$, $\delta > 0$ *are such that*

$$\max \left\{ \int_t^u |dx^1|, \int_t^u |dx^2| \right\} \leq \ell \ \text{ and } \ \int_t^u |dx_r^1 - dx_r^2| \leq \delta;$$

(vi) v *is a bound on* $|V^1|_{\mathrm{Lip}^\gamma}$ *and* $|V^2|_{\mathrm{Lip}^\gamma}$.
Then we have, for some constant $C = C(\gamma)$,

$$\left| \left\{ \pi_{(V^1)} \left(t, y_t^1; x^1 \right)_{t,u} - \pi_{(V^1)} \left(t, \tilde{y}_t^1; x^1 \right)_{t,u} \right\} - \left\{ \pi_{(V^2)} \left(t, y_t^2; x^2 \right)_{t,u} \right. \right.$$
$$\left. \left. - \pi_{(V^2)} \left(t, \tilde{y}_t^2; x^2 \right)_{t,u} \right\} \right|$$
$$\leq Cv\ell \exp (Cv\ell) \left| (y_t^1 - \tilde{y}_t^1) - (y_t^2 - \tilde{y}_t^2) \right|$$
$$+ Cv\ell \exp (Cv\ell) \left(|y_t^1 - \tilde{y}_t^1| + |y_t^2 - \tilde{y}_t^2| \right)^{\min(2,\gamma)-1}$$
$$\times \left(|\tilde{y}_t^1 - \tilde{y}_t^2| + \frac{1}{v} |V^1 - V^2|_{\mathrm{Lip}^{\gamma-1}} + v\delta \right)$$
$$+ C\delta v \exp (Cv\ell) |y_t^2 - \tilde{y}_t^2|.$$

Proof. At the price of replacing γ by $\min (2, \gamma) \in [1, 2]$, we can and will assume that $\gamma \in [1, 2]$. Define for $r \in [t, u]$,

$$y_r^i = \pi_{(V^i)} \left(t, y_t^i; x^i \right)_r, \quad \text{and} \quad \tilde{y}_r^i = \pi_{(V^i)} \left(t, \tilde{y}_t^i; x^i \right)_r \quad \text{with } i = 1, 2.$$

We define

$$e_r = \left| \left\{ y_{t,r}^1 - \tilde{y}_{t,r}^1 \right\} - \left\{ y_{t,r}^2 - \tilde{y}_{t,r}^2 \right\} \right| \quad \text{with } r \in [t, u],$$

and have to estimate e_u. Fom Lemma 10.22, applied with $\gamma \in [1, 2]$, we see for all $r \in [t, u]$,

$$e_r = \left| \int_t^r \left(V^1 (y_s^1) - V^1 (\tilde{y}_s^1) \right) dx_s^1 - \int_t^r \left(V^2 (y_s^2) - V^2 (\tilde{y}_s^2) \right) dx_s^2 \right|$$

$$\leq \begin{cases} v \int_t^r \left| \{ y_s^1 - \tilde{y}_s^1 \} - \{ y_s^2 - \tilde{y}_s^2 \} \right| \cdot |dx_s^1| \\ + \left(|y^1 - \tilde{y}^1|_{\infty;[t,r]} + |y^2 - \tilde{y}^2|_{\infty;[t,r]} \right)^{\gamma-1} \\ \quad \times |x^1|_{1\text{-var};[t,r]} \left(v |\tilde{y}^1 - \tilde{y}^2|_{\infty;[t,r]} + |V^1 - V^2|_{\mathrm{Lip}^{\gamma-1}} \right) \\ + v |y^2 - \tilde{y}^2|_{\infty;[t,r]} |x^1 - x^2|_{1\text{-var};[t,r]}. \end{cases}$$

From Theorem 3.15 we have

$$\left|y^i - \tilde{y}^i\right|_{\infty;[t,r]} \leq c_1 \left|y_t^i - \tilde{y}_t^i\right| \exp\left(c_1 v \ell\right), \quad i = 1, 2,$$

$$\left|\tilde{y}^1 - \tilde{y}^2\right|_{\infty;[t,r]} \leq c_1 \left|\tilde{y}_t^1 - \tilde{y}_t^2\right| \exp\left(c_1 v \ell\right)$$
$$+ c_1 \left|V^1 - V^2\right|_{\mathrm{Lip}^{\gamma-1}} \ell \exp\left(c_1 v \ell\right)$$
$$+ c_1 v \delta \exp\left(c_1 v \ell\right).$$

Hence, we obtain

$$e_r \leq v \int_t^r e_s \left|dx_s^1\right| + v\ell \left|\{y_t^1 - \tilde{y}_t^1\} - \{y_t^2 - \tilde{y}_t^2\}\right|$$
$$+ c_2 \left(\left|y_t^1 - \tilde{y}_t^1\right| + \left|y_t^2 - \tilde{y}_t^2\right|\right)^{\gamma-1} v\ell \exp\left(c_2 v \ell\right)$$
$$\left(\left|\tilde{y}_t^1 - \tilde{y}_t^2\right| + \frac{1}{v}\left|V^1 - V^2\right|_{\mathrm{Lip}^{\gamma-1}} + v\delta\right)$$
$$+ c_1 \left|y_t^2 - \tilde{y}_t^2\right| v\delta \exp\left(c_1 v \ell\right).$$

The proof is then finished by an application of (Gronwall's) Lemma 3.2. ∎

Equipped with these two lemmas, we can prove (under some regularity assumptions) that the map $\mathbf{x} \mapsto \pi_{(V)}\left(0, y_0; \mathbf{x}\right)$ is well-defined and locally Lipschitz continuous in all its parameters (vector fields, initial condition, and driving signal).

Theorem 10.26 ($\overline{\text{Davie}}$) [4] *Assume that*
(i) $V^1 = \left(V_i^1\right)_{1 \leq i \leq d}$ *and* $V^2 = \left(V_i^2\right)_{1 \leq i \leq d}$ *are two collections of* Lip^γ*-vector fields on* $\mathbb{R}^e$ *for* $\gamma > p \geq 1$;
(ii) ω *is a fixed control;* [5]
(iii) $\mathbf{x}^1, \mathbf{x}^2$ *are two weak geometric p-rough paths in* $C^{p\text{-var}}\left([0,T], G^{[p]}\left(\mathbb{R}^d\right)\right)$, *with* $\left\|\mathbf{x}^i\right\|_{p\text{-}\omega} \leq 1$;
(iv) $y_0^1, y_0^2 \in \mathbb{R}^e$ *thought of as time-0 initial conditions;*
(v) v *is a bound on* $\left|V^1\right|_{\mathrm{Lip}^\gamma}$ *and* $\left|V^2\right|_{\mathrm{Lip}^\gamma}$.
Then there exists a unique RDE solution starting at y_0^i *along* V^i *driven by* $\mathbf{x}^i$, *denoted by*

$$y^i \equiv \pi_{(V^i)}\left(0, y_0^i; \mathbf{x}^i\right),$$

for $i = 1, 2$. *Moreover, there exists* $C = C\left(\gamma, p\right)$ *such that* [6]

$$\rho_{p,\omega;[0,T]}\left(y^1, y^2\right) \leq C\left[v\left|y_0^1 - y_0^2\right| + \left|V^1 - V^2\right|_{\mathrm{Lip}^{\gamma-1}} + v\rho_{p,\omega;[0,T]}\left(\mathbf{x}^1, \mathbf{x}^2\right)\right]$$
$$\cdot \exp\left(Cv^p \omega\left(0, T\right)\right).$$

---

[4] The present theorem stands in a similar relation to Davie's lemma as Lemma $\bar{A}$ to A, or Lemma $\bar{B}$ to B.

[5] In view of (iii) one can take $\omega\left(s, t\right) = \sum_{i=1,2} \left\|\mathbf{x}^i\right\|_{p\text{-var};[s,t]}^p$.

[6] Here: $\rho_{p,\omega;[0,T]}\left(y^1, y^2\right) = \left|y^1 - y^2\right|_{p,\omega;[0,T]}$.

Proof. The present regularity assumptions on the vector fields, Lip^γ with $\gamma > p$, are more than enough to guarantee existence of RDE solutions. More precisely, let us pick solutions

$$y^i \in \pi_{(V^i)}\left(0, y_0^i; \mathbf{x}^i\right), \quad i = 1, 2.$$

We may assume, without loss of generality, $p < \gamma < [p] + 1$ so that $[\gamma] = [p]$ and also set

$$\varepsilon := \rho_{p,\omega;[0,T]}\left(\mathbf{x}^1, \mathbf{x}^2\right).$$

For all $s < t$ in $[0, T]$ we can find paths $x^{1,s,t}$ and $x^{2,s,t}$ such that

$$S_{[p]}\left(x^{i,s,t}\right)_{s,t} = \mathbf{x}_{s,t}^i, \quad i = 1, 2,$$

and such that, for a constant $c_1 = c_1\left(p\right)$,

$$\int_s^t \left|dx_r^{i,s,t}\right| \leq c_1 \omega\left(s,t\right)^{1/p},$$

$$\int_s^t \left|dx_r^{1,s,t} - dx_r^{2,s,t}\right| \leq c_1 \varepsilon \omega\left(s,t\right)^{1/p};$$

indeed, this is possible thanks to Proposition 7.64 applied to

$$g_i := \delta_{\frac{1}{\omega(s,t)^{1/p}}} \mathbf{x}_{s,t}^i \in G^{[p]}\left(\mathbb{R}^d\right), \quad i = 1, 2,$$

noting that $\|g_1\|, \|g_2\| \leq 1$ and $|g_1 - g_2|_{T^{[p]}(\mathbb{R}^d)} \leq \varepsilon$. After these preliminaries, let us now fix $s < t < u$ in $[0, T]$ and define

$$x^{i,s,t,u} := x^{i,s,t} \sqcup x^{i,t,u},$$

the concatenation of $x^{i,s,t}$ and $x^{i,t,u}$. Following Davie's lemma, we define

$$\Gamma_{s,t}^i = y_{s,t}^i - \pi_{(V^i)}\left(s, y_s^i; x^{i,s,t}\right)_{s,t}, \quad i = 1, 2,$$

and set $\overline{\Gamma}_{s,t} := \Gamma_{s,t}^1 - \Gamma_{s,t}^2$. From the estimates in (the existence) Theorem 10.14, using the fact that the vector fields are $[p]$-Lipschitz, we have

$$\left|\Gamma_{s,t}^i\right| \leq \frac{1}{2} c_1 \left(\upsilon \omega\left(s,t\right)^{1/p}\right)^{[p]+1}, \quad i = 1, 2, \tag{10.21}$$

and hence

$$\left|\overline{\Gamma}_{s,t}\right| \leq c_1 \left(\upsilon \omega\left(s,t\right)^{1/p}\right)^{[p]+1}.$$

We now proceed similarly as in the proof of Davie's lemma, cf. (10.9). Namely, for $i = 1, 2$, we define

$$A^i \equiv \pi_{(V^i)}\left(s, y_s^i; x^{i,s,t,u}\right)_{s,u} - \pi_{(V^i)}\left(s, y_s^i; x^{i,s,u}\right)_{s,u},$$

noting that $S_{[p]}\left(x^{i,s,t,u}\right) = S_{[p]}\left(x^{i,s,u}\right)$, and

$$
\begin{aligned}
B^i &\equiv \pi_{(V^i)}\left(t, y_t^i; x^{i,t,u}\right)_{t,u} - \pi_{(V^i)}\left(t, \pi_{(V^i)}\left(s, y_s^i; x^{i,s,t}\right)_t; x^{i,t,u}\right)_{t,u} \\
&= \pi_{(V^i)}\left(t, y_t^i; x^{i,t,u}\right)_{t,u} - \pi_{(V^i)}\left(t, y_t^i - \Gamma_{s,t}^i; x^{i,t,u}\right)_{t,u}.
\end{aligned}
$$

We also set $\bar{A} := A^1 - A^2, \bar{B} := B^1 - B^2$ so that $\bar{\Gamma}_{s,u} - \bar{\Gamma}_{s,t} - \bar{\Gamma}_{t,u} = \bar{A} + \bar{B}$ and hence

$$
\left|\bar{\Gamma}_{s,u} - \bar{\Gamma}_{s,t} - \bar{\Gamma}_{t,u}\right| \leq |\bar{A}| + |\bar{B}|.
$$

We now apply *Lemmas* $\overline{A}$ *and* $\overline{B}$ with parameters $\ell := c_1 \omega\left(s, u\right)^{1/p}, \delta := \varepsilon \ell$. *Lemma* $\overline{A}$ was tailor-made to give the estimate

$$
\begin{aligned}
|\bar{A}| &\leq c_2 \left(|y_s^1 - y_s^2| + \frac{1}{\upsilon}|V^1 - V^2|_{\mathrm{Lip}^{\gamma-1}}\right)\left[\upsilon\omega\left(s, u\right)^{1/p}\right]^\gamma \exp\left(c_2 \upsilon\omega\left(s, u\right)^{1/p}\right) \\
&\quad + c_2 \varepsilon \left[\upsilon\omega\left(s, u\right)^{1/p}\right]^{[p]+1} \exp\left(c_2 \upsilon\omega\left(s, u\right)^{1/p}\right).
\end{aligned}
$$

Re-observing that

$$
\max_{i=1,2}\left|\pi_{(V^i)}\left(s, y_s^i; x^{s,t}\right)_t - y_t^i\right| \leq c_1 \left(\upsilon\omega\left(s, t\right)^{1/p}\right)^{[p]+1}, \tag{10.22}
$$

Lemma $\overline{B}$ tells us that

$$
\begin{aligned}
|\bar{B}| &= \left|B^1 - B^2\right| \\
&\leq c_3 \left|\bar{\Gamma}_{s,t}\right| \upsilon\omega\left(s, u\right)^{\frac{1}{p}} \exp\left(c_3 \upsilon\omega\left(s, u\right)^{1/p}\right) \\
&\quad + c_3 \left(|y_t^1 - y_t^2| + \frac{1}{\upsilon}|V^1 - V^2|_{\mathrm{Lip}^{\gamma-1}} + \varepsilon\upsilon\omega\left(s, t\right)^{1/p}\right) \\
&\qquad \cdot \left\{\left[\upsilon\omega\left(s, u\right)^{1/p}\right]^{1+([p]+1)(\min(2,\gamma)-1)} \exp\left(c_3 \upsilon\omega\left(s, u\right)^{1/p}\right)\right\} \\
&\quad + c_3 \varepsilon \left[\upsilon\omega\left(s, u\right)^{1/p}\right]^{[p]+2} \exp\left(c_3 \upsilon\omega\left(s, u\right)^{1/p}\right).
\end{aligned}
$$

Observe that $1+([p]+1)\left(\min\left(2,\gamma\right)-1\right) \geq \min\left([p]+2, 1+\gamma\left(\gamma-1\right)\right) \geq \gamma$, and obviously $[p]+2 \geq \gamma$. Putting things together, we obtain

$$
\begin{aligned}
\left|\bar{\Gamma}_{s,u}\right| &\leq \left|\bar{\Gamma}_{s,t}\right| \exp\left(c_4 \upsilon\omega\left(s, u\right)^{1/p}\right) + \left|\bar{\Gamma}_{t,u}\right| \tag{10.23} \\
&\quad + c_4 \left(\max_{r\in\{s,t\}}|y_r^1 - y_r^2| + \varepsilon + \frac{1}{\upsilon}|V^1 - V^2|_{\mathrm{Lip}^{\gamma-1}}\right) \\
&\qquad \cdot \left\{\left[\upsilon\omega\left(s, u\right)^{1/p}\right]^\gamma \exp\left(c_4 \upsilon\omega\left(s, u\right)^{1/p}\right)\right\}.
\end{aligned}
$$

We also have, from Theorem 3.18, that

$$
\left|\left(y^1 - y^2\right)_{s,t} - \bar{\Gamma}_{s,t}\right| = \left|\pi_{(V^1)}\left(s, y_s^1; x^{1,s,t}\right)_{s,t} - \pi_{(V^2)}\left(s, y_s^2; x^{2,s,t}\right)_{s,t}\right|
$$

is bounded above by

$$c_5 \left(\left| y_s^1 - y_s^2 \right| + \frac{1}{\upsilon} \left| V^1 - V^2 \right|_{\text{Lip}^{\gamma -1}} + \varepsilon \right) \upsilon \omega \, (s,t)^{1/p} \exp \left(c_5 \upsilon \omega \, (s,t)^{1/p} \right).$$
$$(10.24)$$

Thanks to estimates (10.23), (10.24) and (10.21), we can apply Lemma 10.63 to $(s,t) \mapsto \overline{\Gamma}_{s,t}$ and $t \mapsto \left(y_t^1 - y_t^2 \right)$ with the ε parameter in that lemma set to $\frac{1}{\upsilon} \left| V^1 - V^2 \right|_{\text{Lip}^{\gamma -1}} + \varepsilon$. We therefore see that

$$\left| y^1 - y^2 \right|_{\infty;[0,T]} \leq c_6 \left(\left| y_0^1 - y_0^2 \right| + \frac{1}{\upsilon} \left| V^1 - V^2 \right|_{\text{Lip}^{\gamma -1}} + \varepsilon \right) \exp \left(c_6 \upsilon^p \omega \, (0,T) \right),$$

and that for all $s < t$ in $[0,T]$, with $\theta = \gamma/p > 1$,

$$\left| \overline{\Gamma}_{s,t} \right| \leq c_7 \left(\left| y_0^1 - y_0^2 \right| + \varepsilon + \frac{1}{\upsilon} \left| V^1 - V^2 \right|_{\text{Lip}^{\gamma -1}} \right) \upsilon^{p\theta} \omega \, (s,u)^{\theta} \exp \left(c_7 \upsilon^p \omega \, (0,T) \right).$$

These estimates plus (10.24) easily give that, for all $s < t$ in $[0,T]$,

$$\left| y_{s,t}^1 - y_{s,t}^2 \right| \leq c_8 \left(\upsilon \left| y_s^1 - y_s^2 \right| + \left| V^1 - V^2 \right|_{\text{Lip}^{\gamma -1}} + \upsilon \varepsilon \right) \omega \, (s,t)^{1/p} \exp \left(c_8 \upsilon^p \omega \, (0,T) \right)$$

and this implies the claimed Lipschitz estimate. Obtaining uniqueness of the RDE is then easy: take two solutions in $\pi_{(V^1)} \left(0, y_0^1, \mathbf{x}^1 \right)$. The above estimate tells us the supremum distance between these two solutions is 0. $\blacksquare$

We now discuss some corollaries. First observe that a locally Lipschitz estimate in $\rho_{1/p\text{-Höl};[0,T]}$-metric follows immediately from setting $\omega \, (s,t)$ proportional to $(t-s)$. The corresponding result for $\rho_{p\text{-var};[0,T]}$-metric is the content of

Corollary 10.27 *Assume that*
(i) $V^1 = \left(V_i^1 \right)_{1 \leq i \leq d}$ *and* $V^2 = \left(V_i^2 \right)_{1 \leq i \leq d}$ *are two collections of* Lip^{γ}*-vector fields on* $\mathbb{R}^e$ *for* $\gamma > p \geq 1$;
(ii) $\mathbf{x}^1, \mathbf{x}^2$ *are two weak geometric p-rough paths in* $C^{p\text{-var}} \left([0,T], G^{[p]} \left(\mathbb{R}^d \right) \right)$;
(iii) $y_0^1, y_0^2 \in \mathbb{R}^e$ *thought of as time-0 initial conditions;*
(iv) υ *is a bound on* $\left| V^1 \right|_{\text{Lip}^{\gamma}}$ *and* $\left| V^2 \right|_{\text{Lip}^{\gamma}}$ *and* ℓ_p *a bound on* $\left\| \mathbf{x}^1 \right\|_{p\text{-var};[0,T]}$
and $\left\| \mathbf{x}^2 \right\|_{p\text{-var};[0,T]}$.
Then, if $y^i = \pi_{(V^i)} \left(0, y_0^i; \mathbf{x}^i \right)$, *we have for some constant* C *depending only on* γ *and* p,

$$\rho_{p\text{-var};[0,T]} \left(y^1, y^2 \right)$$
$$\leq C \upsilon \ell_p \left[\left(\left| y_0^1 - y_0^2 \right| + \frac{1}{\upsilon} \left| V^1 - V^2 \right|_{\text{Lip}^{\gamma -1}} \right) \right.$$
$$\left. + \rho_{p\text{-var};[0,T]} \left(\delta_{1/\ell_p} \mathbf{x}^1, \delta_{1/\ell_p} \mathbf{x}^2 \right) \right] \exp \left(C \upsilon^p \ell_p^p \right).$$

Proof. Follows from Theorem 10.26 with control ω given by the sum of the controls constructed in Propositions 8.4 and 8.7. (Note that for $p = 1$ we essentially obtain the ODE estimate from Theorem 3.18.) ∎

Thanks to Theorem 8.10 we can "locally uniformly" switch from the inhomogenous path-space metrics $(\rho_{p,\omega}, \rho_{p\text{-var}})$ to the homogenous ones $(d_{p,\omega}, d_{p\text{-var}})$. In fact, we can state the following.

Corollary 10.28 *Let $V = (V_i)_{1 \le i \le d}$ be a collection of Lip^γ-vector fields on $\mathbb{R}^e$ for $\gamma > p \ge 1$. If ω is a control, $p' \ge p$ and $R > 0$, the maps[7]*

$$\mathbb{R}^e \times \left(\left\{ \|\mathbf{x}\|_{p,\omega} \le R \right\}, d_{p',\omega} \right) \quad \to \quad \left(C^{p\text{-}\omega} \left([0,T],\mathbb{R}^e\right), d_{p',\omega} \right)$$

$$(y_0, \mathbf{x}) \quad \mapsto \quad \pi_{(V)}(0, y_0; \mathbf{x})$$

and

$$\mathbb{R}^e \times \left(\left\{ \|\mathbf{x}\|_{p\text{-var}} \le R \right\}, d_{p'\text{-var}} \right) \quad \to \quad \left(C^{p\text{-var}} \left([0,T],\mathbb{R}^e\right), d_{p'\text{-var}} \right)$$

$$(y_0, \mathbf{x}) \quad \mapsto \quad \pi_{(V)}(0, y_0; \mathbf{x})$$

and

$$\mathbb{R}^e \times \left(\left\{ \|\mathbf{x}\|_{p\text{-var}} \le R \right\}, d_\infty \right) \quad \to \quad \left(C^{p\text{-}\omega} \left([0,T],\mathbb{R}^e\right), d_\infty \right)$$

$$(y_0, \mathbf{x}) \quad \mapsto \quad \pi_{(V)}(0, y_0; \mathbf{x})$$

are also uniformly continuous.

Proof. For $p' = p$, this follows from the above Theorem 10.26/Corollary 10.27 combined with the remark that inhomogenous "ρ" path-space metrics are "Hölder equivalent" on bounded sets to the homogenous ones (Theorem 8.10). Recalling the d_0/d_∞-estimate (Proposition 8.15) we only have to consider the metrics $d_{p'\text{-var}}$ for $p' > p$ and d_0 and since

$$d_0\left(\mathbf{x}^1, \mathbf{x}^2\right) \le d_{p'\text{-var}}\left(\mathbf{x}^1, \mathbf{x}^2\right)$$

it suffices to consider the d_0 case. But this is easy: take $\tilde{p} \in (p, \gamma)$ and consider two paths $\mathbf{x}^1$ and $\mathbf{x}^2$ in $\left\{ \|\mathbf{x}\|_{p,\omega} \le R \right\}$ such that $d_0\left(\mathbf{x}^1, \mathbf{x}^2\right) < \varepsilon$. From interpolation,

$$d_{\tilde{p}\text{-}\omega}\left(\mathbf{x}^1, \mathbf{x}^2\right) \le \varepsilon^{1-p/p'} (2R)^{p/p'}$$

and noting $\|\mathbf{x}\|_{\tilde{p}\text{-}\omega} \le cR$ where c may depend on $\omega(0,T)$ it only remains to use (uniform) continuity (on bounded sets) of the Itô–Lyons map, applied with $\tilde{p}$ instead of p. ∎

---

[7] All sets refering to $\mathbf{x}$ are subsets of $C^{p\text{-var}}\left([0,T], G^{[p]}\left(\mathbb{R}^d\right)\right)$.

10.3.5   Convergence of the Euler scheme

Consider an RDE of the form

$$dy = V(y) \, d\mathbf{x}. \tag{10.25}$$

As usual, $\mathbf{x}$ denotes a geometric p-rough path and we assume sufficient regularity on the collection of vector fields V (i.e. that they are $\mathrm{Lip}^{\gamma-1}$, $\gamma > p$) to ensure existence of a solution y. An Euler scheme of order $\geq [p]$ is a natural way to approximate such solutions, at least locally on some (small) time interval $[s,t]$, and we have already seen the error estimate (cf. Corollary 10.15)

$$\left| y_{s,t} - \mathcal{E}_{(V)}(y_s, \mathbf{x}_{s,t}) \right| \lesssim \omega(s,t)^{\gamma/p} \text{ with } \omega(s,t) = \|\mathbf{x}\|_{p\text{-var};[s,t]}^p$$

which was closely related to our existence proof of RDE solutions. We may rewrite the above as

$$y_t = y_s + y_{s,t} \approx y_s + \mathcal{E}_{(V)}(y_s, \mathbf{x}_{s,t}).$$

Iteration of this leads to an approximate solution over the entire time horizon $[0,T]$. We formalize this in

Definition 10.29 (Euler scheme for RDEs) Given $\mathbf{g} \in G^N(\mathbb{R}^d)$ and $V = (V_i)_{1 \leq i \leq d}$, a collection of vector fields in $C^{N-1}(\mathbb{R}^e)$, we write

$$\mathfrak{E}^{\mathbf{g}} y := y + \mathcal{E}_{(V)}(y, \mathbf{g}).$$

Then, given $D = \{0 = t_0 < t_1 < \cdots < t_n = T\}$ and $\mathbf{x} \in C_o^{p\text{-var}}([0,T], G^{[p]}(\mathbb{R}^d))$ and a fixed integer $N \geq p$ we define the "step-N Euler approximation" to (10.25) at time $t_k \in D$ by[8]

$$y_{t_k}^{\mathrm{Euler};D} := \mathfrak{E}^{t_k \longleftarrow t_0} y_0 := \mathfrak{E}^{S_N(\mathbf{x})_{t_{k-1},t_k}} \circ \cdots \circ \mathfrak{E}^{S_N(\mathbf{x})_{t_1,t_0}} y_0.$$

Theorem 10.30 *Assume that*
(i) $\mathbf{x} \in C^{p\text{-var}}([0,T], G^{[p]}(\mathbb{R}^d))$;
(ii) $V = (V_i)_{1 \leq i \leq d}$ *is a collection of vector fields in* $\mathrm{Lip}^\gamma(\mathbb{R}^e)$; $\gamma > p$;
(iii) $D = \{0 = t_0 < t_1 < \cdots < t_n = T\}$ *is a fixed dissection of* $[0,T]$ *and write* $\#D = n$.
Set $N := \lfloor \gamma \rfloor \geq [p]$, *and define the control*

$$\omega(s,t) = |V|_{\mathrm{Lip}^\gamma}^p \|\mathbf{x}\|_{p\text{-var};[s,t]}^p.$$

Then there exists $C = C(p, \gamma)$ *so that, with* $y = \pi_{(V)}(0, y_0; \mathbf{x})$,

$$\left| y_T - y_T^{\mathrm{Euler};D} \right| \leq C e^{C\omega(0,T)} \sum_{k=1}^{\#D} \omega(t_{k-1}, t_k)^\theta, \quad \theta = (N+1)/p > 1.$$

---

[8]$\mathbf{x}$ is a geometric p-rough path and $S_N(\mathbf{x})$ its unique step-N Lyons lift.

If $\mathbf{x} \in C^{1/p\text{-Höl}}\left([0,T], G^{[p]}\left(\mathbb{R}^d\right)\right)$ *and* $t_{k+1} - t_k \equiv |D|$ *for all* k *then*

$$\left| y_T - y_T^{\text{Euler};D} \right| \leq CT \left|V\right|_{\text{Lip}^\gamma}^{N+1} \|\mathbf{x}\|_{1/p\text{-Höl};[0,T]}^{N+1} |D|^{\theta-1}$$

$$\sim |D|^{\theta-1}.$$

Proof. With $D = (t_k)$ let $z^k := \pi_{(V)}\left(t_k, \mathfrak{C}^{t_k \longleftarrow t_0} y_0; \mathbf{x}\right) \in C\left([t_k, T], \mathbb{R}^e\right)$ be the unique RDE solution to $dy = V(y)\,dx$, with time t_k initial condition given by $\mathfrak{C}^{t_k \longleftarrow t_0} y_0$. Note that $z^0 = y_T$ and $z^n = \mathfrak{C}^{T \longleftarrow t_0} y_0 = y_T^{\text{Euler};D}$. Hence

$$\left| y_T - y_T^{\text{Euler};D} \right| \leq \sum_{k=1}^n \left| z_T^k - z_T^{k-1} \right|$$

and since

$$z_T^k = \pi_{(V)}\left(t_k, \mathfrak{C}^{t_{k-1} \longleftarrow t_0} y_0 + \mathcal{E}_{(V)}\left(\mathfrak{C}^{t_{k-1} \longleftarrow t_0} y_0, S_N(\mathbf{x})_{t_{k-1}, t_k}\right); \mathbf{x}\right)_T,$$

$$z_T^{k-1} = \pi_{(V)}\left(t_{k-1}, \mathfrak{C}^{t_{k-1} \longleftarrow t_0} y_0; \mathbf{x}\right)_T$$

$$= \pi_{(V)}\left(t_k, \pi_{(V)}\left(t_{k-1}, \mathfrak{C}^{t_{k-1} \longleftarrow t_0} y_0; \mathbf{x}\right)_{t_k}; \mathbf{x}\right)_T,$$

we can use Lipschitzness of RDE flows (implied a fortiori by theorem 10.26; here $c_1 = c_1(\gamma, p)$) to estimate

$$\left| z_T^k - z_T^{k-1} \right| \leq c_1 \left| z_{t_k}^k - z_{t_k}^{k-1} \right| \exp\left(c_1 \omega(0,T)\right).$$

On the other hand,

$$\left| z_{t_k}^k - z_{t_k}^{k-1} \right| = \left| \mathcal{E}_{(V)}\left(\mathfrak{C}^{t_{k-1} \longleftarrow t_0} y_0, S_N(\mathbf{x})_{t_{k-1}, t_k}\right) \right.$$

$$\left. - \pi_{(V)}\left(t_{k-1}, \mathfrak{C}^{t_{k-1} \longleftarrow t_0} y_0; \mathbf{x}\right)_{t_{k-1}, t_k} \right|$$

$$\leq \left| \pi_{(V)}\left(t_{k-1}, \cdot; \mathbf{x}\right)_{t_{k-1}, t_k} - \mathcal{E}_{(V)}\left(\cdot, S_N(\mathbf{x})_{t_{k-1}, t_k}\right) \right|_\infty$$

and using Euler RDE estimates this is bounded from above by

$$c_2 \omega\left(t_{k-1}, t_k\right)^{(N+1)/p};$$

indeed, $V \in \text{Lip}^\gamma \implies V \in \text{Lip}^{(N+1)-1}$ with $N + 1 = \lfloor \gamma \rfloor + 1 \geq \gamma > p$ and we apply corollary 10.15 with $N + 1$ (instead of γ). ∎

A similar result holds for "geodesic" approximations in the sense of the following definition.

Definition 10.31 (geodesic scheme for RDEs) Given $D = \{0 = t_0 < t_1 < \cdots < t_n = T\}$, $\mathbf{x} \in C_o^{p\text{-var}}\left([0,T], G^{[p]}\left(\mathbb{R}^d\right)\right)$, $V = (V_i)_{1 \leq i \leq d}$ a collection of Lipschitz vector fields on $\mathbb{R}^e$, and a fixed integer $N \geq p$ we define the "step-N geodesic approximation" to (10.25) via any $x^D \in C^{1\text{-var}}\left([0,T],\mathbb{R}^d\right)$ such that

$$S_N\left(x^D\right)_{t_{k-1},t_k} = \mathbf{x}_{t_{k-1},t_k} \quad \text{for all } k=1,\ldots,n;$$

$$\sup_{D\in\mathcal{D}[0,T]} \left\|S_N\left(x^D\right)\right\|_{p\text{-var};[0,T]} \leq K\left\|\mathbf{x}\right\|_{p\text{-var};[0,T]}.$$

The "step-N geodesic approximation" to (10.25) at time $t \in [0,T]$ is then simply defined as

$$y_t^{\text{geo};D} := \pi_{(V)}\left(0,y_0;x^D\right)_t.$$

Remark 10.32 Such x^D always exist and can be constructed as concatenations of geodesics associated to $S_N\left(\mathbf{x}\right)_{t_{k-1},t_k} \in G^N\left(\mathbb{R}^d\right)$. From $\left\|S_N\left(\mathbf{x}\right)\right\|_{p\text{-var};[0,T]} \leq C_{p,N}\left\|\mathbf{x}\right\|_{p\text{-var};[0,T]}$ we can then take $K = 3^{1-1/p}C_{p,N}$.

Proposition 10.33 *Under the assumptions of Theorem 10.30, there exists $C = C(p,\gamma,K)$ such that we have the error estimate (for the step-N geodesic approximation),*

$$\left|y_T - y_T^{\text{geo};D}\right| \leq Ce^{C\omega(0,T)} \sum_{k=1}^{\#D} \omega\left(t_{k-1},t_k\right)^\theta, \quad \theta = (N+1)/p > 1.$$

Proof. With $D = (t_k)$ let $z^k := \pi\left(t_k,y_{t_k},x^D\right) \in C\left([t_k,T],\mathbb{R}^e\right)$ be the unique ODE solution to $dy = V(y)\,dx^D$, with time t_k initial condition given by $y_{t_k} = \pi_{(V)}\left(0,y_0;\mathbf{x}\right)_{t_k}$. Observe that

$$\left|y_T - y_T^{\text{geo};D}\right| \leq \sum_{k=0}^{\#D-1} \left|z_T^{k+1} - z_T^k\right|.$$

Since $z_T^{k+1} = \pi\left(t_{k+1},y_{t_{k+1}},x^D\right)_T$ and $z_T^k = \pi\left(t_{k+1},\pi\left(t_k,y_{t_k},x^D\right)_{t_{k+1}},x^D\right)_T$ we can use Lipschitzness of RDE flows (implied a fortiori by Theorem 10.26) to see that, with $\omega(s,t) = |V|_{\text{Lip}^\gamma}^p \left\|\mathbf{x}\right\|_{p\text{-var};[s,t]}^p$ as earlier,

$$\begin{aligned}\left|z_T^{k+1} - z_T^k\right| &\leq c_1 \exp\left(c_1\omega(t_{k+1},T)\right)\left|y_{t_{k+1}} - \pi_{(V)}\left(t_k,y_{t_k},x^D\right)_{t_{k+1}}\right| \\ &= c_1 \exp\left(c_1\omega(t_{k+1},T)\right)\left|y_{t_k,t_{k+1}} - \pi_{(V)}\left(t_k,y_{t_k},x^D\right)_{t_k,t_{k+1}}\right|\end{aligned}$$

with $c_1 = c_1(p,\gamma,K)$, uniformly over D simply because $\left\|S_N\left(x^D\right)\right\|_{p\text{-var};[t_{k+1},T]} \leq K\left\|\mathbf{x}\right\|_{p\text{-var};[t_{k+1},T]}$ for all k. From the geodesic error estimate it is clear that, with $c_2 = c_2(p,\gamma)$,

$$\left|z_T^{k+1} - z_T^k\right| \leq c_2 \exp\left(c_2\omega(0,T)\right) \times \omega(t_k,t_{k+1})^\theta$$

for $\theta > 1$. In fact, the same argument as in Theorem 10.30 shows that θ can be taken to be $(N+1)/p$. The proof is now easily finished. ∎

10.4 Full RDE solutions

The RDEs we considered in the previous subsection map weak geometric p-rough paths to $\mathbb{R}^e$-valued paths of bounded p-variation. We shall now see that one can construct a "full" solution as a weak geometric p-rough path in its own right. This will allow us to use a solution to a first RDE to be the driving signal for a second RDE. Relatedly, RDE solutions can then be used (as integrators) in rough integrals, cf. Section 10.6 below. This is not only for "functorial" beauty of the theory! It is precisely this reasoning that will enable us later to deal with various derivatives of RDEs such as the Jacobian of an RDE flow. Let us also remark that in Lyons' orginal work [116], existence and uniqueness was established by Picard iteration and so it was a necessity to work with "full" RDE solutions throughout.

10.4.1 Definition

Definition 10.34 Let $\mathbf{x} \in C^{p\text{-var}}\left([0,T], G^{[p]}\left(\mathbb{R}^d\right)\right)$ be a weak geometric p-rough path. We say that $\mathbf{y} \in C\left([0,T], G^{[p]}\left(\mathbb{R}^e\right)\right)$ is a solution to the full rough differential equation (short: a full RDE solution) driven by $\mathbf{x}$ along the vector fields $(V_i)_i$ and started at $\mathbf{y}_0 \in G^{[p]}\left(\mathbb{R}^e\right)$ if there exists a sequence $(x^n)_n$ in $C^{1\text{-var}}\left([0,T], \mathbb{R}^d\right)$ such that (10.13) holds, and ODE solutions $y_n \in \pi_{(V)}\left(0, \pi_1\left(\mathbf{y}_0\right); x^n\right)$ such that

$$\mathbf{y}_0 \otimes S_{[p]}\left(y^n\right) \text{ converges uniformly to } \mathbf{y} \text{ when } n \to \infty.$$

The (formal) equation $d\mathbf{y} = V(y)\,d\mathbf{x}$ is referred to as a full rough differential equation (short: full RDE). □

This definition generalizes immediately to time intervals $[s,T]$ and we define $\boldsymbol{\pi}_{(V)}\left(s, \mathbf{y}_s; \mathbf{x}\right) \subset C\left([s,T], G^{[p]}\left(\mathbb{R}^e\right)\right)$ to be the set of all solutions to the above full RDE starting at $\mathbf{y}_s$ at time s, and in case of uniqueness, $\boldsymbol{\pi}_{(V)}\left(s, \mathbf{y}_s; \mathbf{x}\right)$ is *the* solution of the full RDE.[9]

A key remark about full RDEs (driven by $\mathbf{x}$ along vector fields V) is that they are just RDEs (in the sense of the previous section) driven by $\mathbf{x}$ but along different vector fields as made precise in the next theorem. We have

Theorem 10.35 *Assume that*
(i) $V = (V_i)_{1\le i\le d}$ *is a collection of vector fields in* $\text{Lip}^{\gamma-1}\left(\mathbb{R}^e\right)$*, where* $\gamma > p$*;*
(ii) $\mathbf{x}: [0,T] \to G^{[p]}\left(\mathbb{R}^d\right)$ *is a weak geometric p-rough path;*
(iii) $\mathbf{y}_0 \in G^{[p]}\left(\mathbb{R}^e\right) \subset T^{[p]}\left(\mathbb{R}^e\right) \cong \mathbb{R}^{1+e+\cdots+e^{[p]}}$ *is an initial condition;*
(iv) $\mathbf{y}$ *is a solution of the full RDE driven by $\mathbf{x}$, along (V), started at $\mathbf{y}_0$.*

---

[9]Make sure to distinguish between $\mathbb{R}^e$-valued RDE solutions denoted by $\pi\left(\ldots\right)$ and $G^{[p]}\left(\mathbb{R}^e\right)$-valued full RDE solutions denoted by the bold greek letter $\boldsymbol{\pi}\left(\ldots\right)$.

Then, $u \mapsto z_u := \mathbf{y}_u \in G^{[p]}(\mathbb{R}^e) \subset T^{[p]}(\mathbb{R}^e) \cong \mathbb{R}^{1+e+\cdots+e^{[p]}}$ is a solution to the RDE

$$dz_u = W(z)\,d\mathbf{x}_u$$

driven by $\mathbf{x}$ along $\mathrm{Lip}_{\mathrm{loc}}^{\gamma-1}(\mathbb{R}^e)$-vector fields on $T^{[p]}(\mathbb{R}^e) \cong \mathbb{R}^{1+e+\cdots+e^{[p]}}$ given by

$$W_i(z) = z \otimes V_i(\pi_1(z)), \quad i = 1,\ldots,d.$$

Proof. By definition of full RDEs and RDEs, we can assume that $\mathbf{x} = S_{[p]}(x)$ where x is of bounded variation. Then, if $y = \pi_{(V)}(0, y_0; x)$,

$$
\begin{aligned}
d\mathbf{y}_u &= d\left(\mathbf{y}_0 \otimes S_{[p]}(y)_{0,u}\right) \\
&= \mathbf{y}_0 \otimes S_{[p]}(y)_{0,u} \otimes dy_{0,u} \\
&= \mathbf{y}_u \otimes V(y_u)\,dx_u.
\end{aligned}
$$

■

10.4.2   Existence

Theorem 10.35 is useful as it immediately implies a local existence theorem for full RDEs. For global existence we need to rule out explosion and we do this via the following quantitative estimates for full RDEs.

Theorem 10.36 *Assume that*
(i) $V = (V_i)_{1 \leq i \leq d}$ is a collection of vector fields in $\mathrm{Lip}^{\gamma-1}(\mathbb{R}^e)$, where $\gamma > p$;
(ii) (x_n) is a sequence in $C^{1\text{-var}}([0,T], \mathbb{R}^d)$, and $\mathbf{x}$ is a weak geometric p-rough path such that

$$\lim_{n \to \infty} d_{0;[0,T]}\left(S_{[p]}(x_n), \mathbf{x}\right) = 0 \text{ and } \sup_n \left\| S_{[p]}(x_n) \right\|_{p\text{-var};[0,T]} < \infty;$$

(iii) $\mathbf{y}_0^n \in G^{[p]}(\mathbb{R}^e)$ is a sequence converging to some $\mathbf{y}_0$.
Then, at least along a subsequence, $\mathbf{y}_0^n \otimes S_{[p]}\left(\pi_{(V)}(0, \pi_1(\mathbf{y}_0^n); x_n)\right)$ converges in uniform topology. Moreover, there exists a constant C_1 depending on p, γ, such that for any such limit point $\mathbf{y}$, we have

$$\|\mathbf{y}\|_{p\text{-var};[s,t]} \leq C_1 \left(|V|_{\mathrm{Lip}^{\gamma-1}} \|\mathbf{x}\|_{p\text{-var};[s,t]} \vee |V|_{\mathrm{Lip}^{\gamma-1}}^p \|\mathbf{x}\|_{p\text{-var};[s,t]}^p \right). \tag{10.26}$$

Then, if $x^{s,t} : [s,t] \to \mathbb{R}^d$ is any continuous bounded variation path such that

$$S_{\lfloor \gamma \rfloor}\left(x^{s,t}\right)_{s,t} = S_{\lfloor \gamma \rfloor}(\mathbf{x})_{s,t} \text{ and } \int_s^t \left| dx_u^{s,t} \right| \leq K \|\mathbf{x}\|_{p\text{-var};[s,t]}$$

for some constant $K \geq 1$, we have for all $s < t$ in $[0,T]$ and all $k \in \{1, \ldots, \lfloor \gamma \rfloor\}$,

$$\left| \pi_k \left(S_{\lfloor \gamma \rfloor} (\mathbf{y})_{s,t} - S_{\lfloor \gamma \rfloor} \left(\pi_{(V)} \left(s, \pi_1 (\mathbf{y}_s) ; x^{s,t} \right) \right)_{s,t} \right) \right| \tag{10.27}$$

$$\leq C_2 \left| \left(K \left| V \right|_{\mathrm{Lip}^{\gamma-1}} \|\mathbf{x}\|_{p\text{-var};[s,t]} \right)^{\gamma+k-1} \vee \left(K \left| V \right|_{\mathrm{Lip}^{\gamma-1}} \|\mathbf{x}\|_{p\text{-var};[s,t]} \right)^{pk} \right|$$

where C_2 depends on p and γ. (For $t - s$ small, the term $(\ldots)^{\gamma+k-1}$ dominates.)

Proof. Observe first that it is enough to prove the quantitative estimates for $\mathbf{x} = S_{[p]} (x)$, where x is a bounded variation path, so that $\mathbf{y} = \mathbf{y}_0 \otimes S_{[p]} \left(\pi_{(V)} (0, \pi_1 (\mathbf{y}_0) ; x) \right)$. Define the control ω by[10]

$$\omega (s,t)^{1/p} = K \left| V \right|_{\mathrm{Lip}^{\gamma-1}} \|\mathbf{x}\|_{p\text{-var};[s,t]}$$

and consider the hypothesis $(\mathcal{H}_N)$:

$$\exists c_{1,N} > 0 : \forall \; s < t \text{ in } [0,T] : \; \|S_N (y)\|_{p\text{-var};[s,t]} \leq c_{1,N} \omega (s,t)^{1/p} \vee \omega (s,t).$$

We aim to prove by induction that $(\mathcal{H}_N)$ holds for all $N = 1, \ldots, [p]$. For $N = 1$, $(\mathcal{H}_N)$ follows from (Davie's) Lemma 10.7. We now assume $(\mathcal{H}_N)$ for a fixed $N < [p]$, and aim to prove $(\mathcal{H}_{N+1})$. Fix s and t in $[0,T]$, define $\alpha := \omega (s,t)^{1/p} \vee \omega (s,t)$ and observe that $\|S_N (y)\|_{p\text{-var};[s,t]} \leq c_{1,N} \omega (s,t)^{1/p} \vee \omega (s,t) = c_{1,N} \alpha$ is equivalent to $\left\| S_N \left(\frac{y}{\alpha} \right) \right\|_{p\text{-var};[s,t]} \leq c_{1,N}$. For $u \in [0, t - s]$, define

$$z_u = \alpha . S_{N+1} \left(\frac{1}{\alpha} y \right)_{s,s+u}$$

so that (we could write dx_{s+u} instead of $d\mathbf{x}_{s+u}$ in the next line)

$$dz_u = W_{\alpha, y_s} (z_u) \, d\mathbf{x}_{s+u}, \quad z_0 = (\alpha, 0, \ldots, 0) \in T^{N+1} (\mathbb{R}^e) \tag{10.28}$$

with vector fields W_{α, y_s} on $T^{N+1} (\mathbb{R}^e)$ given by (cf. notation of Theorem 10.35)

$$W_{\alpha, y_s} (z) = \begin{pmatrix} \frac{1}{\alpha} z \otimes V_1 (y_s + \pi_1 (z)) \\ \ldots \\ \frac{1}{\alpha} z \otimes V_d (y_s + \pi_1 (z)) \end{pmatrix} \quad \text{for } z \in T^{N+1} (\mathbb{R}^e).$$

---

[10] For the proof of the first estimate, (10.26), we take $K = 1$.

Observe that $W_{\alpha,y_s}(z)$ only depends on z through $\pi_{0,N}(z)$; that is, it does not depend on $\pi_{N+1}(z)$. Then, for $k = 0, \dots, N$,

$$
\sup_{u \in [0,t-s]} \frac{|\pi_k(z_u)|}{\alpha} = \sup_{u \in [0,t-s]} \left| \pi_k \circ S_N \left(\frac{1}{\alpha} y \right)_{s,s+u} \right|
$$

$$
\leq c_{2,N} \sup_{u \in [0,t-s]} \left\| S_N \left(\frac{1}{\alpha} y \right)_{s,s+u} \right\|^k
$$

$$
\leq c_{3,N} \text{ from the induction hypothesis.}
$$

Define $\Omega = \left\{ z \in T^{N+1}(\mathbb{R}^e), \frac{1}{\alpha} |\pi_{0,N}(z)| < c_{3,N} + 1 \right\}$ and observe that

$$
\{z_u : u \in [0, t-s]\} \subset \Omega.
$$

On the other hand, for $z \in \Omega$,

$$
\left| \frac{1}{\alpha} z \otimes V_i(\dots) \right| \overset{!}{=} \left| \frac{1}{\alpha} \pi_{0,N}(z) \otimes V_i(\dots) \right| \leq (c_{3,N} + 1) |V_i(\dots)|
$$

so that $|W_{\alpha,y_s}|_{\mathrm{Lip}^{\gamma-1}(\Omega)} \leq c_{5,N} |V|_{\mathrm{Lip}^{\gamma-1}}$. We can then find compactly supported vector fields $\tilde{W}_{\alpha,y_s}$ which coincide with W_{α,y_s} on Ω, and such that

$$
\left| \tilde{W}_{\alpha,y_s} \right|_{\mathrm{Lip}^{\gamma-1}} \leq c_4 |W_{\alpha,y_s}|_{\mathrm{Lip}^{\gamma-1}(\Omega)} \leq c_{5,N} |V|_{\mathrm{Lip}^{\gamma-1}}.
$$

Moreover, since $z|_{[0,t-s]}$ remains in Ω, we see that z actually solves

$$
dz_u = \tilde{W}_{\alpha.y_s}(z_u) \, dx_{s+u}, \quad z_0 = (\alpha, 0, \dots, 0).
$$

From Lemma 10.7, we have

$$
|z_{t-s} - \alpha| \leq c_{6,N} \left(\left| \tilde{W}_{y_s} \right|_{\mathrm{Lip}^{\gamma-1}(\Omega)} \|\mathbf{x}\|_{p\text{-var};[s,t]} \vee \left| \tilde{W}_{y_s} \right|_{\mathrm{Lip}^{\gamma-1}(\Omega)}^p \|\mathbf{x}\|_{p\text{-var};[s,t]}^p \right)
$$

$$
\leq c_{7,N} \left(|V|_{\mathrm{Lip}^{\gamma-1}} \|\mathbf{x}\|_{p\text{-var};[s,t]} \vee |V|_{\mathrm{Lip}^{\gamma-1}}^p \|\mathbf{x}\|_{p\text{-var};[s,t]}^p \right) = c_{7,N} \, \alpha.
$$

This reads $\alpha \left| \delta_{\frac{1}{\alpha}} S_{N+1}(y)_{s,t} - 1 \right| \leq c_{7,N} \alpha$ and, using Proposition 7.45, we have

$$
\left\| \delta_{\frac{1}{\alpha}} S_{N+1}(y)_{s,t} \right\| \leq c_{7,N},
$$

which is equivalent to

$$
\left\| S_{N+1}(y)_{s,t} \right\| \leq c_{7,N} \alpha = c_{1,N+1} \omega(s,t)^{1/p} \vee \omega(s,t).
$$

This finishes the induction step and thus the proof of (10.26).

For the proof of (10.27) we proceed similarly. We first write $\alpha.S_{\lfloor\gamma\rfloor}\left(\frac{1}{\alpha}y\right)$ $|_{[0,t-s]}$ as solution to a differential equation of the form (10.28), but now with vector fields W_{α,y_s} on $T^{\lfloor\gamma\rfloor}\left(\mathbb{R}^e\right)$. Applying the "geodesic approximation" error estimate from (Davie's) Lemma 10.7 we see that

$$\left|\alpha.S_{\lfloor\gamma\rfloor}\left(\frac{1}{\alpha}y\right)_{s,t} - \alpha S_{\lfloor\gamma\rfloor}\left(\frac{1}{\alpha}\pi_{(V)}\left(s,y_s;x^{s,t}\right)\right)_{s,t}\right| \le c_1\omega\left(s,t\right)^{\gamma/p},$$

which means that for all $k \in \{1,\ldots,\lfloor\gamma\rfloor\}$,

$$\left|\pi_k\left(S_{\lfloor\gamma\rfloor}\left(y\right)_{s,t} - S_{\lfloor\gamma\rfloor}\left(\pi_{(V)}\left(s,y_s;x^{s,t}\right)\right)_{s,t}\right)\right| \le c_1\alpha^{k-1}\omega\left(s,t\right)^{\gamma/p}.$$

$$(10.29)$$

Recalling that, by definition, $\alpha = \omega\left(s,t\right)^{1/p} \vee \omega\left(s,t\right)$ leads to an estimate of the form (10.27) with right-hand side given by a constant times

$$\left[\omega\left(s,t\right)^{1/p}\right]^{\gamma+k-1} \vee \left[\omega\left(s,t\right)^{1/p}\right]^{\gamma+p(k-1)}.$$

Obviously, the term $\left[\omega\left(s,t\right)^{1/p}\right]^{\gamma+k-1}$ dominates for $\omega\left(s,t\right) \le 1$. For $\omega\left(s,t\right) \ge 1$ it is in fact better to estimate each term on the left-hand side of (10.29) separately, using in particular

$$\left|\pi_k\left(S_{\lfloor\gamma\rfloor}\left(y\right)_{s,t}\right)\right| \le \left\|S_{\lfloor\gamma\rfloor}\left(y\right)_{s,t}\right\|^k \le \|y\|_{p\text{-var};[s,t]}^k,$$

thanks to estimates for the Lyons lift and (10.26). ∎

Remark 10.37 Observe that the (first part of the) above proof shows that for any fixed $s < t$, and $\alpha \ge \omega\left(s,t\right)^{1/p} \vee \omega\left(s,t\right)$ with $\omega\left(s,t\right)^{1/p} = |V|_{\text{Lip}^{\gamma-1}} \|\mathbf{x}\|_{p\text{-var};[s,t]}$

$$r \in [0,t-s] \mapsto z_r = \alpha\left(\delta_{1/\alpha}\mathbf{y}\right)_{s,s+r} \in T^{[p]}\left(\mathbb{R}^d\right)$$

satisfies $dz_r = \tilde{W}_{\alpha,y_s}\left(z_r\right)dx_{s+r}$ started from $z_0 = (\alpha,0,\ldots,0)$ along compactly supported vector fields $\tilde{W}_{\alpha,y_s}$ on $T^{[p]}\left(\mathbb{R}^e\right)$ which satisfy

$$\left|\tilde{W}_{\alpha,y_s}\right|_{\text{Lip}^{\gamma-1}} \le c_{5,N}\,|V|_{\text{Lip}^{\gamma-1}}$$

and $\tilde{W}_{\alpha,y_s}\left(z\right) \equiv \left(\frac{1}{\alpha}z\otimes V_j\left(y_s + \pi_1\left(z\right)\right)\right)_{j=1,\ldots,d}$ on $\left\{z \in T^{[p]}\left(\mathbb{R}^e\right), \frac{1}{\alpha}|z| < c\right\}$ for some c dependent on p. □

10.4.3 Uniqueness and continuity

Theorem 10.26 states uniqueness of RDE solutions, but ignores full RDEs. We observe that full RDE solutions are RDE solutions driven by different vector fields. In particular, if we put ourselves under the same conditions as Theorem 10.26, we automatically obtain uniqueness of full RDEs. What is less obvious is that we have a similar Lipschitz bound for the (full) Itô–Lyons map. Just as in the existence discussion, this estimate on full RDE solutions is actually a consequence of our earlier estimate on RDE solutions.

Theorem 10.38 *Assume that*
(i) $V^1 = \left(V_i^1\right)_{1 \leq i \leq d}$ *and* $V^2 = \left(V_i^2\right)_{1 \leq i \leq d}$ *are two collections of* Lip^γ*-vector fields on* $\mathbb{R}^e$ *for* $\gamma > p \geq 1$;
(ii) ω *is a fixed control;*
(iii) $\mathbf{x}^1, \mathbf{x}^2 \in C^{p\text{-var}}\left([0,T], G^{[p]}\left(\mathbb{R}^d\right)\right)$, *with* $\left\|\mathbf{x}^i\right\|_{p-\omega} \leq 1$;
(iv) $\mathbf{y}_0^1, \mathbf{y}_0^2 \in G^{[p]}\left(\mathbb{R}^e\right)$ *thought of as time-0 initial conditions;*
(v) υ *is a bound on* $\left|V^1\right|_{\mathrm{Lip}^\gamma}$ *and* $\left|V^2\right|_{\mathrm{Lip}^\gamma}$.
For $i = 1, 2$ *we set* $\mathbf{y}^i = \boldsymbol{\pi}_{(V^i)}\left(0, \mathbf{y}_0^i; \mathbf{x}^i\right)$; *that is, the full RDE solutions driven by* $\mathbf{x}^i$, *starting at* y_0, *along the vector fields* $V_1^i, \ldots, V_d^i$. *Then we have the following Lipschitz estimate on the (full) Itô–Lyons map:*[11]

$$\rho_{p,\omega}\left(\mathbf{y}^1, \mathbf{y}^2\right)$$
$$\leq C\left[\upsilon\left|y_0^1 - y_0^2\right| + \left|V^1 - V^2\right|_{\mathrm{Lip}^\gamma-1} + \upsilon\rho_{p,\omega}\left(\mathbf{x}^1, \mathbf{x}^2\right)\right]\exp\left(C\upsilon^p\omega\left(0,T\right)\right),$$

where $C = C\left(\gamma, p\right)$ *and* $y_0^i = \pi_1\left(\mathbf{y}_0^i\right) \in \mathbb{R}^e$.

Proof. Fix $s < t$ in $[0, T]$, and define $\alpha = c_2\omega\left(s, t\right)^{1/p}\exp\left(c_2\omega\left(0,T\right)\right) \in \mathbb{R}$. Define for $i = 1, 2$ and $r \in [0, t - s]$, $z_r^i = \alpha\left(\delta_{1/\alpha}\mathbf{y}_{s,s+r}^i\right)$. As noticed in Remark 10.37, z^i is the solution of the RDE

$$dz_r^i = \tilde{W}_{\alpha, y_s^i}^i\left(z_r^i\right)d\mathbf{x}_{s+r}^i, \qquad z_0^i = \alpha,$$

where $\left|\tilde{W}_{\alpha, y_s}^i\right|_{\mathrm{Lip}^\gamma} \leq c\upsilon$, and $\tilde{W}_{\alpha, y_s}^i\left(z\right) \equiv \left(\frac{1}{\alpha}z \otimes V_j^i\left(y_s + \pi_1\left(z\right)\right)\right)_{j=1,\ldots,d}$ on $\left\{z \in T^{[p]}\left(\mathbb{R}^e\right) : \frac{1}{\alpha}|z| < c\right\}$ for some c dependent on p. Writing $y_t^i = \pi_1\left(\mathbf{y}_t^i\right)$ it is also easy to see that

$$\left|\tilde{W}_{\alpha, y_s^2}^2 - \tilde{W}_{\alpha, y_s^1}^1\right|_{\mathrm{Lip}^\gamma-1} \leq c_1\left|V^2\left(y_s^2 + \cdot\right) - V^1\left(y_s^1 + \cdot\right)\right|_{\mathrm{Lip}^\gamma-1}$$
$$\leq c_1\left|V^2 - V^1\right|_{\mathrm{Lip}^\gamma-1}$$
$$+ c_1\left|V^1\left(y_s^2 + \cdot\right) - V^1\left(y_s^1 + \cdot\right)\right|_{\mathrm{Lip}^\gamma-1}$$
$$\leq c_2\left(\left|V^2 - V^1\right|_{\mathrm{Lip}^\gamma-1} + \left|y_s^2 - y_s^1\right|.\upsilon\right)$$

$$(10.30)$$

---

[11] By definition of a full RDE solution $\mathbf{y}$, any increment $\mathbf{y}_{s,t}$ depends on the starting point $\mathbf{y}_0$ only through $y_0 = \pi_1\left(\mathbf{y}_0\right)$. This explains why we don't have $\left|y_0^1 - y_0^2\right| = \left|\mathbf{y}_0^1 - \mathbf{y}_0^2\right|_{T^{[p]}(\mathbb{R}^e)}$ on the right-hand side.

where we used crucially $V^1 \in \mathrm{Lip}^\gamma$ in the last estimate. Therefore, we obtain from Theorem 10.26 that, with $\varepsilon := \rho_{p,\omega;[0,T]}\left(\mathbf{x}^1,\mathbf{x}^2\right)$,

$$\left|z_{t-s}^1 - z_{t-s}^2\right| \le c_3 \left(\upsilon\left|y_s^1 - y_s^2\right| + \left|V^1 - V^2\right|_{\mathrm{Lip}^{\gamma-1}} + \varepsilon \upsilon\right) \omega\left(s,t\right)^{\frac{1}{p}}$$
$$\exp\left(c_3 \upsilon^p \omega\left(0,T\right)\right).$$

But this says precisely that for all $k = 1, \dots, [p]$ and $s < t$ in $[0,T]$,

$$\left|\pi_k\left(\mathbf{y}_{s,t}^1 - \mathbf{y}_{s,t}^2\right)\right| \le c_3 \left(\upsilon\left|y_s^1 - y_s^2\right| + \left|V^1 - V^2\right|_{\mathrm{Lip}^{\gamma-1}} + \varepsilon \upsilon\right) \omega\left(s,t\right)^{\frac{k}{p}}$$
$$\exp\left(c_3 \upsilon^p \omega\left(0,T\right)\right). \tag{10.31}$$

[We insist that one does *not* have the norm of $\mathbf{y}_s^1 - \mathbf{y}_s^2 \in T^{N+1}\left(\mathbb{R}^e\right)$ on the right-hand side above, but $\left|y_s^1 - y_s^2\right|$ coming from (10.30).] But from Theorem 10.26, noting that $\omega\left(0,s\right)^{1/p}\exp\left(C\upsilon^p\omega\left(0,T\right)\right) \lesssim \upsilon^{-1}\exp\left(c_4\upsilon^p\omega(0,T)\right)$, we have

$$\upsilon\left|y_s^1 - y_s^2\right| \le c_4 \left[\upsilon\left|y_0^1 - y_0^2\right| + \left|V^1 - V^2\right|_{\mathrm{Lip}^{\gamma-1}} + \varepsilon\upsilon\right]\exp\left(c_4\upsilon^p\omega\left(0,T\right)\right).$$

Plugging this inequality into (10.31) gives us the desired result. ∎

The following corollaries can then be proved with almost identical arguments as in the RDE case.

Corollary 10.39 *Assume that*
(i) $V^1 = \left(V_i^1\right)_{1\le i\le d}$ and $V^2 = \left(V_i^2\right)_{1\le i\le d}$ are two collections of Lip^γ-vector fields on $\mathbb{R}^e$ for $\gamma > p \ge 1$;
(ii) $\mathbf{x}^1, \mathbf{x}^2 \in C^{p\text{-var}}\left([0,T], G^{[p]}\left(\mathbb{R}^d\right)\right)$;
(iii) $\mathbf{y}_0^1, \mathbf{y}_0^2 \in G^{[p]}\left(\mathbb{R}^e\right)$ thought of time-0 initial conditions;
(iv) υ is a bound on $\left|V^1\right|_{\mathrm{Lip}^\gamma}$ and $\left|V^2\right|_{\mathrm{Lip}^\gamma}$ and ℓ_p a bound on $\left\|\mathbf{x}^1\right\|_{p\text{-var};[0,T]}$ and $\left\|\mathbf{x}^2\right\|_{p\text{-var};[0,T]}$.
Then, if $\mathbf{y}^i = \boldsymbol{\pi}_{(V^i)}\left(0,\mathbf{y}_0^i;\mathbf{x}^i\right)$, we have

$$\rho_{p\text{-var};[0,T]}\left(\mathbf{y}^1,\mathbf{y}^2\right) \le C\upsilon\ell_p\left[\left(\left|y_0^1 - y_0^2\right| + \frac{1}{\upsilon}\left|V^1 - V^2\right|_{\mathrm{Lip}^{\gamma-1}}\right)\right.$$
$$\left. + \rho_{p\text{-var};[0,T]}\left(\delta_{1/\ell_p}\mathbf{x}^1, \delta_{1/\ell_p}\mathbf{x}^2\right)\right]\exp\left(C\upsilon^p\ell_p^p\right),$$

where $C = C\left(\gamma,p\right)$ and $y_0^i = \pi_1\left(\mathbf{y}_0^i\right) \in \mathbb{R}^e$.

For simplicity of the statement, we once again remove the dependency in the vector fields.

Corollary 10.40 *Let $V = (V_i)_{1\le i\le d}$ be a collection of Lip^γ-vector fields on $\mathbb{R}^e$ for $\gamma > p \ge 1$.*
If ω is a control, $p' \ge p$ and $R > 0$, the maps

$$\mathbb{R}^e \times \left(\left\{\|\mathbf{x}\|_{p,\omega} \le R\right\}, d_{p',\omega}\right) \quad \to \quad \left(C^{p\text{-}\omega}\left([0,T], G^{[p]}\left(\mathbb{R}^e\right)\right), d_{p',\omega}\right)$$
$$(y_0,\mathbf{x}) \quad \mapsto \quad \boldsymbol{\pi}_{(V)}\left(0,y_0;\mathbf{x}\right)$$

and

$$\mathbb{R}^e \times \left(\left\{ \|\mathbf{x}\|_{p\text{-var}} \leq R \right\}, d_{p'\text{-var}} \right) \quad \rightarrow \quad \left(C^{p\text{-var}} \left([0,T], G^{[p]} \left(\mathbb{R}^e \right) \right), d_{p'\text{-var}} \right)$$

$$(y_0, \mathbf{x}) \quad \mapsto \quad \boldsymbol{\pi}_{(V)} \left(0, y_0; \mathbf{x} \right)$$

and

$$\mathbb{R}^e \times \left(\left\{ \|\mathbf{x}\|_{p\text{-}var} \leq R \right\}, d_\infty \right) \quad \rightarrow \quad \left(C^{p\text{-}var} \left([0,T], G^{[p]} \left(\mathbb{R}^e \right) \right), d_\infty \right)$$

$$(y_0, \mathbf{x}) \quad \mapsto \quad \boldsymbol{\pi}_{(V)} \left(0, y_0; \mathbf{x} \right)$$

are also uniformly continuous.

10.5 RDEs under minimal regularity of coefficients

We now show the uniqueness of solutions to RDEs driven by geometric p-rough paths along Lip^p-vector fields. In fact, the driving signal only needs to be of finite $\psi_{p,1}$-variation where we recall from Section 5.4 that

$$\psi_{p,1}(t) = t^p / \left(\ln^* \ln^* 1/t \right) \quad \text{where } \ln^* \equiv \max \left(1, \ln \right).$$

For instance, with probability one, (enhanced) Brownian sample paths have finite $\psi_{2,1}$-variation but are not geometric 2-rough paths (i.e. don't have finite 2-variation), as will be discussed in the later Sections 13.2 and 13.9. The main interest in the refined regularity assumption is that it shows that RDE solutions driven by (enhanced) Brownian motion have unique solutions under Lip^2-regularity assumptions.

Theorem 10.41 *Assume that $p \geq 1$, and that*
(i) $V^1 = \left(V_j^1 \right)_{1 \leq j \leq d}$ and $V^2 = \left(V_j^2 \right)_{1 \leq j \leq d}$ are two collections of Lip^p-vector fields on $\mathbb{R}^e$;
(ii) $\mathbf{x}^1, \mathbf{x}^2 \in C^{\psi_{p,1}\text{-var}} \left([0,T], G^{[p]} \left(\mathbb{R}^d \right) \right)$;
(iii) $y_0^1, y_0^2 \in G^{[p]} \left(\mathbb{R}^e \right)$ thought of as time-0 initial conditions;
(iv) $\mathbf{y}^i \in \boldsymbol{\pi}_{(V^i)} \left(0, y_0^i; \mathbf{x}^i \right)$ for $i = 1,2$ (that is, they are full RDE solutions driven by $\mathbf{x}^i$, starting at y_0^i, along the vector fields V^i);
(v) assume $\left| V^i \right|_{\text{Lip}^p} \leq \upsilon$ and $\left\| \mathbf{x}^i \right\|_{\psi_{p,1}\text{-var};[0,T]} \leq R$ for $i = 1,2$.
Then, $\boldsymbol{\pi}_{(V^i)} \left(0, y_0^i; \mathbf{x}^i \right)$ is a singleton; that is, there exists a unique full RDE solution $\mathbf{y}^i = \boldsymbol{\pi}_{(V^i)} \left(0, y_0^i; \mathbf{x}^i \right)$ starting at y_0^i driven by $\mathbf{x}^i$ along V^i. Moreover, for all $\varepsilon > 0$, there exists $\mu = \mu \left(\varepsilon; p, \upsilon, R \right) > 0$ such that[12]

$$\left| y_0^1 - y_0^2 \right| + \left| V^1 - V^2 \right|_{\text{Lip}^{p-1}} + d_\infty \left(\mathbf{x}^1, \mathbf{x}^2 \right) < \mu$$

implies $d_\infty \left(\mathbf{y}^1, \mathbf{y}^2 \right) < \varepsilon$.

---

[12] For $p = 1$, $\left| V^1 - V^2 \right|_{\text{Lip}^{p-1}}$ is replaced by $2 \left| V^1 - V^2 \right|_\infty$.

Remark 10.42 The theorem applies in particular to geometric p-rough paths, i.e. when $\mathbf{x}^1, \mathbf{x}^2 \in C^{p\text{-var}} \left([0,T], G^{[p]} \left(\mathbb{R}^d \right) \right)$. It is also clear that under Lip^p-regularity, existence of (full) RDE solutions is not an issue here so that the set of RDE solutions $\pi_{(V^i)} \left(0, \mathbf{y}_0^i ; \mathbf{x}^i \right)$ is not empty. $\qquad\square$

Proof. Since we only deal with supremum distance d_∞ and without quantitative estimates, it is enough to prove the above result for RDE rather than full RDE. Running constants $c_1, c_2, \ldots$ may depend on p, υ, R which are kept fixed in this proof. Let $\varepsilon(\mu)$ be the supremum of $d_\infty \left(\mathbf{y}^1, \mathbf{y}^2 \right)$, taken over all RDE solutions $\mathbf{y}^i \in \pi_{(V^i)} \left(0, \mathbf{y}_0^i ; \mathbf{x}^i \right)$, $i = 1, 2$ such that $V^i, \mathbf{x}^i$ satisfy (v) and such that

$$\left| y_0^1 - y_0^2 \right| + \left| V^1 - V^2 \right|_{\mathrm{Lip}^p - 1} + d_\infty \left(\mathbf{x}^1, \mathbf{x}^2 \right) < \mu ;$$

we show that $\varepsilon = \varepsilon(\mu) \to 0$ as $\mu \to 0$.

Construction of $x_{s,t}^i$: For $i = 1, 2$, we define

$$\omega^i(s,t) = \sup_{D \in \mathcal{D}([s,t])} \sum_{t_i \in D} \psi_{p,1} \left[\frac{\left\| \mathbf{x}_{t_i, t_{i+1}}^i \right\|}{\left\| \mathbf{x}^i \right\|_{\psi_{p,1}\text{-var};[0,T]}} \right] ;$$

from Proposition 5.39, ω^i is a control with $\omega^i(0,T) \leq 1$. Define $\omega = \left(\omega^1 + \omega^2 \right)$.

Using the fact that $\psi_{p,1}^{-1}(\cdot) \leq c_1 \psi_{1/p,-1/p}(\cdot) = c_1 \delta(\cdot)^{1/p}$ with $\delta(z) = z \ln^* \ln^* \frac{1}{z}$ (see Lemma 5.48), it follows (again using Proposition 5.39) that for all $s < t$ in $[0,T]$,

$$\left\| \mathbf{x}_{s,t}^i \right\| \leq c_2 \delta \left(\omega(s,t) \right)^{1/p} .$$

As $\delta(z)^{1/p} \leq z^{1/p'}$ for all $p' > p$, we see that $\left\| \mathbf{x}^i \right\|_{p'\text{-var};[s,t]} \leq c_3 \omega(s,t)^{1/p'}$. By interpolation, we obtain that for all $p'' > p'$, there exists g_1, a continuous function 0 at 0 such that, for all $s < t$ in $[0,T]$ and $k = 1, \ldots, [p]$,

$$\left| \pi_k \left(\mathbf{x}_{s,t}^1 - \mathbf{x}_{s,t}^2 \right) \right| \leq g_1(\mu) \omega(s,t)^{k/p''} .$$

This immediately implies that for $\alpha < 1$ (we can take it as close to 1 as we want, and in particular we take it greater than $\{p\}$) and for all $s < t$ in $[0,T]$, and for $k = 1, \ldots, [p]$,

$$\begin{aligned} \left| \pi_k \left(\mathbf{x}_{s,t}^1 - \mathbf{x}_{s,t}^2 \right) \right| &\leq c_3 g_1(\mu) \omega(s,t)^{\frac{k+\alpha-1}{p}} \\ &\leq c_3 g_1(\mu) \delta \left(\omega(s,t) \right)^{\frac{k+\alpha-1}{p}} . \end{aligned}$$

As $\left\| \delta_{\delta(\omega(s,t))^{-1/p}} \mathbf{x}_{s,t}^i \right\| \leq c_2$, and

$$\left| \delta_{\delta(\omega(s,t))^{-1/p}} \mathbf{x}_{s,t}^1 - \delta_{\delta(\omega(s,t))^{-1/p}} \mathbf{x}_{s,t}^2 \right| \leq c_3 g_1(\mu) \delta \left(\omega(s,t) \right)^{(\alpha-1)/p} ,$$

Proposition 7.64 provides us with two paths $x^{1,s,t}$ and $x^{2,s,t}$ such that

$$S_{[p]}\left(x^{i,s,t}\right)_{s,t} = \mathbf{x}^i_{s,t},$$

and

$$\int_s^t d\left|x^{i,s,t}_r\right| \leq c_4 \delta\left(\omega\left(s,t\right)\right)^{1/p},$$

$$\int_s^t d\left|x^{1,s,t}_r - x^{2,s,t}_r\right| \leq c_4 g_1\left(\mu\right) \delta\left(\omega\left(s,t\right)\right)^{\alpha/p}.$$

We define similarly $x^{i,s,u}$ and $x^{i,t,u}$, and then $x^{i,s,t,u}$ to be the concatenation of $x^{i,s,t}$ and $x^{i,t,u}$.

Estimates on Γ: Following closely the pattern of proof of Theorem 10.26, we set for $s < t$,

$$\Gamma^i_{s,t} = y^i_{s,t} - \pi\left(s, y^i_s; x^{i,s,t}\right)_{s,t}, \quad i = 1,2,$$

and $\overline{\Gamma}_{s,t} := \Gamma^1_{s,t} - \Gamma^2_{s,t}$. Theorem 10.14 gives us for $p' \in (p, [p]+1)$,

$$\left|\Gamma^i_{s,t}\right| \leq c_5 \left\|\mathbf{x}^i\right\|_{p'\text{-var};[s,t]}^{[p]+1}.$$

Using Proposition 5.49 and Theorem 5.43, we see that

$$\begin{aligned}
\left\|\mathbf{x}^i\right\|_{p'\text{-var};[s,t]} &\leq c_6 \psi_{p,1}^{-1}\left(\omega^i\left(s,t\right)\right) \\
&\leq c_7 \delta\left(\omega\left(s,t\right)\right)^{1/p}.
\end{aligned}$$

In particular, we have with $\theta = \left([p]+1\right)/p > 1$,

$$\left|\overline{\Gamma}_{s,t}\right| \leq c_7 \delta\left(\omega\left(s,t\right)\right)^{\theta} \leq c_8 \delta\left(\omega\left(s,t\right)\right). \tag{10.32}$$

Define for $i = 1,2$,

$$\begin{aligned}
A^i &:= \pi_{(V^i)}\left(s, y^i_s; x^{i,s,t,u}\right)_{s,u} - \pi_{(V^i)}\left(s, y^i_s; x^{i,s,u}\right)_{s,u} \\
B^i &:= \pi_{(V^i)}\left(t, y^i_t; x^{i,t,u}\right)_{t,u} - \pi_{(V^i)}\left(t, \pi_{(V^i)}\left(s, y^i_s; x^{i,s,t}\right)_t; x^{i,t,u}\right)_{t,u}
\end{aligned}$$

and $\bar{A} := A^1 - A^2, \bar{B} := B^1 - B^2$, we obtain $\overline{\Gamma}_{s,u} - \overline{\Gamma}_{s,t} - \overline{\Gamma}_{t,u} = \tilde{A} + \tilde{B}$ so that

$$\left|\overline{\Gamma}_{s,u} - \overline{\Gamma}_{s,t} - \overline{\Gamma}_{t,u}\right| \leq \left|\bar{A}\right| + \left|\bar{B}\right|.$$

From *Lemma $\overline{A}$*, applied with parameters $\ell := c_4 \delta\left(\omega\left(s,u\right)\right)^{1/p}, \delta := c_4 g_1\left(\mu\right) \delta\left(\omega\left(s,u\right)\right)^{\alpha/p}$, it follows that

$$\begin{aligned}
\left|\bar{A}\right| &\leq c_9 \left(\left|y^1_s - y^2_s\right| + \left|V^1 - V^2\right|_{\text{Lip}^{p}-1}\right) \delta\left(\omega\left(s,u\right)\right) \\
&\quad + c_9 g_1\left(\mu\right) \delta\left(\omega\left(s,u\right)\right)^{\alpha/p} \cdot \delta\left(\omega\left(s,u\right)\right)^{\frac{[p]+\alpha}{p}} \\
&\leq c_9 \left(\left|y^1_s - y^2_s\right| + \left|V^1 - V^2\right|_{\text{Lip}^{p}-1} + g_1\left(\mu\right)\right) \delta\left(\omega\left(s,u\right)\right).
\end{aligned}$$

On the other hand, $\max_{i=1,2}\left|\pi_{(V)}\left(s,y_s^i;x^{s,t}\right)_t - y_t^i\right| = \max_{i=1,2}\left|\Gamma_{s,t}^i\right|$ and with (10.32),

$$\left|\overline{\Gamma}_{s,t}\right| \leq c_8\delta\left(\omega\left(s,t\right)\right)^\theta \leq c_8\delta\left(\omega\left(s,t\right)\right). \tag{10.33}$$

From *Lemma $\overline{B}$* it then follows that

$$\begin{aligned}
\left|\bar{B}\right| \leq\ & c_3\left|\overline{\Gamma}_{s,t}\right|\delta\left(\omega\left(s,u\right)\right)^{\frac{1}{p}} + c_3g_1\left(\mu\right)\delta\left(\omega\left(s,u\right)\right)^{1+\alpha/p} \\
& + c_3\left(\left|y_t^1 - y_t^2\right| + \left|V^1 - V^2\right|_{\mathrm{Lip}^{\gamma}-1} + g_1\left(\mu\right)\delta\left(\omega\left(s,u\right)\right)^{\alpha/p}\right) \\
& \cdot \delta\left(\omega\left(s,u\right)\right)^{\frac{1}{p}\left(1+p\left(\min\left(2,p\right)-1\right)\right)}.
\end{aligned}$$

Observe that $\frac{1}{p}\left(1+p\left(\min\left(2,p\right)-1\right)\right) \geq \frac{1}{p}\min\left(p+1,p^2-2p+1+p\right) \geq 1$, and obviously $1+\alpha/p \geq 1$. Putting things together, we obtain

$$\begin{aligned}
\left|\bar{B}\right| \leq\ & c_{12}\left|\overline{\Gamma}_{s,t}\right|\delta\left(\omega\left(s,u\right)\right)^{\frac{1}{p}} \\
& + c_{12}\left(\left|y_t^1 - y_t^2\right| + \left|V^1 - V^2\right|_{\mathrm{Lip}^p-1} + g_1\left(\mu\right)\right).\delta\left(\omega\left(s,u\right)\right).
\end{aligned}$$

Adding the inequalities on $\left|\bar{A}\right|$ and $\left|\bar{B}\right|$, we obtain that

$$\left|\overline{\Gamma}_{s,u} - \overline{\Gamma}_{s,t} - \overline{\Gamma}_{t,u}\right| \leq c_{13}\left|\overline{\Gamma}_{s,t}\right|\delta\left(\omega\left(s,u\right)\right)^{\frac{1}{p}} \tag{10.34}$$
$$+ c_{13}\left(\max_{r\in\{s,t\}}\left|y_r^1 - y_r^2\right| + \left|V^1 - V^2\right|_{\mathrm{Lip}^p-1} + g_1\left(\mu\right)\right)\delta\left(\omega\left(s,u\right)\right).$$

We also have, from Theorem 3.18, that

$$\begin{aligned}
& \left|\pi_{(V^1)}\left(s,y_s^1;x^{1,s,t}\right) - \pi_{(V^2)}\left(s,y_s^2;x^{2,s,t}\right)\right| \\
& \leq c_{14}\left(\left(\left|y_s^1 - y_s^2\right| + \left|V^1 - V^2\right|_{\mathrm{Lip}^p-1}\right)\delta\left(\omega\left(s,t\right)\right)^{1/p} + g_1\left(\mu\right)\omega\left(s,t\right)^{\alpha/p}\right) \\
& \leq c_{14}\left(\left|y_s^1 - y_s^2\right| + \left|V^1 - V^2\right|_{\mathrm{Lip}^p-1} + g_1\left(\mu\right)\right)\omega\left(s,t\right)^{\alpha/p}. \tag{10.35}
\end{aligned}$$

<u>Conclusion:</u> Thanks to estimates (10.34), (10.35) and (10.33), we can apply Proposition 10.70 (found in Appendix 10.8) to obtain our result. ∎

By interpolation, we obtain the following corollary.

Corollary 10.43 *Let $V = (V_1,\ldots,V_d)$ be a collection of Lip^p-vector fields on $\mathbb{R}^e$. Fix $p'' > p' > p > 1$, a control ω, $R \in (0,\infty]$ and set*[13]

$$\Omega_{p'\text{-}\omega}\left(p,R\right) = \left\{\mathbf{x} : \|\mathbf{x}\|_{p'\text{-}\omega} < R \text{ and } \|\mathbf{x}\|_{\psi_p,1\text{-var}} < R\right\}.$$
$$\Omega\left(p,R\right) = \left\{\mathbf{x} : \|\mathbf{x}\|_{\psi_p,1\text{-var}} < R\right\}.$$

---

[13]In general, $C^{\psi_p,1\text{-var}} \subsetneq C^{p'\text{-}\omega}$.

Then the maps

$$\mathbb{R}^e \times \left(\Omega_{p'\text{-}\omega}\left(p, R\right), d_{p''\text{-}\omega}\right) \rightarrow \left(C^{\psi_{p,1}\text{-var}} \cap C^{p'\text{-}\omega}\left([0, T], G^{[p]}\left(\mathbb{R}^e\right)\right), d_{p''\text{-}\omega}\right)$$

$$(y_0, \mathbf{x}) \mapsto \boldsymbol{\pi}_{(V)}\left(0, y_0; \mathbf{x}\right)$$

and

$$\mathbb{R}^e \times \left(\Omega\left(p, R\right), d_{p'\text{-var}}\right) \rightarrow \left(C^{\psi_{p,1}\text{-var}}\left([0, T], G^{[p]}\left(\mathbb{R}^e\right)\right), d_{p'\text{-var}}\right)$$

$$(y_0, \mathbf{x}) \mapsto \boldsymbol{\pi}_{(V)}\left(0, y_0; \mathbf{x}\right)$$

are uniformly continuous for $R \in (0, \infty)$ and continuous for $R = +\infty$. (This also holds when $d_{p'\text{-var}}$ is replaced by d_∞.)

Proof. Since continuity is a local property, it suffices to consider the case $R < \infty$. Observe that for all $\mathbf{x} \in \Omega\left(p, R\right)$, we have

$$\left\|\boldsymbol{\pi}_{(V)}\left(0, y_0; \mathbf{x}\right)\right\|_{\psi_{p,1}\text{-var}} < c(R),$$

as follows from Corollary 5.44 and Proposition 5.49. Then, we know from Theorem 10.41 that

$$\mathbb{R}^e \times \left(\Omega\left(p, R\right), d_\infty\right) \rightarrow \left(C^{\psi_{p,1}\text{-var}}\left([0, T], G^{[p]}\left(\mathbb{R}^e\right)\right), d_\infty\right)$$

$$(y_0, \mathbf{x}) \mapsto \boldsymbol{\pi}_{(V)}\left(0, y_0; \mathbf{x}\right)$$

is uniformly continuous. In particular, this implies that for some fixed $\varepsilon, R > 0$, there exists $\mu > 0$ such that if $\left|y_0^1 - y_0^2\right| + d_{p'\text{-var}}\left(\mathbf{x}^1, \mathbf{x}^2\right) < \mu$, with $\mathbf{x}^1, \mathbf{x}^2 \in \Omega\left(p, R\right)$, then

$$d_\infty\left(\boldsymbol{\pi}_{(V)}\left(0, y_0^1; \mathbf{x}^1\right), \boldsymbol{\pi}_{(V)}\left(0, y_0^2; \mathbf{x}^2\right)\right) < \varepsilon.$$

Using $\left\|\boldsymbol{\pi}_{(V)}\left(0, y_0^i; \mathbf{x}^i\right)\right\|_{\psi_{p,1}\text{-var}} < c(R)$ we can use interpolation to obtain uniform continuity of

$$\mathbb{R}^e \times \left(\Omega\left(p, R\right), d_{p'\text{-var}}\right) \rightarrow \left(C^{\psi_{p,1}\text{-var}}\left([0, T], G^{[p]}\left(\mathbb{R}^e\right)\right), d_{p'\text{-var}}\right)$$

$$(y_0, \mathbf{x}) \mapsto \boldsymbol{\pi}_{(V)}\left(0, y_0; \mathbf{x}\right).$$

The modulus ω case follows from the same argument, except that for the interpolation step, we need to assume that the p'-ω norm of $\mathbf{x}^i$, $i = 1, 2$ is bounded by R (since, for a given ω, we cannot be sure that the $\psi_{p,1}$-variation of $\mathbf{x}^i$ is controlled by this ω). ∎

10.6 Integration along rough paths

With our main interest in (rough) differential equations we constructed RDEs directly as limits of ODEs. In the same spirit, we now define "rough integrals" as limits of Riemann–Stieltjes integration. Given the work already done, we can take a short-cut and derive existence, uniqueness and continuity properties quickly from the previous RDE results.

Definition 10.44 (rough integrals) Let $\mathbf{x} \in C^{p\text{-var}}\left([0,T], G^{[p]}\left(\mathbb{R}^d\right)\right)$ be a weak geometric p-rough path, and $\varphi = (\varphi_i)_{i=1,\ldots,d}$ a collection of maps from $\mathbb{R}^d$ to $\mathbb{R}^e$. We say that $\mathbf{y} \in C\left([0,T], G^{[p]}\left(\mathbb{R}^e\right)\right)$ is a rough path integral of φ along $\mathbf{x}$, if there exists a sequence (x^n) in $C^{1\text{-var}}\left([0,T], \mathbb{R}^d\right)$ such that

$$\forall n : x_0^n = \pi_1\left(\mathbf{x}_0\right)$$

$$\lim_{n \to \infty} d_{0;[0,T]}\left(S_{[p]}\left(x^n\right), \mathbf{x}\right) = 0$$

$$\sup_n \left\|S_{[p]}\left(x^n\right)\right\|_{p\text{-var};[0,T]} < \infty$$

and

$$\lim_{n \to \infty} d_\infty\left(S_{[p]}\left(\int_0^{\cdot} \varphi\left(x_u^n\right) dx_u^n\right), \mathbf{y}\right) = 0.$$

We will write $\int \varphi\left(x\right) d\mathbf{x}$ for the set of rough path integrals of φ along $\mathbf{x}$. If this set is a singleton, it will denote the rough path integral of φ along $\mathbf{x}$. □

We first note that a classical indefinite Riemann–Stieltjes integral $\int_0^{\cdot} \varphi\left(x\right) dx$ can be written as a (projection to the y-component of the) ODE solution to

$$
\begin{aligned}
dz &= dx, \\
dy &= \varphi\left(z\right) dx, \\
(z, y) &= (x_0, 0).
\end{aligned}
$$

This leads to the key remark that rough integrals can be viewed as (projections of) a solution to (full) RDEs driven by $\mathbf{x}$ along vector fields $V = (V_1, \ldots, V_d)$ given by

$$V_i\left(x, y\right) = \left(e_i, \varphi_i\left(x\right)\right), \tag{10.36}$$

where $(e_1, \ldots, e_d)$ is the standard basis of $\mathbb{R}^d$. Obviously V has the same amount of Lip-regularity as $\varphi = (\varphi_i)_{i=1,\ldots,d}$, viewed as a map

$$\varphi : \mathbb{R}^d \to L\left(\mathbb{R}^d, \mathbb{R}^e\right);$$

in fact, $|V|_{\text{Lip}^{\gamma-1}} \lesssim 1 + |\varphi|_{\text{Lip}^{\gamma-1}}$. Thus, if one is happy with rough integration under Lip^γ-regularity, $\gamma > p$, existence, uniqueness and uniform continuity on bounded sets (in fact: Lipschitz continuity on bounded sets with respect to the correct rough path metric $\rho_{p\text{-var}}$) of

$$\text{Lip}^{\gamma} \times C^{p\text{-var}} \left([0,T], G^{[p]} \left(\mathbb{R}^d \right) \right) \quad \to \quad C^{p\text{-var}} \left([0,T], G^{[p]} \left(\mathbb{R}^e \right) \right)$$

$$(\varphi, \mathbf{x}) \quad \mapsto \quad \int \varphi(x) \, d\mathbf{x}$$

is immediate from the corresponding results on full RDEs in Section 10.4.

The point of the forthcoming theorem is that for rough path integration, one gets away with $\text{Lip}^{\gamma-1}$ regularity. As can be seen already from the next lemma, there is no hope for local Lipschitz continuity of $(\varphi, \mathbf{x}) \mapsto \int \varphi(x) \, d\mathbf{x}$, but uniqueness and uniform continuity on bounded sets hold true.

Lemma 10.45 *Let*

(i) $\left(\varphi_j^1 \right)_{j=1,\dots,d}, \left(\varphi_j^2 \right)_{j=1,\dots,d}$ *be two collections of* $\text{Lip}^{\alpha} \left(\mathbb{R}^d, \mathbb{R}^e \right)$*-maps,* $\alpha \in (0,1]$*, so that* $\max_{i=1,2} \left| \varphi^i \right|_{\text{Lip}^{\alpha}} \leq 1$*;*

(ii) $x^i \in C^{1\text{-var}} \left([s,t], \mathbb{R}^d \right)$ *such that, for some* $\ell \geq 0$ *and* $\varepsilon \in [0,1]$*,*

$$\left| x^i \right|_{1\text{-var};[s,t]} \leq \ell,$$

$$\left| x^2 - x^1 \right|_{1\text{-var};[s,t]} \leq \varepsilon \ell \text{ and } \left| x_s^2 - x_s^1 \right| \leq \varepsilon.$$

Then, for $k \in \{1, 2, \dots\}$*,*

$$\left| \pi_k \left(S_k \left(\int \varphi^2 \left(x^2 \right) dx^2 \right)_{s,t} \right) - \pi_k \left(S_k \left(\int \varphi^1 \left(x^1 \right) dx^1 \right)_{s,t} \right) \right|$$

$$\leq \left(\left| \varphi^2 - \varphi^1 \right|_{\infty} + \varepsilon^{\alpha} \right) \ell^k.$$

Proof. From Proposition 2.8, we easily see that

$$\left| \int \varphi^2 \left(x^2 \right) dx^2 - \int \varphi^1 \left(x^1 \right) dx^1 \right|_{1\text{-var};[s,t]} \leq \left(\left| \varphi^2 - \varphi^1 \right|_{\infty} + \varepsilon^{\alpha} \right) \ell.$$

The conclusion follows then from Proposition 7.63.   ■

Lemma 10.46 (Lemma A_{integral}) *Assume that*

(i) $\left(\varphi_j \right)_{j=1,\dots,d}$ *is a collection of* $\text{Lip}^{\gamma-1} \left(\mathbb{R}^d, \mathbb{R}^e \right)$*-maps where* $\gamma > 1$*;*

(ii) $s < u$ *are some elements in* $[0,T]$*;*

(iii) x *and* $\tilde{x}$ *are some paths in* $C^{1\text{-var}} \left([s,u], \mathbb{R}^d \right)$ *such that* $S_{\lfloor \gamma \rfloor} \left(x \right)_{s,u} = S_{\lfloor \gamma \rfloor} \left(\tilde{x} \right)_{s,u}$ *and* $x_s = \tilde{x}_s$*;*

(iv) $\ell \geq 0$ *is a bound on* $\int_s^u |dx| + \int_s^u |d\tilde{x}|$*.*

Then we have, for some constant $C = C(\gamma)$ *and for all* $k \in \{1, \dots, \lfloor \gamma \rfloor\}$*,*

$$\left| \pi_k \circ S_k \left(\int \varphi(x) \, dx \right)_{s,u} - \pi_k \circ S_k \left(\int \varphi(\tilde{x}) \, d\tilde{x} \right)_{s,u} \right| \leq C \left\| \varphi \right\|_{\text{Lip}^{\gamma-1}}^k \ell^{\gamma+k-1}.$$

Proof. The indefinite $\mathbb{R}^e$-valued integral $\int_s^\cdot \varphi(x)\,dx$ is also a (projection of the) solution to the ODE $y_\cdot = \pi_{(V)}\left(s,(x_s,0);x\right)$ with V given by (10.36). By homogeneity, we may assume $\|\varphi\|_{\mathrm{Lip}^{\gamma-1}} = 1$ so that $|V|_{\mathrm{Lip}^{\gamma-1}} \leq c_1$. We leave it to the reader to provide a self-contained "Riemann–Stieltjes/ODE" proof of the claimed estimate. Somewhat shorter, if more fanciful, we can think of y as a solution (again, strictly speaking, a projection thereof) of an RDE driven by a weak geometric 1-rough path **x**. The claimed estimate is now a consequence of the "geodesics error-estimate" of Theorem 10.36 with $p = 1$ and $K = 1$. Indeed, we take a geodesic $x^{s,u}$ associated with $S_{\lfloor\gamma\rfloor}(x)_{s,u}$ such that

$$S_{\lfloor\gamma\rfloor}\left(x^{s,u}\right)_{s,u} = S_{\lfloor\gamma\rfloor}(x)_{s,u} = S_{\lfloor\gamma\rfloor}(\tilde{x})_{s,u} \quad \text{and}$$

$$\int_s^u |dx^{s,u}| \leq \left\|S_{\lfloor\gamma\rfloor}(x)\right\|_{1\text{-var};[s,u]} = |x|_{1\text{-var};[s,u]}.$$

Since $x_s = \tilde{x}_s$ we see that both $y_\cdot = \pi_{(V)}\left(s,(x_s,0);x\right)$ and $\tilde{y}_\cdot = \pi_{(V)}\left(s,(\tilde{x}_s,0);\tilde{x}\right)$ have the same geodesic approximation given by $\pi_{(V)}(s,(x_s,0);x^{s,u})$. On the "projected" level of the integral this amounts to considering $\int_s^\cdot \varphi(x_s + x_{s,\cdot}^{s,u})\,dx_{s,\cdot}^{s,u})$ and so, for all $s < u$ in $[0,T]$ and for all $k \in \{1,\ldots,\lfloor\gamma\rfloor\}$,

$$\left| \pi_k\left(S_{\lfloor\gamma\rfloor}\left(\int_s^\cdot \varphi(x)\,dx\right)_{s,u} - S_{\lfloor\gamma\rfloor}\left(\int_s^\cdot \varphi\left(x_s + x_{s,\cdot}^{s,u}\right)dx_{\cdot}^{s,u}\right)_{s,u}\right) \right|$$

$$\leq c_2\left(|x|_{1\text{-var};[s,u]}\right)^{\gamma+k-1} \leq c_2 \ell^{\gamma+k-1}.$$

The same estimate holds with x replaced by $\tilde{x}$ and an application of the triangle inequality finishes the proof. ∎

Theorem 10.47 <u>*Existence:*</u> *Assume that*
(i) $\varphi = (\varphi_j)_{j=1,\ldots,d}$ *is a collection of* $\mathrm{Lip}^{\gamma-1}\left(\mathbb{R}^d, \mathbb{R}^e\right)$*-maps where* $\gamma > p \geq 1$;
(ii) **x** *is a geometric p-rough path in* $C^{p\text{-var}}\left([0,T], G^{[p]}\left(\mathbb{R}^d\right)\right)$.
Then, for all $s < t \in [0,T]$, there exists a unique rough-path integral of φ along **x**. *The indefinite integral $\int \varphi(x)\,dx$ is a geometric rough path: there exists a constant C_1 depending only on p and γ such that for all $s < t$ in* $[0,T]$,

$$\left\|\int \varphi(x)\,dx\right\|_{p\text{-var};[s,t]} \leq C_1\,\|\varphi\|_{\mathrm{Lip}^{\gamma-1}}\left(\|\mathbf{x}\|_{p\text{-var};[s,t]} \vee \|\mathbf{x}\|_{p\text{-var};[s,t]}^p\right).$$

Also, if $x^{s,t} : [s,t] \to \mathbb{R}^d$ is any continuous bounded variation path such that

$$x_s^{s,t} = \pi_1(\mathbf{x}_s),\ S_{\lfloor\gamma\rfloor}\left(x^{s,t}\right)_{s,t} = S_{\lfloor\gamma\rfloor}(\mathbf{x})_{s,t} \quad and \quad \int_s^t |dx_u^{s,t}| \leq K\,\|\mathbf{x}\|_{p\text{-var};[s,t]}$$

for some constant K, we have for all $s < t$ in $[0, T]$ with $K \|\mathbf{x}\|_{p\text{-var};[s,t]} < 1$, and all $k \in \{1, \ldots, \lfloor \gamma \rfloor\}$,

$$\left| \pi_k \left\{ \left(\int \varphi(x) \, d\mathbf{x} \right)_{s,t} - S_{\lfloor \gamma \rfloor} \left(\int \varphi\left(x_u^{s,t}\right) \, dx_u^{s,t} \right)_{s,t} \right\} \right|$$
$$\leq C_2 \|\varphi\|_{\mathrm{Lip}^{\gamma-1}}^k \left(K \|\mathbf{x}\|_{p\text{-var};[s,t]} \right)^{\gamma+k-1}, \tag{10.37}$$

where C_2 depends on p and γ.
Uniqueness, continuity: There exists a unique element in $\int \varphi(x) \, d\mathbf{x}$. More precisely, if ω is a fixed control,

$$\max_{i=1,2} \left\{ \left|\varphi^i\right|_{\mathrm{Lip}^{\gamma-1}}, \left\|\mathbf{x}^i\right\|_{p\text{-}\omega;[0,T]} \right\} < R,$$

and

$$\varepsilon = \left| x_0^1 - x_0^2 \right| + \rho_{p\text{-}\omega;[0,T]}\left(\mathbf{x}^1, \mathbf{x}^2\right) + \left|\varphi^2 - \varphi^1\right|_{\mathrm{Lip}^{\gamma-1}},$$

then for some constant $\beta = \beta(\gamma, p) > 0$ and $C = C(R, \gamma, p) \geq 0$,

$$\rho_{p\text{-}\omega;[0,T]}\left(\int \varphi^1\left(x^1\right) \, d\mathbf{x}^1, \int \varphi^2\left(x^2\right) \, d\mathbf{x}^2 \right) \leq C \varepsilon^\beta.$$

Proof. Existence, first proof: For some fixed $r > 0$, consider the collection of vector fields $V^{(r)}$ defined by

$$V_i^{(r)}(x, y) = (e_i, r\varphi_i(x)) \text{ for } x, y \in \mathbb{R}^d \times \mathbb{R}^e,$$

where e_i is the ith standard basis vector of $\mathbb{R}^d$. Observe that $V^{(r)}$ is $(\gamma - 1)$-Lipschitz, and that for a bounded variation path we have

$$\pi_{\left(V^{(r)}\right)}(0, (x_0, 0); x) = \left(x, \int_0^\cdot r\varphi(x_u) \, dx_u \right).$$

In view of $(\gamma - 1)$-Lipschitz regularity of the vector fields we can apply Theorem 10.36; we obtain existence of a rough integral of $\mathbf{x}$ along φ, and that every element $\mathbf{y}^{(r)}$ of $\int r\varphi(x) \, d\mathbf{x}$ satisfies

$$\left\| \mathbf{y}^{(r)} \right\|_{p\text{-var};[s,t]} \leq c_1 \left(\left| V^{(r)} \right|_{\mathrm{Lip}^{\gamma-1}} \|\mathbf{x}\|_{p\text{-var};[s,t]} \vee \left| V^{(r)} \right|_{\mathrm{Lip}^{\gamma-1}}^p \|\mathbf{x}\|_{p\text{-var};[s,t]}^p \right).$$

Now, setting $\mathbf{y} = \mathbf{y}^{(1)}$ it is easy to see that $\mathbf{y}^{(r)} = \delta_r \mathbf{y}$ and $\left| V^{(r)} \right|_{\mathrm{Lip}^{\gamma-1}} \leq c_2 \left(1 + r \left|\varphi\right|_{\mathrm{Lip}^{\gamma-1}} \right)$; thus, picking $r = 1/\left|\varphi\right|_{\mathrm{Lip}^{\gamma-1}}$, we obtain

$$\|\mathbf{y}\|_{p\text{-var};[s,t]} \leq c_3 \left|\varphi\right|_{\mathrm{Lip}^{\gamma-1}} \left(\|\mathbf{x}\|_{p\text{-var};[s,t]} \vee \|\mathbf{x}\|_{p\text{-var};[s,t]}^p \right).$$

The error estimate (10.37) is also a consequence of the corresponding estimate in Theorem 10.36, again applied to full RDEs along vector fields $V^{(r)}$ with $r = 1/ |\varphi|_{\text{Lip}^\gamma - 1}$.

Existence, second proof: We provide a second existence proof, based on by now standard arguments, which is (notationally) helpful for the forthcoming uniquess proof. Similar to the proof of RDE existence, it suffices to establish uniform estimates for $\mathbf{x}$ of the form $\mathbf{x} = S_{[p]}(x)$, followed by a straightforward limiting argument. Let us thus assume $\mathbf{x} = S_{[p]}(x)$ and $\|\mathbf{x}_{s,t}\| \leq \omega(s,t)^{1/p}$ for some control ω. By assumption, there exists $x^{s,t}$, which we think of as an "almost" geodesic path associated with $\mathbf{x}_{s,t}$, and we define

$$\Gamma_{s,t} = S_{\lfloor \gamma \rfloor} \left(\int \varphi(x)\, dx \right)_{s,t} - S_{\lfloor \gamma \rfloor} \left(\int \varphi(x^{s,t})\, dx^{s,t} \right)_{s,t}.$$

Then, for $s < t < u$, we have

$$\Gamma_{s,u} = \Delta_1 - \Delta_2 + \Delta_3$$

where

$$\Delta_1 = S_{\lfloor \gamma \rfloor} \left(\int \varphi(x)\, dx \right)_{s,t} \otimes S_{\lfloor \gamma \rfloor} \left(\int \varphi(x)\, dx \right)_{t,u},$$

$$\Delta_2 = S_{\lfloor \gamma \rfloor} \left(\int \varphi(x^{s,t})\, dx^{s,t} \right)_{s,t} \otimes S_{\lfloor \gamma \rfloor} \left(\int \varphi(x^{t,u})\, dx^{t,u} \right)_{t,u},$$

$$\Delta_3 = S_{\lfloor \gamma \rfloor} \left(\int \varphi(x^{s,t,u})\, dx^{s,t,u} \right)_{s,u} - S_{\lfloor \gamma \rfloor} \left(\int \varphi(x^{s,u})\, dx^{s,u} \right)_{s,u}.$$

First, using Lemma A_{integral} (Lemma 10.46), we obtain that for all $N \in \{1, \ldots, \lfloor \gamma \rfloor\}$

$$|\pi_N(\Delta_3)| \leq c_1 \omega(s,u)^{\frac{\gamma + N - 1}{p}}. \tag{10.38}$$

We also see that

$$\Delta_1 - \Delta_2 = S_{\lfloor \gamma \rfloor} \left(\int \varphi(x)\, dx \right)_{s,t} \otimes \Gamma_{t,u} + \Gamma_{s,t} \otimes S_{\lfloor \gamma \rfloor} \left(\int \varphi(x^{t,u})\, dx^{t,u} \right)_{t,u}$$

so that $\Gamma_{s,u} - \Gamma_{s,t} \otimes \Gamma_{t,u}$ equals

$$S_{\lfloor \gamma \rfloor} \left(\int \varphi(x^{s,t})\, dx^{s,t} \right)_{s,t} \otimes \Gamma_{t,u} + \Gamma_{s,t} \otimes S_{\lfloor \gamma \rfloor} \left(\int \varphi(x^{t,u})\, dx^{t,u} \right)_{t,u} + \Delta_3. \tag{10.39}$$

We now prove by induction in $N \in \{1, \ldots, \lfloor \gamma \rfloor\}$ that

$$\forall s < u \text{ in } [0, T] : |\pi_N(\Gamma_{s,u})| \leq c_2 \omega(s,u)^{\frac{\gamma + N - 1}{p}} \exp\left(c_2 \omega(s,u)^{1/p} \right).$$

For $N = 0$, this is obvious as $\pi_0 \left(\Gamma_{s,t} \right) = 0$. Assume now that

$$\forall s < u \text{ in } [0, T] : \left| \pi_k \left(\Gamma_{s,u} \right) \right| \leq c_2 \omega \left(s, u \right)^{\frac{\gamma + k - 1}{p}} \exp \left(c_2 \omega \left(s, u \right)^{1/p} \right)$$

holds for all $k < N$. Then, from (10.39), we have

$$\pi_N \left(\Gamma_{s,u} \right) - \pi_N \left(\Gamma_{s,t} \right) - \pi_N \left(\Gamma_{t,u} \right) \qquad\qquad (10.40)$$

$$= \pi_N \left(\Delta_3 \right) + \sum_{k=1}^{N-1} \pi_k \left(\Gamma_{s,t} \right) \otimes \pi_{N-k} \left(\Gamma_{t,u} \right)$$

$$+ \sum_{k=1}^{N-1} \pi_k \circ S_{\lfloor \gamma \rfloor} \left(\int \varphi \left(x^{s,t} \right) dx^{s,t} \right)_{s,t} \otimes \pi_{N-k} \left(\Gamma_{t,u} \right)$$

$$+ \sum_{k=1}^{N-1} \pi_k \left(\Gamma_{s,t} \right) \otimes \pi_{N-k} \circ S_{\lfloor \gamma \rfloor} \left(\int \varphi \left(x^{t,u} \right) dx^{t,u} \right)_{t,u},$$

so that, using the induction hypothesis, (10.38) and bounds of the type

$$\left| \pi_k \circ S_{\lfloor \gamma \rfloor} \left(\int \varphi \left(x^{s,t} \right) dx^{s,t} \right)_{s,t} \right| \leq c_2 \omega \left(s, t \right)^{k/p},$$

we have[14]

$$\left| \pi_N \left(\Gamma_{s,u} \right) - \pi_N \left(\Gamma_{s,t} \right) - \pi_N \left(\Gamma_{t,u} \right) \right| \leq c_3 \omega \left(s, u \right)^{\frac{\gamma + N - 1}{p}} \exp \left(c_2 \omega \left(s, u \right)^{1/p} \right).$$
$$(10.41)$$

We can then classically use Lemma 10.59 to obtain that

$$\left| \pi_N \left(\Gamma_{s,u} \right) \right| \leq c_4 \omega \left(s, u \right)^{\frac{\gamma + N - 1}{p}} \exp \left(c_4 \omega \left(s, u \right)^{1/p} \right),$$

which concludes the induction proof. The triangle inequality then leads to, for $k \in \{1, \ldots, \lfloor \gamma \rfloor\}$,

$$\left| \pi_k \left(S_{\lfloor \gamma \rfloor} \left(\int \varphi \left(x \right) dx \right)_{s,t} \right) \right| \leq c_4 \omega \left(s, t \right)^{k/p} \exp \left(c_4 \omega \left(s, t \right)^{1/p} \right),$$

which is equivalent to saying

$$\left\| S_{\lfloor \gamma \rfloor} \left(\int \varphi \left(x \right) dx \right)_{s,t} \right\| \leq c_5 \omega \left(s, t \right)^{1/p} \exp \left(c_5 \omega \left(s, t \right)^{1/p} \right).$$

---

[14] As we shall need (10.41) in the uniqueness/continuity part below, let us point out that (10.41) also follows from the "first" existence proof, namely from the error estimate (10.37).

An application of Proposition 5.10 then gives

$$\left\| S_{\lfloor \gamma \rfloor} \left(\int \varphi(x) \, dx \right)_{s,t} \right\| \leq c_6 \omega(s,t)^{1/p} \vee \omega(s,t).$$

Continuity/uniqueness: Without loss of generality, we assume $\lfloor \gamma \rfloor = [p]$, and also, by simple scaling, that $\max_{i=1,2} |\varphi^i|_{\mathrm{Lip}^{\gamma-1}} \leq 1$ (so that we can use Lemma 10.45). We set

$$\varepsilon = \left| x_0^1 - x_0^2 \right| + \rho_{p\text{-}\omega;[0,T]} \left(\mathbf{x}^1, \mathbf{x}^2 \right) + \left| \varphi^2 - \varphi^1 \right|_{\mathrm{Lip}^{\gamma-1}}$$

and agree that, in this part of the proof, constants may depend on $\omega(0,T)$ and R. Then we define (exactly as in the proof of Theorem 10.26) two paths $x^{1,s,t}$ and $x^{2,s,t}$ such that

$$S_{[p]} \left(x^{i,s,t} \right)_{s,t} = \mathbf{x}_{s,t}^i, \ i = 1, 2,$$

and such that

$$\int_s^t \left| dx_r^{i,s,t} \right| \leq c_7 \omega(s,t)^{1/p}, \ i = 1, 2,$$

$$\int_s^t \left| dx_r^{1,s,t} - dx_r^{2,s,t} \right| \leq c_7 \varepsilon \omega(s,t)^{1/p}.$$

We also define (as usual!)

$$\Gamma_{s,t}^i = \left(\int \varphi^i \left(x_r^i \right) d\mathbf{x}_r^i \right)_{s,t} - S_{[p]} \left(\int \varphi^i \left(x_r^{i,s,t} \right) dx_r^{i,s,t} \right)_{s,t}, \ i = 1, 2,$$

and set $\overline{\Gamma}_{s,t} := \Gamma_{s,t}^1 - \Gamma_{s,t}^2$. From the existence part, we have $\left| \overline{\Gamma}_{s,t} \right| \leq c_8 \omega(s,t)^{\gamma/p}$. Define

$$\overline{\Delta}_3 = \left\{ S_{\lfloor \gamma \rfloor} \left(\int \varphi^1 \left(x^{1,s,t,u} \right) dx^{1,s,t,u} \right)_{s,u} - S_{\lfloor \gamma \rfloor} \left(\int \varphi^1 \left(x^{1,s,u} \right) dx^{1,s,u} \right)_{s,u} \right\}$$

$$- \left\{ S_{\lfloor \gamma \rfloor} \left(\int \varphi^2 \left(x^{2,s,t,u} \right) dx^{2,s,t,u} \right)_{s,u} - S_{\lfloor \gamma \rfloor} \left(\int \varphi^2 \left(x^{2,s,u} \right) dx^{2,s,u} \right)_{s,u} \right\}.$$

Observe that by continuity of the Riemann–Stieltjes integral and the map $S_{\lfloor \gamma \rfloor}$, we have for all integers $k \in \{1, \ldots, \lfloor p \rfloor\}$

$$
\begin{aligned}
&\left| \pi_k \left(\overline{\Delta}_3 \right) \right| \\
&\leq \left| \pi_k \circ S_{\lfloor \gamma \rfloor} \left(\int \varphi^2 \left(x^{2,s,t,u} \right) dx^{2,s,t,u} \right)_{s,u} \right. \\
&\qquad \left. - \pi_k \circ S_{\lfloor \gamma \rfloor} \left(\int \varphi^1 \left(x^{1,s,t,u} \right) dx^{1,s,t,u} \right)_{s,u} \right| \\
&\quad + \left| \pi_k \circ S_{\lfloor \gamma \rfloor} \left(\int \varphi^2 \left(x^{2,s,u} \right) dx^{2,s,u} \right)_{s,u} \right. \\
&\qquad \left. - \pi_k \circ S_{\lfloor \gamma \rfloor} \left(\int \varphi^1 \left(x^{1,s,u} \right) dx^{1,s,u} \right)_{s,u} \right| \\
&\leq 2 \left(\left| \varphi^2 - \varphi^1 \right|_{\mathrm{Lip}^{\gamma-1}} + \rho_{p,\omega} \left(\mathbf{x}^1, \mathbf{x}^2 \right)^{\min\{\gamma-1,1\}} \right) \\
&\qquad \omega \left(s,t \right)^{k/p} \text{ using Lemma 10.45} \\
&\leq \varepsilon^\alpha \omega \left(s,t \right)^{k/p} \text{ for some } \alpha \in (0,1].
\end{aligned}
\tag{10.43}
$$

We now prove by induction on $N \leq \lfloor p \rfloor$ that

$$
\forall k \in \{1, \ldots, N\} : \exists \beta_k > 0 : \left| \pi_k \left(\overline{\Gamma}_{s,t} \right) \right| \leq \varepsilon^{\beta_k} \omega \left(s,t \right).
$$

For $N = 0$, it is obvious, as $\pi_0 \left(\overline{\Gamma}_{s,t} \right) = 0$. Assume now the induction hypothesis is true for all $k < N$. From equation (10.40), we see that $\pi_N \left(\overline{\Gamma}_{s,u} \right) - \pi_N \left(\overline{\Gamma}_{s,t} \right) - \pi_N \left(\overline{\Gamma}_{t,u} \right)$ is equal to $\overline{\Delta}_3 + D_1 + D_2 + D_3$, where

$$
\begin{aligned}
D_1 &= \sum_{k=1}^{N-1} \pi_k \left(\overline{\Gamma}_{s,t} \right) \otimes \pi_{N-k} \left(\Gamma_{t,u}^2 \right) - \sum_{k=1}^{N-1} \pi_k \left(\Gamma_{s,t}^1 \right) \otimes \pi_{N-k} \left(\overline{\Gamma}_{t,u} \right) \\
D_2 &= \sum_{k=1}^{N-1} \pi_k \left(S_{\lfloor \gamma \rfloor} \left(\int \varphi \left(x^{2,s,t} \right) dx^{2,s,t} \right)_{s,t} \right. \\
&\qquad \left. - S_{\lfloor \gamma \rfloor} \left(\int \varphi \left(x^{1,s,t} \right) dx^{1,s,t} \right)_{s,t} \right) \otimes \pi_{N-k} \left(\Gamma_{t,u}^2 \right) \\
&\quad + \sum_{k=1}^{N-1} \pi_k \left(S_{\lfloor \gamma \rfloor} \left(\int \varphi \left(x^{1,s,t} \right) dx^{1,s,t} \right)_{s,t} \right) \otimes \pi_{N-k} \left(\overline{\Gamma}_{t,u} \right)
\end{aligned}
$$

$$D_3 = \sum_{k=1}^{N-1} \pi_k \left(\overline{\Gamma}_{s,t}\right) \otimes \pi_{N-k} \circ S_{\lfloor \gamma \rfloor} \left(\int \varphi\left(x^{2,t,u}\right) dx^{2,t,u}\right)_{t,u}$$

$$- \sum_{k=1}^{N-1} \pi_k \left(\Gamma_{t,u}^1\right) \otimes \pi_{N-k} \left(S_{\lfloor \gamma \rfloor}\left(\int \varphi\left(x^{2,t,u}\right) dx^{2,t,u}\right)_{t,u}\right.$$

$$\left. - S_{\lfloor \gamma \rfloor}\left(\int \varphi\left(x^{1,t,u}\right) dx^{1,t,u}\right)_{t,u}\right).$$

We easily see from the induction hypothesis and Lemma 10.45 that

$$|D_1| + |D_2| + |D_3| \leq c_9 \varepsilon^{\alpha_N} \omega\left(s,t\right)^{N/p}, \quad \text{for some } \alpha_N \in (0, \alpha].$$

In particular, with (10.43), we obtain that

$$\left|\pi_N\left(\overline{\Gamma}_{s,u} - \overline{\Gamma}_{s,t} - \overline{\Gamma}_{t,u}\right)\right| \leq c_{10} \varepsilon^{\alpha_N} \omega\left(s,t\right)^{N/p}.$$

From the existence part, equation (10.41), we also have

$$\left|\pi_N\left(\overline{\Gamma}_{s,u} - \overline{\Gamma}_{s,t} - \overline{\Gamma}_{t,u}\right)\right| \leq \left|\pi_N\left(\Gamma_{s,u}^2 - \Gamma_{s,t}^2 - \Gamma_{t,u}^2\right)\right| + \left|\pi_N\left(\Gamma_{s,u}^1 - \Gamma_{s,t}^1 - \Gamma_{t,u}^1\right)\right|$$

$$\leq c_{11} \omega\left(s,t\right)^{\frac{\gamma+N-1}{p}} \leq c_{12} \omega\left(s,t\right)^{\theta} \quad \text{with } \theta = \gamma/p > 1.$$

These estimates show that assumption (ii) of Lemma 10.61 is satisfied (checking assumption (i) is easy and left to the reader); it then follows that for some $\beta_N \in (0, 1]$,

$$\left|\pi_N\left(\overline{\Gamma}_{s,t}\right)\right| \leq \varepsilon^{\beta_N} \omega\left(s,t\right)$$

and so the induction step is completed. Putting this last inequality and Lemma 10.45 together, we see that for $k \leq [p]$, we have from the triangle inequality

$$\left|\pi_k\left(\left(\int \varphi^2\left(x^2\right) dx^2\right)_{s,t} - \left(\int \varphi^1\left(x^1\right) dx^1\right)_{s,t}\right)\right| \leq \varepsilon^{\min(\beta_N, \gamma-1)} \omega\left(s,t\right)^{k/p}.$$

The proof is now finished. ∎

As a consequence of Theorem 10.47, we have

Corollary 10.48 *Let* $\varphi_i : \mathbb{R}^d \to \mathbb{R}^e$, $i = 1, \dots, d$ *be some* $(\gamma - 1)$*-Lipschitz maps where* $\gamma > p$. *For any fixed control* ω, $R > 0$ *and* $p \leq p'$, *the maps*

$$\text{Lip}^{\gamma-1} \times \left(\left\{\|\mathbf{x}\|_{p\text{-}\omega} \leq R\right\}, d_{p',\omega}\right) \rightarrow \left(C^{p\text{-}\omega}\left([0,T], G^{[p]}\left(\mathbb{R}^e\right)\right), d_{p'\text{-}\omega}\right)$$

$$(\varphi, \mathbf{x}) \mapsto \int \varphi\left(\mathbf{x}\right) d\mathbf{x}$$

and

$$\mathrm{Lip}^{\gamma-1} \times \left(\left\{ \|\mathbf{x}\|_{p\text{-var}} \leq R \right\}, d_{p'\text{-var}} \right) \rightarrow \left(C^{p\text{-var}} \left([0,T], G^{[p]}\left(\mathbb{R}^e\right)\right), d_{p'\text{-var}} \right)$$

$$(\varphi, \mathbf{x}) \mapsto \int \varphi(\mathbf{x}) \, d\mathbf{x}$$

and

$$\mathrm{Lip}^{\gamma-1} \times \left(\left\{ \|\mathbf{x}\|_{p\text{-var}} \leq R \right\}, d_\infty \right) \rightarrow \left(C^{p\text{-var}} \left([0,T], G^{[p]}\left(\mathbb{R}^e\right)\right), d_\infty \right)$$

$$(\varphi, \mathbf{x}) \mapsto \int \varphi(\mathbf{x}) \, d\mathbf{x}$$

are uniformly continuous.

Exercise 10.49 *Extend Theorem 10.47 to* $\varphi_i \in \mathrm{Lip}_{\mathrm{loc}}^{\gamma-1}\left(\mathbb{R}^d, \mathbb{R}^e\right)$, $i = 1, \ldots, d$.

Solution. It suffices to replace φ_i by $\tilde{\varphi}_i \in \mathrm{Lip}^{\gamma-1}\left(\mathbb{R}^d, \mathbb{R}^e\right)$ such that $\varphi \equiv \tilde{\varphi}$ on a ball with radius

$$|x|_{\infty;[0,T]} + 1. \qquad \square$$

10.7 RDEs driven along linear vector fields

In various applications, for instance when studying the "Jacobian of the flow", one encouters RDEs driven along linear vector fields. Since linear vector fields are unbounded, we can, at this stage, only assert uniqueness and local existence (cf. Theorem 10.21). The aim of the present section is then to establish global existence with some precise quantitative estimates.

Exercise 10.50 (linear vector fields) *Let us consider a collection* $V = (V_i)_{i=1,\ldots,d}$ *of linear vector fields of the form*

$$V_i(z) = A_i z + b_i$$

for $e \times e$ *matrices* A_i *and elements* b_i *of* $\mathbb{R}^e$. *Prove (by induction) that*

$$V^{N;i_0,\ldots,i_N}(z) \equiv V_{i_0} \ldots V_{i_N}(z) = A_{i_N} \ldots A_{i_0} z + A_{i_N} \ldots A_{i_1} b_{i_0}.$$

Lemma 10.51 (Lemma A_{linear}) *Assume that*
(i) $V = (V_i)_{1 \leq i \leq d}$ *with* $V_i(z) = A_i z$ *are a collection of linear vector fields, and fix* $N \in \mathbb{N}$;
(ii) s, t *are some elements of* $[0,T]$;
(iii) $x, \tilde{x}$ *are some paths in* $C^{1\text{-var}}\left([s,t], \mathbb{R}^d\right)$ *such that* $S_N(x)_{s,t} = S_N(\tilde{x})_{s,t}$;
(iv) ℓ *is a bound on* $\int_s^t |dx_u|$ *and* $\int_s^t |d\tilde{x}_u|$ *and* υ *is a bound on* $\max_i(|A_i|)$.
Then,

$$\left| \pi_{(V)}(s, y_s; x)_{s,t} - \pi_{(V)}(s, y_s; \tilde{x})_{s,t} \right| \leq C \left[\upsilon \ell \right]^{N+1} \exp\left(C\upsilon\ell\right).$$

Proof. Define for all $r \in [s,t]$, $y_{s,r} = \pi_{(V)}(s, y_s; x)_{s,r}$; we saw in Theorem 3.7 that

$$|y_{s,r}| \leq c(1 + |y_s|) \, v\ell \exp(cv\ell).$$

From Proposition 10.3, we have

$$
\left| \pi_{(V)}(s, y_s; x)_{s,t} - \mathcal{E}_{(V)}\left(y_s, S_N(x)_{s,t}\right) \right|
$$

$$
\leq \sum_{\substack{i_1, \ldots, i_N \\ \in \{1, \ldots, d\}}} \left| \int_s^t (V_{i_1} \ldots V_{i_N} I(y_r) - V_{i_1} \ldots V_{i_N} I(y_s)) \, dx_r^{i_1} \ldots dx_r^{i_N} \right|
$$

$$
\leq \left| A_{i_N} \ldots A_{i_1} \int_s^t y_{s,r} \, dx_r^{i_1} \ldots dx_r^{i_N} \right|
$$

$$
\leq c(1 + |y_s|) [v\ell]^{N+1} \exp(cv\ell).
$$

As $\pi_{(V)}(s, y_s; \tilde{x})_{s,t}$ and $\pi_{(V)}(s, y_s; x)_{s,t}$ share the same Euler approximation, the triangle inequality finishes the proof. ∎

Equipped with this result and the Lipschitzness of the flow for ordinary differential equations (Lemma 12.5), we are ready to obtain a version of Lemma 10.7 for linear vector fields.

Lemma 10.52 (Davie$_{\text{linear}}$) *Let $p \geq 1$. Assume that*
(i) $V = (V_i)_{1 \leq i \leq d}$ is a collection of linear vector fields defined by $V_i(z) = A_i z + b_i$, for some $e \times e$ matrices A_i and elements of $\mathbb{R}^e$ b_i;
(ii) x is a path in $C^{1\text{-var}}([0,T], \mathbb{R}^d)$, and $\mathbf{x} := S_{[p]}(x)$ its canonical lift to a $G^{[p]}(\mathbb{R}^d)$-valued path;
(iii) $y_0 \in \mathbb{R}^e$ is an initial condition;
(iv) v is a bound on $\max_i (|A_i| + |b_i|)$.
*Then there exists a constant C depending on p (but **not** on the 1-variation norm of x) such that for all $s < t$ in $[0,T]$,*

$$
\left| \pi_{(V)}(0, y_0; x)_{s,t} \right| \leq C(1 + |y_0|) \, v \, \|\mathbf{x}\|_{p\text{-var};[s,t]} \exp\left(Cv^p \|\mathbf{x}\|_{p\text{-var};[0,T]}^p \right).
$$

Proof. To simplify the notation we set $y = \pi_{(V)}(0, y_0; x)$ and $N := [p]$, a control function on $[0,T]$ is defined by

$$
\omega(s,t) := v^p \|\mathbf{x}\|_{p\text{-var};[s,t]}^p.
$$

For every $s < t$ in $[0,T]$ we define $x^{s,t}$ as a geodesic path associated with $\mathbf{x}_{s,t} = S_N(x)_{s,t}$, i.e.

$$
S_N\left(x^{s,t}\right)_{s,t} = S_N(x)_{s,t} \quad \text{and} \quad \int_s^t \left| dx^{s,t} \right| = \|\mathbf{x}\|_{p\text{-var};[s,t]}
$$

and we note that, as in (10.7), we have

$$\int_s^t |dx^{s,t}| \le \int_s^t |dx|.$$

(10.45)

Let us now fix $s < t < u$ in $[0, T]$ and define $x^{s,t,u} := x^{s,t} \sqcup x^{t,u}$, the concatenation of $x^{i,s,t}$ and $x^{i,t,u}$. Following the by now classical pattern of proof, first seen in (Davie's) Lemma 10.7, we set

$$\Gamma_{s,t} = \pi_{(V)} (s, y_s; x)_{s,t} - \pi_{(V)} \left(s, y_s; x^{s,t}\right)_{s,t}$$

and observe $\Gamma_{s,u} - \Gamma_{s,t} - \Gamma_{t,u} = A + B$ where

$$
\begin{aligned}
A &= \pi \left(s, y_s; x^{s,t,u}\right)_{s,u} - \pi \left(s, y_s; x^{s,u}\right)_{s,u} \\
B &= \pi \left(t, y_t; x^{t,u}\right)_{t,u} - \pi \left(t, y_t + \Gamma_{s,t}; x^{t,u}\right)_{t,u} .
\end{aligned}
$$

Lemma 10.51 (Lemma A_{linear}) was tailor-made to estimate A and gives

$$|A| \le c_1 \left(1 + |y_s|\right) \omega \left(s, u\right)^{(N+1)/p} \exp \left(c_1 \omega \left(s, u\right)^{1/p}\right).$$

On the other hand, from Theorem 3.8, we have

$$|B| \le c_2 \left|\Gamma_{s,t}\right| \omega \left(t, u\right)^{1/p} \exp \left(c_2 \omega \left(t, u\right)^{1/p}\right)$$

and so

$$\left|\Gamma_{s,u}\right| \le \left|\Gamma_{s,t}\right| e^{2c_2 \omega(s,u)^{1/p}} + \left|\Gamma_{t,u}\right| + c_1 \left(1 + |y_s|\right) \omega \left(s, u\right)^{(N+1)/p} e^{c_1 \omega(s,u)^{1/p}}.$$

(10.46)

Another application of Lemma A_{linear} combined with (10.45) shows that with $\tilde{\omega} \left(s, t\right) := |x|_{1\text{-var};[s,t]}$ we have

$$\lim_{r \to 0} \ \sup_{s,t \text{ s.t } \tilde{\omega}(s,t) \le r} \ \frac{\left|\Gamma_{s,t}\right|}{r} = 0.$$

(10.47)

Also, ODE estimates give that for all $s, t \in [0, T]$, we have

$$\left|\pi_{(V)} \left(s, y_s; x^{s,t}\right)_{s,t}\right| \le c_3 \left(1 + |y_s|\right) \omega \left(s, t\right)^{1/p} e^{c_2 \omega(s,t)^{1/p}}.$$

(10.48)

Inequalities (10.46), (10.47) and (10.48) allow us to use (the analysis) Lemma 10.63 (from the appendix to this chapter) to see that

$$|y|_{\infty;[0,T]} \le c_4 \left(1 + |y_0|\right) \exp \left(c_4 \omega \left(0, T\right)\right),$$

(10.49)

and that for all $s, t \in [0, T]$,

$$\left|\Gamma_{s,t}\right| \le c_4 \left(1 + |y_0|\right) \omega \left(s, t\right)^{\theta} \exp \left(c_4 \omega \left(0, T\right)\right).$$

The triangle inequality then provides that for all $s, t \in [0, T]$,

$$|y_{s,t}| \leq c_5 \left(1 + |y_0|\right) \omega \left(s, t\right)^{1/p} \exp \left(c_5 \omega \left(0, T\right)\right).$$

■

We now give the appropriate extension to (full) RDEs.

Theorem 10.53 *Assume that*
(i) $V = (V_i)_{1 \leq i \leq d}$ *is a collection of linear vector fields defined by* $V_i(z) = A_i z + b_i$, *for some* $e \times e$ *matrices* A_i *and elements of* $\mathbb{R}^e$ b_i;
(ii) $\mathbf{x}$ *is a weak geometric rough path in* $C^{p\text{-var}}\left([0, T], G^{[p]}\left(\mathbb{R}^d\right)\right)$;
(iii) $\mathbf{y}_0 \in G^{[p]}\left(\mathbb{R}^e\right)$ *is an initial condition;*
(iv) v *is a bound on* $\max_i \left(|A_i| + |b_i|\right)$.
Then there exists a unique full RDE solution $\pi_{(V)}\left(0, y_0; \mathbf{x}\right)$ *on* $[0, T]$. *Moreover, there exists a constant* C *depending on* p *such that for all* $s < t$ *in* $[0, T]$, *we have*

$$\left\| \pi_{(V)}\left(0, y_0; \mathbf{x}\right)_{s,t} \right\| \leq C \left(1 + |y_0|\right) v \left\| \mathbf{x} \right\|_{p\text{-var};[s,t]} \exp \left(C v^p \left\| \mathbf{x} \right\|_{p\text{-var};[0,T]}^p\right).$$

Proof. We only have to prove the above estimate for a bounded variation path, the final result following a now classical limiting procedure.

Write $y_0 = \pi_1(\mathbf{y}_0)$, and define $\tilde{y} = \frac{1}{1+|y_0|} \pi_{(V)}\left(0, y_0; x\right)$. Observe that using the linearity of the vector fields, we have

$$\begin{aligned} \tilde{y} &= \pi_{(V)}\left(0, \frac{y_0}{1+|y_0|}; x\right) \\ &= \pi_{(V)}\left(0, \tilde{y}_0; x\right). \end{aligned}$$

From (the proof of) Lemma 10.52 it follows that

$$\begin{aligned} |\tilde{y}|_{\infty;[0,T]} &\leq c_1 \left(1 + |\tilde{y}_0|\right) \exp \left(c_1 v^p \left\| \mathbf{x} \right\|_{p\text{-var};[0,T]}^p\right) \\ &\leq 2c_1 \exp \left(c_1 v^p \left\| \mathbf{x} \right\|_{p\text{-var};[0,T]}^p\right) \equiv R. \end{aligned}$$

For any vector field $\tilde{V}$ such that $V \equiv \tilde{V}$ on $\Omega = \{y \in \mathbb{R}^e : |y| < R + 1\}$ we have

$$\pi_{(V)}\left(0, \tilde{y}_0; x\right) \equiv \pi_{(\tilde{V})}\left(0, \tilde{y}_0; x\right).$$

Moreover, we can (and will) take $\tilde{V}$ such that

$$\left| \tilde{V} \right|_{\text{Lip}^\gamma} \leq c_1 |V|_{\text{Lip}^\gamma(\Omega)} \leq |V|_{\infty;\Omega} + |V'|_{\infty;\Omega} \leq v\left(R + 2\right).$$

It then suffices to use the estimate of Theorem 10.36, applied to the full RDE with vector fields $\tilde{V}$ driven by $S_{[p]}(x)$, to see that

$$\begin{aligned} \left\| S_{[p]}\left(\pi_{(V)}(0, \tilde{y}_0; x)\right)_{s,t} \right\| &\leq c_2 \left(\left| \tilde{V} \right|_{\text{Lip}^{\gamma-1}} \left\| \mathbf{x} \right\|_{p\text{-var};[s,t]} \vee \left| \tilde{V} \right|_{\text{Lip}^{\gamma-1}}^p \left\| \mathbf{x} \right\|_{p\text{-var};[s,t]}^p\right) \\ &\leq c_3 v \left\| \mathbf{x} \right\|_{p\text{-var};[s,t]} \exp \left(c_3 v^p \left\| \mathbf{x} \right\|_{p\text{-var};[0,T]}^p\right). \end{aligned}$$

This implies that

$$\left\| S_{[p]}\left(\pi_{(V)}\left(0,y_0;x\right)\right)_{s,t}\right\| = (1+|y_0|)\left\| S_{[p]}\left(\pi_{(V)}\left(0,\tilde{y}_0;x\right)\right)_{s,t}\right\|$$
$$\leq C(1+|y_0|)\,\upsilon\|\mathbf{x}\|_{p\text{-var};[s,t]}\exp\left(C\upsilon^p\,\|\mathbf{x}\|_{p\text{-var};[0,T]}^p\right).$$

∎

Remark 10.54 This estimate shows in particular that (full) solutions to linear RDEs have growth controlled by

$$\exp\left((\text{const})\times\|\mathbf{x}\|_{p\text{-var};[0,T]}^p\right)$$

which has implications on the integrability of such a solution when the driving signal $\mathbf{x}$ is random. It is therefore interesting to know that this estimate cannot be improved and the reader can find a construction of the relevant examples in [58].   □

Exercise 10.55 *Assume* $\mathbf{x}\in C^{p\text{-var}}\left([0,T],G^{[p]}\left(\mathbb{R}^d\right)\right)$ *drives linear vector fields as in Theorem 10.53 above. If* $\mathbf{x}$ *is controlled by a fixed control* ω *in the sense that*

$$\forall 0\leq s<t\leq T:\upsilon\|\mathbf{x}\|_{p\text{-var};[s,t]}\leq\omega\left(s,t\right)^{1/p},$$

the conclusion of Theorem 10.53 can be written as

$$\left\|\boldsymbol{\pi}_{(V)}\left(0,y_0;\mathbf{x}\right)_{s,t}\right\|\leq C\left(1+|y_0|\right)\omega\left(s,t\right)^{1/p}\exp\left(C\omega\left(0,T\right)\right),$$

valid for all $s<t$ *in* $[0,T]$, *where any dependence on* p,υ *has been included in the constant* C. *Show that an estimate of this exact form remains valid, if the assumption on* $\mathbf{x}$ *is relaxed to*

$$\forall 0\leq s<t\leq T:\upsilon\|\mathbf{x}\|_{p\text{-var};[s,t]}\leq\omega\left(s,t\right)^{1/p}\vee\omega\left(s,t\right).\qquad(10.50)$$

The importance of this exercise comes from the fact (cf. Theorem 10.36) that (10.50) is a typical estimate for solutions of full RDEs, i.e. when $\mathbf{x}$ *itself arises as the solution to a full RDE along* Lip^γ-*vector fields,* $\gamma>p$.

Solution. Assume (10.50) and write $\tilde{\omega}\left(s,t\right)^{1/p}=\omega\left(s,t\right)^{1/p}\vee\omega\left(s,t\right)$. For $s,t:\omega\left(s,t\right)\leq 1$, Theorem 10.53 then gives

$$\left\|\boldsymbol{\pi}_{(V)}\left(0,y_0;\mathbf{x}\right)_{s,t}\right\| \leq C\left(1+|y_s|\right)\omega\left(s,t\right)^{1/p}\exp\left(C\omega\left(s,t\right)\right)$$
$$\leq C\left(1+|y|_{\infty;[0,T]}\right)\omega\left(s,t\right)^{1/p}\exp\left(C\omega\left(0,T\right)\right)$$

and we are done if we can show that

$$|y|_{\infty;[0,T]}\leq c\left(1+|y_0|\right)\exp\left(c\omega\left(0,T\right)\right).$$

From equation (10.49) (now applied with $\tilde{\omega}$!) this estimate follows from (the analysis) Lemma 10.63 but since $\tilde{\omega}\left(0,T\right) = \omega\left(0,T\right)^p$ for large $\omega\left(0,T\right)$, this is not good enough. However, from Remark 10.64 we can do a little better and get

$$|y|_{\infty;[0,T]} \leq c\left(|y_0| + \varepsilon\right)\exp\left(c\sup_{\substack{D=(t_i)\subset[0,T] \text{ such that} \\ \omega(t_i,t_{i+1})\leq 1 \text{ for all } i}} \sum_i \tilde{\omega}\left(t_i,t_{i+1}\right)\right).$$

But since $\tilde{\omega} \equiv \omega$, when $\omega \leq 1$ we can replace $\tilde{\omega}$ by ω, and by superadditivity,

$$|y|_{\infty;[0,T]} \leq c\left(|y_0| + \varepsilon\right)\exp\left(c\omega\left(0,T\right)\right)$$

as required. $\qquad\qquad\qquad\qquad\qquad\qquad\qquad\qquad\qquad\qquad\qquad\qquad\square$

Exercise 10.56 (non-explosion) *Consider* $V = \left(V_i\right)_{1\leq i\leq d}$, *a collection of locally* $\text{Lip}^{\gamma-1}$*-vector fields on* $\mathbb{R}^e$ *for* $\gamma \in (p,[p]+1)$, *such that*
(i) V_i *are Lipschitz continuous;*
(ii) the vector fields $V^{[p]} = \left(V_{i_1}\ldots V_{i_{[p]}}\right)_{i_1,\ldots,i_{[p]}\in\{1,\ldots,d\}}$ *are* $(\gamma - [p])$*-Hölder continuous.*
Show that if **x** *is a geometric p-rough path, and if* y_0, *then* $\pi_{(V)}\left(0,y_0;\mathbf{x}\right)$ *does not explode. Provide a quantitative bound.*

Solution. The argument is the same as for linear RDEs. We only need to extend Lemma A_{linear}, i.e. we need to prove the following: if
(a) s,t are some elements of $[0,T]$;
(b) $x,\tilde{x}$ are some paths in $C^{1\text{-var}}\left([s,t],\mathbb{R}^d\right)$ such that $S_N\left(x\right)_{s,t} = S_N\left(\tilde{x}\right)_{s,t}$;
(c) ℓ is a bound on $\int_s^t |dx_u|$ and $\int_s^t |d\tilde{x}_u|$ and υ is a bound on $\left|V^{[p]}\right|_{(\gamma-[p])\text{-Höl}}^{1/[p]}$ and $\sup_{y,z}|V(y) - V(z)|/|y-z|$.
Then,

$$\left|\pi_{(V)}\left(s,y_s;x\right)_{s,t} - \pi_{(V)}\left(s,y_s;\tilde{x}\right)_{s,t}\right| \leq C\left[\upsilon\ell\right]^\gamma \exp\left(C\upsilon\ell\right).$$

To do so, define for all $r \in [s,t]$, $y_{s,r} = \pi_{(V)}\left(s,y_s;x\right)_{s,r}$; as the vector fields V_i are Lipschitz continuous, Theorem 3.7 gives

$$|y_{s,r}| \leq c_1\left(1 + |y_s|\right)\upsilon\ell\exp\left(c\upsilon\ell\right).$$

From Proposition 10.3,

$$\left| \pi_{(V)}(s, y_s; x)_{s,t} - \mathcal{E}_{(V)}\left(y_s, S_N(x)_{s,t}\right) \right|$$

$$\leq \sum_{\substack{i_1,\dots,i_N \\ \in\{1,\dots,d\}}} \left| \int_s^t \left(V_{i_1}\dots V_{i_{[p]}} I\left(y_r\right) - V_{i_1}\dots V_{i_{[p]}} I\left(y_s\right)\right) dx_r^{i_1}\dots dx_r^{i_{[p]}} \right|$$

$$\leq \sum_{\substack{i_1,\dots,i_N \\ \in\{1,\dots,d\}}} v^{[p]} \int_s^t |y_{s,r}|^{\gamma-[p]} \left|dx_r^{i_1}\right|\dots\left|dx_r^{i_N}\right|$$

$$\leq c_2\left(1+|y_s|\right)^{\gamma-[p]}\left[v\ell\right]^\gamma \exp\left(cv\ell\right)$$

$$\leq c_2\left(1+|y_s|\right)\left[v\ell\right]^\gamma \exp\left(cv\ell\right).$$

As $\pi_{(V)}(s, y_s; \tilde{x})_{s,t}$ and $\pi_{(V)}(s, y_s; x)_{s,t}$ share the same Euler approximation, the triangle inequality finishes the proof.

Equipped with this result, we then prove the exercise by going through the proof of Theorem 10.53. □

10.8   Appendix: p-variation estimates via approximations

Our discussion of the Young–Lóeve inequality was based on some elementary analysis considerations; Lemmas 6.1 and 6.2. We now give the appropriate extensions, still elementary, upon which we base our discussion of rough differential equations.

Lemma 10.57 *Let* $\theta > 1, K, \xi \geq 0, \alpha > 0$. *Assume* $\varrho : [0, R] \to \mathbb{R}^+$ *satisfies*
(i)

$$\lim_{r\to 0} \frac{\varrho(r)}{r} = 0;$$

(ii) for all $r \in [0, R]$,

$$\varrho(r) \leq \left\{2\varrho\left(\frac{r}{2}\right) + \xi r^\theta\right\} \exp\left(Kr^\alpha\right).$$

Then, for all $r \in [0, R]$,

$$\varrho(r) \leq \frac{\xi r^\theta}{1 - 2^{1-\theta}} \exp\left(\frac{2K}{1 - 2^{-\alpha}} r^\alpha\right).$$

Proof. Note that it is enough to prove the final estimate for $r = R$; indeed, given any other $r \in [0, R]$, it suffices to replace the interval $[0, R]$ by $[0, r]$. Assumption (ii) implies that for all $r \in [0, R]$,

$$\varrho(r) \le 2\varrho\left(\frac{r}{2}\right)\exp\left(Kr^\alpha\right) + \hat{\xi}r^\theta$$

with $\hat{\xi} = \xi\exp\left(KR^\alpha\right)$. By induction, we obtain that

$$\varrho(r) \le 2^n\varrho\left(\frac{r}{2^n}\right)\exp\left(Kr^\alpha\sum_{k=0}^{n-1}2^{-k\alpha}\right) + \hat{\xi}r^\theta\sum_{k=0}^{n-1}2^{k(1-\theta)}\exp\left(Kr^\alpha\sum_{j=0}^{k-1}2^{-j\alpha}\right).$$

We bound $\exp\left(Kr^\alpha\sum_{k=0}^{n-1}2^{-k\alpha}\right) \le \exp\left(Kr^\alpha/\left(1 - 2^{-\alpha}\right)\right)$ and so obtain

$$\varrho(r)\exp\left(\frac{-Kr^\alpha}{1 - 2^{-\alpha}}\right) \le 2^n\varrho\left(\frac{r}{2^n}\right) + \hat{\xi}r^\theta\sum_{k=0}^{n-1}2^{k(1-\theta)}.$$

By assumption (i), sending n to ∞ yields

$$\varrho(r) \le \frac{\hat{\xi}r^\theta}{1 - 2^{1-\theta}}\exp\left(\frac{Kr^\alpha}{1 - 2^{-\alpha}}\right).$$

As $\hat{\xi} = \xi\exp\left(KR^\alpha\right) \le \xi\exp\left(KR^\alpha/\left(1 - 2^{-\alpha}\right)\right)$, we then obtain

$$\varrho(r) \le \frac{\xi R^\theta}{1 - 2^{1-\theta}}\exp\left(\frac{2KR^\alpha}{1 - 2^{-\alpha}}\right).$$

∎

As a variation on the theme, let us give

Lemma 10.58 *Let $\theta > 1, K, \xi \ge 0, \alpha > 0$ and $\beta \in [0, 1)$. Assume ϱ :*
$[0, R] \to \mathbb{R}^+$ satisfies
(i)

$$\lim_{r\to 0}\frac{\varrho(r)}{r} = 0;$$

(ii) for all $r \in [0, R]$,

$$\varrho(r) \le \left\{2\varrho\left(\frac{r}{2}\right) + \xi r^\theta \wedge \varepsilon r^\beta\right\}\exp\left(Kr^\alpha\right).$$

Then, for all $r \in [0, R]$, for some constant C depending on θ and β, we have

$$\varrho(r) \le C r \varepsilon^{\frac{\theta-1}{\theta-\beta}}\xi^{\frac{1-\beta}{\theta-\beta}}\exp\left(\frac{2K}{1 - 2^{-\alpha}}r^\alpha\right).$$

Proof. Just as in the previous proof, we only prove the estimate for $r = R$. Defining $\hat{\xi} = \xi \exp(KR^\alpha)$ and $\hat{\varepsilon} = \varepsilon \exp(KR^\alpha)$, we have for all $r \in [0, R]$,

$$\varrho(r) \le 2\varrho\left(\frac{r}{2}\right) \exp(Kr^\alpha) + \hat{\xi}r^\theta \wedge \hat{\varepsilon}r^\beta.$$

By induction, we obtain that

$$\varrho(r) \le 2^n \varrho\left(\frac{r}{2^n}\right) \exp\left(Kr^\alpha \sum_{k=0}^{n-1} 2^{-k\alpha}\right)$$
$$+ \sum_{k=0}^{n-1} \left(\hat{\xi}r^\theta 2^{k(1-\theta)} \wedge \hat{\varepsilon}r^\beta 2^{k(1-\beta)}\right) \exp\left(Kr^\alpha \sum_{j=0}^{k-1} 2^{-j\alpha}\right).$$

We bound $\exp\left(Kr^\alpha \sum_{k=0}^{n-1} 2^{-k\alpha}\right) \le \exp(Kr^\alpha/(1 - 2^{-\alpha}))$ and then let n tend to ∞ to obtain

$$\varrho(r) \exp\left(\frac{-Kr^\alpha}{1 - 2^{-\alpha}}\right) \le \sum_{k=0}^{\infty} \left(\hat{\xi}r^\theta 2^{k(1-\theta)} \wedge \hat{\varepsilon}r^\beta 2^{k(1-\beta)}\right)$$
$$\le \hat{\varepsilon}r^\beta \sum_{0 \le k \le \frac{1}{\theta-\beta}\ln_2\left(\frac{\hat{\xi}}{\hat{\varepsilon}}\right) + \ln_2 r} 2^{k(1-\beta)} + \hat{\xi}r^\theta \sum_{k > \frac{1}{\theta-\beta}\ln_2\left(\frac{\hat{\xi}}{\hat{\varepsilon}}\right) + \ln_2 r} 2^{k(1-\theta)}$$
$$\le c_1 \left(\hat{\varepsilon}r^\beta 2^{\left(\frac{1}{\theta-\beta}\ln_2\left(\frac{\hat{\xi}}{\hat{\varepsilon}}\right) + \ln_2 r\right)(1-\beta)} + \hat{\xi}r^\theta 2^{\left(\frac{1}{\theta-\beta}\ln_2\left(\frac{\hat{\xi}}{\hat{\varepsilon}}\right) + \ln_2 r\right)(1-\theta)}\right)$$
$$\le c_2 r \hat{\varepsilon}^{\frac{\theta-1}{\theta-\beta}} \hat{\xi}^{\frac{1-\beta}{\theta-\beta}},$$

where c_1, c_2 are constants which depend on θ and β. This estimate finishes the proof. ∎

An important consequence of Lemma 10.57 is the following estimate. Typically (e.g. in the proof of Davie's estimate, Lemma 10.7), Γ is the difference between a path y (for which we are trying to bound its p-variation) and a "local" approximation of y which is easier to control.

Lemma 10.59 *Let $\xi > 0$, $\theta > 1, K \ge 0$, $\alpha > 0$ and*

$$\Gamma : \{0 \le s < t \le T\} \equiv \Delta_T \to \mathbb{R}^e$$

be such that:
(i) for some control $\hat{\omega}$,

$$\lim_{r \to 0} \sup_{(s,t) \in \Delta_T : \hat{\omega}(s,t) \le r} \frac{|\Gamma_{s,t}|}{r} = 0; \tag{10.51}$$

(ii) for some control ω we have that, for all $s < t < u$ in $[0, T]$,

$$|\Gamma_{s,u}| \leq \left\{|\Gamma_{s,t}| + |\Gamma_{t,u}| + \xi \omega (s, u)^{\theta}\right\} \exp\left(K\omega\left(s, u\right)^{\alpha}\right). \qquad (10.52)$$

Then, for all $s < t$ in $[0, T]$,

$$|\Gamma_{s,t}| \leq \frac{\xi \omega\left(s, t\right)^{\theta}}{1 - 2^{1-\theta}} \exp\left(\frac{2K}{1 - 2^{-\alpha}} \omega\left(s, t\right)^{\alpha}\right).$$

Remark 10.60 It is important to notice that the control $\hat{\omega}$ is not used in the final estimate.   □

Proof. We assume that $\hat{\omega} \leq \frac{1}{\varepsilon}\omega$ for some $\varepsilon > 0$; otherwise we can replace ω by $\omega + \varepsilon\hat{\omega}$ and let ε tend to 0 at the end. Define for all $r \in [0, \omega\left(0, T\right)]$,

$$\varrho\left(r\right) = \sup_{(s,t)\in\Delta_T : \omega(s,t)\leq r} |\Gamma_{s,t}|.$$

Consider any fixed pair (s, u) with $0 \leq s < u \leq T$ such that $\omega\left(s, u\right) \leq r$. From basic properties of control functions we can then pick t such that $\omega\left(s, t\right)$ and $\omega\left(t, u\right)$ is bounded above by $\omega\left(s, u\right)/2$. It follows that

$$|\Gamma_{s,t}| \leq \varrho\left(r/2\right), \quad |\Gamma_{s,u}| \leq \varrho\left(r/2\right),$$

and by assumption (ii),

$$|\Gamma_{s,u}| \leq \left\{2\varrho\left(\frac{r}{2}\right) + \xi r^{\theta}\right\} \exp\left(Kr^{\alpha}\right).$$

Taking the supremum over all $s < u$ in $[0, T]$ for which $\omega\left(s, u\right) \leq r$ yields that for all $r \in [0, \omega\left(0, T\right)]$,

$$\varrho\left(r\right) \leq \left\{2\varrho\left(\frac{r}{2}\right) + \xi r^{\theta}\right\} \exp\left(Kr^{\alpha}\right).$$

Assumption (i) implies that $\lim_{r \to 0} \varrho\left(r\right)/r = 0$ and by (the previous) Lemma 10.57, we see that for all $r \in [0, \omega\left(0, T\right)]$,

$$\varrho\left(r\right) \leq \frac{\xi r^{\theta}}{1 - 2^{1-\theta}} \exp\left(\frac{2K}{1 - 2^{-\alpha}} r^{\alpha}\right).$$

Obviously, $|\Gamma_{s,t}| \leq \varrho\left(r\right)$ for $r = \omega\left(s, t\right)$ and so the proof is finished.   ∎

The same argument, but using Lemma 10.58 instead of Lemma 10.57, leads to the following estimate which we use in the proof of Theorem 10.47 where we establish continuity of rough integration.

Lemma 10.61 *Let $\theta > 1, K, \xi \geq 0, \alpha > 0, \beta \in [0, 1)$ and*

$$\Gamma : \{0 \leq s < t \leq T\} \equiv \Delta_T \to \mathbb{R}^e$$

be such that:

(i) for some control $\hat{\omega}$,

$$\lim_{r \to 0} \sup_{(s,t) \in \Delta_T \, : \, \hat{\omega}(s,t) \leq r} \frac{|\Gamma_{s,t}|}{r} = 0;$$

(ii) for some control ω we have that, for all $s < t < u$ in $[0,T]$,

$$|\Gamma_{s,u}| \leq \left\{ |\Gamma_{s,t}| + |\Gamma_{t,u}| + \xi \omega (s,u)^{\theta} \wedge \varepsilon \omega (s,u)^{\beta} \right\} \exp \left(K \omega (s,u)^{\alpha} \right).$$

Then, for all $s < t$ in $[0,T]$, for some constant C depending on θ and β, we have

$$|\Gamma_{s,t}| \leq C \omega (s,t)^{\frac{\theta-1}{\theta-\beta}} \xi^{\frac{1-\beta}{\theta-\beta}} \exp \left(\frac{2K}{1-2^{-\alpha}} \omega (s,t)^{\alpha} \right).$$

Remark 10.62 It is worth noting that (ii) is equivalent to saying that, for all $\eta \in [0,1]$,

$$|\Gamma_{s,u}| \leq \left\{ |\Gamma_{s,t}| + |\Gamma_{t,u}| + \xi^{1-\eta} \varepsilon^{\eta} \omega (s,u)^{\theta(1-\eta)+\beta\eta} \right\} \exp \left(K \omega (s,u)^{\alpha} \right),$$

using $|(*)| \leq a \wedge b \Leftrightarrow |(*)| \leq a^{\eta} b^{1-\eta} \; \forall \eta \in [0,1]$, and thus renders Lemma 10.59 applicable. In fact, for any $0 \leq \eta < (\theta - 1) / (\theta - \beta)$ we have $\tilde{\theta} = \beta \eta + \theta (1 - \eta) > 1$; setting also $\tilde{\xi} = \xi^{1-\eta} \varepsilon^{\eta}$, we can apply Lemma 10.59 to get

$$|\Gamma_{s,t}| \leq \frac{1}{1 - 2^{1-\tilde{\theta}}} \omega (s,t)^{\tilde{\theta}} \varepsilon^{\eta} \xi^{1-\eta} \exp \left(\frac{2K}{1-2^{-\alpha}} \omega (s,t)^{\alpha} \right).$$

Although this would be sufficient for our application (namely, the proof of Theorem 10.47) we see that our direct analysis showed that we can take $\eta = (\theta - 1) / (\theta - \beta)$ in the above estimate.   □

As mentioned right above Lemma 10.59, such estimates are typically used when Γ is the difference between a path y and a "local" approximation of y which is easier to control. Sometimes (e.g. in the proof of uniqueness/continuity result for RDEs, Theorem 10.26) the path y itself comes into play; in the sense that (10.52) above has to be replaced by (10.53) below. With some extra information, such as (10.54) below, a similar analysis is possible.

Lemma 10.63 *Let $K \geq 0, \varepsilon > 0, \theta > 1, 1/p > 0$ and*

$$\begin{aligned} \Gamma \quad &: \quad \{0 \leq s < t \leq T\} \equiv \Delta_T \to \mathbb{R}^e, \\ y \quad &: \quad [0,T] \to \mathbb{R}^e \end{aligned}$$

be such that:
(i) for some control $\hat{\omega}$,

$$\lim_{r \to 0} \sup_{(s,t) \in \Delta_T \,:\, \hat{\omega}(s,t) \leq r} \frac{|\Gamma_{s,t}|}{r} = 0;$$

(ii) for some control ω we have that, for all $s < t < u$ in $[0,T]$,

$$|\Gamma_{s,u}| \leq \left\{ |\Gamma_{s,t}| + |\Gamma_{t,u}| + K \left(\varepsilon + \sup_{0 \leq r \leq u} |y_r| \right) \omega(s,u)^\theta \right\} \exp\left(K\omega(s,u)^{1/p} \right);$$

$$(10.53)$$

(iii) for all $s < t$ in $[0,T]$,

$$|y_{s,t} - \Gamma_{s,t}| \leq K \left(\varepsilon + \sup_{r \leq t} |y_r| \right) \omega(s,t)^{\frac{1}{p}} \exp\left(K\omega(s,t)^{1/p} \right). \qquad (10.54)$$

Then we have, for some constant C depending only on K, p and θ,

$$|y|_{\infty;[0,T]} \leq C \exp\left(C\omega(0,T) \right) \left(|y_0| + \varepsilon \right),$$

and for all $s < t$ in $[0,T]$,

$$|\Gamma_{s,t}| \leq C \left(|y_0| + \varepsilon \right) \omega(s,t)^\theta \exp\left(C\omega(0,T) \right).$$

Proof. Fix $v < v'$ in $[0,T]$ and $s < t < u \in [0,v']$. By assumption (ii) we have,

$$|\Gamma_{s,u}| \leq \left\{ |\Gamma_{s,t}| + |\Gamma_{t,u}| + K \left(\varepsilon + |y|_{\infty;[0,v']} \right) \omega(s,u)^\theta \right\} \exp\left(K\omega(s,u)^{1/p} \right).$$

We may thus apply Lemma 10.59 (on the interval $[0,v']$ rather than $[0,T]$ and with parameters $\xi = K \left(\varepsilon + |y|_{\infty;[0,v']} \right)$ and $\alpha = 1/p$). It follows that for all $s < t$ in $[0,v']$,

$$|\Gamma_{s,t}| \leq c_1 \left(\varepsilon + |y|_{\infty;[0,v']} \right) \omega(s,t)^\theta \exp\left(c_1 \omega(s,t)^{1/p} \right)$$

and together with assumption (iii) we see that

$$\sup_{s,t \in [v,v']} |y_{s,t}| \leq c_2 \left(\varepsilon + |y|_{\infty;[0,v']} \right) \omega(v,v')^{1/p} \exp\left(c_2 \omega(v,u)^{1/p} \right).$$

This in turn implies

$$|y|_{\infty;[0,v']} \leq |y|_{\infty;[0,v]} + \sup_{s,t \in [v,v']} |y_{s,t}|$$

$$\leq |y|_{\infty;[0,v]} + \left(\varepsilon + |y|_{\infty;[0,v']} \right) c_2 \omega(s,t)^{1/p} \exp\left(c_2 \omega(v,v')^{1/p} \right).$$

We now pick $v_0 = 0$ and set for $i \in \{0, 1, 2, \dots\}$,

$$
\begin{aligned}
v_{i+1} &= \sup_{r > v_i} \left\{ c_2 \omega \left(v_i, r\right)^{1/p} \exp\left(c_2 \omega\left(v_i, r\right)^{1/p}\right) \leq \frac{1}{2} \right\} \wedge T \\
&= \sup_{r > v_i} \left\{ \omega\left(v_i, r\right) \leq \frac{1}{c_3} \right\} \wedge T,
\end{aligned}
$$

where c_3 was determined from $c_2 \left(1/c_3\right)^{1/p} \exp\left(c_2 \left(1/c_3\right)^{1/p}\right) = 1/2$. It follows that

$$
|y|_{\infty;[0,v_{i+1}]} \leq |y|_{\infty;[0,v_i]} + \frac{1}{2}\left(\varepsilon + |y|_{\infty;[0,v_{i+1}]}\right),
$$

which implies $|y|_{\infty;[0,v_{i+1}]} \leq 2\,|y|_{\infty;[0,v_i]} + \varepsilon$ and then, by induction,

$$
|y|_{\infty;[0,v_i]} \leq 2^i \left(|y_0| + \varepsilon\right).
$$

We claim that $v_N = T$ where $N = [c_3 \omega\left(0, T\right)] + 1$, the first integer strictly greater than $c_3 \omega\left(0, T\right)$. Indeed, $v_N < T$ would imply $\omega\left(v_i, v_{i+1}\right) = 1/c_3$ for all $i < N$, and hence lead to the contradiction

$$
c_3 \omega\left(0, T\right) \geq c_3 \sum_{i=0}^{N-1} \omega\left(v_i, v_{i+1}\right) = N.
$$

We are now able to say that

$$
\begin{aligned}
|y|_{\infty;[0,T]} &\leq 2^{c_3 \omega(0,T)+1} \left(|y_0| + \varepsilon\right) \\
&\leq c_4 \exp\left(c_4 \omega\left(0, T\right)\right)\left(|y_0| + \varepsilon\right).
\end{aligned}
$$

Coming back to inequality (10.53), we obtain that for all $s < t$ in $[0, T]$,

$$
\left|\Gamma_{s,u}\right| \leq \left\{ \left(\left|\Gamma_{s,t}\right| + \left|\Gamma_{t,u}\right|\right) + \exp\left(c_5 \omega\left(0, T\right)\right)\left(|y_0| + \varepsilon\right) \omega\left(s, u\right)^\theta \right\} \\
\exp\left(K \omega\left(s, u\right)^{1/p}\right).
$$

We may thus apply Lemma 10.59 (with parameters $\xi = e^{c_5 \omega(0,T)}\left(|y_0| + \varepsilon\right)$ and $\alpha = 1/p$) once again to obtain that, for all $s < t$ in $[0, T]$,

$$
\left|\Gamma_{s,t}\right| \leq c_6 \left(|y_0| + \varepsilon\right) \omega\left(s, t\right)^\theta \exp\left(c_6 \omega\left(0, T\right)\right).
$$

The proof is now finished. ∎

Remark 10.64 The conclusion of the above lemma can be slightly sharpened to[15]

$$
\begin{aligned}
|y|_{\infty;[0,T]} &\leq C\left(|y_0| + \varepsilon\right) \exp\left(C \sup_{\substack{D=(t_i) \subset [0,T] \text{ such that} \\ \omega(t_i, t_{i+1}) \leq 1 \text{ for all } i}} \sum_i \omega\left(t_i, t_{i+1}\right) \right) \\
&\leq C\left(|y_0| + \varepsilon\right) \exp\left(C \omega\left(0, T\right)\right) \dots \text{by super-additivity of controls.}
\end{aligned}
$$

---

[15] The interest in the sharpening is explained in Exercise 10.55.

To see this, the above arguments remain unchanged until the definition

$$v_{i+1} = \sup_{r > v_i} \left\{ \omega\left(v_i, r\right) \leq \frac{1}{c_3} \right\} \wedge T$$

with c_3 determined from $c_2 \left(1/c_3\right)^{1/p} \exp\left(c_2 \left(1/c_3\right)^{1/p}\right) = 1/2$. Clearly then, by making the preceding constant c_2 bigger if necessary, we may assume that $1/c_3 \leq 1$. As in the proof of Lemma 10.63 we have $|y|_{\infty;[0,v_i]} \leq 2^i \left(|y_0| + \varepsilon\right)$. For any integer N chosen such that $v_N = T$ one then has the conclusion

$$|y|_{\infty;[0,T]} \leq 2^N \left(|y_0| + \varepsilon\right).$$

We claim that

$$N = c_3 \left(\sup_{\substack{D=(t_i) \subset [0,T] \text{ such that} \\ \omega(t_i, t_{i+1}) \leq 1 \text{ for all } i}} \sum_i \omega\left(t_i, t_{i+1}\right) \right) + 1$$

is a valid choice. Assume $v_N < T$. Then $\omega\left(v_i, v_{i+1}\right) = 1/c_3 \leq 1$ for all $i < N$ and so

$$N = c_3 \sum_{i=0}^{N-1} \omega\left(v_i, v_{i+1}\right) \leq c_3 \sup_{\substack{D=(t_i) \subset [0,T] \text{ such that} \\ \omega(t_i, t_{i+1}) \leq 1 \text{ for all } i}} \sum_i \omega\left(t_i, t_{i+1}\right) = N - 1$$

which is a contradiction. $\qquad\qquad\qquad\qquad\qquad\qquad\qquad\qquad\qquad\qquad$ □

In the remainder of this appendix we make the appropriate extensions which are used in Section 10.5 to establish the uniqueness of an RDE solution under minimal regularity assumptions.

Condition 10.65 ($\Psi\left(C, \beta\right)$) We say that an increasing function $\delta : \mathbb{R}^+ \to \mathbb{R}^+$ belongs to the class $\Psi\left(C, \beta\right)$ if
(i) for all $\sigma > 0$, we have $\sum_{n \geq 0} r_n\left(\sigma\right) = \infty$, where

$$r_n\left(\sigma\right) = \inf\left\{ r > 0, \int_1^{2^n} \delta\left(\frac{r}{x}\right) dx \geq \sigma \right\};$$

(ii) there exists $C > 0$, $\beta \in (0, 1)$ such that $\delta\left(x\right) \leq Cx^\beta$ for all $x \in [0, 1]$. □

Exercise 10.66 *Prove that $\delta \in \Psi\left(C, \beta\right)$ for*
(i) $\delta\left(x\right) = x^\theta$ for $\theta > 1$;
(ii) $\delta\left(x\right) = x$;
(iii) $\delta\left(x\right) = x \ln^* \ln^* \frac{1}{x}$.

Exercise 10.67 *Prove that $\delta \notin \Psi\left(C, \beta\right)$ for*
(i) $\delta\left(x\right) = x \ln^* \frac{1}{x}$;
(ii) $\delta\left(x\right) = x^\theta$ for $\theta < 1$.

Lemma 10.68 *Let $K > 0$, $\alpha > 0$. Assume $\varrho : [0, R] \to \mathbb{R}^+$ satisfies for all $r \in [0, R]$,*

$$\varrho(r) \leq 2\varrho\left(\frac{r}{2}\right) \exp\left(K\delta(r)^\alpha\right) + \delta(r),$$

where $\delta \in \Psi(C, \beta)$ for some constants $C, \beta > 0$, and $\alpha > 0$. Then, for all $n \geq 0$, for some constant C_1 depending on K, α, β and C,

$$\varrho(r) \leq \left[2^n \varrho\left(\frac{r}{2^n}\right) + \int_1^{2^n} \delta\left(\frac{r}{x}\right) dx\right] \exp\left(C_1 r^{\alpha\beta}\right).$$

Proof. By induction, we obtain that

$$\varrho(r) \leq 2^n \varrho\left(\frac{r}{2^n}\right) \exp\left(K \sum_{k=0}^{n-1} \delta\left(\frac{r}{2^k}\right)^\alpha\right) + \sum_{k=0}^{n-1} 2^k \exp\left(K \sum_{j=0}^{k-1} \delta\left(\frac{r}{2^j}\right)^\alpha\right) \delta\left(\frac{r}{2^k}\right).$$

We then bound $\exp\left(K \sum_{k=0}^{n-1} \delta\left(\frac{r}{2^k}\right)^\alpha\right)$ by $\exp\left(K C^\alpha r^{\alpha\beta} \sum_{k=0}^{\infty} 2^{-k\alpha\beta}\right) = \exp\left(c r^{\alpha\beta}\right)$ to obtain

$$\varrho(r) \exp\left(-c r^{\alpha\beta}\right) \leq 2^n \varrho\left(\frac{r}{2^n}\right) + \sum_{k=0}^{n-1} 2^k \delta\left(\frac{r}{2^k}\right).$$

By assumption, $x \to \delta\left(\frac{r}{x}\right)$ is a non-increasing function, hence for all k, we have

$$2^k \delta\left(\frac{r}{2^k}\right) \leq \int_{2^k}^{2^{k+1}} \delta\left(\frac{r}{x}\right) dx.$$

Summing up over k, we obtain that for all $n \geq 0$, we have

$$\varrho(r) \exp\left(-c r^{\alpha\beta}\right) \leq 2^n \varrho\left(\frac{r}{2^n}\right) + \int_1^{2^n} \delta\left(\frac{r}{x}\right) dx.$$

∎

Lemma 10.69 *Let ω be a control, $C > 0$, $\theta > 1$ and $\Gamma : \{0 \leq s < t \leq T\} \to \mathbb{R}^e$ a continuous map such that*
(i) for all $s < t$ in $[0, T]$,

$$|\Gamma_{s,t}| \leq C\delta\left(\omega(s,t)\right)^\theta; \tag{10.55}$$

(ii) for all $s < t < u$ in $[0, T]$, and some $\delta \in \Psi(C, \beta)$ with constants $C, \beta > 0$.

$$|\Gamma_{s,u}| \leq (|\Gamma_{s,t}| + |\Gamma_{t,u}|)\exp\left(C\omega(s,u)^{1/p}\right) + \delta\left(\omega(s,u)\right).$$

Then, for all $s < t$ in $[0,T]$, and some constant C_1 depending on C, β and $\omega(0,T)$, we have

$$|\Gamma_{s,t}| \le C_1 \inf_{n \ge 0} \left[2^n \delta \left(\frac{\omega(s,t)}{2^n} \right)^\theta + \int_1^{2^n} \delta \left(\frac{\omega(s,t)}{x} \right) dx \right].$$

Proof. Define

$$\varrho(r) = \sup_{s,t \text{ such that } \omega(s,t) \le r} |\Gamma_{s,t}|,$$

and observe that, as in the proof of Lemma 10.59, we have

$$\varrho(r) \le 2\varrho\left(\frac{r}{2}\right) \exp\left(C\delta(r)^{1/p} \right) + \delta(r).$$

We conclude using Lemma 10.68 and (10.55).   ∎

Proposition 10.70 *Let ω be a control, $C > 0, \mu > 0, \theta > 1, \alpha \in (0,1), \beta \in \left(\frac{1}{\theta}, 1\right)$ and assume that $\delta \in \Psi(C, \beta)$. Let $y : [0,T] \to \mathbb{R}^d$ be a continuous path, and $\Gamma : \{0 \le s < t \le T\} \to \mathbb{R}^e$ a continuous map such that*
(i) for some $\theta > 1$, for all $s < t$ in $[0,T]$,

$$|\Gamma_{s,t}| \le C\delta(\omega(s,t))^\theta ; \tag{10.56}$$

(ii) for all $s < t < u$ in $[0,T]$,

$$|\Gamma_{s,u}| \le (|\Gamma_{s,t}| + |\Gamma_{t,u}|) \exp\left(C\delta(\omega(s,u))^{1/p} \right) + C\left(\mu + \sup_{r \le u} |y_r| \right) \delta(\omega(s,u)) ; \tag{10.57}$$

(iii) for all $s < t$ in $[0,T]$,

$$|y_{s,t} - \Gamma_{s,t}| \le C\left(\mu + \sup_{r \le t} |y_r| \right) \delta(\omega(s,t))^{\alpha/p} .$$

Then, for all $\varepsilon > 0$, there exists $\delta = \delta(\omega(0,T), C, L, K, \theta, p) > 0$ such that $|y_0| + \mu \le \delta$ implies

$$|y|_{\infty;[0,T]} < \varepsilon \text{ and } |y|_{p\text{-}\omega;[0,T]} < \varepsilon.$$

Proof. At the price of replacing p by p/α, we can and will assume $\alpha = 1$. We allow the constant in this proof to depend on $\omega(0,T)$, C, L, K, θ and p. Define $\tau_\gamma = \inf\left\{ t > 0, |y|_{\infty;[0,t]} > \gamma \right\}$. Using Lemma 10.69, we have for all $u, v < \tau_\gamma$ and all $n \ge 0$,

$$\begin{aligned} |\Gamma_{u,v}| &\le c_1 \left[2^n \delta\left(\frac{\omega(u,v)}{2^n} \right)^\theta + (\mu + \gamma) \int_1^{2^n} \delta\left(\frac{\omega(u,v)}{x} \right) dx \right] \\ &\le c_1 \left[2^{n(1-\theta\beta)} \omega(u,v)^{\theta\beta} + (\mu + \gamma) \int_1^{2^n} \delta\left(\frac{\omega(u,v)}{x} \right) dx \right]. \end{aligned}$$

From the triangle inequality, we therefore obtain that for all $u, v < \tau_\alpha$ and all $n \geq 0$,

$$
|y_{u,v}| \leq c_2 (\mu + \gamma) \left(\delta \left(\omega \left(u, v \right) \right)^{1/p} + \int_1^{2^n} \delta \left(\frac{\omega \left(u, v \right)}{x} \right) dx \right)
$$
$$
+ c_2 \omega \left(u, v \right)^{\theta \beta} 2^{n(1-\theta \beta)}.
$$

As $|y|_{\infty;[0,t]} - |y|_{\infty;[0,s]} \leq \sup_{u,v \in [s,t]} |y_{u,v}|$, we have that $|y|_{\infty;[0,t]} - |y|_{\infty;[0,s]}$ is less than or equal to

$$
c_2 \left(\mu + |y|_{\infty;[0,t]} \right) \left(\delta \left(\omega \left(s, t \right) \right)^{1/p} + \int_1^{2^n} \delta \left(\frac{\omega \left(s, t \right)}{x} \right) dx \right) + c_2 \omega \left(s, t \right)^{\theta \beta} 2^{n(1-\theta \beta)}.
$$

$$
(10.58)
$$

Fix $b \in \left(1, 2^{\theta \beta - 1} \right)$ and define

$$
r_k := \inf \left\{ r > 0, c_2 \left(\delta \left(r_k \right)^{1/p} + \int_1^{2^n} \delta \left(\frac{r_k}{x} \right) dx \right) = 1 - \frac{1}{b} \right\} \wedge \omega \left(0, T \right).
$$

By assumption on δ, $\sum_{k=n_0}^{\infty} r_k = +\infty$. In particular, for a fixed n_0 that will be chosen later, we can define n_1 to be the first integer such that

$$
\sum_{k=n_0}^{n_0 + n_1} r_k \geq \omega \left(0, T \right).
$$

We then define the times $(t_i)_{i=0,\ldots,n_1}$ by $t_0 = 0$ and

$$
t_{i+1} = \inf \left\{ t > t_i, \omega \left(t_i, t_{i+1} \right) = r_{n_0 + n_1 - i} \right\} \wedge T.
$$

Observe that by construction, $t_{n_1} = T$. Also, inequality (10.58) gives for all $n \geq 0$,

$$
|y|_{\infty;[0,t_{i+1}]}
$$
$$
\leq |y|_{\infty;[0,t_i]} + c_2 r_{n_0 + n_1 - i}^{\theta \beta} 2^{n(1-\theta \beta)}
$$
$$
+ \left(\mu + |y|_{\infty;[0,t_{i+1}]} \right) c_2 \left(\delta \left(r_{n_0 + n_1 - i} \right)^{1/p} + \int_1^{2^n} \delta \left(\frac{r_{n_0 + n_1 - i}}{x} \right) dx \right).
$$

We will take $n = n_0 + n_1 - i$. By definition of $r_{n_0 + n_1 - i}$,

$$
c_2 \left(\delta \left(r_{n_0 + n_1 - i} \right)^{1/p} + \int_1^{2^{n_0 + n_1 - i}} \delta \left(\frac{r_{n_0 + n_1 - i}}{x} \right) dx \right) = 1 - \frac{1}{b},
$$

so that

$$
|y|_{\infty;[0,t_{i+1}]} \leq b |y|_{\infty;[0,t_i]} + (b-1) \mu + b c_2 r_{n_0 + n_1 - i}^{\theta \beta} 2^{(1-\theta \beta)(n_0 + n_1 - i)}.
$$

As all r_i are bounded by $\omega(0,T)$, we obtain

$$|y|_{\infty;[0,t_{i+1}]} \leq b\left(|y|_{\infty;[0,t_i]} + \frac{(b-1)}{b}\mu + c_3 2^{(1-\theta\beta)(n_0+n_1-i)}\right).$$

An easy induction then gives us that

$$\begin{aligned}
|y|_{\infty;[0,t_k]} &\leq b^k |y_0| + \sum_{j=0}^{k-1} b^{k-j}\left(\frac{(b-1)}{b}\mu + c_3 2^{(1-\theta\beta)(n_0+n_1-j)}\right) \\
&\leq \left(b^k |y_0| + \frac{b^k-1}{b-1}\mu\right) \\
&\quad + c_3 2^{(1-\theta\beta)n_0} \sum_{j=0}^{k-1} 2^{(1-\theta\beta)(n_1-j)} b^{k-j}.
\end{aligned}$$

Applying this to $k = n_1$, we see that

$$\begin{aligned}
|y|_{\infty;[0,T]} &\leq \left(b^{n_1} |y_0| + \frac{b^{n_1}-1}{b-1}\mu\right) \\
&\quad + c_3 2^{(1-\theta\beta)n_0} \sum_{j=0}^{n_1-1} \left(b 2^{(1-\theta\beta)}\right)^{(n_1-j)} \\
&\leq \left(b^{n_1} |y_0| + \frac{b^{n_1}-1}{b-1}\mu\right) \\
&\quad + \frac{c_3}{1-b 2^{(1-\theta\beta)}} 2^{(1-\theta\beta)n_0}.
\end{aligned}$$

For a given $\varepsilon > 0$, we pick n_0 large enough so that $\frac{c_3}{1-b2^{(1-\theta\beta)}} 2^{(1-\theta\beta)n_0} \leq \varepsilon$, to obtain that

$$|y|_{\infty;[0,T]} \leq \left(b^{n_1} |y_0| + \frac{b^{n_1}-1}{b-1}\mu\right) + \varepsilon.$$

Observe that n_1 depends on n_0, which depends on ε. Nonetheless, we see that for $\varepsilon > 0$, there exists $\delta > 0$ such that $|y_0| + \mu \leq \delta$ implies $\left(b^{n_1} |y_0| + \frac{b^{n_1}-1}{b-1}\mu\right) < \varepsilon$ and hence

$$|y|_{\infty;[0,T]} \leq 2\varepsilon.$$

That concludes the proof.   ∎

10.9   Comments

The main result of this chapter, the continuity of the RDE solutions as a function of the driving signal, also known as the *universal limit theorem*, is due to T. Lyons [116, 120], nicely summarized in Lejay [104] and the Saint Flour notes [123]. There have been a number of (re)formulations of rough path theory by other authors, including Davie [37], Feyel and de la Pradelle [50], Gubinelli [74] and Hu and Nualart [86]. Our presentation builds on Friz and Victoir [67] and combines Davie's approach [37] with geometric ideas. It seems to lead to essentially sharp estimates. In particular, we can extend Davie's uniqueness result under Lip^p-regularity, $p < 3$ (compared to $\text{Lip}^{p+\varepsilon}$ in Lyons' uniqueness proof via Picard iteration) to the case of arbitrary $p \geq 1$. In this case, the flow need not be Lipschitz continuous (A. M. Davie, personal communication). Convergence of Euler schemes for rough differential equations is established in Davie [37] for $[p] = 1, 2$; the general case, Section 10.3.5, is new.

Some of our estimates appear as special cases in previous works, for instance in Hu and Nualart [87] in the "Young" case of $1/p$-Hölder paths with $p \in [1, 2)$. Lipschitz estimates for rough integration or differential equations, at least for $p \in [2, 3)$, appear in Hu and Nualart [86] and Lyons and Qian [120]; see also Gubinelli [74].

11

RDEs: smoothness

We remain in the RDE setting of the previous chapter; that is, we consider rough differential equations of the form

$$dy = V(y)\, d\mathbf{x}, \quad y(0) = y_0,$$

where $\mathbf{x} = \mathbf{x}(t)$ is a weak geometric p-rough path. In the present chapter we investigate various smoothness properties of the solution, in particular as a function of y_0 and $\mathbf{x}$. In particular, we shall see that RDE solutions induce flows of diffeomorphisms which depend continuously on $\mathbf{x}$. As an application, we consider a class of parabolic partial differential equations with "rough" coefficients in a transport term.

11.1 Smoothness of the Itô–Lyons map

Assuming $x \in C^{1\text{-var}}([0,T], \mathbb{R}^d)$ we saw in Chapter 4 (cf. Remark 4.5) that the $\mathbb{R}^e$-valued ODE solution $y = \pi_{(V)}(0, y_0, x)$ together with its directional derivative $z = \frac{d}{d\varepsilon}\{\pi_{(V)}(0, y_0 + \varepsilon v; x + \varepsilon h)\}_{\varepsilon=0}$ satisfies the system

$$dy = V(y)\, dx, \quad dz = (DV(y)\, dx) \cdot z + V(y)\, dh$$

started at $(y_0, v) \in \mathbb{R}^e \oplus \mathbb{R}^e$. In particular, we may write $(y, z) = \pi_{(W)}(0, (y_0, v); (x, h))$ where (W) are the induced vector fields on $\mathbb{R}^e \oplus \mathbb{R}^e$. In this formulation, the extension to a rough path setting is easy. Assume at first that $V \in \mathrm{Lip}^{\gamma+1}$ with $\gamma > p$ so $W \in \mathrm{Lip}^{\gamma}_{loc}$, and assume furthermore that there exists a $G^{[p]}(\mathbb{R}^d \oplus \mathbb{R}^d)$-valued geometric p-rough path $\boldsymbol{\chi}$ which projects onto the $G^{[p]}(\mathbb{R}^d)$-valued geometric p-rough paths $\mathbf{x}$ and $\mathbf{h}$. After a localization argument (exploiting the structure of (y, z) and in particular the fact that linear RDEs do not explode; cf. the argument in Section 11.1.1 below) we may assume $W \in \mathrm{Lip}^{\gamma}$ and have existence/uniqueness/continuity properties of the RDE $\pi_{(W)}(0, (y_0, v); \boldsymbol{\chi})$ with values in $\mathbb{R}^e \oplus \mathbb{R}^e$. Projection to the second component gives (at least a candidate for) the directional derivative

$$z = \frac{d}{d\varepsilon}\{\pi_{(V)}(0, y_0 + \varepsilon v; \mathbf{plus} \circ \delta_{1,\varepsilon}(\boldsymbol{\chi}))\}_{\varepsilon=0}.$$

(Recall from Section 7.5.6 that **plus** resp. $\delta_{1,\varepsilon}$ is defined as the unique extension of

$$(x,h) \in \mathbb{R}^{2e} \mapsto (x+h) \in \mathbb{R}^e \text{ resp. } (x,h) \in \mathbb{R}^{2e} \mapsto (x,\varepsilon h) \in \mathbb{R}^{2e}$$

to a homomorphism between the respective free nilpotent groups.) When h enjoys complementary Young regularity to $\mathbf{x}$, say $h \in C^{q\text{-var}}\left([0,T],\mathbb{R}^d\right)$ with $1/p + 1/q > 1$, we naturally take $\boldsymbol{\chi} = S_{[p]}(\mathbf{x},h)$, the Young pairing of $\mathbf{x}$ and h, in which case

$$\mathbf{plus} \circ \delta_{1,\varepsilon}(\boldsymbol{\chi}) = T_{\varepsilon h}\mathbf{x}.$$

(The translation operator T was introduced in Section 9.4.6.) The proof that z is not only a candidate but indeed is the directional derivative can then be done by passing to limit in the corresponding ODE statements, using both continuity of the Lyons–Itô maps and "closedness of the derivative operator" (Proposition B.7). Unfortunately, this reasoning requires one degree too much regularity (our discussion above started with $V \in \mathrm{Lip}^{\gamma+1}$). With a little extra effort we can prove differentiability for $V \in \mathrm{Lip}^\gamma, \gamma > p$. The argument exploits, of course, the specific structure of (y,z) and in particular the fact that

$$DV \in \mathrm{Lip}^{\gamma-1}$$

only appears in a rough integration procedure. (Recall from Section 10.6 that existence/uniqueness/continuity for rough integrals holds under $\mathrm{Lip}^{\gamma-1}$-regularity, $\gamma > p$.)

11.1.1 *Directional derivatives*

All smoothness properties under consideration will be *local*. On the other hand, the differential equation satisfied by these derivatives (and higher derivatives) naturally exhibits growth beyond the standard conditions for global existence. To make (iterated) localization arguments transparent, we make the following

Definition 11.1 Let $V = (V_1, \ldots, V_d)$ be a collection of vector fields on $\mathbb{R}^e$. We say that V satisfies the p non-explosion condition if for all $R > 0$, there exists $M > 0$ such that if $(y_0,\mathbf{x}) \in \mathbb{R}^e \times C^{p\text{-var}}\left([0,T];G^{[p]}\left(\mathbb{R}^d\right)\right)$ with $\|\mathbf{x}\|_{p\text{-var};[0,T]} + |y_0| < R$,

$$\left\|\boldsymbol{\pi}_{(V)}(0,y_0,\mathbf{x})\right\|_{\infty;[0,T]} < M. \qquad \square$$

Following our usual convention we agree that, in the case of non-uniqueness, $\boldsymbol{\pi}_{(V)}(0,y_0;\mathbf{x})$ stands for any full RDE solutions driven by $\mathbf{x}$ along vector fields V started at y_0. For example, a collection of $\mathrm{Lip}^{\gamma-1}(\mathbb{R}^e)$-vector fields, with $\gamma > p$, satisfies the p non-explosion condition.

In what follows, we fix a collection $V = (V_1, \ldots, V_d)$ of $\mathrm{Lip}^\gamma_{loc}\,(\mathbb{R}^e)$-vector fields that satisfies the p non-explosion condition. Motivated by the presentation of directional derivatives of ODE solutions established in Theorem 4.4, we make the following definitions.

1. Consider the ODE

$$d\begin{pmatrix} x \\ h \\ y \end{pmatrix} = \begin{pmatrix} dx \\ dh \\ V(y_u)\,dx_u \end{pmatrix} \equiv \tilde{V}\begin{pmatrix} x \\ h \\ y \end{pmatrix} d\chi, \qquad (11.1)$$

where $\chi_t = (x_t, h_t) \in \mathbb{R}^d \oplus \mathbb{R}^d$ and $\tilde{V} \in \mathrm{Lip}^\gamma_{loc}$ is defined by the last equality. We then define the map

$$f_{1,(V)} : \mathbb{R}^e \times C^{p\text{-var}}\left([0,T], G^d\left(\mathbb{R}^d \oplus \mathbb{R}^d\right)\right) \to$$
$$C^{p\text{-var}}\left([0,T], G^{[p]}\left(\mathbb{R}^d \oplus \mathbb{R}^d \oplus \mathbb{R}^e\right)\right)$$

by[1]

$$f_{1,(V)} : (y_0, \chi) \mapsto \pi_{(\tilde{V})}\left(0, \exp\left((0, 0, y_0)\right); \chi\right).$$

2. Consider the (Riemann–Stieltjes) integral

$$\begin{pmatrix} M \\ H \end{pmatrix} = \int_0^\cdot \begin{pmatrix} DV \\ V \end{pmatrix}(y_t)\, d\begin{pmatrix} x_t \\ h_t \end{pmatrix} \equiv \int_0^\cdot \varphi(w_t)\,dw_t, \qquad (11.2)$$

where $w = (x, h, y) \in \mathbb{R}^d \oplus \mathbb{R}^d \oplus \mathbb{R}^e$ and $\varphi \in \mathrm{Lip}^{\gamma-1}_{loc}$ is defined by the last equality. We then define the map

$$f_{2,(V)} : C^{p\text{-var}}\left([0,T], G^{[p]}\left(\mathbb{R}^d \oplus \mathbb{R}^d \oplus \mathbb{R}^e\right)\right) \to$$
$$C^{p\text{-var}}\left([0,T], G^{[p]}\left(\mathbb{R}^{e\times e} \oplus \mathbb{R}^e\right)\right)$$

as the rough integral

$$f_{2,(V)} : \mathbf{w} \mapsto \int_0^\cdot \varphi(\mathbf{w})\,d\mathbf{w}.$$

3. Consider the linear ODE

$$dz_t = dM_t \cdot z_t + dH_t \equiv A(z_t)\,d\begin{pmatrix} M_t \\ H_t \end{pmatrix} \qquad (11.3)$$

---

[1]There is some irrelevant freedom in choosing the starting point. Any $\mathbf{y}_0 \in G^{[p]}\left(\mathbb{R}^d \oplus \mathbb{R}^d \oplus \mathbb{R}^e\right)$ with the property that the last component of $\pi_1(\mathbf{y}_0) \in \mathbb{R}^d \oplus \mathbb{R}^d \oplus \mathbb{R}^e$ is equal to $y_0 \in \mathbb{R}^e$ will do.

where A is a collection of linear (strictly speaking: affine-linear) vector fields. We then define the map

$$f_{3,(V)} : \mathbb{R}^e \times C^{p\text{-var}}\left([0,T], G^{[p]}\left(\mathbb{R}^{e\times e} \oplus \mathbb{R}^e\right)\right) \to C^{p\text{-var}}\left([0,T], \mathbb{R}^e\right)$$

as a solution to the corresponding (linear) RDE, namely

$$f_{3,(V)} : (z_0, \boldsymbol{\xi}) \mapsto \pi_{(A)}(0, z_0; \boldsymbol{\xi}).$$

Remark 11.2 Observe that if $(y_0, x), (v, h) \in \mathbb{R}^e \times C^{1\text{-var}}\left([0,T], \mathbb{R}^d\right)$, then we proved in Theorem 4.4 that the derivative of the ODE map $\pi_{(V)}(0, y_0; x)$ in (y_0, x) is given by

$$D_{(v,h)}\pi_{(V)}(0, y_0; x) = f_{3,(V)}\left(v, f_{2,(V)} \circ f_{1,(V)}\left(y_0, S_{[p]}(x \oplus h)\right)\right). \quad \square$$

We are now ready for our main theorem:

Theorem 11.3 *(directional derivatives in starting point and perturbation)*
(i) $\gamma > p$, and $V = (V_1, \ldots, V_d)$ is a collection of $\text{Lip}^\gamma_{\text{loc}}(\mathbb{R}^e)$-vector fields that satisfies the p non-explosion condition;
(ii) $\mathbf{x} \in C^{p\text{-var}}\left([0,T], G^{[p]}(\mathbb{R}^d)\right)$ is a weak geometric p-rough path;
(iii) $v \in \mathbb{R}^e$ and $h \in C^{q\text{-var}}\left([0,T], \mathbb{R}^d\right)$ with $1/p + 1/q > 1$ and $q \le p$.
Then $\varepsilon \mapsto \pi_{(V)}(0, y_0 + \varepsilon v; T_{\varepsilon h}(\mathbf{x}))$ is differentiable in $C^{p\text{-var}}\left([0,T], \mathbb{R}^e\right)$, and its derivative at 0 is given by

$$f_{3,(V)}\left(v, f_{2,(V)} \circ f_{1,(V)}\left(y_0, S_{[p]}(\mathbf{x} \oplus h)\right)\right).$$

Proof. Without loss of generality, we can assume that the vector fields are in $\text{Lip}^\gamma(\mathbb{R}^e)$. We then consider any sequence of paths $(x_n, h_n) \in C^{1\text{-var}}\left([0,T], \mathbb{R}^d\right)$ such that

$$\sup_n \left(\left\|S_{[p]}(x_n)\right\|_{p\text{-var};[0,T]} + \|h_n\|_{q\text{-var};[0,T]}\right) < \infty$$

$$\lim_{n\to\infty} d_\infty\left(S_{[p]}(x_n), \mathbf{x}\right) + d_\infty\left(S_{[p]}(h_n), \mathbf{h}\right) = 0.$$

From basic continuity properties of the Young pairing of such rough paths (cf. Remark 9.32) this implies that

$$\sup_n \left\|S_{[p]}(x_n \oplus h_n)\right\|_{p\text{-var};[0,T]} < \infty$$

$$\lim_{n\to\infty} d_\infty\left(S_{[p]}(x_n \oplus h_n), S_{[p]}(\mathbf{x} \oplus h)\right) = 0.$$

Let us use the notations $\mathbb{Y}^p \equiv C^{p\text{-var}}\left([0,T], \mathbb{R}^d\right)$ and $\mathbb{Y}^\infty \equiv C\left([0,T], \mathbb{R}^d\right)$; they are Banach spaces when equipped with p-variation and ∞-norm. For any fixed $n \in \mathbb{N}$, the map $\theta \in \mathbb{R} \mapsto \pi_{(V)}\left(0, y_0 + \theta v; S_{[p]}\left(x_n + \theta h_n\right)\right) \in \mathbb{Y}^1$ is continuously differentiable in $\mathbb{Y}^1$ with derivative given by

$$g_n\left(\theta\right) \equiv f_{3,(V)}\left(v, f_{2,(V)} \circ f_{1,(V)}\left(y_0 + \theta v, S_{[p]}\left(x_n + \theta h_n\right)\right)\right),$$

a consequence of our smoothness results on ODE solution maps, Theorem 4.4. By the fundamental theorem of calculus (in a Banach setting, cf Section B.1 in Appendix B), for all $\varepsilon \in [0,1]$ and $n \in \mathbb{N}$ we then have

$$\pi_{(V)}\left(0, y_0 + \varepsilon v; S_{[p]}\left(x_n + \varepsilon h_n\right)\right) - \pi_{(V)}\left(0, y_0; S_{[p]}\left(x_n\right)\right) = \int_0^\varepsilon g_n\left(\theta\right) d\theta,$$

$$(11.4)$$

as an equation in $\mathbb{Y}^1$ (by which we mean in particular that the integral appearing on the right-hand side is the limit in $\mathbb{Y}^1$ of its Riemann-sum approximations). From the continuous embedding $\mathbb{Y}^1 \hookrightarrow \mathbb{Y}^\infty$ we can view (11.4) as an equation in $\mathbb{Y}^\infty$ and as such we now try to send $n \to \infty$ in (11.4). By continuity of the translation operator and the Itô–Lyons map (Theorem 9.33) we have

$$\pi_{(V)}\left(0, y_0 + \varepsilon v; S_{[p]}\left(x_n + \varepsilon h_n\right)\right) \to \pi_{(V)}\left(0, y_0 + \varepsilon v; T_{\varepsilon h}\left(\mathbf{x}\right)\right) \text{ in } \mathbb{Y}^\infty$$

(even with uniform p-variation bounds) for any ε (including $\varepsilon = 0$) which justifies the passage to the limit in the left-hand side of (11.4) to

$$\pi_{(V)}\left(0, y_0 + \varepsilon v; T_{\varepsilon h}\left(\mathbf{x}\right)\right) - \pi_{(V)}\left(0, y_0; \mathbf{x}\right)$$

(which is actually an element of $\mathbb{Y}^p \equiv C^{p\text{-var}}\left([0,T], \mathbb{R}^d\right)$, not only $\mathbb{Y}^\infty$). On the other hand, from Theorem 10.47, we see that

$$\sup_{\theta \in [0,1]} \left|g_n\left(\theta\right) - g\left(\theta\right)\right|_{\mathbb{Y}^\infty} \to 0 \text{ as } n \to \infty$$

where $g\left(\theta\right) \equiv f_{3,(V)}\left(v, f_{2,(V)} \circ f_{1,(V)}\left(y_0 + \theta v, T_{\theta h}\left(\mathbf{x}\right)\right)\right)$. Clearly then

$$\left|\int_0^\varepsilon \left(g_n\left(\theta\right) - g\left(\theta\right)\right) d\theta\right|_{\mathbb{Y}^\infty} \leq \int_0^\varepsilon \left|g_n\left(\theta\right) - g\left(\theta\right)\right|_{\mathbb{Y}^\infty} d\theta \to 0 \text{ as } n \to \infty$$

which justifies the passage to the limit in the right-hand side of (11.4) and we obtain

$$\pi_{(V)}\left(0, y_0 + \varepsilon v; T_{\varepsilon h}\left(\mathbf{x}\right)\right) - \pi_{(V)}\left(0, y_0; \mathbf{x}\right) = \int_0^\varepsilon g\left(\theta\right) d\theta,$$

as an equation in $\mathbb{Y}^\infty$. Now, $\pi_{(V)}\left(0, y_0; \mathbf{x}\right), \pi_{(V)}\left(0, y_0 + \varepsilon v; T_{\varepsilon h}\left(\mathbf{x}\right)\right) \in \mathbb{Y}^p \subsetneq \mathbb{Y}^\infty$ and for the integrand on the right-hand side we even have

$$\{\theta \mapsto g\left(\theta\right)\} \in C\left([0,1], \mathbb{Y}^p\right).$$

To prove this, from the continuity of the Itô map (Theorem 10.26 and its corollaries), it is enough to prove that

$$\theta \to T_{\theta h}\left(\mathbf{x}\right) = \mathbf{plus} \circ \delta_{1,\theta}\left(S_{[p]}\left(\mathbf{x} \oplus h\right)\right)$$

is a continuous function from $[0,1]$ into $\mathbb{Y}^p$. But this is easily implied by Proposition 8.11. By a simple fact of Banach calculus (Proposition B.1 in Appendix B) it then follows that $\varepsilon \mapsto \pi_{(V)}\left(0, y_0 + \varepsilon v; T_{\varepsilon h}\left(\mathbf{x}\right)\right)$ is continuously differentiable in $\mathbb{Y}^p$ and the proof is then finished. ∎

We can generalize the previous theorem by perturbing the driving rough path in a more general way. Indeed, after replacing in the proof above $T_{\varepsilon h}\left(\mathbf{x}\right)$ by $\mathbf{plus} \circ \delta_{1,\varepsilon}\left(S_{[p]}\left(\mathbf{x}, h\right)\right)$ (the two are equal), we observe that all we need to translate a rough path $\mathbf{x}$ is a path $\boldsymbol{\chi} \in C^{p\text{-var}}\left([0,T], G^{[p]}\left(\mathbb{R}^d \oplus \mathbb{R}^d\right)\right)$ that projects onto $\mathbf{x}$, i.e. such that $\mathbf{plus} \circ \delta_{1,0}\left(\boldsymbol{\chi}\right) = \mathbf{x}$. We obtain the following result.

Proposition 11.4 *(i)* $\gamma > p$, *and* $V = (V_1, \ldots, V_d)$ *is a collection of locally* $\mathrm{Lip}^\gamma\left(\mathbb{R}^e\right)$*-vector fields that satisfies the p non-explosion condition;*
(ii) $\boldsymbol{\chi} \in C^{p\text{-var}}\left([0,T], G^{[p]}\left(\mathbb{R}^d \oplus \mathbb{R}^d\right)\right)$;
(iii) $v \in \mathbb{R}^e$.
Then $\varepsilon \mapsto \pi_{(V)}\left(0, y_0 + \varepsilon v; \mathbf{plus} \circ \delta_{1,\varepsilon}\left(\boldsymbol{\chi}\right)\right)$ *is differentiable in* $C^{p\text{-var}}$ $\left([0,T], \mathbb{R}^e\right)$, *and its derivative at 0 is given* $f_{3,(V)}\left(v, f_{2,(V)} \circ f_{1,(V)}\left(y_0, \boldsymbol{\chi}\right)\right)$.

If $\boldsymbol{\chi} = S_{[p]}\left(\mathbf{x}, h\right)$, the proposition is exactly the previous theorem. If we want to differentiate $\pi_{(V)}\left(0, y_0; \mathbf{x}\right)$ in the direction of a path $\mathbf{h} \in C^{p\text{-var}}\left([0,T], G^{[p]}\left(\mathbb{R}^d\right)\right)$ which is not of finite q-variation with $q^{-1} + p^{-1} > 1$, then the previous proposition tells us to construct a rough path $\boldsymbol{\chi} \in C^{p\text{-var}}\left([0,T], G^{[p]}\left(\mathbb{R}^d \oplus \mathbb{R}^d\right)\right)$ that projects on both $\mathbf{x}$ and $\mathbf{h}$ (in the sense that $\mathbf{plus} \circ \delta_{1,0}\left(\boldsymbol{\chi}\right) = \mathbf{x}$ and $\mathbf{plus} \circ \delta_{0,1}\left(\boldsymbol{\chi}\right) = \mathbf{h}$). This would allow us, for instance, to differentiate the solution of an SDE in the direction of another Brownian motion, or in the direction of (Lévy-)area perturbations. That said, we shall not pursue these directions here and return to the Young perturbation setting of Theorem 11.3.

We now address the question of higher directional derivatives.

Proposition 11.5 *Assume that*
(i) $\gamma > p$, $k \geq 1$, *and* $V = (V_1, \ldots, V_d)$ *is a collection of* $\mathrm{Lip}_{\mathrm{loc}}^{k-1+\gamma}\left(\mathbb{R}^e\right)$*-vector fields that satisfies the p non-explosion condition;*
(ii) $\mathbf{x} \in C^{p\text{-var}}\left([0,T], G^{[p]}\left(\mathbb{R}^d\right)\right)$ *is a weak geometric p-rough path;*
(iii) $v_1, \ldots v_k \in \mathbb{R}^e$ *and* $h_1, \ldots, h_k \in C^{q\text{-var}}\left([0,T], \mathbb{R}^d\right)$ *with* $1/p + 1/q > 1$ *and* $q \leq p$.
Then the following directional derivatives exist in $C^{p\text{-var}}\left([0,T], \mathbb{R}^e\right)$ *for all*

$j \in \{1, \ldots, k\}$,

$$D_{(v_1, \ldots, v_j; h_1, \ldots, h_j)} \pi_{(V)}(0, y_0; \mathbf{x})$$

$$= \left\{ \frac{\partial^j}{\partial \varepsilon_1 \ldots \partial \varepsilon_j} \pi_{(V)} \left(0, y_0 + \sum_{i=1}^{j} \varepsilon_i v_i, T_{\left(\sum_{i=1}^{j} \varepsilon_i h_i \right)}(\mathbf{x}) \right) \right\}_{\varepsilon_1 = \ldots \varepsilon_j = 0}$$

and the ensemble of these derivatives satisfies the RDE obtained by formal differentiation.

Proof. The argument is the same as in the ODE case (cf. Proposition 4.6). All we need to observe is that $f_{3,(V)}\left(v, f_{2,(V)} \circ f_{1,(V)}\left(y_0, S_{[p]}(\mathbf{x}, h)\right)\right)$ can be obtained by the projection of the solution of an RDE driven along locally $\mathrm{Lip}^{k-2+\gamma}$ vector fields satisfying the p non-explosion condition. ∎

11.1.2 Fréchet differentiability

Theorem 11.6 *Assume that*
(i) $\gamma > p$, $k \geq 1$, and $V = (V_1, \ldots, V_d)$ is a collection of $\mathrm{Lip}^{k-1+\gamma}(\mathbb{R}^e)$-vector fields that satisfies the p non-explosion condition;
(ii) $\mathbf{x} \in C^{p\text{-var}}\left([0, T], G^{[p]}(\mathbb{R}^d)\right)$ is a geometric p-rough path.
Then, the map

$$(y_0, h) \in \mathbb{R}^e \times C^{q\text{-var}}\left([0, T], \mathbb{R}^d\right) \mapsto \pi_{(V)}(0, y_0; T_h(\mathbf{x})) \in C^{p\text{-var}}\left([0, T], \mathbb{R}^e\right)$$

is C^k-Fréchet.

Proof. Once again, the proof is identical to the ODE case (cf. Proposition 4.8). The map $(y_0, \mathbf{x}), (v_i, h_i)_{1 \leq i \leq k} \mapsto D^k_{(v_i, h_i)_{1 \leq i \leq k}} \pi_{(V)}(0, y_0; \mathbf{x})$ is uniformly continuous on bounded sets because of (i) uniform continuity on bounded sets of the Itô–Lyons map and (ii) uniform continuity on bounded sets of rough integration. We can now appeal to Corollary B.11 in Appendix B. The proof is then finished. ∎

Corollary 11.7 (Li–Lyons) *Let $k \in \{1, 2, \ldots\}$ and $p \in [1, 2)$ and consider the (Young) differential equation*

$$dy = V(y)\,dx$$

along Lip^{γ}-vector fields on $\mathbb{R}^e$ for $\gamma > p - 1 + k$ with (unique) solution $y = \pi_{(V)}(0, y_0; x)$. Then

$$(y_0, x) \in \mathbb{R}^e \times C^{p\text{-var}}\left([0, T], \mathbb{R}^d\right) \mapsto \pi_{(V)}(0, y_0; x) \in C^{p\text{-var}}\left([0, T], \mathbb{R}^e\right)$$

is C^k in the Fréchet sense.

Proof. Apply the previous theorem with $p = q \in [1, 2)$. In this case, the driving signal is $\mathbb{R}^d$-valued, say x, and since then $T_h(x) = x + h$ we can take $x \equiv 0$. The previous theorem shows C^k-Fréchetness of the map from (y_0, h) to the $C^{p\text{-var}}([0, T], \mathbb{R}^e)$-valued solution of

$$dy = V(y) d(0 + h).$$

Replacing the letter h by x leads to the claimed statement. ∎

Exercise 11.8 (Kusuoka) *(i) Assume that $\alpha \in (1/4, 1/2)$ and set $p = 1/\alpha$. Let $\gamma > p$, $k \geq 1$, and $V = (V_1, \ldots, V_d)$ be a collection of $\mathrm{Lip}^{k-1+\gamma}(\mathbb{R}^e)$-vector fields that satisfies the p non-explosion condition;*
(ii) $\mathbf{x} \in C^{\alpha\text{-Höl}}([0, T], G^{[p]}(\mathbb{R}^d))$ is a geometric α-Hölder rough path. Take $\delta \in (1/2, 1/2 + \alpha)$ and let us consider the fractional Sobolev (or Besov) space $W_0^{\delta, 2}$ (which is strictly bigger than the usual Cameron–Martin space $W_0^{1,2}$). Show that the map

$$(y_0, k) \in \mathbb{R}^e \times W_0^{\delta, 2}([0, T], \mathbb{R}^d) \mapsto \pi_{(V)}(0, y_0; T_k(\mathbf{x})) \in C^{\alpha\text{-Höl}}([0, T], \mathbb{R}^e)$$

is C^k-Fréchet.

Solution. Using the p-variation properties of Besov spaces, as discussed in Example 5.16, the Young pairing

$$(\mathbf{x}, h) \mapsto S_{[p]}(\mathbf{x} \oplus h)$$

is continuous from $C^{\alpha\text{-Höl}} \times W_0^{\delta, 2} \to C^{\alpha\text{-Höl}}$ and this is the only modification needed in the arguments of this section. □

Exercise 11.9 (Duhamel's principle) *Write $J_{t \leftarrow s}^{y_s, \mathbf{x}}$ for the derivative ("Jacobian") of $y_t \equiv \pi_{(V)}(s, \cdot; \mathbf{x})_t : \mathbb{R}^e \to \mathbb{R}^e$ at some point $y_s \in \mathbb{R}^e$. Establish the formula*

$$D_{(v,h)} \pi_{(V)}(0, y_0; \mathbf{x})_t = J_{t \leftarrow 0}^{y_0, \mathbf{x}} \cdot v$$
$$+ \sum_{i=1}^{d} \int_0^t J_{t \leftarrow s}^{y_s, \mathbf{x}} \cdot V_i\left(\pi_{(V)}(0, y_0; \mathbf{x})_s\right) dh_s^i.$$

Detail all assumptions.

Exercise 11.10 *Assume $V \in \mathrm{Lip}^\gamma(\mathbb{R}^e)$ and write $J_{\cdot \leftarrow 0}^{y_0, \mathbf{x}}$ for the (Fréchet) derivative of $y_t \equiv \pi_{(V)}(0, \cdot; \mathbf{x}) : \mathbb{R}^e \to C^{p\text{-var}}([0, T], \mathbb{R}^d)$ at some point $y_0 \in \mathbb{R}^e$. Noting that $J_{\cdot \leftarrow 0}^{y_0, \mathbf{x}}$ can be viewed as an element in $C^{p\text{-var}}([0, T], \mathbb{R}^{e \times e})$, show that*

$$\left|J_{\cdot \leftarrow 0}^{y_0, \mathbf{x}}\right|_{p\text{-var};[0,T]} \leq C \exp\left(C \|\mathbf{x}\|_{p\text{-var};[0,T]}^p\right)$$

with a suitable constant C depending on p, γ and $|V|_{\mathrm{Lip}^\gamma}$.

Solution. One proceeds as in Exercise 10.55, noting that $J_{\cdot \leftarrow 0}^{y_0, \mathbf{x}}$ satisfies a linear RDE starting at I, the identity map in $\mathbb{R}^e$. Note the constant C can be chosen independent of y_0 thanks to translation invariance of the Lip^γ-norm, i.e. $|V(y_0 + \cdot)|_{\mathrm{Lip}^\gamma} = |V|_{\mathrm{Lip}^\gamma}$. □

11.2   Flows of diffeomorphisms

We saw that $\text{Lip}^{\gamma+k-1}$-regularity on the vector fields V implies that

$$y_0 \in \mathbb{R}^e \mapsto \pi_{(V)}(0, y_0; \mathbf{x}) \in C^{p\text{-var}}([0, T], \mathbb{R}^e)$$

is C^k-Fréchet. Relatedly, under the same regularity assumptions, we now show that the map $(t, y_0) \mapsto \pi(0, y_0; \mathbf{x})_t$ is a flow of C^k-diffeomorphisms, i.e. an element in the space $\mathcal{D}_k(\mathbb{R}^e)$ defined as

$$\mathcal{D}_k(\mathbb{R}^e) := \left\{ \begin{array}{c} \phi : [0, T] \times \mathbb{R}^e \to \mathbb{R}^e : (t, y) \mapsto \phi_t(y) \text{ such that} \\ \forall t \in [0, T] : \phi_t \text{ is a } C^k\text{-diffeomorphism of } \mathbb{R}^e \\ \forall \alpha : |\alpha| \le k : \partial_\alpha \phi_t(y), \partial_\alpha \phi_t^{-1}(y) \text{ are continuous in } (t, y) \end{array} \right\}.$$

$$(11.5)$$

Proposition 11.11 *Let $p \ge 1$ and $k \in \{1, 2, \ldots\}$ and assume $V = (V_1, \ldots, V_d)$ is a collection of $\text{Lip}^{\gamma+k-1}$-vector fields on $\mathbb{R}^e$ for $\gamma > p$. Assume $\mathbf{x} \in C^{p\text{-var}}([0, T], G^{[p]}(\mathbb{R}^d))$. Then, the map*

$$\phi : (t, y) \in [0, T] \times \mathbb{R}^e \mapsto \pi_{(V)}(0, y; \mathbf{x})_t \in \mathbb{R}^e$$

is a flow of C^k-diffeomorphisms. Moreover, for any multi-index α with $1 \le |\alpha| \le k$, the maps

$$(t, y) \in [0, T] \times \mathbb{R}^e \mapsto \partial_\alpha \phi_t(y), \partial_\alpha \phi_t^{-1}(y)$$

are bounded by a constant only depending on $p, \gamma, k, \|\mathbf{x}\|_{p\text{-var};[0,T]}$ and $|V|_{\text{Lip}^{\gamma+k-1}}$.

Proof. We proceed as in the ODE case (Corollary 4.9). Clearly, $y_0 \in \mathbb{R}^e \mapsto \pi_{(V)}(0, y_0; \mathbf{x})_t$ is in $C^k(\mathbb{R}^e, \mathbb{R}^e)$. We then argue that $\pi_{(V)}(0, \cdot; \mathbf{x})_t^{-1} = \pi_{(V)}(0, \cdot; \overleftarrow{\mathbf{x}})_t$, where $\overleftarrow{\mathbf{x}}(\cdot) = \mathbf{x}(t - \cdot) \in C^{p\text{-var}}([0, t], G^{[p]}(\mathbb{R}^d))$. Indeed, we have seen that this holds (cf. the proof of Corollary 4.9) in the ODE case, i.e. when $\mathbf{x}$ is replaced by some continuous, bounded variation path x. A simple limit argument (in fact: our definition of an RDE solution combined with uniqueness) then shows that this identity remains valid in the RDE setting. It follows that $\pi_{(V)}(0, \cdot; \mathbf{x})_t$ is a bijection whose inverse is also in $C^k(\mathbb{R}^e, \mathbb{R}^e)$. This finishes the proof that $\pi_{(V)}(0, \cdot; \mathbf{x})_t$ is a C^k-diffeomorphism of $\mathbb{R}^e$. At last, each ∂_α-derivative of $\pi_{(V)}(0, \cdot; \mathbf{x})_t$ resp. $\pi_{(V)}(0, \cdot; \mathbf{x})_t^{-1}$ can be represented via (non-explosive) RDE solutions which plainly implies joint continuity in t and y_0. This also yields the claimed boundedness, since for a fixed y say

$$\partial_\alpha \pi_{(V)}(0, y; \mathbf{x}) = \partial_\alpha \pi_{(\tilde{V})}(0, 0; \mathbf{x})$$

where $\tilde{V} = V(y + \cdot)$; it is then clear that $\sup_{t \in [0,T]} |\partial_\alpha \pi_{(\tilde{V})}(0,0;\mathbf{x})_t|$ will be bounded by a constant depending only on $k, p, \gamma, \|\mathbf{x}\|_{p\text{-var};[0,T]}$ and

$$\left|\tilde{V}\right|_{\text{Lip}^{\gamma+k-1}} = |V|_{\text{Lip}^{\gamma+k-1}},$$

thanks to translation invariance of Lip-norms.   ∎

The following statement is a first limit theorem for RDE flows. The uniformity in $y_0 \in \mathbb{R}^e$ is a consequence of the invariance of the Lip^γ-norm under translation,

$$\forall y_0 \in \mathbb{R}^e, \gamma \geq 1 : |V|_{\text{Lip}^\gamma} = |V(y_0 + \cdot)|_{\text{Lip}^\gamma}. \tag{11.6}$$

Theorem 11.12 *Let $p \geq 1$ and $k \in \{1, 2, \dots\}$ and assume $V = (V_1, \dots, V_d)$ is a collection of $\text{Lip}^{\gamma+k-1}$-vector fields on $\mathbb{R}^e$ for $\gamma > p$. Write $\alpha = (\alpha_1, \dots, \alpha_e) \in \mathbb{N}^e$ and $|\alpha| = \alpha_1 + \cdots + \alpha_e \leq k$. Then the ensemble*

$$\left\{\partial_\alpha \pi_{(V)}(0, y_0; \mathbf{x}) : |\alpha| \leq k\right\}$$

depends continuously on $\mathbf{x} \in C^{p\text{-var}}\left([0,T], G^{[p]}(\mathbb{R}^d)\right)$. More precisely, for all $\varepsilon, R > 0$ there exists δ (depending also on p, γ, k and $|V|_{\text{Lip}^{\gamma+k-1}}$) such that for all $\mathbf{x}^1, \mathbf{x}^2$ with $\max_{i=1,2} \|\mathbf{x}^i\|_{p\text{-var};[0,T]} \leq R$ and

$$d_{p\text{-var};[0,T]}\left(\mathbf{x}^1, \mathbf{x}^2\right) < \delta$$

we have

$$\sup_{y_0 \in \mathbb{R}^e} \left|\partial_\alpha \pi_{(V)}\left(0, y_0; \mathbf{x}^1\right) - \partial_\alpha \pi_{(V)}\left(0, y_0; \mathbf{x}^2\right)\right|_{p\text{-var};[0,T]} < \varepsilon. \tag{11.7}$$

If $\mathbf{x}$ is a geometric $1/p$-Hölder rough path, we may replace p-variation by $1/p$-Hölder throughout.

Proof. We show for all $\varepsilon > 0$ that there exists δ such that $d_{p\text{-var};[0,T]}\left(\mathbf{x}^1, \mathbf{x}^2\right) < \delta$ implies

$$\sup_{y_0 \in \mathbb{R}^e} \left|\partial_\alpha \pi_{(V)}\left(0, y_0; \mathbf{x}^1\right) - \partial_\alpha \pi_{(V)}\left(0, y_0; \mathbf{x}^2\right)\right|_{p\text{-var};[0,T]} < \varepsilon$$

where α is an arbitrary multi-index with $|\alpha| = \alpha_1 + \cdots + \alpha_e \leq k$. The main observation is that we can take $y_0 = 0$ at the price of replacing V by $V(y_0 + \cdot)$. Thus, thanks to (11.6), uniformity in $y_0 \in \mathbb{R}^e$ will come for free provided our choice of δ depends on V only through $|V|_{\text{Lip}^{\gamma+k-1}}$.

Case 1: Assume $k = 1$ so that $V \in \text{Lip}^\gamma$. In this case, $\partial_\alpha \pi_{(V)}(0, y_0; \mathbf{x})$ corresponds to a directional derivative in one of the basis directions of $\mathbb{R}^e$, say e_j. From Theorem 11.3 we can write $\partial_\alpha \pi_{(V)}(0, 0; \mathbf{x})$ as a composition of the form

$$f_{3,(V)}\left(e_j, f_{2,(V)} \circ f_{1,(V)}(0, \mathbf{x})\right).$$

Inspection of the respective definitions of these maps shows continuous dependence in $\mathbf{x}$ with modulus of continuity only depending on $|V|_{\text{Lip}^\gamma}$, as required. More precisely, $f_{1,(V)}$ was defined as a full RDE solution and the continuity estimate for full RDE solutions, Corollary 10.39, clearly shows that the modulus of continuity only depends on $|V|_{\text{Lip}^\gamma}$. Similar remarks apply to $f_{2,(V)}$ and $f_{3,(V)}$ after inspection of the continuity estimates for rough integrals and solutions of RDEs with linear vector fields.

Case 2: Now assume $V \in \text{Lip}^{\gamma+k-1}$ for $k > 1$. We have already pointed out that the ensemble

$$\left\{ \partial_\alpha \pi_{(V)} (0,0;\mathbf{x}) : |\alpha| \le k-1 \right\}$$

can be written as a solution to an RDE along Lip^γ_{loc}-vector fields (satisfying the p non-explosion condition). After localization, it can be written as an RDE solution along genuine Lip^γ-vector fields where we insist that the Lip^γ-norm of these localized vector fields only depends on $\|\mathbf{x}\|_{p\text{-var};[0,T]}$ and $|V|_{\text{Lip}^{\gamma+k-1}}$. We can now appeal to case 1 and the proof is finished.

(The adaptation to the Hölder case is left to the reader as a simple exercise.) ∎

Theorem 11.13 *The conclusion of Theorem 11.12 holds with (11.7) replaced by*[2]

$$\sup_{y_0 \in \mathbb{R}^e} \left| \partial_\alpha \pi_{(V)} \left(0, y_0; \mathbf{x}^1\right)^{-1} - \partial_\alpha \pi_{(V)} \left(0, y_0; \mathbf{x}^2\right)^{-1} \right|_{p\text{-var};[0,T]} < \varepsilon. \quad (11.8)$$

If $\mathbf{x}$ is a geometric $1/p$-Hölder rough path, we may replace p-var by $1/p$-Höl throughout.

Proof. We proceed as in the proof of Theorem 11.12 and observe that we can take $y_0 = 0$ at the price of replacing V by $V(y_0 + \cdot)$. Secondly, we only consider the case $|\alpha| = k = 1$ so that $V \in \text{Lip}^\gamma$. (The general case is reduced to this one as in the proof of Theorem 11.12, case 2.)

It helps to note that with $\hat{y} = \pi(s,y;\mathbf{x})_t$ we have

$$\pi_{(V)} \left(0, y; \mathbf{x}(s-\cdot)\right)_s = \pi_{(V)} \left(0, \hat{y}; \mathbf{x}(t-\cdot)\right)_t.$$

We now consider the (inverse) flow of two RDEs driven by $\mathbf{x}^i$, $i = 1,2$ respectively, so that

$$\left(\pi_{(V)} \left(0, y^i; \mathbf{x}^i\right)^{-1}\right)_{s,t}$$
$$= \pi_{(V)} \left(0, y^i; \mathbf{x}^i(t-\cdot)\right)_t - \pi_{(V)} \left(0, \hat{y}^i; \mathbf{x}^i(t-\cdot)\right)_t$$
$$= \underbrace{\left(y^i - \hat{y}^i\right)}_{=\pi(s,y^i;\mathbf{x}^i)_{s,t}} \cdot \underbrace{\int_0^1 \nabla \pi \left(0, \hat{y}^i + \tau \left(y^i - \hat{y}^i\right); \mathbf{x}^i(t-\cdot)\right)_t d\tau}_{=:F(\mathbf{x}^i)}.$$

---

[2] $\partial_\alpha \pi_{(V)} (0, y_0; \mathbf{x})^{-1}$ denotes the path $t \mapsto \partial_\alpha \pi_{(V)} (0, \cdot; \mathbf{x})_t^{-1}|_{\cdot=y_0} \in \mathbb{R}^e$.

We can write

$$\left| \left(\pi_{(V)} \left(0, y^1; \mathbf{x}^1\right)^{-1} \right)_{s,t} - \left(\pi_{(V)} \left(0, y^2; \mathbf{x}^2\right)^{-1} \right)_{s,t} \right|$$

$$\leq \left| \left(\pi \left(s, y^1; \mathbf{x}^1\right)_{s,t} - \pi \left(s, y^2; \mathbf{x}^2\right)_{s,t} \right) \right| \left| F\left(\mathbf{x}^1\right) \right|$$

$$+ \left| \pi \left(s, y^2; \mathbf{x}^2\right)_{s,t} \right| \left| F\left(\mathbf{x}^1\right) - F\left(\mathbf{x}^2\right) \right|,$$

which leaves us with estimating four terms. First, from Corollary 10.27,

$$\left| \left(\pi \left(s, y^1; \mathbf{x}^1\right)_{s,t} - \pi \left(s, y^2; \mathbf{x}^2\right)_{s,t} \right) \right| \leq c_3 \left[\rho_{p\text{-var};[s,t]} \left(\mathbf{x}^1, \mathbf{x}^2\right) + \left| y^1 - y^2 \right| \right]$$

where c_3 may depend on $R \geq \max_{i=1,2} \left\| \mathbf{x}^i \right\|_{p\text{-var};[0,T]}$ and $\left| V \right|_{\text{Lip}^\gamma}$. Second, it is easy to see that

$$\left| F\left(\mathbf{x}^1\right) \right| \leq \sup_{t \in [0,T]} \sup_{y_0 \in \mathbb{R}^e} \nabla \pi \left(0, y_0; \mathbf{x}^i \left(t - \cdot\right)\right)_t$$

$$\leq c_4 = c_4 \left(R, \left|V\right|_{\text{Lip}^\gamma}\right).$$

Third, for $c_5 = c_5 \left(R, \left|V\right|_{\text{Lip}^\gamma}\right)$ we have $\left| \pi \left(s, y^2; \mathbf{x}^2\right)_{s,t} \right| \leq c_5 \left\| \mathbf{x}^2 \right\|_{p\text{-var};[s,t]}$ by Theorem 10.14. At last, we easily see that $\left| F\left(\mathbf{x}^1\right) - F\left(\mathbf{x}^2\right) \right|$ is bounded from above by

$$\sum_{\alpha : |\alpha| = 1} \sup_{y_0 \in \mathbb{R}^e} \left| \partial_\alpha \pi_{(V)} \left(0, y_0; \mathbf{x}^1\right) - \partial_\alpha \pi_{(V)} \left(0, y_0; \mathbf{x}^2\right) \right|_{p\text{-var};[0,T]} < \varepsilon,$$

for an arbitrary fixed $\varepsilon > 0$ which is possible by Theorem 11.12 provided $\mathbf{x}^1, \mathbf{x}^2$ satisfy $\max_{i=1,2} \left\| \mathbf{x}^i \right\|_{p\text{-var};[0,T]} \leq R$ and $d_{p\text{-var};[0,T]} \left(\mathbf{x}^1, \mathbf{x}^2\right) < \delta$ for some $\delta = \delta\left(\varepsilon, R\right)$. Putting things together, and taking $y^1 = y^2 = y_0 = 0$, we see that

$$\left| \left(\pi_{(V)} \left(0, 0; \mathbf{x}^1\right)^{-1} \right)_{s,t} - \left(\pi_{(V)} \left(0, 0; \mathbf{x}^2\right)^{-1} \right)_{s,t} \right|$$

$$\leq c_3 c_4 \left| \rho_{p\text{-var};[s,t]} \left(\mathbf{x}^1, \mathbf{x}^2\right) \right| + c_5 \left\| \mathbf{x}^2 \right\|_{p\text{-var};[s,t]} \left| F\left(\mathbf{x}^1\right) - F\left(\mathbf{x}^2\right) \right|$$

$$\leq c_6 d_{p\text{-var};[s,t]} \left(\mathbf{x}^1, \mathbf{x}^2\right) + c_6 \left\| \mathbf{x}^2 \right\|_{p\text{-var};[s,t]} \varepsilon \quad \text{(thanks to Theorem 8.10)}$$

and, by super-additivity of $d_{p\text{-var};[\cdot,\cdot]} \left(\mathbf{x}^1, \mathbf{x}^2\right)$ resp. $\left\| \mathbf{x}^2 \right\|_{p\text{-var};[\cdot,\cdot]}$, this becomes

$$\left| \pi_{(V)} \left(0, 0; \mathbf{x}^1\right)^{-1} - \pi_{(V)} \left(0, 0; \mathbf{x}^2\right)^{-1} \right|_{p\text{-var};[0,T]} \leq c_6 \delta + c_6 R \varepsilon$$

with $c_6 = c_6 \left(R, \left|V\right|_{\text{Lip}^\gamma}\right)$. This estimate is more than enough to finish the proof; e.g. replace δ by $\min\left(\delta, R\varepsilon\right)$ and start the argument with $\varepsilon / \left(c_6 R\right)$

instead of ε. (The adaption to the Hölder case is left to the reader as a simple exercise.) ∎

We already pointed out that $\pi_{(V)}(0,\cdot;\mathbf{x})$ can be viewed as an element in $\mathcal{D}_k(\mathbb{R}^e)$, the space of flows of C^k-diffeomorphisms. For any bounded set $\mathfrak{K} \subset \mathbb{R}^e$ one can define

$$|\phi|_{(0,k);\mathfrak{K}} := \sup_{\substack{t\in[0,T],\\ \alpha:|\alpha|\le k,\\ y\in\mathfrak{K}}} |\partial_\alpha \phi_t(y)|$$

and, setting $\mathfrak{K}^n := \{y \in \mathbb{R}^e : |y| \le n\}$,

$$d_{\mathcal{D}_k(\mathbb{R}^e)}(\phi,\psi) = \sum_{n=1}^{\infty} \frac{1}{2^n} \frac{|\phi-\psi|_{(0,k);\mathfrak{K}^n}}{1+|\phi-\psi|_{(0,k);\mathfrak{K}^n}}$$

and also $\tilde{d}_{\mathcal{D}_k(\mathbb{R}^e)}(\phi,\psi) = d_{\mathcal{D}_k(\mathbb{R}^e)}(\phi,\psi)+d_{\mathcal{D}_k(\mathbb{R}^e)}(\phi^{-1},\psi^{-1})$. One can check that $\mathcal{D}_k(\mathbb{R}^e)$ is a Polish space under $\tilde{d}_{\mathcal{D}_k(\mathbb{R}^e)}$ and that convergence $\tilde{d}_{\mathcal{D}_k(\mathbb{R}^e)}$ $(\phi^n,\phi) \to 0$ is equivalent to

$$\sup_{\substack{t\in[0,T],\\ y\in\mathfrak{K},\\ \alpha:|\alpha|\le k}} |\partial_\alpha \phi_t^n(y) - \partial_\alpha \phi_t(y)| + \left|\partial_\alpha (\phi_t^n)^{-1}(y) - \partial_\alpha (\phi_t)^{-1}(y)\right| \to 0$$

for all compact subsets $\mathfrak{K} \subset \mathbb{R}^e$. We then have the following "limit theorems for flows of diffeomorphisms", as an immediate consequence of Theorems 11.12 and 11.13.

Corollary 11.14 *Under the assumptions of Theorem 11.12, the map*

$$\mathbf{x} \in C^{p\text{-var}}\left([0,T], G^{[p]}(\mathbb{R}^d)\right) \mapsto \pi_{(V)}(0,\cdot;\mathbf{x}) \in \mathcal{D}_k(\mathbb{R}^e)$$

is (uniformly) continuous (on bounded sets).

Exercise 11.15 *Establish continuity of* $\mathbf{x} \in C^{p\text{-var}}\left([0,T], G^{[p]}(\mathbb{R}^d)\right) \mapsto \pi_{(V)}(0,\cdot;\mathbf{x}) \in \mathcal{D}_k^{p\text{-var}}(\mathbb{R}^e)$ *where* $\mathcal{D}_k^{p\text{-var}}$ *is constructed as* $\mathcal{D}_k(\mathbb{R}^e)$ *but with the semi-norm* $|\phi|_{(0,k);\mathfrak{K}}$ *replaced by*

$$\sup_{\substack{\alpha:|\alpha|\le k,\\ y\in\mathfrak{K}}} \left(\sup_{(t_i)\subset[0,T]} \sum \left|\partial_\alpha \phi_{t_i,t_{i+1}}(y)\right|^p\right)^{1/p}.$$

Conduct a similar dicussion in the $1/p$-Hölder context.

11.3   Application: a class of rough partial differential equations

Let S^n denote the set of symmetric $n \times n$ matrices and consider the partial differential equations of parabolic type

$$\dot{u} = F\left(t, x, Du, D^2 u\right), \tag{11.9}$$

$$u\left(0, \cdot\right) = u_0 \in \mathrm{BUC}\left(\mathbb{R}^n\right), \tag{11.10}$$

where $F = F\left(t, x, p, X\right) \in C\left([0,T], \mathbb{R}^n, \mathbb{R}^n, S^n\right)$ is assumed to be *degenerate elliptic*[3] and $u = u\left(t, x\right) \in \mathrm{BUC}\left([0,T] \times \mathbb{R}^n\right)$ is a real-valued function of time and space.[4] Equation (11.9) will be interpreted in *viscosity* sense and we recall[5] that this means that u is a viscosity sub- (and super-) solution to $\partial_t - F = 0$; that is, if $\psi \in C^2\left((0,T) \times \mathbb{R}^n\right)$ is such that $\left(\bar{t}, \bar{x}\right)$ is a maximum (resp. minimum) of $u - \psi$ then

$$\dot{\psi}\left(\bar{t}, \bar{x}\right) \leq (\text{resp. } \geq) \; F\left(\bar{t}, \bar{x}, D\psi\left(\bar{t}, \bar{x}\right), D^2\psi\left(\bar{t}, \bar{x}\right)\right).$$

The aim of this section is to allow for some "rough" perturbation of the form

$$Du \cdot V\left(x\right) dz_t \equiv \sum_{i,j} \partial_i u\left(t, x\right) V_j^i\left(x\right) dz_t^j$$

where $z = \left(z^1, \ldots, z^d\right) : [0,T] \to \mathbb{R}^d$ and, as usual, $V = \left(V_1, \ldots, V_d\right)$ denotes a collection of sufficiently nice vector fields on $\mathbb{R}^e$. As pointed out in [111], classical (deterministic) second-order viscosity theory can deal at best with $z \in W^{1,1}\left([0,T], \mathbb{R}^d\right)$, i.e. measurable dependence in time of dz/dt. Any "rough" partial differential equation of the form

$$du = F\left(t, x, Du, D^2 u\right) dt - Du\left(t, x\right) \cdot V\left(x\right) dz_t$$

where z enjoys only "Brownian" regularity of z (i.e. just below $1/2$-Hölder), or less, falls dramatically outside the scope of the deterministic theory. However, one can give meaning to this equation (and then establish existence, uniqueness, stability, etc.) via ideas from rough path theory; that is, by accepting that z should be replaced by a geometric p-rough path $\mathbf{z}$. The main result of this section is[6]

---

[3] This means $F\left(\ldots, X\right) \geq F\left(\ldots, Y\right)$ whenever $X \geq Y$ in the sense of symmetric matrices.

[4] BUC denotes the space of bounded, uniformly continuous functions, equipped with local-uniform topology.

[5] See the "User's guide" [34] or Fleming and Soner's textbook [53].

[6] Unless otherwise stated, BUC-spaces will be equipped with the topology of locally uniform convergence.

Theorem 11.16 *Let $p \geq 1$ and $(z^\varepsilon) \subset C^\infty ([0,T], \mathbb{R}^d)$ so that*

$$S_{[p]} (z^\varepsilon) \to \mathbf{z} \in C^{0,p\text{-var}} \left([0,T], G^{[p]} (\mathbb{R}^d)\right).$$

Assume that

$$u_0^\varepsilon \in \text{BUC} (\mathbb{R}^n) \to u_0 \in \text{BUC} (\mathbb{R}^n),$$

locally uniformly as $\varepsilon \to 0$. Let $F = F(t,x,p,X)$ be continuous, degenerate elliptic, and assume that $\partial_t - F$ satisfies $\Phi^{(3)}$-invariant comparison (cf. definition 11.19 below). Assume that $V = (V_1, \ldots, V_d)$ is a collection of $\text{Lip}^{\gamma+2} (\mathbb{R}^n)$ vector fields with $\gamma > p$. Consider (necessarily unique[7]) viscosity solutions $u^\varepsilon \in \text{BUC} ([0,T] \times \mathbb{R}^n)$ to

$$du^\varepsilon = F\left(t,x,Du^\varepsilon, D^2 u^\varepsilon\right) dt - Du^\varepsilon \cdot V(x) dz^\varepsilon (t) = 0, \quad (11.11)$$
$$u^\varepsilon (0, \cdot) = u_0^\varepsilon, \quad\quad\quad\quad\quad\quad\quad\quad\quad\quad\quad\quad\quad\quad (11.12)$$

and assume that the resulting family $(u^\varepsilon : \varepsilon > 0)$ is locally uniformly bounded[8]. Then

(i) there exists a unique $u \in \text{BUC} ([0,T] \times \mathbb{R}^n)$, only dependent on $\mathbf{z}$ and u_0 but not on the particular approximating sequences, such that $u^\varepsilon \to u$ locally uniformly. We write (formally)

$$du = F\left(t,x,Du, D^2 u\right) dt - Du \cdot V(x) d\mathbf{z}(t) = 0, \quad (11.13)$$
$$u(0, \cdot) = u_0; \quad\quad\quad\quad\quad\quad\quad\quad\quad\quad\quad\quad\quad\quad (11.14)$$

and also $u = u^{\mathbf{z}}$ when we want to indicate the dependence on $\mathbf{z}$;

(ii) we have the contraction property

$$\left| u^{\mathbf{z}} - \hat{u}^{\mathbf{z}} \right|_{\infty; \mathbb{R}^n \times [0,T]} \leq |u_0 - \hat{u}_0|_{\infty; \mathbb{R}^n}$$

where $\hat{u}^{\mathbf{z}}$ is defined as the limit of $\hat{u}^\eta$, defined as in (11.11) with u^ε replaced by $\hat{u}^\eta$ throughout;

(iii) the solution map $(\mathbf{z}, u_0) \mapsto u^{\mathbf{z}}$ from

$$C^{p\text{-var}} \left([0,T], G^{[p]} (\mathbb{R}^d)\right) \times \text{BUC} (\mathbb{R}^n) \to \text{BUC} ([0,T] \times \mathbb{R}^n)$$

is continuous.

Let us recall that *comparison* (for BUC-solutions of $\partial_t - F = 0$) means that, whenever $u, v \in \text{BUC} ([0,T] \times \mathbb{R}^n)$ are viscosity sub- (resp. super-) solutions to (11.9) with respective BUC-initial datas $u_0 \leq v_0$, then

$$u \leq v \text{ on } [0,T] \times \mathbb{R}^n.$$

---

[7] This follows from the first five lines in the proof of this theorem.

[8] A simple sufficient condition is boundedness of $F(\cdot, \cdot, 0, 0)$ on $[0,T] \times \mathbb{R}^n$, and the assumption that $u_0^\varepsilon \to u_0$ uniformly, as can be seen by comparison.

Given $F \in C\left([0,T], \mathbb{R}^n, \mathbb{R}^n, S^n\right)$, a sufficient condition for comparison[9] is the following technical[10]

Condition 11.17 There exists a function $\theta : [0,\infty] \to [0,\infty]$ with $\theta\left(0+\right) = 0$, such that for each fixed $t \in [0,T]$,

$$F\left(t,x,\alpha\left(x-\tilde{x}\right),X\right) - F\left(t,\tilde{x},\alpha\left(x-\tilde{x}\right),Y\right) \le \theta\left(\alpha\left|x-\tilde{x}\right|^2 + \left|x-\tilde{x}\right|\right)$$

$$(11.15)$$

for all $\alpha > 0$, $x,\tilde{x} \in \mathbb{R}^n$ and $X,Y \in S^n$ (the space of $n \times n$ symmetric matrices) satisfy

$$-3\alpha \begin{pmatrix} I & 0 \\ 0 & I \end{pmatrix} \le \begin{pmatrix} X & 0 \\ 0 & -Y \end{pmatrix} \le 3\alpha \begin{pmatrix} I & -I \\ -I & I \end{pmatrix}.$$

Furthermore, we require $F = F\left(t,x,p,X\right)$ to be uniformly continuous on sets $B_R := \left\{(t,x,p,X) \in [0,T] \times \mathbb{R}^n \times \mathbb{R}^n \times S^n : |p| \le R, |X| \le R\right\} \forall R < \infty$.

Remark 11.18 Using the elementary inequalities,

$$\left|\sup\left(.\right) - \sup\left(*\right)\right|, \left|\inf\left(.\right) - \inf\left(*\right)\right| \le \sup\left|.-*\right|$$

one immediately sees that if $F_\gamma, F_{\gamma,\beta}$ satisfy (11.15) for γ, β in some index set with a common modulus θ, then $\inf_\gamma F_\gamma, \sup_\beta \inf_\gamma F_{\beta,\gamma}$, etc. again satisfy (11.15). Similar remarks apply to the uniform continuity property; provided there exists, for any $R < \infty$, a common modulus of continuity σ_R for all $F_\gamma, F_{\gamma,\beta}$ restricted to B_R.

To state our key assumption on F we need some preliminary remark on the transformation behaviour of

$$Du = \left(\partial_1 u, \ldots, , \partial_n u\right), \ D^2 u = \left(\partial_{ij}u\right)_{i,j=1,\ldots,n}$$

under change of coordinates on $\mathbb{R}^n$ where $u = u\left(t,\cdot\right)$, for fixed t. Let us allow the change of coordinates to depend on t, say $v\left(t,\cdot\right) := u\left(t,\phi_t\left(\cdot\right)\right)$ where $\phi_t : \mathbb{R}^n \to \mathbb{R}^n$ is a diffeomorphism. Differentiating $v\left(t,\phi_t^{-1}\left(\cdot\right)\right) = u\left(t,\cdot\right)$ twice, followed by evaluation at $\phi_t\left(y\right)$, we have, with summation over repeated indices,

$$\begin{aligned}
\partial_i u\left(t,\phi_t\left(x\right)\right) &= \partial_k v\left(t,x\right) \partial_i \phi_t^{-1;k}|_{\phi_t(x)} \\
\partial_{ij} u\left(t,\phi_t\left(x\right)\right) &= \partial_{kl} v\left(t,x\right) \partial_i \phi_t^{-1;k}|_{\phi_t(x)} \partial_j \phi_t^{-1;l}|_{\phi_t(x)} \\
&\quad + \partial_k v\left(t,x\right) \partial_{ij} \phi_t^{-1;k}|_{\phi_t(x)}.
\end{aligned}$$

---

[9]... which, en passant, implies degenerate ellipticity, cf. page 18 in [34].
[10]See e.g. [34, (3.14) and Section 8], [53, Section V.7, V.8] or the appendix of [24].

We shall write this, somewhat imprecisely[11] but conveniently, as

$$Du|_{\phi_t(x)} = \left\langle Dv|_x, D\phi_t^{-1}|_{\phi_t(x)} \right\rangle, \tag{11.16}$$

$$D^2u|_{\phi_t(x)} = \left\langle D^2v|_x, D\phi_t^{-1}|_{\phi_t(x)} \otimes D\phi_t^{-1}|_{\phi_t(x)} \right\rangle + \left\langle Dv|_x, D^2\phi_t^{-1}|_{\phi_t(x)} \right\rangle.$$

Let us now introduce $\Phi^{(k)}$ as the class of all flows of C^k-diffeomorphisms of $\mathbb{R}^n$, $\phi = (\phi_t : t \in [0,T])$, such that $\phi_0 = \text{Id } \forall \phi \in \Phi^{(k)}$ and such that ϕ_t and ϕ_t^{-1} have k bounded derivatives, uniformly in $t \in [0,T]$. Since $\Phi^{(k)} \subset \mathcal{D}_k(\mathbb{R}^n)$ we inherit a natural notion of convergence: $\phi(n) \to \phi$ in $\Phi^{(k)}$ iff for all multi-indices α with $|\alpha| \leq k$,

$$\partial_\alpha \phi(n) \to \partial_\alpha \phi_t, \ \partial_\alpha \phi(n)^{-1} \to \partial_\alpha \phi_t^{-1} \text{ locally uniformly in } [0,T] \times \mathbb{R}^n.$$

Definition 11.19 ($\Phi^{(k)}$-invariant comparison) Let $k \geq 2$ and

$$F^\phi\left((t,x,p,X)\right) := F\left(t, \phi_t(x), \left\langle p, D\phi_t^{-1}|_{\phi_t(x)} \right\rangle,\right.$$
$$\left. \left\langle X, D\phi_t^{-1}|_{\phi_t(x)} \otimes D\phi_t^{-1}|_{\phi_t(x)} \right\rangle + \left\langle p, D^2\phi_t^{-1}|_{\phi_t(x)} \right\rangle \right). \tag{11.17}$$

We say that $\partial_t = F$ satisfies $\Phi^{(k)}$-invariant comparison if, for every $\phi \in \Phi^{(k)}$, comparison holds for BUC solutions of $\partial_t - F^\phi = 0$. □

Example 11.20 (F linear) Suppose that $\sigma(t,x) : [0,T] \times \mathbb{R}^n \to \mathbb{R}^{n \times n'}$ and $b(t,x) : [0,T] \times \mathbb{R}^n \to \mathbb{R}^n$ are bounded, continuous in t and Lipschitz continuous in x, uniformly in $t \in [0,T]$. If $F(t,x,p,X) = \text{Tr}\left[\sigma(t,x)\sigma(t,x)^T X\right] + b(t,x) \cdot p$, then $\Phi^{(3)}$-invariant comparison holds. Although this is a special case of the following example, let us point out that F^ϕ is of the same form as F with σ, b replaced by

$$\sigma^\phi(t,x)_m^k = \sigma_m^i(t, \phi_t(x)) \partial_i \phi_t^{-1;k}|_{\phi_t(x)}, \ k = 1, \ldots, n; m = 1, \ldots, n'$$

$$b^\phi(t,x)^k = \left[b^i(t, \phi_t(x)) \partial_i \phi_t^{-1;k}|_{\phi_t(x)}\right]$$
$$+ \sum_{i,j}\left(\sigma_m^i \sigma_m^j \partial_{ij} \phi_t^{-1;k}|_{\phi_t(y)}\right), \ k = 1, \ldots, n.$$

By defining properties of flows of diffeomorphisms, $t \mapsto \partial_i \phi_t^{-1;k}|_{\phi_t(x)}$, $\partial_{ij}\phi_t^{-1;k}|_{\phi_t(y)}$ is continuous and the C^3-boundedness assumption inherent in our definition of $\Phi^{(3)}$ ensures that σ^ϕ, b^ϕ are Lipschitz in x, uniformly in $t \in [0,T]$. It is then easy to see (cf. the argument of [53, Lemma 7.1]) that F^ϕ satisfies condition 11.17 for every $\phi \in \Phi^{(3)}$. This implies that $\Phi^{(3)}$-invariant comparison holds for BUC solutions of $\partial_t - F^\phi = 0$.

---

[11] Strictly speaking, one should view $\left(Du, D^2u\right)|.$ as a *second-order* cotangent vector, the pull-back of $\left(Dv, D^2v\right)|_x$ under ϕ_t^{-1}.

Exercise 11.21 (F quasi-linear) *Let*

$$F(t, x, p, X) = \mathrm{Tr}\left[\sigma(t, x, p)\,\sigma(t, x, p)^T\, X\right] + b(t, x, p). \qquad (11.18)$$

(i) Assume that $b = b(t, x, p) : [0, T] \times \mathbb{R}^n \times \mathbb{R}^n \to \mathbb{R}$ is bounded, continuous and Lipschitz continuous in x and p, uniformly in $t \in [0, T]$.
(ii) Assume that $\sigma = \sigma(t, x, p) : [0, T] \times \mathbb{R}^n \times \mathbb{R}^n \to \mathbb{R}^{n \times n'}$ is a continuous, bounded map such that $\sigma(t, \cdot, p)$ is Lipschitz continuous, uniformly in $(t, p) \in [0, T] \times \mathbb{R}^n$; assume also existence of a constant $c > 0$, such that

$$\forall p, q \in \mathbb{R}^n : |\sigma(t, x, p) - \sigma(t, x, q)| \le c \frac{|p - q|}{1 + |p| + |q|} \qquad (11.19)$$

for all $t \in [0, T]$ and $x \in \mathbb{R}^n$. Show that under these assumptions $\Phi^{(3)}$-invariant comparison holds for $\partial_t = F$.

Example 11.22 (F of Hamilton–Jacobi–Bellman type) From Example 11.20 and Remark 11.18, we see that $\Phi^{(3)}$-invariant comparison holds when F is given by

$$F(t, x, p, X) = \inf_{\gamma \in \Gamma} \left\{ \mathrm{Tr}\left[\sigma(t, x; \gamma)\,\sigma(t, x; \gamma)^T\, X\right] + b(t, x; \gamma) \cdot p \right\},$$

the usual non-linearity in the Hamilton–Jacobi–Bellman equation, whenever the conditions in Example 11.20 are satisfied uniformly with respect to $\gamma \in \Gamma$. More generally, one can take the infimum of quasi-linear F_γ, provided the conditions in Exercise 11.21 are satisfied uniformly.

Example 11.23 (F of Isaac type) Similarly, $\Phi^{(3)}$-invariant comparison holds for

$$F(t, x, p, X) = \sup_\beta \inf_\gamma \left\{ \mathrm{Tr}\left[\sigma(t, x; \beta, \gamma)\,\sigma(t, x; \beta, \gamma)^T\, X\right] + b(t, x; \beta, \gamma) \cdot p \right\}$$

(such non-linearities arise in the Isaac equation in the theory of differential games), and more generally

$$\begin{aligned} F(t, x, p, X) \\ = \sup_\beta \inf_\gamma \left\{ \mathrm{Tr}\left[\sigma(t, x, p; \beta, \gamma)\,\sigma(t, x, p; \beta, \gamma)^T \cdot X\right] + b(t, x, p; \beta, \gamma) \right\} \end{aligned}$$

whenever the conditions in Examples 11.20 and 11.21 are satisfied uniformly with respect to $\beta \in \mathcal{B}$ and $\gamma \in \Gamma$, where $\mathcal{B}$ and Γ are arbitrary index sets. □

Lemma 11.24 *Let $z : [0, T] \to \mathbb{R}^d$ be smooth and assume that we are given Lip^γ-vector fields $V = (V_1, \ldots, V_d)$ with $\gamma > 3$. Then the ODE*

$$dy_t = V(y_t)\,dz_t, \quad t \in [0, T]$$

has a unique solution flow $\phi = \phi^z \in \Phi^{(3)}$.

Proof. This follows directly from Proposition 11.11 applied with $p = 1$. A direct ODE proof, building on Corollary 4.9 and then arguing as in the proof of Proposition 11.11 is also not difficult (and actually shows that C^3-boundedness of V is enough here). ∎

Proposition 11.25 *Let z, V and ϕ be as in Lemma 11.24. Then u is a viscosity sub- (resp. super-) solution (always assumed BUC) of*

$$\dot{u}(t,x) = F\left(t, x, Du, D^2 u\right) - Du(t,x) \cdot V(x) \dot{z}_t \qquad (11.20)$$

if and only if $v(t,x) := u(t, \phi_t(x))$ is a viscosity sub- (resp. super-) solution of

$$\dot{v}(t,x) = F^\phi\left(t, x, Dv, D^2 v\right) \qquad (11.21)$$

where F^ϕ was defined in (11.17).

Proof. Set $y = \phi_t(x)$. When u is a classical sub-solution, it suffices to use the chain rule and definition of F^ϕ to see that

$$
\begin{aligned}
\dot{v}(t,x) &= \dot{u}(t,y) + Du(t,y) \cdot \dot{\phi}_t(x) = \dot{u}(t,y) + Du(t,y) \cdot V(y) \dot{z}_t \\
&\leq F\left(t, y, Du(t,y), D^2 u(t,y)\right) = F^\phi\left(t, x, Dv(t,x), D^2 v(t,x)\right).
\end{aligned}
$$

The case when u is a viscosity sub-solution of (11.20) is not much harder: suppose that $(\bar{t}, \bar{x})$ is a maximum of $v - \xi$, where $\xi \in C^2\left((0,T) \times \mathbb{R}^n\right)$ and define $\psi \in C^2\left((0,T) \times \mathbb{R}^n\right)$ by $\psi(t,y) = \xi\left(t, \phi_t^{-1}(y)\right)$. Set $\bar{y} = \phi_{\bar{t}}(\bar{x})$ so that

$$F\left(\bar{t}, \bar{y}, D\psi(\bar{t}, \bar{y}), D^2 \psi(\bar{t}, \bar{y})\right) = F^\phi\left(\bar{t}, \bar{x}, D\xi(\bar{t}, \bar{x}), D^2 \xi(\bar{t}, \bar{x})\right).$$

Obviously, $(\bar{t}, \bar{y})$ is a maximum of $u - \psi$, and since u is a viscosity sub-solution of (11.20) we have

$$\dot{\psi}(\bar{t}, \bar{y}) + D\psi(\bar{t}, \bar{y}) V(\bar{y}) \dot{z}(\bar{t}) \leq F\left(\bar{t}, \bar{y}, D\psi(\bar{t}, \bar{y}), D^2 \psi(\bar{t}, \bar{y})\right).$$

On the other hand, $\xi(t,x) = \psi(t, \phi_t(x))$ implies $\dot{\xi}(\bar{t}, \bar{x}) = \dot{\psi}(\bar{t}, \bar{y}) + D\psi(\bar{t}, \bar{y}) V(\bar{y}) \dot{z}(\bar{t})$ and putting things together we see that

$$\dot{\xi}(\bar{t}, \bar{x}) \leq F^\phi\left(\bar{t}, \bar{x}, D\xi(\bar{t}, \bar{x}), D^2 \xi(\bar{t}, \bar{x})\right),$$

which says precisely that v is a viscosity sub-solution of (11.21). Replacing maximum by minimum and $\leq$ by $\geq$ in the preceding argument, we see that if u is a super-solution of (11.20), then v is a super-solution of (11.21). Conversely, the same arguments show that if v is a viscosity sub- (resp. super-) solution for (11.21), then $u(t,y) = v\left(t, \phi^{-1}(y)\right)$ is a sub- (resp. super-) solution for (11.20). ∎

We can now give the proof of the main result.

Proof. (Theorem 11.16) Using Lemma 11.24, we see that $\phi^\varepsilon \equiv \phi^{z^\varepsilon}$, the solution flow to $dy = V(y) \, dz^\varepsilon$, is an element of $\Phi \equiv \Phi^{(3)}$. Set $F^\varepsilon := F^{\phi^\varepsilon}$. From Proposition 11.25, we know that u^ε is a solution to

$$du^\varepsilon = F\left(t, y, Du^\varepsilon, D^2 u^\varepsilon\right) dt - Du^\varepsilon(t, y) \cdot V(y) \, dz_t^\varepsilon, \quad u^\varepsilon(0, \cdot) = u_0^\varepsilon$$

if and only if v^ε is a solution to $\partial_t - F^\varepsilon = 0$. By assumption of Φ-invariant comparison,

$$\left| v^\varepsilon - \hat{v}^\varepsilon \right|_{\infty; \mathbb{R}^n \times [0,T]} \leq \left| v_0 - \hat{v}_0 \right|_{\infty; \mathbb{R}^n},$$

where $v^\varepsilon, \hat{v}^\varepsilon$ are viscosity solutions to $\partial_t - F^\varepsilon = 0$. Let $\phi^{\mathbf{z}}$ denote the solution flow to the rough differential equation

$$dy = V(y) \, d\mathbf{z}.$$

Thanks to $\mathrm{Lip}^{\gamma+2}$-regularity of the vector fields $\phi^{\mathbf{z}} \in \Phi$, and in particular a flow of C^3-diffeomorphisms. Set $F^{\mathbf{z}} = F^{\phi^{\mathbf{z}}}$. The "universal" limit theorem [120] holds; in fact, on the level of flows of diffeomorphisms (see [119] and the rest of this chapter for more details) it tells us that, since z^ε tends to $\mathbf{z}$ in rough path sense,

$$\phi^\varepsilon \to \phi^{\mathbf{z}} \text{ in } \Phi$$

so that, by continuity of F (more precisely: uniform continuity on compacts), we easily deduce that

$$F^\varepsilon \to F^{\mathbf{z}} \text{ locally uniformly.}$$

From the "Barles–Perthame" method of semi-relaxed limits (Lemma 6.1 and Remarks 6.2, 6.3 and 6.4 in [34], see also [53]), the pointwise (relaxed) limits

$$\bar{v} \quad := \quad \limsup{}^* v^\varepsilon,$$
$$\underline{v} \quad := \quad \liminf{}_* v^\varepsilon$$

are viscosity (sub- resp. super-) solutions to $\partial_t - F^{\mathbf{z}} = 0$, with identical initial data. As the latter equation satisfies comparison, one has trivially uniqueness and hence $v := \bar{v} = \underline{v}$ is the unique (and continuous, since $\bar{v}, \underline{v}$ are upper resp. lower semi-continuous) solution to

$$\partial_t v = F^{\mathbf{z}} v, \quad v(0, \cdot) = u_0(\cdot).$$

Moreover, using a simple Dini-type argument (e.g. [34, p. 35]) one sees that this limit must be uniform on compacts. It follows that v is the unique solution to

$$\partial_t v = F^{\mathbf{z}} v, \quad v(0, \cdot) = u_0(\cdot)$$

(hence does not depend on the approximating sequence to $\mathbf{z}$) and the proof of (i) is finished by setting

$$u^{\mathbf{z}}(t, x) := v\left(t, (\phi_t^{\mathbf{z}})^{-1}(x)\right).$$

(ii) The comparison $|u^{\mathbf{z}} - \hat{u}^{\mathbf{z}}|_{\infty;[0,T] \times \mathbb{R}^n} \leq |u_0 - \hat{u}_0|_{\infty;\mathbb{R}^n}$ is a simple consequence of comparison for $v, \hat{v}$ (solutions to $\partial_t v = F^{\mathbf{z}} v$). At last, to see (iii), we argue in the very same way as in (i), starting with

$$F^{\mathbf{z}_n} \to F^{\mathbf{z}} \text{ locally uniformly}$$

to see that $v^n \to v$ locally uniformly, i.e. uniformly on compacts. ∎

11.4 Comments

Flows of RDE solutions were first studied by Lyons and Qian [119], see also [120]; perturbations in the driving signal in [118]. Theorem 11.6 appears to be new; Corollary 11.7 was established by Li and Lyons in [109]. Exercise 11.8 is the rough path generalization of an SDE regularity result of Kusuoka [98]; his result is recovered upon taking the driving rough path to be enhanced Brownian motion. The limit Theorem 11.12 is the rough path generalization of the corresponding limit theorems for stochastic flows as discussed in Ikeda and Watanabe [88], Kusuoka [98] or Malliavin [124].

Our definition of $\mathcal{D}_k(\mathbb{R}^e)$, equation (11.5), follows Ben Arous and Castell [11]; see also Kusuoka [98]. Corollary 11.14 is somewhat cruder than Theorem 11.12 but helpful in making the link to various works on stochastic flows, including Kusuoka [98] and Li and Lyons [109]. Section 11.3 on rough partial differential equations

$$du = F\left(t, x, Du, D^2 u\right) dt + H\left(x, Du\right) d\mathbf{z},$$

with F fully non-linear but $H = (H_1, \ldots, H_d)$ linear in Du is taken from Caruana *et al.* [24]; the case when F and H are both linear (with respect to the derivatives of u) was considered in Caruana and Friz [23]. From the works of Lions and Souganidis [111–113] we conjecture that the present results extend to sufficiently smooth but non-linear H. Other classes of rough partial differential equations have been studied; in Gubinelli *et al.* [77] the authors consider the evolution problem $dY = -AY dt + B(Y) dX$ where $-A$ is the generator of an analytic semi-group; the solution is understood in *mild* sense, with the integrals involved being of Young type. An extension to a genuine rough setting (i.e. beyond the Young setting) is discussed in Gubinelli and Tindel [76].

12

RDEs with drift and other topics

In the last two chapters we discussed various properties of rough differential equations of the form

$$dy = V(y)d\mathbf{x},$$

where $V = (V_1, \ldots, V_d)$ denotes, as usual, a collection of vector fields. In applications, the term $V(y)d\mathbf{x}$ may model a state-dependent perturbation of the classical ODE $\dot{y} = W(y)$. This leads to differential equations of the form

$$dy = V(y)d\mathbf{x} + W(y)\,dt$$

where $W(dy)\,dt$ is viewed as a drift term. To some extent, no new theory is required here. It suffices to replace V by $\tilde{V} = (V_1, \ldots, V_d, W)$ and the geometric p-rough path $\mathbf{x}$ by the "space-time" rough path $\tilde{\mathbf{x}} : t \mapsto S_{[p]}(\mathbf{x}, t)$, as discussed in Section 9.4. The downside of this approach is that one has to impose the *same* regularity assumptions on V and W, which is wasteful.[1] We shall see in this chapter that the regularity assumptions on W can be significantly weakened. Moreover, our estimates will be important in their own right as they will lead us to a deterministic understanding of McShane-type approximation results.[2]

12.1   RDEs with drift terms

It is helpful to consider drift terms of the more general form $W(y)\,dh$ where $W = (W_1, \ldots, W_{d'})$ and $h \in C^{q\text{-var}}\left([0, T], \mathbb{R}^{d'}\right)$. It is natural to assume that the drift term signal h has better regularity than x; which is to say that $q \leq p$. A well-defined Young pairing $S_{[p]}(\mathbf{x}, h)$ is still necessary and so we assume that

$$1/p + 1/q > 1.$$

It follows that $q \in [1, 2)$ and hence $[q] = 1$. This implies that h is actually a geometric q-rough path, say

$$\mathbf{h} \in C^{q\text{-var}}\left([0, T], G^{[q]}\left(\mathbb{R}^{d'}\right)\right).$$

---

[1] From ODE theory, one expects that $W \in \text{Lip}^1$ will suffice for uniqueness, in contrast to $V \in \text{Lip}^p$ needed for RDE uniqueness ...

[2] To be discussed in Sections 12.2 and 13.3.4.

In fact, we shall find it convenient *not* to impose that $q \leq p$, since this will allow us to keep the symmetry between $\mathbf{x}$ and $\mathbf{h}$. The object of study is then the rough differential equation of the form

$$dy = V(y)d\mathbf{x} + W(y)d\mathbf{h}, \tag{12.1}$$

where $\mathbf{x}$ is a weak geometric p-rough path, and $\mathbf{h}$ is a weak geometric q-rough path.[3] In many applications, $h_t = t$ or h is of bounded variation, i.e. $q = 1$. We note again that (12.1) can be rewritten as a standard RDE driven by the geometric p-rough path $S_{[p]}(\mathbf{x} \oplus \mathbf{h})$, along the vector fields (V, W), at the price of suboptimal regularity assumptions on V and W. Our direct analysis of (12.1) starts with the following

Definition 12.1 Let $p, q \geq 1$ such that $1/p + 1/q > 1$. Let $\mathbf{x} \in C^{p\text{-var}}$ $\left([0,T], G^{[p]}\left(\mathbb{R}^d\right)\right)$ be a weak geometric p-rough path, and $\mathbf{h} \in C^{q\text{-var}}$ $\left([0,T], G^{[q]}\left(\mathbb{R}^{d'}\right)\right)$ be a weak geometric q-rough path. We say that $y \in C\left([0,T], \mathbb{R}^e\right)$ is a solution to the rough differential equation (short: an RDE solution) driven by $(\mathbf{x}, \mathbf{h})$ along the collection of $\mathbb{R}^e$-vector fields $\left((V_i)_{1 \leq i \leq d}, (W_j)_{1 \leq j \leq d'}\right)$ and started at y_0 if there exists a sequence (x^n, h^n) $\subset C^{1\text{-var}}\left([0,T], \mathbb{R}^d\right) \times C^{1\text{-var}}\left([0,T], \mathbb{R}^{d'}\right)$ such that

$$\sup_n \left\| S_{[p]}(x^n) \right\|_{p\text{-var};[0,T]} + \left\| S_{[q]}(h^n) \right\|_{q\text{-var};[0,T]} < \infty,$$

$$\lim_{n \to \infty} d_{0;[0,T]}(S_{[p]}(x^n), \mathbf{x}) = 0 \text{ and } \lim_{n \to \infty} d_{0;[0,T]}\left(S_{[q]}(h^n), \mathbf{h}\right) = 0$$

and ODE solutions $y^n \in \pi_{(V,W)}(0, y_0; (x^n, h^n))$ such that

$$y^n \to y \text{ uniformly on } [0,T] \text{ as } n \to \infty.$$

The (formal) equation $dy = V(y)d\mathbf{x} + W(y)d\mathbf{h}$ is referred to as a rough differential equation with drift (short: an RDE with drift). $\qquad \square$

This definition generalizes immediately to time intervals $[s,T]$ and we define $\pi_{(V,W)}(s, y_s; (\mathbf{x}, h)) \subset C\left([s,T], \mathbb{R}^e\right)$ to be the set of all solutions to the above RDE with drift starting at y_s at time s, and in case of uniqueness, $\pi_{(V,W)}(s, y_s; (\mathbf{x}, \mathbf{h}))$ is *the* solution of the RDE with drift.

We will also be interested in full RDE solutions with drift. Let us define this concept.

Definition 12.2 Let $p, q \geq 1$ such that $1/p + 1/q > 1$. Let $\mathbf{x} \in C^{p\text{-var}}\left([0,T], G^{[p]}\left(\mathbb{R}^d\right)\right)$ be a weak geometric p-rough path, and $\mathbf{h} \in C^{q\text{-var}}\left([0,T], G^{[q]}\left(\mathbb{R}^{d'}\right)\right)$ be a weak geometric q-rough path. We say that $\mathbf{y} \in C\left([0,T],\right.$

---

[3]In the case that $q > p$, one sees that $1/p + 1/q > 1$ implies $p \in [1,2)$ and so $V(y)d\mathbf{x}$ plays the role of the drift term.

$G^{\lfloor \max(p,q) \rfloor} \left(\mathbb{R}^e \right)$) is a solution to the full rough differential equation (short: a full RDE solution) driven by $(\mathbf{x}, \mathbf{h})$ along the collection of $\mathbb{R}^e$-vector fields $\left((V_i)_{1 \le i \le d} , (W_j)_{1 \le j \le d'} \right)$ and started at y_0 if there exists a sequence $(x^n, h^n)_n$ in $C^{1\text{-var}} \left([0,T] , \mathbb{R}^d \right) \times C^{1\text{-var}} \left([0,T] , \mathbb{R}^{d'} \right)$ such that

$$\sup_n \left\| S_{[p]} \left(x^n \right) \right\|_{p\text{-var}} + \left\| S_{[q]} \left(h^n \right) \right\|_{q\text{-var}} < \infty,$$

$$\lim_{n \to \infty} d_0 (S_{[p]} \left(x^n \right), \mathbf{x}) = 0 \text{ and } \lim_{n \to \infty} d_0 \left(S_{[q]} \left(h^n \right), \mathbf{h} \right) = 0$$

and ODE solutions $y_n \in \pi_{(V,W)} \left(0, \pi_1 \left(y_0 \right); (x^n, h^n) \right)$ such that

$$y_0 \otimes S_{[\max(p,q)]} \left(y^n \right) \to \mathbf{y} \text{ uniformly on } [0,T] \text{ as } n \to \infty .$$

The (formal) equation $d\mathbf{y} = V \left(\mathbf{y} \right) d\mathbf{x} + W \left(\mathbf{y} \right) d\mathbf{h}$ is referred to as a full rough differential equation with drift (short: a full RDE with drift).   □

This definition generalizes immediately to time intervals $[s,T]$ and we define $\pi_{(V,W)} \left(s, y_s; (\mathbf{x}, \mathbf{h}) \right) \subset C \left([s,T], G^{[\max(p,q)]} \left(\mathbb{R}^e \right) \right)$ to be the set of all solutions to the above full RDE with drift starting at y_s at time s, and in case of uniqueness, $\pi_{(V,W)} \left(s, y_s; (\mathbf{x}, \mathbf{h}) \right)$ is *the* solution of the full RDE with drift.

12.1.1   Existence

We start by comparing ODE solutions with drift with their counterparts where we remove the drift.

Lemma 12.3 *Assume that*
(i) $V = (V_i)_{1 \le i \le d}$ is a collection of vector fields in $\mathrm{Lip}^{\gamma-1} (\mathbb{R}^e)$ with $\gamma > 1$;
(i bis) $W = (W_j)_{1 \le j \le d'}$ is a collection of vector fields in $\mathrm{Lip}^{\beta-1} (\mathbb{R}^e)$ with $\beta > 1$;
(ii) s,t are some elements of $[0,T]$;
(iii) $y_s \in \mathbb{R}^e$ is an initial condition;
(iv) x and h are two paths in $C^{1\text{-var}} \left([s,t], \mathbb{R}^d \right)$ and $C^{1\text{-var}} \left([s,t], \mathbb{R}^{d'} \right)$;
(v) $\ell_x \ge 0$ is a bound on $|V|_{\mathrm{Lip}^{\gamma-1}} \int_s^t |dx_r|$ and $\ell_h \ge 0$ is a bound on $|W|_{\mathrm{Lip}^{\beta-1}} \int_s^t |dh_r|$.
Then we have, for some constant $C = C (\gamma, \beta)$,

$$\left| \pi_{(V,W)} \left(s, y_s; (x,h) \right)_{s,t} - \pi_{(V)} \left(s, y_s; x \right)_{s,t} - \pi_{(W)} \left(s, y_s; h \right)_{s,t} \right|$$
$$\le C \left(\ell_h^{\beta} + \ell_x \ell_h^{\gamma-1} + \ell_x \ell_h + \ell_x^{\beta-1} \ell_h + \ell_x^{\gamma} \right) \exp \left(C \left(\ell_x + \ell_h \right) \right).$$

In the case when $\gamma \geq 2$ and $\beta \geq 2$, we have

$$\left| \pi_{(V,W)} \left(s, y_s; (x, h)\right)_{s,t} - \pi_{(V)} \left(s, y_s; x\right)_{s,t} - \pi_{(W)} \left(s, y_s; h\right)_{s,t} \right|$$
$$\leq C \ell_x \ell_h \exp\left(C \left(\ell_x + \ell_h\right)\right).$$

Proof. Without loss of generality, we assume $(s, t) = (0, 1)$.

<u>Case 1:</u> We first assume that $\gamma \geq 2$ and $\beta \geq 2$. Define, for $u \in [0, 1]$, $z_u = \pi_{(V,W)} \left(0, y_0; (x, h)\right)_u$, $y_u^x = \pi_{(V)} \left(0, y_0; x\right)_u$ and $y_u^h = \pi_{(W)} \left(0, y_0; h\right)_u$. Define also

$$e_u = \left| z_{0,u} - \left(y_{0,u}^x + y_{0,u}^h\right) \right|.$$

First observe that

$$\left| y_{0,u}^x \right| \leq c \ell_x \text{ and } \left| y_{0,u}^h \right| \leq c \ell_h.$$

Then, by definition of y and z, for $u \in [0, 1]$,

$$
\begin{aligned}
e_u &= \left| \int_0^u \left\{ V\left(z_r\right) - V\left(y_r^x\right) \right\} dx_r + \int_0^u \left\{ W\left(z_r\right) - W\left(y_r^h\right) \right\} dh_r \right| \\
&\leq c_2 \left| V \right|_{\mathrm{Lip}^{\gamma}-1} \int_0^u \left| z_r - y_r^x \right| \cdot \left| dx_r \right| + c_2 \left| W \right|_{\mathrm{Lip}^{\beta}-1} \int_0^u \left| z_r - y_r^h \right| \cdot \left| dh_r \right| \\
&\leq c_2 \int_0^u \left| z_{0,r} - y_{0,r}^x - y_{0,r}^h \right| \cdot \left(\left| V \right|_{\mathrm{Lip}^{\gamma}-1} \left| dx_r \right| + \left| W \right|_{\mathrm{Lip}^{\beta}-1} \left| dh_r \right| \right) \\
&\quad + c_2 \left| V \right|_{\mathrm{Lip}^{\gamma}-1} \int_0^u \left| y_{0,r}^h \right| \cdot \left| dx_r \right| + c_2 \left| W \right|_{\mathrm{Lip}^{\beta}-1} \int_0^u \left| y_{0,r}^x \right| \cdot \left| dh_r \right| \\
&\leq c_2 \int_0^u e_r \left(\left| V \right|_{\mathrm{Lip}^{\gamma}-1} \left| dx_r \right| + \left| W \right|_{\mathrm{Lip}^{\beta}-1} \left| dh_r \right| \right) + c_4 \ell_x \ell_h.
\end{aligned}
$$

We conclude the proof by using Gronwall's inequality.

<u>Case 2:</u> We still assume $\gamma \geq 2$, but $\beta < 2$. As $\left| \pi_{(W)} \left(0, y_0; h\right)_{s,t} - W\left(y_0\right) h_{s,t} \right|$ $\leq c_1 \ell_h^{\beta}$, we see that we can replace $\pi_{(W)} \left(s, y_s; h\right)_{s,t}$ by $W\left(y_0\right) h_{s,t}$. Define this time $e_u = \left| z_u - y_u - W\left(y_0\right) h_{0,u} \right|$, and observe that $\left| z_{0,1} \right| \leq$

$c_1 \left(\ell_x + \ell_h \right)$. Then, by definition of y and z, for $u \in [0,1]$,

$$
\begin{aligned}
e_u &= \left| \int_0^u \{V(z_r) - V(y_r)\} \, dx_r + \int_0^u \{W(z_r) - W(z_0)\} \, dh_r \right| \\
&\leq c_2 \left| V \right|_{\mathrm{Lip}^{\gamma - 1}} \int_0^u |z_r - y_r| \cdot |dx_r| + c_2 \left| W \right|_{\mathrm{Lip}^{\beta - 1}} \sup_r |z_{0,r}|^{\beta - 1} \int_0^u |dh_r| \\
&\leq c_2 \left| V \right|_{\mathrm{Lip}^{\gamma - 1}} \int_0^u e_r |dx_r| + c_2 \left| W \right|_{\mathrm{Lip}^{\beta - 1}} \sup_r |z_{0,r}|^{\beta - 1} \int_0^u |dh_r| \\
&\quad + c_2 \left(|V|_{\mathrm{Lip}^{\gamma - 1}} \int_0^u |dx_r| \right) \left(|W|_{\mathrm{Lip}^{\beta - 1}} \int_0^u |dh_r| \right) \\
&\leq c_3 \left| V \right|_{\mathrm{Lip}^{\gamma - 1}} \int_0^u e_r |dx_r| + c_3 \ell_h \left((\ell_x + \ell_h)^{\beta - 1} + \ell_x \right) \\
&\leq c_3 \left| V \right|_{\mathrm{Lip}^{\gamma - 1}} \int_0^u e_r |dx_r| + c_4 \left(\ell_h^\beta + \ell_h \ell_x^{\beta - 1} \right) \exp\left(c_4 \ell_x \right).
\end{aligned}
$$

We conclude this case with Gronwall's lemma once again.

The case $\gamma < 2$ and $\beta \geq 2$ is of course the symmetric case.

<u>Case 3:</u> We finally consider the case $\gamma < 2$ and $\beta < 2$, which is the simplest case. As

$$
\left| \pi_{(W)} (0, y_0; h)_{0,1} - W(y_0) h_{0,1} \right| \leq c_1 \ell_h^\beta,
$$

$$
\left| \pi_{(V)} (0, y_0; x)_{0,1} - V(y_0) x_{0,1} \right| \leq c_1 \ell_x^\gamma,
$$

we see that we can replace $\pi_{(W)} (0, y_0; h)_{0,1}$ by $W(y_0) h_{0,1}$ and $\pi_{(V)} (0, y_0; x)_{0,1}$ by $V(y_0) x_{0,1}$. But

$$
\begin{aligned}
&\left| \pi_{(V,W)} (0, y_0; (x,h))_{0,1} - V(y_0) x_{0,1} - W(y_0) h_{0,1} \right| \\
&\leq \left| \int_0^1 [V(z_u) - V(z_0)] \, dx_u \right| \\
&\quad + \left| \int_0^1 [W(z_u) - W(z_0)] \, dh_u \right| \\
&\leq c_4 (\ell_x + \ell_h)^{\gamma - 1} \ell_x + c_4 (\ell_x + \ell_h)^{\beta - 1} \ell_h \\
&\leq c_5 \left(\ell_h^\beta + \ell_x \ell_h^{\gamma - 1} + \ell_x^{\beta - 1} \ell_h + \ell_x^\gamma \right).
\end{aligned}
$$

That concludes the proof. ∎

It is now easy to give the following generalization of Lemma 10.5.

Lemma 12.4 (Lemma A_{drift}) *Assume that*

(i) $V = (V_i)_{1 \leq i \leq d}$ *is a collection of vector fields in* $\mathrm{Lip}^{\gamma - 1} (\mathbb{R}^e)$ *with* $\gamma > 1$;

(i bis) $W = (W_j)_{1 \leq j \leq d'}$ *is a collection of vector fields in* $\mathrm{Lip}^{\beta - 1} (\mathbb{R}^e)$ *with* $\beta > 1$;

(ii) $s < u$ *are some elements of* $[0, T]$;

(iii) $x, \tilde{x}$ are two paths in $C^{1\text{-var}}\left([s,u], \mathbb{R}^d\right)$ such that $S_{\lfloor\gamma\rfloor}(x)_{s,u} = S_{\lfloor\gamma\rfloor}(\tilde{x})_{s,u}$;
(iii bis) $h, \tilde{h}$ are two paths in $C^{1\text{-var}}\left([s,u], \mathbb{R}^{d'}\right)$ such that $S_{\lfloor\beta\rfloor}(h)_{s,u} = S_{\lfloor\beta\rfloor}\left(\tilde{h}\right)_{s,u}$;
(iv) $\ell_x \geq 0$ is a bound on $|V|_{\text{Lip}^{\gamma-1}} \max\left\{\int_s^u |dx|, \int_s^u |d\tilde{x}|\right\}$ and $\ell_h \geq 0$ is a bound on $|W|_{\text{Lip}^{\beta-1}} \max\left\{\int_s^u |dh|, \int_s^u \left|d\tilde{h}\right|\right\}$.
We then have, for some constant $C = C(\gamma, \beta)$,

$$\left|\pi_{(V,W)}(s, y_s; (x,h))_{s,u} - \pi_{(V,W)}\left(s, y_s; \left(\tilde{x}, \tilde{h}\right)\right)_{s,u}\right|$$
$$\leq C\left(\ell_h^\beta + \ell_x \ell_h^{\gamma-1} + \ell_x \ell_h + \ell_x^{\beta-1}\ell_h + \ell_x^\gamma\right) \exp\left(C\left(\ell_x + \ell_h\right)\right).$$

Proof. Write $\pi_{(V,W)}(s, y_s; (x,h))_{s,u} - \pi_{(V,W)}\left(s, y_s; \left(\tilde{x}, \tilde{h}\right)\right)_{s,u}$ as $\Delta_1 + \Delta_2 + \Delta_3$, where

$$\Delta_1 = \pi_{(V,W)}(s, y_s; (x,h))_{s,u} - \pi_{(V)}(s, y_s; x)_{s,u} - \pi_{(W)}(s, y_s; h)_{s,t}$$
$$- \left(\pi_{(V,W)}(s, y_s; (x,h))_{s,u} - \pi_{(V,W)}(s, y_s; \tilde{x})_{s,u} - \pi_{(W)}\left(s, y_s; \tilde{h}\right)_{s,t}\right),$$

and

$$\Delta_2 = \pi_{(V)}(s, y_s; x)_{s,u} - \pi_{(\tilde{V})}(s, y_s; \tilde{x})_{s,u}$$
$$\Delta_3 = \pi_{(W)}(s, y_s; h)_{s,t} - \pi_{(W)}\left(s, y_s; \tilde{h}\right)_{s,t}.$$

Lemma 12.3 gives

$$|\Delta_1| \leq c_1\left(\ell_h^\beta + \ell_x \ell_h^{\gamma-1} + \ell_x \ell_h + \ell_x^{\beta-1}\ell_h + \ell_x^\gamma\right) \exp\left(c_1 \ell_x\right),$$

Lemma 10.5 (which we called "Lemma A") gives

$$|\Delta_2| \leq c_2 \ell_x^\gamma \exp\left(c_2 \ell_x\right),$$

and

$$|\Delta_3| \leq c_3 \ell_h^\beta \exp\left(c_3 \ell_h\right).$$

The triangle inequality then finishes the proof. ∎

Lemma 12.5 (Lemma B_{drift}) *Assume that*
(i) $V = (V_i)_{1 \leq i \leq d}$ is a collection of vector fields in $\text{Lip}^{\gamma-1}(\mathbb{R}^e)$ with $\gamma > 1$;
(i bis) $W = (W_j)_{1 \leq j \leq d'}$ is a collection of vector fields in $\text{Lip}^{\beta-1}(\mathbb{R}^e)$ with $\beta > 1$;
(ii) $t < u$ are some elements of $[0, T]$;
(iii) $y_t, \tilde{y}_t \in \mathbb{R}^e$ (thought of as "time-t" initial conditions);
(iv) x is a path in $C^{1\text{-var}}\left([t, u], \mathbb{R}^d\right)$;

(iv bis) h is a path in $C^{\text{1-var}}\left([t,u],\mathbb{R}^{d'}\right)$;

(v) $\ell_x \geq 0$ is a bound on $|V|_{\text{Lip}^{\gamma-1}} \int_t^u |dx_r|$ and $\ell_h \geq 0$ is a bound on $|W|_{\text{Lip}^{\beta-1}} \int_t^u |dh_r|$.

Then, if $\pi_{(V)}(t,\cdot;x)$ denotes the unique solution to $dy = V(y)\,dx$ from some time-t initial condition, we have for some $C = C(\gamma,\beta)$,

$$\left| \pi_{(V,W)}(t,y_t;(x,h))_{t,u} - \pi_{(V,W)}(t,\tilde{y}_t;(x,h))_{t,u} \right|$$

$$\leq \begin{array}{l} C\,|y_t - \tilde{y}_t| \cdot \left(\ell_x^{\gamma-1} + \ell_h^{\beta-1}\right) \exp\left(C\left(\ell_x + \ell_h\right)\right) \\ + \left(\ell_h^\beta + \ell_x \ell_h^{\gamma-1} + \ell_x \ell_h + \ell_x^{\beta-1}\ell_h + \ell_x^\gamma\right) \exp\left(C\left(\ell_x + \ell_h\right)\right). \end{array}$$

Proof. Write once again $\pi_{(V,W)}(t,y_t;(x,h))_{t,u} - \pi_{(V,W)}(t,\tilde{y}_t;(x,h))_{t,u}$ as $\Delta_1 + \Delta_2 + \Delta_3$, where

$$\Delta_1 = \pi_{(V,W)}(t,y_t;(x,h))_{t,u} - \pi_{(V)}(t,y_t;x)_{t,u} - \pi_{(W)}(t,y_t;h)_{t,u}$$
$$- \left(\pi_{(V,W)}(t,\tilde{y}_t;(x,h))_{t,u} - \pi_{(V)}(t,\tilde{y}_t;x)_{t,u} - \pi_{(W)}(t,\tilde{y}_t;h)_{t,u}\right),$$

and

$$\Delta_2 = \pi_{(V)}(t,y_t;x)_{t,u} - \pi_{(V)}(t,\tilde{y}_t;x)_{t,u}$$
$$\Delta_3 = \pi_{(W)}(t,y_t;h)_{t,u} - \pi_{(W)}(t,\tilde{y}_t;h)_{t,u}.$$

Lemma 12.3 gives

$$|\Delta_1| \leq c_1 \left(\ell_h^\beta + \ell_h \ell_x^{\beta-1} + \ell_h^{\gamma-1}\ell_\beta + \ell_x^\gamma\right) \exp\left(c_1 \ell_x\right).$$

Then remark that Lemma 12.5 (Lemma B) in the case $\gamma \geq 2$ gives

$$|\Delta_2| \leq c_2 \left(|y_t - \tilde{y}_t| . \ell_x\right) \exp\left(c_2 \ell_x\right),$$

while inequality (10.12) in Exercise 10.12 easily leads to

$$|\Delta_2| \leq c_2 \left(|y_t - \tilde{y}_t| . \ell_x^{\gamma-1} + \ell_x^\gamma\right) \exp\left(c_2 \ell_x\right)$$

in the case $\gamma < 2$. We obtain similarly

$$|\Delta_2| \leq c_2 \left(|y_t - \tilde{y}_t| . \ell_h^{\beta-1} + \ell_h^\beta\right) \exp\left(c_2 \ell_h\right).$$

∎

We are now ready for our existence theorem:

Theorem 12.6 *Assume that $p,q,\gamma,\beta \in [1,\infty)$ are such that*

$$1/p + 1/q > 1 \tag{12.2}$$
$$\gamma > p \text{ and } \beta > q \tag{12.3}$$
$$\frac{\gamma-1}{q} + \frac{1}{p} > 1 \text{ and } \frac{1}{q} + \frac{\beta-1}{p} > 1. \tag{12.4}$$

(i) $V = (V_i)_{1 \leq i \leq d}$ *is a collection of vector fields in* $\mathrm{Lip}^{\gamma-1}(\mathbb{R}^e)$;

(i bis) $W = (W_i)_{1 \leq i \leq d'}$ *is a collection of vector fields in* $\mathrm{Lip}^{\beta-1}(\mathbb{R}^e)$;

(ii) (x_n) *is a sequence in* $C^{1\text{-var}}([0,T], \mathbb{R}^d)$, *and* $\mathbf{x}$ *is a weak geometric p-rough path such that*

$$\lim_{n \to \infty} d_{0;[0,T]}(S_{[p]}(x_n), \mathbf{x}) \text{ and } \sup_n \left\| S_{[p]}(x_n) \right\|_{p\text{-var};[0,T]} < \infty;$$

(ii bis) (h_n) *is a sequence in* $C^{1\text{-var}}([0,T], \mathbb{R}^{d'})$, *and* $\mathbf{h}$ *is a weak geometric q-rough path such that*

$$\lim_{n \to \infty} d_{0;[0,T]}(S_{[q]}(h_n), \mathbf{h}) \text{ and } \sup_n \left\| S_{[q]}(h_n) \right\|_{q\text{-var};[0,T]} < \infty;$$

(iii) $\mathbf{y}_0^n \in G^{[\max(p,q)]}(\mathbb{R}^e)$ *is a sequence converging to some* $\mathbf{y}_0$;

(iv) ω *is the control defined by*

$$\omega(s,t) = \left(|V|_{\mathrm{Lip}^{\gamma-1}} \|\mathbf{x}\|_{p\text{-var};[s,t]} \right)^p + \left(|W|_{\mathrm{Lip}^{\beta-1}} \|\mathbf{h}\|_{q\text{-var};[s,t]} \right)^q.$$

Then, at least along a subsequence, $\mathbf{y}_0^n \otimes S_{[\max(p,q)]}\left(\pi_{(V,W)}(0, \pi_1(\mathbf{y}_0^n); (x_n, h_n)) \right)$ *converges in uniform topology, and there exists a constant* C_1 *depending on* $p, q, \gamma,$ *and* β *such that for any limit point* $\mathbf{y}$, *and all* $s < t$ *in* $[0,T]$,

$$\|\mathbf{y}\|_{\max(p,q)\text{-var};[s,t]} \leq C_1 \left(\omega(s,t)^{1/\max(p,q)} \vee \omega(s,t) \right).$$

Finally, if $x^{s,t} : [s,t] \to \mathbb{R}^d$ *and* $h^{s,t} : [s,t] \to \mathbb{R}^{d'}$ *are two continuous paths of bounded variation such that*

$$S_{[p]}\left(x^{s,t} \right)_{s,t} = \mathbf{x}_{s,t} \text{ and } \int_s^t \left| dx_u^{s,t} \right| \leq K \|\mathbf{x}\|_{p\text{-var};[s,t]}$$

$$S_{[q]}\left(h^{s,t} \right) = \mathbf{h}_{s,t} \text{ and } \int_s^t \left| dh_u^{s,t} \right| \leq K \|\mathbf{h}\|_{q\text{-var};[s,t]}$$

for some constant K *then, for all* $s < t : \omega(s,t) \leq 1$, *there exists* C_2 *and* $\theta > 1$,

$$\left| \mathbf{y}_{s,t} - S_{[\max(p,q)]}\left(\pi_{(V,W)}(s, \pi_1(\mathbf{y}_s); (x^{s,t}, h^{s,t})) \right)_{s,t} \right| \leq C_2 \omega(s,t)^\theta,$$

$$\tag{12.5}$$

where C_2 *and* θ *depend on* p, q, γ, β *and* K.

Proof. The argument follows exactly the line of the proof of Lemma 10.7, Theorem 10.14, and Theorem 10.36. We adapt Lemma 10.7 by first assuming $\mathbf{x}$ and $\mathbf{h}$ to be the lift of smooth paths and concentrating on RDEs rather than full RDEs. We then define geodesics $h^{s,t}, x^{s,t}$ corresponding to the elements $\mathbf{h}_{s,t}$ and $\mathbf{x}_{s,t}$, and $\Gamma_{s,t}$ to be the difference of the ODE solution driven by $h^{s,t}, x^{s,t}$ and $\mathbf{h}, \mathbf{x}$. Then we bound $\Gamma_{s,u} - \Gamma_{s,t} - \Gamma_{t,u}$ using Lemma A_{drift} and Lemma B_{drift}, and conclude this first part with Lemma 10.59. When using Lemma A_{drift} and Lemma B_{drift}, the variables ℓ_x and ℓ_h are set to $\omega(s,u)^{1/p}$ and $\omega(s,u)^{1/q}$, respectively. To be able to use Lemma 10.59, all the powers in the expression

$$\omega(s,u)^{\beta/q} + \omega(s,u)^{\frac{\gamma-1}{q}+\frac{1}{p}} + \omega(s,u)^{1/q+1/p} + \omega(s,u)^{\frac{1}{q}+\frac{\beta-1}{p}} + \omega(s,u)^{\gamma/p}$$

(which comes to $\ell_h^\beta + \ell_x \ell_h^{\gamma-1} + \ell_x \ell_h + \ell_x^{\beta-1}\ell_h + \ell_x^\gamma$) must be strictly greater than 1. That explains the Conditions 12.2, 12.3, and 12.4. Equipped with the equivalent of Lemma 10.7, a limit argument similar to the one in Theorem 10.14 allows us to prove the current theorem, for the case of RDEs (rather than full RDEs). We then use similar arguments as the one in the proof of Theorem 10.36 to conclude the proof. ∎

Remark 12.7 The conditions on p, q, γ, β in Theorem 12.6 may look surprisingly complicated and it helps to play through a few special cases. To this end, let us note that the conditions have been stated in such a way as to emphasize the symmetric roles of these parameters. We may break this symmetry by assuming, without loss of generality, that $p \geq q$ in which case the first condition in (12.4) is seen to be redundant since

$$\frac{\gamma-1}{q} + \frac{1}{p} \geq \frac{\gamma-1}{p} + \frac{1}{p} = \frac{\gamma}{p} > 1.$$

The conditions on p, q, γ, β then reduce to

$$1/p + 1/q > 1, \quad \gamma > p \text{ and } \beta > \max\left\{q, 1 + p\left(1 - \frac{1}{q}\right)\right\}.$$

Let us look at three special cases:
(i) The case $q = 1$, frequently encountered in applications. The condition reduces to

$$\gamma > p \text{ and } \beta > q = 1,$$

which is natural when compared with the regularity assumptions for RDE existence.
(ii) The case $q = p$. The conditions now reduce to

$$p < 2, \quad \gamma > p \text{ and } \beta > p.$$

and we effectively consider RDEs (or better: Young ODEs) driven by $t \mapsto (x_t, h_t)$ of finite p-variation.
(iii) A good understanding of the term "$\max\left\{q, 1 + p\left(1 - \frac{1}{q}\right)\right\}$" is the content of the forthcoming Remark 12.12. □

The following corollary will be important as it often allows us to identify RDEs with drift. Examples will be seen in Section 12.2 below.

Corollary 12.8 (Euler estimate, RDE with drift) *In the setting of the previous Theorem 12.6, but focusing for simplicity on RDE solutions rather than full RDE solutions, we have for $M > 0$, for all s, t such that $\omega(s, t) \leq M$,*

$$\left| \pi_{(V,W)}(s, y_s; ((\mathbf{x}, \mathbf{h})))_{s,t} - \pi_{(V)}(s, y_s; x^{s,t})_{s,t} - \pi_{(W)}(s, y_s; h^{s,t})_{s,t} \right| \leq C_1 \omega(s, t)^\theta$$

for some $\theta > 1$ and $C_1 = C(M, p, q, \gamma, \beta)$. We also have for some constant $C_2 = C(M, p, q, \gamma, \beta)$,

$$\left| \pi_{(V,W)}(s, y_s; ((\mathbf{x}, \mathbf{h})))_{s,t} - \mathcal{E}_{(V)}(y_s, \mathbf{x}_{s,t}) - \mathcal{E}_{(W)}(y_s, \mathbf{h}_{s,t}) \right| \leq C_2 \omega(s, t)^\theta . \tag{12.6}$$

Proof. We assume $M = 1$; the general case follows the same lines. By the triangle inequality and inequality (12.5), it suffices to prove that

$$\left| \pi_{(V,W)}\left(s, y_s; \left(x^{s,t}, h^{s,t}\right)\right)_{s,t} - \pi_{(V)}\left(s, y_s; x^{s,t}\right)_{s,t} - \pi_{(W)}\left(s, y_s; h^{s,t}\right)_{s,t} \right|$$

is bounded by $C \omega(s, t)^\theta$. But this follows from Lemma 12.3.

The second inequality follows easily from the first one. ∎

12.1.2 Uniqueness and continuity

For existence of RDE with drift, we started by comparing $\pi_{(V,W)}(s, y_s; (x, h))_{s,t}$ to $\pi_{(V)}(s, y_s; x)_{s,t} + \pi_{(W)}(s, y_s; h)_{s,t}$. We now look at the continuity of the difference between those two terms.

Lemma 12.9 *Assume that*
(i) $V = (V_i)_{1 \leq i \leq d}$ and $\tilde{V} = (V_i)_{1 \leq i \leq d}$ are two collections of vector fields in $\mathrm{Lip}^\gamma(\mathbb{R}^e)$, $\gamma \geq 1$;
(i bis) $W = (W_j)_{1 \leq j \leq d'}$ and $\tilde{W} = (W_i)_{1 \leq i \leq d}$ are two collections of vector fields in $\mathrm{Lip}^\beta(\mathbb{R}^e)$, $\beta \geq 1$;
(ii) $s < t$ are some elements of $[0, T]$;
(iii) $y_s \in \mathbb{R}^e$ is an initial condition;
(iv) $(x, \tilde{x})$ and $\left(h, \tilde{h}\right)$ are two pairs in $C^{1\text{-var}}\left([s, t], \mathbb{R}^d\right)^{\times 2}$ and $C^{1\text{-var}}\left([s, t], \mathbb{R}^{d'}\right)^{\times 2}$;
(v) ℓ_x and ℓ_h are such that

$$\max\left\{ |V|_{\mathrm{Lip}^\gamma} \int_s^t |dx_r|, \left|\tilde{V}\right|_{\mathrm{Lip}^\gamma} \int_s^t |d\tilde{x}_r| \right\} \leq \ell_x,$$

$$\max\left\{ |W|_{\mathrm{Lip}^\beta} \int_s^t |dh_r|, \left|\tilde{W}\right|_{\mathrm{Lip}^\beta} \int_s^t \left|d\tilde{h}_r\right| \right\} \leq \ell_h;$$

(vi) δ_x *and* δ_h *are such that*

$$\max\left(\left|V\right|_{\mathrm{Lip}^\gamma},\left|\tilde{V}\right|_{\mathrm{Lip}^\gamma}\right)\int_s^t|dx_r-d\tilde{x}_r|\leq\delta_x,$$

$$\max\left(\left|W\right|_{\mathrm{Lip}^{\beta-1}},\left|\tilde{W}\right|_{\mathrm{Lip}^{\beta-1}}\right)\int_s^t\left|dh_r-d\tilde{h}_r\right|\leq\delta_h;$$

(vii) ε_V *and* ε_W *are such that*

$$\frac{1}{\max\left(\left|V\right|_{\mathrm{Lip}^\gamma},\left|\tilde{V}\right|_{\mathrm{Lip}^\gamma}\right)}\left|V-\tilde{V}\right|_{\mathrm{Lip}^{\gamma-1}}\leq\varepsilon_V,$$

$$\frac{1}{\max\left(\left|W\right|_{\mathrm{Lip}^\beta},\left|\tilde{W}\right|_{\mathrm{Lip}^\beta}\right)}\left|W-\tilde{W}\right|_{\mathrm{Lip}^{\beta-1}}\leq\varepsilon_W.$$

Then, if Δ *is defined by*

$$\Delta=\left(\pi_{(V,W)}\left(s,y_s;(x,h)\right)_{s,t}-\pi_{(V)}\left(s,y_s;x\right)_{s,t}-\pi_{(W)}\left(s,y_s;h\right)_{s,t}\right)$$

$$-\left(\pi_{(\tilde{V},\tilde{W})}\left(s,\tilde{y}_s;\left(\tilde{x},\tilde{h}\right)\right)_{s,t}-\pi_{(\tilde{V})}\left(s,\tilde{y}_s;\tilde{x}\right)_{s,t}-\pi_{(\tilde{W})}\left(s,\tilde{y}_s;\tilde{h}\right)_{s,t}\right),$$

we have for some constant $C=C\left(\gamma,\beta\right),$

$$\Delta\leq C\left(|y_0-\tilde{y}_0|+\varepsilon_V+\varepsilon_W\right)\left(\ell_h\ell_x+\ell_h\ell_x^{\beta-1}+\ell_h^{\gamma-1}\ell_x\right)\exp\left(C\left(\ell_x+\ell_h\right)\right)$$

$$+C\left(\delta_x\left(\ell_h+\ell_h^{\gamma-1}\ell_x\right)+\delta_h\left(\ell_c+\ell_x^{\beta-1}\ell_h\right)\right)\exp\left(C\left(\ell_x+\ell_h\right)\right).$$

Proof. Without loss of generality, we assume $(s,t)=(0,1)$. Define for $u\in[0,1]$,

$$z_u=\pi_{(V,W)}\left(0,y_0;(x,h)\right)_u,\ y_u^x=\pi_{(V)}\left(0,y_0;x\right)_u\text{ and }y_u^h=\pi_{(W)}\left(0,y_0;h\right)_u;$$

$$\tilde{z}_u=\pi_{(\tilde{V},\tilde{W})}\left(0,\tilde{y}_0;\left(\tilde{x},\tilde{h}\right)\right)_u,\ \tilde{y}_u^{\tilde{x}}=\pi_{(\tilde{V})}(0,\tilde{y}_0;\tilde{x})_u\text{ and }y_u^{\tilde{h}}=\pi_{(\tilde{W})}\left(0,\tilde{y}_0;\tilde{h}\right)_u.$$

We also set

$$e_u=\left(z_{0,u}-y_{0,u}^x-y_{0,u}^h\right)-\left(\tilde{z}_{0,u}-\tilde{y}_{0,u}^{\tilde{x}}-\tilde{y}_{0,u}^{\tilde{h}}\right).$$

We obtain by definition of $y,z,\tilde{y},\tilde{z}$ for $u\in[0,1]$ that $e_u=\Delta_u^1+\Delta_u^2$, where

$$\Delta_u^1=\int_0^u\{V\left(z_r\right)-V\left(y_r^x\right)\}dx_r-\int_0^u\left\{\tilde{V}\left(\tilde{z}_r\right)-\tilde{V}\left(\tilde{y}_r^{\tilde{x}}\right)\right\}d\tilde{x}_r,$$

$$\Delta_u^2=\int_0^u\{W\left(z_r\right)-W\left(y_r^h\right)\}dh_r-\int_0^u\left\{\tilde{W}\left(\tilde{z}_r\right)-\tilde{W}\left(\tilde{y}_r^{\tilde{h}}\right)\right\}d\tilde{h}_r.$$

Lemma 10.22 implies that

$$\left|\Delta_u^1\right| \le |V|_{\mathrm{Lip}^\gamma} \int_0^u \left|z_r - y_r^x - \tilde{z}_r - \tilde{y}_r^{\tilde{x}}\right| . |dx_r|$$
$$+ \left(|z - y^x|_{\infty;[0,1]} + |\tilde{z} - \tilde{y}^{\tilde{x}}|_{\infty;[0,1]}\right)^{\gamma-1} \left(|y^x - \tilde{y}^{\tilde{x}}|_{\infty;[0,1]} + \varepsilon_V\right) \ell_x$$
$$+ \left|\tilde{z} - \tilde{y}^{\tilde{x}}\right| . \delta_x.$$

There are a few terms in here we know how to bound: first from Lemma 12.3 (that we have to use with Lipschitz parameters $\gamma + 1$ and $\beta + 1$ which are greater than 2), we have

$$\begin{aligned}
|z - y^x|_{\infty;[0,1]} &\le \left|y_{0,.}^h\right|_{\infty;[0,1]} + \left|z - y^x - y_{0,.}^h\right|_{\infty;[0,1]} \\
&\le c_1 \ell_h + c_1 \ell_x \ell_h \exp\left(c_1\left(\ell_x + \ell_h\right)\right) \\
&\le c_2 \ell_h \exp\left(c_2\left(\ell_x + \ell_h\right)\right).
\end{aligned}$$

Similarly, we have

$$\left|\tilde{z} - \tilde{y}^{\tilde{x}}\right|_{\infty;[0,1]} \le c_2 \ell_h \exp\left(c_2\left(\ell_x + \ell_h\right)\right).$$

Then, Theorem 3.15 provides

$$\left|y^x - \tilde{y}^{\tilde{x}}\right|_{\infty;[0,1]} \le c_3 \left(|y_0 - \tilde{y}_0| + \delta_x + \varepsilon_V \ell_x\right).$$

Hence, we obtain

$$\begin{aligned}
\left|\Delta_u^1\right| \le\ & |V|_{\mathrm{Lip}^\gamma} \int_0^u |e_r| . |dx_r| + |V|_{\mathrm{Lip}^\gamma} \int_0^u \left|y_{0,u}^h - \tilde{y}_{0,u}^{\tilde{h}}\right| . |dx_r| \\
&+ c_4 \left(|y_0 - \tilde{y}_0| + \varepsilon_V\left(1 + \ell_x\right)\right) . \ell_h^{\gamma-1} \ell_x \exp\left(c_4\left(\ell_x + \ell_h\right)\right) \\
&+ c_4 \delta_x \left(\ell_h + \ell_h^{\gamma-1} \ell_x\right) \exp\left(c_4\left(\ell_x + \ell_h\right)\right).
\end{aligned}$$

Theorem 3.18 also provides

$$\left|y_{0,.}^h - \tilde{y}_{0,.}^{\tilde{h}}\right|_{\infty;[0,1]} \le c_5 \left(|y_0 - \tilde{y}_0| \ell_h + \delta_h + \varepsilon_W \ell_h\right) \exp\left(c_5\left(\ell_x + \ell_h\right)\right).$$

That gives our final bound on Δ_u^1, namely

$$\begin{aligned}
\left|\Delta_u^1\right| \le\ & |V|_{\mathrm{Lip}^\gamma} \int_0^u |e_r| . |dx_r| \\
&+ c_6 \left(|y_0 - \tilde{y}_0| \left(\ell_h \ell_x + \ell_h^{\gamma-1} \ell_x\right)\right. \\
&+ \left. \varepsilon_V \ell_h^{\gamma-1} \ell_x + \varepsilon_W \ell_h \ell_x\right) \exp\left(c_6\left(\ell_x + \ell_h\right)\right) \\
&+ c_6 \left(\delta_h \ell_x + \delta_x \left(\ell_h + \ell_h^{\gamma-1} \ell_x\right)\right) \exp\left(c_6\left(\ell_x + \ell_h\right)\right).
\end{aligned}$$

By symmetry, we obtain

$$
\begin{aligned}
\left|\Delta_u^2\right| \;\leq\; & |W|_{\mathrm{Lip}^\beta} \int_0^u |e_r| \cdot |dh_u| \\
& + c_7 \left(|y_0 - \tilde{y}_0| \left(\ell_h \ell_x + \ell_h \ell_x^{\beta-1}\right)\right. \\
& \left.+ \varepsilon_W \, \ell_h \ell_x^{\beta-1} + \varepsilon_V \, \ell_h \ell_x \right) \exp\left(c_7 \left(\ell_x + \ell_h\right)\right) \\
& + c_7 \left(\delta_x \ell_h + \delta_h \left(\ell_c + \ell_x^{\beta-1} \ell_h\right)\right) \exp\left(c_6 \left(\ell_x + \ell_h\right)\right).
\end{aligned}
$$

In particular, we obtain that

$$
\begin{aligned}
|e_u| \;\leq\; & \int_0^u |e_r| \cdot \left(|W|_{\mathrm{Lip}^\beta} |dh_r| + |V|_{\mathrm{Lip}^\gamma} |dx_r|\right) \\
& + c_8 \left(|y_0 - \tilde{y}_0| + \varepsilon_V + \varepsilon_W\right) \left(\ell_h \ell_x + \ell_h \ell_x^{\beta-1} + \ell_h^{\gamma-1} \ell_x\right) \exp\left(c_6 \left(\ell_x + \ell_h\right)\right) \\
& + c_8 \left(\delta_x \left(\ell_h + \ell_h^{\gamma-1} \ell_x\right) + \delta_h \left(\ell_c + \ell_x^{\beta-1} \ell_h\right)\right) \exp\left(c_6 \left(\ell_x + \ell_h\right)\right).
\end{aligned}
$$

We conclude with Gronwall's lemma.   ∎

In the existence part, we used Lemma 12.3 to extend Lemma A and Lemma B to be able to generalize the RDE existence theorem to the RDE with drift existence theorem. Here, with Lemma 12.9, we can do the same, and generalize the RDE continuity theorems to RDE with drift continuity theorems. Without further details, we therefore present the uniqueness/continuity theorem for RDE with drift:

Theorem 12.10 *Assume that* $p, q, \gamma, \beta \in [1, \infty)$ *are such that*

$$
1/p + 1/q > 1
$$
$$
\gamma > p \ \text{and} \ \beta > q
$$
$$
\frac{\gamma - 1}{q} + \frac{1}{p} > 1 \ \text{and} \ \frac{1}{q} + \frac{\beta - 1}{p} > 1;
$$

(i) $V^1 = \left(V_i^1\right)_{1 \leq i \leq d}$ *and* $V^2 = \left(V_i^2\right)_{1 \leq i \leq d}$ *are two collections of vector fields in* $\mathrm{Lip}^\gamma\left(\mathbb{R}^e\right)$;
(i bis) $W^1 = \left(W_i^1\right)_{1 \leq i \leq d'}$ *and* $W^2 = \left(W_i^2\right)_{1 \leq i \leq d'}$ *are two collections of vector fields in* $\mathrm{Lip}^\beta\left(\mathbb{R}^e\right)$;
(ii) ω *is a fixed control;*
(iii) $\mathbf{x}^1, \mathbf{x}^2$ *are two weak geometric p-rough paths in* $C^{p\text{-var}}\left([0, T], G^{[p]}\left(\mathbb{R}^d\right)\right)$, *with* $\left\|\mathbf{x}^i\right\|_{p-\omega} \leq 1$;
(iii bis) $\mathbf{h}^1, \mathbf{h}^2$ *are two weak geometric q-rough paths in* $C^{q\text{-var}}\left([0, T], G^{[q]}\left(\mathbb{R}^{d'}\right)\right)$, *with* $\left\|\mathbf{h}^i\right\|_{q-\omega} \leq 1$;
(iv) $\mathbf{y}_0^1, \mathbf{y}_0^2 \in G^{[\max(p,q)]}\left(\mathbb{R}^e\right)$ *thought of as time-0 initial conditions;*
(v) υ *is a bound on* $\left|V^1\right|_{\mathrm{Lip}^\gamma}, \left|V^2\right|_{\mathrm{Lip}^\gamma}, \left|W^1\right|_{\mathrm{Lip}^\beta}$ *and* $\left|W^2\right|_{\mathrm{Lip}^\beta}$.
Then, $\boldsymbol{\pi}_{\left(V^i, W^i\right)}\left(0, \mathbf{y}_0^i; \left(\mathbf{x}^i, \mathbf{h}^i\right)\right)$ *is a singleton; that is, there exists a unique full RDE solution* $\mathbf{y}^i = \boldsymbol{\pi}_{\left(V^i, W^i\right)}\left(0, \mathbf{y}_0^i; \left(\mathbf{x}^i, \mathbf{h}^i\right)\right)$ *started at* $\mathbf{y}_0^i$ *driven by*

$(\mathbf{x}^i, \mathbf{h}^i)$ *along* (V^i, W^i). *Moreover,*

$$\rho_{\max(p,q)\text{-}\omega}\left(\mathbf{y}^1, \mathbf{y}^2\right) \le C\varepsilon \exp\left(Cv\omega\left(0, T\right)\right)$$

where $C = C\left(\gamma, \beta, p, q\right)$ *and*

$$
\begin{aligned}
\varepsilon &= v\left|y_0^1 - y_0^2\right| + \left|V^1 - V^2\right|_{\text{Lip}^{\gamma-1}} + \left|W^1 - W^2\right|_{\text{Lip}^{\beta-1}} \\
&\quad + v\left(\rho_{p\text{-}\omega}\left(\mathbf{x}^1, \mathbf{x}^2\right) + \rho_{q\text{-}\omega}\left(\mathbf{h}^1, \mathbf{h}^2\right)\right).
\end{aligned}
$$

Remark that the metric $\rho_{\max(p,q)\text{-}\omega}$, unlike $d_{\infty;[0,T]}$ in the next statement, only measures the distance between the increments of two paths. Then, in the above definition of ε, it is really $\left|y_0^1 - y_0^2\right| = \left|y_0^1 - y_0^2\right|_{\mathbb{R}^e}$ rather than $\left|y_0^1 - y_0^2\right|_{T^{[p\vee q]}(\mathbb{R}^e)}$. We now state the refined uniqueness theorem, which also extends to the drift case without difficulties.

Theorem 12.11 *Assume that* $p, q, \gamma, \beta \in [1, \infty)$ *are such that*

$$1/p + 1/q > 1$$

$$\gamma \ge p \text{ and } \beta \ge q$$

$$\frac{\gamma - 1}{q} + \frac{1}{p} \ge 1 \text{ and } \frac{1}{q} + \frac{\beta - 1}{p} \ge 1.$$

(i) $\left(V_j^1\right)_{1 \le j \le d}$ *and* $\left(V_j^2\right)_{1 \le j \le d}$ *are two collections of vector fields in* $\text{Lip}^p\left(\mathbb{R}^e\right)$;
(i bis) $\left(W_j^1\right)_{1 \le j \le d'}$ *and* $\left(W_j^2\right)_{1 \le j \le d'}$ *are two collections of vector fields in* $\text{Lip}^q\left(\mathbb{R}^e\right)$;
(ii) $\mathbf{x}^1, \mathbf{x}^2 \in C^{\psi_p\text{-var}}\left([0, T], G^{[p]}\left(\mathbb{R}^d\right)\right)$, *with* $\left\|\mathbf{x}^i\right\|_{\psi_p\text{-var}} \le R$;
(ii bis) $\mathbf{h}^1, \mathbf{h}^2 \in C^{\psi_q\text{-var}}\left([0, T], G^{[q]}\left(\mathbb{R}^{d'}\right)\right)$, *with* $\left\|\mathbf{h}^i\right\|_{\psi_q\text{-var}} \le R$;
(iii) $y_0^1, y_0^2 \in G^{[p \vee q]}\left(\mathbb{R}^e\right)$ *thought of as time-0 initial conditions;*
(iv) $\mathbf{y}^i$ *are some arbitrary elements of* $\boldsymbol{\pi}_{(V^i)}\left(0, y_0^i; \mathbf{x}^i\right)$ *(that is, they are RDE solutions driven by* $\mathbf{x}^i$, *starting at* y_0^i, *along the vector fields* V^i);
(v) v *is a bound on* $\left|V^1\right|_{\text{Lip}^p}, \left|V^2\right|_{\text{Lip}^p}, \left|W^1\right|_{\text{Lip}^q}$ *and* $\left|W^2\right|_{\text{Lip}^q}$.
Then, $\boldsymbol{\pi}_{(V^i, W^i)}\left(0, y_0^i; (\mathbf{x}^i, \mathbf{h}^i)\right)$ *is a singleton; that is, there exists a unique full RDE solution* $\mathbf{y}^i = \boldsymbol{\pi}_{(V^i, W^i)}\left(0, y_0^i; (\mathbf{x}^i, \mathbf{h}^i)\right)$ *started at* y_0^i *driven by* $(\mathbf{x}^i, \mathbf{h}^i)$ *along* (V^i, W^i). *Moreover, for all* $\varepsilon > 0$, *there exists* $\mu = \mu\left(\varepsilon; p, q, v, R\right) > 0$ *such that*

$$\left|y_0^1 - y_0^2\right| + \left|V^1 - V^2\right|_{\text{Lip}^{p-1}} + \left|W^1 - W^2\right|_{\text{Lip}^{q-1}} + d_\infty\left(\mathbf{x}^1, \mathbf{x}^2\right) < \mu$$

implies that

$$d_{\infty;[0,T]}\left(\mathbf{y}^1, \mathbf{y}^2\right) < \varepsilon.$$

Remark 12.12 The conditions on γ, β, p, q above already appeared in the existence theorem for RDEs with drift and all comments made then (Remark 12.7) remain valid. In particular, assuming $p \ge q$ without loss of generality, the conditions reduce to

$$1/p + 1/q > 1, \ \gamma > p \text{ and } \beta > \max\left\{q, 1 + p\left(1 - \frac{1}{q}\right)\right\}.$$

In Section 12.2 we shall see that $q := p/[p] \geq 1$ and $\beta := \gamma - [p] + 1$ arises naturally when perturbing the (centre of the) driving geometric p-rough path. (In fact, the drift vector fields, then consist of $[p] - 1$ iterated Lie brackets of the original Lip^γ-vector fields which explains the choice of β.) An elementary computation then gives

$$\max\left\{q, 1 + p\left(1 - \frac{1}{q}\right)\right\} = p - [p] + 1 < \beta,$$

which shows that this condition is natural after all.   □

Of course, Corollaries 10.39 and 10.40 also extend to the drift case and we leave the details to the reader.

We conclude this section with an exercise in which the reader is invited to implement the so-called Doss–Sussmann method for RDEs with drift. For simplicity, we only deal with $\mathbb{R}^e$-valued RDE solutions. Let us also note that it does not (seem to) lead to optimal regularity assumptions.

Exercise 12.13 (Doss–Sussmann) *Let* $\mathbf{x} \in C^{p\text{-var}}\left([0,T], G^{[p]}\left(\mathbb{R}^d\right)\right)$,

$$V_0 \in \text{Lip}^1\left(\mathbb{R}^e\right), V = (V_1, \dots, V_d) \in \text{Lip}^{\gamma+1}\left(\mathbb{R}^e\right)$$

with $\gamma > p$. *Let* $J^{\mathbf{x}}_{0 \leftarrow t}$ *be the Jacobian of* $\pi_{(V)}(0, \cdot; \mathbf{x})_t^{-1} : \mathbb{R}^e \to \mathbb{R}^e$ *and set*

$$W(t, y) \equiv J^{\mathbf{x}}_{0 \leftarrow t}(y) \cdot V_0\left(\pi_{(V)}(0, y; \mathbf{x})_t\right). \tag{12.7}$$

(i) Show that the ordinary, time-inhomogenous ODE

$$\dot{z}_t = W(t, z_t), \ z(0) = y_0 \tag{12.8}$$

admits a unique, non-explosive solution on $[0, T]$.
(ii) Show that the solution to the RDE with drift $dy = V(y)\,d\mathbf{x} + V_0(y)\,dt$, *started at* y_0, *is given by*

$$y = \pi_{(V)}(0, z_t; \mathbf{x})_t, \ t \in [0, T]. \tag{12.9}$$

(iii) Deduce an Euler estimate for RDEs with drift of form (12.6).

12.2 Application: perturbed driving signals and impact on RDEs

12.2.1 *(Higher-)area perturbations and modified drift terms*

We consider a driving rough path $\mathbf{x}$ and consider what happens if we perturb it on some "higher-area" level such as

$$\mathcal{V}^N\left(\mathbb{R}^d\right) := \mathfrak{g}^N\left(\mathbb{R}^d\right) \cap \left(\mathbb{R}^d\right)^{\otimes N},$$

the centre of the Lie algebra $\mathfrak{g}^N\left(\mathbb{R}^d\right)$; for example, $\mathcal{V}^2\left(\mathbb{R}^d\right) = so\,(d)$, the space of anti-symmetric $d \times d$ matrices. Unless otherwise stated, $\mathcal{V}^N\left(\mathbb{R}^d\right)$ will be equipped with the Euclidean metric.

Theorem 12.14 (centre perturbation) *Let* $p,r \geq 1$ *and* $N \in \mathbb{N}$ *such that*

$$[r] = N \geq [p]\,.$$

Given a weak geometric p-rough path $\mathbf{x} : [0,T] \to G^{[p]}\left(\mathbb{R}^d\right)$ *and*

$$\varphi \in C^{\frac{r}{[r]}\text{-var}}\left([0,T],\mathcal{V}^N\left(\mathbb{R}^d\right)\right)$$

we define the perturbation

$$\mathbf{x}^\varphi := \exp\left(\log\left(S_N\left(\mathbf{x}\right)\right) + \varphi\right). \tag{12.10}$$

Then $\mathbf{x}^\varphi$ *is a weak geometric* $\max\,(p,r)$-*rough path. Assume* $V \in \mathrm{Lip}^\gamma$ *with* $\gamma > \max\,(p,r)$, *so that* $dy = V\,(y)\,d\mathbf{x}^\varphi$, $y\,(0) = y_0$ *has a unique RDE solution. Then there is a unique solution to the RDE with drift,*

$$dz = V\,(z)\,d\mathbf{x} + W\,(z)\,d\varphi,\ z\,(0) = y_0$$

where W is the collection of vector fields given by

$$\left(\left[V_{i_1},\left[\ldots,\left[V_{i_N-1},V_{i_N}\right]\right]\ldots\right]\right)_{i_1,\ldots,i_N\in\{1,\ldots,d\}}$$

and

$$y = \pi_{(V)}\left(0,y_0;\mathbf{x}^\varphi\right) \equiv \pi_{(V,W)}\left(0,y_0;(\mathbf{x},\varphi)\right) = z.$$

We prepare the proof with

Lemma 12.15 *Let* $k\in\mathbb{N}$. *Given a multi-index* $\alpha=(\alpha_1,\ldots,\alpha_k)\in\{1,\ldots,d\}^k$ *and* Lip^{k-1} *vector fields* $V_1,\ldots,V_k$ *on* $\mathbb{R}^e$, *define*

$$V_\alpha = \left[V_{\alpha_k},\left[V_{\alpha_{k-1}},\ldots,\left[V_{\alpha_2},V_{\alpha_1}\right]\right]\right].$$

Further, let $e_1,\ldots,e_d$ *denote the canonical basis of* $\mathbb{R}^d$. *Then* $\mathfrak{g}^n\left(\mathbb{R}^d\right)$, *the step-n free Lie algebra, is generated by elements of the form*

$$e_\alpha = \left[e_{\alpha_k},\left[e_{\alpha_{k-1}},\ldots,\left[e_{\alpha_2},e_{\alpha_1}\right]\right]\right] \in \left(\mathbb{R}^d\right)^{\otimes k},\ k \leq n$$

with $[u,v] = u \otimes v - v \otimes u$ *and*[4]

$$\sum_{i_1,\ldots,i_k\in\{1,\ldots,d\}} V_{i_k}\cdots V_{i_1}\left(e_\alpha\right)^{i_k,\ldots,i_1} = V_\alpha.$$

---

[4] A k-tensor $u \in \left(\mathbb{R}^d\right)^{\otimes k}$ is written as $u = \sum_{i_1,\ldots,i_k\in\{1,\ldots,d\}} u^{i_k,\ldots,i_1}\, e_{i_k} \otimes \ldots \otimes e_{i_1}$.

Proof. It is clear that $\mathfrak{g}^n\left(\mathbb{R}^d\right)$ is generated by the e_α. We prove the second statement by induction: a straightforward calculation shows that it holds for $k = 2$. Now suppose it holds for $k - 1$ and denote $V_{\tilde{\alpha}} = \left[V_{\alpha_{k-1}}, \ldots, [V_{\alpha_2}, V_{\alpha_1}]\right]$. Then (using summation convention),

$$
\begin{aligned}
V_{i_k}\ldots V_{i_1}\left(e_\alpha\right)^{i_k,\ldots,i_1} &= V_{i_k}\ldots V_{i_1}\left(e_{\alpha_k}\otimes\left[e_{\alpha_{k-1}},\ldots,[e_{\alpha_2},e_{\alpha_1}]\right]\right)^{i_k,\ldots,i_1}\\
&\quad -V_{i_k}\ldots V_{i_1}\left(\left[e_{\alpha_{k-1}},\ldots,[e_{\alpha_2},e_{\alpha_1}]\right]\otimes e_{\alpha_k}\right)^{i_k,\ldots,i_1}\\
&= V_{i_k}\ldots V_{i_1}\delta^{\alpha_k,i_k}\otimes\left[e_{\alpha_{k-1}},\ldots,[e_{\alpha_2},e_{\alpha_1}]\right]^{i_{k-1},\ldots,i_1}\\
&\quad -V_{i_k}\ldots V_{i_1}\left[e_{\alpha_{k-1}},\ldots,[e_{\alpha_2},e_{\alpha_1}]\right]^{i_k,\ldots,i_2}\otimes\delta^{\alpha_k,i_1}\\
&= V_{\alpha_k}V_{i_{k-1}}\ldots V_{i_1}\left[e_{\alpha_{k-1}},\ldots,[e_{\alpha_2},e_{\alpha_1}]\right]^{i_{k-1},\ldots,i_1}\\
&\quad -V_{i_k}\ldots V_{i_2}\left[e_{\alpha_{k-1}},\ldots,[e_{\alpha_2},e_{\alpha_1}]\right]^{i_k,\ldots,i_2}V_{\alpha_k}\\
&= V_{\alpha_k}V_{\tilde{\alpha}}-V_{\tilde{\alpha}}V_{\alpha_k} = \left[V_{\alpha_k},\left[V_{\alpha_{k-1}},\ldots,[V_{\alpha_2},V_{\alpha_1}]\right]\right],
\end{aligned}
$$

where we set $\tilde{\alpha} = (\alpha_{k-1},\ldots,\alpha_1)$ and use the induction hypothesis that $V_{\tilde{\alpha}}$ equals

$$
\begin{aligned}
V_{i_{k-1}}\ldots V_{i_1}&\left[e_{\alpha_{k-1}},\ldots,[e_{\alpha_2},e_{\alpha_1}]\right]^{i_{k-1},\ldots,i_1}\\
&= V_{i_k}\ldots V_{i_2}\left[e_{\alpha_{k-1}},\ldots,[e_{\alpha_2},e_{\alpha_1}]\right]^{i_k,\ldots,i_2}.
\end{aligned}
$$

∎

Proof of Theorem 12.14. Remark that W,φ satisfy the regularity condition of Theorem 12.10 (cf. Remark 12.12), and so RDEs of type $dz = V(z)\,d\mathbf{x}+W(z)\,d\varphi$ have unique solutions. It suffices to show that $y_T = z_T$. Take a dissection $D = (t_i)$ of $[0,T]$ and define

$$
z_t^i = \pi_{(V,W)}\left(t_i,y_{t_i};(\mathbf{x},\varphi)\right)_t \text{ for } t\in[t_i,T].
$$

Note that $z_T^0 = z_T$ and $z_T^{|D|} = y_T$, hence

$$
|z_T - y_T| \le \sum_{i=1}^{|D|}\left|z_T^i - z_T^{i-1}\right|.
$$

Now,

$$
\begin{aligned}
\left|z_T^i - z_T^{i-1}\right| &= \left|\pi_{(V,W)}\left(t_i,y_{t_i};(\mathbf{x},\varphi)\right)_T - \pi_{(V,W)}\left(t_{i-1},y_{t_{i-1}};(\mathbf{x},\varphi)\right)_T\right|\\
&= \left|\pi_{(V,W)}\left(t_i,y_{t_i};(\mathbf{x},\varphi)\right)_T\right.\\
&\quad \left. -\pi_{(V,W)}\left(t_i,\pi_{(V,W)}\left(t_{i-1},y_{t_{i-1}};(\mathbf{x},\varphi)\right)_{t_i};(\mathbf{x},\varphi)\right)_T\right|\\
&\lesssim \left|y_{t_i} - \pi_{(V,W)}\left(t_{i-1},y_{t_{i-1}};(\mathbf{x},\varphi)\right)_{t_i}\right|
\end{aligned}
$$

thanks to Lipschitzness of the flow (which was established in Theorem 12.10). By subtracting/adding $\mathcal{E}_{(V)}\left(y_{t_{i-1}},\mathbf{x}_{t_{i-1},t_i}\right) + \mathcal{E}_{(W)}\left(y_{t_{i-1}},\varphi_{t_{i-1},t_i}\right)$

we estimate $\left| z_T^i - z_T^{i-1} \right| \leq \Delta_1 + \Delta_2$ where

$$\Delta_1 = \left| y_{t_i} - \mathcal{E}_{(V)} \left(y_{t_{i-1}}, \mathbf{x}_{t_{i-1},t_i} \right) - \mathcal{E}_{(W)} \left(y_{t_{i-1}}, \varphi_{t_{i-1},t_i} \right) \right|,$$

$$\Delta_2 = \left| \pi_{(V,W)} \left(t_{i-1}, y_{t_{i-1}}; (\mathbf{x},\varphi) \right)_{t_i} - \mathcal{E}_{(V)} \left(y_{t_{i-1}}, \mathbf{x}_{t_{i-1},t_i} \right) - \mathcal{E}_{(W)} \left(y_{t_{i-1}}, \varphi_{t_{i-1},t_i} \right) \right|.$$

Thanks to Lemma 12.15 and $\mathcal{E}_{(W)} \left(y_{t_{i-1}}, \varphi_{t_{i-1},t_i} \right) = W \left(y_{t_{i-1}} \right) \cdot \varphi_{t_{i-1},t_i}$ we have

$$\mathcal{E}_{(V)} \left(y_{t_{i-1}}, \mathbf{x}_{t_{i-1},t_i} \right) + \mathcal{E}_{(W)} \left(y_{t_{i-1}}, \varphi_{t_{i-1},t_i} \right) = \mathcal{E}_{(V)} \left(y_{t_{i-1}}, \mathbf{x}_{t_{i-1},t_i}^{\varphi} \right)$$

and hence, from the Euler estimate for RDEs, Corollary 10.15, $\Delta_1 \leq c_1 \omega$ $(t_{i-1}, t_i)^{\theta}$ for some control ω and some $\theta > 1$. On the other hand, our Euler estimates for RDEs with drift as stated in Corollary 12.8 imply that, similarly, $\Delta_2 \leq c_2 \omega (t_{i-1}, t_i)^{\theta}$. It follows that, with $c_3 = c_1 + c_2$,

$$\left| z_T^i - z_T^{i-1} \right| \leq c_3 \omega (t_{i-1}, t_i)^{\theta}$$

and so $|z_T - y_T| \leq c_3 \sum_{i=1}^{|D|} \omega (t_{i-1}, t_i)^{\theta} \to 0$ as $|D| \to 0$. ∎

In Theorem 12.14 we have studied the impact of *level-N perturbation* of a driving signal. More precisely, given a weak geometric p-rough path $\mathbf{x}$ with $[p] \leq N$ and a sufficiently regular map

$$\varphi^{(N)} : [0, T] \to \mathcal{V}^N \left(\mathbb{R}^d \right)$$

we defined in (12.10) a perturbation of $\mathbf{x}$, which we now denote by

$$T_{\varphi^{(N)}} \mathbf{x} := \exp[\log S_N \left(\mathbf{x} \right) + \left(0, \ldots, 0, \varphi^{(N)} \right)],$$

and then saw that RDEs driven along vector fields $V = (V_1, \ldots, V_d)$ by $d\mathbf{x}$ versus $d \left(T_{\varphi^{(N)}} \mathbf{x} \right)$ effectively differ by a drift term of the form

$$\left(\left[V_{i_1}, \left[\ldots, \left[V_{i_{N-1}}, V_{i_N} \right] \right] \ldots \right] \right) d\varphi^{N; i_1, \ldots, i_N}$$

with summation over iterated indices. On the other hand, the "natural" *level-1 perturbation* of a geometric p-rough path $\mathbf{x}$, in the direction of a sufficiently regular path $\varphi^{(1)} : [0, T] \to \mathbb{R}^d$, is given by (cf. Section 9.4.6) the translation operator $T_{\varphi^{(1)}}$ acting on $\mathbf{x}$. The respective RDEs driven by $d\mathbf{x}$ and $d\mathbf{x}^{\varphi^{(1)}}$ obviously differ by a drift term of the form

$$V_i d\varphi^{1; i}.$$

These two perturbations can be thought of as special cases of a general perturbation. To this end, we now consider a general perturbation

$$\varphi = \left(\varphi^{(1)}, \ldots, \varphi^{(N)} \right) : [0, T] \to \mathfrak{g}^N \left(\mathbb{R}^d \right),$$

assumed (for simplicity) to be of bounded variation with respect to the Euclidean metric on $\mathfrak{g}^N\left(\mathbb{R}^d\right)$. Let us also assume, at first, $\mathbf{x} = S_N(x)$ where $x \in C^{\text{1-var}}\left([0,T],\mathbb{R}^d\right)$. We then define, inductively,

$$T_{\varphi^{(1)}} x := x + \varphi^{(1)} \in C^{\text{1-var}}\left([0,T],\mathbb{R}^d\right) \tag{12.11}$$

$$T_{\left(\varphi^{(1)},\varphi^{(2)}\right)} x := \exp[\log S_2\left(T_{\varphi^{(1)}} x\right) + \left(0,\varphi^{(2)}\right) \in C^{\text{2-var}}\left([0,T],G^2\left(\mathbb{R}^d\right)\right)$$

$$\vdots$$

$$T_{\left(\varphi^{(1)},\dots,\varphi^{(N)}\right)} x := \exp[\log S_N\left(T_{\left(\varphi^{(1)},\dots,\varphi^{(N-1)}\right)} x\right)$$
$$+ \left(0,\dots,0,\varphi^{(N)}\right) \in C^{\text{N-var}}\left([0,T],G^N\left(\mathbb{R}^d\right)\right)$$

and note that, even though x was assumed to be of bounded variation, $T_{\left(\varphi^{(1)},\dots,\varphi^{(N)}\right)} x$ is a genuine (weak) geometric N-rough path.

Theorem 12.16 (general perturbation) *(i) Let $p \geq 1$ and $[p] \leq N$. Given $\mathbf{x} \in C_o^{p\text{-var}}\left([0,T],G^{[p]}\left(\mathbb{R}^d\right)\right)$ and $\varphi : [0,T] \to \mathfrak{g}^N\left(\mathbb{R}^d\right)$ of bounded variation with respect to the Euclidean metric on $\mathfrak{g}^N\left(\mathbb{R}^d\right)$, there exists a unique*

$$T_\varphi \mathbf{x} := T_{\left(\varphi^{(1)},\dots,\varphi^{(N)}\right)} \mathbf{x} \in \left\{ \begin{array}{ll} C^{p\text{-var}}\left([0,T],G^N\left(\mathbb{R}^d\right)\right) & \text{if } [p] = N \\ C^{N\text{-var}}\left([0,T],G^N\left(\mathbb{R}^d\right)\right) & \text{if } [p] > N \end{array} \right.$$

with the property that, whenever $S_{[p]}(x^n) \to \mathbf{x}$ uniformly and $\sup_n \left\|S_{[p]}(x^n)\right\|_{p\text{-var}} < \infty$ then

$$T_{\left(\varphi^{(1)},\dots,\varphi^{(N)}\right)} x^n \to T_{\left(\varphi^{(1)},\dots,\varphi^{(N)}\right)} \mathbf{x}$$

uniformly and with uniform p- (resp. N-) variation bounds.
(ii) Assume $V \in \text{Lip}^\gamma, \gamma > \max(p,N)$. Then

$$y \equiv \pi_{(V)}(0,y_0;T_\varphi \mathbf{x}) \quad \text{equals} \quad z \equiv \pi_{(V,*V)}(0,y_0;(\mathbf{x};\varphi))$$

where y is the RDE solution to $dy = V(y)\,d\left(T_{\left(\varphi^{(1)},\dots,\varphi^{(N)}\right)}\mathbf{x}\right)$ and z the solution of the following RDE with drift,

$$dz = V(z)\,d\mathbf{x} + (*V)(z)\,d\varphi, \quad z(0) = y_0$$

where

$$(*V)(\cdot)\,d\varphi = \sum_{k=1}^N \sum_{i_1,\dots,i_k} \left(\left[V_{i_1},\left[\dots,\left[V_{i_{k-1}},V_{i_k}\right]\right]\dots\right]\right)|.d\varphi^{(k);i_1,\dots,i_k}.$$

Proof. When $\mathbf{x} = S_N(x)$ for $x \in C^{\text{1-var}}\left([0,T],\mathbb{R}^d\right)$, we can use (12.11) and apply iteratively Theorem 12.14 to see that

$$\pi_{(V)}(0,y_0;T_\varphi \mathbf{x}) \quad \text{equals} \quad z \equiv \pi_{(V,*V)}(0,y_0;(\mathbf{x};\varphi)).$$

For the general case, we need to properly define $T_\varphi \mathbf{x}$. To this end, let $C = (\partial_1, \ldots, \partial_d)$ be the collection of coordinate vector fields on $\mathbb{R}^d$. The full RDE solution $\mathbf{y} = \boldsymbol{\pi}_{(C)}(0, \mathbf{x}_0; \mathbf{x})$ is identically equal to the input signal $\mathbf{x}$, which suggests *defining*

$$T_\varphi \mathbf{x} := \boldsymbol{\pi}_{(V, *V)}(0, o; (\mathbf{x}; \varphi))$$

as an RDE with drift. We can now use continuity results for RDEs with drift to see that $\mathbf{x} \mapsto T_\varphi \mathbf{x}$ has the required continuity properties, as stated in part (i). ∎

12.2.2 Limits of Wong–Zakai type with modified area

The next theorem describes a situation in which a piecewise linear approximation is twisted in such a way as to lead to a centre perturbation. In view of the forthcoming examples (Section 13.3.4) we state the following results only for geometric Hölder rough paths.

Definition 12.17 Let $\alpha \in (0, 1]$, $\mathbf{x} \in C^{\alpha\text{-Höl}}([0, T], G^{[1/\alpha]}(\mathbb{R}^d))$ and write $x = \pi_1(\mathbf{x})$ for its projection to a path with values in $\mathbb{R}^d$.
(i) Assume $[1/\alpha] \le N \in \mathbb{N}$ and let $(D_n) = (t_i^n : i)$ be a sequence of dissections of $[0, T]$ such that[5]

$$\sup_{n \in \mathbb{N}} \left\| S_{[1/\alpha]}(x^{D_n}) \right\|_{\alpha\text{-Höl}} = M < \infty \text{ and } d_\infty\left(S_{[1/\alpha]}(x^{D_n}), \mathbf{x}\right) \to_{n \to \infty} 0.$$

If $(x^n) \subset C^{1\text{-Höl}}([0, T], \mathbb{R}^d)$ is such that[6]

$$\mathbf{p}_t^n := S_N(x^n)_{0,t} \otimes S_N(x^{D_n})_{0,t}^{-1}$$

takes values in the centre of $G^N(\mathbb{R}^d)$ whenever $t \in D_n$ then we say that (x^n) is an approximation on (D_n) with perturbations $(\mathbf{p}^n)$ on level N to $\mathbf{x}$.
(ii) Let $\beta \in (0, 1]$ such that

$$[1/\beta] = N \ge [1/\alpha]. \tag{12.12}$$

We say that an approximation (x^n) on (D_n) with perturbations $(\mathbf{p}^n)$ on level N to $\mathbf{x}$ is $\min(\alpha, \beta)$-Hölder comparable (with constants c_1, c_2, c_3) if for all $t_i^n, t_{i+1}^n \in D_n$

$$|x^n|_{1\text{-Höl};[t_i^n, t_{i+1}^n]} \le c_1 |x^{D_n}|_{1\text{-Höl};[t_i^n, t_{i+1}^n]} + c_2 |t_{i+1}^n - t_i^n|^{\beta-1} \text{ and}$$

$$\|\mathbf{p}_{s,t}^n\| \le c_3 |t - s|^\beta \text{ for all } s, t \in D_n.$$

□

---

[5] We recall that x^D is the piecewise linear approximation to x based on the dissection D.

[6] It is *not* assumed that $\mathbf{p}_t^n \in$ centre of $G^N(\mathbb{R}^d)$ when $t \notin D_n$.

Although at first sight technical, these definitions are fairly natural: firstly, we restrict our attention to Hölder rough paths $\mathbf{x}$ which are the limit of "their (lifted) piecewise linear approximations". As we shall see in Part III, this covers the bulk of stochastic processes which admit a lift to a rough path. Assumption (ii) in the above definition then guaranteess that (x^n) remains, at $\min(\alpha,\beta)$-Hölder scale, comparable to the piecewise linear approximations. In particular, the assumption on $|x^n|_{1\text{-Höl};\left[t_i^n,t_{i+1}^n\right]} = |\dot{x}^n|_{\infty;\left[t_i^n,t_{i+1}^n\right]}$ will be easy to verify in all examples (cf. below). The intuition is that, if we assume x^n runs at constant speed over any interval $I = \left[t_i^n, t_{i+1}^n\right]$, $D_n = (t_i^n)$, it is equivalent to saying that

$$\text{length}\left(x^n|_I\right) \leq c_1 \text{length}\left(x^{D_n}|_I\right) + c_2 |I|^\beta$$
$$(\quad = \quad c_1 \left|x_{t_i^n,t_{i+1}^n}\right| + c_2 \left|t_{i+1}^n - t_i^n\right|^\beta).$$

Theorem 12.18 *Let $\alpha,\beta \in (0,1]$ and assume $[1/\beta] = N \geq [1/\alpha]$. Assume $\mathbf{x} \in C^{\alpha\text{-Höl}}\left([0,T], G^{[1/\alpha]}\left(\mathbb{R}^d\right)\right)$ and let (x^n) be an approximation on some sequence (D_n) of dissections of $[0,T]$ with perturbations $(\mathbf{p}^n)$ on level N to $\mathbf{x}$.*
(i) If the approximation is $\min(\alpha,\beta)$-Hölder comparable (with constants c_1,c_2,c_3) then there exists a constant $C = C(\alpha,\beta,c_1,c_2,M,T,N)$ such that

$$\sup_{n\in\mathbb{N}} \|S_N(x^n)\|_{\min(\alpha,\beta)\text{-Höl}} \leq C \left(\sup_{n\in\mathbb{N}} \|S_{[1/\alpha]}(x^{D_n})\|_{\alpha\text{-Höl}} + c_3 + 1\right) < \infty.$$

(ii) If $\mathbf{p}_t^n \to \mathbf{p}_t$ for all $t \in \cup_n D_n$ and $\cup_n D_n$ is dense in $[0,T]$ then $\mathbf{p}$ is a β-Hölder continuous path with values in the centre of $G^N\left(\mathbb{R}^d\right)$ and for every $t \in [0,T]$,

$$d\left(S_N(x^n)_{0,t}, S_N(\mathbf{x})_{0,t} \otimes \mathbf{p}_{0,t}\right) \leq d\left(S_N(x^{D_n})_{0,t}, S_N(\mathbf{x})_{0,t}\right) + d\left(\mathbf{p}_{0,t}^n, \mathbf{p}_{0,t}\right)$$
$$\to 0 \text{ as } n \to \infty.$$

(iii) If the assumptions of both (i) and (ii) are met then, for all $\gamma < \min(\alpha,\beta)$,

$$d_{\gamma\text{-Höl}}\left(S_N(x^n), \mathbf{x}^\varphi\right) \to_{n\to\infty} 0$$

where $\varphi := \log\mathbf{p} \in \mathcal{V}^N\left(\mathbb{R}^d\right)$ and $\mathbf{x}^\varphi = \exp\left(\log(S_N(\mathbf{x})) + \varphi\right)$.

Proof. (i) Take $s < t$ in $[0, T]$. If $s, t \in \left[t_i^n, t_{i+1}^n\right]$ we have by our assumption on $|x^n|_{1\text{-Höl};[t_i, t_{i+1}]}$

$$
\begin{aligned}
\left\| S_N\left(x^n\right)_{s,t} \right\| &\leq |t - s| \left\| S_N\left(x^n\right) \right\|_{1\text{-Höl};\left[t_i^n, t_{i+1}^n\right]} \\
&= |t - s| \, |x^n|_{1\text{-Höl};\left[t_i^n, t_{i+1}^n\right]} \\
&\leq |t - s| \left\{ c_1 \left| x_{t_i^n, t_{i+1}^n} \right| / \left| t_{i+1}^n - t_i^n \right| + c_2 \left| t_{i+1}^n - t_i^n \right|^{\beta - 1} \right\} \\
&\leq |t - s| \left\{ c_1 \, |x|_{\alpha\text{-Höl}} \left| t_{i+1}^n - t_i^n \right|^{\alpha - 1} + c_2 \left| t_{i+1}^n - t_i^n \right|^{\beta - 1} \right\} \\
&\leq c_4 \, |t - s|^{\min(\alpha, \beta)},
\end{aligned}
$$

with suitable constant c_4. Otherwise we can find $t_i^n \leq t_j^n$ so that $s \leq t_i^n \leq t_j^n \leq t$ and

$$
\left\| S_N\left(x^n\right)_{s,t} \right\| \leq 2 c_4 \, |t - s|^\gamma + \left\| S_N\left(x^n\right)_{t_i^n, t_j^n} \right\|.
$$

Estimates for the Lyons lift $\mathbf{x} \mapsto S_N(\mathbf{x})$, Proposition 9.3, then guarantee existence of a constant c_5 such that

$$
\begin{aligned}
\left\| S_N\left(x^n\right)_{t_i^n, t_j^n} \right\| &\leq \left\| S_N\left(x^{D_n}\right)_{t_i^n, t_j^n} \right\| + \left\| \mathbf{p}_{t_i^n, t_j^n}^n \right\| \\
&\leq c_5 \left\| S_{[1/\alpha]}\left(x^{D_n}\right) \right\|_{\alpha\text{-Höl}} \left| t_j^n - t_i^n \right|^\alpha + c_3 \left| t_j^n - t_i^n \right|^\beta \\
&\leq \left(c_5 \left\| S_{[1/\alpha]}\left(x^{D_n}\right) \right\|_{\alpha\text{-Höl}} + c_3 \right) |t - s|^{\min(\alpha, \beta)}
\end{aligned}
$$

and, since $\sup_n \left\| S_{[1/\alpha]}\left(x^{D_n}\right) \right\|_{\alpha\text{-Höl}} < \infty$ by assumption, the proof of the uniform Hölder bound is finished.

(ii) By assumption, $\mathbf{p}^n$ is uniformly β-Hölder. By a standard Arzela–Ascoli-type argument, it is clear that every pointwise limit (if only on the dense set $\cup_n D_n$) is a uniform limit and β-Hölder regularity is preserved in this limit, i.e. $\mathbf{p}$ is β-Hölder itself. For every $t \in \cup_n D_n$, $\mathbf{p}_t^n$ takes values in the centre of $G^N\left(\mathbb{R}^d\right)$ and hence (density of $\cup_n D_n$, continuity of $\mathbf{p}$) it is easy to see that $\mathbf{p}$ takes values in the centre for all $t \in [0, T]$. Now take $t \in D_n$. Since elements in the centre commute with all elements in $G^N\left(\mathbb{R}^d\right)$, we have

$$
\begin{aligned}
&d\left(S_N\left(x^n\right)_{0,t}, S_N(\mathbf{x})_{0,t} \otimes \mathbf{p}_{0,t} \right) \\
={} &\left\| S_N\left(x^n\right)_{0,t}^{-1} \otimes S_N\left(x^{D_n}\right)_{0,t} \otimes \mathbf{p}_{0,t}^n \otimes S_N\left(x^{D_n}\right)_{0,t}^{-1} \right. \\
&\left. \otimes S_N(\mathbf{x})_{0,t} \otimes \left(\mathbf{p}_{0,t}^n\right)^{-1} \otimes \mathbf{p}_{0,t} \right\| \\
={} &\left\| S_N\left(x^{D_n}\right)_{0,t}^{-1} \otimes S_N(\mathbf{x})_{0,t} \otimes \left(\mathbf{p}_{0,t}^n\right)^{-1} \otimes \mathbf{p}_{0,t} \right\| \\
\leq{} &d\left(S_N\left(x^{D_n}\right)_{0,t}, S_N(\mathbf{x})_{0,t} \right) + d\left(\mathbf{p}_{0,t}^n, \mathbf{p}_{0,t} \right).
\end{aligned}
$$

On the other hand, given an arbitrary element $t \in [0, T]$ we can take t^n to be the closest neighbour in D_n and so

$$
\begin{aligned}
&d \left(S_N \left(x^n \right)_{0,t}, S_N \left(\mathbf{x} \right)_{0,t} \otimes \mathbf{p}_{0,t} \right) \\
= \quad &d \left(S_N \left(x^{D_n} \right)_{0,t}, S_N \left(\mathbf{x} \right)_{0,t} \right) + 2d \left(\mathbf{p}_{0,t}^n, \mathbf{p}_{0,t} \right) \\
&+ d \left(S_N \left(\mathbf{x} \right)_{0,t}^{-1} \otimes S_N \left(x^n \right)_{0,t}, S \left(\mathbf{x} \right)_{0,t^n}^{-1} \otimes S_N \left(x^n \right)_{0,t^n} \right).
\end{aligned}
$$

From the assumptions and Hölder (resp. uniform Hölder) continuity of $S \left(\mathbf{x} \right)$ (resp. $S_N \left(x^n \right)$), we see that $d \left(S_N \left(x^n \right)_{0,t}, S_N \left(\mathbf{x} \right)_{0,t} \otimes \mathbf{p}_{0,t} \right) \to 0$, as required.

(iii) Uniform $\min \left(\alpha, \beta \right)$-Hölder bounds imply equivalence of pointwise and uniform convergence; convergence with Hölder exponent $\gamma < \min \left(\alpha, \beta \right)$ then follows by interpolation. Observe also that

$$
S_N \left(\mathbf{x} \right)_{0,t} \otimes \mathbf{p}_{0,t} = \mathbf{x}^{\varphi} = \exp \left(\log \left(S_N \left(\mathbf{x} \right) \right) + \log \mathbf{p} \right)
$$

is a simple consequence of $\mathbf{p}_{0,t}$ taking values in the centre.   ∎

12.3   Comments

The present exposition of RDEs with drift is new. A detailed study of RDEs with drift was previously carried out in Lejay and Victoir [107]. Exercise 12.13 goes back to Doss–Sussmann [44, 166] and is taken from Friz and Oberhauser [58], as is the bulk of material in Section 12.2 which can be used to prove optimality of various rough-path estimates for RDEs and linear RDEs obtained in Chapter 10.

Part III

Stochastic processes lifted to rough paths

13

Brownian motion

We discuss how Brownian motion can be enhanced, essentially by adding Lévy's stochastic area, to a process ("enhanced Brownian motion", EBM) with the property that almost every sample path is a geometric rough path ("Brownian rough path"). Various approximation results are studied, followed by a discussion of large deviations and support descriptions in rough path topology.

13.1 Brownian motion and Lévy's area

13.1.1 Brownian motion

We start with the following fundamental

Definition 13.1 (Brownian motion) A real-valued stochastic process $(\beta_t : t \geq 0)$ is a (1-dimensional) Brownian motion if it has the properties

(i) $\beta_0(\omega) = 0$ for all ω;
(ii) the map $t \mapsto \beta_t(\omega)$ is a continuous function of $t \in \mathbb{R}^+$ for all ω;
(iii) for every $t, h \geq 0, \beta_{t,t+h} \equiv \beta_{t+h} - \beta_t$ is independent of $(\beta_u : 0 \leq u \leq t)$, and has Gaussian distribution with mean 0 and variance h.

An $\mathbb{R}^d$-valued stochastic process $(B_t : t \geq 0)$ is a (d-dimensional) Brownian motion if it has independent components $(B^1, \ldots, B^d)$, each of which is a 1-dimensional Brownian motion. A realization of Brownian motion is called a Brownian path. □

It is an immediate corollary of properties (i)–(iii) that Brownian motion has *stationary increments*, that is

$$(B_{s,s+t} : t \geq 0) \overset{\mathcal{D}}{=} (B_t : t \geq 0),$$

as for the *Brownian scaling* property,

$$\forall \lambda > 0 : (B_{\lambda^2 t} : t \geq 0) \overset{\mathcal{D}}{=} (\lambda B_t : t \geq 0)$$

and

$$(B_t : t \geq 0) \overset{\mathcal{D}}{=} (t B_{1/t} : t \geq 0). \tag{13.1}$$

We trust the reader is familiar with the following basic facts concerning Brownian motion. (Some references are given in the comments section at the end of this chapter.)

- **Existence**. More precisely, there exists a unique Borel probability measure $\mathbb{W}$ on $C\left([0,\infty),\mathbb{R}^d\right)$ so that the coordinate function $B_t\left(\omega\right) = \omega_t$ defines a Brownian motion. The aforementioned measure is known as the (d-dimensional) *Wiener measure*.

- Brownian motion is a **martingale**. In fact, a theorem of P. Lévy states that if (M_t) denotes any $\mathbb{R}^d$-valued continuous martingale started at zero, such that

$$M_t \otimes M_t - t \times I$$

(where I is the ($d \times d$)-identiy matrix) is also a martingale, then (M_t) must be a d-dimensional Brownian motion.

- Brownian motion is a zero-mean **Gaussian process** with covariance function[1]

$$(s,t) \mapsto \mathbb{E}\left(B_s \otimes B_t\right) = (s \wedge t) \times I.$$

As for every continuous Gaussian process, mean and covariance fully determine the law of the process.

- Brownian motion is a (time-homogenous) **Markov process**. Its transition density – also known as *heat-kernel* – is given by

$$p_t\left(x,y\right) = \frac{1}{(2\pi t)^{d/2}} e^{-\frac{|x-y|^2}{2t}},$$

where $|\cdot|$ denotes the Euclidean norm on $\mathbb{R}^d$.

- Brownian sample paths are of unbounded variation, i.e. for any $T > 0$

$$|B|_{1\text{-var};[0,T]} = +\infty \text{ a.s.}$$

In fact, Brownian sample paths have unbounded p-variation for any $p \leq 2$ and the reader can find a self-contained proof in Section 13.9.

13.1.2 Lévy's area: definition and exponential integrability

Given two independent Brownian motions, say β and $\tilde{\beta}$, we define their *Lévy's area* as the stochastic Itô integral[2]

$$t \in [0,\infty) \mapsto \frac{1}{2}\left(\int_0^t \beta_s d\tilde{\beta}_s - \tilde{\beta}_s d\beta_s\right). \tag{13.2}$$

---

[1] $s \wedge t = \min\left(s,t\right).$

[2] Since $\left\langle\beta,\tilde{\beta}\right\rangle = \left\langle\tilde{\beta},\beta\right\rangle$ $(=0)$ it would not make a difference here to use Stratonovich integration.

We recall that (Itô) stochastic integrals are limits of their (left-point) Riemann–Stieltjes approximations, uniformly on compact time intervals. Indeed, given any sequence of dissections[3] (D_n) of $[0, T]$ with mesh $|D_n| \to 0$, one has[4]

$$\mathbb{E}\left[\sup_{t \in [0,T]} \left| \int_0^t \beta_s \, d\tilde{\beta}_s - \sum_{s_i \in D_n} \beta_{s_i} \left(\tilde{\beta}_{s_{i+1} \wedge t} - \tilde{\beta}_{s_i \wedge t} \right) \right|^2 \right] \to 0 \text{ as } n \to \infty.$$

(13.3)

Since uniform limits of continuous functions are continuous, (13.3) implies in particular that (13.2) can be taken to be continuous in t, with probability one.

Definition 13.2 (Lévy's area) Given a d-dimensional Brownian motion $B = \left(B^1, \dots, B^d \right)$, we define the d-dimensional Lévy area $A = \left(A^{i,j} : i, j \in \{1, \dots, d\} \right)$ as the continuous process

$$t \mapsto A_t^{i,j} = \frac{1}{2} \left(\int_0^t B_s^i \, dB_s^j - B_s^j \, dB_s^i \right).$$

We also define the Lévy area increments as, for any $s < t$ in $[0, T]$,

$$\begin{aligned} A_{s,t}^{i,j} &= A_t^{i,j} - A_s^{i,j} - \frac{1}{2} \left(B_s^i B_{s,t}^j - B_s^j B_{s,t}^i \right) \\ &= \frac{1}{2} \left(\int_s^t B_{s,r}^i \, dB_r^j - B_{s,r}^j \, dB_r^i \right). \end{aligned}$$

$\square$

We note that $A_t = A_{0,t}$ and more generally, $A_{s,t}$ take values in $so(d) \equiv \left[\mathbb{R}^d, \mathbb{R}^d \right]$, the space of anti-symmetric $d \times d$ matrices, and it suffices to consider $i \neq j$ (or $i < j$ if you wish). As a consequence of basic properties of Brownian motion, we have that

$$\forall \lambda > 0 : (A_{\lambda t} : t \geq 0) \overset{\mathcal{D}}{=} (\lambda A_t : t \geq 0),$$

$$\forall 0 \leq s < t < \infty : A_{s,t} \overset{\mathcal{D}}{=} A_{0,t-s}.$$

We now establish that the Lévy area has exponential integrability and note that by scaling it suffices to consider $t = 1$. There are many ways to see this integrability (including integrability properties of the Wiener–Itô chaos and heat-kernel estimates), but we have chosen a fairly elementary one.

---

[3] Unless otherwise stated, dissections are assumed to be deterministic.

[4] This can be taken as the very definition of the stochastic integral $\int_0^{\cdot} \beta \, d\beta$. Alternatively, only accepting that $\int_0^{\cdot} \beta_s \, d\tilde{\beta}_s$ is an L^2-martingale, one employs Doob's L^2-inequality to "get rid" of the sup inside the expectation, followed by Itô's L^2-isometry to establish (13.3).

Lemma 13.3 *Let B be a d-dimensional Brownian motion. Then, for all $\eta < 1/2$, we have*

$$\mathbb{E}\left[\exp\left(\frac{\eta}{T}\,|B|_{\infty;[0,T]}^2\right)\right] < \infty.$$

Proof. It suffices to consider $d = 1$, and $T = 1$. From $B \overset{\mathcal{D}}{=} (-B)$ and the *reflection principle* for Brownian motion,[5] we see that

$$
\begin{aligned}
\mathbb{P}\left[|B|_\infty \geq M\right] &\leq 2\mathbb{P}\left[\sup_{0 \leq t \leq 1} B_t \geq M\right] \\
&= 4\mathbb{P}\left(B_1 \geq M\right).
\end{aligned}
$$

The result follows from the usual tail behaviour of $B_1 \sim N(0,1)$.   ∎

Proposition 13.4 *Let B be a d-dimensional Brownian motion, and A its Lévy area. There exists $\eta > 0$ such that for any $0 \leq s < t \leq T$,*

$$\mathbb{E}\left[\exp\left(\eta\frac{|A_{s,t}|}{|t-s|}\right)\right] < \infty.$$

Proof. Since $A_{s,t}/|t-s| \overset{\mathcal{D}}{=} A_{0,1}$ it is enough to prove exponential integrability of Lévy's area at time 1. To this end, it suffices to consider a "building block" of Lévy's area of form $\int_0^1 \beta d\tilde{\beta}$. We observe that, *conditional on* $\beta(.)$, we can view $\int_0^1 \beta d\tilde{\beta}$ as if the integrand β were deterministic and from a very basic form of Itô's isometry,

$$\int_0^1 \beta_s d\tilde{\beta}_s \sim N\left(0, \int_0^1 \beta_s^2 ds\right).$$

It follows that, conditional on $\mathfrak{F}^\beta = \sigma\left(\beta_s : 0 \leq s \leq 1\right)$,

$$
\begin{aligned}
&\mathbb{E}\left[\left. e^{\eta\left|\int_0^1 \beta_s d\tilde{\beta}_s\right|}\right| \mathfrak{F}^\beta\right] \\
&= \mathbb{E}\left[\left. e^{\eta|Z|}\right| \mathfrak{F}^\beta\right] \text{ with } Z \sim N\left(0, \int_0^1 \beta_s^2 ds\right) \\
&\leq 2\mathbb{E}\left[\left. e^{\eta Z}\right| \mathfrak{F}^\beta\right] = 2\exp\left(\frac{\eta^2}{2}\int_0^1 \beta_s^2 ds\right) \\
&\leq 2\exp\left(\frac{\eta^2}{2}\,|\beta|_{\infty;[0,1]}^2\right)
\end{aligned}
$$

and, after taking expectations,

$$\mathbb{E}\left[e^{\eta\left|\int_0^1 \beta_s d\tilde{\beta}_s\right|}\right] \leq 2\mathbb{E}\exp\left(\frac{\eta^2}{2}\,|\beta|_{\infty;[0,1]}^2\right) < \infty$$

for $\eta > 0$ small enough, thanks to Lemma 13.3. The proof is now finished.   ∎

---

[5] For example, [143, p. 105].

Exercise 13.5 (Lévy's construction of Lévy's area) *Let β and $\tilde{\beta}$ be two independent Brownian motions on $[0,T]$. Consider a sequence of dyadic dissections $D_n = \{iT2^{-n} : i = 0, \ldots, 2^n\}$ and write $\beta^{D_n}, \tilde{\beta}^{D_n}$ for the resulting piecewise linear approximations. Show that*

$$A_n(\omega) := \frac{1}{2}\left(\int_0^T \beta_s^{D_n}\, d\tilde{\beta}_s^{D_n} - \tilde{\beta}_s^{D_n}\, d\beta_s^{D_n}\right)$$

is a (discrete-time) martingale with respect to the filtration

$$\mathfrak{F}_n := \sigma\left(\left(\beta_t, \tilde{\beta}_t\right) : t \in D_n\right)$$

which converges in $L^2(\mathbb{P})$. Identify the limit as Lévy's area (at time T), as defined in (13.2).

Exercise 13.6 *Consider a d-dimensional Brownian motion with its associated, so (d)-valued, Lévy area process A. Let $(0 = t_0 < t_1 < \cdots < t_n = T)$ be a dissection of $[0,T]$. Show that*

$$\left|\sum_{i=1}^n A_{t_{i-1},t_i}\right|_{L^q(\mathbb{P})} \le cq \left|\sum_{i=1}^n A_{t_{i-1},t_i}\right|_{L^2(\mathbb{P})}$$

where c is a constant, independent of n and q. (Remark that this estimate is an immediate consequence of integrability properties of Wiener–Itô chaos as discussed in Section D.4 in Appendix D. The point of this exercise is to give an elementary proof.)

Solution. Without loss of generality, we take $T = 1$. If $X_i := A_{t_{i-1},t_i}$ and $S_n = X_1 + \cdots + X_n$, a sum of independent random variables, it suffices to show existence of $\eta, M > 0$, independent of n and q, such that $\mathbb{E}\left(e^{\eta |S_n|/|S_n|_{L^2}}\right) < M < \infty$. From $S_n \overset{\mathcal{D}}{=} -S_n$ and $e^{|x|} \le e^x + e^{-x}$ it is enough to estimate

$$\mathbb{E}\left(e^{\eta S_n/|S_n|_{L^2}}\right) = \Pi_{i=1}^n \mathbb{E}\left(e^{\eta X_i/|S_n|_{L^2}}\right).$$

Note that, for all λ small enough, $\mathbb{E}\left(e^{\lambda A_{0,1}}\right) = \mathbb{E}\left(e^{\lambda^2 Z}\right)$ for some random variable Z with an exponential tail. (This can be seen from the identity $\mathbb{E}\exp\left(\lambda \int_0^1 \beta d\tilde{\beta}\right) = \mathbb{E}\exp\left(\frac{\lambda^2}{2}\int_0^1 \beta^2 dt\right)$ obtained by conditioning on β, exactly as in the proof of Proposition 13.4.) Note also that by scaling properties of Lévy's area,

$$X_i \overset{\mathcal{D}}{=} |t_i - t_{i-1}|\, A_{0,1} = c|X_i|_{L^2}\, A_{0,1} \quad \text{with } c = 1/|A_{0,1}|_{L^2}.$$

It follows that, for η small enough, and with $\theta_i = |X_i|_{L^2}^2 / |S_n|_{L^2}^2$,

$$\mathbb{E}\left(\exp\left(\eta\, X_i/|S_n|_{L^2}\right)\right) = \mathbb{E}\left(\exp\left(\eta c \frac{|X_i|_{L^2}}{|S_n|_{L^2}} A_{0,1}\right)\right)$$

$$= \mathbb{E}\left(\exp\left(\eta^2 c^2 \theta_i Z\right)\right) \leq \mathbb{E}\left(\exp\left(\eta^2 c^2 Z\right)\right)^{\theta_i}$$

by Jensen's inequality. As a result, $\mathbb{E}\left(e^{\eta\, S_n/|S_n|_{L^2}}\right) \leq \mathbb{E}\left(\exp\left(\eta^2 c^2 Z\right)\right) < \infty$
for η small enough.   □

13.1.3   Lévy's area as time-changed Brownian motion

The following result will only be used in Section 13.8 on the support theorem in its conditional form.

Proposition 13.7 *Let B be a d-dimensional Brownian motion and fix two distinct components $\beta = B^i, \tilde{\beta} = B^j$ where $i, j \in \{1, \ldots, d\}$ and $i \neq j$. Set*

$$A(t) := \frac{1}{2}\left(\int_0^t \beta_s d\tilde{\beta}_s - \tilde{\beta}_s d\beta_s\right) \quad and \quad a(t) := \frac{1}{4}\int_0^t \left(\beta_s^2 + \tilde{\beta}_s^2\right) ds.$$

Then $\left(A\left(a^{-1}(t)\right) : t \geq 0\right)$ is a (1-dimensional) Brownian motion, independent of the process $\left(\beta_s^2 + \tilde{\beta}_s^2 : s \geq 0\right)$, and hence independent of the radial process $\{|B_s| : s \geq 0\}$ where $|\cdot|$ denotes Euclidean norm on $\mathbb{R}^d$.

Proof. Set $r_t(\omega) \equiv r_t \equiv \sqrt{\beta_t^2 + \tilde{\beta}_t^2}$. By Itô's formula,

$$\frac{r_t^2}{2} = \int_0^t r_s d\gamma_s + t \tag{13.4}$$

where

$$\gamma_t(\omega) = \int_0^t \frac{\beta_s}{r_s} d\beta_s + \int_0^t \frac{\tilde{\beta}_s}{r_s} d\tilde{\beta}_s.$$

(Note that dr and $d\gamma$ differ by a drift differential.) Clearly, the system of martingales (A, γ) satisifes the bracket relations $\langle\gamma\rangle_t = t$, $\langle\gamma, A\rangle_t = 0$ and

$$\langle A\rangle_t = \frac{1}{4}\int_0^t r_s^2 ds.$$

Let $\tilde{\gamma}_t = A(\phi_t)$ where $\phi_t = a^{-1}(t)$. By Lévy's characterization, γ and $\tilde{\gamma}$ are two mutually independent Brownian motions. Moreover, (13.4) shows that r_t is the pathwise unique solution to an SDE driven by $\tilde{\gamma}$ and, in particular, $\sigma[r_s : s \leq t] \subset \sigma[\gamma_s : s \leq t]$. Consequently, the processes $(\tilde{\gamma}_t)$ and (r_t) are independent and we arrive at the representation

$$A(t) = \tilde{\gamma}\left(\frac{1}{4}\int_0^t r_s^2 ds\right)$$

where $\tilde{\gamma}$ is a Brownian motion independent of the process (r_t). This concludes the proof. ∎

Exercise 13.8 *Derive the characteristic function of* $\frac{1}{2}\left(\int_0^t \beta_s d\tilde{\beta}_s - \tilde{\beta}_s d\beta_s\right)$.

13.2 Enhanced Brownian motion

13.2.1 Brownian motion lifted to a $G^2\left(\mathbb{R}^d\right)$-valued path

Recall that exp denotes the exponential map from $g^2\left(\mathbb{R}^d\right) \cong \mathbb{R}^d \oplus so\left(d\right)$ $\rightarrow G^2\left(\mathbb{R}^d\right)$. Its inverse $\log\left(\cdot\right)$ can be viewed as a global chart for $G^2\left(\mathbb{R}^d\right)$, which is therefore diffeomorphic to a Euclidean space of dimension $d + d\left(d-1\right)/2$. (As far as the geometry is concerned, it cannot get much simpler!) If $x : [0,T] \rightarrow \mathbb{R}^d$ is a smooth path started at 0, then its step-2 lift satisfies

$$S_2\left(x\right)_t = \exp\left(x_t + a_t\right) \in G^2\left(\mathbb{R}^d\right)$$

with area

$$a_t^{i,j} = \frac{1}{2}\left(\int_0^t x_s^i dx_s^j - x_s^j dx_s^i\right).$$

Recall also that

$$S_2\left(x\right)_{s,t} = S_2\left(x\right)_s^{-1} \otimes S_2\left(x\right)_t = \exp\left(x_{s,t} + a_{s,t}\right),$$

where $x_{s,t} = x_t - x_s$ and $a_{s,t} \in so\left(d\right)$ is given by

$$
\begin{aligned}
a_{s,t}^{i,j} &= \frac{1}{2}\left(\int_s^t x_{s,r}^i dx_r^j - x_{s,r}^j dx_r^i\right) \\
&= a_t^{i,j} - a_s^{i,j} - \frac{1}{2}\left(x_s^i x_{s,t}^j - x_s^j x_{s,t}^i\right).
\end{aligned}
$$

This motivates us to define the lift Brownian motion to a process with values in $G^2\left(\mathbb{R}^d\right)$ as follows.

Definition 13.9 (enhanced Brownian motion, or EBM) Let B and A denote a d-dimensional Brownian motion and its Lévy area process. The continuous $G^2\left(\mathbb{R}^d\right)$-valued process $\mathbf{B}$, defined by

$$\mathbf{B}_t := \exp\left[B_t + A_t\right], \quad t \geq 0,$$

is called **enhanced Brownian motion**; if we want to stress the underlying process we call $\mathbf{B}$ the **natural lift** of B. Sample path realizations of $\mathbf{B}$ are called **Brownian rough paths**. (This terminology is motivated by the forthcoming Corollary 13.14). □

We also write $\mathbf{B}_{s,t} = \mathbf{B}_s^{-1} \otimes \mathbf{B}_t \in G^2(\mathbb{R}^d)$ and observe that this is consistent with

$$\mathbf{B}_{s,t} = \exp[B_{s,t} + A_{s,t}],$$

where $B_{s,t} = B_t - B_s$ (as usual) and $A_{s,t} \in so(d)$ is given by

$$\begin{aligned} A_{s,t}^{i,j} &= A_t^{i,j} - A_s^{i,j} - \frac{1}{2}\left(B_s^i B_{s,t}^j - B_s^j B_{s,t}^i\right) \\ &= \frac{1}{2}\left(\int_s^t B_{s,r}^i dB_r^j - B_{s,r}^j dB_r^i\right) \quad \text{a.s.} \end{aligned}$$

Why? We have just recalled that all this holds for smooth paths, where one can write out all iterated integrals as Riemann–Stieltjes integrals. This is still true for the Brownian case but now convergence is only in the L^2-sense, see (13.3), and L^2-limits are only defined up to null-sets, hence the a.s. above.

Exercise 13.10 *(i) Check that, almost surely,*

$$\mathbf{B}_t = \left(1, B_t, \int_0^t B \otimes \circ dB\right) \in G^2(\mathbb{R}^d),$$

where $\circ dB$ denotes Stratonovich integration.
(ii) Show that

$$\hat{\mathbf{B}}_t = \left(1, B_t, \int_0^t B \otimes dB\right) \in T_1^2(\mathbb{R}^d),$$

where dB denotes Itô integration, does <u>not</u> yield a geometric rough path. Hint: Consider $i = j$ and compute the expectation.

The following proposition should be compared with our definition of Brownian motion, Definition 13.1. It identifies enhanced Brownian motion as a special case of a left-invariant Brownian motion on a Lie group.

Proposition 13.11 *Enhanced Brownian motion $\mathbf{B}$ is a left-invariant Brownian motion on the Lie group $\left(G^2(\mathbb{R}^d), \otimes, ^{-1}, 1\right)$ in the sense that*
(i) $\mathbf{B}_0(\omega) = 1$ for all ω;
(ii) the map $t \mapsto \mathbf{B}_t(\omega)$ is a continuous function of $t \in \mathbb{R}^+$ for all ω;
(iii) for every $t, h \geq 0$, $\mathbf{B}_{t,t+h} = \mathbf{B}_t^{-1} \otimes \mathbf{B}_{t+h}$ is independent of $\sigma(\mathbf{B}_u : u \leq t)$;
(iv) it has stationary increments,

$$(\mathbf{B}_{s,s+t} : t \geq 0) \overset{\mathcal{D}}{=} (\mathbf{B}_t : t \geq 0).$$

Proof. (i),(ii) are trivial. For (iii) observe that, since A_r is measurable function of $\{B_u : u \leq r\}$,

$$\sigma(\mathbf{B}_r : r \leq s) = \sigma(B_r, A_r : r \leq s) = \sigma(B_r : r \leq s).$$

On the other hand, $(B_{s,s+t}, A_{s,s+t} : t \geq 0)$ is measurably determined by

$$\sigma(B_{s,r} : r \geq s) = \sigma(B_{s,s+t} : t \geq 0),$$

see (13.5) in particular. From defining properties of Brownian motion, $\sigma\left(B_r : r \leq s\right)$ and $\sigma\left(B_{s,s+t} : t \geq 0\right)$ are indepedent and this finishes the proof.

(iv) Recall that for Brownian motion, for all $s \geq 0$,

$$\left(B_{s,s+t} : t \geq 0\right) \overset{\mathcal{D}}{=} \left(B_t : t \geq 0\right).$$

Then, for s fixed,

$$
\begin{aligned}
\left(A_{s,s+t}^{i,j}\right)_{t\geq 0} &= \frac{1}{2}\left(\int_s^{s+t} B_{s,r}^i \, dB_r^j - B_{s,r}^j \, dB_r^i\right)_{t\geq 0} \\
&= \frac{1}{2}\left(\int_s^{s+t} B_{s,r}^i \, dB_{s,r}^j - B_{s,r}^j \, dB_{s,r}^i\right)_{t\geq 0} \\
&\overset{\mathcal{D}}{=} \frac{1}{2}\left(\int_0^t B_r^i \, dB_r^j - B_r^j \, dB_r^i\right)_{t\geq 0}
\end{aligned}
\tag{13.5}
$$

and the same holds for the pair

$$\left(B_{s,t}, A_{s,s+t}\right)_{t\geq 0} \overset{\mathcal{D}}{=} \left(B_t, A_t\right)_{t\geq 0}.$$

∎

Recall that the Lie group $\left(G^2\left(\mathbb{R}^d\right), \otimes, {}^{-1}, e\right)$ has the additional structure of *dilation*. As we shall now see, it fits together perfectly with scaling properties of enhanced Brownian motion.

Lemma 13.12 (EBM, scaling) *Let* **B** *be an enhanced Brownian motion. For all* $\lambda > 0$ *we have*

$$\left(\mathbf{B}_{\lambda^2 t} : t \geq 0\right) \overset{\mathcal{D}}{=} \left(\delta_\lambda \mathbf{B}_t : t \geq 0\right),$$

where δ *is the dilatation operator on* $G^2\left(\mathbb{R}^d\right)$.

Proof. From Brownian scaling, for any $\lambda > 0$ we have

$$\left(B_{\lambda^2 t}\right)_{t\geq 0} \overset{\mathcal{D}}{=} \left(\lambda B_t\right)_{t\geq 0}.$$

That is, speeding up time by a factor of λ^2 is, in law, equivalent to spatial scaling by a factor of λ. Since A is determined as the limit of a homogenous polynomial of degree 2 in terms of Brownian increments, see (13.3), the scaling factor λ appears twice and one has

$$\left(B_{\lambda^2 t}, A_{\lambda^2 t}\right)_{t\geq 0} \overset{\mathcal{D}}{=} \left(\lambda B_t, \lambda^2 A_t\right)_{t\geq 0}.$$

Now apply $\exp: \mathbb{R}^d \oplus so\left(d\right) \to G^2\left(\mathbb{R}^d\right)$.

∎

13.2.2 Rough path regularity

As shown in Theorem 13.69 in the appendix to this chapter, Brownian motion has infinite 2-variation[6] and hence infinite q-variation for any $q \leq 2$. Therefore, our only chance to construct a "Brownian" rough path is to look for a geometric p-rough path, $p > 2$, which lifts B. In other words, we look for a process $\mathbf{B}$ with values in $G^{[p]}\left(\mathbb{R}^d\right)$, with finite p-variation (or $1/p$-Hölder regularity) with respect to the Carnot–Caratheodory distance, such that

$$\pi_1\left(\mathbf{B}\right) = B.$$

We shall see that an enhanced Brownian motion, i.e. the $G^2\left(\mathbb{R}^d\right)$-valued process $t \mapsto \mathbf{B}_t \equiv \exp\left(B_t, A_t\right)$, has in fact a.s. finite α-Hölder regularity for any $a < 1/2$. In particular, there is no cost in assuming $\alpha \in \left(1/3, 1/2\right)$ so that $[p] = [1/\alpha] = 2$, which confirms that a.e. realization of $\mathbf{B} = \mathbf{B}\left(\omega\right)$ is a geometric p-rough path (in fact a geometric $1/p$-Hölder rough path), $p \in \left(2, 3\right)$, in the sense of Definition 9.15.

In order to establish that $\mathbf{B}$ is a.s. a geometric α-Hölder rough path we need to show that, for some $\alpha \in \left(1/3, 1/2\right)$, the path $t \mapsto \mathbf{B}_t$ is Hölder regular with respect to $d\left(g, h\right) = \left\|g^{-1} \otimes h\right\|$, the Carnot–Caratheodory norm. Using equivalence of homogenous norms, all this boils down to the question of whether

$$d\left(\mathbf{B}_s, \mathbf{B}_t\right) = d\left(e^{B_s + A_s}, e^{B_t + A_t}\right) = \left\|e^{B_{s,t} + A_{s,t}}\right\| \sim |B_{s,t}| \vee |A_{s,t}|^{1/2}$$

is bounded by $C\left(\omega\right)|t - s|^\alpha$, uniformly for s, t on a finite interval such as $[0, 1]$.

Obviously, this is true for the Brownian increment $B_{s,t}$ and we are only left with the question

"Does there exist $\alpha > 1/3$ such that $\displaystyle \sup_{s,t \in [0,1]} \frac{|A_{s,t}|}{|t - s|^{2\alpha}} < \infty$ a.s. ?"

$$(13.6)$$

To fully appreciate the forthcoming Corollary 13.14, the reader is urged to think for a moment about how to prove this! To avoid misunderstandings, let us point out two things:

(i) α-Hölder regularity, for any $\alpha < 1/2$, of $t \mapsto A_t\left(\omega\right) \in so\left(d\right)$ is a straightforward application of a suitable version of Kolmogorov's regularity criterion, applied to a process with values in the Euclidean space $so\left(d\right)$. It also follows from Proposition 13.7 that we can represent Lévy area as (α-Hölder continuous) Brownian motion run at a (Lipschitz continuous) random clock.

---

[6] This is not to be confused with the important fact that Brownian motion has *finite quadratic variation* in the sense of Theorem 13.70.

(ii) The *cancellation* on the right-hand side of

$$A_{s,t} = (A_t - A_s) - \frac{1}{2} [B_s, B_{s,t}] \qquad (13.7)$$

is essential. Thoughtless application of the triangle inequality only shows

$$|A_{s,t}| \leq |A_t - A_s| + \frac{1}{2} |B|_\infty |B_{s,t}|$$
$$\lesssim C(\omega) |t - s|^\alpha,$$

which is *not* the positive answer to (13.6) which we seek. In fact, this only shows that sample paths $t \mapsto \mathbf{B}_t(\omega)$ are a.s. Hölder of exponent less than $1/4$ and there is *a world of difference* between $\alpha < 1/4$ and $\alpha > 1/3$. The first is a sample path property of limited interest, the latter implies that almost every realization $\mathbf{B}(\omega)$ is a geometric rough path to which all of the theory of rough paths is applicable!

Theorem 13.13 *Write* $\mathbf{B}$ *for a* $G^2(\mathbb{R}^d)$*-valued enhanced Brownian motion on* $[0, T]$*. Then there exists* $\eta > 0$*, not dependent on* T*, such that*

$$\sup_{s,t \in [0,T]} \mathbb{E}\left[\exp\left(\eta \frac{d(\mathbf{B}_s, \mathbf{B}_t)^2}{|t - s|} \right) \right] < \infty. \qquad (13.8)$$

Proof. From scaling properties of enhanced Brownian motion, $\mathbf{B}_{s,t} \overset{\mathcal{D}}{=} \delta_{(t-s)^{1/2}} \mathbf{B}_{0,1}$ so that

$$d(\mathbf{B}_s, \mathbf{B}_t)^2 = \|\mathbf{B}_{s,t}\|^2 \overset{\mathcal{D}}{=} (t - s) \|\mathbf{B}_1\|^2.$$

Hence, it suffices to find η small enough so that

$$\mathbb{E}\left[\exp\left(\eta \|\mathbf{B}_1\|^2 \right) \right] < \infty.$$

By equivalence of homogenous norms, $\|\mathbf{B}_1\|^2 \sim |B_1|^2 + |A_1|$ where B_1 (resp. A_1) denotes d-dimensional Brownian motion (resp. Lévy area) at time 1. Thus, everything boils down to (trivial) Gaussian integrabilty of $B_1 \sim N(0,1)$ and exponential integrability of Lévy area, which was established in Proposition 13.4. ∎

Thanks to (13.8), we can appeal to general regularity results for stochastic processes (as discussed in Section A.4 in Appendix A).

Corollary 13.14 *Write* $\mathbf{B}$ *for a* $G^2(\mathbb{R}^d)$*-valued enhanced Brownian motion on* $[0, T]$*.*
(i) Let $\alpha \in [0, 1/2)$*. Then there exists* $\eta > 0$*, not dependent on* T*, such that*

$$\mathbb{E}\left[\exp\left(\frac{\eta}{T^{1-2\alpha}} \|\mathbf{B}\|_{\alpha\text{-Höl};[0,T]}^2 \right) \right] < \infty.$$

(ii) Assume that φ is a fixed increasing function such that $\varphi(h) = \sqrt{h \log(1/h)}$ (Lévy modulus) in a positive neighbourhood of 0. Then there exists $\eta = \eta(T) > 0$ such that

$$\mathbb{E}\left[\exp\left(\eta \|\mathbf{B}\|^2_{\varphi\text{-Höl};[0,T]}\right)\right] < \infty$$

where $\|\mathbf{B}\|_{\varphi\text{-Höl};[0,T]} = \sup_{s<t \text{ in}[0,T]} d(\mathbf{B}_s, \mathbf{B}_t)/\varphi(t-s)$.

Proof. (i) By scaling, $T^{2\alpha-1}\|\mathbf{B}\|^2_{\alpha\text{-Höl};[0,T]}$ has the same distribution as $\|\mathbf{B}\|^2_{\alpha\text{-Höl};[0,1]}$. We then apply Theorem A.19 with modulus function $h \mapsto h^{1/\alpha}$.

(ii) A direct application of Theorem A.19 with modulus function $h \mapsto \varphi(h)$. ∎

Let us remark that we may take $\alpha \in (1/3, 1/2)$ in the previous corollary, which therefore implies a fortiori that $\mathbf{B}$ is a.s. a geometric α-Hölder rough path. Since α-Hölder regularity implies p-variation regularity with $p = 1/\alpha \in (2,3)$, we trivially see that $\mathbf{B}$ is a.s. a geometric p-rough path. Similarly, φ-Hölder regularity implies φ^{-1}-variation regularity where

$$\varphi^{-1}(h) \sim \frac{h^2}{\log(1/h)}.$$

In fact, more is true and the general results of Section A.4 (Theorem A.24 to be precise) show that (13.8) implies

Theorem 13.15 (exact variation for EBM) *Write $\mathbf{B}$ for a $G^2(\mathbb{R}^d)$-valued enhanced Brownian motion on $[0,T]$. Let[7]*

$$\psi_{2,1}(h) = \frac{h^2}{\ln^* \ln^*(1/h)},$$

where $\ln^* = \max(1, \ln)$. Then there exists $\eta > 0$, not dependent on T, such that

$$\mathbb{E}\left[\exp\left(\frac{\eta}{T}\|\mathbf{B}\|^2_{\psi_{2,1}\text{-var};[0,T]}\right)\right] < \infty$$

where the reader is reminded that

$$\|\mathbf{B}\|_{\psi\text{-var};[0,T]} = \inf\left\{M > 0, \sup_{D \in \mathcal{D}[0,T]} \sum_{t_i \in D} \psi\left[d(\mathbf{B}_{t_i}, \mathbf{B}_{t_{i+1}})/M\right] \leq 1\right\}.$$

Enhanced Brownian motion also satisfies a law of iterated logarithms. We first recall *Khintchine's law of iterated logarithms* for a Brownian motion[8]

---

[7] This is one instance of a (generalized) variation function, as introduced in Definition 5.45.

[8] See McKean's classical text [125, p. 12] or [94, Theorem 9.23], for instance; it can also be obtained as a consequence of Schilder's theorem, to be discussed in Section 13.6.

which states that, for a 1-dimensional Brownian motion β,

$$\mathbb{P}\left[\limsup_{h\to 0} \frac{|\beta_h|}{\varphi(h)} = c\right] = 1, \tag{13.9}$$

where $c \in (0,\infty)$ is a deterministic constant (equal to $\sqrt{2}$, in fact) and $\varphi(h) = \sqrt{h \ln^* \ln^* (1/h)}$. (Observe that φ is Lipschitz equivalent to the inverse of $\psi_{2,1}$, see Lemma 5.48).

Proposition 13.16 (law of the iterated logarithm for EBM) *Write* **B** *for* $G^2\left(\mathbb{R}^d\right)$*-valued enhanced Brownian motion on* $[0,T]$*. Let* $\varphi(h) = \sqrt{h \ln^* \ln^* (1/h)}$. *Then there exists a deterministic constant* $c \in (0,\infty)$ *such that*

$$\mathbb{P}\left[\limsup_{h\to 0} \frac{\|\mathbf{B}\|_{0;[0,h]}}{\varphi(h)} = c\right] = 1.$$

Proof. From general principles (Theorem A.21 in Section A.4) we see that (13.8) implies

$$L := \limsup_{h\to 0} \frac{\|\mathbf{B}\|_{0;[0,h]}}{\varphi(h)}$$

defines an almost surely finite random variable, i.e. $L(\omega) < \infty$ almost surely. On the other hand, by the classical law of iterated logarithms for Brownian motion, it is clear that

$$\limsup_{h\to 0} \frac{\|\mathbf{B}\|_{0;[0,h]}}{\varphi(h)} \geq \limsup_{h\to 0} \frac{|\beta_h|}{\varphi(h)} = \tilde{c} > 0 \text{ a.s.}$$

where $\tilde{c} = \sqrt{2}$ is the constant from Khintchine's law of iterated logarithms. It follows that

$$0 < \tilde{c} \leq L(\omega) < \infty \text{ a.s.}$$

By construction of enhanced Brownian motion, $\|\mathbf{B}\|_{0;[0,h]}$ is $\sigma(B_t : t \in [0,h])$ measurable where $B = \pi_1(\mathbf{B})$ denotes the underlying d-dimensional Brownian motion. It now follows from Blumenthal's zero–one law for Brownian motion[9] that L equals, almost surely, a deterministic constant. ∎

13.3   Strong approximations

We discuss a number of approximation results in which enhanced Brownian motion arises as an *almost-sure limit* or *limit in probability*, always in the appropriate rough path metrics. The interest in these results is that either convergence is preserved under continuous maps; applied to the Itô–Lyons

---

[9] See, for example, [143, Chapter III] or [94, Theorem 7.17].

map in rough path topology all convergence results discussed then translate immediately to strong convergence results in which the limit of certain (random ODEs/RDEs) is identified as an RDE solution driven by **B**, i.e. as the solution to a Stratonovich SDE.

Our list of approximations is not exhaustive and several strong convergence results (including convergence of non-nested piecewise linear approximations and Karhunen–Loéve approximations) are left to a later chapter on Gaussian processes (Chapter 15) which provides the natural framework for these convergence results.

13.3.1 Geodesic approximations

Let us fix $\alpha < 1/2$. From the last section, we know that enhanced Brownian motion **B** has sample paths with

$$\mathbf{B}\left(\omega\right) \in C_0^{\alpha\text{-Höl}}\left(\left[0,T\right], G^2\left(\mathbb{R}^d\right)\right)$$

almost surely. From general interpolation results, it follows that for every $\alpha < 1/2$ we also have

$$\mathbf{B}\left(\omega\right) \in C_0^{0,\alpha\text{-Höl}}\left(\left[0,T\right], G^2\left(\mathbb{R}^d\right)\right)$$

almost surely. From the very definition of the space $C_0^{0,\alpha\text{-Höl}}$ it then follows that almost every $\mathbf{B}\left(\omega\right)$ is the $d_{\alpha\text{-Höl}}$-limit of smooth paths lifted to $G^2\left(\mathbb{R}^d\right)$. (When $\alpha \in (1/2, 1/3)$ this is precisely the difference between weak geometric α-Hölder rough paths and (genuine) geometric α-Hölder rough paths.) The important remarks here are that (i) these approximations are based on entirely deterministic facts and applied to almost every ω and (ii) they rely on all the information contained in $\mathbf{B}\left(\omega\right)$, that is on the underlying Brownian path $B\left(\omega\right) = \pi_1\left(\mathbf{B}\right)$ and the Lévy area $A\left(\omega\right) = \text{Anti}\left(\pi_2\left(\mathbf{B}\right)\right)$. This is in strict contrast to all *probabilistic* approximations discussed in the following sections. These are only based on the $\mathbb{R}^d$-valued Brownian motion $B = \pi_1\left(\mathbf{B}\right)$ and frequently (but not always!) give rise to the standard Lévy area which underlies our definition of enhanced Brownian motion.

13.3.2 Nested piecewise linear approximations

As earlier, $\mathbf{B} = \exp\left(B + A\right)$ denotes an enhanced Brownian motion, the natural lift of a d-dimensional Brownian motion B. We now consider a sequence (D_n) of *nested* dissections, that is $D_n \subset D_{n+1}$ for all n, such that $|D_n|$, the mesh of D_n, tends to zero as $n \to \infty$. The reason for this assumption is that then

$$\mathfrak{F}_n := \sigma\left(B_t : t \in D_n\right)$$

forms a family of σ-algebras increasing in n. In other words, $(\mathfrak{F}_n)$ is a filtration and this will allow us to use elegant martingale arguments. (What we will <u>not</u> use here is the fact that $t \mapsto B_t$ is a martingale.) Define $B^n = B^{D_n}(\omega)$ as the piecewise linear approximation based on the dissection D_n. We consider the step-2 lift and write, as usual,

$$\mathbf{B}^n := S_2\left(B^n\right) = \exp\left(B^n + A^n\right).$$

Proposition 13.17 *For fixed t in $[0, T]$ the convergence $\mathbf{B}_t^n \to \mathbf{B}_t$ holds almost surely and in $L^2(\mathbb{P})$.*

Proof. The statement is $d\left(\mathbf{B}_t^n, \mathbf{B}_t\right) \to 0$ (a.s. and in L^2). This is equivalent to

$$\text{(a)} \quad |B_t^n - B_t| \to 0 \quad \text{and} \quad \text{(b)} \quad |A_t^n - A_t| \to 0.$$

Ad (a), Since $\{D_n\}$ is nested, $\mathfrak{F}_n := \sigma\left(B_t : t \in D_n\right)$ forms a filtration. We claim that a.s.

$$\mathbb{E}\left[B_t | \mathfrak{F}_n\right] = B_t^n \text{ and } \mathbb{E}\left[A_t | \mathfrak{F}_n\right] = A_t^n.$$

Using the Markov property of B,

$$\mathbb{E}\left[B_t | \mathfrak{F}_n\right] = \mathbb{E}\left[B_t | B_{t_i}, B_{t_{i+1}}\right]$$

where t_i, t_{i+1} are two neighbours in D_n with $t \in [t_i, t_{i+1}]$. It is a simple exercise of Gaussian conditioning[10] to see that

$$\mathbb{E}\left[B_t | B_{t_i}, B_{t_{i+1}}\right] = \frac{t_{i+1} - t}{t_{i+1} - t_i} B_{t_i} + \frac{t - t_i}{t_{i+1} - t_i} B_{t_{i+1}}$$

and this is precisely equal to B_t^n. Mesh $|D_n| \to 0$ implies that B_t is $(\vee_n \mathfrak{F}_n)$-measurable and martingale convergence shows that[11]

$$B_t^n = \mathbb{E}\left[B_t | \mathfrak{F}_n\right] \to B_t \text{ a.s and in } L^2.$$

Ad (b). We first fix n and show $\mathbb{E}\left[A_t | \mathfrak{F}_n\right] = A_t^n$. It simplifies things to set $\beta = B^i, \tilde{\beta} = B^j, i \neq j$ and consider $\int_0^t \beta d\tilde{\beta}$. Let $\{\tilde{D}^m\}$ be a dissection of $[0, t]$, with t fixed, and mesh $\left|\tilde{D}^m\right| \to 0$. By L^2-continuity of $\mathbb{E}\left[\cdot | \mathfrak{F}_n\right]$ and

---

[10] The reader might be familiar with $\mathbb{E}\left[B_t | B_T\right] = (t/T) B_T$.

[11] There are more elementary arguments for $B_t^n \to B$ but this one extends to the area level.

(13.3),

$$\mathbb{E}\left[\int_0^t \beta d\tilde{\beta}\Big|\mathfrak{F}_n\right] = \lim_{m\to\infty}\mathbb{E}\left[\sum_{t_i\in\tilde{D}^m}\beta_{t_i}\tilde{\beta}_{t_i,t_{i+1}}\Big|\mathfrak{F}_n\right]$$

$$= \lim_{m\to\infty}\sum_{t_i\in\tilde{D}^m}\mathbb{E}\left[\beta_{t_i}\tilde{\beta}_{t_i,t_{i+1}}\Big|\mathfrak{F}_n\right]$$

$$= \lim_{m\to\infty}\sum_{t_i\in\tilde{D}^m}\beta_{t_i}^n\tilde{\beta}_{t_i,t_{i+1}}^n \qquad (\text{use } \beta\perp\tilde{\beta} \text{ and part (a)})$$

$$= \int_0^t \beta^n d\tilde{\beta}^n,$$

by definition of the Riemann–Stieltjes integral applied to the (bounded variation!) integrator $\tilde{\beta}^n$. After exchanging the roles of β and $\tilde{\beta}$ and subtraction, we find $\mathbb{E}[A_t|\mathfrak{F}_n] = A_t^n$ as claimed. The final reasoning is as above: A_t is $(\vee_n\mathfrak{F}_n)$-measurable, this follows from (13.3), and by martingale convergence

$$A_t^n = \mathbb{E}[A_t|\mathfrak{F}_n] \to A_t \text{ a.s and in } L^2.$$

∎

Theorem 13.18 *For every $\alpha\in[0,1/2)$, there exists a positive random variable M with Gaussian tails, in particular $M<\infty$ a.s., such that*

$$\sup_{n=1,\dots,\infty}\|\mathbf{B}^n\|_{\alpha\text{-Höl};[0,T]} \le M$$

where $\mathbf{B}^\infty\equiv\mathbf{B}$.

Proof. We keep the notation of the last proof, where we established

$$B_t^n = \mathbb{E}[B_t|\mathfrak{F}_n] \text{ and } A_t^n = \mathbb{E}[A_t|\mathfrak{F}_n].$$

Simple algebra (attention $A_{s,t}\ne A_t - A_s$!) yields

$$B_{s,t}^n = \mathbb{E}[B_{s,t}|\mathfrak{F}_n] \text{ and } A_{s,t}^n = \mathbb{E}[A_{s,t}|\mathfrak{F}_n]. \qquad (13.10)$$

We focus on one component in the matrix $A_{s,t}$, say $A_{s,t}^{i,j}$ with $i\ne j$. Clearly,

$$\left|A_{s,t}^{i,j}\right| \le |A_{s,t}| \le \left(|B_{s,t}|\vee|A_{s,t}|^{1/2}\right)^2 \sim \|\mathbf{B}_{s,t}\|^2,$$

where $\sim$ is a reminder of the Lipschitz equivalence of homogenous norms on $G^2(\mathbb{R}^d)$. From Corollary 13.14, $\|\mathbf{B}_{s,t}\|^2 \le M_1(t-s)^{2\alpha}$ for a non-negative r.v. M_1 with Gauss tail. In particular, $|M_1|_{L^q}<\infty$ for all $q<\infty$. (More precisely, the Gauss tail is captured in $|M_1|_{L^q} = O(q^{1/2})$ for q large.) We then have

$$-M_1(t-s)^{2\alpha} \le A_{s,t}^{i,j} \le M_1(t-s)^{2\alpha}$$

and conditioning with respect to $\mathfrak{F}_n$ yields

$$-M_2 \left(t - s\right)^{2\alpha} \leq \mathbb{E}[A_{s,t}^{i,j}|\mathfrak{F}_n] \leq M_2 \left(t - s\right)^{2\alpha}$$

where $M_2 = \sup\{\mathbb{E}[M_1|\mathfrak{F}_n] : n \geq 1\}$ has its L^q-norm controlled by Doob's maximal inequality

$$|M_2|_{L^q} \leq \frac{q}{q-1} |M_1|_{L^q} = O\left(q^{1/2}\right) \text{ as } q \to \infty.$$

(The square-root growth implies that M_2 has a Gauss tail.) From (13.10) we have

$$-M_2 \left(t - s\right)^{2\alpha} \leq A_{s,t}^{n;i,j} \leq M_2 \left(t - s\right)^{2\alpha},$$

where M_2 is independent of n. If necessary, replace M_2 by $d^2 M_2$ to obtain the estimate

$$\sup_n \left|A_{s,t}^n\right| \leq M_2 \left(t - s\right)^{2\alpha}.$$

The same reasoning, easier in fact, shows that

$$\sup_n \left|B_{s,t}^n\right| \leq M_2 \left(t - s\right)^{\alpha}.$$

Putting everything together

$$\left\|\mathbf{B}_{s,t}^n\right\| \sim \left|B_{s,t}^n\right| \vee \left|A_{s,t}^n\right|^{1/2} \leq M_2 \left(t - s\right)^{\alpha},$$

which is precisely the required estimate on $\|\mathbf{B}^n\|_{\alpha\text{-Höl}}$, uniform over $n \geq 1$. Setting $M = M_1 + M_2$ finishes the proof. ∎

With the uniform bounds of Theorem 13.18, a simple argument (interpolation plus Hölder's inequality) leads to

Corollary 13.19 *Let* (D_n) *be a sequence of nested dissections of* $[0, T]$, *that is* $D_n \subset D_{n+1}$ *for all* n, *such that mesh* $|D_n| \to 0$ *as* $n \to \infty$. *Then*

$$d_{\alpha\text{-Höl};[0,T]} \left(S_2 \left(B^{D_n}\right), \mathbf{B}\right) \to 0$$

almost surely and in $L^q \left(\mathbb{P}\right)$ *for all* $q \in [1, \infty)$.

13.3.3 General piecewise linear approximations

We saw that martingale arguments lead to a quick proof of convergence of (lifted) piecewise linear approximations to enhanced Brownian motion, along a *nested* sequence of dissections. Dealing with an arbitrary sequence (D_n) requires a direct analysis. We first establish pointwise L^q-estimates (only here we use the specifics of *piecewise linear* approximations) followed by a general Besov–Hölder-type embedding which implies the corresponding rough path estimates.

Proposition 13.20 *Let D be a dissection of $[0, T]$ and $1/r \in [0, 1/2]$. Then there exists $C = C(T)$ such that for $k = 1, 2$ and all $0 \le s < t \le T$, $q \ge 1$,*

$$\left| \pi_k \left(\mathbf{B}_{s,t} - S_2 \left(B^D \right)_{s,t} \right) \right|_{L^q(\mathbb{P})} \le C |D|^{1/2 - 1/r} \left(\sqrt{q} \, |t - s|^{1/r} \right)^k.$$

Proof. We write A for the Lévy area, i.e. $\mathbf{B} = \exp(B + A)$.
Step 1: We consider first $s, t \in D$. In this case, the level 1 estimate is trivial as $B^D_{s,t} = B_{s,t}$. For level 2, observe that if $s = t_m$ and $t = t_n$ for some $m < n$, we have

$$\mathbf{B}_{s,t} - S_2 \left(B^D \right)_{s,t} = \sum_{i=m}^{n-1} A_{t_l, t_{l+1}}.$$

From Exercise 13.6,

$$\left| \pi_2 \left(S_2 \left(B^D \right)_{s,t} - \mathbf{B}_{s,t} \right) \right|_{L^q} \le c_1 q \left| \sum_{i=m}^{n-1} A_{t_i, t_{i+1}} \right|_{L^2} \le c_2 q \sqrt{\sum_{i=m}^{n-1} (t_{i+1} - t_i)^2}.$$

Since $\sum_{i=m}^{n-1} (t_{i+1} - t_i)^2 \le |t - s| \min(|D|, |t - s|)$, we have

$$\left| \pi_2 \left(S_2 \left(B^D \right)_{s,t} - \mathbf{B}_{s,t} \right) \right|_{L^q} \le c_2 q \left(|D| \wedge |t - s| \right)^{1/2} \cdot |t - s|^{1/2}.$$

Step 2: (small intervals) Consider the case $s_D \le s < t \le s^D$. Then

$$
\begin{aligned}
\left| B^D_{s,t} - B_{s,t} \right|_{L^q} &\le \frac{t - s}{s^D - s_D} \left| B_{s_D, s^D} \right|_{L^q} + \left| B_{s,t} \right|_{L^q} \\
&\le c_3 q^{1/2} |t - s|^{1/2} = c_3 q^{1/2} (|D| \wedge |t - s|)^{1/2},
\end{aligned}
$$

which settles level 1. For level 2, we estimate

$$
\begin{aligned}
\left| \pi_2 \left(S_2 \left(B^D \right)_{s,t} - \mathbf{B}_{s,t} \right) \right|_{L^q} &\le \left| \pi_2 \circ S_2 \left(B^D \right)_{s,t} \right|_{L^q} + \left| \pi_2 \left(\mathbf{B}_{s,t} \right) \right|_{L^q} \\
&\le c_4 q \left(\frac{|t - s|^2}{|s^D - s_D|} + |t - s| \right) \\
&\le 2 c_4 q |t - s| \\
&= 2 c_4 q \left(|D| \wedge |t - s| \right)^{1/2} \cdot |t - s|^{1/2}.
\end{aligned}
$$

Step 3: (arbitrary intervals) It remains to deal with $s < t$ such that $s \le s^D \le t_D \le t$. The level 1 estimate follows immediately from the level 1 estimate of step 2; indeed,

$$
\begin{aligned}
\left| B^D_{s,t} - B_{s,t} \right|_{L^q} &\le \left| B^D_{s,s^D} - B_{s,s^D} \right|_{L^q} + \left| B^D_{t_D, t} - B_{t_D, t} \right|_{L^q} \\
&\le 2 c_3 q^{1/2} (|D| \wedge |t - s|)^{1/2}.
\end{aligned}
$$

For the level 2 estimate, we note the algebraic identity in $T^2\left(\mathbb{R}^d\right)$,

$$S_2\left(B^D\right)_{s,t} - \mathbf{B}_{s,t} = \left(S_2\left(B^D\right)_{s,s^D} - \mathbf{B}_{s,s^D}\right) \otimes S_2\left(B^D\right)_{s^D,t} \quad (13.11)$$

$$+\mathbf{B}_{s,s^D} \otimes \left(S_2\left(B^D\right)_{s^D,t_D} - \mathbf{B}^D_{s^D,t_D}\right)$$

$$\otimes S_2\left(B^D\right)_{t_D,t} \quad (13.12)$$

$$+\mathbf{B}_{s,t_D} \otimes \left(S_2\left(B^D\right)_{t_D,t} - \mathbf{B}_{t_D,t}\right). \quad (13.13)$$

Projection to level 2 yields an expression of $\pi_2\left(S_2\left(B^D\right)_{s,t} - \mathbf{B}_{s,t}\right)$ in terms of the first *and* second level of all involved terms. For instance, the L_q-norm of (13.12) projected to level 2 is readily estimated by

$$\left|\pi_2\left(\mathbf{B}_{s,s^D}\right)\right|_{L^q} + \left|\pi_2\left(S_2\left(B^D\right)_{s^D,t_D} - \mathbf{B}^D_{s^D,t_D}\right)\right|_{L^q} + \left|\pi_2\left(S_2\left(B^D\right)_{t_D,t}\right)\right|_{L^q}$$

$$+ \left(\left|\mathbf{B}_{s,s^D}\right|_{L^q} + \left|B_{t_D,t}\right|_{L^q}\right) . \underbrace{\left|B^D_{s^D,t_D} - B_{s^D,t_D}\right|_{L^q}}_{=0} + \left|\mathbf{B}_{s,s^D}\right|_{L^q} . \left|B_{t_D,t}\right|_{L^q}$$

which, by the previous steps, is bounded by a constant times q times

$$\left|s^D - s\right| + \left(|D| \wedge \left|t_D - s^D\right|\right)^{1/2} . \left|t_D - s^D\right|^{1/2} + \left|t - t_D\right| + \left|s^D - s\right|^{1/2} . \left|t - t_D\right|^{1/2}$$

$$\leq 3(|D| \wedge |t - s|)^{1/2} |t - s|^{1/2}.$$

The estimates for the L_q-norm of (13.11), (13.13) projected to level 2 are very similar and also lead to bounds of the form $O\left(q|D| \wedge |t - s|\right)^{1/2}$ $|t - s|^{1/2}$. We omit the details.

Step 4: The estimates of steps 1–3 can be summarized in

$$\left|\pi_k\left(S_2\left(B^D\right)_{s,t} - \mathbf{B}_{s,t}\right)\right|_{L^q} \leq c_5\left(q^{\frac{k}{2}}(|D| \wedge |t - s|)^{1/2} . |t - s|^{\frac{k-1}{2}}\right),$$

valid for $k = 1, 2$ and all $s < t$ in $[0, T]$. By geometric interpolation, using $2/r \in [0, 1]$, we also have

$$\left|\pi_k\left(S_2\left(B^D\right)_{s,t} - \mathbf{B}_{s,t}\right)\right|_{L^q} \leq c_5\left(q^{\frac{k}{2}}\left(|D|^{1-2/r} \wedge |t - s|^{2/r}\right)^{1/2} . |t - s|^{\frac{k-1}{2}}\right)$$

$$\leq c_6\left(q^{\frac{k}{2}}|D|^{1/2-1/r} . |t - s|^{k/r}\right).$$

∎

We then obtain the following *quantitative estimates* in both (homogenous, inhomogenous) Hölder rough path metrics.

Corollary 13.21 *Let $0 \leq \alpha < 1/2$. Then, for every $\eta \in (0, 1/2 - \alpha)$, there exists a constant $C = C(\alpha, \eta, T)$ such that, for all $q \in [1, \infty)$,*

$$\left| d_{\alpha\text{-Höl}} \left(S_2 \left(B^D \right), \mathbf{B} \right) \right|_{L^q(\mathbb{P})} \leq C q^{1/2} \left| D \right|^{\eta/2}$$

and also, for $k \in \{1, 2\}$,

$$\left| \rho^{(k)}_{\alpha\text{-Höl};[0,T]} \left(\mathbf{B}, S_2 \left(B^D \right) \right) \right|_{L^q(\mathbb{P})} \leq C q^{\frac{k}{2}} \left| D \right|^{\eta}.$$

In particular, $S_2 \left(B^D \right) \to \mathbf{B}$ in $L^q(\mathbb{P})$ for all $q < \infty$ as $|D| \to 0$, with respect to either α-Hölder rough path metric.

Proof. Define r by $1/r := 1/2 - \eta$ and note that $\alpha < 1/r < 1/2$. Write $c_1, c_2, \ldots$ for constants which may depend on T (and tacitly on d). We have, for any $q \in [1, \infty), 0 \leq s < t \leq T$ and dissection D of $[0, T]$,

$$\left| \pi_k \left(\mathbf{B}_{s,t} - S_2 \left(B^D \right)_{s,t} \right) \right|_{L^q(\mathbb{P})} \leq c_1 \left| D \right|^{1/2 - 1/r} \left(\sqrt{q} \left| t - s \right|^{1/r} \right)^k, \quad k = 1, 2,$$

by the previous result (Proposition 13.20). Also,

$$\mathbb{E} \left(\|\mathbf{B}_{s,t}\|^q \right)^{1/q} \leq c_2 \left(\sqrt{q} \left| t - s \right|^{1/2} \right) \leq c_3 \left(\sqrt{q} \left| t - s \right|^{\gamma} \right)$$

from basic scaling and integrability of enhanced Brownian motion and both together easily imply that

$$\mathbb{E} \left(\left\| S_2 \left(B^D \right)_{s,t} \right\|^q \right)^{1/q} \leq c_4 \left(\sqrt{q} \left| t - s \right|^{\gamma} \right).$$

We can then appeal to Theorem A.13 in Appendix A to see that

$$\left| \sup_{s,t \in [0,T]} \frac{\left| \pi_k \left(\mathbf{B}_{s,t} - S_2 \left(B^D \right)_{s,t} \right) \right|}{|t - s|^{k\alpha}} \right|_{L^q(\mathbb{P})} \leq c_5 q^{\frac{k}{2}} \left| D \right|^{1/2 - 1/r} = c_5 q^{\frac{k}{2}} \left| D \right|^{\eta}$$

and also

$$\left| d_{\alpha\text{-Höl}} \left(S_2 \left(B^D \right), \mathbf{B} \right) \right|_{L^q(\mathbb{P})} \leq c_6 q^{1/2} \left| D \right|^{\eta/2}.$$

∎

Exercise 13.22 *Let $(D_n) \subset \mathcal{D}([0, T])$ be a sequence of dissections of $[0, T]$. Show that, $S_2 \left(B^{D_n} \right) \to \mathbf{B}$ almost surely with respect to α-Hölder rough path topology, $\alpha \in [0, 1/2)$, provided $\text{mesh}(D_n) \to 0$ fast enough. Hint: Borel-Cantelli.*

13.3.4 Limits of Wong–Zakai type with modified Lévy area

We formulate the following result for random rough paths which are the limits of their "piecewise linear approximations". Although the example we have in mind here is enhanced Brownian motion (in which case $\alpha \in (1/3, 1/2)$, $N \in \{2, 3, \dots\}$ and $L^q(\mathbb{P})/\alpha$-Hölder convergence was established in the previous section), it applies to the bulk of stochastic processes which admit a lift to rough path.

Definition 13.23 Let $\alpha \in (0, 1]$ and assume $\mathbf{X} = \mathbf{X}(\omega)$ has sample paths in $C_0^{\alpha\text{-Höl}}\left([0, T], G^{[1/\alpha]}(\mathbb{R}^d)\right)$; write $X = \pi_1(\mathbf{X})$ for its projection to a process with values in $\mathbb{R}^d$.

(i) Let $N \geq [1/\alpha]$ and let $(D_n) = (t_i^n : i)$ be a sequence of dissections of $[0, T]$ such that

$$\forall q \in \mathbb{N} : \sup_{n \in \mathbb{N}} \left| \left\| S_{[1/\alpha]}\left(X^{D_n}\right) \right\|_{\alpha\text{-Höl};[0,T]} \right|_{L^q(\mathbb{P})} < \infty$$

$$d_\infty\left(S_{[1/\alpha]}\left(X^{D_n}\right), \mathbf{X}\right) \to 0 \text{ in probability as } n \to \infty.$$

If $(X^n(\omega)) \subset C^{1\text{-Höl}}\left([0, T], \mathbb{R}^d\right)$ such that, for all ω,[12]

$$\mathbf{P}_t^n(\omega) := S_N\left(X^n(\omega)\right)_{0,t} \otimes S_N\left(X^{D_n}(\omega)\right)_{0,t}^{-1}$$

takes values in the centre of $G^N(\mathbb{R}^d)$ whenever $t \in D_n$ then we say that (X^n) is an approximation on (D_n) with perturbations $(\mathbf{P}^n)$ on level N to the random rough path $\mathbf{X}$.

(ii) Let $\beta \in (0, 1]$ and $[1/\beta] = N \geq [1/\alpha]$. We say that an approximation (X^n) on (D_n) with perturbations $(\mathbf{P}^n)$ on level N to the random rough path $\mathbf{X} \in C^{\alpha\text{-Höl}}\left([0, T], G^{[1/\alpha]}(\mathbb{R}^d)\right)$ is $\min(\alpha, \beta)$-Hölder comparable (with constants c_1, c_2, c_3) if for all $t_i^n, t_{i+1}^n \in D_n$, all ω and all $q \in [1, \infty)$,

$$\left| X^n \right|_{1\text{-Höl};[t_i^n, t_{i+1}^n]} \leq c_1 \left| X^{D_n} \right|_{1\text{-Höl};[t_i^n, t_{i+1}^n]} + c_2 \left| t_{i+1}^n - t_i^n \right|^{\beta-1}$$

$$\left| \left\| \mathbf{P}_{s,t}^n \right\| \right|_{L^q(\mathbb{P})} \leq c_3 \left| t - s \right|^\beta \text{ for all } s, t \in [0, T].$$

$\square$

Theorem 13.24 *Let $\alpha, \beta \in (0, 1]$ and $[1/\beta] = N \geq [1/\alpha]$. Assume $\mathbf{X} = \mathbf{X}(\omega)$ has sample paths in $C_0^{\alpha\text{-Höl}}\left([0, T], G^{[1/\alpha]}(\mathbb{R}^d)\right)$ and write $X = \pi_1(\mathbf{X})$ for its projection to a process with values in $\mathbb{R}^d$; let (X^n) be an approximation on (D_n) with perturbations $(\mathbf{P}^n)$ on level N to $\mathbf{X}$.*

(i) If the approximation is $\min(\alpha, \beta)$-Hölder comparable (with constants c_1, c_2, c_3) then for all $\gamma < \min(\alpha, \beta)$,

$$\forall q \in [1, \infty) : \sup_{n \in \mathbb{N}} \left| \left\| S_N(x^n) \right\|_{\gamma\text{-Höl};[0,T]} \right|_{L^q(\mathbb{P})} < \infty.$$

---

[12]It is *not* assumed that $\mathbf{P}_t^n(\omega) \in$ centre of $G^N(\mathbb{R}^d)$ when $t \notin D_n$.

(ii) If $\mathbf{P}_t^n \to \mathbf{P}_t$ *in probability for all* $t \in \cup_{n \in \mathbb{N}} D_n$ *dense in* $[0,T]$ *then, for all such* t,

$$d\left(S_N\left(X^n\right)_{0,t}, S_N\left(\mathbf{X}\right)_{0,t} \otimes \mathbf{P}_{0,t}\right) \to 0 \text{ in probability.}$$

(iii) If the assumptions of both (i) and (ii) are met then, for all $\gamma < \min(\alpha, \beta)$,

$$d_{\gamma\text{-Höl};[0,T]}\left(S_N\left(X^n\right), \mathbf{X}^\varphi\right) \to 0 \text{ in } L^q \text{ for all } q \in [1, \infty)$$

where $\varphi := \log \mathbf{P} \in \mathcal{V}^N\left(\mathbb{R}^d\right) \equiv \mathfrak{g}^N\left(\mathbb{R}^d\right) \cap \left(\mathbb{R}^d\right)^{\otimes N}$ *and* $\mathbf{X}^\varphi = \exp\left(\log\left(S_N\right.\right.$ $\left.\left(\mathbf{X}\right)\right) + \varphi)$.

Proof. (i) By a standard Garsia–Rodemich–Rumsey or Kolmogorov argument, the assumption on $\left\|\left\|\mathbf{P}_{s,t}^n\right\|\right\|_{L^q(\mathbb{P})}$ implies, for any $\tilde{\beta} < \beta$, the existence of $C_3 \in L^q$ for all $q \in [1, \infty)$ so that

$$\forall s < t \text{ in } [0,T] : \left\|\mathbf{P}_{s,t}^n\right\| \leq C_3\left(\omega\right)|t-s|^{\tilde{\beta}}.$$

We can pick $\tilde{\beta}$ large enough so that $\left[1/\tilde{\beta}\right] = [1/\beta] = N$ and $\gamma < \min\left(\alpha, \tilde{\beta}\right)$. We can then apply Theorem 12.18 with $\tilde{\beta}$ instead of β and learn that there exists a deterministic constant c such that

$$\sup_n \left\|S_N\left(X^n\right)\right\|_{\min(\alpha, \tilde{\beta})\text{-Höl}} \leq c\left(\sup_n \left\|S_{[1/\alpha]}\left(X^{D_n}\right)\right\|_{\alpha\text{-Höl}} + 1 + C_3\right).$$

Taking L^q-norms finishes the uniform L^q-bound.

(ii) From Theorem 12.18

$$d\left(S_N\left(X^n\right)_{0,t}, S_N\left(\mathbf{X}\right)_{0,t} \otimes \mathbf{P}_{0,t}\right) \leq d\left(S_N\left(X^{D_n}\right)_{0,t}, S_N\left(\mathbf{X}\right)_{0,t}\right)$$
$$+ d\left(\mathbf{P}_{0,t}^n, \mathbf{P}_{0,t}\right)$$

which, from the assumptions, obviously converges to 0 (in probability) for every fixed $t \in \cup_n D_n$.

(iii) General facts, about L^q-convergence of rough paths (cf. Section A.3.2 in Appendix A; inspection of the proofs shows that convergence in probability for all t in a dense set of $[0,T]$ is enough), implies the claimed convergence. ∎

Remark 13.25 The assumptions on X^n and $\mathbf{P}^n$ guarantee that the (X^n) remain, at $\min(\alpha, \beta)$-Hölder scale, comparable to the piecewise linear approximations. In particular, the assumption on $\left|X^n\right|_{1\text{-Höl};\left[t_i^n, t_{i+1}^n\right]} = \left|\dot{X}^n\right|_{\infty;\left[t_i^n, t_{i+1}^n\right]}$ is easy to verify in all examples below. The intuition is that, if we assume that X^n runs at constant speed over any interval $I = [t_i^n, t_{i+1}^n]$, $D_n = (t_i^n)$, it is equivalent to saying that

$$\text{length}\,(X^n|_I) \;\leq\; c_1 \text{length}\,\left(X^{D_n}|_I\right) + c_2\,|I|^{\beta}$$

$$(\quad = \quad c_1 \left|X_{t_i^n, t_{i+1}^n}\right| + c_2 \left|t_{i+1}^n - t_i^n\right|^{\beta}).$$

$\square$

Remark 13.26 In both examples below we have $\beta = 1/N$. It is Theorem 12.18, from which Theorem 13.24, was essentially obtained as a corollary, which suggests the need for the slightly looser condition $[1/\beta] = N$. $\quad\square$

Example 13.27 (Sussmann) Take any sequence of dissection of $[0, T]$, say (D_n) with mesh $|D_n| \to 0$ and $\mathbf{X}(\omega)$ such as in Theorem 13.24. The piecewise linear approximation X^{D_n} is nothing but the repeated concatentation of linear chords connecting the points $(X_t : t \in D_n)$. For some fixed $\mathbf{v} \in \mathcal{V}^N\left(\mathbb{R}^d\right)$, $N \in \{2, 3, \dots\}$ we now construct *Sussmann's nonstandard approximation* X^n as (repeated) concatenation of linear chords and "geodesic loops". First, we require $X^n(t) = X(t)$ for all $t \in D_n = (t_i^n : i)$. For intermediate times, i.e. $t \in \left(t_{i-1}^n, t_i^n\right)$ for some i, we proceed as follows: For $t \in [t_{i-1}^n, \left(t_{i-1}^n + t_i^n\right)/2]$ we run linearly and at constant speed from $X\left(t_{i-1}^n\right)$ so as to reach $X\left(t_i^n\right)$ by time $\left(t_{i-1}^n + t_i^n\right)/2$. (This is the usual linear interpolation between $X\left(t_{i-1}^n\right)$ and $X\left(t_i^n\right)$ but run at double speed.) This leaves us with the interval $[\left(t_{i-1}^n + t_i^n\right)/2, t_i^n]$ for other purposes and we run, starting at $x\left(t_i^n\right) \in \mathbb{R}^d$, through a "geodesic" ξ : $[\left(t_{i-1}^n + t_i^n\right)/2, t_i^n] \to \mathbb{R}^d$ associated with $\exp\left(\mathbf{v}/\left|t_i^n - t_{i-1}^n\right|\right) \in G^N\left(\mathbb{R}^d\right)$. Since $N > 1$, $\pi_1\left(\exp\left(\mathbf{v}/\left|t_i^n - t_{i-1}^n\right|\right)\right) = 0$ and so this geodesic path returns to its starting point in $\mathbb{R}^d$; in particular,

$$X^n\left(\left(t_{i-1}^n + t_i^n\right)/2\right) = X^n\left(t_i^n\right) = X\left(t_i^n\right).$$

It is easy to see (via Chen's theorem) that this approximation satisfies the assumptions of Theorem 13.24 with

$$\mathbf{P}_{s,t}^n := S_N\left(X^n\right)_{s,t} \otimes S_N\left(X^{D_n}\right)_{s,t}^{-1} = e^{\mathbf{v}(t-s)} \;\forall s, t \in D_n$$

(so that $\left.|\mathbf{P}_{s,t}^n|\right|_{L^q} = |\mathbf{P}_{s,t}^n| \lesssim |t - s|^{1/N}$ first for all $s, t \in D_n$ and then, easy to see, for all s, t) and deterministic limit $\mathbf{P}_{0,t} = e^{\mathbf{v}t}$, $\beta = 1/N$. Indeed, the length of x^n over any interval $I = [t_{i-1}^n, t_i^n]$ is obviously bounded by the length of the corresponding linear chord plus the length of the geodesic associated with $\exp(|I|\,v)$, which is precisely equal to

$$\|\exp(|I|\,v)\| = |I|^{1/N}\,\|\exp(\mathbf{v})\| =: c_2\,|I|^{1/N}\,.$$

An application of Theorem 13.24, applied to $\mathbf{X} = \mathbf{B}$, i.e. enhanced Browian motion, gives the following convergence result. For any $\gamma < 1/N$ we have

$$d_{\gamma\text{-Höl}}\left(S_N\left(B^n\right), S_N\left(\mathbf{B}\right) \otimes e^{\mathbf{v}\cdot}\right) \to 0 \text{ in } L^q \text{ for all } q \in [1, \infty).$$

Observe that γ for $1/\gamma \in [N, N+1)$ is a genuine rough path convergence. In Section 12.2 we have identified RDEs driven by

$$T_{(0,\dots,0,\mathbf{v}\cdot)} S_N (\mathbf{B}) = S_N (\mathbf{B}) \otimes e^{\mathbf{v}\cdot}$$

as RDEs driven by $\mathbf{B}$ with an additional drift. $\qquad\qquad\qquad\qquad\square$

Example 13.28 (McShane) Given $x \in C ([0,T], \mathbb{R}^2)$, an *interpolation function* $\phi = (\phi^1, \phi^2) \in C^1 ([0,1], \mathbb{R}^2)$ with $\phi(0) = (0,0)$ and $\phi(1) = (1,1)$ and a fixed dissection $D = (t_i)$ of $[0,T]$ we define the *McShane interpolation* $\tilde{x}^D \in C ([0,T], \mathbb{R}^2)$ componentwise by

$$\tilde{x}_t^{D;i} := x_{t_D}^i + \phi^{\Delta(t,i)} \left(\frac{t - t_D}{t^D - t_D} \right) x_{t_D, t^D}^i, \quad i = 1, 2.$$

The points $t_D, t^D \in D$ denote the left, resp. right, neighbouring points of t in the dissection and

$$\Delta(t, i) := \begin{cases} i, & \text{if } x_{t_D, t^D}^1 x_{t_D, t^D}^2 \geq 0 \\ 3 - i, & \text{if } x_{t_D, t^D}^1 x_{t_D, t^D}^2 < 0. \end{cases}$$

As a simple consequence of this definition, for $u < v$ in $[t_i, t_{i+1}]$

$$S_2 (\tilde{x}^D)_{u,v} = \exp \left(\tilde{x}_{u,v}^D + \left| x_{t_i, t_{i+1}}^1 \right| \left| x_{t_i, t_{i+1}}^2 \right| A^\phi \left(\frac{u - t_i}{t_{i+1} - t_i}, \frac{v - t_i}{t_{i+1} - t_i} \right) \right)$$

where $A^\phi (u, v) \equiv A_{u,v}^\phi$ is the area increment of ϕ over $[u, v] \subset [0, 1]$. Consider now $\mathbf{X}(\omega) = \mathbf{B}(\omega) = \exp (B + A) \in C_0^{\alpha\text{-Höl}} ([0,1], G^{[1/\alpha]} (\mathbb{R}^2))$ with $\alpha \in (1/3, 1/2)$ and take any $(D_n)_{n \in \mathbb{N}}$ with $|D_n| \to 0$. (We know from Section 13.3.3 that $S_2 (B^{D_n})$ converges to $\mathbf{B}$ in α-Hölder rough path topology and in L^q for all q.) It is easy to see (via Chen's theorem) that McShane's approximation to 2-dimensional Brownian motion satisfies the assumptions of Theorem 13.24 with $\beta = 1/2$, $N = 2$. Indeed, writing

$$B^n := \tilde{B}^{D_n}$$

for McShane's approximations, it is clear that for any $s < t$

$$\mathbf{P}_{s,t}^n = \exp \left(\left| x_{t_D, t^D}^1 \right| \left| x_{t_D, t^D}^2 \right| \times A^\phi \left(\frac{s - t_D}{t^D - t_D}, \frac{t - t_D}{t^D - t_D} \right) \right) \text{ with } D = D_n$$

and for two points $t_i < t_j$ in D_n the relevant increment is given by

$$\mathbf{P}_{t_i, t_j}^n = \exp \left(A_{0,1}^\phi \sum_{k=i+1}^{j} \left| B_{t_k, t_{k+1}}^1 \right| \left| B_{t_k, t_{k+1}}^2 \right| \right).$$

It is easy to see that $\sum_{k=i+1}^{j} \left| B^1_{t_k,t_{k+1}} \right| \left| B^2_{t_k,t_{k+1}} \right|$ converges, in L^2 say, to its mean

$$\frac{2}{\pi} \sum_{k=i+1}^{j} (t_{k+1} - t_k) = \frac{2}{\pi} |t_j - t_i|,$$

while $\left\| \mathbf{P}^n_{t_i,t_j} \right\|_{L^q} \leq \tilde{c}_q |t_j - t_i|^{1/2}$ follows directly from

$$\left\| \mathbf{P}^n_{t_i,t_j} \right\| \sim \left(\sum_{k=i+1}^{j} \left| B^1_{t_k,t_{k+1}} \right| \left| B^2_{t_k,t_{k+1}} \right| \right)^{1/2}$$

and

$$\left| \sum_{k=i+1}^{j} \left| B^1_{t_k,t_{k+1}} \right| \left| B^2_{t_k,t_{k+1}} \right| \right|_{L^q} \leq \sum_{k=i+1}^{j} \left| B^1_{t_k,t_{k+1}} \right|_{L^q} \left| B^2_{t_k,t_{k+1}} \right|_{L^q} = c_q |t_j - t_i|.$$

In fact, $\left\| \left| \mathbf{P}^n_{s,t} \right| \right\|_{L^q} \leq \tilde{c}_q |t - s|^{1/2}$ for all s, t since for $u < v$ in $[t_i, t_{i+1}]$

$$
\begin{aligned}
\left\| \mathbf{P}^n_{u,v} \right\|_{L^q} &= \left(\mathbb{E} \left| \left| x^1_{t_i,t_{i+1}} \right| \left| x^2_{t_i,t_{i+1}} \right| A^\phi \left(\frac{v - t_i}{t_{i+1} - t_i}, \frac{u - t_i}{t_{i+1} - t_i} \right) \right|^{q/2} \right)^{1/q} \\
&= c_{\phi,q} (t_{i+1} - t_i)^{1/2} \left| \frac{u - v}{t_{i+1} - t_i} \right| \leq c_{\phi,q} |u - v|^{1/2}.
\end{aligned}
$$

At last, for any $t_i \in D_n$, we have $|B^n|_{1\text{-Höl};[t_i,t_{i+1}]} \leq |\phi'|_\infty |B^{D_n}|_{1\text{-Höl};[t_i,t_{i+1}]}$. This shows that all assumptions of Theorem 13.24 are satisfied and we have, for all $\alpha \in [0, 1/2)$,

$$d_{\alpha\text{-Höl}} \left(S_2 (B^n), \exp (B_t + A_t + t\mathbf{\Gamma}) \right) \to 0 \text{ in } L^q \text{ for all } q \in [1, \infty)$$

where A_t is the usual $so(2)$-valued Lévy's area and

$$\mathbf{\Gamma} = \begin{pmatrix} 0 & \frac{2}{\pi} A^\phi_{0,1} \\ -\frac{2}{\pi} A^\phi_{0,1} & 0 \end{pmatrix} \in so(2).$$ $\qquad \square$

13.3.5 Convergence of 1D Brownian motion and its ε-delay

A real-valued Brownian motion β and its ε-delay $\beta^\varepsilon \equiv \beta(\cdot - \varepsilon)$ give rise to the $\mathbb{R}^2$-valued process

$$t \mapsto (\beta^\varepsilon_t, \beta_t) := (\beta_{t-\varepsilon}, \beta_t).$$

We shall assume $\varepsilon > 0$. On a sufficiently small time interval (of length $\leq \varepsilon$), it is clear that β^ε and β have independent Brownian increments so that

$$\left(\beta^\varepsilon_{s,t}, \beta_{s,t} : t \in [s, s + \varepsilon] \right)$$

has the distribution of a 2-dimensional standard Brownian motion $(B_t : t \in [0, \varepsilon])$. This suggests defining the stochastic area increments of $(\beta^\varepsilon, \beta)$ as

$$
\begin{aligned}
A^\varepsilon_{s,t} &= \frac{1}{2} \left(\int_s^t \beta^\varepsilon_{s,.} . d\beta - \beta_{s,.} . d\beta^\varepsilon \right) \\
&= \int_s^t \beta^\varepsilon_{s,.} . d\beta - \frac{1}{2} \beta^\varepsilon_{s,t} \beta_{s,t}.
\end{aligned}
$$

In particular, we can define the $G^2(\mathbb{R}^2)$-valued continuous process $(\mathbf{X}^\varepsilon_t : t \geq 0)$ as $(\beta^\varepsilon_{0,.}, \beta_{0,.})$ enhanced with the area process $A^\varepsilon_{0,.}$ so that

$$
\log \mathbf{X}^\varepsilon_t = \begin{pmatrix} \beta^\varepsilon_{0,t} \\ \beta_{0,t} \end{pmatrix} + \begin{pmatrix} 0 & A^\varepsilon_{0,t} \\ -A^\varepsilon_{0,t} & 0 \end{pmatrix}.
$$

It is left for the reader to check, as a simple consequence of Chen's relation, that the area-component of $\mathbf{X}^\varepsilon_{s,t} = (\mathbf{X}^\varepsilon_s)^{-1} \otimes \mathbf{X}^\varepsilon_t$ is indeed given by $A^\varepsilon_{s,t}$. We then have

Lemma 13.29 *There exists $\eta > 0$ such that*

$$
\sup_{\varepsilon \in (0,1]} \sup_{s,t \in [0,T]} \mathbb{E} \left[\exp \left(\eta \frac{d(\mathbf{X}^\varepsilon_s, \mathbf{X}^\varepsilon_t)^2}{|t - s|} \right) \right] < \infty.
$$

Proof. We estimate

$$
\begin{aligned}
|A^\varepsilon_{s,t}|_{L^q(\mathbb{P})} &\leq \left| \int_s^t \beta^\varepsilon_{s,.} . d\beta \right|_{L^q(\mathbb{P})} + \frac{1}{2} |\beta^\varepsilon_{s,t} \beta_{s,t}|_{L^q(\mathbb{P})} \\
&\leq \left| \int_s^t \beta^\varepsilon_{s,.} . d\beta \right|_{L^q(\mathbb{P})} + \frac{1}{2} |\beta_{s-\varepsilon,t-\varepsilon}|_{L^{2q}(\mathbb{P})} |\beta_{s,t}|_{L^{2q}(\mathbb{P})} \\
&\leq \left| \int_s^t \beta^\varepsilon_{s,.} . d\beta \right|_{L^q(\mathbb{P})} + c_1 |t - s| q
\end{aligned}
$$

since $|\beta_{s,t}|_{L^{2q}(\mathbb{P})} / |t - s|^{1/2} = |\beta_{0,1}|_{L^{2q}(\mathbb{P})} = O(q^{1/2})$, cf. Lemma A.17. Thus it will be enough to show that

$$
\left| \int_s^t \beta^\varepsilon_{s,.} . d\beta \right|_{L^q(\mathbb{P})} = O((t - s) q).
$$

To this end, we first observe that by stationarity of Brownian increments we may replace (s, t) by $(0, t - s)$. In other words, it suffices to estimate L^q-moments of the continuous martingale

$$
M_t = \int_0^t \beta^\varepsilon_{0,.} . d\beta.
$$

Noting $\langle M \rangle_t = \int_0^t \left| \beta_{-\varepsilon, s-\varepsilon} \right|^2 ds$, the exponential martingale inequality gives

$$
\begin{aligned}
\mathbb{P}\left(M_t > tx\right) &\leq \mathbb{P}\left(M_t > tx, \langle M \rangle_t \leq xt^2\right) + \mathbb{P}\left(\langle M \rangle_t > xt^2\right) \\
&\leq \exp\left(-\frac{1}{2}\frac{(tx)^2}{xt^2}\right) + \mathbb{P}\left(t\left|\beta\right|^2_{\infty;[0,t]} > xt^2\right) \\
&= \exp\left(-\frac{1}{2}x\right) + \mathbb{P}\left(\left|\beta\right|^2_{\infty;[0,1]} > x\right).
\end{aligned}
$$

The same argument applies to $-M_t$ and we see that $\left|M_t\right|/t$ has an exponential tail. Equivalently,

$$
\left| \frac{1}{t}\int_0^t \beta^\varepsilon_{0,\cdot}\,d\beta \right|_{L^q(\mathbb{P})} = O\left(q\right),
$$

which is what we wanted to show. ∎

An appeal to general regularity results for stochastic processes (see Section A.4 in Appendix A) then gives

Proposition 13.30 *Let* $\alpha \in [0, 1/2)$. *Then there exists* $\eta > 0$ *such that*

$$
\sup_{\varepsilon \in (0,1]} \mathbb{E}\left[\exp \eta \left\|\mathbf{X}^\varepsilon\right\|^2_{\alpha\text{-Höl};[0,T]}\right] < \infty
$$

and

$$
\sup_{\varepsilon \in (0,1]} \mathbb{E}\left[\exp \eta \left\|\mathbf{X}^\varepsilon\right\|^2_{\psi_{2,1}\text{-var};[0,T]}\right] < \infty.
$$

Theorem 13.31 *Let* β *be a* 1-*dimensional Brownian motion with* ε-*delay* $\beta^\varepsilon \equiv \beta(\cdot - \varepsilon)$, *lifted to* $G^2\left(\mathbb{R}^2\right)$-*valued geometric* α-*Hölder rough path,* $\alpha \in (1/3, 1/2)$ *given by*

$$
\mathbf{X}^\varepsilon_t = \exp\left(\left(\beta^\varepsilon_{0,t}, \beta_{0,t}\right); A^\varepsilon_{0,t}\right).
$$

Set also

$$
\tilde{\mathbf{X}}_t := \exp\left(\left(\beta_t, \beta_t\right); -t/2\right).
$$

Then, for any $q \in [1, \infty)$ *we have*

$$
\left| d_{\alpha\text{-Höl};[0,T]}\left(\mathbf{X}^\varepsilon, \tilde{\mathbf{X}}\right) \right|_{L^q(\mathbb{P})} \to 0 \text{ as } \varepsilon \to 0.
$$

Proof. Thanks to Proposition A.15 of Appendix A, in the presence of uniform α-Hölder bounds (which we established in Proposition 13.30), it suffices to show that

$$
\log \mathbf{X}^\varepsilon_t \equiv \left(\beta_{0,t-\varepsilon}, \beta_{0,t}; A^\varepsilon_{0,t}\right) \to \left(\beta_t, \beta_t, -t/2\right) \text{ as } \varepsilon \to 0
$$

in probability and pointwise, i.e. for fixed $t \in [0, T]$. Clearly it is enough to focus on the area. Using

$$\int_0^t \beta_{0,\cdot}^\varepsilon . d\beta \to \int_0^t \beta_{0,\cdot} . d\beta$$

(in probability and pointwise as $\varepsilon \to 0$), easily seen from Itô's isometry, we have

$$
\begin{aligned}
A_{0,t}^\varepsilon &= \int_0^t \beta_{0,\cdot}^\varepsilon . d\beta - \frac{1}{2} \beta_{0,t}^\varepsilon \beta_{0,t} \\
&\to \int_0^t \beta_{0,\cdot} . d\beta - \frac{1}{2} \left(\beta_{0,t} \right)^2 \\
&= \frac{1}{2} \left(\left(\beta_{0,t} \right)^2 - t \right) - \frac{1}{2} \left(\beta_{0,t} \right)^2 = -\frac{1}{2} t
\end{aligned}
$$

and the proof is finished. ∎

13.4   Weak approximations

We now turn to weak approximations of enhanced Brownian motion **B** and prove a Donsker-type theorem in the Brownian rough-path setting. In later chapters on Gaussian resp. Markov processes we shall encounter other weak convergence results which may be applied to enhanced Brownian motion and thus deserve to be mentioned briefly here: from general principles on Gaussian rough paths we have, for instance, that enhanced fractional Brownian motion $\mathbf{B}^H$ converges weakly to **B** as $H \to 1/2$ (recall that Brownian motion is fractional Brownian with Hurst parameter $H = 1/2$). Similarly, a sequence of Markov processes (X^{a_n}) on $\mathbb{R}^d$ with (uniformly elliptic) generator of divergence form $\nabla \cdot (a_n \nabla)$, enhanced with suitable stochastic area to a $G^2 \left(\mathbb{R}^d \right)$-valued process $\mathbf{X}^{a_n}$, will be seen to converge weakly to **B** provided that $2a_n \to I$, the $d \times d$ identity matrix. In all these cases weak convergence holds with respect to a rough path metric (namely, α-Hölder topology with any $\alpha < 1/2$). The interest in such results is that weak convergence is preserved under continuous maps; applied to the Itô–Lyons map in rough path topology all these weak convergence results translate immediately to weak convergence results in which the limit of certain (random ODEs/RDEs) is identified as an RDE solution driven by **B**, i.e. as the solution to a Stratonovich SDE.

13.4.1   Donsker's theorem for enhanced Brownian motion

Consider a random walk in $\mathbb{R}^d$, given by the partial sums of a sequence of independent random variables $(\xi_i : i = 1, 2, 3, \dots)$, identically distributed,

$\xi_i \overset{\mathcal{D}}{=} \xi$ with zero mean and unit covariance matrix, $\mathbb{E}\left(\xi \otimes \xi\right) = I$. *Donsker's theorem* (e.g. [143]) states that the rescaled, piecewise linearly connected random walk

$$W_t^{(n)} = \frac{1}{n^{1/2}}\left(\xi_1 + \cdots + \xi_{[tn]} + (nt - [nt])\,\xi_{[nt]+1}\right)$$

converges weakly to standard Brownian motion, on $C\left([0,1],\mathbb{R}^d\right)$ with sup topology. It was observed by Lamperti in [100] that this convergence takes place in α-Hölder topology, for $\alpha < (p-1)/2p$ provided $\mathbb{E}\left(|\xi|^{2p}\right) < \infty$, $p > 1$; and this is essentially sharp. In particular, for convergence in α-Hölder topology for any $\alpha < 1/2$ one needs finite moments of any order. We now extend this to a rough-path setting. More precisely, we show weak convergence in homogenous α-Hölder norm of the lifted rescaled random walk to $G^2\left(\mathbb{R}^d\right)$-valued enhanced Brownian motion $\mathbf{B}$. Observe that this implies a weak Wong–Zakai-type theorem: ODEs driven by $W^{(n)}$ converge weakly (in α-Hölder topology) to the corresponding Stratonovich SDE solution.

Theorem 13.32 (Donsker's theorem for EBM) *Assume ξ has zero mean, unit covariance and $\mathbb{E}\left(|\xi|^p\right) < \infty$ for all $p \in [1,\infty)$ and $\alpha < 1/2$. Then $S_2(W_\cdot^{(n)})$ converges weakly to $\mathbf{B}$, in $C^{0,\alpha\text{-H\"ol}}\left([0,1],G^2\left(\mathbb{R}^d\right)\right)$.*

We shall, in fact, prove a more general theorem that deals with random walks on groups. More precisely, Chen's theorem implies

$$S_2\left(W^{(n)}\right)_t = \delta_{n^{-1/2}}\left(e^{\xi_1} \otimes \cdots \otimes e^{\xi_{[nt]}} \otimes e^{(nt-[nt])\xi_{[nt]+1}}\right)$$

where δ denotes dilation on $G^2\left(\mathbb{R}^d\right)$ and $e^v = \left(1, v, \frac{v^{\otimes}}{2}\right)$, the usual step-2 exponential map. Observe that $\left(\boldsymbol{\xi}_i\right) = \left(e^{\xi_i}\right)$ is an independent, identically distributed sequence of $G^2\left(\mathbb{R}^d\right)$-valued random variables centred in the sense that

$$\mathbb{E}\left(\pi_1\left(\boldsymbol{\xi}_i\right)\right) = \mathbb{E}\xi_i = 0$$

(π_1 is the projection from $G^2\left(\mathbb{R}^d\right) \to \mathbb{R}^d$). Let us also observe that the shortest path which connects the unit element $1 \in G^2\left(\mathbb{R}^d\right)$ with e^{ξ_i} is precisely $e^{t\xi_i}$, so that piecewise linear interpolation on $\mathbb{R}^d$ lifts to geodesic interpolation on $G^2\left(\mathbb{R}^d\right)$. This suggests the following Donsker-type theorem:

Theorem 13.33 *Let $\left(\boldsymbol{\xi}_i\right)$ be a centred sequence of independent and identically distributed $G^2\left(\mathbb{R}^d\right)$-valued random variables with finite moments of all orders,*

$$\forall q \in [1,\infty) : E\left(\|\boldsymbol{\xi}_i\|^q\right) < \infty,$$

such that $\pi_1(\xi_i)$ has zero mean and unit variance, and consider the rescaled random walk defined by $\mathbf{W}_0^{(n)} = 1$ and

$$\mathbf{W}_t^{(n)} = \delta_{n^{-1/2}} \left(\boldsymbol{\xi}_1 \otimes \cdots \otimes \boldsymbol{\xi}_{[tn]} \right) \text{ for } nt = [nt]$$

for $t \in \{0, \frac{1}{n}, \frac{2}{n}, \dots \}$, piecewise geodesically connected in between (i.e. $\mathbf{W}_t^{(n)}|_{[\frac{i}{n}, \frac{i+1}{n}]}$ is a geodesic connecting $\mathbf{W}_{i/n}^{(n)}$ and $\mathbf{W}_{(i+1)/n}^{(n)}$). Then, for any $\alpha < 1/2$, $\mathbf{W}^{(n)}$ converges weakly to $\mathbf{B}$, in $C^{0,\alpha\text{-Höl}} \left([0,1], G^2 \left(\mathbb{R}^d \right) \right)$.

Proof. Following a standard pattern of proof, weak convergence follows from convergence of the finite-dimensional distributions and tightness (here in α-Hölder topology).

Step 1: (Convergence of the finite-dimensional distributions) This is an immediate consequence of a *central limit theorem* on free nilpotent groups (see comments to this chapter).

Step 2: (Tightness) We need to find positive constants a, b, c such that for all $u, v \in [0,1]$,

$$\sup_n \mathbb{E} \left[d \left(\mathbf{W}_v^{(n)}, \mathbf{W}_u^{(n)} \right)^a \right] \le c |v - u|^{1+b},$$

so that we can apply Kolmogorov's tightness criterion (Corollary A.11 in Appendix A) to obtain tightness in γ-Hölder rough path topology, for any $\gamma < b/a$. Using basic properties of geodesic interpolation, we see that it is enough to consider $u, v \in \{0, \frac{1}{n}, \frac{2}{n}, \dots \}$ and then, of course, there is no loss of generality in taking $[u, v] = [0, k/n]$ for some $k \in \{0, \dots, n\}$. It follows that what has to be established reads

$$\frac{1}{n^{a/2}} E \left[\| \boldsymbol{\xi}_1 \otimes \cdots \otimes \boldsymbol{\xi}_k \|^a \right] \le c_1 \left| \frac{k}{n} \right|^{1+b},$$

uniformly over all $n \in \mathbb{N}$ and $0 \le k \le n$, and with b/a arbitrarily close to $1/2$. To this end, it is enough to show that for all $p \in \{1, 2, \dots \}$,

$$(*) : \mathbb{E} \left[\| \boldsymbol{\xi}_1 \otimes \cdots \otimes \boldsymbol{\xi}_k \|^{4p} \right] = O \left(k^{2p} \right),$$

since we can then take $a = 4p$, $b = 2p - 1$ and of course $b/a = (2p-1)/(4p) \uparrow 1/2$ as $p \uparrow \infty$. Thus, the proof is finished once we show $(*)$ and this is the content of the last step of this proof.

Step 3: Let P be a polynomial function on $G^2(\mathbb{R}^d)$, i.e. a polynomial in $a^{1;i}, a^{2;ij}$ where

$$a = \left(a^{1;i}, a^{2;ij}; 1 \le i \le d, 1 \le i < j \le d \right) \in \mathfrak{g}^2 \left(\mathbb{R}^d \right)$$

is the log-chart of $G^2(\mathbb{R}^d)$, $g \mapsto a = \log(g)$. We define the degree $d^\circ P$ by agreeing that monomials of the form

$$\left(a^{1;i} \right)^{\alpha_i} \left(a^{2;ij} \right)^{\alpha_{i,j}}$$

have degree $\sum \alpha_i + 2 \sum \alpha_{i,j}$. An easy application of the Campbell–Baker–Hausdorff formula reveals that

$$TP : g \mapsto \mathbb{E}\left(P\left(g \otimes \boldsymbol{\xi}\right)\right) - P\left(g\right)$$

is also a polynomial function, of degree $\leq d^\circ P - 2$. For instance,

$$T\left(a \mapsto \left(a^{2;ij}\right)^m\right)$$

is seen to contain terms $\left(a^{2;ij}\right)^{m-1}$ and $\left(a^{2;ij}\right)^{m-2}\left(a^{1;k}\right)^2$, etc. (all of which are of degree $2m - 2$). Now, for any $p \in \{1, 2, \ldots\}$,

$$\begin{aligned}
\|e^a\|^{4p} &\sim \sum_i \left|a^{1;i}\right|^{4p} + \sum_{i<j} \left|a^{1;ij}\right|^{2p} \\
&= \sum_i \left(a^{1;i}\right)^{4p} + \sum_{i<j} \left(a^{1;ij}\right)^{2p} =: P\left(e^a\right)
\end{aligned}$$

where P is a polynomial of degree $4p$. Recalling the definition of the operator T and using independence, we have

$$\begin{aligned}
\mathbb{E}\left[P(\boldsymbol{\xi}_1 \otimes \cdots \otimes \boldsymbol{\xi}_k)\right] &= \mathbb{E}\left[\mathbb{E}\left[P((\boldsymbol{\xi}_1 \otimes \cdots \otimes \boldsymbol{\xi}_{k-1}) \otimes \boldsymbol{\xi}_k) \mid \boldsymbol{\xi}_1, \ldots, \boldsymbol{\xi}_{k-1}\right]\right] \\
&= \mathbb{E}\left[TP(\boldsymbol{\xi}_1 \otimes \cdots \otimes \boldsymbol{\xi}_{k-1}) + P(\boldsymbol{\xi}_1 \otimes \cdots \otimes \boldsymbol{\xi}_{k-1})\right] \\
&= \cdots = \\
&= (T+1)^k P(1) \\
&= \sum_{l \geq 0} \binom{k}{l} T^l P(1),
\end{aligned}$$

but the function $TP : g \mapsto \mathbb{E}(P(g \otimes \boldsymbol{\xi})) - P(g)$ is a polynomial function of degree at most $d^\circ P - 2 = 4p - 2$. Hence, $d^\circ T^l P \leq d^\circ P - 2l = 2(2p - l)$ and the above sum contains only a finite number of terms, more precisely

$$\mathbb{E}\left[P(\boldsymbol{\xi}_1 \otimes \cdots \otimes \boldsymbol{\xi}_k)\right] = \sum_{l=0}^{2p} \binom{k}{l} T^l P(1).$$

Since each of these terms is $O(k^{2p})$, as $k \to \infty$, we are done. ∎

Exercise 13.34 *Generalize Theorem 13.32 to a random walk with* $\mathbb{E}\left(\xi\right) = 0$ *and arbitrary non-degenerate covariance matrix.*

13.5 Cameron–Martin theorem

For the reader's convenience, we state a general fact of Gaussian analysis, Theorem D.2 in Appendix D, in a Brownian context. Recall that the

Cameron–Martin space for d-dimensional Brownian motion is given by (cf. Section 1.4.1)

$$\mathcal{H} = W_0^{1,2}\left([0,T],\mathbb{R}^d\right) = \left\{ \int_0^{\cdot} \dot{h}_t dt : \ \dot{h} \in L^2\left[0,1\right],\mathbb{R}^d \right\} \tag{13.14}$$

and has Hilbert structure given by $\langle h, g \rangle_{\mathcal{H}} = \left\langle \dot{h}, \dot{g} \right\rangle_{L^2}$.

Theorem 13.35 *Let B be a d-dimensional Brownian motion on $[0,T]$. Let $h \in \mathcal{H}$ be a Cameron–Martin path. Then the law of B is equivalent to the law of $T_h(B) \equiv B + h$. (These laws are viewed as Borel measures on $C_0\left([0,T],\mathbb{R}^d\right)$, denoted by $\mathbb{W}$ and $(T_h)_* \mathbb{W} \equiv \mathbb{W}^h$ respectively.) In fact,*

$$\frac{d\mathbb{W}^h}{d\mathbb{W}} = \exp\left(\int_0^T \dot{h} dB - \frac{1}{2} \int_0^T \left| \dot{h}_t \right|^2 dt \right).$$

Almost surely, enhanced Brownian motion $\mathbf{B}$ has sample paths in $C^{\alpha\text{-Höl}}\left([0,T],G^2\left(\mathbb{R}^d\right)\right)$ for any $\alpha \in [0,1/2)$. By interpolation, we also have that a.s. $\mathbf{B}$ takes values in the Polish space $C^{0,\alpha\text{-Höl}}\left([0,T],G^2\left(\mathbb{R}^d\right)\right)$ and in fact in the closed subspace of paths starting at the unit element of G^2. We view $\mathbf{B}$ as a $C^{0,\alpha\text{-Höl}}\left([0,T],G^2\left(\mathbb{R}^d\right)\right)$-valued random variable; the law of $\mathbf{B}$ is then a Borel probability measure on $C^{0,\alpha\text{-Höl}}\left([0,T],G^2\left(\mathbb{R}^d\right)\right)$.

Theorem 13.36 *Let $\mathbf{B}$ be a $G^2\left(\mathbb{R}^d\right)$-valued enhanced Brownian motion on $[0,T]$. Let $h \in \mathcal{H}$ be a Cameron–Martin path. Then the law of $T_h(\mathbf{B})$ is equivalent to the law of $\mathbf{B}$.*

Proof. We can assume that the underlying probability space is a Wiener space $C_0\left([0,T],\mathbb{R}^d\right)$ equipped with Wiener measure $\mathbb{W}$. In particular, Brownian motion is the coordinate map $B(t,\omega) = \omega(t)$. The law of B is $\mathbb{W}$ and we write $\mathbb{W}^h$ for the law of $B + h$. From the Cameron–Martin theorem, we know that the measures are equivalent, $\mathbb{W} \sim \mathbb{W}^h$. Now, $\mathbf{B}$ is a measurable map from $C\left([0,T],\mathbb{R}^d\right) \to C^{0,\alpha\text{-Höl}}\left([0,T],G^2\left(\mathbb{R}^d\right)\right)$. It is easy to see that (using Stratonovich calculus or, more elementary, the L^2-convergent Riemann–Stieltjes sum for the area) that

$$\mathbf{B}(\cdot + h) = T_h \mathbf{B} \text{ a.s.} \tag{13.15}$$

and hence the law of $T_h \mathbf{B}$ is $\mathbf{B}_* \mathbb{W}^h$, the usual short notation for the image measure of $\mathbb{W}^h$ under $\mathbf{B}$. On the other hand, the law of $\mathbf{B}$ is $\mathbf{B}_* \mathbb{W}$. Equivalence of measures implies equivalence of image measures, and we find $\mathbf{B}_* \mathbb{W} \sim \mathbf{B}_* \mathbb{W}^h$. The proof is now easily finished. ∎

Let us elaborate a bit further on property (13.15).

Proposition 13.37 Let $B = B(\omega)$ be $\mathbb{R}^d$-valued Brownian motion, realized as coordinate map on Wiener space, and $\mathbf{B}$ be the corresponding $G^2(\mathbb{R}^d)$-valued enhancement, realized as the limit of lifted piecewise linear approximations, say $d_{\alpha\text{Höl};[0,T]}\left(\mathbf{B}, S_2\left(B^{D_n}\right)\right) \to 0$ in probability. Then

$$\mathbb{P}\left(\{\omega : \mathbf{B}(\omega + h) \equiv T_h\mathbf{B}(\omega) \text{ for all } h \in \mathcal{H}\}\right) = 1.$$

Proof. As in the proof of Theorem 13.36, we assume that Brownian motion is realized as the coordinate map on Wiener space, $B(t, \omega) = \omega(t)$, under Wiener measure $\mathbb{W}$. It is clear that

$$S_2\left(B^{D_n}(\omega + h)\right) = S_2\left(\omega^{D_n} + h^{D_n}\right) = T_{h^{D_n}}\left(S_2\left(\omega^{D_n}\right)\right), \qquad (13.16)$$

where ω^D, h^D etc denote the piecewise linear approximations of the respective paths based on some dissection D. By passing to a subsequence, if necessary, we may assume that

$$\lim_{n\to\infty} S_2\left(B^{D_n}(\omega)\right)$$

(with respect to $d_{\alpha\text{-Höl}}$) exists for $\mathbb{W}$-almost surely ω, the limit being, by definition, the geometric α-Hölder rough path $\mathbf{B}(\omega)$. Fixing such an ω, chosen from a set of full $\mathbb{W}$-measure, we note that, for any $h \in \mathcal{H}$, the sequence

$$\left(S_2\left(B^{D_n}(\omega + h)\right) : n = 1, 2, \dots\right)$$

is also convergent. Indeed, from (13.16) and basic continuity properties of the translation operator $(h, \mathbf{x}) \mapsto T_h\mathbf{x}$ we see that, always in α-Hölder rough path topology,

$$S_2\left(B^{D_n}(\omega + h)\right) \to T_h\left(\mathbf{B}(\omega)\right) \text{ as } n \to \infty.$$

On the other hand, we have

$$\mathbf{B}(\omega + h) = \lim S_2\left(B^{D_n}(\omega + h)\right),$$

thanks to the existence of the limit on the right-hand side and the very realization of $\mathbf{B}$ as the limit of lifted piecewise linear approximations. This, of course, allows us to identify

$$\mathbf{B}(\omega + h) = T_h\left(\mathbf{B}(\omega)\right)$$

and we stress the fact that ω was chosen in a set of full measure independent of h. This concludes the proof. ∎

13.6   Large deviations

Let B denote d-dimensional standard Brownian motion. It is rather obvious that $\varepsilon B \to 0$ in distribution as $\varepsilon \to 0$. The same can be said for enhanced

Brownian motion **B** provided scalar multiplication by ε on $\mathbb{R}^d$ is replaced by dilation δ_ε on $G^2\left(\mathbb{R}^d\right)$, i.e. $\delta_\varepsilon \mathbf{B} \to o$ in distribution as $\varepsilon \to 0$. It turns out that, to leading order, the speed of this convergence can be computed very precisely. This is a typical example of a *large deviations* statement for sample paths. We assume in this section that the reader is familiar with the rudiments of large deviations as collected in the Appendix. Adopting standard terminology, the goal of this section is prove a *large deviation principle* for enhanced Brownian motion **B** in suitable rough path metrics. There is an obvious motivation for all this. The *contraction principle* will imply – by continuity of the Itô–Lyons maps and without any further work – a large deviation principle for rough differential equations driven by enhanced Brownian motion. Combined with the fact that RDEs driven by enhanced Brownian motion are exactly Stratonovich stochastic differential equations, this leads directly to large deviations for SDEs, better known as Freidlin–Wentzell estimates.

13.6.1 *Schilder's theorem for Brownian motion*

Let B denote d-dimensional standard Brownian motion on $[0,T]$. If $P_\varepsilon \equiv (\varepsilon B)_* \mathbb{P}$ denotes the law of εB, viewed as a Borel measure on $C_0\left([0,T],\mathbb{R}^d\right)$, the next theorem can be summarized in saying that $(P_\varepsilon)_{\varepsilon>0}$ satisfies a large deviation principle on the space $C_0\left([0,T],\mathbb{R}^d\right)$ with rate function I. (When no confusion arises, we shall simply say that $(\varepsilon B)_{\varepsilon>0}$ satisfies a large deviation principle.) All subsequent large deviation statements will involve the *good rate function* (cf. Exercise 13.39)

$$I\left(h\right) = \left\{ \begin{array}{ll} \frac{1}{2}\left\langle h,h\right\rangle_{\mathcal{H}} & \text{if } h \in \mathcal{H} \\ +\infty & \text{otherwise} \end{array} \right.$$

where $\mathcal{H}$ denotes the Cameron–Martin space for B as defined in (13.14). We now show that $(\varepsilon B)_{\varepsilon>0}$ satisfies a large deviation principle in uniform topology with good rate function I. This is nothing other than a special case of the general large deviation result for Gaussian measures on Banach spaces, see Section D.2 in Appendix D. However, in an attempt to keep the present chapter self-contained, we include the following classical proof based on Fernique estimates.[13]

Theorem 13.38 (Schilder) *Let B be a d-dimensional Brownian motion on $[0,T]$. For any measurable $A \subset C_0\left([0,T],\mathbb{R}^d\right)$ we have*

$$-I\left(A^\circ\right) \leq \liminf_{\varepsilon \to 0} \varepsilon^2 \log \mathbb{P}\left[\varepsilon B \in A\right] \leq \limsup_{\varepsilon \to 0} \varepsilon^2 \log \mathbb{P}\left[\varepsilon B \in A\right] \leq -I\left(\bar{A}\right).$$

$$(13.17)$$

Here, A° and $\bar{A}$ denote the interior and closure of A with respect to uniform topology.

---

[13] $I(A) = \inf(I(h)\colon h \in A)$.

Proof. For simplicity of notation we assume $T = 1$. We write $C_0\left([0,1]\right)$ instead of $C_0\left([0,1],\mathbb{R}^d\right)$ and assume $d = 1$ since the extension to $d > 1$ only involves minor notational changes.

(Upper bound) Write x for a generic path in $C_0\left([0,1]\right)$ and let x^m denote the piecewise linear approximation of x interpolated at points in $D_m = \{i/m : i = 0,\ldots,m\}$.

Step 1: We define $U_m := \langle x^m, x^m \rangle_{\mathcal{H}}$. In other words,

$$U_m = \int_0^1 |\dot{x}_t^m|^2 \, dt = m \sum_{i=1}^m \left|x_{i/m} - x_{(i-1)/m}\right|^2.$$

Under $P_\varepsilon = (\varepsilon B)_* \mathbb{P}$, the random variable U_m is distributed like $\varepsilon^2 \chi^2$ with m degrees of freedom and so

$$P_\varepsilon\left[U_m \geq l\right] = \frac{1}{2^{m/2}\,\Gamma(m)} \int_{l/\varepsilon^2}^\infty e^{-u/2} u^{m/2-1} \, du.$$

Therefore, for arbitrary m, l we have

$$\limsup_{\varepsilon \to 0} \varepsilon^2 \log P_\varepsilon\left(U_m > l\right) \leq -l/2.$$

For G open and non-empty, $l := \inf\left\{\langle h, h\rangle_{\mathcal{H}} : h \in G \cap \mathcal{H}\right\} < \infty$ and so

$$P_\varepsilon\left[x^m \in G\right] = P_\varepsilon\left[x^m \in G \cap \mathcal{H}\right] \leq P_\varepsilon\left[\langle x^m, x^m\rangle_{\mathcal{H}} \geq l\right].$$

From the preceding tail estimate on $U_m = \langle x^m, x^m\rangle_{\mathcal{H}}$ it plainly follows that

$$\limsup_{\varepsilon \to 0} \varepsilon^2 \log P_\varepsilon\left[x^m \in G\right] \leq -l/2 = -\frac{1}{2} I(G).$$

Step 2: We fix $\alpha \in (0, 1/2)$. From our Fernique estimates (Corollary 13.14), $Z := 2 |B|_{\alpha\text{-Höl};[0,1]}$ has a Gauss tail and so there exists $c_1 > 0$ such that

$$P_\varepsilon\left[|x^m - x|_{\infty;[0,1]} \geq \delta\right] = \mathbb{P}\left[|B^m - B|_{\infty;[0,1]} \geq \delta/\varepsilon\right] \leq m\mathbb{P}\left[Z\left(\frac{1}{m}\right)^\alpha \geq \delta/\varepsilon\right]$$

$$\leq m\mathbb{P}\left[Z \geq m^\alpha \delta/\varepsilon\right] \leq \frac{m}{c_1} \exp\left(-c_1\left(m^\alpha \delta/\varepsilon\right)^2\right).$$

This shows that piecewise linear approximations are *exponentially good* in the sense

$$\limsup_{\varepsilon \to 0} \varepsilon^2 \log P_\varepsilon\left[|x^m - x|_{\infty;[0,1]} \geq \delta\right] \leq -c_1 m^{2\alpha}\delta^2 \to -\infty \text{ as } m \to \infty. \tag{13.18}$$

Step 3: Write $B(y,\delta) \equiv \{x \in C_0\left([0,1]\right) : |x - y|_\infty < \delta\}$. Given a closed set F, its open δ-neighbourhood F^δ is defined as $\cup\{B(y,\delta) : y \in F\}$. Clearly,

$$P_\varepsilon(F) \leq P_\varepsilon\left[x^m(.) \in F^\delta\right] + P_\varepsilon\left[|x^m - x|_{\infty;[0,1]} \geq \delta\right]$$

and by combining the estimates obtained in the first two steps we see that

$$\limsup_{\varepsilon \to 0} \varepsilon^2 \log P_\varepsilon (F) \le \max \left(-I \left(F^\delta \right), -c_1 m^{2\alpha} \delta \right).$$

Now let $m \to \infty$ and then $\delta \to 0$. The convergence $I \left(F^\delta \right) \to I (F)$ is standard, see Lemma C.1.

(**Lower bound**) It is enough to consider an open ball of fixed radius δ centred at some $h \in \mathcal{H}$. Define $Z = B - \varepsilon^{-1} h$ and $A_\varepsilon = \{ B \in B \left(0, \delta \varepsilon^{-1} \right) \}$. By the Cameron–Martin theorem, Theorem 13.35,

$$
\begin{aligned}
\mathbb{P} \left[\varepsilon B \in B (h, \delta) \right] &= \mathbb{P} \left[Z \in B \left(0, \delta \varepsilon^{-1} \right) \right] \\
&\quad \cdot \mathbb{E} \left(\exp \left[-\frac{1}{\varepsilon} \int_0^1 \dot{h}_t t dB_t - \frac{1}{2\varepsilon^2} \int_0^1 \left| \dot{h}_t \right|^2 dt \right]; A_\varepsilon \right) \\
&= e^{-I(h)/\varepsilon^2} \mathbb{E} \left(\exp \left\{ -\frac{1}{\varepsilon} \int_0^1 \dot{h}_t dB_t \right\}; A_\varepsilon \right) \\
&= e^{-I(h)/\varepsilon^2} \mathbb{E} \left(\exp \left[-\frac{1}{\varepsilon} \int_0^1 \dot{h}_t dB_t \right] \middle| A_\varepsilon \right) \mathbb{P} \left(A_\varepsilon \right) \\
&\ge e^{-I(h)/\varepsilon^2} \mathbb{P} \left(A_\varepsilon \right) = e^{-I(h)/\varepsilon^2} \left(1 + o(1) \right).
\end{aligned}
$$

In the last line we used symmetry (B and $-B$ having identical distributions implies $\mathbb{E} \left(\int_0^1 g dB \middle| A_\varepsilon \right) = 0$ for all deterministic integrands g such as $-\varepsilon^{-1} \dot{h}$), and Jensen's inequality

$$\mathbb{E} \left(\exp \left[\int_0^1 g dB_t \right] \middle| A_\varepsilon \right) \ge \exp \mathbb{E} \left(\int_0^1 g dB \middle| A_\varepsilon \right) = 1.$$

∎

Exercise 13.39 *Show that*

$$h \mapsto \begin{cases} \frac{1}{2} \langle h, h \rangle_{\mathcal{H}} & \text{if } h \in \mathcal{H} \\ +\infty & \text{otherwise} \end{cases}$$

is a good rate function. (Hint: Compactness of level sets follows from equicontinuity and Arzela–Ascoli).

13.6.2 Schilder's theorem for enhanced Brownian motion

Let $\Phi_m : x \in C \left([0, T], \mathbb{R}^d \right) \mapsto x^{D_m}$ denote the piecewise linear approximation map along the dissection D_m given by $\{ iT/m : i = 0, \ldots, m \}$. Clearly, $\Phi_m (\varepsilon B)$ satisfies a large deviation principle, as can be seen from elementary m-dimensional Gaussian considerations, or Schilder's theorem and the contraction principle applied to the continuous (linear) map Φ_m. We have seen in Section 13.3.3 that, for any $\alpha \in [0, 1/2)$, there exist positive constants $C = C_{\alpha,T}, \eta > 0$ such that for all $q \in [1, \infty)$,

$$\left| d_{\alpha\text{-Höl};[0,T]} \left(S_2 \left(\Phi_m \left(B \right) \right), \mathbf{B} \right) \right|_{L^q(\mathbb{P})} \leq C \sqrt{q} \left| \frac{1}{m} \right|^{\eta}. \qquad (13.19)$$

As an almost immediate consequence, we see that piecewise linear approximations are exponentially good in the following sense.

Lemma 13.40 *For any $\delta > 0$ and $\alpha \in [0, 1/2)$ we have*

$$\lim_{m \to \infty} \limsup_{\varepsilon \to 0} \varepsilon^2 \log \mathbb{P} \left(d_{\alpha\text{-Höl};[0,T]} \left(S_2 \circ \Phi_m \left(\varepsilon B \right), \delta_\varepsilon \mathbf{B} \right) > \delta \right) = -\infty.$$

Proof. We define $\alpha_m = C \left| \frac{1}{m} \right|^{\eta}$. Using inequality (13.19), we estimate

$$\mathbb{P} \left(d_{\alpha\text{-Höl};[0,1]} \left(S_2 \left(\Phi_m \left(\varepsilon B \right) \right), \delta_\varepsilon \mathbf{B} \right) > \delta \right) = \mathbb{P} \left(d_{\alpha\text{-Höl};[0,T]} \left(S_2 \left(\Phi_m \left(B \right) \right), \mathbf{B} \right) > \frac{\delta}{\varepsilon} \right)$$

$$\leq \left(\frac{\delta}{\varepsilon} \right)^{-q} \sqrt{q}^q \alpha_m^q$$

$$\leq \exp \left[q \log \left(\frac{\varepsilon}{\delta} \alpha_m \sqrt{q} \right) \right],$$

and after choosing $q = 1/\varepsilon^2$ we obtain, for ε small enough,

$$\varepsilon^2 \log \mathbb{P} \left(d_{\alpha\text{-Höl};[0,1]} \left(S_2 \left(\Phi_m \left(\varepsilon B \right) \right), \delta_\varepsilon \mathbf{B} \right) > \delta \right) \leq \log \left(\frac{\alpha_m}{\delta} \right).$$

Now it suffices to take the lim sup with $\varepsilon \to 0$ and note that $\log \left(\alpha_m / \delta \right) \to -\infty$ as $m \to \infty$. ∎

We also need the following (uniform) continuity property on level sets of the rate function. As will be seen in the proof below, this is a consequence of

$$\left| h \right|_{1\text{-var};[s,t]} \leq \left| t - s \right|^{1/2} \left| h \right|_{\mathcal{H}} \qquad (13.20)$$

and general continuity properties of the lifting map S_N in variation metrics.

We recall that the good rate function I is defined by

$$I \left(h \right) = \begin{cases} \frac{1}{2} \left\langle h, h \right\rangle_{\mathcal{H}} & \text{if } h \in \mathcal{H} \\ +\infty & \text{otherwise.} \end{cases}$$

Lemma 13.41 *For all $\Lambda > 0$ and $\alpha \in [0, 1/2)$ we have*

$$\sup_{\{h : I(h) \leq \Lambda\}} d_{\alpha\text{-Höl};[0,T]} \left(S_2 \left(\Phi_m \left(h \right) \right), S_2 \left(h \right) \right) \to 0 \text{ as } m \to \infty.$$

Proof. Without loss of generality, we take $T = 1$. First observe that

$$
\begin{aligned}
\|S_2 \left(\Phi_m \left(h\right)\right)\|_{1\text{-var};[s,t]} &\leq |\Phi_m \left(h\right)|_{1\text{-var};[s,t]} \\
&\leq |h|_{1\text{-var};[s,t]} \\
&\leq \sqrt{2\Lambda} \, |t - s|^{1/2} .
\end{aligned}
$$

Hence, we see that interpolation allows us to restrict ourselves to the case $\alpha = 0$. Furthermore, Proposition 8.15 allows us to actually replace $d_{\alpha\text{-Höl}}$ by d_∞. Then, we easily see that

$$
\begin{aligned}
d_\infty \left(S_2 \left(\Phi_m \left(h\right)\right), S_2 \left(h\right)\right) &\leq \max_{i=0}^{m-1} d \left(S_2 \left(\Phi_m \left(h\right)\right)_{\frac{i}{m}}, S_2 \left(h\right)_{\frac{i}{m}}\right) \\
&\quad + \max_{i=0}^{m-1} \left(\|S_2 \left(\Phi_m \left(h\right)\right)\|_{0;\left[\frac{i}{m},\frac{i+1}{m}\right]} \right. \\
&\quad \left. + \|S_2 \left(h\right)\|_{0;\left[\frac{i}{m},\frac{i+1}{m}\right]}\right).
\end{aligned}
$$

Clearly,

$$
\begin{aligned}
\|S_2 \left(\Phi_m \left(h\right)\right)\|_{0;\left[\frac{i}{m},\frac{i+1}{m}\right]} &\leq \|S_2 \left(\Phi_m \left(h\right)\right)\|_{1\text{-var};\left[\frac{i}{m},\frac{i+1}{m}\right]} \\
&\leq \sqrt{2\Lambda} m^{-1/2},
\end{aligned}
$$

and similarly,

$$
\|S_2 \left(h\right)\|_{0;[t_i,t_{i+1}]} \leq \sqrt{2\Lambda} m^{-1/2}.
$$

Then, because $\Phi_m \left(h\right)_{\frac{i}{m}} = h_{\frac{i}{m}}$, using equivalence of homogenous norms we have

$$
\begin{aligned}
\max_{i=0}^{m-1} d \left(S_2 \left(\Phi_m \left(h\right)\right)_{\frac{i}{m}}, S_2 \left(h\right)_{\frac{i}{m}}\right) &\leq c_1 \max_{i=0}^{m-1} \left|\sum_{j=0}^{i-1} \pi_2 \circ S_2 \left(h\right)_{\frac{j}{m},\frac{j+1}{m}}\right|^{1/2} \\
&\leq c_1 \left(\sum_{j=0}^{m-1} \left|\pi_2 \circ S_2 \left(h\right)_{\frac{j}{m},\frac{j+1}{m}}\right|\right)^{1/2} \\
&\leq c_2 \left(\sum_{j=0}^{m-1} |h|^2_{1\text{-var};\left[\frac{j}{m},\frac{j+1}{m}\right]}\right)^{1/2} \\
&\leq c_2 |h|^{1/2}_{1\text{-var};[0,1]} \left(\max_{i=0}^{m-1} |h|_{1\text{-var};\left[\frac{j}{m},\frac{j+1}{m}\right]}\right)^{1/2} \\
&\leq c_3 \Lambda^{1/2} m^{-1/4}.
\end{aligned}
$$

In particular, we see that

$$
\sup_{\{h: I(h) \leq \Lambda\}} d_\infty \left(S_2 \left(\Phi_m \left(h\right)\right), S_2 \left(h\right)\right) \leq c_4 \Lambda^{1/2} m^{-1/4},
$$

which concludes the proof.   ∎

Theorem 13.42 *For any $\alpha \in [0, 1/2)$, the family $(\delta_\varepsilon \mathbf{B} : \varepsilon > 0)$ satisfies a large deviation in homogenous α-Hölder topology. More precisely, viewing $\mathbf{P}_\varepsilon := (\delta_\varepsilon \mathbf{B})_* \mathbb{P}$ as a Borel measure on the Polish space $(C_0^{0,\alpha\text{-Höl}}([0,T], G^2(\mathbb{R}^d)), d_{\alpha\text{-Höl}})$, the family $(\mathbf{P}_\varepsilon : \varepsilon > 0)$ satisfies a large deviation principle on this space with good rate function, defined for $x \in C_0^{0,\alpha\text{-Höl}}([0,T], G^2(\mathbb{R}^d))$, given by*

$$J(x) = \frac{1}{2} \langle \pi_1(x), \pi_1(x) \rangle_{\mathcal{H}} \quad \text{if } \pi_1(y) \in \mathcal{H}.$$

Proof. We once again assume $T = 1$ without loss of generality. We know from Section 13.3.2 that $\mathbf{B}$ is the almost sure $d_{\alpha\text{-Höl}}$ limit of the lifted piecewise linear approximation, based on the dyadics dissections $D_n = (i/2^n : i = 0, \dots, 2^n)$ for instance. We may assume that the underlying probability space is the usual d-dimensional Wiener space $C_0([0,1], \mathbb{R}^d)$, so that $\mathbb{P}$ is a Wiener measure and $B(\omega) = \omega$. At the price of modifying $\mathbf{B}$ on a set of probability zero, we can and will assume that

$$\mathbf{B}(\omega) := \lim_{n \to \infty} S_2(B^{D_n}(\omega)) \text{ with respect to } d_{\alpha\text{-Höl}}$$

(arbitrarily defined on the null set where this limit does not exist) so that $\mathbf{B}$ is well-defined on $\mathcal{H} \subset C^{1\text{-var}}$ and coincides with the map $h \mapsto S_2(h)$, based on Riemann–Stieltjes integration.[14] We approximate the measurable map $\mathbf{B}(\cdot)$ by

$$\omega \in C_0([0,1], \mathbb{R}^d) \mapsto S_2(\Phi_m(\omega)) \in C_0^{0,\alpha\text{-Höl}}([0,1], G^2(\mathbb{R}^d)),$$

which is a *continuous* map (for fixed m) as is easily seen from continuity of the two maps

$$\omega \;\in\; C_0([0,1], \mathbb{R}^d) \mapsto \Phi_m(\omega) \in C_0^{0,1\text{-Höl}}([0,1], \mathbb{R}^d),$$
$$x \;\in\; C_0^{1\text{-Höl}}([0,1], \mathbb{R}^d) \mapsto S_2(x) \in C_0^{1\text{-Höl}}([0,1], G^2(\mathbb{R}^d)).$$

The *extended contraction principle*, Section C.2 in Appendix C, implies the required large deviation principle for enhanced Brownian motion provided we check (i) exponential goodness of these approximations and (ii) a (uniform) continuity property on level sets of the rate function. But these properties were the exact content of Lemmas 13.40 and 13.41 above. ∎

It should be noted that the proof of Theorem 13.42 uses few specifics of (enhanced) Brownian motion and only relies on reasonably good ("Gaussian") estimates of piecewise linear approximations and some regularity of

---

[14]More generally, the map $h \mapsto \lim_{n \to \infty} S_N(h^{D_n})$ is well-defined on $C^{\rho\text{-var}}$, $\rho < 2$, and coincides with the step-N Young lift of h.

the Cameron–Martin space: (13.19), (13.20). Indeed, as will be discussed in Section 15.7, an almost identical proof carries through in a general Gaussian (rough path) setting.

We also note that it would be sufficent to prove Theorem 13.42 in uniform topology, i.e. for $\alpha = 0$, by appealing to the so-called *inverse contraction principle*, Section C.2 in Appendix C. We have

Proposition 13.43 *Assume Theorem 13.42 holds for $\alpha = 0$. Then it also holds for any $\alpha \in [0, 1/2)$.*

Proof. By the inverse contraction principle all we have to do is check that $\{\delta_\varepsilon \mathbf{B}\}$ is exponentially tight in α-Hölder topology. But this follows from the compact embedding of

$$C^{\alpha'\text{-Höl}}\left([0,T], G^2\left(\mathbb{R}^d\right)\right) \hookrightarrow C^{0,\alpha\text{-Höl}}\left([0,T], G^2\left(\mathbb{R}^d\right)\right)$$

and Gauss tails of $\|\mathbf{B}\|_{\alpha'\text{-Höl}}$ where $\alpha < \alpha' < 1/2$, i.e.

$$\exists c > 0 : \mathbb{P}\left[\|\mathbf{B}\|_{\alpha'\text{-Höl}} > l\right] \leq \exp\left(-cl^2\right).$$

Indeed, defining the following precompact sets in α-Hölder topology,

$$K_M = \left\{x : |x|_{\alpha'\text{-Höl}} \leq \sqrt{M/c}\right\},$$

exponential tightness follows from

$$\begin{aligned}
\varepsilon^2 \log\left[\mathbf{P}_\varepsilon\left(K_M^c\right)\right] &= \varepsilon^2 \log \mathbb{P}\left[\|\delta_\varepsilon \mathbf{B}\|_{\alpha'\text{-Höl}} > \sqrt{M/c}\right] \\
&= \varepsilon^2 \log \mathbb{P}\left[\|\mathbf{B}\|_{\alpha'\text{-Höl}} > \sqrt{\frac{M}{c\varepsilon^2}}\right] \leq -M.
\end{aligned}$$

∎

Exercise 13.44 (Schilder for EBM via Itô calculus) *The purpose of this exercise is to give a direct proof of Theorem 13.42 using martingale techniques. Thanks to Proposition 13.43, we only need to consider the uniform topology.*
(i) Define the so(d)-valued approximations to Lévy's area process,

$$A_t^m := \frac{1}{2}\int_0^t B_{[ms]/m} \otimes dB_s - \int_0^t dB_s \otimes B_{[ms]/m}.$$

Use the fact that $t \mapsto A_t$ is a martingale to show that they give rise to exponentially good approximations to $\{\delta_\varepsilon \mathbf{B}\}$:

$$\lim_{m \to \infty} \overline{\lim}_{\varepsilon \to 0} \varepsilon^2 \log \mathbb{P}\left[d_{\infty;[0,T]}\left(\exp\left(\varepsilon B + \varepsilon^2 A^m\right), \delta_\varepsilon \mathbf{B}\right) \geq \delta\right] = -\infty.$$

(ii) Define $A(h)_t^m = \frac{1}{2} \int_0^t h_{[ms]/m} \otimes dh_s - \int_0^t dh_s \otimes h_{[ms]/m}$ for any $h \in \mathcal{H}$. Show that for all $\Lambda > 0$,

$$\lim_{m \to \infty} \sup_{\{h \in \mathcal{H} : I(h) \leq \Lambda\}} d_{\infty;[0,T]} (\exp(h + A(h)^m), S_2(h)) = 0.$$

(iii) Deduce a large deviation principle for enhanced Brownian motion in uniform topology. More precisely, show that $\mathbf{P}_\varepsilon = (\delta_\varepsilon \mathbf{B})_* \mathbb{P}$ viewed as a Borel measure on the Polish space $(C_0([0,T], G^2(\mathbb{R}^d)), d_\infty)$ satisfies a large deviation principle with good rate function $J(\mathbf{y}) = I(\pi_1(\mathbf{y}))$.

Exercise 13.45 The purpose of this exercise is to give a direct proof of Theorem 13.42 using Markovian techniques. Again, thanks to Proposition 13.43, it suffices to consider the uniform topology.

Let $\mathbf{p}(t, x, y)$ denote the transition density for enhanced Brownian motion seen as a Markov process on $G^2(\mathbb{R}^d)$. Use Varadhan's formula (cf. Section E.5 in Appendix E

$$\lim_{\varepsilon \to 0} 2\varepsilon \log \mathbf{p}(\varepsilon, x, y) = -d(x, y)^2$$

and the fact that $G^2(\mathbb{R}^d)$ is a geodesic space to establish a large deviation principle for enhanced Brownian motion in uniform topology.

Exercise 13.46 (Strassen's law) Let $\mathbf{B}$ denote enhanced Brownian motion $\mathbf{B}$ on $[0,1]$. Establish the following functional version of the law of iterated logarithm for $\mathbf{B}$ in α-Hölder (rough path) topology, $\alpha < 1/2$: let $\varphi(h) = \sqrt{h \ln \ln(1/h)}$ for h small enough, and show that

$$t \in [0,1] \mapsto \delta_{\frac{1}{\varphi(h)}} \mathbf{B}_{h\cdot}(\omega)$$

is relatively compact as a random variable with values in $C^{0,\alpha\text{-Höl}}([0,1], G^2(\mathbb{R}^d))$ with the compact set of limit points as $h \to 0$ given by $S_2(\mathcal{K})$ where $\mathcal{K} = \{h \in \mathcal{H} : |h|_{\mathcal{H}} \leq \sqrt{2}\}$.

13.7 Support theorem

13.7.1 Support of Brownian motion

Almost surely, the d-dimensional Brownian motion $B \in C^{\alpha\text{-Höl}}([0,T], \mathbb{R}^d)$ for $\alpha \in [0, 1/2)$, and hence we also have that B belongs almost surely to the Polish space $C^{0,\alpha\text{-Höl}}([0,T], \mathbb{R}^d)$, and in fact in the closed subspace of paths started at 0, $C_0^{0,\alpha\text{-Höl}}([0,T], \mathbb{R}^d)$. B can then be viewed as a $C_0^{0,\alpha\text{-Höl}}$-valued random variable and its law of B is a Borel probability measure on $C_0^{0,\alpha}([0,T], \mathbb{R}^d)$.

Definition 13.47 Let μ be a Borel probability measure on some Polish space (E, d). The (topological) support of μ is the smallest closed set of full measure.   □

We recall that $\mathcal{H} = W_0^{1,2}\left([0,T], \mathbb{R}^d\right)$ denotes the Cameron–Martin space for Brownian motion. Let us also recall (cf. Theorem 13.35) that, for any $h \in \mathcal{H}$, the law of $T_h(B) = B + h$ is equivalent to the law of B. Let us record some simple properties of T_h. Thanks to $\mathcal{H} \hookrightarrow C^{1/2\text{-Höl}}$, and $\alpha < 1/2$, it is a continuous map of $C_0^{0,\alpha}\left([0,T], \mathbb{R}^d\right)$ into itself and bijective with inverse T_{-h}. In particular, the image of any open sets under T_h is again open.

Corollary 13.48 *Let h be a Cameron–Martin path and $x \in C^{0,\alpha}\left([0,T], \mathbb{R}^d\right)$. Then if x belongs to the support of the law of B, so does $T_h(x)$.*

Proof. Write $\mathcal{N}(x)$ for all open neighbourhoods of x. To show that $T_h(x)$ is in the support, it suffices to show that

$$\forall V \in \mathcal{N}(T_h(x)) : \mathbb{P}(B \in V) > 0.$$

Fix $V \in \mathcal{N}(T_h(x))$. By continuity, there exists $U \in \mathcal{N}(x)$ so that $T_h(U) \subset V$. From the above remark, $T_h(U) \in \mathcal{N}(T_h(x))$. Thus

$$\begin{aligned} \mathbb{P}(B \in V) &\geq \mathbb{P}(B \in T_h(U)) \\ &= \mathbb{P}(T_{-h}B \in U) \end{aligned}$$

and from Cameron–Martin the last expression is positive if and only if $\mathbb{P}(B \in U)$ is positive. But this is true since $U \in \mathcal{N}(x)$ and x is in the support.   ■

Theorem 13.49 *Let $\alpha \in (0, 1/2)$. The topological support of the law of Brownian motion on $[0,T]$ in α-Hölder topology is precisely $C_0^{0,\alpha}\left([0,T]; \mathbb{R}^d\right)$.*

Proof. Almost surely, $B(\omega) \in C_0^{0,\alpha}\left([0,T]; \mathbb{R}^d\right)$ which is closed in α-Hölder topology. Therefore, the support of the law of B is included in $C_0^{0,\alpha}\left([0,T]; \mathbb{R}^d\right)$.

Vice-versa, the support contains (trivially!) one point, say

$$x \in C_0^{0,\alpha}\left([0,T]; \mathbb{R}^d\right).$$

From the (defining) properties of the space $C_o^{0,\alpha}$, there are smooth paths $\{x^n\}$ with $x^n(0) = 0$ so that

$$x - x^n = T_{-x^n}(x) \to 0 \text{ in } \alpha\text{-Hölder topology.}$$

Any such x^n is a Cameron–Martin path and so $T_{-x^n}(x) \in$ support for all n. By definition, the support is closed (in α-Hölder topology) and therefore $0 \in$ support. But then any translate $T_h(0) = h$ belongs to the support, for all Lipschitz (in fact, Cameron–Martin) paths h. Since Lipschitz paths are dense in $C^{0,\alpha}\left([0,T];\mathbb{R}^d\right)$, taking the closure yields

$$C^{0,\alpha}\left([0,T];\mathbb{R}^d\right) \subset \operatorname{supp}\left(\text{law of } B\right).$$

∎

13.7.2 Recalls on translations of rough paths

We just used the translation map $T_h\left(x\right) = x + h$ for $\mathbb{R}^d$-valued paths x and h. Assume both x and h are Lipschitz, started at 0, and consider the step-2 lift: $\mathbf{x} \equiv S_2\left(x\right)$, and $T_h\left(\mathbf{x}\right) \equiv S_2\left(T_h\left(x\right)\right)$. From definition of S_2,

$$T_h\left(\mathbf{x}\right) = 1 + \left(\mathbf{x}^1 + h\right) + \left(\mathbf{x}^2 + \int_0^\cdot x \otimes dh + \int_0^\cdot h \otimes dx + S_2\left(h\right)\right).$$

The following proposition is an easy consequence of the results of Section 9.4.6.

Proposition 13.50 *Let $\alpha \in (1/3, 1/2]$. The map $\mathbf{x} \mapsto T_h\left(\mathbf{x}\right)$ can be extended to a continuous map of $C^{0,\alpha}\left([0,1], G^2\left(\mathbb{R}^d\right)\right)$ into itself and T_h also denotes this extension. It is bijective with inverse T_{-h}. In particular, the image of any open set under T_h is again open.*

13.7.3 Support of enhanced Brownian motion

Following Section 13.5 we recall that $\mathbf{B}_*\mathbb{P}$, the law of $\mathbf{B}$, can be viewed as a Borel probability measure on $C^{0,\alpha}\left([0,T], G^2\left(\mathbb{R}^d\right)\right)$, $\alpha \in [0, 1/2)$. Moreover, we saw that the law of $T_h\left(\mathbf{B}\right)$ is equivalent to the law of $\mathbf{B}$ when $h \in \mathcal{H}$ is a Cameron–Martin path. As a consequence, we have

Proposition 13.51 *Let $h \in \mathcal{H}$ be a Cameron–Martin path and $\mathbf{x} \in \operatorname{supp}\left(\mathbf{B}_*\mathbb{P}\right)$. Then $T_h\left(\mathbf{x}\right) \in \operatorname{supp}\left(\mathbf{B}_*\mathbb{P}\right)$.*

Proof. With the properties of $\mathbf{x} \mapsto T_h\left(\mathbf{x}\right)$ we established in Proposition 13.50 and

$$\left(\text{law of } \mathbf{B}\right) \sim \left(\text{law of } T_h\left(\mathbf{B}\right)\right),$$

the proof given earlier for Brownian motion (Corollary 13.48) adapts with no changes. ∎

Lemma 13.52 *Let $\alpha \in (0, 1/2)$. There exist $\mathbf{x} \in \operatorname{supp}\left(\mathbb{P}_*\mathbf{B}\right)$ and $\left(x^n\right) \subset \mathcal{H}$ so that*

$$\left\|T_{-x^n}\mathbf{x}\right\|_{\alpha\text{-Höl};[0,T]} \to 0 \text{ as } n \to \infty.$$

Remark 13.53 Note that $\left\|T_{-B^n}\mathbf{B}\right\|_{\alpha\text{-Höl};[0,T]} \to 0$ does not follow as a deterministic consequence from $d_{\alpha\text{-Höl};[0,T]}\left(\mathbf{B}, S_2\left(B_n\right)\right) \to 0$. □

Proof. If B^n denotes the piecewise linear approximation based on a nested sequence of dissections, we saw that

$$S(B^n) \to \mathbf{B} \text{ a.s. (pointwise)}$$

with uniform α-Hölder bounds. In fact, the essential observation was that

$$\mathbb{E}\left[\int_0^t \beta d\tilde{\beta}|\mathfrak{F}_n\right] = \int_0^t \beta^n d\tilde{\beta}^n$$

where $\mathfrak{F}_n = \sigma\left(\beta_t, \tilde{\beta}_t : t \in D_n\right)$. The arguments given in Section 13.3.2 also give

$$\mathbb{E}\left[\int_0^t \beta d\tilde{\beta}\,\middle|\,\sigma\left(\beta_t : t \in D_n\right) \vee \sigma(\tilde{\beta}_t : t \in [0,T])\right] = \int_0^t \beta^n d\tilde{\beta},$$

$$\mathbb{E}\left[\int_0^t \beta d\tilde{\beta}\,\middle|\,\sigma\left(\beta_t : t \in [0,T]\right) \vee \sigma(\tilde{\beta}_t : t \in D_n)\right] = \int_0^t \beta d\tilde{\beta}^n.$$

Both integrals on the right-hand side make sense as Riemann–Stieltjes integrals, and

$$T_{-B^n}\mathbf{B} \to 0 \text{ a.s. (pointwise)}$$

with uniform α-Hölder bounds. The usual interpolation finishes the proof. Indeed, we could have started with $\tilde{\alpha} \in (\alpha, 1/2)$, got uniform $\tilde{\alpha}$-bounds and used interpolation to obtain $T_{-B^n}\mathbf{B} \to 0$ in α-Hölder topology. This statement holds a.s. and we can take any $\mathbf{x} = \mathbf{B}(\omega)$ for ω in a set of full measure. ∎

Theorem 13.54 *Let $\alpha \in (0, 1/2)$. The topological support of the law of $G^2(\mathbb{R}^d)$-valued enhanced Brownian motion on $[0,T]$ with respect to $d_{\alpha\text{-Höl}}$ is precisely $C_0^{0,\alpha}([0,T]; G^2(\mathbb{R}^d))$.*

Proof. Thanks to Proposition 13.51 and Lemma 13.52, the argument is the same as that for d-dimensional Brownian motion, as given in the proof of Theorem 13.49. ∎

13.8 Support theorem in conditional form

13.8.1 Brownian motion conditioned to stay near the origin

We want to condition d-dimensional standard Brownian motion B to stay ε-close to the origin over the time interval $[0,1]$. In other words, we want to condition with respect to the event

$$\left\{|B|_{\infty;[0,1]} < \varepsilon\right\} = \left\{\sup_{t \in [0,1]} \sqrt{\sum_{i=1}^d (B_t^i)^2} < \varepsilon\right\}. \tag{13.21}$$

Despite the equivalence of norms on $\mathbb{R}^d$, Brownian motion *does* care how it is confined and it is important that we use the Euclidean norm on $\mathbb{R}^d$. (See Proposition 13.7, which we shall use below.) From Theorem 13.49, we know that $\left\{ |B|_{\infty;[0,1]} < \varepsilon \right\}$ has positive probability, but the next lemma gives a precise quantitative bound.

Lemma 13.55 *Let $\lambda > 0$ denote the lowest eigenvalue of $-\frac{1}{2}\Delta$ with Dirichlet boundary conditions on $\partial B(0,1)$, the boundary of the Euclidean unit ball. Then there exists a constant $C > 0$ such that*

$$\frac{1}{C} \exp\left(-\lambda \frac{t}{\varepsilon^2}\right) \leq \mathbb{P}\left[|B|_{\infty;[0,t]} < \varepsilon \right]. \tag{13.22}$$

Proof. By Brownian scaling it suffices to consider $\varepsilon = 1$. Let $p_t(x,y)$ denote the Dirichlet heat-kernel for $B(0,1)$. Then,

$$\mathbb{P}\left[|B|_{\infty;[0,t]} < 1 \right] = \int_{B(0,1)} p_t(0,y)\,dy.$$

Recall that the lowest eigenvalue is simple and that the (up to multiplicative constants unique) eigenfunction $\psi(\cdot)$ corresponding to λ can be taken positive,[15] continuous (in fact, smooth[16]) and L^2-normalized so that

$$\int_{B(0,1)} \psi^2(z)\,dz = 1.$$

In particular,

$$
\begin{aligned}
\psi(y) &= e^\lambda \int_{B(0,1)} p_1(y,z)\,\psi(z)\,dz \\
&\leq e^\lambda \sqrt{\int_{B(0,1)} p_1(y,z)^2\,dy} \sqrt{\int_{B(0,1)} \psi^2(z)\,dz} \\
&\leq e^\lambda \sqrt{p_2(y,y)} \leq e^\lambda (4\pi)^{-d/4}
\end{aligned}
$$

and the proof is finished with the estimate

$$
\begin{aligned}
0 \; &< \; \psi(0) = e^{\lambda t} \int_{B(0,1)} p_t(0,y)\,\psi(y)\,dy \\
&\leq \; (4\pi)^{-d/4}\, e^{\lambda(t+1)} \int_{B(0,1)} p_t(0,y)\,dy.
\end{aligned}
$$

∎

We shall need to complement Lemma 13.55 with an upper estimate and write $\mathbb{P}_x$ to indicate that Brownian motion B is started at $B(0) = x$.

---

[15] See, for example, [71, Theorem 8.38].
[16] This follows from standard elliptic regularity theory.

Lemma 13.56 *Let $\lambda > 0$ denote the lowest eigenvalue of $-\frac{1}{2}\Delta$ with Dirich-let boundary conditions on the Euclidean unit ball $B(0,1)$. Then there exists a constant $C > 0$ such that*

$$\sup_{x \in B(0,1)} \mathbb{P}_x \left[|B|_{\infty;[0,t]} < \varepsilon \right] < C \exp\left(-\lambda \frac{t}{\varepsilon^2} \right). \tag{13.23}$$

Proof. Again, by Brownian scaling it suffices to consider $\varepsilon = 1$. Let $x \in B(0,1)$, the Euclidean ball $B(0,1) \subset \mathbb{R}^d$. From symmetry considerations (cf. Exercise 13.57 below) we see that

$$\mathbb{P}_x \left[|B|_{\infty;[0,t]} < 1 \right] \leq \mathbb{P}_0 \left[|B|_{\infty;[0,t]} < 1 \right] = \int_{B(0,1)} p_t(0,x)\, dx,$$

where $p_t(\cdot,\cdot)$ denotes the Dirichlet heat-kernel for $B(0,1)$. Using $p_t(x,y) = p_t(y,x)$ and writing P_t for the associated semi-group on $L^2(B(0,1))$, we have

$$\int_{B(0,1)} p_t(0,x)\, dx = \int_{B(0,1)} \left(\int_{B(0,1)} p_{t-1}(x,y)\, p_1(0,y)\, dy \right) dx$$

$$\leq \sqrt{|B(0,1)|} \left| \left(\int_{B(0,1)} p_{t-1}(\cdot,y)\, p_1(0,y)\, dy \right) \right|_{L^2(B(0,1))}$$

$$\leq \sqrt{|B(0,1)|}\, |P_{t-1} p_1(0,\cdot)|_{L^2(B(0,1))}$$

$$\leq \sqrt{|B(0,1)|}\, e^{-(t-1)\lambda}\, |p_1(0,\cdot)|_{L^2(B(0,1))}$$

$$= \sqrt{|B(0,1)|}\, e^{-(t-1)\lambda}\, \sqrt{p_2(0,0)},$$

as required.   ∎

Exercise 13.57 *Let B denote Brownian motion on $\mathbb{R}^d$ equipped with Euclidean distance. Show that for all $x \in B(0,1)$,*

$$\mathbb{P}_x \left[|B|_{\infty;[0,t]} < 1 \right] \leq \mathbb{P}_0 \left[|B|_{\infty;[0,t]} < 1 \right].$$

(Hint: Use symmetry).

We can now define the conditional probabilities

$$\mathbb{P}^\varepsilon(\bullet) \doteq \mathbb{P}\left(\bullet \mid |B|_{\infty;[0,1]} < \varepsilon \right).$$

(Since the conditioning event has positive probability, this notion is elementary.)

Lemma 13.58 (increments over small times) *There exists $C > 0$ such that for all $R > 0$ and $0 < \varepsilon < 1$,*

$$\mathbb{P}^\varepsilon \left(\exists\, 0 \leq s < t \leq 1,\, |t-s| < \varepsilon^2 : \frac{\|\mathbf{B}_{s,t}\|}{|t-s|^\alpha} > R \right) \leq C \exp\left[-\frac{1}{C} \left(\frac{R}{\varepsilon^{1-2\alpha}} \right)^2 \right].$$

Proof. Step 1: Suppose there exists a pair of times $s, t \in [0, 1]$ such that

$$s < t, \ |t - s| < \varepsilon^2 \text{ and } \frac{\|\mathbf{B}_{s,t}\|}{|t - s|^\alpha} > R.$$

Then there exists a $k \in \{1, \ldots, \lceil 1/\varepsilon^2 \rceil\}$ so that $[s, t] \subset [(k-1)\varepsilon^2, (k+1)\varepsilon^2]$. In particular, the probability that such a pair of times exists is at most

$$\sum_{k=1}^{\lceil 1/\varepsilon^2 \rceil} \mathbb{P}^\varepsilon \left(\|\mathbf{B}\|_{\alpha\text{-Höl};[(k-1)\varepsilon^2,(k+1)\varepsilon^2]} > R \right). \tag{13.24}$$

We will see in step 2 below that each term in this sum is exponentially small with ε, namely bounded by

$$\mathbb{P}^\varepsilon \left(\|\mathbf{B}\|_{\alpha\text{-Höl};[(k-1)\varepsilon^2,(k+1)\varepsilon^2]} > R \right) \leq C \exp\left[-\frac{1}{C} \left(\frac{R}{\varepsilon^{1-2\alpha}} \right)^2 \right] \tag{13.25}$$

for some positive constant C. Since there are only $\lceil 1/\varepsilon^2 \rceil$ terms in this sum, it suffices to make C slightly bigger to control the entire sum and this finishes the proof of Lemma 13.58, subject to proving (13.25).

Step 2: We now show that for any $T_1 < T_2$ in $[0, 1]$ with $T_2 - T_1 \leq 2\varepsilon^2$ we have

$$\mathbb{P}^\varepsilon \left(\|\mathbf{B}\|_{\alpha;[T_1,T_2]} > R \right) < C \exp\left[-\frac{1}{C} \left(\frac{R}{\varepsilon^{1-2\alpha}} \right)^2 \right].$$

(Applied to $T_1 = (k-1)\varepsilon^2, T_2 = (k+1)\varepsilon^2$ we will then obtain the estimate (13.25), as desired.) Writing out the very definition of $\mathbb{P}^\varepsilon$ leads immediately to

$$\mathbb{P}^\varepsilon \left(\|\mathbf{B}\|_{\alpha;[T_1,T_2]} > R \right)$$

$$\leq \frac{\mathbb{P}_0 \left(\|\mathbf{B}\|_{\alpha\text{-Höl};[T_1,T_2]} > R; |B|_{0;[0,T_1]} < \varepsilon; |B|_{0;[T_2,1]} < \varepsilon \right)}{\mathbb{P}_0 \left[|B|_{0;[0,1]} < \varepsilon \right]}$$

By using the Markov property, this equals

$$\frac{\mathbb{E}_0 \left[\mathbb{P}_{B(T_2)} \left(|B|_{0;[0,1-T_2]} < \varepsilon \right); \|\mathbf{B}\|_{\alpha\text{-Höl};[T_1,T_2]} > R; |B|_{0;[0,T_1]} < \varepsilon \right]}{\mathbb{P}_0 \left[B_{0;[0,1]} < \varepsilon \right]}$$

$$\leq c_1 e^{\lambda \varepsilon^{-2}} \mathbb{E}_0 \left[e^{-\lambda(1-T_2)\varepsilon^{-2}}; \|\mathbf{B}\|_{\alpha\text{-Höl};[T_1,T_2]} > R; |B|_{0;[0,T_1]} < \varepsilon \right],$$

where c_1 is the product of the respective multiplicative constants of Lemmas 13.55 and 13.56. Using independence of (enhanced) Brownian increments, see Proposition 13.11, the last equation line is

$$
= \quad c_1 e^{\lambda T_2 \varepsilon^{-2}} \mathbb{P}_0 \left(\|\mathbf{B}\|_{\alpha\text{-Höl};[0,T_2-T_1]} > R \right) \mathbb{P}_0 \left(|B|_{0;[0,T_1]} < \varepsilon \right)
$$
$$
\leq \quad c_2 e^{\lambda(T_2-T_1)\varepsilon^{-2}} \mathbb{P}_0 \left(\|\mathbf{B}\|_{\alpha\text{-Höl};[0,T_2-T_1]} > R \right)
$$

as another application of Lemma 13.55. Using $T_2 - T_1 \leq 2\varepsilon^2$ this expression is bounded by

$$
\leq \quad c_2 e^{2\lambda} \mathbb{P}_0 \left(\|\mathbf{B}\|_{\alpha\text{-Höl};[0,2\varepsilon^2]} > R \right)
$$
$$
\leq \quad c_2 e^{2\lambda} c_3 \exp \left[-\frac{1}{c_3} \frac{R^2}{(2\varepsilon^2)^{1-2\alpha}} \right]
$$

where we used scaling and Fernique estimates for enhanced Brownian motion in the last step. The proof is now finished. ∎

Lemma 13.59 (increments over large times) *There exists $C > 0$ such that for all $R > 0$ and $0 < \varepsilon$ small enough, namely such that (13.27) is satisfied,*

$$
\mathbb{P}^\varepsilon \left(\exists 0 \leq s < t \leq 1, |t-s| \geq \varepsilon^2 : \frac{\|\mathbf{B}_{s,t}\|}{|t-s|^\alpha} > R \right) \leq C \exp \left(-\frac{1}{C} \left(\frac{R^2}{\varepsilon^{1-2\alpha}} \right)^2 \right).
$$

Proof. Let us first recall Lipschitz equivalence of homogenous norms,

$$
\|\mathbf{B}_{s,t}\| \sim |B_t - B_s| \vee \sqrt{|A_{s,t}|}.
$$

We can thus establish Lemma 13.59 by estimating

$$
\mathbb{P}^\varepsilon \left(\exists s,t \in [0,1], |t-s| \geq \varepsilon^2 : \frac{|B_t - B_s|}{|t-s|^\alpha} \vee \frac{\sqrt{|A_{s,t}|}}{|t-s|^\alpha} > R \right)
$$
$$
\leq \quad \mathbb{P}^\varepsilon \left(\exists s,t \in [0,1], |t-s| \geq \varepsilon^2 : 2\varepsilon^{1-2\alpha} \vee \frac{\sqrt{|A_{s,t}|}}{|t-s|^\alpha} > R \right). \quad (13.26)
$$

Two observations are of assistance. First, upon assuming ε small enough, namely such that

$$
2\varepsilon^{1-2\alpha} < R, \quad (13.27)
$$

we have

$$
2\varepsilon^{1-2\alpha} \vee \frac{\sqrt{|A_{s,t}(\omega)|}}{|t-s|^\alpha} > R \quad \Longleftrightarrow \quad \frac{\sqrt{|A_{s,t}(\omega)|}}{|t-s|^\alpha} > R.
$$

Second, $|B|_{\infty;[0,1]} \leq \varepsilon$ implies that the area increments become "almost" additive. More precisely, cf. (13.7),

$$
A_{s,t} = (A_t - A_s) - \frac{1}{2}[B_s, B_{s,t}] \quad \Longrightarrow \quad |A_{s,t}| \leq |A_t - A_s| + 2\varepsilon^2
$$

and it follows that

$$\frac{|A_{s,t}|}{|t-s|^{2\alpha}} \leq \frac{|A_t - A_s|}{|t-s|^{2\alpha}} + 2\varepsilon^{2-4\alpha}$$

$$\leq \frac{|A_t - A_s|}{\varepsilon^{2\alpha}|t-s|^{\alpha}} + \frac{R^2}{2}$$

where we used $|t-s| \geq \varepsilon^2$ and (13.27). Putting things together shows that (13.26) is

$$= \mathbb{P}^{\varepsilon}\left(\exists s, t \in [0,1], |t-s| \geq \varepsilon^2 : \frac{|A_{s,t}|}{|t-s|^{2\alpha}} > R^2\right)$$

$$\leq \mathbb{P}^{\varepsilon}\left(\exists s, t \in [0,1] : \frac{|A_t - A_s|}{\varepsilon^{2\alpha}|t-s|^{\alpha}} > \frac{R^2}{2}\right)$$

and it is of course enough to consider

$$\mathbb{P}^{\varepsilon}\left(\exists s, t \in [0,1] : \frac{|Z_t - Z_s|}{\varepsilon^{2\alpha}|t-s|^{\alpha}} > \frac{R^2}{2}\right)$$

$$= \mathbb{P}^{\varepsilon}\left(|Z|_{\alpha\text{-Höl};[0,1]} > \frac{R^2}{2}\varepsilon^{2\alpha}\right) \tag{13.28}$$

where Z is one component of the Lévy's area,

$$Z_t \equiv A_t^{i,j} = \frac{1}{2}\left(\int_0^t B^i dB^j - \int_0^t B^j dB^i\right)$$

for fixed $i \neq j$ in $\{1,\dots,d\}$. Proposition 13.7 tells us that there exists a 1-dimensional Brownian motion, say W, such that

$$Z_t = W(a(t)) \text{ with } a(t) := \frac{1}{4}\int_0^t \left((B_s^i)^2 + (B_s^j)^2\right) ds$$

and W is independent of the process $(B^i)^2 + (B^j)^2$ and so independent of $|B|_{\infty;[0,1]}$. In order to control the Hölder norm of Z we use a basic fact about composition of Hölder functions,

$$|f \circ g|_{(\alpha\beta)\text{-Höl}} \leq |f|_{\alpha\text{-Höl}}\left(|g|_{\beta\text{-Höl}}\right)^{\alpha},$$

with the remark that $|f|_{\alpha\text{-Höl}}$ can be replaced by the α-Hölder norm of f restricted to the range of g. Applying this to $W \circ a$ yields

$$|Z|_{\alpha\text{-Höl};[0,1]} \leq |W|_{\alpha\text{-Höl};[0,a(1)]}\left(|a|_{\text{Lip};[0,1]}\right)^{\alpha}.$$

On the conditioning event $\left\{ |B|_{\infty;[0,1]} \leq \varepsilon \right\}$ we have both a (1) and $|a|_{\text{Lip};[0,1]}$ $\leq \varepsilon^2/4 \leq \varepsilon^2$ and so we can continue to estimate (13.28):

$$
\begin{aligned}
\mathbb{P}^\varepsilon \left(|Z|_{\alpha\text{-Höl};[0,1]} > \frac{R^2}{2} \varepsilon^{2\alpha} \right) &\leq \mathbb{P}^\varepsilon \left(|W|_{\alpha\text{-Höl};[0,\varepsilon^2]} > \frac{R^2}{2} \right) \\
&= \mathbb{P} \left(|W|_{\alpha\text{-Höl};[0,\varepsilon^2]} > \frac{R^2}{2} \right) \\
&= \mathbb{P} \left(\varepsilon^{1-2\alpha} |W|_{\alpha\text{-Höl};[0,1]} > \frac{R^2}{2} \right).
\end{aligned}
$$

From the second to the third line above, when replacing $\mathbb{P}^\varepsilon$ by $\mathbb{P}$, we crucially used that W is independent of $(B^i)^2 + (B^j)^2$ and so independent of the radial process $(B.)$ and in particular of $|B|_{\infty;[0,1]}$. The proof of Lemma 13.59 is then finished with Fernique estimates, i.e. Gaussian integrability for the α-Hölder norm of W. ∎

Remark 13.60 Fix any $s < t$ in $[0,1]$ with the property that $t - s \geq \varepsilon^2$. Then

$$
\begin{aligned}
\mathbb{P}^\varepsilon \left(\frac{\|\mathbf{B}_{s,t}\|}{|t-s|^{1/2}} > R \right) &\leq \mathbb{P}^\varepsilon \left(\frac{\|\mathbf{B}_{s,t}\|}{|t-s|^\alpha} > R\varepsilon^{1-2\alpha} \right) \\
&\leq C \exp \left(-\frac{1}{C} R^4 \right) \quad \text{for } R > 2.
\end{aligned}
$$

The last estimate comes from Lemma 13.59, applied with $R\varepsilon^{1-2\alpha}$ instead of R and noting that condition (13.27) is satisfied for $R > 2$. □

We are now able to state the main result of this section.

Theorem 13.61 *Let* $\alpha \in [0,1/2)$. *Then, for any* $\delta > 0$,

$$
\lim_{\varepsilon \to 0} \mathbb{P}^\varepsilon \left(\|\mathbf{B}\|_{\alpha\text{-Höl};[0,1]} > \delta \right) = 0.
$$

Proof. An obvious consequence of Lemmas 13.58 and 13.59. ∎

Exercise 13.62 (i) *Let* φ *be a fixed increasing function such that* $\varphi(h) = \sqrt{h \log(1/h)}$ *in a positive neighbourhood of* 0. *Prove that there exists* $\eta > 0$ *such that*

$$
\sup_{\varepsilon \in (0,1)} \mathbb{E}^\varepsilon \exp \left(\eta \|\mathbf{B}\|_{\varphi\text{-Höl},[0,1]} \right) < \infty,
$$

where $\|\mathbf{B}\|_{\varphi\text{-Höl};[0,T]} = \sup_{s,t\in[0,T]} d(\mathbf{B}_s, \mathbf{B}_t) / \varphi(t-s)$.
(ii) *Show that there exists* $\eta > 0$ *such that*

$$
\sup_{\varepsilon \in (0,1)} \mathbb{E}^\varepsilon \exp \left(\eta \|\mathbf{B}\|^2_{\psi_{2,1}\text{-var};[0,1]} \right) < \infty.
$$

Hint: It suffices to establish

$$\exists \eta > 0 : \sup_{\varepsilon \in (0,1]} \sup_{s,t \in [0,T]} \mathbb{E}^{\varepsilon} \left(\exp \left(\eta \frac{\|\mathbf{B}_{s,t}\|^2}{|t-s|} \right) \right) < \infty.$$

The case $t - s \leq \varepsilon^2$ follows from the argument in the proof of Lemma 13.58, line by line with s,t instead of T_1, T_2 and $\|\mathbf{B}\|_{\alpha\text{-Höl};[0,T_2-T_1]}$ replaced by $\|\mathbf{B}_{s,t}\| / |t-s|^{1/2}$; also noting that

$$\mathbb{P}^{\varepsilon} \left(\frac{\|\mathbf{B}_{s,t}\|}{|t-s|^{1/2}} > R \right) \leq (\text{const}) e^{\lambda} \mathbb{P} \left(\frac{\|\mathbf{B}_{s,t}\|}{|t-s|^{1/2}} > R \right).$$

13.8.2 *Intermezzo on rough path distances*

Recall that S_2 maps a sufficiently regular $\mathbb{R}^d$-valued path started at 0 into the $G^2 \left(\mathbb{R}^d \right)$-valued path $1 + h(\cdot) + \int_0^{\cdot} h \otimes dh$. We then have

Proposition 13.63 *Let $\mathbf{X} = \exp(X + A) \in C_0^{\alpha\text{-Höl}} \left([0,1], G^2 \left(\mathbb{R}^d \right) \right)$ and $h \in C_0^{1\text{-var}} \left([0,1], \mathbb{R}^d \right)$. Then there exists a constant C such that*

$$\left| d \left(\mathbf{X}_{s,t}, S_2 \left(h \right)_{s,t} \right) - \left\| \left(T_{-h} \mathbf{X} \right)_{s,t} \right\| \right| \leq C \sqrt{|X - h|_{\alpha\text{-Höl}} \, |h|_{1\text{-var};[s,t]}} \, |t-s|^{\alpha}.$$

In particular, when $h \in \mathcal{H} \equiv W_0^{1,2} \left([0,1], \mathbb{R}^d \right)$ this implies

$$\left| d_{\alpha\text{-Höl}} \left(\mathbf{X}, S_2 \left(h \right) \right) - \| T_{-h} \mathbf{X} \|_{\alpha\text{-Höl}} \right| \leq C \sqrt{|h|_{\mathcal{H}} \, |X - h|_{\alpha\text{-Höl}}}.$$

Proof. By symmetry of the Carnot–Caratheodory norm and the triangle inequality it follows that

$$\big| \|a\| - \|b\| \big| = \big| \| a^{-1} \| - \|b\| \big| \leq d \left(a^{-1}, b \right) = \| a \otimes b \|.$$

We apply this with (A^h denotes the area associated to h)

$$a = \mathbf{X}_{s,t}^{-1} \otimes (S_2(h))_{s,t} = \exp \left(\overbrace{-X_{s,t}^1 + h_{s,t}} - A_{s,t} + A_{s,t}^h - \frac{1}{2} \left[X_{s,t}^1 - h_{s,t}, h_{s,t} \right] \right)$$

and

$$b = (T_{-h} \mathbf{X})_{s,t} = \exp \left(\overbrace{X_{s,t}^1 - h_{s,t}} + A_{s,t} - A_{s,t}^h - \frac{1}{2} \int_s^t [h_{s,\cdot}, d(X - h)] \right.$$
$$\left. - \frac{1}{2} \int_s^t [X_{s,\cdot} - h_{s,\cdot}, dh] \right).$$

By the Campbell–Baker–Hausdorff formula, noting the cancellation of the indicated terms,

$$a \otimes b = \exp\left(-\frac{1}{2}\left[X_{s,t}^1 - h_{s,t}, h_{s,t}\right] - \frac{1}{2}\int_s^t [h_{s,\cdot}, d\,(X-h)] - \frac{1}{2}\int_s^t [X_{s,\cdot} - h_{s,\cdot}, dh]\right)$$

and it follows that $\left| d\left(\mathbf{X}_{s,t}, S_2\,(h)_{s,t}\right) - \left\|(T_{-h}\mathbf{X})_{s,t}\right\|\right|$ is less than or equal to a constant times

$$\left|\left[X_{s,t}^1 - h_{s,t}, h_{s,t}\right]\right|^{1/2} + \left|\int_s^t [h_{s,\cdot}, d\,(X-h)]\right|^{1/2} + \left|\int_s^t [X_{s,\cdot} - h_{s,\cdot}, dh]\right|^{1/2}.$$

The first term is estimated via

$$\left|\left[X_{s,t}^1 - h_{s,t}, h_{s,t}\right]\right| \leq 2\,|X - h|_{\alpha\text{-H\"ol}}\,|h|_{1\text{-var};[s,t]}\,|t - s|^\alpha =: \Delta,$$

for the third we note that

$$\left|\int_s^t (X_{s,\cdot} - h_{s,\cdot}) \otimes dh_r\right| \leq |h|_{1\text{-var};[s,t]}\,\sup_{r\in[s,t]}|X_{s,r} - h_{s,r}|$$
$$\leq |h|_{1\text{-var};[s,t]}\,|X - h|_{\alpha\text{-H\"ol}}\,|t - s|^\alpha$$

so that $\int_s^t [X_{s,\cdot} - h_{s,\cdot}, dh]$ is also bounded by Δ and a similar bound is obtained for the middle term after integration by parts. The final statement comes from

$$|h|_{1\text{-var};[s,t]} \leq |h|_{\mathcal{H}}\,|t - s|^{\frac{1}{2}}.$$

∎

13.8.3 *Enhanced Brownian motion under conditioning*

We now condition d-dimensional standard Brownian motion B to stay ε-uniformly close to a given path h over the time interval $[0, 1]$ and ask what happens to enhanced Brownian motion $\mathbf{B}$ when $\varepsilon \to 0$. To this end, let us write out the conditioning event in more detail,

$$\left\{\,|B - h|_{\infty;[0,1]} < \varepsilon\right\} = \left\{\sup_{t\in[0,1]}\sqrt{\sum_{i=1}^d \left(B_t^i - h_t^i\right)^2} < \varepsilon\right\}$$

and introduce the notation

$$\mathbb{P}^{\varepsilon,h}\,(\bullet) = \mathbb{P}\left(\,\bullet\,\middle|\,\,|B - h|_{\infty;[0,1]} < \varepsilon\right).$$

We assume that $h \in \mathcal{H} \equiv W_0^{1,2}\left([0, 1], \mathbb{R}^d\right)$, the Cameron–Martin space.

Lemma 13.64 *Given $h \in \mathcal{H}$ and $\alpha < 1/2$ then, for any $\delta > 0$,*

$$\lim_{\varepsilon \to 0} \mathbb{P}^{\varepsilon,h}\big(d_{\alpha\text{-Höl};[0,1]}(\mathbf{B}, S_2(h)) > \delta\big) = 0 \ \textit{iff} \ \lim_{\varepsilon \to 0} \mathbb{P}^{\varepsilon,h}\big(\|T_{-h}\mathbf{B}\|_{\alpha\text{-Höl};[0,1]} > \delta\big) = 0.$$

Proof. Immediate from Proposition 13.63, noting that $|B - h|_{\alpha\text{-Höl}}$ is dominated by either quantity

$$d_{\alpha\text{-Höl};[0,1]}(\mathbf{B}, S_2(h)), \ \|T_{-h}\mathbf{B}\|_{\alpha\text{-Höl};[0,1]}.$$

∎

Lemma 13.65 *Given $h \in \mathcal{H}$ and $\alpha < 1/2$ then, for any $\delta > 0$,*

$$\mathbb{P}^{\varepsilon,h}\big(\|T_{-h}\mathbf{B}\|_{\alpha\text{-Höl};[0,1]} > \delta\big) \le \sqrt{\mathbb{P}^{\varepsilon}\big[\|\mathbf{B}\|_{\alpha\text{-Höl};[0,1]} > \delta\big]}\sqrt{\mathbb{E}^{\varepsilon}\big[\exp\big(-2\,I[\dot{h}]\big)\big]}$$

where we write $I\big[\dot{h}\big] \equiv \int_0^1 \dot{h}_t dB_t = \sum_{i=1}^d \int_0^1 \dot{h}_t^i dB_t^i$.

Proof. From Proposition 13.37, $(T_{-h}\mathbf{B})(\omega) = \mathbf{B}(\omega - h)$ a.s. and we proceed by the Cameron–Martin therorem. A drift term $-h$ corresponds to the Radon–Nikodym density

$$R_h = \exp\left(-I[\dot{h}] - \frac{1}{2}\int_0^1 \big|\dot{h}_t\big|^2 dt\right)$$

which will allow us to write out $\mathbb{P}^{\varepsilon,h}$ in terms of $\mathbb{E}^{\varepsilon,0}$. We first note that by symmetry,

$$\frac{\mathbb{E}\left[\int_0^1 \dot{h}_t dB_t, \left\{\ |B|_{\infty;[0,1]} < \varepsilon\right\}\right]}{\mathbb{P}\left[|B|_{\infty;[0,1]} < \varepsilon\right]} = \frac{\mathbb{E}\left[-\int_0^1 \dot{h}_t dB_t, \left\{\ |{-B}|_{\infty;[0,1]} < \varepsilon\right\}\right]}{\mathbb{P}\left[|{-B}|_{\infty;[0,1]} < \varepsilon\right]},$$

which shows that $\mathbb{E}^{\varepsilon}\big(I[\dot{h}]\big) = 0$. From Jensen's inequality we then have

$$\mathbb{E}^{\varepsilon}\big(e^{-I[\dot{h}]}\big) \ge \exp \mathbb{E}^{\varepsilon}\big(-I[\dot{h}]\big) = 1.$$

After these short preparations, we can write

$$\mathbb{P}^{\varepsilon,h}\big(\|T_{-h}\mathbf{B}\|_{\alpha\text{-Höl};[0,1]} > \delta\big)$$

$$= \frac{\mathbb{E}\left[R_h, \left\{\|\mathbf{B}\|_{\alpha\text{-Höl};[0,1]} > \delta\right\} \cap \left\{\ |B|_{\infty;[0,1]} < \varepsilon\right\}\right]}{\mathbb{E}\left[R_h, \left\{\ |B|_{\infty;[0,1]} < \varepsilon\right\}\right]}$$

$$= \frac{\mathbb{E}^{\varepsilon}\left[e^{-I[\dot{h}]}, \left\{\|\mathbf{B}\|_{\alpha\text{-Höl};[0,1]} > \delta\right\}\right]}{\mathbb{E}^{\varepsilon}\left[e^{-I[\dot{h}]}\right]}$$

$$\le \ \mathbb{E}^{\varepsilon}\left[e^{-I[\dot{h}]}, \left\{\|\mathbf{B}\|_{\alpha\text{-Höl};[0,1]} > \delta\right\}\right].$$

Cauchy–Schwarz finishes the proof.   ■

We can now state the main result of this section.

Theorem 13.66 *Given* $h \in C_0^2\left([0,1],\mathbb{R}^d\right)$ *and* $\alpha < 1/2$ *then, for any* $\delta > 0$,

$$\lim_{\varepsilon \to 0} \mathbb{P}^{\varepsilon,h}\left(d_{\alpha\text{-H\"ol};[0,1]}\left(\mathbf{B}, S_2\left(h\right)\right) > \delta\right) = 0. \tag{13.29}$$

Proof. Combining the previous two lemmas gives the desired conclusion *provided*

$$\varlimsup_{\varepsilon \to 0} \mathbb{E}^\varepsilon\left[\exp\left(-2I[\dot{h}]\right)\right] < \infty.$$

When $h \in C_0^2\left([0,1],\mathbb{R}^d\right)$ this is very easy; it suffices to write $g := -2\dot{h} \in C^1$ so that

$$|I[g]| = \left|\int_0^1 g\,dB\right| = \left|g_1 B_1 - \int_0^1 B\,dg\right| \leq 2\,|B|_{\infty;[0,1]} \times |\dot{g}|_{\infty;[0,1]}\,.$$

■

The reader may suspect that the restriction to $h \in C_0^2\left([0,1],\mathbb{R}^d\right) \subsetneqq \mathcal{H}$ in the previous statement was unnecessary. The extension to $h \in \mathcal{H}$ turns out to be rather subtle and is discussed in the following

Exercise 13.67 *Show that for all* $g \in L^2\left([0,1],\mathbb{R}^d\right)$,

$$\mathbb{E}^\varepsilon\left[\exp\left(I\left[g\right]\right)\right] \leq \mathbb{E}\left[\exp\left(|I\left[g\right]|\right)\right].$$

Hint: A classical correlation inequality [35, Theorem 2.1] states that for any i.i.d. family of standard Gaussian (X_i) *and any convex, symmetric set* $C_k \subset R^k$,

$$\mathbb{P}\left[|X_1| < \eta,\, (X_1,\ldots,X_k) \in C_k\right] \leq \mathbb{P}\left[|X_1| < \eta\right]\mathbb{P}\left[(X_1,\ldots,X_k) \in C_k\right]. \tag{13.30}$$

Solution. It suffices to show that for any $\eta > 0$ and $\varepsilon > 0$,

$$\mathbb{P}^\varepsilon\left(|I\left(g\right)| < \eta\right) \geq \mathbb{P}\left(|I\left(g\right)| < \eta\right) \tag{13.31}$$

since then

$$
\begin{aligned}
\mathbb{E}^\varepsilon\left(\exp I\left(g\right)\right) &\leq \mathbb{E}^\varepsilon\left(\exp|I\left(g\right)|\right) = \int_0^\infty e^x \mathbb{P}^\varepsilon\left(|I\left(g\right)| > x\right)dx \\
&\leq \int_0^\infty e^x \mathbb{P}\left(|I\left(g\right)| > x\right)dx = \mathbb{E}\left(\exp|I\left(g\right)|\right).
\end{aligned}
$$

To show inequality (13.31), let (g_i) denote an orthonormal basis for $L^2\left([0,1],\mathbb{R}^d\right)$ such that $g_1 = g$. Set $X_i := \int_0^1 g_i\,dB$ and denote by X the infinite vector whose components are X_i. Note that the X_i are standard, i.i.d. normal random variables. Let C be a convex, symmetric set in R^∞ and denote by

C_k its projection on R^k. Clearly, C_k is convex and symmetric and (13.30) applies to it. By dominated convergence,

$$\mathbb{P}\left[|X_1| < \eta,\, X \in C\right] \leq \mathbb{P}\left[|X_1| < \eta\right]\mathbb{P}\left[X \in C\right].$$

Therefore, choosing $C := \left\{\, |B|_{\infty;[0,1]} < \varepsilon\right\}$ and noting that C is both symmetric and convex, we obtain inequality (13.31). □

13.9   Appendix: infinite 2-variation of Brownian motion

Let B denote d-dimensional standard Brownian motion. We now show that $|B|_{2\text{-var};[0,T]} = +\infty$ a.s. The importance of this statement is that it rules out a stochastic integration based on Young integrals.

Lemma 13.68 (Vitali covering) *Assume a set $E \subset [0,1]$ admits a "Vitali cover"; that is, a (possibly uncountable) family $\mathcal{I} = \{I_\alpha\}$ of closed intervals (with non-empty interior) in $(0,1)$ so that for every $t \in E, \eta > 0$ there exists an interval $I \in \mathcal{I}$ with length $|I| < \eta$ and $t \in I$. Then, for every $\varepsilon > 0$, there exist disjoint intervals $I_1, \ldots, I_n \in \mathcal{I}$ such that*[17]

$$|E \backslash (I_1 \cup \cdots \cup I_n)| < \varepsilon. \tag{13.32}$$

Proof. We start by taking any interval $I_1 \in \mathcal{I}$ and assume that $I_1, \ldots, I_k$ have been defined. If these intervals cover E, the construction is trivially finished. Otherwise, we set $r_k := \sup\{|I| : I \in \mathcal{I},\, I \text{ disjoint of } I_1 \cup \cdots \cup I_k\}$ and note $r_k \in (0,1]$. We can then pick $I_{k+1} \in \mathcal{I}$, disjoint of $I_1 \cup \cdots \cup I_k$, with $|I_{k+1}| > r_k/2$. Assuming the construction does not finish trivially, we obtain a family $(I_k : k \in \mathbb{N})$ of closed, disjoint intervals in $(0,1)$. Clearly,

$$\sum_{k=1}^{\infty} |I_k| \leq |\cup_{k=1}^{\infty} I_{k+1}| \leq 1 \tag{13.33}$$

and we can pick $n \in \mathbb{N}$ such that $\sum_{k>n}^{\infty} |I_k| < \varepsilon/5$; with this choice of n we now verify (13.32). To this end, take any $t \in E \backslash (I_1 \cup \cdots \cup I_n)$ and then $I \in \mathcal{I}$, disjoint of $I_1 \cup \cdots \cup I_n$, with $t \in I$. There exists an integer $l \geq n+1$ such that I is disjoint of $I_1 \cup \cdots \cup I_{l-1}$ but $I \cap I_l \neq \varnothing$; otherwise $|I| \leq r_{k-1} < 2|I_k|$ for all k, in contradiction to (13.33). We also have $|I| \leq r_{l-1} < 2|I_l|$ and thinking of I_l as a "ball" of radius $|I_l/2|$ it is

---

[17]We write $|\cdot|$ for the Lebesgue measure. If E is not measurable, the statement and proof remain valid if $|\cdot|$ is understood as an outer Lebesgue measure.

then clear from $t \in I$, $I \cap I_l \neq \varnothing$ that t is also contained in a "ball" with the same centre but radius $5 |I_l/2|$. In other words, t is contained in some interval J_l with $|J_l| = 5 |I_l|$. By our choice of n we then see that $|E \backslash (I_1 \cup \cdots \cup I_n)| \leq \sum_{l=n+1}^{\infty} |J_l| < \varepsilon$. ∎

Theorem 13.69 *Let B denote d-dimensional Brownian motion on $[0,T]$. Assume that for some function ψ, defined in a positive neighbourhood of 0,*

$$\frac{h^2 / (\log \log 1/h)}{\psi(h)} \to 0 \ \text{as } h \downarrow 0.$$

Then $\sup_D \sum_{t_i \in D} \psi \left(\left| B_{t_i, t_{i+1}} \right| \right) = +\infty$ *where the sup runs over all dissections of $[0,T]$. In particular, for any $q \leq 2$ we have $|B|_{q\text{-var};[0,T]} = +\infty$ with probability one.*

Proof. Without loss of generality, we may take $T = 1$ and argue on 1-dimensional Brownian motion β. From Khintchine's law of iterated logarithms, see (13.9), there exists a deterministic constant $c \in (0, \infty)$ such that, with probability one,

$$\limsup_{t \downarrow 0} \frac{|\beta_t|}{\bar{\varphi}(t)} = c$$

where $\bar{\varphi}(h) = \sqrt{h \log \log 1/h}$ is well-defined for h small enough. (The fact that $c = 2^{1/2}$ is irrelevant for the argument.) For every fixed t, $\left(\beta_{t,t+h} : h \geq 0 \right)$ is a Brownian motion and so it is clear that (for fixed t) with probability one,

$$\limsup_{h \downarrow 0} \frac{|\beta_{t,t+h}|}{\bar{\varphi}(h)} = c.$$

Noting that $\bar{\psi}(h) := h^2 / (\log \log 1/h)$ is the asympotic inverse of $\bar{\varphi}$ at 0, this implies $\mathbb{P}(t \in E_\delta) = 1$ where

$$E_\delta = \left\{ t \in (0,1) : \bar{\psi} \left(|\beta_{t,t+h}| \right) > \frac{c}{2} h \ \text{for some } h \in (0, \delta) \right\}.$$

A Fubini argument applied to the product of $\mathbb{P}$ and Lebesgue measure $|\cdot|$ on $(0,1)$ shows that $\mathbb{P}(|E_\delta|) = 1$. But then $E := \cap_{\delta > 0} E_\delta = \cap_n E_{1/n}$ also satisfies $\mathbb{P}(|E|) = 1$ and since for each $t \in E$ there are arbitrarily small intervals $[t, t+h]$ such that $\bar{\psi} \left(|\beta_{t,t+h}| \right) > ch/2$, this family of all such intervals forms a Vitali cover of E. We can fix $\delta > 0$, and discarding all intervals of length $\geq \delta$ still leaves us with a Vitali cover of E. By Vitali's covering lemma, there are disjoint intervals $[t_i, t_i + h_i]$ for $i = 1, \ldots, n$, with $h_i < \delta$, of total length $\sum_{i=1}^{d} h_i$ arbitrarily close to 1 and in particular $\geq 1/2$, say. We can complete the endpoints of these disjoint intervals to a dissection $D^\delta = (s_j)$ of $[0,1]$ with mesh $\leq \delta$ and

$$\sum_j \bar{\psi}\left(\left|\beta_{s_j,s_{j+1}}\right|\right) \;\geq\; \sum_i \bar{\psi}\left(\left|\beta_{t_i,t_i+h_i}\right|\right)$$

$$\geq\; \sum_i ch_i/2$$

$$\geq\; c/4.$$

On the other hand, writing $\Delta\left(\delta,\omega\right) = \inf_{s_j \in D^\delta}\psi\left(\left|\beta_{s_j,s_{j+1}}\right|\right)/\bar{\psi}\left(\left|\beta_{s_j,s_{j+1}}\right|\right)$,

$$\sup_{D=(r_j)}\sum_j \psi\left(\left|\beta_{r_j,r_{j+1}}\right|\right) \;\geq\; \sum_{s_j \in D^\delta}\psi\left(\left|\beta_{s_j,s_{j+1}}\right|\right)$$

$$\geq\; \Delta\left(\delta,\omega\right)\sum_j \bar{\psi}\left(\left|\beta_{s_j,s_{j+1}}\right|\right)$$

$$\geq\; \Delta\left(\delta,\omega\right)c/4.$$

It is now an easy consequence of (uniform) continuity of β on $[0,T]$ and the assumption $\bar{\psi}/\psi \to 0$ that $\Delta\left(\delta,\omega\right) \to +\infty$ as $\delta \to 0$. This finishes the proof. ∎

The case $q = 2$ should be compared with *finite quadratic variation* of Brownian motion in the commonly used sense of semi-martingale theory. The reader can find a proof in [143, Section II.2.12].

Theorem 13.70 *Let β denote Brownian motion on $[0,T]$. If (D_n) is a sequence of nested (i.e. $D_n \subset D_{n+1}$) dissections of $[0,T]$ such that $|D_n| \to 0$ as $n \to \infty$, then*

$$\lim_{n\to\infty}\sum_{t_i \in D_n}\left|\beta_{t_i,t_{i+1}}\right|^2 = T \qquad \text{almost surely.}$$

If we drop the nestedness assumption, convergence holds in probability.

13.10 Comments

Section 13.1: The definition and basic properties of Brownian motion are classical. Among the many good textbooks, let us mention the classics: Ikeda and Watanabe [88], Karatzas and Shreve [94], Revoz and Yor [143], Rogers and Williams [144] and Stroock [159]. The definition of Lévy area based on stochastic integration appears in Ikeda and Watanabe [88, p. 128] or Malliavin [124, p. 216]. Lévy's original (martingale) construction, discussed in Exercise 13.5, only uses discrete-time martingale techniques and corresponds to a Karhunen–Lóeve-type convergence result which extends to other Gaussian processes (see Section 15.5.3). A completely different

Markovian construction of Lévy area will be given in the later chapter on Markovian processes, starting with Section 16.1. At last, Section 13.1.3 follows Ikeda and Watanabe [88].

Section 13.2: The name *enhanced Brownian motion* first appears in [115]. Once it is identified as a special case of a left-invariant Brownian motion on a (free, nilpotent) Lie group, properties such as those given in Proposition 13.11 are well known (e.g. Rogers and Williams [145] or the works of Baldi, Ben Arous). Rough path regularity of enhanced Brownian motion was established in unpublished thesis work of Sipiläinen [156]. Following the monograph of Lyons and Qian [120], it follows from showing the dyadic piecewise linear approximations converge in p-variation (rough path) metric. Our exposition here is a simple abstraction of Friz and Victoir [62], based on general Besov–Hölder embedding-type results for paths with values in metric spaces. A Fernique-type estimate for rough path norms of enhanced Brownian motion (Corollary 13.14) was also established by Inahama [90].

Section 13.3: Geodesic approximations were introduced in the rough path context in Friz and Victoir [63]; the resulting convergence results are trivial but worth noting. Rough path convergence of dyadic piecewise linear approximations to Brownian motion was established in unpublished thesis work of Sipiläinen [156], see also Lyons and Qian [120], and underwent several simplifications, notably Friz [69]. Non-standard approximations to Brownian motion were pioneered by McShane [126] and Sussmann [167]; the corresponding subsection is taken from Friz and Oberhauser [58].

Section 13.4: The discussion of Donsker's theorem for enhanced Brownian motion is taken from Breuillard *et al.* [16]. For the central limit theorem on free nilpotent groups, see Crépel and Raugi [36].

Section 13.5: The Cameron–Martin theorem for Brownian motion is classical. See Stroock [159], for instance. The proper abstract setting is for Gaussian measures on Banach spaces and (cf. Appendix D and the references therein). Theorem 13.36 is a simple observation and appears in Friz and Victoir [62]; see also Inahama [91].

Section 13.6: Large deviations for Brownian motion in uniform topology were obtained by Schilder, See Kusuoka [98]. Extensions to Hölder topology are discussed in Baldi *et al.* [5]; one can even do without any topology, Ledoux [102]. Large deviations for EBM were first established in p-variation rough path topology, Ledous *et al.* [101]; the $1/p$-Hölder case was obtained in Friz and Victoir [62]. Proposition 13.43 is taken from Friz and Victoir [65]. Exercise 13.44 follows the usual martingale arguments of the Freidlin–Wentzell estimates; Exercise 13.45 is a special case of the large deviation principle established in Section 16.7. Exercise 13.46 follows from the same arguments as in the Brownian motion case, Baldi *et al.* [5].

Section 13.7: The support description of Brownian motion itself is a trivial consequence of the Cameron–Martin theorem. The support description of EBM is subtle because of Lévy's area. Based on correlation inequalities, it was first obtained in Ledoux *et al.* [101] in p-variation topology. The arguments were simplified and strengthened to $1/p$-Hölder (resp. Lévy modulus) topology in Friz [69] (resp. Friz and Victoir [62]); the present discussion is a further streamlining.

Section 13.8: The discussion in this section follows closely Friz *et al.* [57]. The subtle gap between C^2 and $\mathcal{H}$ as discussed in Exercise 13.67 was noted in the Onsager–Machlup context, see Shepp and Zeitouni [154], but seems new in the support context.

Appendix: The infinite 2-variation of Brownian motion, not to be confused with the finite "quadratic variation" of Brownian motion, is well known, e.g. Freedman [56, Chapter 1]; our slightly sharper result is taken from Taylor [168].

14

Continuous (semi-)martingales

We have seen in the previous chapter that Brownian motion B can be enhanced to a stochastic process $\mathbf{B} = \mathbf{B}(\omega)$ for which almost every realization is a geometric $1/p$-Hölder rough path (and hence a geometric p-rough path), $p \in (2,3)$. In this chapter, we show that any continuous, d-dimensional semi-martingale, say $S = M + V$ where M is a continuous local martingale and V a continuous path of bounded variation on any compact time interval, admits a similar enhancement with p-variation rough path regularity, $p \in (2,3)$.

In fact, it suffices to construct a lift of M, denoted by $\mathbf{M}$, since then the lift of S is given deterministically via the translation operator, $\mathbf{S} = T_V \mathbf{M}$. Note that convergence of lifted piecewise linear approximations, in the sense that[1]

$$d_{p\text{-var};[0,T]}\left(S_2\left(S^{D_n}\right), \mathbf{S}\right) \to_{n\to\infty} 0 \text{ in probability},$$

is readily reduced to showing the convergence

$$d_{p\text{-var};[0,T]}\left(S_2\left(M^{D_n}\right), \mathbf{M}\right) \to_{n\to\infty} 0 \text{ in probability}.$$

Indeed, since $\left|V^{D_n} - V\right|_{(1+\varepsilon)\text{-var};[0,T]} \to_{n\to\infty} 0$ follows readily from Proposition 1.28 plus interpolation, it suffices to use basic continuiuty properties of the translation operator $(h, \mathbf{x}) \mapsto T_h \mathbf{x}$ as a map from

$$C^{q\text{-var}}\left([0,T], \mathbb{R}^d\right) \times C^{p\text{-var}}\left([0,T], G^2\left(\mathbb{R}^d\right)\right) \mapsto C^{p\text{-var}}\left([0,T], G^2\left(\mathbb{R}^d\right)\right),$$

valid for $1/q + 1/p > 1$. After these preliminary remarks we can and will focus our attention on continuous local martingales. We assume the reader is familiar with the basic aspects of this theory.

14.1 Enhanced continuous local martingales

We write $\mathcal{M}^c_{0,\mathrm{loc}}\left([0,\infty), \mathbb{R}^d\right)$ or $\mathcal{M}^c_{0,\mathrm{loc}}\left(\mathbb{R}^d\right)$ for the class of $\mathbb{R}^d$-valued continuous local martingales $M : [0,\infty) \to \mathbb{R}^d$ null at 0, defined on some filtered probability space $(\Omega, \mathfrak{F}, (\mathfrak{F}_t), \mathbb{P})$. We define the process $\langle M \rangle : \Omega \times [0,\infty) \to \mathbb{R}^d$ componentwise, with the ith component being defined as the

---

[1] As usual, S^{D_n} denotes the piecewise linear approximation to a path S based on some dissection $D_n \in \mathcal{D}[0,T]$.

"usual" bracket (or quadratic variation) process $\langle M^i \rangle \equiv \langle M^i, M^i \rangle$ of (the real-valued, continuous local martingale) M^i; that is, the unique real-valued continuous increasing process such that $\left(M^i \right)^2 - \langle M^i \rangle$ is a continuous local martingale null at zero.[2]

The area-process $A : \Omega \times [0, \infty) \to so\,(d)$ is defined by Itô or Stratonovich stochastic integration

$$
\begin{aligned}
A_t^{i,j} &= \frac{1}{2} \left(\int_0^t M_r^i dM_r^j - \int_0^t M_r^j dM_r^i \right) \\
&= \frac{1}{2} \left(\int_0^t M_r^i \circ dM_r^j - \int_0^t M_r^j \circ dM_r^i \right), \quad i,j \in \{1, \ldots, d\};
\end{aligned}
$$

the equality being a consequence of the fact that the covariation $\langle M^i, M^j \rangle$ is symmetric in i, j. (As is well known, Itô and Stratonovich integrals differ by $1/2$ in the covariation between integrand and integrator.[3]) We note that the area-process is a vector-valued continuous local martingale. By disregarding a null set we can and will assume that M and A are continuous.

Definition 14.1 If M is an $\mathbb{R}^d$-valued continuous local martingale, define $\mathbf{M} := \exp\,(M + A)$ to be its lift, and observe that $\mathbf{M}$ has sample paths in $C\,([0, \infty),\ G^2\,(\mathbb{R}^d))$. The resulting class of enhanced (continuous, local) martingales is denoted by $\mathcal{M}_{0,\mathrm{loc}}^c\,(G^2\,(\mathbb{R}^d))$. □

The lift is compatible with the stopping and time-changes.

Lemma 14.2 *Let M be an $\mathbb{R}^d$-valued continuous local martingale, and $\mathbf{M}$ its lift.*
(i) Let τ be a stopping time. Then, $\mathbf{M}^\tau : t \to \mathbf{M}_{t \wedge \tau}$ is the lift of $M^\tau : t \to M_{t \wedge \tau}$.
(ii) Let ϕ be a time-change; that is, a family ϕ_s, $s \geq 0$, of stopping times such that the maps $s \mapsto \phi_s$ are a.s. increasing and right-continuous. If M is constant on each interval $[\phi_{t-}, \phi_t]$, then $M \circ \phi$ is a continuous local martingale and its lift is $\mathbf{M} \circ \phi$.

Proof. Stopped processes are special cases of time-changed processes (take $\phi_t = t \wedge \tau$) so it suffices to show the second statement. This follows from the compatibility of a time change ϕ and stochastic integration with respect to a continuous local martingale, constant on each interval $[\phi_{t-}, \phi_t]$.[4] The lift is of course a special case of stochastic integration. ∎

The lift is also compatible with respect to scaling and concatenation of (local martingale) paths.

---

[2] See, for example, [143, Chapter 4, Theorem 1.8] or [145, p. 54].
[3] See, for example, [160].
[4] See Proposition V.1.5. (ii) of [143], for example.

Lemma 14.3 *Let* M *be an* $\mathbb{R}^d$*-valued continuous local martingale, and* $\mathbf{M}$ *its lift. If* $\delta_c : G^2\left(\mathbb{R}^d\right) \to G^2\left(\mathbb{R}^d\right)$ *is the dilation operator (see Definition 7.13), then* $\delta_c\mathbf{M}$ *is the lift of the martingale* cM.

Proof. Left to the reader. ■

14.2   The Burkholder–Davis–Gundy inequality

Definition 14.4 $F : \mathbb{R}^+ \to \mathbb{R}^+$ *is* moderate *if*
(i) F *is continuous and increasing,*
(ii) $F\left(x\right) = 0$ *if and only if* $x = 0$, *and*
(iii) *for some (and then for every)* $\alpha > 1$,

$$\sup_{x>0} \frac{F\left(\alpha x\right)}{F\left(x\right)} < \infty.$$
□

A few properties of moderate functions are collected in the following lemma.

Lemma 14.5 *(i)* $x \mapsto F\left(x\right)$ *is moderate if and only if* $x \mapsto F\left(x^{1/2}\right)$ *is moderate.*
(ii) Given $c, A, B > 0 : c^{-1}A \le B \le cA \implies \exists C = C\left(c, F\right):$

$$C^{-1}F\left(A\right) \le F\left(B\right) \le CF\left(A\right).$$

(iii) $\exists C : \forall x, y > 0 : F\left(x + y\right) \le C\left[F\left(x\right) + F\left(y\right)\right].$

Proof. (i),(ii) are left to the reader. Ad (iii). Without loss of generality, we assume $x < y$; then $F\left(x + y\right) \le F\left(2y\right) \le c_1 F\left(y\right)$ by moderate growth of F. ■

We now recall the classical Burkholder–Davis–Gundy inequality for continuous local martingales.[5]

Theorem 14.6 (Burkholder–Davis–Gundy) *Let* F *be a moderate function,* $M \in \mathcal{M}^c_{0,\mathrm{loc}}\left(\mathbb{R}^d\right)$ *some continuous local martingale. Then there exists a constant* $C = C\left(F, d\right)$ *such that*

$$C^{-1}\mathbb{E}\left(F\left(|\langle M\rangle_\infty|^{1/2}\right)\right) \le \mathbb{E}\left(F\left(\sup_{s\ge 0}|M_s|\right)\right) \le C\mathbb{E}\left(F\left(|\langle M\rangle_\infty|^{1/2}\right)\right).$$

Observe that if one knows the above statement only for $\mathbb{R}$-valued martingales, then using the norm on $\mathbb{R}^d$, $|a| = \max\left\{|a^1|, \dots, |a^d|\right\}$, the $\mathbb{R}^d$ Burkholder–Davis–Gundy inequality is a simple consequence of the

---

[5] A proof can be found in [145, p. 93], for instance.

Burkholder–Davis–Gundy inequality for $\mathcal{M}^c_{0,\mathrm{loc}}(\mathbb{R})$, applied component-wise. Lemma 14.5 shows that one can switch to Lipschitz equivalent norms.

In Section 14.5 we shall need a Burkholder–Davis–Gundy-type upper bound for real-valued discrete-time martingales. To state this, let us first define the p-variation of a discrete-time martingale (Y_n) as

$$|Y|_{p\text{-var}} \equiv \left[\sup_{(n_k)\nearrow} \sum_k |Y_{n_{k+1}} - Y_{n_k}|^p \right]^{1/p}.$$

A proof of the following lemma can be found in [108, Proposition 2b] for $d = 1$. The extension to dimension $d > 1$ is straightforward.

Lemma 14.7 *Let F be moderate, and $Y : \mathbb{N} \to \mathbb{R}^d$ a discrete martingale. If $1 < q < p \le 2$ or $1 = q = p$, there exists a constant $C = C(F,d)$,*

$$\mathbb{E}\left(F\left(|Y|_{p\text{-var}} \right) \right) \le C \mathbb{E}\left[F\left(\left[\sum_n |Y_{n+1} - Y_n|^q \right]^{1/q} \right) \right].$$

We now derive the Burkholder–Davis–Gundy inequality for *enhanced* (continuous, local) martingales.

Theorem 14.8 (BDG for enhanced martingales) *Let F be a moderate function, $\mathbf{M} \in \mathcal{M}^c_{0,\mathrm{loc}}\left(G^2\left(\mathbb{R}^d \right) \right)$ be the lift of some local martingale M. Then there exists a constant $C = C(F,d)$ so that*

$$C^{-1}\mathbb{E}\left(F\left(|\langle M \rangle_\infty|^{1/2} \right) \right) \le \mathbb{E}\left(F\left(\sup_{s,t\ge 0} \|\mathbf{M}_{s,t}\| \right) \right) \le C \mathbb{E}\left(F\left(|\langle M \rangle_\infty|^{1/2} \right) \right).$$

Proof. The lower bound comes from $\|\mathbf{M}_{s,t}\| \ge |M_{s,t}|$, monotonicity of F and the classical Burkholder–Davis–Gundy lower bound. For the upper bound we note that $\sup_{u,v\ge 0} \|\mathbf{M}_{u,v}\| \le 2 \sup_{t\ge 0} \|\mathbf{M}_t\|$. By the equivalence of homogenous norms,

$$\|\mathbf{M}_t\| \le c_1 \left(|M_t| + |A_t|^{1/2} \right)$$

and using "$F(x+y) \lesssim F(x) + F(y)$", combined with the classical Burkholder–Davis–Gundy upper bound, it suffices to show that

$$\mathbb{E}\left(F\left(\sup_{t\ge 0} |A_t|^{1/2} \right) \right) \le c_2 \mathbb{E}\left(F\left(|\langle M \rangle_\infty|^{1/2} \right) \right).$$

But this is easy using the fact that $F\left((\cdot)^{1/2} \right)$ is moderate and A itself is a martingale with bracket

$$|\langle A^{i,j} \rangle_t| =$$

$$\left| \frac{1}{4}\left(\int_0^t |M^i|^2 \, d\langle M^j \rangle + \int_0^t |M^j|^2 \, d\langle M^i \rangle \right) - \frac{1}{2}\int_0^t M^i M^j d\langle M^i, M^j \rangle \right|$$

$$\le c_1 \sup_{t\ge 0} |M_t| \cdot |\langle M \rangle_t|$$

to which we can apply the Burkholder–Davis–Gundy inequality:

$$
\begin{aligned}
\mathbb{E}\left(F\left(\sup_{t \geq 0} |A_t|^{1/2} \right) \right)
&\leq c_2 \mathbb{E}\left(F\left(|\langle A \rangle_\infty|^{1/4} \right) \right) \\
&\leq c_3 \mathbb{E}\left(F\left(\sup_{t \geq 0} |M_t|^{1/2} \times |\langle M \rangle_\infty|^{1/4} \right) \right) \\
&\leq c_4 \mathbb{E}\left(F\left(\sup_{t \geq 0} |M_t| + |\langle M \rangle_\infty|^{1/2} \right) \right) \\
&\leq c_5 \mathbb{E}\left(F\left(\sup_{t \geq 0} |M_t| \right) \right) + \mathbb{E}\left(F\left(|\langle M \rangle_\infty|^{1/2} \right) \right) \\
&\leq c_6 \mathbb{E}\left(F\left(|\langle M \rangle_\infty|^{1/2} \right) \right).
\end{aligned}
$$

Here, we used "$F(xy) \leq F(x^2 + y^2) \lesssim F(x^2) + F(y^2)$" and, of course, the classical Burkholder–Davis–Gundy upper bound in the last step. ∎

14.3 p-Variation rough path regularity of enhanced martingales

We now show that every $\mathbf{M} \in \mathcal{M}_{0,\mathrm{loc}}^c\left(G^2\left(\mathbb{R}^d \right) \right)$ is a geometric p-rough path for $p \in (2,3)$. In other words, for every $T > 0$,

$$
\|\mathbf{M}\|_{p\text{-var};[0,T]} < \infty \text{ a.s.} \tag{14.1}
$$

The Burkholder–Davis–Gundy inequality on the group allows for an elegant proof of this.

Proposition 14.9 (enhanced martingale, p-variation regularity)
Let $p > 2$ and $\mathbf{M} \in \mathcal{M}_{0,\mathrm{loc}}^c\left(G^2\left(\mathbb{R}^d \right) \right)$. Then, for every $T > 0$,

$$
\|\mathbf{M}\|_{p\text{-var};[0,T]} < \infty \quad a.s.
$$

Proof. There exists a sequence of stopping times τ_n that converges to ∞ almost surely, such that M^{τ_n} and $\langle M^{\tau_n} \rangle$ are bounded (for instance, $\tau_n = \inf\{t : |M_t| > n \text{ or } |\langle M \rangle_t| > n\}$ will do). Since

$$
\mathbb{P}\left(\|\mathbf{M}\|_{p\text{-var};[0,T]} \neq \|\mathbf{M}\|_{p\text{-var};[0,T \wedge \tau_n]} \right) \leq \mathbb{P}\left(\tau_n < T \right) \to 0 \text{ as } n \to \infty
$$

it suffices to consider the lift of a bounded continuous martingale with bounded quadratic variation. We can work with the l^1-norm on $\mathbb{R}^d$, $|a| = \sum_{i=1}^d |a_i|$. The time-change $\phi(t) := \inf\{s : |\langle M \rangle_s| > t\}$ may have jumps, but continuity of $|\langle M \rangle|$ ensures that $|\langle M \rangle_{\phi(t)}| = t$. From the definition of

ϕ and the Burkholder–Davis–Gundy inequality on the group, both $\langle M \rangle$ and $\mathbf{M}$ are constant on the intervals $[\phi_{t-}, \phi_t]$. It follows that $\mathbf{X}_t = \mathbf{M}_{\phi(t)}$ defines a continuous[6] path from $[0, |\langle M \rangle_T|]$ to $G^2(\mathbb{R}^d)$ and from the invariance of the p-variation with respect to time-changes, we have

$$\|\mathbf{X}\|_{p\text{-var};[0,|\langle M \rangle_T|]} = \|\mathbf{M}\|_{p\text{-var};[0,T]} .$$

As argued at the beginning of the proof, we may assume that $|\langle M \rangle_T| \leq R$ for some deterministic R large enough. Therefore,

$$\mathbb{P}\left(\|\mathbf{M}\|_{p\text{-var};[0,T]} > K \right) \qquad (14.2)$$

$$= \mathbb{P}\left(\|\mathbf{X}\|_{p\text{-var};[0,|\langle M \rangle_T|]} > K, |\langle M \rangle_T| \leq R \right)$$

$$\leq \mathbb{P}\left(\|\mathbf{X}\|_{p\text{-var};[0,R]} > K \right) .$$

We go on to show that $\mathbf{X}$ is in fact Hölder continuous. For $0 \leq s \leq t \leq R$, we can use the Burkholder–Davis–Gundy inequality on the Group, Theorem 14.8, to obtain

$$\mathbb{E}\left(\|\mathbf{X}_{s,t}\|^{2q} \right) = \mathbb{E}\left(\|\mathbf{M}_{\phi(s),\phi(t)}\|^{2q} \right) \leq c_q \mathbb{E}\left(\left| \langle M \rangle_{\phi(t)} - \langle M \rangle_{\phi(s)} \right|^q \right) .$$

Observe that

$$\left| \langle M \rangle_{\phi(t)} - \langle M \rangle_{\phi(s)} \right| = \sum_i \left(\langle M^i \rangle_{\phi(t)} - \langle M^i \rangle_{\phi(s)} \right)$$

$$= \left| \langle M \rangle_{\phi(t)} \right| - \left| \langle M \rangle_{\phi(s)} \right| = t - s.$$

Thus, for all $q < \infty$ there exists a constant c_q such that

$$\mathbb{E}\left(\|\mathbf{X}_{s,t}\|^{2q} \right) \leq c_q |t - s|^q .$$

We can now apply Theorem A.10 to see that $\|\mathbf{X}\|_{1/p\text{-Höl};[0,R]} \in L^q$ for all $q \in [1, \infty)$ and

$$\mathbb{P}\left(\|\mathbf{X}\|_{p\text{-var};[0,R]} > K \right) \leq \frac{\mathbb{E}\left(\|\mathbf{X}\|_{p\text{-var};[0,R]} \right)}{K} \leq \frac{\mathbb{E}\left(\|\mathbf{X}\|_{1/p\text{-Höl};[0,R]} \cdot R^{1/p} \right)}{K}$$

tends to zero as $K \to \infty$. Together with (14.2) we see that $\|\mathbf{M}\|_{p\text{-var};[0,T]} < \infty$ with probability 1, as claimed. ∎

---

[6]From Lemma 14.2, $\mathbf{X}$ is the lift of $M \circ \phi$, which is a continuous local martingale. This is another way to see continuity of $\mathbf{X}$.

14.4   Burkholder–Davis–Gundy with p-variation rough path norm

Following a classical approach to Burkholder–Davis–Gundy-type inequalities, we first prove a Chebyshev-type estimate.

Lemma 14.10 *There exists a constant A such that for all continuous local martingales M, for all $\lambda > 0$,*

$$\mathbb{P}\left(\|\mathbf{M}\|_{p\text{-var};[0,\infty)} > \lambda\right) \leq A\frac{\mathbb{E}\left(|\langle M\rangle_\infty|\right)}{\lambda^2}, \qquad (14.3)$$

where $\mathbf{M}$ denotes the lift of M.

Proof. It suffices to prove the statement when $\lambda = 1$ (the general case follows by considering M/λ with lift $\delta_{1/\lambda}\mathbf{M}$). The statement then reduces to

$$\exists A : \forall M : \mathbb{P}\left[\|\mathbf{M}\|_{p\text{-var};[0,\infty)} > 1\right] \leq A\,\mathbb{E}\left(|\langle M\rangle_\infty|\right).$$

Assume this is false. Then for every A, and in particular for $A(k) \equiv k^2$, there exists $M \equiv M^{(k)}$ with lift $\mathbf{M}^{(k)}$ such that the condition is violated, i.e. we have:

$$k^2\,\mathbb{E}\left[\left|\left\langle M^{(k)}\right\rangle_\infty\right|\right] < \mathbb{P}\left[\left\|\mathbf{M}^{(k)}\right\|_{p\text{-var};[0,\infty)} > 1\right].$$

Set $u_k = \mathbb{P}\left[\left\|\mathbf{M}^{(k)}\right\|_{p\text{-var};[0,\infty)} > 1\right]$, $n_k = [1/u_k + 1] \in \mathbb{N}$ and note that $1 \leq n_k u_k \leq 2$. Observe that

$$n_k k^2 \mathbb{E}\left[\left|\left\langle M^{(k)}\right\rangle_\infty\right|\right] \leq n_k \mathbb{P}\left[\left\|\mathbf{M}^{(k)}\right\|_{p\text{-var};[0,\infty)} > 1\right] = n_k u_k \leq 2.$$

We now "expand" the sequence $\left(M^{(k)} : k = 1, 2, \dots\right)$ by replacing each $M^{(k)}$ with n_k independent copies of $M^{(k)}$. This yields another sequence of continuous local martingales, say $\left(N^{(k)} : k = 1, 2, \dots\right)$. Writing $\mathbf{N}^{(k)}$ for the lift of $N^{(k)}$ we clearly see that

$$\sum_k \mathbb{P}\left[\left\|\mathbf{N}^{(k)}\right\|_{p\text{-var};[0,\infty)} > 1\right] = \sum_k n_k u_k = +\infty,$$

while

$$\sum_k \mathbb{E}\left[\left|\left\langle N^{(k)}\right\rangle_\infty\right|\right] = \sum_k n_k \mathbb{E}\left[\left|\left\langle M^{(k)}\right\rangle_\infty\right|\right] \leq \sum_k \frac{2}{k^2} < \infty.$$

Thus, if the claimed statement is false, there exists a sequence of martingales N^k with lift $\mathbf{N}^k$ each defined on some filtered probability space

$\left(\Omega^k, \left(\mathcal{F}_t^k\right), \mathbb{P}^k\right)$ with the two properties

$$\sum_k \mathbb{P}^k \left[\left\| \mathbf{N}^{(k)} \right\|_{p\text{-var};[0,\infty)} > 1 \right] = +\infty \text{ and } \sum_k \mathbb{E}^k \left[\left| \left\langle N^{(k)} \right\rangle_\infty \right| \right] < \infty.$$

Define the probability space $\Omega = \bigotimes_{k=1}^\infty \Omega^k$, the probability $\mathbb{P} = \bigotimes_{k=1}^\infty \mathbb{P}^k$, and the filtration $(\mathcal{F}_t)$ on Ω given by

$$\mathcal{F}_t = \left(\bigotimes_{i=1}^{k-1} \mathcal{F}_\infty^i \right) \otimes \mathcal{F}_{g(k-t)}^k \otimes \left(\bigotimes_{j=k+1}^\infty \mathcal{F}_0^k \right) \quad \text{for } k-1 \le t < k,$$

where $g(u) = 1/u - 1$ maps $[0,1] \to [0,\infty]$. Then, a continuous martingale on $(\Omega, (\mathcal{F}_t), \mathbb{P})$ is defined by concatenation,

$$N_t = \sum_{i=1}^{k-1} N_\infty^{(i)} + N_{g(k-t)}^{(k)} \quad \text{for} \quad k-1 \le t < k,$$

and hence its lift $\mathbf{N}$ satisfies

$$\mathbf{N}_t = \left(\bigotimes_{i=1}^{k-1} \mathbf{N}_\infty^{(i)} \right) \otimes \mathbf{N}_{g(k-t)}^{(k)}.$$

We also observe that, again for $k-1 \le t < k$,

$$\langle N \rangle_t = \sum_{i=1}^{k-1} \left\langle N^{(i)} \right\rangle_\infty + \left\langle N^{(k)} \right\rangle_{g(k-t)}.$$

In particular, $\langle N \rangle_\infty = \sum_k \left\langle N^{(k)} \right\rangle_\infty$ and, using the second property of the martingale sequence, $\mathbb{E}\left(|\langle N \rangle_\infty| \right) < \infty$. Define the events

$$A_k = \left\{ \| \mathbf{N} \|_{p\text{-var};[k-1,k]} > 1 \right\}.$$

Then, using the first property of the martingale sequence,

$$\sum_k \mathbb{P}(A_k) = \sum_k \mathbb{P}^k \left(\left\| \mathbf{N}^{(k)} \right\|_{p\text{-var};[0,\infty)} > 1 \right) = \infty.$$

Since the events $\{A_k : k \ge 1\}$ are independent, the Borel–Cantelli lemma implies that

$$\mathbb{P}(A_k \text{ infinitely often}) = 1.$$

Thus, almost surely, for all $K > 0$ there exist a finite number of increasing times $t_0, \ldots, t_n \in [0,\infty)$ so that

$$\sum_{i=1}^n \left\| \mathbf{N}_{t_{i-1},t_i} \right\| > K$$

and $\|\mathbf{N}\|_{p\text{-var};[0,\infty)}$ must be equal to $+\infty$ with probability one. We now define a martingale X by time-change, namely via $f(t) = t/(1-t)$ for $0 \le t < 1$ and $f(t) = \infty$ for $t \ge 1$,

$$X : t \mapsto N_{f(t)}.$$

Note that $\mathbb{E}\left(|\langle N \rangle_\infty|\right) < \infty$ so that N can be extended to a (continuous) martingale indexed by $[0,\infty]$ and X is indeed a continuous martingale with lift $\mathbf{X}$. Since lifts interchange with time-changes, $\|\mathbf{X}\|_{p\text{-var};[0,1]} = \|\mathbf{N}\|_{p\text{-var};[0,\infty)} = +\infty$ with probability one. But this contradicts the p-variation regularity of enhanced martingales. ∎

The passage from the above Chebyshev-type estimate to the full Burkholder–Davis–Gundy inequality is made possible by the following lemma. The proof can be found in [145, p. 94].

Lemma 14.11 (good λ inequality) *Let X, Y be non-negative random variables, and suppose there exists $\beta > 1$ such that for all $\lambda > 0, \delta > 0$,*

$$\mathbb{P}\left(X > \beta\lambda, Y < \delta\lambda\right) \le \psi\left(\delta\right) \mathbb{P}\left(X > \lambda\right)$$

where $\psi\left(\delta\right) \searrow 0$ when $\delta \searrow 0$. Then, for each moderate function F, there exists a constant C depending only on β, ψ, F such that

$$\mathbb{E}\left(F\left(X\right)\right) \le C\mathbb{E}\left(F\left(Y\right)\right).$$

We now derive the Burkholder–Davis–Gundy inequality for enhanced (continuous, local) martingales in homogenous p-variation norm.

Theorem 14.12 *(BDG inequality in homogenous p-variation norm)* *Let F be a moderate function, $\mathbf{M} \in \mathcal{M}_{0,\text{loc}}^c\left(G^2\left(\mathbb{R}^d\right)\right)$ the lift of some continuous local martingale M, and $p > 2$. Then there exists a constant $C = C\left(p, F, d\right)$ so that*

$$C^{-1}\mathbb{E}\left(F\left(|\langle M \rangle_\infty|^{1/2}\right)\right) \le \mathbb{E}\left(F\left(\|\mathbf{M}\|_{p\text{-var};[0,\infty)}\right)\right) \le C\mathbb{E}\left(F\left(|\langle M \rangle_\infty|^{1/2}\right)\right).$$

Proof. Only the upper bound requires a proof. Fixing $\lambda, \delta > 0$ and $\beta > 1$, we define the stopping times

$$S_1 = \inf\left\{t > 0, \|\mathbf{M}\|_{p\text{-var};[0,t]} > \beta\lambda\right\},$$

$$S_2 = \inf\left\{t > 0, \|\mathbf{M}\|_{p\text{-var};[0,t]} > \lambda\right\},$$

$$S_3 = \inf\left\{t > 0, |\langle M \rangle_t|^{1/2} > \delta\lambda\right\},$$

with the convention that the infimum of the empty set is ∞. Define the local martingale $N_t = M_{S_3 \wedge S_2, (t+S_2) \wedge S_3}$ and its lift $\mathbf{N}$; note that $N_t \equiv 0$ on $\{S_2 = \infty\}$. It is easy to see that

$$\|\mathbf{M}\|_{p\text{-var};[0,S_3]} \le \|\mathbf{M}\|_{p\text{-var};[0,S_3 \wedge S_2]} + \|\mathbf{N}\|_{p\text{-var}},$$

where $\|\mathbf{N}\|_{p\text{-var}} \equiv \|\mathbf{N}\|_{p\text{-var};[0,\infty)}$. By definition of the relevant stopping times,

$$\mathbb{P}\left(\|\mathbf{M}\|_{p\text{-var}} > \beta\lambda, |\langle M\rangle_\infty|^{1/2} \le \delta\lambda\right) = \mathbb{P}\left(S_1 < \infty, S_3 = \infty\right).$$

On the event $\{S_1 < \infty, S_3 = \infty\}$ one has

$$\|\mathbf{M}\|_{p\text{-var};[0,S_3]} > \beta\lambda$$

and, since $S_2 \le S_1$, one also has $\|\mathbf{M}\|_{p\text{-var};[0,S_3 \wedge S_2]}$. Hence, on $\{S_1 < \infty, S_3 = \infty\}$,

$$\|\mathbf{N}\|_{p\text{-var}} \ge \|\mathbf{M}\|_{p\text{-var};[0,S_3]} - \|\mathbf{M}\|_{p\text{-var};[0,S_3 \wedge S_2]} \ge (\beta - 1)\lambda.$$

Therefore, using (14.3),

$$\mathbb{P}\left(\|\mathbf{M}\|_{p\text{-var}} > \beta\lambda, |\langle M\rangle_\infty|^{1/2} \le \delta\lambda\right) \le \mathbb{P}\left(\|\mathbf{N}\|_{p\text{-var}} \ge (\beta - 1)\lambda\right)$$

$$\le \frac{A}{(\beta - 1)^2 \lambda^2} \mathbb{E}\left(|\langle N\rangle_\infty|\right).$$

From the definition of N, for every $t \in [0,\infty]$,

$$\langle N\rangle_t = \langle M\rangle_{S_3 \wedge S_2, (t+S_2) \wedge S_3}.$$

On $\{S_2 = \infty\}$ we have $\langle N\rangle_\infty = 0$ while on $\{S_2 < \infty\}$ we have, from definition of S_3,

$$|\langle N\rangle_\infty| = \left|\langle M\rangle_{S_3 \wedge S_2, S_3}\right| = \left|\langle M\rangle_{S_3} - \langle M\rangle_{S_3 \wedge S_2}\right| \le 2\left|\langle M\rangle_{S_3}\right| = 2\delta^2\lambda^2.$$

It follows that

$$\mathbb{E}\left(|\langle N\rangle_\infty|\right) \le 2\delta^2\lambda^2 \mathbb{P}\left(S_2 < \infty\right) = 2\delta^2\lambda^2 \mathbb{P}\left(\|\mathbf{M}\|_{p\text{-var}} > \lambda\right)$$

and we have the estimate

$$\mathbb{P}\left(\|\mathbf{M}\|_{p\text{-var}} > \beta\lambda, |\langle M\rangle_\infty|^{1/2} \le \delta\lambda\right) \le \frac{2A\delta^2}{(\beta - 1)^2} \mathbb{P}\left(\|\mathbf{M}\|_{p\text{-var}} > \lambda\right).$$

An application of the good λ-inequality finishes the proof. ∎

14.5 Convergence of piecewise linear approximations

Recall that x^D denotes the piecewise linear approximation to some continuous $\mathbb{R}^d$-valued path x, based on some dissection D of $[0,T]$. Given $M \in \mathcal{M}_{0,\text{loc}}^c\left(\mathbb{R}^d\right)$ the same notation applies (path-by-path) and we write $M^D = M^D(\omega)$. The next lemma involves no probabilty.

Lemma 14.13 *Let $p \geq 1$ and $\mathbf{x} : [0, T] \to G^2\left(\mathbb{R}^d\right)$ be a weak geometric p-rough path. Set $x = \pi_1\left(\mathbf{x}\right)$ and let D be a dissection of $[0, T]$. Then there exists a constant $C = C\left(p\right)$ such that*

$$\left\| S_2\left(x^D\right) \right\|_{p\text{-var};[0,T]} \leq C \left\| \mathbf{x} \right\|_{p\text{-var};[0,T]}$$

$$+ C \left(\max_{(s_k) \subset D} \sum_k d\left(\mathbf{x}_{s_k, s_{k+1}}, S_2\left(x^D\right)_{s_k, s_{k+1}} \right)^p \right)^{1/p}.$$

Proof. We first note that $\left\| S_2\left(x^D\right)_{s,t} \right\|^p \leq 3^{p-1} \left[\left| x^D_{s,s^D} \right|^p + \left\| S_2\left(x^D\right)_{s^D, t_D} \right\|^p \right.$ $\left. + \left| x^D_{t_D, t} \right|^p \right]$. Now let (u_k) be a dissection of $[0, T]$, unrelated to D. Recall that u^D resp. u_D refers to the right- resp. left-neighbours of u in D.

$$3^{1-p} \sum_k \left\| S_2\left(x^D\right)_{u_k, u_{k+1}} \right\|^p$$

$$\leq \sum_k \left[\left| x^D_{u_k, u_k^D} \right|^p + \left| x^D_{u_{k+1,D}, u_k} \right|^p \right] + \sum_k \left\| S_2\left(x^D\right)_{u_k^D, u_{k+1,D}} \right\|^p$$

$$\leq 2 \left| x^D \right|^p_{p\text{-var};[0,T]} + \max_{(s_j) \subset D} \sum_j \left\| S_2\left(x^D\right)_{s_j, s_{j+1}} \right\|^p$$

$$\leq 2c_1 \left| x \right|^p_{p\text{-var};[0,T]} + \max_{(s_j) \subset D} \sum_j \left\| S_2\left(x^D\right)_{s_j, s_{j+1}} \right\|^p.$$

Trivially, $\left| x \right|_{p\text{-var};[0,T]} \leq \left\| \mathbf{x} \right\|_{p\text{-var};[0,T]}$. On the other hand, using $(a+b)^p \leq 2^{p-1}\left(a^p + b^p\right)$ when $a, b > 0$, the triangle inequality gives

$$2^{1-p} \max_{(s_k) \subset D} \sum_k \left\| S_2\left(x^D\right)_{s_k, s_{k+1}} \right\|^p$$

$$\leq \max_{(s_k) \subset D} \sum_k d\left(\mathbf{x}_{s_k, s_{k+1}}, S_2\left(x^D\right)_{s_k, s_{k+1}} \right)^p + \left\| \mathbf{x} \right\|^p_{p\text{-var};[0,T]}.$$

Taking the supremum over all possible subdivisions (u_k) finishes the proof. ∎

Lemma 14.14 *Let F be a moderate function, $\mathbf{M} \in \mathcal{M}^c_{0,\text{loc}}\left(G^2\left(\mathbb{R}^d\right)\right)$ the lift of some continuous local martingale M. Assume $2 < p' < p \leq 4$. Then there exists a constant $C = C\left(p, p', F\right)$ so that for all dissections $D = \{t_l\}$ of $[0, T]$,*

$$\mathbb{E}\left[F\left(\left(\max_{(s_k) \subset D} \sum_k d\left(\mathbf{M}_{s_k, s_{k+1}}, S_2\left(M^D\right)_{s_k, s_{k+1}} \right)^p \right)^{1/p} \right) \right]$$

$$\leq C\mathbb{E}\left[F\left(\left(\sum_l \left\| \mathbf{M}_{t_l, t_{l+1}} \right\|^{p'} \right)^{1/p'} \right) \right].$$

Proof. For fixed k, there are $i < j$ so that $s_k = t_i$ and $s_{k+1} = t_j$. Then

$$\mathbf{M}_{s_k, s_{k+1}} = \bigotimes_{l=i}^{j-1} \exp\big(M_{t_l, t_{l+1}} + A_{t_l, t_{l+1}}\big), \; S_2\left(M^D\right)_{s_k, s_{k+1}} = \bigotimes_{l=i}^{j-1} \exp\big(M_{t_l, t_{l+1}}\big).$$

From equivalence of homogenous norms we have

$$d\left(\mathbf{M}_{s_k, s_{k+1}}, S_2\left(M^D\right)_{s_k, s_{k+1}}\right) = \left\|\mathbf{M}_{s_k, s_{k+1}}^{-1} \otimes S_2\left(M^D\right)_{s_k, s_{k+1}}\right\| \quad (14.4)$$

$$= \left\|\exp\left(\sum_{l=i}^{j-1} A_{t_l, t_{l+1}}\right)\right\|$$

$$\leq c_1 \left|\sum_{l=i}^{j-1} A_{t_l, t_{l+1}}\right|^{1/2}.$$

The idea is to introduce the (vector-valued) discrete-time martingale

$$Y_j = \sum_{l=0}^{j-1} A_{t_l, t_{l+1}} \in so\,(d)$$

so that

$$\max_{(s_k) \subset D} \sum_k d\left(\mathbf{M}_{s_k, s_{k+1}}, S_2\left(M^D\right)_{s_k, s_{k+1}}\right)^p$$

$$\leq c_1 \max_{\{i_1, \dots, i_n\} \subset \{1, \dots, \#D\}} \sum_k \left|Y_{i_{k+1}} - Y_{i_k}\right|^{p/2}$$

which can be rewritten as

$$\left(\max_{(s_k) \subset D} \sum_k d\left(\mathbf{M}_{s_k, s_{k+1}}, S_2\left(M^D\right)_{s_k, s_{k+1}}\right)^p\right)^{1/p} \leq c_1^{1/p} \sqrt{|Y|_{p/2\text{-var}}}.$$

Noting that $F \circ \sqrt{\cdot}$ is moderate and that $1 < p'/2 < p/2 \leq 2$, Lemma 14.7 yields

$$\mathbb{E}\left[F \circ \sqrt{\cdot}\left(|Y|_{p/2\text{-var}}\right)\right] \leq c_2 \mathbb{E}\left[F \circ \sqrt{\cdot}\left(\left(\sum_l |Y_{l+1} - Y_l|^{p'/2}\right)^{2/p'}\right)\right]$$

$$= c_2 \mathbb{E}\left[F \circ \sqrt{\cdot}\left(\left(\sum_l |A_{t_l, t_{l+1}}|^{p'/2}\right)^{2/p'}\right)\right]$$

$$\leq c_3 \mathbb{E}\left[F\left(\left(\sum_l \|\mathbf{M}_{t_l, t_{l+1}}\|^{p'}\right)^{1/p'}\right)\right].$$

∎

Theorem 14.15 *Let F be a moderate function, $\mathbf{M} \in \mathcal{M}^c_{0,\text{loc}}\left(G^2\left(\mathbb{R}^d\right)\right)$ the lift of a continuous local martingale M. Then there exists a constant $C = C\left(p, F, d\right)$ so that for all dissections D of $[0, T]$,*

$$\mathbb{E}\left(F\left(\left\|S_2\left(M^D\right)\right\|_{p\text{-var};[0,T]}\right)\right) \leq C\mathbb{E}\left(F\left(\left|\langle M\rangle_T\right|^{1/2}\right)\right).$$

Proof. From Lemma 14.13,

$$\left\|S_2\left(M^D\right)\right\|_{p\text{-var};[0,T]} \leq c_1 \left\|\mathbf{M}\right\|_{p\text{-var};[0,T]}$$

$$+ c_1 \left(\max_{(s_k) \subset D} \sum_k d\left(\mathbf{M}_{s_k, s_{k+1}}, S_2\left(M^D\right)_{s_k, s_{k+1}}\right)^p\right)^{1/p}.$$

Using "$F\left(x + y\right) \lesssim F\left(x\right) + F\left(y\right)$" and the above lemma, with $p' = 1 + p/2$ for instance, we obtain

$$\mathbb{E}\left[F\left(\left\|S_2\left(M^D\right)\right\|_{p\text{-var}}\right)\right] \leq c_2 \mathbb{E}\left[F\left(\left\|\mathbf{M}\right\|_{p\text{-var}}\right)\right]$$

$$+ c_2 \mathbb{E}\left[F\left(\left(\sum_l \left\|\mathbf{M}_{t_l, t_{l+1}}\right\|^{p'}\right)^{1/p'}\right)\right]$$

$$\leq c_3 \mathbb{E}\left[F\left(\left\|\mathbf{M}\right\|_{p\text{-var}}\right)\right] + c_3 \mathbb{E}\left[F\left(\left\|\mathbf{M}\right\|_{p'\text{-var}}\right)\right].$$

The proof is now finished with the Burkholder–Davis–Gundy inequality on the group in p- (resp. p'-)variation norm. ∎

Theorem 14.16 *Assume that M is a continuous local martingale with lift $\mathbf{M} \in \mathcal{M}^c_{0,\text{loc}}\left(G^2\left(\mathbb{R}^d\right)\right)$. If*

$$\left|M\right|_{\infty;[0,T]} \in L^q\left(\Omega\right) \text{ for some } q \geq 1, \tag{14.5}$$

then $d_{p\text{-var};[0,T]}\left(S_2\left(M^D\right), \mathbf{M}\right)$ converges to 0 in L^q. If M is a continuous local martingale, then convergence holds in probability.

Remark 14.17 *If $q > 1$, Doob's maximal inequality implies that (14.5) holds for any L^q-martingale.* □

Proof. Observe first that when $t = t_j \in D$, as in the last lemma,

$$d\left(\mathbf{M}_t, S_2\left(M^D\right)_t\right) \leq c_1 \left|\sum_{l=0}^{j-1} A_{t_l, t_{l+1}}\right|^{1/2}.$$

The path M^D restricted to $[t_i, t_{i+1}]$ is a straight line with no area, hence

$$S_2\left(M^D\right)_{t_i, t} = \exp\left(\frac{t - s}{t_{i+1} - t_i} M_{t_i, t_{i+1}}\right)$$

and

$$d_\infty \left(\mathbf{M}, S_2(M^D)\right) = \max_i \sup_{t \in [t_i, t_{i+1}]} d\left(\mathbf{M}_{t_i} \otimes \mathbf{M}_{t_i, t}, S_2(M^D)_{t_i} \otimes S_2(M^D)_{t_i, t}\right)$$

$$= \sup_{t \in [0,T]} \left\| S_2(M^D)_{t_i, t}^{-1} \otimes S_2(M^D)_{t_i}^{-1} \otimes \mathbf{M}_{t_i} \otimes \mathbf{M}_{t_i, t} \right\|$$

$$\leq \max_i \sup_{t \in [t_i, t_{i+1}]} \left\| \left(S_2(M^D)_{t_i, t} \right) \right\|$$

$$+ \max_i \sup_{t \in [t_i, t_{i+1}]} \left(\left\| S_2(M^D)_{t_i}^{-1} \otimes \mathbf{M}_{t_i} \right\| + \left\| \mathbf{M}_{t_i, t} \right\| \right)$$

$$\leq 2 \sup_{0 < v - u \leq |D|} \|\mathbf{M}_{u,v}\| + c_1 \max_{i,j} \left| \sum_{l=i}^{j-1} A_{t_l, t_{l+1}} \right|^{1/2}$$

$$\leq 2 \sup_{0 < v - u \leq |D|} \|\mathbf{M}_{u,v}\| + 2c_1 \max_j \left| \sum_{l=0}^{j-1} A_{t_l, t_{l+1}} \right|^{1/2}.$$

Now, using the classical Burkholder–Davis–Gundy inequality, we have

$$E\left(\max_j \left| \sum_{l=0}^{j-1} A_{t_l, t_{l+1}} \right|^{q/2} \right) \leq c_2 E\left(\left(\sum_{l=0}^{j-1} \left| A_{t_l, t_{l+1}} \right|^2 \right)^{q/4} \right)$$

$$\leq c_3 E\left(\left(\sum_{l=0}^{j-1} \left\| \mathbf{M}_{t_l, t_{l+1}} \right\|^4 \right)^{q/4} \right)$$

$$\leq c_3 E\left(\left(\max_l \left\| \mathbf{M}_{t_l, t_{l+1}} \right\|^{q/4} \right) \right.$$

$$\left. \left(\sum_{l=0}^{j-1} \left\| \mathbf{M}_{t_l, t_{l+1}} \right\|^3 \right)^{q/4} \right).$$

Hölder's inequality, Theorem 14.12 and Theorem 14.8 then lead us to

$$E\left(\max_j \left| \sum_{l=0}^{j-1} A_{t_l, t_{l+1}} \right|^{q/2} \right)$$

$$\leq c_3 E\left(\max_l \left\| \mathbf{M}_{t_l, t_{l+1}} \right\|^q \right)^{1/4} E\left(\left(\sum_{l=0}^{j-1} \left\| \mathbf{M}_{t_l, t_{l+1}} \right\|^3 \right)^{q/3} \right)^{3/4}$$

$$\leq c_3 E\left(\max_l \left\| \mathbf{M}_{t_l, t_{l+1}} \right\|^q \right)^{1/4} E\left(\|\mathbf{M}\|_{3\text{-var};[0,T]}^q \right)^{3/4}$$

$$\leq c_4 E\left(\sup_{0 < v - u \leq |D|} \|\mathbf{M}_{u,v}\|^q \right)^{1/4} E\left(\|\mathbf{M}\|_\infty^q \right)^{3/4}.$$

This proves that

$$E\left(d_\infty\left(\mathbf{M}, S_2\left(M^D\right)\right)^q\right) \tag{14.6}$$

$$\leq c_5 E\left(\sup_{0<v-u\leq|D|}\|\mathbf{M}_{u,v}\|^q\right)^{1/4} E\left(\|\mathbf{M}\|_\infty^q\right)^{3/4}$$

$$+ c_5 E\left(\sup_{0<v-u\leq|D|}\|\mathbf{M}_{u,v}\|^q\right).$$

Since $\mathbf{M}$ is almost surely continuous, and hence uniformly continuous on $[0,T]$,

$$\sup_{0<v-u\leq|D|}\|\mathbf{M}_{u,v}\| \to 0 \text{ a.s. with } |D| \to 0;$$

by dominated convergence (with $\|\mathbf{M}\|_\infty \in L^q$, seen by (14.5) and Burkholder–Davis–Gundy inequality on the group), this convergence also holds in L^q. Hence, using (14.6), we see that $d_{\infty;[0,T]}\left(\mathbf{M}, S_2\left(M^D\right)\right) \to 0$ in L^q. Recall from Proposition 8.15 that

$$d_{0;[0,T]}\left(\mathbf{M}, S_2\left(M^D\right)\right) \leq c_6 d_{\infty;[0,T]}\left(\mathbf{M}, S_2\left(M^D\right)\right)$$

$$+ c_6 \left|\|\mathbf{M}\|_\infty d_{\infty;[0,T]}\left(\mathbf{M}, S_2\left(M^D\right)\right)\right|^{1/2}.$$

It suffices to use Cauchy–Schwarz,

$$\mathbb{E}\left(\left|\|\mathbf{M}\|_\infty d_{\infty;[0,T]}\left(\mathbf{M}, S_2\left(M^D\right)\right)\right|^{q/2}\right)$$

$$\leq \mathbb{E}\left(\|\mathbf{M}\|_\infty^q\right)^{1/2}\mathbb{E}\left(d_{\infty;[0,T]}\left(\mathbf{M}, S_2\left(M^D\right)\right)^q\right)^{1/2}$$

to see that $d_0\left(\mathbf{M}, S_2\left(M^D\right)\right) \to 0$ in L^q. We then use interpolation (Lemma 8.16) to see that for $2 < p' < p$,

$$d_{p\text{-var};[0,T]}\left(\mathbf{M}, S_2\left(M^D\right)\right)$$

$$\leq c_7 d_0\left(\mathbf{M}, S_2\left(M^D\right)\right)^{1-\frac{p'}{p}}\left(\|\mathbf{M}\|_{p'\text{-var};[0,T]}^{\frac{p'}{p}} + \left\|S_2\left(M^D\right)\right\|_{p'\text{-var};[0,T]}^{\frac{p'}{p}}\right).$$

Hence,

$$\mathbb{E}\left(d_{p\text{-var};[0,T]}\left(S_2\left(M^D\right), \mathbf{M}\right)^q\right)$$

$$\leq c_7^q \mathbb{E}\left(\left(\|\mathbf{M}\|_{p'\text{-var};[0,T]}^{q\frac{p'}{p}} + \left\|S_2\left(M^D\right)\right\|_{p'\text{-var};[0,T]}^{q\frac{p'}{p}}\right) d_0\left(\mathbf{M}, S_2\left(M^D\right)\right)^{q\left(1-\frac{p'}{p}\right)}\right).$$

Using Hölder's inequality with conjugate exponents $1/(p'/p)$ and $1/(1-p'/p)$ gives

$$\mathbb{E}\left(d_{p\text{-var}}\left(S_2\left(M^D\right), \mathbf{M}\right)^q\right) \leq c_8 \mathbb{E}\left(\|\mathbf{M}\|_{p'\text{-var}}^q + \left\|S_2\left(M^D\right)\right\|_{p'\text{-var}}^q\right)^{p'/p}$$

$$\times \left[\mathbb{E}\left(d_0\left(\mathbf{M}, S_2\left(M^D\right)\right)^q\right)\right]^{1-p'/p}.$$

But now it suffices to remark, using our Burkholder–Davis–Gundy estimates (Theorems 14.12 and 14.15), that

$$\max\left[\mathbb{E}\left(\|\mathbf{M}\|_{p'\text{-var};[0,T]}^{q}\right), \sup_{D\in\mathcal{D}[0,T]} \mathbb{E}\left(\|S_2\left(M^D\right)\|_{p'\text{-var};[0,T]}^{q}\right)\right]$$
$$\leq c_9 \mathbb{E}\left(|\langle M\rangle_T|^{q/2}\right) \leq c_{10}\mathbb{E}\left(\left|\,|M|_{\infty;[0,T]}\right|^{q}\right),$$

and the last term is finite by assumption. We proved that $d_{p\text{-var}}\left(S_2\left(M^D\right), \mathbf{M}\right) \to 0$ in L^q for any martingale M s.t. $|M|_{\infty;[0,T]} \in L^q$. At last, if M is a local martingale one obtains convergence in probability by a simple localization argument. ∎

14.6 Comments

Local (and semi)martingales, including the classical Burkholder–Davis–Gundy inequality, are discussed in many textbooks, see e.g. Revuz and Yor [143], Rogers and Williams [145], and Stroock [160]. Proposition 14.7 was strengthened in Pisier and Xu [137], Theorem 2.1 (ii) to $1 \leq p = q < 2$.

Rough path regularity of enhanced martingales and certain convergence results were first established in Coutin and Lejay [31]. The proof of the Burkholder–Davis–Hundy inequality for enhanced martingales in p-variation rough path norm follows closely Lépingle [108] and is taken from Friz and Victoir [66], as is the rough path convergence of piecewise linear approximation.

An interesting recent application of rough paths to semi-martingale theory was given in Feng and Zhao [49]: the authors construct a stochastic area between the local time $x \mapsto L_t^x$ of a real-valued semi-martingale and a deterministic function $g = g(x)$ of finite q-variation, $q < 3$; as an application, to obain a generalization of the Tanaka–Meyer formula.

A large deviation principle for square-integrable martingales over Brownian filtration is discussed in forthcoming work by Z. Qian and C. Xu.

15

Gaussian processes

We have seen in a previous chapter that d-dimensional Brownian motion B can be enhanced to a stochastic process $\mathbf{B} = \mathbf{B}(\omega)$ for which almost every realization is a geometric $1/p$-Hölder rough path (and hence a geometric p-rough path), $p \in (2, 3)$. Now, B is a continuous, centred Gaussian process, with independent components $(B^1, \ldots, B^d)$, whose law is fully determined by its *covariance function*

$$\begin{aligned} R(s,t) &= \mathbb{E}(B_s \otimes B_t) \\ &= \mathrm{diag}(s \wedge t, \ldots, s \wedge t). \end{aligned}$$

Let us note that this covariance function, $R = R(s,t)$, has finite 1-variation (in 2D sense, where the variation of R is based on its rectangular increments, cf. Section 5.5).

In the present chapter, our aim is to replace Brownian motion by a d-dimensional, continuous, centred Gaussian process X with independent components $(X^1, \ldots, X^d)$; again, its law is fully determined by its covariance function. In particular, we want to construct a reasonable lifted process $\mathbf{X}$ with geometric rough (sample) paths (in short: a *Gaussian rough path*) and study its probabilistic properties. We shall see that this is possible whenever the covariance function has finite ρ-variation (in 2D sense), for some $\rho \in [1, 2)$, so that $\mathbf{X}$ is a geometric p-rough path for $p > 2\rho$. This also leaves considerable room to deal with Gaussian processes (with sample path behaviour) worse than Brownian motion.

The main tools in this chapter are 2D Young theory (cf. Section 6.4) and then integrability of Gaussian chaos and L^2-expansions (the essentials of which are collected in Appendix D).

15.1 Motivation and outlook

Let $X = (X_t^1, \ldots, X_t^d : t \in [0, T])$ be a d-dimensional, continuous and centred Gaussian process with independent components. By a trivial reparametrization, $t \mapsto X_{tT}$, we can and will take $T = 1$. The law of such a process is fully characterized by its covariance function,

$$R(s,t) = \mathrm{diag}\left(\mathbb{E}(X_s^1 X_t^1), \ldots, \mathbb{E}(X_s^d X_t^d)\right), \quad s, t \in [0, 1].$$

To explain the main idea, assume at first that X has smooth sample paths so that X can be lifted canonically via iterated integration. With focus on the first set of iterated integrals, assuming $X_0 = 0$ for simpler notation, we can write

$$\mathbb{E}\left(\left|\int_0^t X_u^i \, dX_u^j\right|^2\right) = \int_{[0,t]^2} R_i(u,v) \frac{\partial^2}{\partial u \partial v} R_j(u,v) \, du dv$$

$$\equiv \int_{[0,t]^2} R_i(u,v) \, dR_j(u,v),$$

where R_i is the covariance function of X^i. The integral which appears on the right-hand side above can be viewed as a 2-dimensional (2D) Young integral. From the Young–Lóeve–Towghi inequality (Theorem 6.18) we see that under the assumption $\rho < 2$ we have, for $0 \le t \le 1$,

$$\mathbb{E}\left(\left|\int_0^t X_u^i \, dX_u^j\right|^2\right) \le (\text{const}) \times |R_i|_{\rho\text{-var};[0,1]^2} \, |R_j|_{\rho\text{-var};[0,1]^2}$$

$$\le (\text{const}) \times |R|^2_{\rho\text{-var};[0,1]^2} \, .$$

This gives us *uniform* L^2-bounds provided the covariance of the process has finite ρ-variation in 2D sense with $\rho \in [1,2)$. It is then relatively straightforward to define

$$\int X \otimes dX$$

(... and then a "natural" lift of X to a geometric rough path $\mathbf{X}$ ...) in L^2-sense, as long as X has covariance with finite ρ-variation in 2D sense. Recall that the latter means[1]

$$|R|^\rho_{\rho\text{-var};[0,1]^2} = \sup \sum_{i,j} \left| R\left(\begin{matrix} t_i, t_{i+1} \\ t'_j, t'_{j+1} \end{matrix} \right) \right|^\rho$$

$$= \sup \sum_{i,j} \left| \mathbb{E}\left[(X_{t_{i+1}} - X_{t_i}) (X_{t'_{j+1}} - X_{t'_j}) \right] \right|^\rho < \infty.$$

One should note that the assumption $R \in C^{\rho\text{-var}}\left([0,1]^2\right)$ really encodes some information about the *decorrelation* of the increments of X. In the extreme case of uncorrelated increments (example: Brownian motion or L^2-martingales) the double-sum reduces to the summation over $i = j$. (In particular, one sees that the covariance of Brownian motion has finite $\rho = 1$ variation in 2D sense.)

The Gaussian nature of X starts to play a role when turning L^2-esimates into L^q-estimates for all $q < \infty$, an essentially free consequence of Wiener–Itô chaos integrability. This will be seen to imply (rough path) regularity

---

[1] The sup runs over all dissections $(t.) = D, (t'.) = D'$ of $[0,1]$. If one takes the sup only over D, so that both t_i and $t'_j = t_j$ are taken from D, then this will still suffice to control the ρ-variation of R; see Lemma 5.54.

for **X** and also "Fernique" estimates, by which we mean Gaussian tails of homogenous rough path norms of **X**. Another useful consequence is the fact that, assuming finite ρ-variation of R, the Cameron–Martin space $\mathcal{H}$ is continuously embedded in the space of finite ρ-variation paths. When $\rho \in [1,2)$, the standing assumption in a Gaussian rough path setting, we see that Cameron–Martin paths are fully accessible to Young theory. This in turn is crucial for various results, including rough path convergence of Karhunen–Loève approximations, large deviation and support statements. (Further applications towards Malliavin calculus will be discussed in a later chapter.)

15.2   One-dimensional Gaussian processes

Throughout this section, X will be a real-valued centred Gaussian process on $[0,1]$ with continuous sample paths and (continuous) covariance $R = R(s,t) = \mathbb{E}(X_s X_t)$. We note that the law of X induces a Gaussian measure on the Banach space $C([0,1],\mathbb{R})$.

15.2.1 ρ-Variation of the covariance

For a covariance function, as a function of two variables $(s,t) \in [0,1]^2 \mapsto R(s,t)$, we have a well-defined concept of ρ-variation (in the 2D sense) as discussed in Section 5.5. We start with some simple examples.

Example 15.1 (Brownian motion) Standard Brownian motion on $[0,1]$ has covariance

$$R_{\mathrm{BM}}(s,t) = \min(s,t).$$

Trivially, $(s,t) \in [0,1]^2 \mapsto R_{\mathrm{BM}}(s,t)$ has finite ρ-variation with $\rho = 1$, controlled by

$$\omega([s,t] \times [u,v]) := |(s,t) \cap (u,v)| = \int_{[s,t] \times [u,v]} \delta_{x=y}(dxdy),$$

where δ is the Dirac mass. Since $\omega\left([s,t]^2\right) = |t-s|$, we see that ω is a Hölder-dominated 2D control (in the sense of Definition 5.51).   □

Example 15.2 (Gaussian martingales) Any continuous Gaussian martingale M has a deterministic bracket.[2] Since

$$(M(t) : t \geq 0) \overset{\mathcal{D}}{=} \left(B_{\langle M \rangle_t} : t \geq 0\right)$$

---

[2]See, for example, [143, Chapter IV, (1.35)].

we see that

$$R\left(s,t\right) = \min\left\{\langle M\rangle_s, \langle M\rangle_t\right\} = \langle M\rangle_{\min(s,t)}.$$

But the notion of ρ-variation is invariant under reparametrization and it follows that R has finite 1-variation since R_{BM} has finite 1-variation.[3] $\quad\square$

Exercise 15.3 (Gaussian bridge processes) *Gaussian bridge processes are immediate generalizations of the Brownian bridge: given a continuous, centred Gaussian process X on $[0,1]$ with covariance R of finite ρ-variation, the corresponding bridge is defined as*

$$X_{\mathrm{Bridge}}\left(t\right) := X\left(t\right) - tX\left(1\right)$$

with covariance R_{Bridge}.

Prove that R_{Bridge} has finite ρ-variation, and that if R has its ρ-variation controlled by a Hölder-dominated 2D control then the same is true for R_{Bridge}.

Exercise 15.4 (Ornstein–Uhlenbeck) *Show that the usual (real-valued) Ornstein–Uhlenbeck (stationary or started at a fixed point) has covariance of finite 1-variation, controlled by a Hölder-dominated 2D control.*

We now turn to *fractional Brownian motion* β^H on $[0,1]$ with Hurst parameter $H \in (0,1)$. It is a zero-mean Gaussian process with covariance

$$R^H\left(s,t\right) = \mathbb{E}\left(\beta_s^H \beta_t^H\right) = \frac{1}{2}\left(t^{2H} + s^{2H} - |t-s|^{2H}\right).$$

For Hurst parameter $H > 1/2$, fractional Brownian motion has Hölder sample paths with exponent greater than $1/2$ which is, in the context of rough paths, a trivial case. We shall therefore make the standing assumption

$$H \leq 1/2$$

noting that this covers Brownian motion with $H = 1/2$.

Proposition 15.5 (fractional Brownian motion) *Let β^H be fractional Brownian motion with Hurst parameter $H \in (0,1/2]$. Then, its covariance*

---

[3] One should note that L^2-martingales (without assuming a Gaussian structure) have orthogonal increments, i.e.

$$\mathbb{E}\left(X_{s,t}X_{u,v}\right) = 0 \text{ if } s < t < u < v$$

and this alone will take care of the (usually difficult to handle) off-diagonal part in the variation of the covariance $(s,t) \mapsto \mathbb{E}\left(X_s X_t\right)$.

is of finite $1/(2H)$-variation, controlled by

$$\omega_H\left(\cdot,\cdot\right) := \left|R^H\right|_{1/(2H)\text{-var};[\cdot,\cdot]\times[\cdot,\cdot]}^{1/(2H)} . \tag{15.1}$$

Moreover, there exists a constant $C = C\left(H\right)$ such that, for all $s < t$ in $[0,1]$,

$$\left|R^H\right|_{1/(2H)\text{-var};[s,t]^2} \leq C_H \left|t-s\right|^{\frac{1}{2H}}$$

so that ω_H is a Hölder-dominated control.

Proof. Let $D = \{t_i\}$ be a dissection of $[s,t]$, and let us look at

$$\sum_{i,j}\left|\mathbb{E}\left(\beta_{t_i,t_{i+1}}^H \beta_{t_j,t_{j+1}}^H\right)\right|^{\frac{1}{2H}} .$$

For a fixed i and $i \neq j$, as $H \leq \frac{1}{2}$, $\mathbb{E}\left(\beta_{t_i,t_{i+1}}^H \beta_{t_j,t_{j+1}}^H\right)$ is negative, hence,

$$\sum_{j}\left|\mathbb{E}\left(\beta_{t_i,t_{i+1}}^H \beta_{t_j,t_{j+1}}^H\right)\right|^{\frac{1}{2H}}$$

$$\leq \sum_{j\neq i}\left|\mathbb{E}\left(\beta_{t_i,t_{i+1}}^H \beta_{t_j,t_{j+1}}^H\right)\right|^{\frac{1}{2H}} + \mathbb{E}\left(\left|\beta_{t_i,t_{i+1}}^H\right|^2\right)^{\frac{1}{2H}}$$

$$\leq \left|\mathbb{E}\left(\sum_{j\neq i}\beta_{t_i,t_{i+1}}^H \beta_{t_j,t_{j+1}}^H\right)\right|^{\frac{1}{2H}} + \mathbb{E}\left(\left|\beta_{t_i,t_{i+1}}^H\right|^2\right)^{\frac{1}{2H}}$$

$$\leq \left(2^{\frac{1}{2H}-1}\left|\mathbb{E}\left(\sum_{j}\beta_{t_i,t_{i+1}}^H \beta_{t_j,t_{j+1}}^H\right)\right|^{\frac{1}{2H}} + 2^{\frac{1}{2H}-1}\mathbb{E}\left(\left|\beta_{t_i,t_{i+1}}^H\right|^2\right)^{\frac{1}{2H}}\right)$$

$$+ \mathbb{E}\left(\left|\beta_{t_i,t_{i+1}}^H\right|^2\right)^{\frac{1}{2H}}$$

$$\leq C_H\left|\mathbb{E}\left(\beta_{t_i,t_{i+1}}^H \beta_{s,t}^H\right)\right|^{\frac{1}{2H}} + C_H\mathbb{E}\left(\left|\beta_{t_i,t_{i+1}}^H\right|^2\right)^{\frac{1}{2H}} .$$

Hence,

$$\sum_{i,j}\left|\mathbb{E}\left(\beta_{t_i,t_{i+1}}^H \beta_{t_j,t_{j+1}}^H\right)\right|^{\frac{1}{2H}} \leq C_H\sum_{i}\mathbb{E}\left(\left|\beta_{t_i,t_{i+1}}^H\right|^2\right)^{\frac{1}{2H}}$$

$$+ C_H\sum_{i}\left|\mathbb{E}\left(\beta_{t_i,t_{i+1}}^H \beta_{s,t}^H\right)\right|^{\frac{1}{2H}} .$$

The first term is equal to $C_H |t - s|$, so we just need to prove that

$$\sum_i \left| \mathbb{E} \left(\beta^H_{t_i,t_{i+1}} \beta^H_{s,t} \right) \right|^{\frac{1}{2H}} \leq C_H |t - s|. \tag{15.2}$$

To achieve this, it will be enough to prove that for $[u, v] \subset [s, t]$,

$$\left| \mathbb{E} \left(\beta^H_{u,v} \beta^H_{s,t} \right) \right| \leq C_H |v - u|^{2H}.$$

First recall that as $2H < 1$, if $0 < x < y$, then $(x + y)^{2H} - x^{2H} \leq y^{2H}$. Hence, using this inequality and the triangle inequality,

$$\begin{aligned}
\left| \mathbb{E} \left(\beta^H_{u,v} \beta^H_{s,t} \right) \right| &= c_H \left| (t - v)^{2H} + (u - s)^{2H} - (v - s)^{2H} - (t - u)^{2H} \right| \\
&\leq c_H \left((t-u)^{2H} - (t-v)^{2H} \right) + c_H \left((v - s)^{2H} - (u - s)^{2H} \right) \\
&\leq 2c_H (v - u)^{2H}.
\end{aligned}$$

∎

Exercise 15.6 *We say that a real-valued Gaussian process X on $[0, 1]$ satisfies the Coutin–Qian conditions if, for some $H \in (0, 1)$, $c_H > 0$ and all s, t*

$$\mathbb{E} \left(|X_{s,t}|^2 \right) \leq c_H |t - s|^{2H}, \tag{15.3}$$

$$\left| \mathbb{E} \left(X_{s,s+h} X_{t,t+h} \right) \right| \leq c_H |t - s|^{2H-2} h^2, \text{ for } 0 < h < t - s. \tag{15.4}$$

Let ω_H be the 2D control for the covariance of fBM, as defined in (15.1). Show that, for all $s \leq t$ and $u \leq v$ in $[0, 1]$,

$$\left| \mathbb{E} \left(X_{s,t} X_{u,v} \right) \right| \leq C_H \omega_H \left([s, t] \times [u, v] \right)^{2H},$$

and conclude that the covariance of X has finite $1/(2H)$-variation, controlled by a Hölder-dominated 2D control.

Solution. Working as in Lemma 5.54, at the price of a factor $3^{\frac{1}{2H} - 1}$, we can restrict ourselves to the cases $s = u \leq t = v$ and $s \leq t \leq u \leq v$. The first case is given by assumption (15.3), so let us focus on the second one. Assume first we can write $t - s = nh$, $v - u = mh$ and that $u - t > h$. Then,

$$\mathbb{E} \left(X_{s,t} X_{u,v} \right) = \sum_{k=0}^{n-1} \sum_{l=0}^{m-1} \mathbb{E} \left(X_{s+kh,s+(k+1)h} X_{t+lh,t+(l+1)h} \right).$$

Using the triangle inequality and our assumption,

$$
\begin{aligned}
\left|\mathbb{E}\left(X_{s,t} X_{u,v}\right)\right| &= \sum_{k=0}^{n-1}\sum_{l=0}^{m-1}\left|\mathbb{E}\left(X_{s+kh,s+(k+1)h} X_{u+lh,u+(l+1)h}\right)\right| \\
&\leq C_H \sum_{k=0}^{n-1}\sum_{l=0}^{m-1}\left|(u+lh)-(s+kh)\right|^{2H-2} h^2 \\
&\leq C_H \sum_{k=0}^{n-1}\sum_{l=0}^{m-1}\int_{u+(l-1)h}^{u+lh}\int_{s+kh}^{s+(k+1)h}\left|y-x\right|^{2H-2} dx dy \\
&\leq C_H \int_{u-h}^{v-h}\int_{s}^{t}\left|y-x\right|^{2H-2} dx dy \\
&\leq C_H \left|\mathbb{E}\left(\beta_{u-h,v-h}^{H}\beta_{s,t}^{H}\right)\right|.
\end{aligned}
$$

Letting h tend to 0, by continuity, we easily see that

$$
\left|\mathbb{E}\left(X_{s,t}X_{u,v}\right)\right| \leq C_H \left|\mathbb{E}\left(\beta_{u,v}^{H}\beta_{s,t}^{H}\right)\right|,
$$

which implies our statement for $s \leq t \leq u \leq v$. That concludes the proof. □

15.2.2 *A Cameron–Martin/variation embedding*

As in the last section, X is a real-valued centred Gaussian process on $[0,1]$ with continuous sample paths and hence induces a Gaussian measure on the Banach space $C\left([0,1],\mathbb{R}\right)$. From general principles (see Appendix D) the associated *Cameron–Martin space*[4] $\mathcal{H} \subset C\left([0,1],\mathbb{R}\right)$ consists of paths $t \mapsto h_t = \mathbb{E}\left(ZX_t\right)$ where Z is an element of the L^2-closure of span $\{X_t : t \in [0,1]\}$, a Gaussian random variable. We recall that if $\tilde{h} = E\left(\tilde{Z}X.\right)$ denotes another element in $\mathcal{H}$, the inner product $\langle h, h'\rangle_{\mathcal{H}} = \mathbb{E}\left(ZZ'\right)$ makes $\mathcal{H}$ a Hilbert space. The following embedding theorem will prove crucial in our later applications to support theorems and large deviations.

Proposition 15.7 *Assume the covariance* $R : (s,t) \mapsto \mathbb{E}\left(X_s X_t\right)$ *is of finite* ρ-*variation (in 2D sense) for* $\rho \in [1,\infty)$. *Then* $\mathcal{H}$ *is continuously embedded in the space of continuous paths of finite* ρ-*variation. More precisely, for all* $h \in \mathcal{H}$ *and all* $s < t$ *in* $[0,1]$,

$$
\left|h\right|_{\rho\text{-var};[s,t]} \leq \sqrt{\langle h,h\rangle_{\mathcal{H}}}\sqrt{R_{\rho\text{-var};[s,t]^2}}.
$$

---

[4]Equivalently: *reproducing kernel Hilbert space*.

Proof. Let $h = \mathbb{E}(ZX.)$ and assume, without loss of generality, that $\langle h, h \rangle_{\mathcal{H}}^{1/2} = |Z|_{L^2} = 1$. Let (t_j) be a dissection of $[s, t]$. Let ρ' be the Hölder conjugate of ρ. Using duality for l^ρ-spaces, we have[5]

$$\left(\sum_j \left| h_{t_j, t_{j+1}} \right|^\rho \right)^{1/\rho}$$

$$= \sup_{\beta, |\beta|_{l^{\rho'}} \leq 1} \sum_j \beta_j h_{t_j, t_{j+1}} = \sup_{\beta, |\beta|_{l^{\rho'}} \leq 1} \mathbb{E} \left(Z \sum_j \beta_j X_{t_j, t_{j+1}} \right)$$

$$\leq \sup_{\beta, |\beta|_{l^{\rho'}} \leq 1} \sqrt{ \sum_{j,k} \beta_j \beta_k \mathbb{E} \left(X_{t_j, t_{j+1}} X_{t_k, t_{k+1}} \right) } \quad \text{(Cauchy–Schwarz)}$$

$$\leq \sup_{\beta, |\beta|_{l^{\rho'}} \leq 1} \sqrt{ \left(\sum_{j,k} |\beta_j|^{\rho'} |\beta_k|^{\rho'} \right)^{\frac{1}{\rho'}} \left(\sum_{j,k} \left| \mathbb{E} \left(X_{t_j, t_{j+1}} X_{t_k, t_{k+1}} \right) \right|^\rho \right)^{\frac{1}{\rho}} }$$

$$\leq \left(\sum_{j,k} \left| \mathbb{E} \left(X_{t_j, t_{j+1}} X_{t_k, t_{k+1}} \right) \right|^\rho \right)^{1/(2\rho)} \leq \sqrt{ R_{\rho\text{-var};[s,t]^2} } .$$

The proof is then finished by taking the supremum over all $(t_j) \in \mathcal{D}[s, t]$. ∎

Remark 15.8 Assume that the ρ-variation of R is controlled by a Hölder-dominated control, i.e.

$$\forall s < t \text{ in } [0, 1] : R^\rho_{\rho\text{-var};[s,t]^2} \leq K \, |t - s| .$$

Then Proposition 15.7 implies that

$$|h_{s,t}| \leq |h|_{\rho\text{-var};[s,t]} \leq |h|_{\mathcal{H}} \, K^{1/2} \, |t - s|^{1/(2\rho)}$$

which tells us that $\mathcal{H}$ is continuously embedded in the space of $1/(2\rho)$-Hölder continuous paths (which can also be seen directly from $h_{s,t} = \mathbb{E}(ZX_{s,t})$ and Cauchy–Schwarz). The point is that $1/(2\rho)$-Hölder only implies 2ρ-variation regularity, in contrast to the sharper result of Proposition 15.7. □

Remark 15.9 Let $\mathcal{H}_{\text{BM}}$ denote the Cameron–Martin space of real-valued Brownian motion $(\beta_t : t \in [0, 1])$; defined as the set of all paths $t \mapsto h_t = \mathbb{E}(Z\beta_t)$ where Z is in the L^2-closure of span $\{\beta_t : t \in [0, 1]\}$. As is well known (see Example D.3), $\mathcal{H}_{\text{BM}}$ is identified with the Sobolev space $W_0^{1,2}$

---

[5] The case $\rho = 1$ may be seen directly by taking $\beta_j = sgn\left(h_{t_j, t_{j+1}} \right)$.

$([0,1], \mathbb{R})$. It is worth noting that Proposition 15.7 implies $|h|_{1\text{-var};[s,t]} \leq M |t - s|^{1/2}$ with $M = \sqrt{\langle h, h \rangle_{\mathcal{H}_{\mathrm{BM}}}}$; and this property alone, using

$$|h|_{W^{1,2}} = \sup_{(t_i) \in \mathcal{D}[0,1]} \left(\sum_i |h_{t_i, t_{i+1}}|^2 / |t_{i+1} - t_i| \right)^{1/2}$$

implies the (important) estimate $|h|_{W^{1,2}} \leq \sqrt{\langle h, h \rangle_{\mathcal{H}_{\mathrm{BM}}}}$. $\qquad\square$

Remark 15.10 Consider $\mathcal{H}_{\mathrm{fBM}}^H \equiv \mathcal{H}^H$, the Cameron–Martin space of fractional Brownian motion with Hurst parameter H. It can be useful to know that smooth paths started at the origin are contained in $\mathcal{H}^H$. In fact, one even has (e.g. [40, 63])

$$C_0^1 \left([0, T], \mathbb{R}^d \right) \subset \mathcal{H}^H. \tag{15.5}$$

Let us now focus on the interesting regime $H \in (0, 1/2]$. Proposition 15.7 immediately gives

$$\mathcal{H}^H \hookrightarrow C^{1/(2H)\text{-var}}$$

which shows that fractional Cameron–Martin paths have finite $q \in [1, 2)$-variation as long as $H > H^* = 1/4$. In fact, one can do a little better and show that for any $\delta \in (1/2, 1/2 + H)$,

$$\mathcal{H}^H \hookrightarrow W_0^{\delta,2} \hookrightarrow C^{1/\delta\text{-var}}.$$

The first embedding is well known: from [40] and the references therein we know that $\mathcal{H}^H$ is continuously embedded in the potential space $I_{H+1/2,2}^+$, which we need not define here, and from [51, 40] one has $I_{H+1/2,2}^+ \subset W^{\delta,2}$; a direct proof can be found in [64]. The second embedding is a Besov-variation embedding, see Corollary A.3 in Appendix A. $\qquad\square$

15.2.3 *Covariance of piecewise linear approximations*

Let X be a centred real-valued continuous Gaussian process on $[0, 1]$ with covariance $R = R_X$ assumed to be of finite ρ-variation, dominated by some 2D control function ω. We now discuss what happens to (the ρ-variation of) the covariance of piecewise linear approximations to X. To this end, let $D = (\tau_i), \tilde{D} = (\tilde{\tau}_j)$ be dissections of $[0, 1]$ and write X^D for the piecewise linear approximation to X, i.e. $X_t^D = X_t$ for $t \in D$ and X^D is linear between two successive points of D. If $(s, t) \times (u, v) \subset (\tau_i, \tau_{i+1}) \times (\tilde{\tau}_j, \tilde{\tau}_{j+1})$ we set, consistent with Definition 5.59,

$$R^{D, \tilde{D}} \begin{pmatrix} s, t \\ u, v \end{pmatrix} \quad : \quad = \mathbb{E} \left(\int_s^t \dot{X}_r^D \, dr \int_u^v \dot{X}_r^{\tilde{D}} \, dr \right)$$

$$= \frac{t - s}{\tau_{i+1} - \tau_i} \times \frac{v - u}{\tilde{\tau}_{j+1} - \tilde{\tau}_j} R \begin{pmatrix} \tau_i, \tau_{i+1} \\ \tau_j, \tau_{j+1} \end{pmatrix}.$$

In particular, $R^D := R^{D,D}$ is then precisely R_{X^D} i.e. the covariance of X^D.

Proposition 15.11 (covariance of piecewise, linear approximations)
Let X be a continuous, centred, real-valued Gaussian process on $[0,1]$ with covariance R assumed to be of finite ρ-variation. Let $D, \tilde{D} \in \mathcal{D}[0,1]$. Then $\left(X^D, X^{\tilde{D}}\right)$ is jointly Gaussian with covariance

$$R_{\left(X^D, X^{\tilde{D}}\right)} : \begin{pmatrix} s \\ t \end{pmatrix} \mapsto \begin{pmatrix} \mathbb{E}\left(X_s^D X_t^D\right) & \mathbb{E}\left(X_s^D X_t^{\tilde{D}}\right) \\ \mathbb{E}\left(X_s^{\tilde{D}} X_t^D\right) & \mathbb{E}\left(X_s^{\tilde{D}} X_t^{\tilde{D}}\right) \end{pmatrix}$$

of finite ρ-variation. Moreover,

$$\left|R_{\left(X^D, X^{\tilde{D}}\right)}\right|_{\rho\text{-var};[0,1]^2} \leq 4.9^{1-\frac{1}{\rho}} |R|_{\rho\text{-var};[0,1]^2} .$$

Proof. It is easy to check that $\left(X^D, X^{\tilde{D}}\right)$ is jointly Gaussian. Observe that, using the notation of Definition 5.59,

$$R_{\left(X^D, X^{\tilde{D}}\right)} = \begin{pmatrix} R^{D,D} & R^{D,\tilde{D}} \\ R^{\tilde{D},D} & R^{\tilde{D},\tilde{D}} \end{pmatrix}.$$

It then follows from Proposition 5.60 that each component of this matrix has finite ρ-variation in 2D sense, controlled by $9^{\rho-1} |f|_{\rho\text{-var}}^\rho$. ∎

We now go a bit further in our analysis of piecewise linear approximation and show that Hölder-domination of the ρ-variation (on the "diagonal" $[s,t]^2$) remains valid when switching from $R = R_X$ to $R_{(X^D,X)}$ (this will only be used in Section 15.5.1 for establishing Hölder convergence of piecewise linear approximations). As usual, given $s \in [0,1]$, we write s_D for the greatest element of D such that $s_D \leq s$, and s^D the smallest element of D such that $s < s^D$.

Lemma 15.12 *Let X be a continuous, centred, real-valued Gaussian process on $[0,1]$ with covariance R. Then*
(i) for all $u_1, v_1, u_2, v_2 \in D$,

$$\left|R^D\right|_{\rho\text{-var};[u_1,v_1]\times[u_2,v_2]} \leq 9^{1-\frac{1}{\rho}} |R|_{\rho\text{-var};[u_1,v_1]\times[u_2,v_2]} ;$$

(ii) for all $s,t \in [0,1]$, with $s_D \leq s, t \leq s^D$, for all $u,v \in D$,

$$\left|R^D\right|_{\rho\text{-var};[s,t]\times[u,v]} \leq 9^{1-\frac{1}{\rho}} \left|\frac{t-s}{s^D-s_D}\right| \mathbb{E}\left(\left|X_{s_D,s^D}\right|^2\right)^{1/2} |R|_{\rho\text{-var};[u,v]^2}^{1/2} ;$$

(iii) for all $s_1, t_1, s_2, t_2 \in [0,1]$, with $s_{1,D} \leq s_1, t_1 \leq s_1^D$, $s_{2,D} \leq s_2, t_2 \leq s_2^D$,

$$\left|R^D\right|_{\rho\text{-var};[s_1,t_1]\times[s_2,t_2]} \leq \left|\frac{t_1-s_1}{s_1^D-s_{1,D}}\right|\left|\frac{t_2-s_2}{s_2^D-s_{2,D}}\right| \left|\mathbb{E}\left(X_{s_{1,D},s_1^D} X_{s_{2,D},s_2^D}\right)\right|.$$

Proof. (i) This follows from Proposition 15.11; indeed, there is no difference in the argument between working with $[0,1]^2$ or rectangles whose interval endpoints are elements of D. (ii) The second estimate is a bit more subtle. Take $s, t \in [0,1]$, with $s_D \leq s, t \leq s^D$, $u, v \in D$, (s_i) and (t_j) subdivisions of $[s,t]$ and $[u,v]$. Then, if $h_t^{i,D} = \mathbb{E}\left(X_{s_i,s_{i+1}}^D X_t^D\right)$, we know from Proposition 15.7 that

$$
\begin{aligned}
\left|h^{i,D}\right|_{\rho\text{-var};[u,v]} &\leq \left|R^D\right|_{\rho\text{-var};[u,v]}^{1/2} \mathbb{E}\left(\left|X_{s,t}^D\right|^2\right)^{1/2} \\
&\leq 9^{\rho-1}\frac{s_{i+1}-s_i}{s^D-s_D}\left|R\right|_{\rho\text{-var};[u,v]^2}^{1/2} \mathbb{E}\left(\left|X_{s_D,s^D}\right|^2\right)^{1/2}.
\end{aligned}
$$

Hence, for a fixed i,

$$
\begin{aligned}
\sum_j \left|\mathbb{E}\left(X_{s_i s_{i+1}}^D X_{t_j,t_{j+1}}^D\right)\right|^\rho &\leq \left|h^{i,D}\right|_{\rho\text{-var};[u,v]}^\rho \\
&\leq \left(9^{\rho-1}\frac{s_{i+1}-s_i}{s^D-s_D}\left|R\right|_{\rho\text{-var};[u,v]^2}^{1/2}\mathbb{E}\left(\left|X_{s_D,s^D}\right|^2\right)^{1/2}\right)^\rho.
\end{aligned}
$$

Summing over i and taking the supremum over all dissections ends the proof of the second estimate. We leave the easy proof of the third estimate to the reader. ∎

Proposition 15.13 (Hölder estimate in piecewise linear case) *Let X be a continuous, centred, real-valued Gaussian process on $[0,1]$ with covariance R assumed to be of finite ρ-variation. Then*

$$
\left|R_X\right|_{\rho\text{-var};[s,t]^2}^\rho \leq K\left|t-s\right| \quad \text{for all } s < t \text{ in } [0,1]
$$

implies, for some constant $C = C(\rho)$,

$$
\left|R_{(X,X^D)}\right|_{\rho\text{-var};[s,t]^2}^\rho \leq CK\left|t-s\right| \quad \text{for all } s < t \text{ in } [0,1].
$$

Proof. We need to estimate the ρ-variation of all the entries of $R_{(X,X^D)}$,

$$
\begin{pmatrix} s \\ t \end{pmatrix} \mapsto \begin{pmatrix} \mathbb{E}\left(X_s X_t\right) & \mathbb{E}\left(X_s X_t^D\right) \\ \mathbb{E}\left(X_s^D X_t\right) & \mathbb{E}\left(X_s^D X_t^D\right) \end{pmatrix}
$$

and focus on the lower-right entry, which is precisely R_{X^D}. By scaling we assume, without loss of generality, that $K = 1$. Then, by an argument similar to the proof of Proposition 5.60 (or Exercise 5.11 for the analogous 1-dimensional case), we may estimate its ρ-variation over some $[s,t]^2$ (with the property that $s^D \leq t_D$) in terms of the ρ-variation of R^D over smaller rectangles, namely

$$\frac{1}{9^{\rho-1}}\left|R^D\right|^\rho_{\rho\text{-var};[s,t]^2} \leq \left|R^D\right|^\rho_{\rho\text{-var};[s,s^D]^2} + \left|R^D\right|^\rho_{\rho\text{-var};[s,s^D]\times[s^D,t_D]}$$
$$+ \left|R^D\right|^\rho_{\rho\text{-var};[s,s^D]\times[t_D,t]} + \left|R^D\right|^\rho_{\rho\text{-var};[s^D,t_D]\times[s,s^D]}$$
$$+ \left|R^D\right|^\rho_{\rho\text{-var};[s^D,t_D]\times[s^D,t_D]} + \left|R^D\right|^\rho_{\rho\text{-var};[s^D,t_D]\times[t_D,t]}$$
$$+ \left|R^D\right|^\rho_{\rho\text{-var};[t_D,t]\times[s,s^D]} + \left|R^D\right|^\rho_{\rho\text{-var};[t_D,t]\times[s^D,t_D]}$$
$$+ \left|R^D\right|^\rho_{\rho\text{-var};[t_D,t]^2} \cdot$$

The proof is then easily finished with Lemma 15.12 and the fact that, for $s_D \leq s, t \leq s^D$, we have estimates of the form

$$\left|\frac{t-s}{s^D-s_D}\right| \mathbb{E}\left(\left|X_{s_D,s^D}\right|^2\right)^{1/2} \leq \left|\frac{t-s}{s^D-s_D}\right|\left|s^D-s_D\right|^{1/(2\rho)}$$
$$= \left|\frac{t-s}{s^D-s_D}\right|^{1-1/2\rho} |t-s|^{1/(2\rho)}$$
$$\leq |t-s|^{1/(2\rho)} .$$

Similar arguments apply to the ρ-variation of $(s,t)\mapsto \mathbb{E}\left(X_s X_t^D\right), \mathbb{E}\left(X_s^D X_t\right)$ with details left to the reader. The proof is then finished. ∎

15.2.4 Covariance of mollifier approximations

Let X be a continuous centred, real-valued Gaussian process on $[0,1]$ with covariance $R = R_X$ assumed to be of finite ρ-variation, dominated by some 2D control function ω. We now consider *mollifier* approximations. To this end, let us first extend X_t from $[0,1]$ to $(-\infty,\infty)$ by setting $X_t \equiv X_0$ for $t < 0$ and $X_t \equiv X_1$ for $t > 1$. As a simple consequence of this, for any rectangle $Q \subset \mathbb{R}^2$,

$$\left|R_X\right|_{\rho\text{-var};Q} = \left|R_X\right|_{\rho\text{-var};Q\cap[0,1]^2} . \tag{15.6}$$

Then, given a "mollifier" probability measure μ on $\mathbb{R}$, compactly supported, we define

$$X_t^\mu = \int X_{t-u}\,d\mu(u);$$

we also recall the notation (cf. Proposition 5.64)

$$\omega^{\mu,\tilde{\mu}}\left(\begin{array}{c}s,t\\u,v\end{array}\right) = \int\int \omega\left(\begin{array}{c}s-a,t-a\\u-b,v-b\end{array}\right)d\mu(a)\,d\tilde{\mu}(b).$$

Proposition 15.14 (covariance of mollifier approximations) *Let X be a continuous, centred, real-valued Gaussian process on $[0,1]$ with covariance R assumed to be of finite ρ-variation, controlled by*

$$\omega(Q) = |R|^\rho_{\rho\text{-var};Q} \text{ for any rectangle } Q \subset \mathbb{R}^2.$$

Let μ be a compactly supported probability measure on $\mathbb{R}$. Then X^μ is a Gaussian process with covariance of finite ρ-variation controlled by $\omega^{\mu,\mu}$. Moreover, $\omega^{\mu_n,\mu_n} \to \omega$ (pointwise) along any sequence $\mu_n \longrightarrow \delta_0$, the Dirac measure at zero.[6] If $\tilde{\mu}$ denotes another compactly supported probability measure on $\mathbb{R}$, then $\left(X^\mu, X^{\tilde{\mu}}\right)$ is jointly Gaussian with covariance

$$R_{(X^\mu, X^{\tilde{\mu}})} : \begin{pmatrix} s \\ t \end{pmatrix} \mapsto \begin{pmatrix} \mathbb{E}\left(X_s^\mu X_t^\mu\right) & \mathbb{E}\left(X_s^\mu X_t^{\tilde{\mu}}\right) \\ \mathbb{E}\left(X_s^{\tilde{\mu}} X_t^\mu\right) & \mathbb{E}\left(X_s^{\tilde{\mu}} X_t^{\tilde{\mu}}\right) \end{pmatrix}$$

of finite ρ-variation, controlled by a 2D control $\hat{\omega}$ which satisfies

$$\left| R_{(X^\mu, X^{\tilde{\mu}})} \right|_{\rho\text{-var};[0,1]^2} \leq \hat{\omega}\left([0,1]^2\right)^{1/\rho} \leq 4.4^{1/\rho} \omega\left([0,1]^2\right).$$

Remark 15.15 We shall apply $\mu, \tilde{\mu}$ given by μ^δ, μ^η where $d\mu^\delta := \frac{1}{\delta}\varphi\left(\frac{u}{\delta}\right)$ and $\varphi \in C^\infty\left(\mathbb{R}, \mathbb{R}^+\right)$, supported on $[-1,1]$ with total mass $\int \varphi(t)\,dt = 1$. Note that μ^δ converges to the Dirac measure at zero, as $\delta \to 0$. $\square$

Proof. We leave it to the reader to check that X^μ, and then $\left(X^\mu, X^{\tilde{\mu}}\right)$, are Gaussian processes. Proposition 5.64 then implies all statements, noting that

$$R_{(X^\mu, X^{\tilde{\mu}})} = \begin{pmatrix} R^{\mu,\tilde{\mu}} & R^{\mu,\tilde{\mu}} \\ R^{\tilde{\mu},\mu} & R^{\tilde{\mu},\tilde{\mu}} \end{pmatrix}$$

has finite ρ-variation controlled by $\hat{\omega}^{\mu,\tilde{\mu}} = 4^\rho\left(\omega^{\mu,\mu} + \omega^{\mu,\tilde{\mu}} + \omega^{\tilde{\mu},\mu} + \omega^{\tilde{\mu},\tilde{\mu}}\right)$. ∎

15.2.5 *Covariance of Karhunen–Loève approximations*

Let X be a continuous centred, real-valued Gaussian process on $[0,1]$ with covariance $R = R_X$ assumed to be of finite ρ-variation, dominated by some 2D control function ω. We now consider *Karhunen–Loève* approximations, also known as L^2-approximations. The situation here is more subtle than for piecewise linear or mollifier approximations. We will focus on the important case $\rho \in [1,2)$, although we only obtain *uniform* 2-variation bounds rather than uniform ρ-variation bounds (there is a world of difference, as will be seen in the next section: $\rho < 2$ allows for many uniform estimates which do not hold with $\rho = 2$). For a precise statement, we need some notation: $\mathcal{H} \subset C\left([0,1], \mathbb{R}\right)$ denotes the Cameron–Martin space, for which we fix an orthonormal basis $\left(h^k : k \in \mathbb{N}\right)$. From general principles $\mathcal{H}$ embeds isometrically into a (Gaussian) subspace of $L^2\left(\mathbb{P}\right)$,

$$h \in \mathcal{H} \mapsto \xi(h) \in L^2\left(\mathbb{P}\right)$$

and there is an L^2-expansion/approximation of the following type, where ξ is the Paley–Wiener map (cf. Section D.3 in Appendix D).

---

[6]That is, for all bounded continuous functions f, $\lim_{n\to\infty} \int f d\mu_n = f(0)$.

Definition 15.16 (Karhunen–Loève approximation) For a fixed orthonormal basis $(h^k : k \in \mathbb{N})$ in $\mathcal{H} \subset C([0,1], \mathbb{R})$ consider the L^2-expansion of X,

$$X = \sum_{k \in \mathbb{N}} Z_k h^k, \quad \text{convergent a.s. and in } L^2(\mathbb{P}),$$

where $Z_k := \xi(h^k)$, $k \in \mathbb{N}$, is a sequence of independent standard normal random variables. For a fixed set $A \subset \mathbb{N}$, define

$$\mathcal{F}_A = \sigma(Z_k, k \in A) \quad \text{and} \quad X_t^A = \mathbb{E}[X_t | \mathcal{F}_A].$$

The sequence $(X^{\{1,\dots,n\}} : n \in \mathbb{N})$ is then called a Karhunen–Loève approximation to X. □

Remark 15.17 Observe that $t \mapsto X_t^A$ is a Gaussian process in its own right with covariance function

$$R^A(s,t) := R_{X^A}(s,t) := \mathbb{E}[X_s^A X_t^A] = \sum_{i \in A} h_s^i h_t^i.$$ □

Lemma 15.18 *Let X be a continuous, centred, real-valued Gaussian process on $[0,1]$ with covariance R assumed to be of finite ρ-variation, for some $\rho \geq 1$. Then, for all subsets A of $\mathbb{N}$,*

$$|R_{X^A}|_{\rho\text{-var};[s,t]^2} \leq (1 + \min\{|A|, |A^c|\}) |R_X|_{\rho\text{-var};[s,t]^2},$$
$$|R_{X^A}|_{2\text{-var};[s,t]^2} \leq |R|_{2\text{-var};[s,t]^2}.$$

Proof. We first prove the first inequality, assuming that $|A^c| < \infty$ so that

$$R_{X^A} = \sum_{k \in A^c} h_k \otimes h_k.$$

It is then clear from Proposition 15.7, using $|h_k|_{\mathcal{H}} = 1$, that

$$|h_k \otimes h_k|_{\rho\text{-var};[s,t]^2} \leq |h_k|_{\rho\text{-var};[s,t]}^2 \leq |R|_{\rho\text{-var};[s,t]^2},$$

and it follows from the triangle inequality that

$$|R_{X^A}|_{\rho\text{-var};[s,t]^2} \leq |R|_{\rho\text{-var};[s,t]^2} + \sum_{k \in A^c} |h_k \otimes h_k|_{\rho\text{-var};[s,t]^2}$$
$$\leq (1 + |A^c|) |R|_{\rho\text{-var};[s,t]^2}.$$

The case $|A| < \infty$ is similar but easier and left to the reader. We now turn to the proof of the second inequality: let $D = (t_i)$ be a dissection of $[s,t]$ and set $X_i^A = X_{t_i, t_{i+1}}^A$. Let $\beta = (\beta_{i,j})$ be a positive symmetric matrix, and let us estimate $\left| \sum_{i,j} \beta_{i,j} \mathbb{E}(X_i^A X_j^A) \right|$. To this end, note

$$\mathbb{E}(X_i^A X_j^A) = \sum_{k \in A} \mathbb{E}(Z_k X_i) \mathbb{E}(Z_k X_j) = \frac{1}{2} \sum_{k \in A} \mathbb{E}\left((Z_k^2 - \mathbb{E}(Z_k^2)) X_i X_j \right),$$

so that

$$\sum_{i,j} \beta_{i,j} \mathbb{E}\left(X_i^A X_j^A\right) = \frac{1}{2} \sum_{k \in A} \mathbb{E}\left(\left(Z_k^2 - \mathbb{E}\left(Z_k^2\right)\right) \sum_{i,j} \beta_{i,j} X_i X_j\right).$$

As β is symmetric, we can write $\beta = P^T D P$, with PP^T the identity matrix, and D a diagonal matrix which contains the (non-negative) eigenvalues (d_i) of β. Simple linear algebra gives

$$\sum_{i,j} \beta_{i,j} X_i X_j = (PX)^T D (PX) = \sum_i d_i (PX)_i^2,$$

and we can compute

$$
\begin{aligned}
\sum_{i,j} \beta_{i,j} \mathbb{E}\left(X_i^A X_j^A\right) &= \sum_{k \in A} \sum_i d_i \frac{1}{2} \mathbb{E}\left(\left(Z_k^2 - \mathbb{E}\left(Z_k^2\right)\right)(PX)_i^2\right) \\
&= \sum_i d_i \sum_{k \in A} \mathbb{E}\left(Z_k (PX)_i\right)^2 \\
&\leq \sum_i d_i \mathbb{E}\left((PX)_i^2\right) \qquad \text{(Parseval inequality)} \\
&= \mathbb{E}\left((PX)^T D (PX)\right) \\
&= \sum_{i,j} \beta_{i,j} \mathbb{E}\left(X_i X_j\right) \\
&\leq |\beta|_{l^2} |R|_{2\text{-var};[s,t]^2} \qquad \text{(Hölder inequality).}
\end{aligned}
$$

(Note that finite $\rho \in [1,2)$-variation of R implies finite 2-variation of R.) We now apply this estimate with $\beta_{i,j} = \mathbb{E}\left(X_i^A X_j^A\right)$ and find

$$\sqrt{\sum_{i,j} \left|\mathbb{E}\left(X_i^A X_j^A\right)\right|^2} \leq |R|_{2\text{-var}}.$$

The proof is finished by taking the supremum over all dissections of $[s,t]$. ∎

15.3 Multidimensional Gaussian processes

Any $\mathbb{R}^d$-valued centred Gaussian process $X = \left(X^1, \ldots, X^d\right)$ with continuous sample paths gives rise to an abstract Wiener space $(E, \mathcal{H}, \mathbb{P})$ with $E = C\left([0,1], \mathbb{R}^d\right)$ and $\mathcal{H} \subset C\left([0,1], \mathbb{R}^d\right)$. If $\mathcal{H}^i$ denotes the Cameron–Martin space associated with the 1-dimensional Gaussian process X^i and all $\left\{X^i : i = 1, \ldots, d\right\}$ are independent, then $\mathcal{H} \cong \oplus_{i=1}^d \mathcal{H}^i$. Recall that $\mathcal{H}$ embeds isometrically into a (Gaussian) subspace of $L^2(\mathbb{P})$,

$$h \in \mathcal{H} \mapsto \xi(h) \in L^2(\mathbb{P}).$$

15.3.1 Wiener chaos

From Section D.4 of Appendix D, there is an (orthogonal) decomposition of the form

$$L^2(\mathbb{P}) = \oplus_{n=0}^{\infty} \mathcal{W}^{(n)}(\mathbb{P}).$$

The subspaces $\mathcal{W}^n(\mathbb{P})$ are known as *homogenous Wiener chaos* of order n and $\mathcal{C}^n(\mathbb{P}) := \oplus_{j=0}^n \mathcal{W}^{(i)}(\mathbb{P})$ denotes the *Wiener chaos* (or non-homogenous chaos) of order n. Our interest in Wiener chaos comes from the fact that $\mathcal{C}^n(\mathbb{P})$ is precisely the closure (in probability, say) of polynomials of degree less than or equal to n in the variables $\xi(h_k)$ where $(h_k) \subset \mathcal{H}$ is any fixed orthonormal basis. In particular, any polynomial in $X_{t_k}^{i_k}$ for finitely many $i_k \in \{1,\dots,d\}$ and $t_k \in [0,1]$ is an element of $\mathcal{C}^n(\mathbb{P})$ for sufficiently large n.

Proposition 15.19 *Assume the $\mathbb{R}^d$-valued continuous centred Gaussian process $X = (X^1,\dots,X^d)$ has sample paths of finite variation and let $S_N(X) \equiv \mathbf{X}$ denote its natural lift to a process with values in $G^N(\mathbb{R}^d) \subset T^N(\mathbb{R}^d)$. Then, for $n = 1,\dots,N$ and any $s,t \in [0,1]$ the random variable $\pi_n(\mathbf{X}_{s,t})$ is an element in the nth (in general, not homogenous) Wiener chaos.[7]*

Proof. $\pi_n(\mathbf{X})$ is given by n iterated integrals which can be written out in terms of (a.s. convergent) Riemann–Stieltjes sums. Each such Riemann–Stieltjes sum is a polynomial of degree at most n and with variables of the form $X_{s,t}$. It now suffices to remark that the (not necessarily homogenous) nth Wiener chaos contains all such polynomials and is closed under convergence in probability. ∎

As a special case of the Wiener chaos integrability, see (D.5) in Section D.4, we have

Lemma 15.20 *Let $n \in \mathbb{N}$ and $Z \in \mathcal{C}^n(\mathbb{P})$. Then, for $q > 2$,*

$$|Z|_{L^2} \le |Z|_{L^q} \le |Z|_{L^2}(n+1)(q-1)^{n/2}.$$

A simple but useful consequence is that, for random variables $Z, W \in \mathcal{C}^n(\mathbb{P})$ we have

$$|WZ|_{L^2} \le C(n)|W|_{L^2}|Z|_{L^2}. \tag{15.7}$$

(There is nothing special about L^2 here, but this is how we usually use it.) We now discuss some more involved corollaries.

Corollary 15.21 *Let g be a random element of $G^N(\mathbb{R}^d)$ such that for all $1 \le n \le N$ the projection $\pi_n(g)$ is an element of the nth Wiener chaos. Let δ be a positive real. Then the following statements are equivalent:*

---

[7] Strictly speaking, the $(\mathbb{R}^d)^{\otimes n}$-valued chaos.

(i) there exists a constant $C_1 > 0$ such that for all $n = 1, \ldots, N$, there exists $q = q(n) \in [1, \infty) : |\pi_n(g)|_{L^q} \le C_1 \delta^n$;

(ii) there exists a constant $C_2 > 0$ such that for all $n = 1, \ldots, N$ and for all $q \in [1, \infty) : |\pi_n(g)|_{L^q} \le C_2 q^{\frac{n}{2}} \delta^n$;

(iii) there exists a constant $C_3 > 0$ and there exists $q \in [1, \infty) : \mathbb{E}(\|g\|^q)^{1/q} \le C_3 \delta$;

(iv) there exists a constant $C_4 > 0$ such that for all $q \in [1, \infty) : \mathbb{E}(\|g\|^q)^{1/q} \le C_4 q^{\frac{1}{2}} \delta$.

When switching from the ith to the jth statement, the constant C_j depends only on C_i and N.

Proof. Clearly, (iv)$\Longrightarrow$(iii), (ii)$\Longrightarrow$(i), and Lemma 15.20 shows (i)$\Longrightarrow$(ii). It is therefore enough to prove (ii)$\Longrightarrow$(iv), (iii)$\Longrightarrow$(i),

(ii)$\Longrightarrow$(iv): By equivalence of homogenous norms, we have

$$\|g\| \le c_1 \max_{n=1}^{N} |\pi_n(g)|^{1/n},$$

so that,

$$
\begin{aligned}
\mathbb{E}(\|g\|^q)^{1/q} &\le c_2 \max_{n=1}^{N} \left(\mathbb{E}\left(|\pi_n(g)|^{q/n} \right) \right)^{1/q} \\
&\le c_3 \left(\max_{n=1}^{N} \left(q^{\frac{n}{2}} \delta^n \right)^{q/n} \right)^{1/q} \\
&\le c_4 q^{\frac{1}{2}} \delta.
\end{aligned}
$$

(iii)$\Longrightarrow$(i): By equivalence of homogenous norms, we have

$$|\pi_n(g)|^{1/n} \le c_5 \|g\|.$$

Hence,

$$\mathbb{E}\left(|\pi_n(g)|^{q_0/n} \right)^{n/q_0} \le c_5^n \mathbb{E}(\|g\|^{q_0})^{n/q_0} \le c_6 \delta^n.$$

■

Proposition 15.22 *Let $\mathbf{X}$ be a continuous $G^N(\mathbb{R}^d)$-valued stochastic process and ω a control function on $[0,1]$. Assume that for all $s < t$ in $[0,1]$ and $n = 1, \ldots, N$, the projection $\pi_n(\mathbf{X}_{s,t})$ is an element in the nth Wiener chaos and that, for some constant C,*

$$|\pi_n(\mathbf{X}_{s,t})|_{L^2} \le C\omega(s,t)^{\frac{n}{2\rho}}. \tag{15.8}$$

Then,
(i) there exists a constant $C' = C'(\rho, N)$ such that for all $s < t$ in $[0,1]$ and $q \in [1, \infty)$,

$$|d(\mathbf{X}_s, \mathbf{X}_t)|_{L^q} \le C' \sqrt{q} \omega(s,t)^{\frac{1}{2\rho}}; \tag{15.9}$$

(ii) if $p > 2\rho$ then $\|\mathbf{X}\|_{p\text{-var};[0,1]}$ has a Gaussian tail. More precisely, if $\omega(0,1) \le K$ then there exists $\eta = \eta(p,\rho,N,K) > 0$ such that

$$\mathbb{E}\exp\left(\eta\|\mathbf{X}\|^2_{p\text{-var};[0,1]}\right) < \infty; \qquad (15.10)$$

(iii) if $\omega(s,t) \le K|t-s|$ for all $s < t$ in $[0,1]$ we may replace $\|\mathbf{X}\|_{p\text{-var};[0,1]}$ in (15.10) by $\|\mathbf{X}\|_{1/p\text{-Höl};[0,1]}$.

Proof. (i) is a clear consequence of Corollary 15.21 and (iii) follows from a (probabilistic) Besov–Hölder embedding, Theorem A.12, in Appendix A. At last, (ii) follows from (iii) by reparametrization. Indeed, assuming without loss of generality that $\omega(0,1) > 0$, super-additivity of controls implies

$$\forall s < t \text{ in } [0,T]: \omega(s,t) \le \omega(0,1)\left[\frac{\omega(0,t)}{\omega(0,1)} - \frac{\omega(0,s)}{\omega(0,1)}\right]$$

and we may define $\left(\tilde{\mathbf{X}}_t : 0 \le t \le 1\right)$ by requiring that

$$\tilde{\mathbf{X}}_{\omega(0,t)/\omega(0,1)} = \mathbf{X}_t.$$

(Note that $\omega(0,s)/\omega(0,1) = \omega(0,t)/\omega(0,1) \implies \omega(s,t) = 0 \implies \mathbf{X}|_{[s,t]} \equiv \mathbf{X}_s$ a.s. from (15.8) and $\tilde{\mathbf{X}}$ is indeed well-defined.) Clearly, $\tilde{\mathbf{X}}$ satisfies the assumptions for (iii) with $K = \omega(0,1)$ and we conclude with invariance of variation norms under reparametrization,

$$\|\mathbf{X}\|_{p\text{-var};[0,1]} = \left\|\tilde{\mathbf{X}}\right\|_{p\text{-var};[0,1]} \le \left\|\tilde{\mathbf{X}}\right\|_{1/p\text{-Höl};[0,1]}.$$

∎

Remark 15.23 (Lévy modulus and exact variation) In the setting of Proposition 15.22 and under a Hölder assumption on ω, i.e.

$$\forall s < t \text{ in } [0,1]: \omega(s,t) \le K|t-s|$$

it is immediate from (15.8), cf. Lemma A.17, that there exists $\eta = \eta(\rho,N) > 0$ so that

$$\sup_{s,t\in[0,1]} \mathbb{E}\exp\left(\eta\frac{d(\mathbf{X}_s,\mathbf{X}_t)^2}{|t-s|^{1/\rho}}\right) < \infty.$$

In the language of Section A.4, Appendix A, this shows that $\mathbf{X}$ satisfies the "Gaussian integrability condition (2ρ)". From the general results of that appendix it then follows that $\mathbf{X}$ has a.s. Lévy modulus-type regularity and also finite $\psi_{2\rho,\rho}$-variation. In fact, the same reparametrization argument that was used in the proof of Proposition 15.22 shows that finite $\psi_{2\rho,\rho}$-variation holds without the Hölder assumption on ω. We note that, at least for $\rho = 1$, the interest in generalized variation regularity comes from Section 10.5. □

Proposition 15.24 *Let* $\mathbf{X}, \mathbf{Y}$ *be two continuous* $G^N\left(\mathbb{R}^d\right)$-*valued stochastic processes and* ω *a control function on* $[0, 1]$. *Assume that for all* $s < t$ *in* $[0, 1]$ *and* $n = 1, \ldots, N$, $\pi_n\left(\mathbf{X}_{s,t}\right)$ *and* $\pi_n\left(\mathbf{Y}_{s,t}\right)$ *are elements of the* nth *Wiener chaos and that, for some* $C > 0$ *and* $\varepsilon > 0$,

$$\left|\pi_n\left(\mathbf{X}_{s,t}\right)\right|_{L^2} \leq C\omega\left(s,t\right)^{\frac{n}{2\rho}} \text{ and } \left|\pi_n\left(\mathbf{Y}_{s,t}\right)\right|_{L^2} \leq C\omega\left(s,t\right)^{\frac{n}{2\rho}}, \quad (15.11)$$

$$\left|\pi_n\left(\mathbf{Y}_{s,t} - \mathbf{X}_{s,t}\right)\right|_{L^2} \leq C\varepsilon\omega\left(s,t\right)^{\frac{n}{2\rho}}. \quad (15.12)$$

Then,

(i) there exists a constant $C' = C'\left(C, \rho, N\right)$ *such that for all* $q \in [1, \infty)$

$$\left|\pi_n\left(\mathbf{Y}_{s,t} - \mathbf{X}_{s,t}\right)\right|_{L^q} \leq C' q^{\frac{n}{2}} \varepsilon\omega\left(s,t\right)^{\frac{n}{2\rho}};$$

(ii) if $p > 2\rho$ *there exists a constant* $C'' = C''\left(C, p, \rho, N\right)$ *such that*

$$\left|d_{p\text{-var};[0,1]}\left(\mathbf{X}, \mathbf{Y}\right)\right|_{L^q} \leq C'' \max\left\{\varepsilon^{1/N}, \varepsilon\right\} \sqrt{q} \quad (15.13)$$

and, for all $n = 1, \ldots, N$, *we have*

$$\left|\rho_{p\text{-var};[0,T]}^{(n)}\left(\mathbf{X}, \mathbf{Y}\right)\right|_{L^q} \leq C'' q^{\frac{n}{2}} \varepsilon; \quad (15.14)$$

(iii) if $\omega\left(s,t\right) \leq K\left|t - s\right|$ *for all* $s < t$ *in* $[0, 1]$ *then we may replace* $d_{p\text{-var};[0,1]}, \rho_{p\text{-var};[0,T]}^{(n)}$ *in (15.13), (15.14) by* $d_{1/p\text{-Höl};[0,1]}, \rho_{1/p\text{-Höl};[0,T]}^{(n)}$ *respectively.*

Proof. (i) is a clear consequence of Corollary 15.21 and (iii) follows from a (probabilistic) Besov–Hölder "distance" comparison, Theorem A.13. Case (ii) then follows from (iii) by the same reparametrization argument that we used in the proof of Proposition 15.22. ∎

Remark 15.25 Recall from Definition 8.6 that the inhomogenous p-variation distance was given by

$$\rho_{p\text{-var};[0,1]}\left(\mathbf{X}, \mathbf{Y}\right) = \max_{n=1,\ldots,N} \rho_{p\text{-var};[0,1]}^{(n)}\left(\mathbf{X}, \mathbf{Y}\right)$$

with

$$\rho_{p\text{-var};[0,1]}^{(n)}\left(\mathbf{X}, \mathbf{Y}\right) = \sup_{(t_i) \in \mathcal{D}[0,T]} \left(\sum_i \left|\pi_k\left(\mathbf{X}_{t_i,t_{i+1}} - \mathbf{Y}_{t_i,t_{i+1}}\right)\right|^{p/k}\right)^{k/p}$$

so that in the context of Proposition 15.24 one has, for $\varepsilon \in (0, 1], q \in [1, \infty)$,

$$\left|\rho_{p\text{-var};[0,1]}\left(\mathbf{X}, \mathbf{Y}\right)\right|_{L^q} \leq C'' q^{\frac{N}{2}} \varepsilon,$$

$$\left|d_{p\text{-var};[0,1]}\left(\mathbf{X}, \mathbf{Y}\right)\right|_{L^q} \leq C'' \varepsilon^{1/N} \sqrt{q}.$$

(Similarly with $1/p$-Hölder distances provided $\omega(s,t) \leq (\text{const}) \times |t-s|$.)
In other words, working with an "inhomogenous" p-variation distance yields
linear estimates in ε (which is useful, since by Theorem 10.38 the Itô–Lyons
map is locally Lipschitz continuous in $\rho_{p\text{-var};[0,1]}$) while working with "homogenous" p-variation distance $d_{p\text{-var}}$ has the advantage that the random
variable $d_{p\text{-var};[0,1]}(\mathbf{X},\mathbf{Y})$ has a Gaussian tail (which will be useful to establish "exponential goodness" of certain approximations in the context of
large deviations, cf. Section 15.7). □

15.3.2 Uniform estimates for lifted Gaussian processes

As in the previous section, we consider an $\mathbb{R}^d$-valued continuous centred
Gaussian process X with independent components $X^1, \ldots, X^d$. We shall
assume that the sample paths $(X_t(\omega) : t \in [0,1])$ are of finite variation.
This implies that iterated integrals of X are well-defined as Riemann–
Stieltjes integrals and we shall see that their second moments are controlled
uniformly (i.e. with constants not depending on the finite-variation sample
path assumption!) using the estimates for 2D Young integrals (from Section
6.4) of the respective covariances, which explains the standing assumption

$$\exists \rho \in [1,2) : |R_X|_{\rho\text{-var};[0,1]^2} < \infty.$$

(Recall that $R_X(s,t) = \mathbb{E}(X_s \otimes X_t) = \text{diag}(R_{X^1}, \ldots, R_{X^d})$ is the $(\mathbb{R}^d \otimes \mathbb{R}^d)$-valued covariance function of X.) We shall also control the difference
between (the iterated integrals of) a pair of Gaussian processes (X,Y), in
which case we will make the stronger assumption

$$\exists \rho \in [1,2) : |R_{(X,Y)}|_{\rho\text{-var};[0,1]^2} < \infty.$$

The following exercise shows that the above assumption indeed implies that
$X, Y, X-Y$, etc. have covariance of finite ρ-variation and we shall use this
without further notice.

Exercise 15.26 *Let $Z = (Z_1, \ldots, Z_n)$ be a centred, n-dimensional Gaussian process, with covariance R_Z of finite ρ-variation controlled by some
2D control ω. Let α be a linear map from $\mathbb{R}^n$ into $\mathbb{R}^d$, then the covariance
of αZ also has finite ρ-variation controlled by $C\omega$ where $C = C(\alpha)$.*

In a typical application below, (X,Y) is a $(2d)$-dimensional, centred
Gaussian process in which all coordinate pairs $(X^1, Y^1), \ldots, (X^d, Y^d)$ are
independent (think of Y as the coordinate-wise piecewise linear or mollifier
approximation to X) which allows us to reduce parts of the analysis to
$d=1$. We will need

Lemma 15.27 *Let (X,Y) be a 2-dimensional centred Gaussian process
with covariance R of finite ρ-variation controlled by ω. Then, for fixed*

$s < t$ in $[0,1]$, *the function*

$$(u,v) \in [s,t]^2 \mapsto f(u,v) := \mathbb{E}\left(X_{s,u}Y_{s,u}X_{s,v}Y_{s,v}\right)$$

satisfies $f(s,\cdot) = f(\cdot,s) = 0$ *and has finite ρ-variation. More precisely, there exists a constant $C = C(\rho)$ such that*

$$|f|^\rho_{\rho\text{-var};[s,t]^2} \le C\omega\left([s,t]^2\right)^2.$$

Proof. We fix $u < u'$, $v < v'$, all in $[s,t]$. Using

$$X_{s,u'}Y_{s,u'} - X_{s,u}Y_{s,u} = X_{u,u'}Y_{s,u'} + X_{s,u}Y_{u,u'},$$

we bound $|\mathbb{E}\left((X_{s,u'}Y_{s,u'} - X_{s,u}Y_{s,u})(X_{s,v'}Y_{s,v'} - X_{s,v}Y_{s,v})\right)|$ by

$$|\mathbb{E}\left(X_{u,u'}Y_{s,u'}X_{v,v'}Y_{s,v'}\right)| + |\mathbb{E}\left(X_{s,u}Y_{u,u'}X_{v,v'}Y_{s,v'}\right)|$$
$$+ |\mathbb{E}\left(X_{u,u'}Y_{s,u'}X_{s,v}Y_{v,v'}\right)| + |\mathbb{E}\left(X_{s,u}Y_{u,u'}X_{s,v}Y_{v,v'}\right)|.$$

To estimate the second expression, for example, we use a well-known identity for the product of Gaussian random variables,[8]

$$\mathbb{E}\left(X_{s,u}Y_{u,u'}X_{v,v'}Y_{s,v'}\right) = \mathbb{E}\left(X_{s,u}Y_{u,u'}\right)\mathbb{E}\left(X_{v,v'}Y_{s,v'}\right)$$
$$+ \mathbb{E}\left(X_{s,u}X_{v,v'}\right)\mathbb{E}\left(Y_{u,u'}Y_{s,v'}\right)$$
$$+ \mathbb{E}\left(X_{s,u}Y_{s,v'}\right)\mathbb{E}\left(X_{v,v'}Y_{u,u'}\right),$$

to obtain

$$\frac{1}{C_\rho}|\mathbb{E}\left(X_{s,u}Y_{u,u'}X_{v,v'}Y_{s,v'}\right)|^\rho \le \omega\left([s,u]\times[u,u']\right)\omega\left([v,v']\times[s,v']\right)$$
$$+ \omega\left([s,u]\times[v,v']\right)\omega\left([u,u']\times[s,v']\right)$$
$$+ \omega\left([s,u]\times[s,v']\right)\omega\left([u,u']\times[v,v']\right)$$
$$\le \omega\left([s,t]\times[u,u']\right)\omega\left([v,v']\times[s,t]\right)$$
$$+ \omega\left([s,t]\times[v,v']\right)\omega\left([u,u']\times[s,t]\right)$$
$$+ \omega\left([s,t]\times[s,t]\right)\omega\left([u,u']\times[v,v']\right).$$

Working similarly with all terms, we obtain that this last expression controls the ρ-variation of $(u,v) \in [s,t]^2 \to \mathbb{E}\left(X_{s,u}Y_{s,u}X_{s,v}Y_{s,v}\right)$, and the bound on the ρ-variation on $[s,t]^2$. ∎

Proposition 15.28 *Assume that*
(i) $X = \left(X^1,\ldots,X^d\right)$ *is a centred continuous Gaussian process with independent components and with bounded variation sample paths;*

---

[8]This is a consequence of the so-called Wick formula for Gaussian random variables; see also [120, Lemma 4.5.1].

(ii) the covariance of X is of finite ρ-variation dominated by a 2D control ω, for some $\rho \in [1, 2)$;

(iii) $\mathbf{X} = S_3(X)$.

There exists $C = C(\rho)$ such that for all $s < t$ in $[0, 1]$, for $n = 1, 2, 3,$;

$$|\pi_n(\mathbf{X}_{s,t})|_{L^2(\mathbb{P})} \leq C\omega\left([s, t]^2\right)^{\frac{n}{2\rho}}.$$

Proof. From Proposition 15.61 in the appendix to this chapter, it is enough to prove

$$(a) \quad \mathbb{E}\left(\left|X_{s,t}^i\right|^2\right) \leq \omega\left([s, t]^2\right)^{1/\rho} \quad \text{for all } i;$$

$$(b) \quad \mathbb{E}\left(\left|\mathbf{X}_{s,t}^{i,j}\right|^2\right) \leq C\omega\left([s, t]^2\right)^{2/\rho} \quad \text{for } i, j \text{ distinct};$$

$$(c) \quad \mathbb{E}\left(\left|\mathbf{X}_{s,t}^{i,i,j}\right|^2\right) \leq C\omega\left([s, t]^2\right)^{3/\rho} \quad \text{for } i, j \text{ distinct};$$

$$(d) \quad \mathbb{E}\left(\left|\mathbf{X}_{s,t}^{i,j,k}\right|^2\right) \leq C\omega\left([s, t]^2\right)^{3/\rho} \quad \text{for } i, j, k \text{ distinct}.$$

The level-one estimate (a) is obvious. For the level-two estimate (b), we fix $i \neq j$ and $s < t$, $s' < t'$. Then, using independence of X^i and X^j,

$$
\begin{aligned}
\mathbb{E}\left(\mathbf{X}_{s,t}^{i,j}\mathbf{X}_{s',t'}^{i,j}\right) &= \mathbb{E}\left(\int_s^t \int_{s'}^{t'} X_{s,u}^i X_{s',v}^i dX_u^j dX_v^j\right) \\
&= \int_s^t \int_{s'}^{t'} \mathbb{E}\left(X_{s,u}^i X_{s',v}^i\right) d\mathbb{E}\left(X_u^j X_v^j\right) \\
&= \int_s^t \int_{s'}^{t} R_{X^i}\begin{pmatrix} s, u \\ s', v \end{pmatrix} dR_{X^j}\begin{pmatrix} u \\ v \end{pmatrix} \\
&\leq C\omega\left([s, t] \times [s', t']\right)^{2/\rho} \quad \text{by Young 2D estimate.}
\end{aligned}
$$

(b) follows trivially from setting $s = s', t = t'$ (the general result will be used in the level-three estimates, see step 2 below). We break up the level-three estimates into several steps. Throughout, the indices i, j (and then k) are assumed to be distinct.

Step 1: For fixed $s < t, s' < t', t' < u'$ we claim that

$$\mathbb{E}\left(\mathbf{X}_{s,t}^{i,j}X_{s',t'}^i X_{t',u'}^j\right) \leq C\omega\left([s, t] \times [s', t']\right)^{1/\rho}\omega\left([s, t] \times [t', u']\right)^{1/\rho}.$$

Indeed, with $d\mathbb{E}\left(X_{t',u'}^j X_u^j\right) \equiv \mathbb{E}\left(X_{t',u'}^j \dot{X}_u^j\right) du$ we have

$$
\begin{aligned}
\mathbb{E}\left(\mathbf{X}_{s,t}^{i,j}X_{s',t'}^i X_{t',u'}^j\right) &= \mathbb{E}\left(\int_s^t X_{s,u}^i X_{s',t'}^i X_{t',u'}^j dX_u^j\right) \\
&= \int_{u=s}^t \mathbb{E}\left(X_{s,u}^i X_{s',t'}^i\right) d\mathbb{E}\left(X_{t',u'}^j X_u^j\right).
\end{aligned}
$$

Since the 1D ρ-variation of $u \mapsto \mathbb{E}\left(X^i_{s,u} X^i_{s',t'}\right)$ is controlled by $(u,v) \mapsto \omega\left([u,v] \times [s',t']\right)$, and similarly for $u \mapsto \mathbb{E}\left(X^j_{t',u'} X^j_u\right)$, the (classical 1D) Young estimate gives

$$\left| \int_{u=s}^{t} \mathbb{E}\left(X^i_{s,u} X^i_{s',t'}\right) d\mathbb{E}\left(X^j_{t',u'} X^j_u\right) \right|$$
$$\leq C\omega\left([s,t] \times [s',t']\right)^{1/\rho} \omega\left([s,t] \times [t',u']\right)^{1/\rho}.$$

Step 2: For fixed $s < t$, we claim that the 2D map $(u,v) \in [s,t]^2 \mapsto \mathbb{E}\left(\mathbf{X}^{i,j}_{s,u} \mathbf{X}^{i,j}_{s,v}\right)$ has finite ρ-variation controlled by

$$[u_1,u_2] \times [v_1,v_2] \mapsto C\omega\left([s,t]^2\right) \omega\left([u_1,u_2] \times [v_1,v_2]\right).$$

Indeed, using the level-two estimate and step 1, for $u_1 < u_2$, $v_1 < v_2$ all in $[s,t]$,

$$\mathbb{E}\left(\left(\mathbf{X}^{i,j}_{s,u_2} - \mathbf{X}^{i,j}_{s,u_1}\right)\left(\mathbf{X}^{i,j}_{s,v_2} - \mathbf{X}^{i,j}_{s,v_1}\right)\right)$$
$$= \mathbb{E}\left(\left(\mathbf{X}^{i,j}_{u_1,u_2} + X^i_{s,u_1} X^j_{u_1,u_2}\right)\left(\mathbf{X}^{i,j}_{v_1,v_2} + X^i_{s,v_1} X^j_{v_1,v_2}\right)\right)$$
$$= \mathbb{E}\left(\mathbf{X}^{i,j}_{u_1,u_2} \mathbf{X}^{i,j}_{v_1,v_2}\right)$$
$$+ \mathbb{E}\left(\mathbf{X}^{i,j}_{u_1,u_2} X^i_{s,v_1} X^j_{v_1,v_2}\right)$$
$$+ \mathbb{E}\left(X^i_{s,u_1} X^j_{u_1,u_2} \mathbf{X}^{i,j}_{v_1,v_2}\right)$$
$$+ \mathbb{E}\left(X^i_{s,u_1} X^i_{s,v_1}\right)\mathbb{E}\left(X^j_{u_1,u_2} X^j_{v_1,v_2}\right)$$
$$\leq \omega\left([u_1,u_2] \times [v_1,v_2]\right)^{2/\rho}$$
$$+ \omega\left([u_1,u_2] \times [s,v_1]\right)^{1/\rho} \omega\left([u_1,u_2] \times [v_1,v_2]\right)^{1/\rho}$$
$$+ \omega\left([s,u_1] \times [v_1,v_2]\right)^{1/\rho} \omega\left([u_1,u_2] \times [v_1,v_2]\right)^{1/\rho}$$
$$+ \omega\left([s,u_1] \times [s,v_1]\right)^{1/\rho} \omega\left([u_1,u_2] \times [v_1,v_2]\right)^{1/\rho}$$
$$\leq 4\left\{\omega\left([s,t]^2\right) \omega\left([u_1,u_2] \times [v_1,v_2]\right)\right\}^{1/\rho}.$$

(Here we used the fact that ω can be taken symmetric.)
Step 3: We now establish the actual estimates and start with (d). For i,j,k distinct, we have

$$\mathbb{E}\left(\left|\int_s^t \mathbf{X}^{i,j}_{s,u} dX^k_u\right|^2\right) = \int\int_{[s,t]^2} \mathbb{E}\left(\mathbf{X}^{i,j}_{s,u} \mathbf{X}^{i,j}_{s,v}\right) dR_k(u,v).$$

By Young's 2D estimate, combined with ρ-variation regularity of the integrand established in step 2, we obtain

$$\mathbb{E}\left(\left|\int_s^t \mathbf{X}^{i,j}_{s,u} dX^k_u\right|^2\right) \leq C\omega\left([s,t]^2\right)^{3/\rho},$$

as desired. The estimate (c) follows from

$$\mathbb{E}\left(\left|\int_s^t \left(X^i_{s,u}\right)^2 dX^k_u\right|^2\right) = \int\int_{[s,t]^2} \mathbb{E}\left(\left(X^i_{s,u}\right)^2 \left(X^i_{s,v}\right)^2\right) dR_k(u,v)$$

and Young's 2D estimate, combined with ρ-variation regularity of the integrand which follows as a special case of Lemma 15.27 (the full generality will be used in the next section). ∎

Corollary 15.29 *Let* $\mathbf{X}, \rho, \omega$ *be as in the last proposition. Then*
(i) there exists a constant $C = C(\rho, N)$ *such that for all* $s < t$ *in* $[0, 1]$ *and* $q \in [1, \infty)$,

$$|d(\mathbf{X}_s, \mathbf{X}_t)|_{L^q} \le C\sqrt{q}\,\omega(s, t)^{\frac{1}{2\rho}};$$ (15.15)

(ii) if $p > 2\rho$ *then* $\|\mathbf{X}\|_{p\text{-var};[0,1]}$ *has a Gaussian tail. More precisely, if* $\omega(0, 1) \le K$ *then there exists* $\eta = \eta(p, \rho, N, K) > 0$ *such that*

$$\mathbb{E}\exp\left(\eta\,\|\mathbf{X}\|^2_{p\text{-var};[0,1]}\right) < \infty;$$ (15.16)

(iii) if $\omega(s, t) \le K|t - s|$ *for all* $s < t$ *in* $[0, 1]$ *we may replace* $\|\mathbf{X}\|_{p\text{-var};[0,1]}$ *in (15.16) by* $\|\mathbf{X}\|_{1/p\text{-Höl};[0,1]}$.

Proof. An immediate consequence of the estimates of Propositions 15.28 and 15.22, applied with (1D) control $(s, t) \mapsto \omega\left([s, t]^2\right)$. ∎

Our next task is to establish suitable moment estimates for the difference of the (first three) iterated integrals of two nice Gaussian processes.

Proposition 15.30 *Let*
(i) $(X, Y) = (X^1, Y^1, \ldots, X^d, Y^d)$ *be a centred continuous Gaussian process with bounded variation sample paths, such that* (X^i, Y^i) *is independent of* (X^j, Y^j) *when* $i \ne j$;
(ii) the covariance of (X, Y) *is of finite* ρ-variation dominated by a 2D control ω, for some $\rho \in [1, 2)$;
(iii) $\mathbf{X} = S_3(X)$ *and* $\mathbf{Y} = S_3(X)$;
(iv) $\varepsilon > 0$ *such that for all* $s < t$ *in* $[0, 1]$,

$$|R_{X-Y}|_{\rho\text{-var};[s,t]^2} \le \varepsilon^2 \omega\left([s, t]^2\right)^{1/\rho}.$$

Then, for $\omega\left([0, 1]^2\right) \le K$, *there exists a constant* $C = C(\rho, K)$ *such that for all* $s < t$ *in* $[0, 1]$ *and* $n = 1, 2, 3$, *we have*

$$|\pi_n(\mathbf{X}_{s,t} - \mathbf{Y}_{s,t})|_{L^2(\mathbb{P})} \le C\varepsilon\omega\left([s, t]^2\right)^{\frac{n}{2\rho}}.$$

Proof. From Proposition 15.62 in the appendix to this chapter, it is enough to prove

(a) $\mathbb{E}\left(\left|\mathbf{X}_{s,t}^{i}-\mathbf{Y}_{s,t}^{i}\right|^{2}\right) \leq \varepsilon\omega\left([s,t]^{2}\right)^{1/\rho}$ for all i;

(b) $\mathbb{E}\left(\left|\mathbf{X}_{s,t}^{i,j}-\mathbf{Y}_{s,t}^{i,j}\right|^{2}\right) \leq C\varepsilon\omega\left([s,t]^{2}\right)^{2/}$ for i,j distinct;

(c) $\mathbb{E}\left(\left|\mathbf{X}_{s,t}^{i,i,j}-\mathbf{Y}_{s,t}^{i,i,j}\right|^{2}\right) \leq C\varepsilon\omega\left([s,t]^{2}\right)^{3/\rho}$ for i,j distinct;

(d) $\mathbb{E}\left(\left|\mathbf{X}_{s,t}^{i,j,k}-\mathbf{Y}_{s,t}^{i,j,k}\right|^{2}\right) \leq C\varepsilon\omega\left([s,t]^{2}\right)^{3/\rho}$ for i,j,k distinct.

The level-one estimate (a) is obvious from

$$\mathbb{E}\left(\left|X_{s,t}^{i}-Y_{s,t}^{i}\right|^{2}\right) \leq \left|R_{X_i-Y_i}\right|_{\rho\text{-var};[s,t]^2} \leq \left|R_{X-Y}\right|_{\rho\text{-var};[s,t]^2} \leq \varepsilon\omega\left([s,t]^{2}\right)^{1/\rho}.$$

For the level-two estimate (b) we fix $i \neq j$. By inserting/subtracting $\int_s^t X_{s,u}^i dY_u^j$ we have

$$\left|\mathbf{X}_{s,t}^{i,j}-\mathbf{Y}_{s,t}^{i,j}\right|_{L^2}^{2} \leq 2\left|\mathbf{X}_{s,t}^{i,j}-\int_s^t X_{s,u}^i dY_u^j\right|_{L^2}^{2} + 2\left|\int_s^t X_{s,u}^i dY_u^j - \mathbf{Y}_{s,t}^{i,j}\right|_{L^2}^{2}$$

$$\leq 2\varepsilon\left\{\left|\int_s^t X_{s,u}^i d\left(\frac{X_u^j - Y_u^j}{\sqrt{\varepsilon}}\right)\right|_{L^2}^{2}\right.$$

$$\left. + \left|\int_s^t \left(\frac{X_{s,u}^i - Y_{s,u}^i}{\sqrt{\varepsilon}}\right) dY_u^j\right|_{L^2}^{2}\right\}$$

$$\leq 2c_1\omega\left([s,t]^{2}\right)^{2/\rho}$$

where the last estimate comes from application of Proposition 15.28 to a (2-dimensional) Gaussian process of the form $\tilde{X} = \left(X^i, \left(X^j - Y^j\right)/\sqrt{\varepsilon}\right)$. (Note that $\tilde{X}$ has indeed independent components and covariance of finite ρ-variation, controlled by ω.) The level-three estimate (d), on the variance of $\mathbf{X}_{s,t}^{i,j,k}-\mathbf{Y}_{s,t}^{i,j,k}$, with i,j,k distinct and fixed, is proved in a similar fashion: after adding/subtracting $\int_{[s,t]} \mathbf{X}_{s,\cdot}^{i,j} dY^k$ we are left with an integral of the form $\int \mathbf{X}_{s,\cdot}^{i,j} d\left(X-Y\right)^k$ and a second one of the form

$$\int \left(\mathbf{X}_{s,u}^{i,j}-\mathbf{Y}_{s,u}^{i,j}\right) dY_u^k = \iint X_{s,\cdot}^i d\left(X^j-Y^j\right) dY^k + \iint \left(X_{s,\cdot}^i-Y_{s,\cdot}^i\right) dY^j dY^k.$$

It then suffices to apply Proposition 15.28 to a (3-dimensional) Gaussian process of the form $\left(X^i, X^j, \left(X^k - Y^k\right)/\sqrt{\varepsilon}\right)$.

It remains to prove the other level-three estimate (c) and we keep $i \neq j$ fixed throughout. We have

$$\left| \mathbf{X}_{s,t}^{i,i,j} - \mathbf{Y}_{s,t}^{i,i,j} \right|_{L^2}^2 \leq 2 \left| \int_s^t \left(X_{s,u}^i \right)^2 d \left(X_u^j - Y_u^j \right) \right|_{L^2}^2$$

$$+ 2 \left| \int_s^t \left\{ \left(X_{s,u}^i \right)^2 - \left(Y_{s,u}^i \right)^2 \right\} dY_u^j \right|_{L^2}^2.$$

The variance of $\int_s^t \left(X_{s,u}^i \right)^2 d \left(X_u^j - Y_u^j \right)$ can be written as a 2D Young integral and by Lemma 15.27 and 2D Young estimates we obtain

$$\left| \int_s^t \left(X_{s,u}^i \right)^2 d \left(X_u^j - Y_u^j \right) \right|_{L^2}^2 \leq c_2 \varepsilon \omega \left(s, t \right)^{3/\rho'}.$$

To deal with the other term, we first note that, again using Lemma 15.27, the ρ-variation of

$$(u, v) \mapsto g(u,v) \equiv \frac{1}{\varepsilon} \mathbb{E} \left[\left\{ \left(X_{s,u}^i \right)^2 - \left(Y_{s,u}^i \right)^2 \right\} \left\{ \left(X_{s,v}^i \right)^2 - \left(Y_{s,v}^i \right)^2 \right\} \right]$$

$$= \mathbb{E} \left[\left(\frac{X_{s,u}^i - Y_{s,u}^i}{\sqrt{\varepsilon}} \right) \left(X_{s,u}^i + Y_{s,u}^i \right) \left(\frac{X_{s,v}^i - Y_{s,v}^i}{\sqrt{\varepsilon}} \right) \left(X_{s,v}^i + Y_{s,v}^i \right) \right]$$

over $[s,t]^2$ is controlled by a constant times $\omega \left([s,t]^2 \right)^2$; then, again using 2D Young estimates with $1/\rho + 1/\rho > 1$, we see that

$$\left| \int_s^t \left\{ \left(X_{s,u}^i \right)^2 - \left(Y_{s,u}^i \right)^2 \right\} dY_u^j \right|_{L^2}^2 = \varepsilon \int_{[s,t]^2} g(u,v) dR_{Y^j}(u,v)$$

$$\leq c_4 \varepsilon \omega \left([s,t]^2 \right)^{3/\rho}.$$

The proof is then finished. ∎

Corollary 15.31 *Let* $\mathbf{X} = S_3(X), \mathbf{Y} = S_3(X), \omega, K, \rho$ *as in the previous proposition and in particular*

$$\left| R_{X-Y} \right|_{\rho\text{-var};[s,t]^2} \leq \varepsilon^2 \omega \left([s,t]^2 \right)^{1/\rho}. \tag{15.17}$$

Then
(i) there exists a constant $C = C(\rho, K)$ *such that for all* $s < t$ *in* $[0,1]$, $q \in [1, \infty)$ *and* $n = 1, 2, 3$,

$$\left| \pi_n \left(\mathbf{Y}_{s,t} - \mathbf{X}_{s,t} \right) \right|_{L^q(\mathbb{P})} \leq C q^{\frac{n}{2}} \varepsilon \omega \left([s,t]^2 \right)^{\frac{n}{2\rho}};$$

(ii) if $p > 2\rho$ *then there exists a constant* $C' = C'(p, \rho, K)$ *such that*

$$\left| d_{p\text{-var};[0,1]} \left(\mathbf{X}, \mathbf{Y} \right) \right|_{L^q(\mathbb{P})} \leq C' \max \left\{ \varepsilon^{1/3}, \varepsilon \right\} \sqrt{q} \tag{15.18}$$

and for $n = 1, 2, 3$ we have

$$\left| \rho^{(n)}_{p\text{-var};[0,1]} (\mathbf{X}, \mathbf{Y}) \right|_{L^q(\mathbb{P})} \leq C' q^{\frac{n}{2}} \varepsilon; \qquad (15.19)$$

(iii) if $\omega(s,t) \leq K |t - s|$ for all $s < t$ in $[0,1]$ then $d_{p\text{-var};[0,1]}, \rho^{(n)}_{p\text{-var},[0,1]}$ in (15.18), (15.19) may be replaced by $d_{1/p\text{-Höl};[0,1]}, \rho^{(n)}_{1/p\text{-Höl};[0,1]}$ respectively.

Proof. An immediate consequence of the estimates of Propositions 15.30 and 15.24, applied with (1D) control $(s,t) \mapsto \omega\left([s,t]^2\right)$ and $\rho' \in (\rho, 2)$. For (ii), (iii) we may take $\rho' = \rho + \min(p, 4)/2$ (so that $2\rho < 2\rho' < p$) so that C' has no explicit dependence on ρ'. ∎

Remark 15.32 Assume the covariance of (X, Y) is of finite ρ-variation dominated by a 2D control ω, for some $\rho \in [1, 2)$. Then, by interpolation, for all $\rho' > \rho$

$$\begin{aligned}
|R_{X-Y}|_{\rho'\text{-var};[s,t]^2} &\leq |R_{X-Y}|_\infty^{1-\rho/\rho'} \cdot |R_{X-Y}|_{\rho\text{-var};[s,t]^2}^{\rho/\rho'} \\
&\leq |R_{X-Y}|_\infty^{1-\rho/\rho'} \, \omega\left([s,t]^2\right)^{1/\rho'} .
\end{aligned}$$

But ω also controls the ρ'-variation of the covariance of (X, Y); indeed,

$$\begin{aligned}
\left| R_{(X,Y)} \right|_{\rho'\text{-var};[s,t] \times [uv]}^{\rho'} &\leq \left| R_{(X,Y)} \right|_{\rho\text{-var};[s,t] \times [uv]}^{\rho'} \\
&\leq \left| R_{(X,Y)} \right|_{\rho\text{-var};[0,1]^2}^{\rho'-\rho} \left| R_{(X,Y)} \right|_{\rho\text{-var};[s,t] \times [u,v]}^{\rho}
\end{aligned}$$

and hence, with $c = K^{\rho'/\rho - 1}$, where $\left| R_{(X,Y)} \right|_{\rho\text{-var};[0,1]^2}^{\rho} \leq \omega\left([0,1]^2\right) \leq K$,

$$\left| R_{(X,Y)} \right|_{\rho'\text{-var};[s,t] \times [uv]}^{\rho'} \leq c\omega([s,t] \times [u,v]) .$$

It follows that Corollary 15.31 may be applied with parameter ρ', control $c\omega$ and

$$\varepsilon^2 = |R_{X-Y}|_\infty^{1-\rho/\rho'} . \qquad \square$$

15.3.3 Enhanced Gaussian process

The *uniform* estimates of the previous section, proved under the assumption of bounded variation sample paths, allow for a simple passage to the limit. Indeed, given a (d-dimensional) continuous Gaussian process X, whose sample paths are not of bounded variation, but whose covariance has finite ρ-variation, $\rho \in [1, 2)$, we may consider suitably smooth approximations (X^n) for which

$$\sup_{n,m} \omega^{n,m}\left([0,1]^2\right) \leq K$$

where $\omega^{n,m}$ is 2D control which controls the ρ-variation of $R_{(X^n, X^m)}$. (In fact, we have already seen that the above supremum bound is satisfied for either piecewise linear or mollifier approximations.) Corollary 15.31 then implies that $S_3(X_n)$ is Cauchy-in-probability in $C^{0,p\text{-var}}\left([0,1], G^3(\mathbb{R}^d)\right)$, for any $p > 2\rho$, which leads us to the following result.

Theorem 15.33 (enhanced Gaussian process) *Assume* $X = (X^1, \ldots, X^d)$ *is a centred continuous Gaussian process with independent components. Let* $\rho \in [1,2)$ *and assume the covariance of* X *is of finite* ρ-*variation dominated by a 2D control* ω *with* $\omega\left([0,1]^2\right) \leq K$. *Then, there exists a unique continuous* $G^3(\mathbb{R}^d)$-*valued process* $\mathbf{X}$, *such that:*
(i) $\mathbf{X}$ *"lifts" the Gaussian process* X *in the sense* $\pi_1(\mathbf{X}_t) = X_t - X_0$;
(ii) there exists $C = C(\rho)$ *such that for all* $s < t$ *in* $[0,1]$ *and* $q \in [1, \infty)$,

$$|d(\mathbf{X}_s, \mathbf{X}_t)|_{L^q} \leq C\sqrt{q}\,\omega\left([s,t]^2\right)^{\frac{1}{2\rho}}; \tag{15.20}$$

(iii) (Fernique-estimates) for all $p > 2\rho$ *and* $\omega\left([0,1]^2\right) \leq K$, *there exists* $\eta = \eta(p, \rho, K) > 0$, *such that*

$$\mathbb{E}\left(\exp\left(\eta\,\|\mathbf{X}\|^2_{p\text{-var};[0,1]}\right)\right) < \infty$$

and if $\omega\left([s,t]^2\right) \leq K|t-s|$ *for all* $s < t$ *in* $[0,1]$, *then we may replace* $\|\mathbf{X}\|_{p\text{-var};[0,1]}$ *by* $\|\mathbf{X}\|_{1/p\text{-Höl};[0,1]}$;
(iv) the lift $\mathbf{X}$ *is natural in the sense that it is the limit of* $S_3(X^n)$ *where* X^n *is any sequence of piecewise linear or mollifier approximations to* X *such that* $d_\infty(X^n, X)$ *converges to 0 almost surely.*

Definition 15.34 A $G^3(\mathbb{R}^d)$-valued process $\mathbf{X}$ as constructed above is called an **enhanced Gaussian process**; if we want to stress the underlying Gaussian process we call $\mathbf{X}$ the **natural lift** of X. Sample path realizations of $\mathbf{X}$ are called **Gaussian rough paths**, as is motivated by
(i) for $\rho \in [1, 3/2)$, we see that $\mathbf{X}$ has almost surely finite p-variation, for any $p \in (2\rho, 3)$, and hence so does its projection to $G^2(\mathbb{R}^d)$, which is therefore almost surely a geometric p-rough path;
(ii) for $\rho \in [3/2, 2)$, we see that $\mathbf{X}$ has almost surely finite p-variation, for any $p \in (2\rho, 4)$, and is therefore almost surely a geometric p-rough path. □

Remark 15.35 With the notation of the above theorem, if X has a.s. sample paths of finite $[1,2)$-variation, $\mathbf{X}$ coincides with the canonical lift obtained by iterated Young integration of X. If

$$\tilde{\mathbf{X}} = (1, \pi_1(\mathbf{X}), \pi_2(\mathbf{X})) \in C_0^{0,p\text{-var}}\left([0,1], G^2(\mathbb{R}^d)\right) \text{ a.s.}$$

for $p < 3$ then $\tilde{\mathbf{X}}$ is a geometric p-rough path and $\mathbf{X}$ coincides with the Lyons lift of $\tilde{\mathbf{X}}$. Let us also observe that only point (iv) guarantees the uniqueness of the lift. For $p \in [3,4)$, if $e_1, \ldots, e_d$ is a basis of $\mathbb{R}^d$,

$$\tilde{\mathbf{X}}_t := \mathbf{X}_t \otimes \exp\left(t\left[e_1, [e_1, e_2]\right]\right)$$

would also satisfy conditions (i) to (iii). Similarly, for $p \in [2,3)$, the Lyons lift of the projection to $G^2\left(\mathbb{R}^d\right)$ of $t \mapsto \mathbf{X}_t \otimes \exp\left(t\left[e_1, e_2\right]\right)$ would also satisfy conditions (i) to (iii). □

Proof. Fix a mollifier function $\varphi\left(\cdot\right)$ and set $d\mu_n\left(u\right) = n\varphi\left(nu\right)du$. Define smooth approximations to X, componentwise by convolution against $d\mu_n$; that is,[9]

$$t \mapsto X_t^n = \int X_{t-u}\, d\mu_n\left(u\right),$$

so that X_t^n is a smooth function in t. From Proposition 15.14 there exists $c_1 = c_1\left(\rho\right)$ so that

$$\sup_{n,m}\left|R_{\left(X^n, X^m\right)}\right|^\rho_{\rho\text{-var};[0,1]^2} \le c_1\left|R\right|^\rho_{\rho\text{-var};[0,1]^2} =: c_1 K$$

and from Corollary 15.31 (plus Remark 15.32) we see that there exists $\theta > 0$ and $c_2 = c_2\left(p, q, K, \theta\right)$ so that

$$\left|d_{p\text{-var};[0,1]}\left(S_3\left(X_n\right), S_3\left(X_m\right)\right)\right|_{L^q\left(\mathbb{P}\right)} \le c_1\left|R_{X_n - X_m}\right|^\theta_\infty.$$

It follows that $S_3\left(X_n\right)$ is Cauchy-in-probability as a sequence of $C_o^{0,p\text{-var}}$-valued random variables[10] and so there exists $\mathbf{X} \in C_o^{0,p\text{-var}}\left([0,1], G^3\left(\mathbb{R}^d\right)\right)$ such that $d_{p\text{-var}}\left(S_3\left(X^{D_n}\right), \mathbf{X}\right) \to 0$ in probability and from the uniform estimates from Corollary 15.29 also in L^q for all $q \in [1, \infty)$. From Corollary 15.29 we have the estimate

$$\left|d\left(S_3\left(X_n\right)_s, S_3\left(X_n\right)_t\right)\right|_{L^q} \le C\sqrt{q}\omega^n\left(\left[s,t\right]^2\right)^{\frac{1}{2\rho}} \tag{15.21}$$

for any 2D control ω^n which controls the ρ-variation of R_{X_n} and in particular for $\omega^n = \omega^{\mu_n, \mu_n}$, the "$\mu_n$-convolution of ω" from Proposition 15.14. Sending $n \to \infty$ then shows that

$$\left|d\left(\mathbf{X}_s, \mathbf{X}_t\right)\right|_{L^q} \le C\sqrt{q}\omega\left(\left[s,t\right]^2\right)^{\frac{1}{2\rho}}.$$

Obviously, the increments $\mathbf{X}_{s,t} = \mathbf{X}_s^{-1} \otimes \mathbf{X}_t$ are limits (in probability, say) of $S_3\left(X^n\right)_{s,t}$ and so, from Proposition 15.19 and closedness of the

---

[9] We could also use piecewise linear (instead of mollifier) approximations.

[10] A Cauchy criterion for convergence in probability of r.v.s with values in a Polish space is an immediate generalization of the corresponding real-valued case.

Wiener–Itô chaos under convergence in probability, $\pi_n\left(\mathbf{X}_{s,t}\right)$ is indeed an element of the nth (not necessarily homogenous) Wiener–Itô chaos. The statements of (ii),(iii) then follow directly from Proposition 15.22, applied with 1D control $s,t \mapsto \omega\left(\left[s,t\right]^2\right)$.

For (iv), as of yet, our construction of $\mathbf{X}$ may depend on the particular mollifier function φ. Assume now that $\left(X^1,Y^1,\ldots,X^d,Y^d\right)$ is Gaussian with independent $\left\{\left(X^i,Y^i\right):i=1,\ldots,d\right\}$ such that Y has bounded variation sample paths. Then

$$\left|R_{(X^n,Y)}\right|_{\rho\text{-var};[0,1]^2} \leq \left|R_{(X^n,X)}\right|_{\rho\text{-var};[0,1]^2} + \left|R_{(X,Y)}\right|_{\rho\text{-var};[0,1]^2}$$

which is finite, whenever $R_X \in C^{\rho\text{-var}}$, uniformly in n and uniformly over all Y given by (componentwise) piecewise linear or mollifier approximations to X. (This follows from Propositions 15.11 and 15.14 respectively.) We can therefore, as in part (i), pass to the limit in

$$\left|d_{p\text{-var};[0,1]}\left(S_3\left(X_n\right),S_3\left(Y\right)\right)\right|_{L^q(\mathbb{P})} \leq c_4\left|R_{X_n-Y}\right|_\infty^\theta$$

to learn that

$$\left|d_{p\text{-var};[0,1]}\left(\mathbf{X},S_3\left(Y\right)\right)\right|_{L^q(\mathbb{P})} \leq c_4\left|R_{X-Y}\right|_\infty^\theta.$$

When applied to $Y = X^{D_n}$ with $|D_n| \to 0$ resp. $Y = X^{\mu_n}$ for any $\mu_n \to \delta_0$. the right-hand side above tends to zero and the proof of (iv) is finished. ∎

Theorem 15.33 asserts in particular that d-dimensional Brownian motion can be naturally lifted to an enhanced Gaussian process, easily identified as enhanced Brownian motion (in view of (iv) and the results of Section 13.3.3). Other examples are obtained by considering d independent (continuous, centred) Gaussian processes, each of which satisfies the condition that its covariance is of finite ρ-variation, for some $\rho < 2$. For example (cf. Proposition 15.5) one may take d independent copies of fractional Brownian motion: the resulting $\mathbb{R}^d$-valued fractional Brownian motion B^H can be lifted to an enhanced Gaussian process ("enhanced fractional Brownian motion", $\mathbf{B}^H$) provided $H > 1/4$. Further examples are constructed by consulting the list of Gaussian processes in Section 15.2.

Exercise 15.36 *In the context of Theorem 15.33,*
(i) show that there exists $\eta = \eta\left(\rho,K\right) > 0$ such that

$$\sup_{0 \leq s < t \leq 1} \mathbb{E}\left(\exp\left(\eta\left[\frac{d\left(\mathbf{X}_s,\mathbf{X}_t\right)}{\omega\left(s,t\right)^{\frac{1}{2\rho}}}\right]^2\right)\right) < \infty;$$

(ii) define a deterministic time-change from $[0,1]$ onto itself, given by

$$\tau\left(t\right) = \left|R_X\right|_{\rho\text{-var};[0,t]^2}^\rho / \left|R_X\right|_{\rho\text{-var};[0,1]^2}^\rho$$

and define the Gaussian process $\left(\tilde{X}_t : 0 \leq t \leq 1 \right)$ *by requiring that* $\tilde{X}_{\tau(t)} = X_t$. *Show that* $\tilde{X}$ *admits a natural lift* $\tilde{\mathbf{X}}$ *so that*

$$\tilde{\mathbf{X}}_{\tau(t)} = \mathbf{X}_t$$

and such that

$$\sup_{0 \leq s < t \leq 1} \mathbb{E} \left(\exp \left(\eta \left[\frac{d \left(\tilde{\mathbf{X}}_s, \tilde{\mathbf{X}}_t \right)}{|t - s|^{\frac{1}{2\rho}}} \right]^2 \right) \right) < \infty;$$

(iii) deduce from the results of Section A.4, Appendix A, that for a suitable constant c,

$$\mathbb{E} \left(\exp \left(\eta c \left\| \tilde{\mathbf{X}} \right\|^2_{\psi_{2\rho,\rho}\text{-var};[0,1]} \right) \right) < \infty$$

and then[11]

$$\mathbb{E} \left(\exp \left(\eta c \left\| \mathbf{X} \right\|^2_{\psi_{2\rho,\rho}\text{-var};[0,1]} \right) \right) < \infty.$$

Solution. (i) is a consequence of (15.20), cf. Lemma A.17. (ii) Assume $\omega(0,1) = 1$ for simplicity. By definition of τ and $\tilde{X}$ we see that

$$\begin{aligned} |R_{\tilde{X}}|^\rho_{\rho\text{-var};[\tau(s),\tau(t)]^2} &\leq |R_{\tilde{X}}|^\rho_{\rho\text{-var};[0,\tau(t)]^2} - |R_{\tilde{X}}|^\rho_{\rho\text{-var};[0,\tau(s)]^2} \\ &= |R_X|^\rho_{\rho\text{-var};[0,t]^2} - |R_X|^\rho_{\rho\text{-var};[0,s]^2} \\ &= |R_X|^\rho_{\rho\text{-var};[0,1]^2} \left(\tau(t) - \tau(s) \right), \end{aligned}$$

which implies that $\tilde{X}$ has finite ρ-variation controlled by $\tilde{\omega} = |R_{\tilde{X}}|^\rho_{\rho\text{-var};[\cdot,\cdot]\times[\cdot,\cdot]}$. Clearly then, $\tilde{\omega} \left([s,t]^2 \right) \leq K |t - s|$ and the claimed estimate follows from (15.20), applied with $\tilde{\omega}$.

(iii) This is a straightforward consequence of the results of Section A.4, Appendix A, and invariance of (generalized) variation norms under reparametrization. □

Theorem 15.37 *Let* $(X,Y) = \left(X^1, Y^1, \ldots, X^d, Y^d \right)$ *be a centred continuous Gaussian process such that* $\left(X^i, Y^i \right)$ *is independent of* $\left(X^j, Y^j \right)$ *when* $i \neq j$. *Let* $\rho \in [1,2)$ *and assume the covariance of* (X,Y) *is of finite* ρ-*variation, controlled by a 2D control* ω *with* $\omega \left([0,1]^2 \right) \leq K$, *and write* $\mathbf{X}$ *and* $\mathbf{Y}$ *for the natural lift of* X *and* Y. *Assume also that*

$$|R_{X-Y}|_{\rho\text{-var};[s,t]^2} \leq \varepsilon^2 \omega \left([s,t]^2 \right)^{1/\rho}.$$

---

[11] This sharpens the statement on finite p-variation, $p > 2\rho$, and is relevant (at least when $\rho = 1$) as it allows for unique RDE solutions driven by $\mathbf{X}$ along Lip^2-vector fields.

Then
(i) there exists a constant $C = C(\rho, K)$ such that for all $s < t$ in $[0,1]$, $q \in [1, \infty)$ and $n = 1, 2, 3$,

$$\left| \pi_n \left(\mathbf{Y}_{s,t} - \mathbf{X}_{s,t} \right) \right|_{L^q(\mathbb{P})} \leq C q^{\frac{n}{2}} \varepsilon \omega \left([s,t]^2 \right)^{\frac{n}{2\rho}} ; \qquad (15.22)$$

(ii) if $p > 2\rho$ then there exists a constant $C' = C'(p, \rho, K)$ such that

$$\left| d_{p\text{-var};[0,1]} \left(\mathbf{X}, \mathbf{Y} \right) \right|_{L^q(\mathbb{P})} \leq C' \max \left\{ \varepsilon^{1/3}, \varepsilon \right\} \sqrt{q} \qquad (15.23)$$

and for $n = 1, 2, 3$ we have

$$\left| \rho^{(n)}_{p\text{-var};[0,1]} \left(\mathbf{X}, \mathbf{Y} \right) \right|_{L^q(\mathbb{P})} \leq C' q^{\frac{n}{2}} \varepsilon; \qquad (15.24)$$

(iii) if $\omega(s,t) \leq K |t - s|$ for all $s < t$ in $[0,1]$ then $d_{p\text{-var};[0,1]}, \rho^{(n)}_{p\text{-var};[0,1]}$ in (15.18), (15.19) may be replaced by $d_{1/p\text{-Höl};[0,1]}, \rho^{(n)}_{1/p\text{-Höl};[0,1]}$ respectively.

Proof. The statements are precisely those of Corollary 15.31 but without assuming that $\mathbf{X}, \mathbf{Y}$ are the step-three lift of processes with bounded-variation sample paths. The proof is then completed with the same passage to limit, along the lines of the previous proof. $\blacksquare$

Remark 15.38 As already noted in Remark 15.32, estimates (15.22), (15.23), (15.24) of Theorem 15.37 apply in particular after replacing ρ by $\rho' \in (\rho, 2)$, such that $2\rho < 2\rho' < p$, and after replacing ε^2 by $|R_{X-Y}|_\infty^{1-\rho/\rho'}$. In particular, there exist positive constants θ, C depending only on ρ, ρ', K, p such that

$$\left| d_{p\text{-var};[0,1]} \left(\mathbf{X}, \mathbf{Y} \right) \right|_{L^q(\mathbb{P})} \leq C |R_{X-Y}|_\infty^\theta \sqrt{q}. \qquad \square$$

15.4 The Young–Wiener integral

Given a suitable d-dimensional Gaussian process X, assuming in particular finite ρ-variation of the covariance for some $\rho \in [1, 2)$, we have constructed a Gaussian rough path $\mathbf{X}$ of finite p-variation, for any $p > 2\rho$. At the same time we have seen that for any $h \in \mathcal{H}$, the associated Cameron–Martin space has finite ρ-variation. Clearly, integrals of the form $\int h \otimes dh$ are well-defined Young integrals. However, cross integrals of the form

$$\int h dX$$

are only well-defined as Young integrals if $1/\rho + 1/p > 1$, which would require $\rho \in [1, 3/2)$. However, we can define the integral probabilistically, say in L^2-sense, and it will suffice to look at the scalar-valued case. Let us remark that such cross integrals arise if we consider perturbations of the random variable $\mathbf{X}(\cdot)$ in Cameron–Martin directions or when dealing with non-centred Gaussian processes. We have

Proposition 15.39 (Young–Wiener integral) *Assume X is a continuous, centred Gaussian with covariance R of finite ρ-variation. Let $h \in C^{q\text{-var}}([0,1],\mathbb{R})$, with $q^{-1} + \rho^{-1} > 1$. Then, for any piecewise linear or mollifier approximation (X^n) to X, the indefinite integral*

$$\int_0^t h\,dX^n$$

converges, for each $t \in [0,1]$, in L^2 and its common limit is denoted by $\int_0^t h\,dX$. For all $s < t$ in $[0,1]$, we have the Young–Wiener isometry

$$\mathbb{E}\left(\left|\int_s^t h_u\,dX_u\right|^2\right) = \int_{[s,t]^2} h_u h_v\,dR(u,v),$$

and if $h(s) = 0$ we have the Young–Wiener estimate

$$\mathbb{E}\left(\left|\int_s^t h_u\,dX_u\right|^2\right) \le C_{\rho,q}\,|h|_{q\text{-var};[s,t]}^2\,|R|_{\rho\text{-var};[s,t]^2}. \tag{15.25}$$

At last, the process $t \mapsto \int_0^t h\,dX$ admits a continuous version with sample path of finite p-variation, for any $p > 2\rho$.

Proof. When X has (piecewise) smooth sample paths, the Young–Wiener isometry is obvious from

$$\mathbb{E}\left(\left|\int_s^t h_u\,dX_u\right|^2\right) = \mathbb{E}\left(\int_s^t \int_s^t h_u h_v\,dX_u\,dX_v\right) = \int_{[s,t]^2} h_u h_v\,dR(u,v).$$

Finite q-variation of h implies that $h \otimes h$ also has finite q-variation (now in 2D sense) and from the Young 2D estimates it follows that

$$
\begin{aligned}
\mathbb{E}\left(\left|\int_s^t h_u\,dX_u\right|^2\right) &\le\ c_1\,|h \otimes h|_{q\text{-var};[s,t]}\,|R|_{\rho\text{-var};[s,t]^2} \\
&\le\ c_2\,|h|_{q\text{-var};[s,t]}^2\,|R|_{\rho\text{-var};[s,t]^2}
\end{aligned}
$$

where c_1, c_2 depend on p, ρ. Replace X by $X^n - X^m$, then piecewise linear or mollifier approximation yields

$$\sup_{t\in[0,1]} \mathbb{E}\left(\left|\int_0^t h\,dX^n - \int_0^t h\,dX^m\right|^2\right) \le c_2\,|h|_{q\text{-var};[0,1]}^2\,|R_{X^n-X^m}|_{\rho\text{-var};[0,1]^2}.$$

In fact, by choosing $\rho' > \rho$ small enough (so that $1/q + 1/\rho' > 1$) we can use interpolation to see that (constants may now also depend on h and ρ')

$$\sup_{t \in [0,1]} \mathbb{E} \left(\left| \int_0^t h \, dX^n - \int_0^t h \, dX^m \right|^2 \right)$$

$$\leq \quad c_3 \left| R_{X^n - X^m} \right|_{\rho'\text{-var};[0,1]}^2$$

$$\leq \quad c_4 \left| R_{X^n - X^m} \right|_{\infty;[0,1]^2}^{1-\rho/\rho'} \sup_{n,m} \left| R_{(X^n,X^m)} \right|_{\rho\text{-var};[0,1]^2}^{\rho/\rho'}$$

$$\leq \quad c_5 \left| R_{X^n - X^m} \right|_{\infty;[0,1]^2}^{1-\rho/\rho'},$$

where the last estimate is justified exactly as in step (i) of the proof of Theorem 15.33; that is, by means of Proposition 15.14. It follows that $\left(\int_0^t h \, dX^n : n \in \mathbb{N} \right)$ is Cauchy in $L^2 (\mathbb{P})$ and hence convergent. Then, similar to step (iii) of the aforementioned proof, one sees that this limit does not depend on a particular approximation. At last, the p-variation regularity is the content of Exercise 15.41 below.   ∎

Remark 15.40 When X is Brownian motion, $dR = \delta_{\{s=t\}}$ and we recover the Itô isometry for Itô–Wiener integrals.   □

Exercise 15.41 *In the context of Proposition 15.39, assuming in particular that X has covariance of finite ρ-variation controlled by some 2D control ω, show that $\int h \, dX$ admits a version which has finite p-variation for any $p > 2\rho$.*

Solution. Since $I_t - I_s := \int_s^t h \, dX$ is Gaussian, we have

$$\left| I_t - I_s \right|_{L^r(\mathbb{P})} \leq \quad \left| \int_s^t h_{s,u} \, dX_u \right|_{L^r(\mathbb{P})} + |h|_\infty |X_{s,t}|_{L^r(\mathbb{P})}$$

$$\leq \quad c_1 \left[\left| \int_s^t h_{s,u} \, dX_u \right|_{L^2(\mathbb{P})} + |h|_\infty |X_{s,t}|_{L^2(\mathbb{P})} \right]$$

$$\leq \quad c_2 \left(|h|_{q\text{-var};[s,t]} + |h|_\infty \right) |R|_{\rho\text{-var};[s,t]^2}^{1/2}$$

where c_1, c_2 may depend on r, ρ, q. Setting $\omega (s,t) := |R|_{\rho\text{-var};[s,t]^2}^{\rho}$ yields

$$\left| I_t - I_s \right|_{L^r(\mathbb{P})} = O \left(|\omega (0,t) - \omega (0,s)|^{\frac{1}{2\rho}} \right)$$

so that by Kolmogorov's criterion $J = I \circ \omega (0, \cdot)^{-1}$ is Hölder continuous with any exponent less than $1/2\rho$. It follows that J (and then I) have the claimed p-variation regularity, $p > 2\rho$.   □

15.5   Strong approximations

15.5.1   *Piecewise linear approximations*

We now establish the rate of convergence for piecewise linear approximations with focus. Those results are here for clarity, as we only need to put pieces together to obtain them.

Theorem 15.42 *Assume that $X = \left(X^1, \ldots, X^d\right)$ is a centred continuous Gaussian process with independent components and covariance R of finite ρ-variation, $\rho \in [1, 2)$, controlled by some 2D control ω. Fix an arbitrary $p \in (2\rho, 4)$, $\eta \in \left(0, \frac{1}{2\rho} - \frac{1}{p}\right)$ and write $\mathbf{X}$ for the natural lift of X. Then*

(i) if $\omega\left([0,1]^2\right) \leq K$, there exists some constant $C_1 = C\left(\rho, p, K, \theta\right)$, such that for all $D \in \mathcal{D}[0,1]$ and $q \in [1, \infty)$,

$$\left|d_{p\text{-var};[0,1]}\left(\mathbf{X}, S_3\left(X^D\right)\right)\right|_{L^q(\mathbb{P})} \leq C_1 \sqrt{q} \max_{t_i \in D} \omega\left([t_i, t_{i+1}]^2\right)^{\eta/3}, \quad (15.26)$$

and also

$$\forall n \in \{1,2,3\}: \left|\rho^{(n)}_{p\text{-var};[0,1]}\left(\mathbf{X}, S_3\left(X^D\right)\right)\right|_{L^q(\mathbb{P})} \leq C_1 q^{\frac{n}{2}} \max_{t_i \in D} \omega\left([t_i, t_{i+1}]^2\right)^{\eta};$$

(ii) if $\omega(s,t) \leq K|t-s|$ for all $s < t$ in $[0,1]$ then $d_{p\text{-var};[0,1]}, \rho^{(n)}_{p\text{-var};[0,1]}$ in the above estimates may be replaced by $d_{1/p\text{-Höl};[0,1]}, \rho^{(n)}_{1/p\text{-Höl};[0,1]}$ respectively.

Remark 15.43 If $\rho \in [1, 3/2)$ we can take $p \in (2\rho, 3)$ and then only need a step-2 lift. Since the power $1/3$ in (15.26) is readily traced back to (15.13) we see that, in the case $\rho \in [1, 3/2)$, we have

$$\left|d_{1/p\text{-Höl};[0,1]}\left(\mathbf{X}, S_3\left(X^D\right)\right)\right|_{L^q(\mathbb{P})} \leq C\sqrt{q}\,|D|^{\eta/2}.$$

In particular, the above estimates applied to enhanced Brownian motion are in precise agreement with those obtained earlier (Corollary 13.21) by direct computation in a Brownian context. $\quad\Box$

Proof. Pick $\rho' \in (\rho, p/2)$ and note that, following Remark 15.32,

$$\omega_D(A) := \left|R_{(X,X^D)}\right|^{\rho}_{\rho\text{-var};A} \quad \text{(any rectangle } A \subset [0,1]^2\text{)}$$

also controls the ρ'-variation of $\left(X, X^D\right)$, while interpolation gives

$$\left|R_{X-X^D}\right|_{\rho'\text{-var};[s,t]^2} \leq c_2 \left|R_{X-X^D}\right|^{1-\rho/\rho'}_{\infty} \omega_D\left([s,t]^2\right)^{1/\rho'}$$

where we note that $|R_{X-X^D}|_\infty = \sup_{s,t\in[0,1]} \mathbb{E}\left[\left(X_s - X_s^D\right)\left(X_t - X_t^D\right)\right]$ is bounded by

$$\sup_{t\in[0,1]} \mathbb{E}\left[\left(X_t - X_t^D\right)^2\right] \leq 2\max_{t_i\in D} \mathbb{E}\left(\left|X_{t_i,t_{i+1}}\right|^2\right)$$

$$\leq 2\max_{t_i\in D} \omega_D\left(\left[t_i, t_{i+1}\right]^2\right)^{1/\rho}.$$

Proposition 15.13, applied with ρ' instead of ρ and $\varepsilon^2 = c_3 \max_{t_i\in D} \omega$ $\left(\left[t_i, t_{i+1}\right]^2\right)^{\frac{1}{\rho}-\frac{1}{\rho'}}$, yields

$$\left|d_{p\text{-var}}\left(\mathbf{X}, S_3\left(X^D\right)\right)\right|_{L^q(\mathbb{P})} \leq c_1 q^{1/2} \max_{t_i\in D} \omega_D\left(\left[t_i, t_{i+1}\right]^2\right)^{\frac{1}{6\rho}-\frac{1}{6\rho'}},$$

and for $k = 1, 2, 3$,

$$\left|\rho_{p\text{-var}}^{(k)}\left(\mathbf{X}, S_3\left(X^D\right)\right)\right|_{L^q(\mathbb{P})} \leq c_1 q^{k/2} \max_{t_i\in D} \omega_D\left(\left[t_i, t_{i+1}\right]^2\right)^{\frac{1}{2\rho}-\frac{1}{2\rho'}}.$$

We conclude the p-variation estimates by observing that $\omega_D\left(\left[t_i, t_{i+1}\right]^2\right) \leq c_2\omega\left(\left[t_i, t_{i+1}\right]^2\right)$ (see Proposition 15.11 and Lemma 15.12). At last, the Hölder estimate is obtained similarly. ∎

Exercise 15.44 *Assume $D_n = \left\{\frac{k}{2^n}, 0 \leq k \leq 2^n\right\}$. Show that under the assumptions of Theorem 15.42, part (ii),*

$$d_{1/p\text{-Höl};[0,1]}\left(\mathbf{X}, S_3\left(X^D\right)\right) \to 0 \text{ a.s.}$$

Solution. From Theorem 15.42, there exists $\theta > 0$ such that

$$\left|d_{1/p\text{-Höl};[0,1]}\left(\mathbf{X}, S_3\left(X^D\right)\right)\right|_{L^q(\mathbb{P})} \leq C 2^{-n\theta}\sqrt{q}.$$

A standard Borell–Cantelli argument finishes the proof. □

15.5.2 Mollifier approximations

Theorem 15.45 *Assume that X is a centred $\mathbb{R}^d$-valued continuous Gaussian process with independent components and covariance $R = R_X$ of finite ρ-variation, $\rho \in [1,2)$, so that there exists a natural lift $\mathbf{X}$, with p-variation sample paths for any $p \in (2\rho, 4)$. Fix a mollifier function $\varphi(\cdot) : \mathbb{R} \to \mathbb{R}$, set $d\mu_n(u) = n\varphi(nu)\,dt$ and define (componentwise) approximations by*

$$t \mapsto X_t^n = \int X_{t-u}\,d\mu_n(u).$$

Then

$$\sup_{q\in[1,\infty)} \frac{\left|d_{p\text{-var};[0,1]}\left(\mathbf{X}, S_3\left(X^n\right)\right)\right|_{L^q}}{\sqrt{q}} \to 0 \text{ as } n \to \infty.$$

Proof. Similar to the arguments of step 1 in Theorem 15.33. The details are left to the reader. ∎

15.5.3 Karhunen–Loève approximations

Any $\mathbb{R}^d$-valued centred Gaussian process $X = \left(X^1, \ldots, X^d\right)$ with continuous sample paths gives rise to an abstract Wiener space $(E, \mathcal{H}, \mathbb{P})$ with $E = C\left([0,1], \mathbb{R}^d\right)$ and $\mathcal{H} \subset C\left([0,1], \mathbb{R}^d\right)$. From general principles (cf. Section D.3 of Appendix D), for any fixed orthonormal basis $\left(h^k : k \in \mathbb{N}\right) \subset \mathcal{H}$, there is a Karhunen–Loève expansion (a.s. and L^2-convergent)

$$X = \sum_{k \in \mathbb{N}} Z_k h^k \,,$$

where Z_k, the image of h_k under the Payley–Wiener map, is a sequence of independent standard normal random variables. With our standing assumptions of independence of its component processes, each component gives rise to an abstract Wiener space on $C\left([0,1], \mathbb{R}\right)$ with Cameron–Martin space $\mathcal{H}^i$ and $\mathcal{H} \cong \oplus_{i=1}^d \mathcal{H}^i$. The 1-dimensional considerations of Section 15.2.5 then apply without changes to the d-dimensional setting (with d independent components) and we have from Lemma 15.18, setting again

$$X^A = \mathbb{E}\left[X.|\mathcal{F}_A\right] \text{ where } \mathcal{F}_A = \sigma\left(Z_k, k \in A\right) \text{ and } A \subset \mathbb{N},$$

that for any $\rho \geq 1$ and $A \subset \mathbb{N}$,

$$\left|R_{X^A}\right|_{\rho\text{-var}} \leq \left(1 + \min\left\{|A|, |A^c|\right\}\right) |R|_{\rho\text{-var};[s,t]^2} \qquad (15.27)$$

and

$$\left|R_{X^A}\right|_{2\text{-var};[s,t]^2} \leq |R|_{2\text{-var};[s,t]^2} \,. \qquad (15.28)$$

We now assume that R has finite ρ-variation for some $\rho \in [1,2)$ dominated by some 2D control ω. For fixed $A \subset \mathbb{N}$, finite or with finite complement, it follows from (15.27) that $X^A = \left(X^{A,1}, \ldots, X^{A,d}\right)$ admits a natural $G^3\left(\mathbb{R}^d\right)$-valued lift, denoted by $\mathbf{X}^A$. Of course, $\mathbf{X}^{\mathbb{N}} = \mathbf{X}$, the natural lift of X.

Lemma 15.46 *Assume that*
(i) $X = \left(X^1, \ldots, X^d\right)$ is a centred continuous Gaussian process with independent components;
(ii) X has Karhunen–Loève expansion $\sum_{k \in \mathbb{N}} Z_k h^k$ where $(h^k = (h^{k;1}, \ldots, h^{k;d}))$ is an orthonormal basis for $\mathcal{H}$;
(iii) the covariance of X is of finite ρ-variation, for some $\rho \in [1,2)$, controlled by some 2D control ω;
(iv) $A \subset \mathbb{N}$ so that $\min\left\{|A|, |A^c|\right\} < \infty$;

Then,
(a) for all $s < t$ in $[0,1]$, for all i, j, k distinct in $\{1, \ldots, d\}$, we have[12]

$$\mathbb{E}\left(X_{s,t}^i | \mathcal{F}_A\right) = X_{s,t}^{A,i} \tag{15.29}$$

$$\mathbb{E}\left(\mathbf{X}_{s,t}^{i,j} | \mathcal{F}_A\right) = \mathbf{X}_{s,t}^{A,i,j} \tag{15.30}$$

$$\mathbb{E}\left(\mathbf{X}_{s,t}^{i,j,k} | \mathcal{F}_A\right) = \mathbf{X}_{s,t}^{A,i,j,k} \tag{15.31}$$

$$\mathbb{E}\left(\mathbf{X}_{s,t}^{i,i,j} | \mathcal{F}_A\right) = \mathbf{X}_{s,t}^{A,i,i,j} + \frac{1}{2}\int_s^t \mathbb{E}\left(\left|X_{s,u}^{A^c;i}\right|^2\right) dX_u^{A;j}; \tag{15.32}$$

(b) for all $s < t$ in $[0,1]$ and $n \in \{1, 2, 3, \ldots\}$, we have

$$\sup_{A \subset \mathbb{N}, \min\{|A|,|A^C|\} < \infty} \mathbb{E}\left(\left|\pi_n\left(\mathbf{X}_{s,t}^A\right)\right|^2\right) \le C\omega\left([s,t]^2\right)^{n/\rho}$$

where C depends on ρ.

Proof. (a) Equality (15.29) is essentially the definition of X^A. Equality (15.30) is also easy: one just needs to note that $\mathbb{E}\left(\cdot | \mathcal{F}_A\right)$ is a projection in L^2 and hence L^2-continuous; since both $\mathbf{X}$ and $\mathbf{X}^A$ are L^2-limits of their respective lifted piecewise linear approximations (a general feature of enhanced Gaussian processes), the claim follows.

The proof of equality (15.31) follows the same argument, while (15.32) is a consequence of

$$\mathbb{E}\left(\left|X_{s,u}^i\right|^2 \Big| \mathcal{F}_A\right) - \left|X_{s,u}^{A,i}\right|^2 = \mathbb{E}\left(\left|X_{s,u}^{A^c;i}\right|^2\right).$$

From the L^2-projection property of $\mathbb{E}\left(\cdot | \mathcal{F}_A\right)$ we then see that, for i, j, k distinct,

$$\mathbb{E}\left(\left|\mathbf{X}_{s,t}^{A,i}\right|^2\right) \le \mathbb{E}\left(\left|\mathbf{X}_{s,t}^i\right|^2\right) \le c_1\omega\left([s,t]^2\right)^{1/\rho}$$

$$\mathbb{E}\left(\left|\mathbf{X}_{s,t}^{A,i,j}\right|^2\right) \le \mathbb{E}\left(\left|\mathbf{X}_{s,t}^{i,j}\right|^2\right) \le c_1\omega\left([s,t]^2\right)^{2/\rho}$$

$$\mathbb{E}\left(\left|\mathbf{X}_{s,t}^{A,i,j,k}\right|^2\right) \le \mathbb{E}\left(\left|\mathbf{X}_{s,t}^{i,j,k}\right|^2\right) \le c_1\omega\left([s,t]^2\right)^{3/\rho}$$

$$\mathbb{E}\left(\left|\mathbb{E}\left(\mathbf{X}_{s,t}^{i,i,j} | \mathcal{F}_A\right)\right|^2\right) \le \mathbb{E}\left(\left|\mathbf{X}_{s,t}^{i,i,j}\right|^2\right) \le c_1\omega\left([s,t]^2\right)^{3/\rho};$$

for some constant $c_1 = c_1(\rho)$, thanks to (15.20). Thus, to prove (b), it only remains to prove that

$$\mathbb{E}\left(\left|\int_s^t \mathbb{E}\left(\left|X_{s,u}^{A^c;i}\right|^2\right) dX_u^{A;j}\right|^2\right) \le c_2\omega\left([s,t]^2\right)^{3/\rho}.$$

---

[12] $A^c = \mathbb{N}\backslash A$.

To this end, observe that, thanks to $i \neq j$,

$$\int_s^t \mathbb{E}\left(\left|X_{s,u}^{A^c;i}\right|^2\right) dX_u^{A;j} = \mathbb{E}\left(\left.\int_s^t \mathbb{E}\left(\left|X_{s,u}^{A^c;i}\right|^2\right) dX_u^j \right| \mathcal{F}_A\right),$$

and hence

$$\mathbb{E}\left(\left|\int_s^t \mathbb{E}\left(\left|X_{s,u}^{A^c;i}\right|^2\right) dX_u^{A;j}\right|^2\right) \leq \mathbb{E}\left(\left|\int_s^t \mathbb{E}\left(\left|X_{s,u}^{A^c;i}\right|^2\right) dX_u^j\right|^2\right).$$

We define $f(u) := \mathbb{E}(|X_{s,u}^{A^c;i}|^2)$, noting that $f(s) = 0$, and for $u < t$ in $[s,t]$,

$$f_{u,v} = R_{X^{A^c;i}}\begin{pmatrix} u,v \\ u,v \end{pmatrix} + R_{X^{A^c;i}}\begin{pmatrix} s,u \\ u,v \end{pmatrix} + R_{X^{A^c;i}}\begin{pmatrix} u,v \\ s,u \end{pmatrix}$$

so that

$$|f_{u,v}|^2 \leq |R_{X^{A^c;i}}|_{2\text{-var};[u,v]^2}^2 + |R_{X^{A^c;i}}|_{2\text{-var};[u,v]\times[s,t]}^2 + |R_{X^{A^c;i}}|_{2\text{-var};[s,t]\times[u,v]}^2.$$

As the right-hand side above is super-additive in $[u,v]$, it follows from the uniform 2-variation estimates (15.28) that

$$|f|_{2\text{-var};[s,t]}^2 \leq 3\,|R_{X^{A^c;i}}|_{2\text{-var};[s,t]^2}^2 \leq 3\,|R_{X^i}|_{2\text{-var};[s,t]^2}^2 \leq 3\omega\left([s,t]^2\right)^{2/\rho}$$

and we conclude with the Young–Wiener estimate of Proposition 15.39. ∎

Theorem 15.47 *Assume that*
(i) $X = \left(X^1, \ldots, X^d\right)$ *is a centred continuous Gaussian process with independent components;*
(ii) X *has Karhunen–Loève expansion* $\sum_{k\in\mathbb{N}} Z_k h^k$ *where* $(h^k = (h^{k;1}, \ldots, h^{k;d}))$ *is an orthonormal basis for* $\mathcal{H}$;
(iii) *the covariance of* X *is of finite* ρ-*variation, for some* $\rho \in [1,2)$, *controlled by some 2D control* ω *with* $\omega\left([0,1]^2\right) \leq K$;
(iv) $p > 2\rho$ *and* $A_n := \{1, \ldots, n\}$.
Then, there exists a constant $\eta = \eta(p,\rho,K) > 0$

$$\sup_{n\in\mathbb{N}} \mathbb{E}\exp\left(\eta\left\|\mathbf{X}^{A_n}\right\|_{p\text{-var};[0,1]}^2\right) < \infty \tag{15.33}$$

and, for all $q \in [1,\infty)$,

$$d_{p\text{-var};[0,1]}\left(\mathbf{X}^{A_n}, \mathbf{X}\right) \quad \rightarrow \quad 0 \text{ in } L^q(\mathbb{P}) \text{ as } n \to \infty, \tag{15.34}$$

$$\left\|\mathbf{X}^{A_n^c}\right\|_{p\text{-var};[0,1]} \quad \rightarrow \quad 0 \text{ in } L^q(\mathbb{P}) \text{ as } n \to \infty. \tag{15.35}$$

If ω *is Hölder dominated, i.e.* $\omega(s,t) \leq K|t-s|$ *for all* $s < t$ *in* $[0,1]$, *then (15.33), (15.34), (15.35) also hold in* $1/p$-*Hölder sense.*

Proof. Inequality (15.33) follows from Lemma 15.46 and Proposition 15.22. Let us observe that the proof of (15.34) can be reduced to pointwise convergence

$$d\left(\mathbf{X}_t^{A_n}, \mathbf{X}_t\right) \to 0 \text{ in probability} \tag{15.36}$$

under the Hölder assumption "$\omega\left([s,t]^2\right) \leq K |t - s|$". Indeed, assuming this Hölder domination on ω, this follows directly from Proposition A.15 whereas the general case is reduced to the Hölder one by considering $\tilde{X} := X \circ [(\omega(0, \cdot)/\omega([0,1]^2)]^{-1}$ and noting that both natural lift and Karhunen–Loève expansions commute with a deterministic, continuous time-change,

$$d_{p\text{-var};[0,1]}\left(\mathbf{X}^{A_n}, \mathbf{X}\right) = d_{p\text{-var};[0,1]}\left(\tilde{\mathbf{X}}^{A_n}, \tilde{\mathbf{X}}\right) \leq d_{1/p\text{-Höl};[0,1]}\left(\tilde{\mathbf{X}}^{A_n}, \tilde{\mathbf{X}}\right).$$

We thus turn to the proof of (15.36). From Proposition 15.62, it will be enough to prove that for i, j, k distinct,

$$\left|\mathbf{X}_t^{A_n,i} - \mathbf{X}_t^i\right| \to 0 \text{ in } L^2(\mathbb{P}),$$

$$\left|\mathbf{X}_t^{A_n,i,j} - \mathbf{X}_t^{i,j}\right| \to 0 \text{ in } L^2(\mathbb{P}),$$

$$\left|\mathbf{X}_t^{A_n,i,j,k} - \mathbf{X}_t^{i,j,k}\right| \to 0 \text{ in } L^2(\mathbb{P}),$$

$$\left|\mathbf{X}_t^{A_n,i,i,j} - \mathbf{X}_t^{i,i,j}\right| \to 0 \text{ in } L^2(\mathbb{P}).$$

The first three convergence results are pure martingale convergence results. For the last one, in view of (15.32) we also need to prove that $\int_s^t \mathbb{E}\left(\left|X_{s,u}^{A_n^c;i}\right|^2\right) dX_u^{A_n;j}$ converges to 0 in L^2. To this end we note that, again by a martingale argument,

$$\sup_{u\in[s,t]} \mathbb{E}\left(\left|X_{s,u}^{A_n^c;i}\right|^2\right) = \sup_{u\in[s,t]} \mathbb{E}\left(\left|X_{s,u}^i - X_{s,u}^{A_n;i}\right|^2\right) \to 0 \text{ as } n \to \infty.$$

On the other hand, we saw in the proof of Lemma 15.46, that the 2-variation of $u \in [s,t] \to \mathbb{E}\left(\left|X_{s,u}^{A_n^c;i}\right|^2\right)$ is bounded by $c_1\omega\left([s,t]^2\right)$. By interpolation, this means that for $\varepsilon > 0$, its $(2+\varepsilon)$-variation converges to 0 when n tends to ∞. We pick ε such that $\frac{1}{2+\varepsilon} + \frac{1}{2\rho} > 1$. After recalling that

$$\mathbb{E}\left(\left|\int_s^t \mathbb{E}\left(\left|X_{s,u}^{A_n^c;i}\right|^2\right) dX_u^{A_n;j}\right|^2\right) \leq \mathbb{E}\left(\left|\int_s^t \mathbb{E}\left(\left|X_{s,u}^{A_n^c;i}\right|^2\right) dX_u^{j}\right|^2\right),$$

we therefore obtain, using the Young–Wiener integral bounds (Proposition 15.39), that

$$\mathbb{E}\left(\left|\int_s^t \mathbb{E}\left(\left|X_{s,u}^{A_n^c;i}\right|^2\right) dX_u^{A_n;j}\right|^2\right) \leq c_2 \left|\mathbb{E}\left(\left|X_{s,\cdot}^{A_n^c;i}\right|^2\right)\right|_{(2+\varepsilon)\text{-var};[s,t]}^2 |R|_{\rho\text{-var};[s,t]},$$

and hence $\mathbb{E}\left(\left|\int_s^t \mathbb{E}\left(\left|X_{s,u}^{A_n^c;i}\right|^2\right)dX_u^{A_n;j}\right|^2\right) \to 0$ as n tends to ∞. It only remains to prove (15.35) which is reduced, as above, to pointwise convergence (in probability or L^2) of the $i, (i,j), (i,j,k), (i,i,j)$-coordinates. By the backward martingale convergence theorem and Kolmogorov's 0–1 law, for i, j, k distinct,

$$X_t^{A_n^c,i} = \mathbb{E}\left(X_t^{;i}\,\middle|\,\mathcal{F}_{A^c}\right) \to \mathbb{E}\left(X_t^{;i}\,\middle|\,\cap_k \mathcal{F}_{A^c}\right) = \mathbb{E}\left(X_t^{;i}\right) = 0$$

(with convergence in L^2, as $n \to \infty$) and similarly, using the fact that i, j, k are distinct,

$$\left|\mathbf{X}_t^{A_n^c,i,j}\right|, \left|\mathbf{X}_t^{A_n^c,i,j,k}\right| \to 0 \text{ in } L^2\left(\mathbb{P}\right),$$

so that we are only left to show that $\left|\mathbf{X}_t^{A_n^c,i,i,j}\right| \to 0$ in $L^2\left(\mathbb{P}\right)$, which in view of (15.32) requires us to prove that $\left|\int_s^t \mathbb{E}\left(\left|X_{s,u}^{A_n;i}\right|^2\right)dX_u^{A_n^c;j}\right|^2 \to 0$ in L^2 as $n \to \infty$. From

$$\lim_{n\to\infty} \mathbb{E}\left(\left|\int_s^t \left[\mathbb{E}\left(\left|X_{s,u}^{A_n;i}\right|^2\right) - \mathbb{E}\left(\left|X_{s,u}^i\right|^2\right)\right]dX_u^{A_n^c;j}\right|^2\right) = 0,$$

this can be reduced to L^2-convergence of $\left|\int_s^t \mathbb{E}\left(\left|X_{s,u}^i\right|^2\right)dX_u^{A_n^c;j}\right|^2 \to 0$, which follows, thanks to

$$\int_s^t \mathbb{E}\left(\left|X_{s,u}^i\right|^2\right)dX_u^{A_n^c;j} = \mathbb{E}\left(\int_s^t \mathbb{E}\left(\left|X_{s,u}^i\right|^2\right)dX_u^j\,\middle|\,\mathcal{F}_{A_n^c}\right),$$

from backward martingale convergence. The proof is then finished. ∎

Exercise 15.48 *In the context of Theorem 15.47, show that*

$$\exists \eta > 0 : \sup_{n \in \mathbb{N}} \mathbb{E}\exp\left(\eta\left\|\mathbf{X}^{A_n}\right\|_{\psi_{\rho,\rho/2}\text{-var};[0,1]}^2\right) < \infty.$$

15.6 Weak approximations

15.6.1 Tightness

Proposition 15.49 *Assume that*
(i) ω is a 2D control;
(ii) (X^n) is a sequence of centred, d-dimensional, continuous Gaussian processes with independent components;
(iii) for $\rho \in [1,2)$ and for some constant C and for all $s < t$ in $[0,1]$,

$$\sup_n |R_{X^n}|_{\rho\text{-var};[s,t]^2}^\rho \le C\omega\left([s,t]^2\right);$$

(iv) $\mathbf{X}^n$ *denotes the natural lift of* X^n *with sample paths in* $C_o^{0,p\text{-var}}\left([0,1],\right.$
$\left.G^3\left(\mathbb{R}^d\right)\right)$, *for some* $p > 2\rho$.
Then the family $(\mathbb{P}_*\mathbf{X}^n)$, *i.e. the laws of* $\mathbf{X}^n$ *viewed as Borel measures on the Polish space* $C_o^{0,p\text{-var}}\left([0,1],G^3\left(\mathbb{R}^d\right)\right)$, *are tight. If* ω *is Hölder dominated then tightness holds in* $C_o^{0,1/p\text{-Höl}}\left([0,1],G^3\left(\mathbb{R}^d\right)\right)$.

Proof. Let us fix $p' \in (2\rho, p)$ and consider first the case of Hölder-dominated ω. Define

$$K_R = \left\{ \mathbf{x}: \ \|\mathbf{x}\|_{1/p'\text{-Höl}} \leq R \right\}$$

and note that K_R is a relatively compact set in $C_0^{0,1/p\text{-Höl}}\left([0,1],G^3\left(\mathbb{R}^d\right)\right)$, which is a simple consequence from Arzela–Ascoli and interpolation (see Proposition 8.17). From the Fernique estimates in Theorem 15.33, there exists a constant c such that

$$\sup_n \mathbb{P}\left(\mathbf{X}^n \in K_R\right) \leq c e^{-R^2/c}$$

and the tightness result follows. The general case is a time-changed version of the Hölder case, using relative compactness of

$$\left\{ \mathbf{x}: d\left(\mathbf{x}_s, \mathbf{x}_t\right) \leq R \left(\omega\left([0,t]^2\right) - \omega\left([0,s]^2\right) \right)^{1/p'} \right\}$$

in $C_o^{0,p\text{-var}}\left([0,1],G^3\left(\mathbb{R}^d\right)\right)$. We leave the details to the reader. ∎

15.6.2 Convergence

We now turn to convergence. By Prohorov's theorem,[13] tightness already implies existence of weak limits and so it only remains to see that there is one and only one limit point; the classical way to see this is by checking convergence of the finite-dimensional distributions. We need a short lemma concerning the interchanging of limits.

Lemma 15.50 *Let* (E,d) *a Polish space and* $\left\{ Z^{m,n}: m \in \bar{\mathbb{N}}, n \in \bar{\mathbb{N}} \right\}$ *a collection of* E-*valued random variables. Assume* $Z^{m,n}$ *converges weakly to* $Z^{m,\infty}$ *as* $n \to \infty$ *for every* $m \in \mathbb{N}$. *Assume also* $Z^{m,n} \to Z^{\infty,n}$ *in probability, uniformly in* n; *that is,*

$$\forall \delta > 0: \sup_{n \in \bar{\mathbb{N}}} \mathbb{P}\left(d\left(Z^{m,n}, Z^{\infty,n}\right) > \delta\right) \to 0 \ as \ m \to \infty.$$

Then $Z^{\infty,n}$ *converges weakly to* $Z^{\infty,\infty}$.

---

[13] For example, [13].

Proof. By the Portmanteau theorem,[14] it suffices to show that for every $f : E \to \mathbb{R}$, bounded and uniformly continuous,

$$\mathbb{E}f\left(Z^{\infty,n}\right) \to \mathbb{E}f\left(Z^{\infty,\infty}\right).$$

To see this, fix $\varepsilon > 0$ and $\delta = \delta\left(\varepsilon\right) > 0$ such that $d\left(x,y\right) < \delta$ implies $|f\left(x\right) - f\left(y\right)| < \varepsilon$. By assumption we can take $m = m\left(\varepsilon\right)$ large enough such that

$$\sup_{0 \leq n \leq \infty} \mathbb{P}\left(d\left(Z^{m,n}, Z^{\infty,n}\right) > \delta\right) < \varepsilon.$$

Hence,

$$\sup_{0 \leq n \leq \infty} |\mathbb{E}f\left(Z^{\infty,n}\right) - \mathbb{E}f\left(Z^{m,n}\right)|$$

$$\begin{aligned}
&\leq \sup_{0 \leq n \leq \infty} |\mathbb{E}\left[|f\left(Z^{\infty,n}\right) - f\left(Z^{m,n}\right)|\, ; d\left(Z^{\infty,n}, Z^{m,n}\right) \geq \delta\right]| \\
&\quad + \sup_{0 \leq n \leq \infty} |\mathbb{E}\left[|f\left(Z^{\infty,n}\right) - f\left(Z^{m,n}\right)|\, ; d\left(Z^{\infty,n}, Z^{m,n}\right) < \delta\right]| \\
&\leq 2\left|f\right|_{\infty} \sup_{0 \leq n \leq \infty} \mathbb{P}\left(d_{\infty}\left(\mathbf{X}_n, S_3\left(X_n^D\right)\right) \geq \delta\right) + \varepsilon \\
&\leq \left(2\left|f\right|_{\infty} + 1\right)\varepsilon.
\end{aligned}$$

On the other hand, for $n \geq n_0\left(m, \varepsilon\right) = n_0\left(\varepsilon\right)$ large enough, we also have

$$|\mathbb{E}f\left(Z^{m,n}\right) - \mathbb{E}f\left(Z^{m,\infty}\right)| \leq \varepsilon$$

and the proof is then finished with the triangle inequality,

$$\begin{aligned}
|\mathbb{E}f\left(Z^{\infty,n}\right) - \mathbb{E}f\left(Z^{\infty,\infty}\right)| &\leq |\mathbb{E}f\left(Z^{\infty,n}\right) - \mathbb{E}f\left(Z^{m,\infty}\right)| \\
&\quad + |\mathbb{E}f\left(Z^{m,\infty}\right) - \mathbb{E}f\left(Z^{\infty,\infty}\right)| \\
&\quad + |\mathbb{E}f\left(Z^{m,n}\right) - \mathbb{E}f\left(Z^{m,\infty}\right)| \\
&\leq \left(2\left|f\right|_{\infty} + 1\right)2\varepsilon + \varepsilon.
\end{aligned}$$

∎

Theorem 15.51 *Assume that*
(i) $\left(X^n\right)_{0 \leq n \leq \infty}$ *is a sequence of centred, d-dimensional, continuous Gaussian processes on* $[0,1]$ *with independent components;*
(ii) the covariances of X^n *are of finite ρ-variation, $\rho \in [1,2)$, uniformly controlled by some 2D control ω;*
(iii) $\mathbf{X}^n$ *denotes the natural lift of* X^n *with sample paths in* $C_o^{0,p\text{-var}}\left([0,1],\right.$ $\left. G^3\left(\mathbb{R}^d\right)\right)$, *for some $p > 2\rho$;*
(iv) R_{X^n} *converges pointwise on* $[0,1]^2$, *to* R_{X^∞}.
Then, for any $p > 2\rho$, $\mathbf{X}^n$ *converges weakly to* $\mathbf{X}^\infty$ *with respect to p-variation topology. If ω is Hölder-dominated, then convergence holds with respect to $1/p$-Hölder topology.*

---

[14] For example, [13].

Proof. Tightness was established in Proposition 15.49 so we only need weak convergence of the finite-dimensional distributions:

$$(\mathbf{X}_t^n : t \in S) \implies (\mathbf{X}_t^\infty : t \in S) \quad \text{for any } S \in \mathcal{D}[0,1].$$

By assumption (iv) this holds on level one, meaning that

$$(X_t^n : t \in S) \implies (X_t^\infty : t \in S).$$

Now, given a continuous path $x \in C([0,1], \mathbb{R}^d)$ it is easy to see that

$$(x : t \in S) \mapsto S_3(x^D) \in C([0,1], G^3(\mathbb{R}^d))$$

is continuous and so it is clear that

$$\left(S_3(X^{n,D})_t : t \in S\right) \implies \left(S_3(X^{\infty,D})_t : t \in S\right).$$

On the other hand, it follows from Theorem 15.42 that, along any sequence $(D_m) \subset \mathcal{D}[0,1]$ with mesh tending to zero, $S_3(X^{n,D_m}) \to \mathbf{X}^n$, pointwise and in probability (much more was shown!), and also uniformly in n, thanks to the explicit estimates of Theorem 15.42. It then suffices to apply Lemma 15.50 with $Z^{m,n} = \left(S_3(X^{n,D_m})_t : t \in S\right)$ with state-space $E = G^3(\mathbb{R}^d)^{\times(\#S)}$. ∎

Example 15.52 Set $R(s,t) = \min(s,t)$. The covariance of fractional Brownian motion is given by

$$R^H(s,t) = \frac{1}{2}\left(s^{2H} + t^{2H} - |t-s|^{2H}\right).$$

Take a sequence $H_n \to 1/2$. It is easy to see that $R^{H_n} \to R$ pointwise and from our discussion of fractional Brownian motion, for any $\rho > 1$,

$$\limsup_{n \to \infty} \frac{\left|R^{H_n}\right|_{\rho\text{-var};[s,t]^2}}{|t-s|^{1/\rho}} < \infty. \qquad \square$$

15.7 Large deviations

As in previous sections, $X = (X^1, \ldots, X^d)$ denotes a centred continuous Gaussian process on $[0,1]$, with independent components, each with covariance of finite ρ-variation for some $\rho \in [1,2)$ and dominated by some 2D control ω. We write $\mathcal{H}$ for its associated Cameron–Martin space. Recall from Section 15.3.3 that X admits a natural lift to a $G^3(\mathbb{R}^d)$-valued process $\mathbf{X}$, obtained as the limit of lifted piecewise linear approximations along dissections D with mesh $|D| \to 0$,

$$d_{p\text{-var};[0,1]}\left(\mathbf{X}, S_3(X^D)\right) \to 0 \text{ in } L^q(\mathbb{P}) \text{ for all } q \in [1, \infty).$$

Since the law of X induces a Gaussian measure on $C\left([0,1],\mathbb{R}^d\right)$, it follows from general principles (see Section D.2 in Appendix D) that $(\varepsilon X : \varepsilon > 0)$ satisfies a large deviation principle with good rate function I in uniform topology, where I is given by

$$I(x) = \begin{cases} \frac{1}{2}\langle x,x\rangle_{\mathcal{H}} & \text{if } x \in \mathcal{H} \subset C\left([0,1],\mathbb{R}^d\right) \\ +\infty & \text{otherwise.} \end{cases}$$

We write Φ_m for the piecewise linear approximations along the dissection $D_m = \{i/m : i = 0,\dots,m\}$. It is clear that

$$S_3 \circ \Phi_m : \left(C\left([0,1],\mathbb{R}^d\right),|\cdot|_{\infty}\right) \to C\left([0,1],G^3\left(\mathbb{R}^d\right),d_{\infty}\right)$$

is continuous. By the contraction principle, $S_3\left(\varepsilon\Phi_m\left(X\right)\right)$ satisfies a large deviation principle with good rate function

$$J_m\left(y\right) = \inf\left\{I\left(x\right),\ x \text{ such that } S_3\left(\Phi_m\left(x\right)\right) = y\right\},$$

the infimum of the empty set being $+\infty$. Essentially, a large deviation principle for $\delta_{\varepsilon}\mathbf{X}$ is obtained by sending m to infinity. To this end we now prove that $S_3\left(\Phi_m\left(X\right)\right)$ is an exponentially good approximation to $\mathbf{X}$.

Lemma 15.53 *Let $\delta > 0$ fixed. Then, for $p > 2\rho$, we have*

$$\lim_{m\to\infty}\overline{\lim_{\varepsilon\to0}}\,\varepsilon^2\log\mathbb{P}\left(d_{p\text{-var}}\left(S_3\left(\Phi_m\left(\varepsilon X\right)\right),\delta_{\varepsilon}\mathbf{X}\right) > \delta\right) = -\infty.$$

If ω is Hölder-dominated, then

$$\lim_{m\to\infty}\overline{\lim_{\varepsilon\to0}}\,\varepsilon^2\log\mathbb{P}\left(d_{1/p\text{-Höl}}\left(S_3\left(\Phi_m\left(\varepsilon X\right)\right),\delta_{\varepsilon}\mathbf{X}\right) > \delta\right) = -\infty.$$

Proof. First observe that

$$d_{p\text{-var}}\left(S_3\left(\Phi_m\left(\varepsilon X\right)\right),\delta_{\varepsilon}\mathbf{X}\right) = \varepsilon d_{p\text{-var}}\left(S_3\left(\Phi_m\left(X\right)\right),\mathbf{X}\right).$$

Clearly, for $\theta > 0$, $\alpha_m \equiv |R_{X-X^{D_m}}|_{\infty}^{\theta} \to 0$ as $m \to \infty$ and from Theorem 15.42,

$$\left|d_{p\text{-var}}\left(S_3\left(\Phi_m\left(X\right)\right),\mathbf{X}\right)\right|_{L^q} \equiv C\sqrt{q}\alpha_m \to 0. \tag{15.37}$$

We then estimate

$$\mathbb{P}\left(d_{p\text{-var}}\left(S_3\left(\Phi_m\left(\varepsilon X\right)\right),\delta_{\varepsilon}\mathbf{X}\right) > \delta\right) = \mathbb{P}\left(d_{p\text{-var}}\left(S_3\left(\Phi_m\left(X\right)\right),\mathbf{X}\right) > \frac{\delta}{\varepsilon}\right)$$

$$\leq \left(\frac{\delta}{\varepsilon}\right)^{-q}\sqrt{q}^q\alpha_m^q$$

$$\leq \exp\left[q\log\left(\frac{\varepsilon}{\delta}\alpha_m\sqrt{q}\right)\right],$$

and after choosing $q = 1/\varepsilon^2$ we obtain, for ε small enough,

$$\varepsilon^2\log\mathbb{P}\left(d_{p\text{-var}}\left(S_3\left(\Phi_m\left(\varepsilon X\right)\right),\delta_{\varepsilon}\mathbf{X}\right) > \delta\right) \leq \log\left(\frac{\alpha_m}{\delta}\right).$$

Now take the limits $\overline{\lim}_{\varepsilon \to 0}$ and $\lim_{m \to \infty}$ to finish the proof, for the $d_{p\text{-var}}$ case. The proof is (almost) identical for the $1/p$-Hölder case. ∎

From our embedding of the Cameron–Martin space into the space of paths of finite ρ-variation, we obtain

Lemma 15.54 *For all* $\Lambda > 0$, *and* $p > 2\rho$, *we have*

$$\lim_{m \to \infty} \sup_{\{h:I(h) \leq \Lambda\}} d_{p\text{-var}} \left[(S_3 \circ \Phi_m)(h), S_3(h) \right] = 0. \tag{15.38}$$

If ω *is Hölder-dominated, then*

$$\lim_{m \to \infty} \sup_{\{h:I(h) \leq \Lambda\}} d_{1/p\text{-Höl}} \left[(S_3 \circ \Phi_m)(h), S_3(h) \right] = 0. \tag{15.39}$$

Proof. First, let us observe that for $s < t$ in $[0,T]$, we have, as $\rho < 2$, from Theorem 9.5 and Proposition 5.20,

$$\begin{aligned}
\| (S_3 \circ \Phi_m)(h) \|_{\rho\text{-var};[s,t]} &\leq c_1 \| \Phi_m(h) \|_{\rho\text{-var};[s,t]} \\
&\leq c_2 \| h \|_{\rho\text{-var};[s,t]}.
\end{aligned}$$

Now using Proposition 15.7, we obtain for h with $I(h) \leq \Lambda$,

$$\| (S_3 \circ \Phi_m)(h) \|_{\rho\text{-var};[s,t]} \leq c_3 \Lambda^{1/2} \omega \left([s,t]^2 \right)^{1/2\rho}.$$

In particular, we see that

$$\sup_{m \geq 0} \sup_{\{h:I(h) \leq \Lambda\}} \| (S_3 \circ \Phi_m)(h) \|_{2\rho\text{-var};[0,1]} \leq \sup_m \sup_{\{h:I(h) \leq \Lambda\}} \| (S_3 \circ \Phi_m)(h) \|_{\rho\text{-var};[0,1]}$$
$$< \infty$$

and, if ω is Holder-dominated,

$$\sup_m \sup_{\{h:I(h) \leq \Lambda\}} \| (S_3 \circ \Phi_m)(h) \|_{1/2\rho\text{-Höl};[0,1]} < \infty.$$

In particular, we first see that by interpolation, to prove (15.38) and (15.39), it is enough to prove that

$$\lim_{m \to \infty} \sup_{\{h:I(h) \leq \Lambda\}} d_0 \left[(S_3 \circ \Phi_m)(h), S_3(h) \right] = 0.$$

We will actually prove the stronger statement

$$\lim_{m \to \infty} \sup_{\{h:I(h) \leq \Lambda\}} d_{\rho'\text{-var};[0,1]} \left[(S_3 \circ \Phi_m)(h), S_3(h) \right] = 0,$$

for $\rho' \in (\rho, 2)$. But, as we picked $\rho' < 2$, we can use the uniform continuity on bounded sets of the map S_3 (Corollary 9.11) to see that it only remains to prove

$$\lim_{m \to \infty} \sup_{\{h:I(h) \leq \Lambda\}} d_{\rho'\text{-var};[0,1]} \left[\Phi_m(h), h \right] = 0.$$

Using interpolation once again, it is enough to prove that

$$\lim_{m \to \infty} \sup_{\{h: I(h) \le \Lambda\}} d_{\infty\text{-var};[0,1]} \left[\Phi_m \left(h\right), h\right] = 0.$$

This follows from

$$d_{\infty\text{-var};[0,1]} \left[\Phi_m \left(h\right), h\right] \le \max_{i=0}^{m-1} |h|_{\rho\text{-var};\left[\frac{i}{m}, \frac{i+1}{m}\right]}$$

$$\le (2\Lambda)^{1/2} \max_{i=0}^{m-1} \omega \left(\left[\frac{i}{m}, \frac{i+1}{m}\right]^2\right)^{1/2\rho}.$$

That concludes the proof.   ∎

We are now in a position to state the main theorem of this section:

Theorem 15.55 *Assume that*
(i) $X = \left(X^1, \ldots, X^d\right)$ *is a centred continuous Gaussian process on* $[0,1]$ *with independent components;*
(ii) $\mathcal{H}$ *denotes the Cameron–Martin space associated with* X*;*
(iii) the covariance of X *is of finite* ρ*-variation dominated by some 2D control* ω*, for some* $\rho \in [1, 2)$*;*
(iv) $\mathbf{X}$ *denotes the natural lift of* X *to a* $G^3 \left(\mathbb{R}^d\right)$*-valued process.*
Then, for any $p \in (2\rho, 4)$*, the family* $(\delta_\varepsilon \mathbf{X})_{\varepsilon > 0}$ *satisfies a large deviation principle in* p*-variation topology with good rate function, defined for* $\mathbf{x} \in C_o^{0,p\text{-var}} \left([0,1], G^3 \left(\mathbb{R}^d\right)\right)$*, given by*

$$J\left(\mathbf{x}\right) = \frac{1}{2} \left\langle \pi_1 \left(\mathbf{x}\right), \pi_1 \left(\mathbf{x}\right) \right\rangle_{\mathcal{H}} \quad \text{if } \pi_1 \left(\mathbf{x}\right) \in \mathcal{H}.$$

If ω *is Hölder-dominated then the large deviation principle holds in* $1/p$*-Hölder topology.*

Proof. The proof is the same as in the Brownian motion case: after (re)stating the large deviation principle satisfied by $S_3 \left(\varepsilon \Phi_m \left(X\right)\right)$, we only need to use the extended contraction principle and Lemmas 15.53 and 15.54 that $(\delta_\varepsilon \mathbf{X})_{\varepsilon > 0}$.   ∎

15.8 Support theorem

We recall the standing assumptions. Under some probability measure $\mathbb{P}$ we have a d-dimensional Gaussian process X on $[0,1]$, always assumed to be centred, continuous, with independent components. We write $\mathcal{H}$ for the associated Cameron–Martin space. Under the assumption that X has covariance of finite ρ-variation for some $\rho \in [1, 2)$, we have seen in Section 15.3.3 that X admits a natural lift to a $G^3 \left(\mathbb{R}^d\right)$-valued process $\mathbf{X}$ whose sample paths are, almost surely, geometric p-rough paths, $p \in (2\rho, 4)$. We can and will assume that $\mathbb{P}$ is a Gaussian measure on $C\left([0,1],\mathbb{R}^d\right)$ so that $X\left(\omega\right) = \omega_t$ is realized as a coordinate process. $\mathbf{X}$ can then be viewed as a measurable map from $C\left([0,1],\mathbb{R}^d\right)$ into the Polish space $\Omega := C_0^{0,p\text{-var}}\left([0,1],G^3\left(\mathbb{R}^d\right)\right)$, resp. $C_0^{0,1/p\text{-Höl}}\left([0,1],G^3\left(\mathbb{R}^d\right)\right)$, almost surely defined as

$$\mathbf{X}\left(\omega\right) = \lim_{n\to\infty} S_3\left(\omega^{D_n}\right),$$

in probability say, where ω^{D_n} denotes the piecewise linear approximation based on any sequence of dissections (D_n) with mesh $|D_n|$ tending to zero. The law of $\mathbf{X}$ is viewed as a Borel measure on Ω. We now introduce the assumption of complementary Young regularity.

Condition 15.56 There exists $q \geq 1$ with $1/p + 1/q > 1$ so that

$$\mathcal{H} \hookrightarrow C^{q\text{-var}}\left([0,T],\mathbb{R}^d\right).$$

We say that $\mathcal{H}$ **has complementary Young regularity to** X. $\qquad\square$

Thanks to Proposition 15.7, Condition 15.56 is satisfied when X has covariance of finite ρ-variation for some $\rho \in [1,3/2)$; indeed, this follows from considering

$$\frac{1}{\rho} + \frac{1}{p} > 1$$

where the critical value $\rho^* = 3/2$ is obtained by replacing p by (its lower bound) 2ρ and "greater than 1" by "equal to 1".

Remark 15.57 An application of Proposition 15.5 shows that fractional Brownian motion ("$\rho = 1/\left(2H\right)$") satisfies Condition 15.56 for Hurst parameter $H > 1/3$. One can actually do better: it follows from Remark 15.10 that for any $H > 1/4$ complementary Young regularity holds. $\qquad\square$

Lemma 15.58 *Assume complementary Young regularity. Then,*
(i) for $\mathbb{P}$-almost every ω we have

$$\forall h \in \mathcal{H} : \mathbf{X}\left(\omega + h\right) = T_h\mathbf{X}\left(\omega\right)$$

where T denotes the translation operator for geometric rough paths;
(ii) for every $h \in \mathcal{H}$ the laws of $\mathbf{X}$ and $T_h\mathbf{X}$ are equivalent.

Proof. The arguments are essentially identical to those employed for Brownian motion (Theorem 13.36 and Proposition 13.37):

Ad (i). By switching to a subsequence if needed, we may assume that $\mathbf{X}\left(\omega\right)$ is defined as $\lim_{n\to\infty} S_3\left(\omega^{D_n}\right)$ whenever this limit exists (and arbitrarily on the remaining null-set N). Now fix $h \in \mathcal{H}$; using complementary Young regularity, we have

$$S_3\left(\omega^{D_n} + h^{D_n}\right) = T_{h^{D_n}} S_3\left(\omega^{D_n}\right) \to T_h\mathbf{X}\left(\omega\right) \text{ as } n \to \infty$$

and thus see that $\mathbf{X}(\omega + h) = T_h \mathbf{X}(\omega)$ for all h and $\omega \notin N$.

Ad (ii). By Cameron–Martin, the laws of X and $X + h$, as Borel measures on $C([0,1], \mathbb{R}^d)$, are equivalent. It follows that the image measures under the measurable map $\mathbf{X}(\cdot)$, Borel measures on Ω, are equivalent. But this says precisely that the laws of $\mathbf{X}$ and $\mathbf{X}(\cdot + h)$ are equivalent and the proof is finished since $\mathbf{X}(\cdot + h) = T_h \mathbf{X}$ almost surely. ∎

Although elementary, let us spell out the following in its natural generality.

Lemma 15.59 *Let S, S' be two Polish spaces and μ a Borel measure on S. Assume $x \in \operatorname{supp}[\mu]$ and f is continuous at x. Then $f(x) \in \operatorname{supp}[f_*\mu]$. If, in addition, $S' = S$ and $f_*\mu \sim \mu$ then $f(x) \in \operatorname{supp}[\mu]$.*

Proof. Write $B_\delta(x)$ for an open ball, centred at x of radius $\delta > 0$. For every $\varepsilon > 0$ there exists δ such that $B_\delta(x) \subset f^{-1}(B_\varepsilon(f(x)))$ and hence $0 < \mu(B_\delta(x)) \le (f_*\mu)(B_\varepsilon(f(x)))$ so that $f(x) \in \operatorname{supp} f_*\mu$. If $f_*\mu \sim \mu$ then and $0 < (f_*\mu)(B_\varepsilon(f(x))) \implies 0 < \mu(B_\varepsilon(f(x)))$ and so $f(x) \in \operatorname{supp}[\mu]$. ∎

We are now ready to state the main result in this section.

Theorem 15.60 *Let $\mathbf{X}_*\mathbb{P}$ denote the law of $\mathbf{X}$, a Borel measure on the Polish space $C_0^{0,p\text{-var}}([0,1], G^3(\mathbb{R}^d))$ where $p > 2\rho$. Assume that complementary Young regularity holds. Then*

$$\operatorname{supp}[\mathbf{X}_*\mathbb{P}] = \overline{S_3(\mathcal{H})},$$

where support and closure are with respect to p-variation topology. If ω is Hölder-dominated, i.e. $\omega\left([s,t]^2\right) \le K|t - s|$ for some constant K, we can use $1/p$-Hölder topology instead of p-variation topology.

Proof. As a preliminary remark, note that $S_3(\mathcal{H})$ is meaningful since any $h \in \mathcal{H}$ has finite ρ-variation (Proposition 15.7) and hence lifts canonically to a $G^3(\mathbb{R}^d)$-valued path (of finite ρ-variation) by iterated Young integration (or more precisely, as an application of Theorem 9.5).

Step 1: $\subset$-inclusion. Since $X^{\{1,\ldots,n\}} := \mathbb{E}\left[X. | \mathcal{F}_{\{1,\ldots,n\}}\right] \in \mathcal{H}$ almost surely and converges to $\mathbf{X}$ in the respective rough path metrics, the first inclusion is clear.

Step 2: $\supset$-inclusion. The idea is to find at least one fixed $\hat\omega \in C([0,1], \mathbb{R}^d)$ such that $\mathbf{X}(\hat\omega) \in \operatorname{supp}[\mathbf{X}_*\mathbb{P}]$ and such that there exists a (deterministic!) sequence $(g_n) \subset \mathcal{H}$, which can and will depend on $\hat\omega$, such that $T_{-g_n}\mathbf{X}(\hat\omega) = \mathbf{X}(\hat\omega - g_n) \to \mathbf{X}(0) = S_3(0)$ in rough path metric. Having found such an element $\hat\omega$ (with suitable sequence g_n) we can apply Lemma 15.59 with μ as the law of $\mathbf{X}$, a Borel measure on $S = C_0^{0,p\text{-var}}([0,1], G^3(\mathbb{R}^d))$ resp. $C_0^{0,1/p\text{-Höl}}([0,1], G^3(\mathbb{R}^d))$, $S' = S$ and *continuous* function $f : S \to S$ given by $f : \mathbf{x} \mapsto T_{-g_n}\mathbf{x}$; using the fact that the law of $T_h\mathbf{X}$ is equivalent to the law of **X**, cf. Lemma 15.58, we conclude that $T_{-g_n}\mathbf{X}(\hat{\omega}) \in \text{supp}[\mathbf{X}_*\mathbb{P}]$. This holds true for all n and by closedness of the support, the limit $\mathbf{X}(0) = S_3(0)$ must be in the support. The same argument shows that any further translate $T_h S_3(0) = S_3(h)$ must be in the support and thus

$$\text{supp}[\mathbf{X}_*\mathbb{P}] \supset S_3(\mathcal{H}).$$

Passing the (p-variation resp. $1/p$-Hölder rough path) closure on both sides then finishes the proof. It remains to see how to find $\hat{\omega}$ with the required properties. Since $\mathbf{X}(\omega) \in \text{supp}[\mathbf{X}_*\mathbb{P}]$ and $T_{-g_n}\mathbf{X}(\omega) = \mathbf{X}(\omega - g_n)$ holds true for almost every ω, there is a null-set N_1 so that any $\omega \notin N_1$ will have these properties. Furthermore, Theorem 15.47, tells us that there is another null-set N_2 so that we can pick $\hat{\omega} \notin (N_2 \cup N_1)$ and

$$\mathbf{X}(\hat{\omega}) = \lim_{m \to \infty} S_3\left(\sum_{i=1}^{m} \xi(h_k)|_{\hat{\omega}} h_k(\cdot)\right) = \lim_{m \to \infty} \mathbf{X}^{\{1,\dots,m\}}(\hat{\omega})$$

$$\mathbf{X}^{\{n+1,n+2,\dots\}}(\hat{\omega}) \to S_3(0).$$

It now suffices to set $g_n(\cdot) = \sum_{i=1}^{n} \xi(h_k)|_{\hat{\omega}} h_k(\cdot) \in \mathcal{H} \hookrightarrow C^{q\text{-var}}$; we then see that

$$
\begin{aligned}
\mathbf{X}(\hat{\omega} - g_n) &= T_{-g_n}\mathbf{X}(\hat{\omega}) = \lim_{m \to \infty} T_{-g_n}\mathbf{X}^{\{1,\dots,m\}}(\hat{\omega}) \\
&= \lim_{m \to \infty} \mathbf{X}^{\{n+1,\dots,m\}}(\hat{\omega}) \\
&= \mathbf{X}^{\{n+1,n+2,\dots\}}(\hat{\omega}) \to \mathbf{X}(0) = S_3(0),
\end{aligned}
$$

as required, and this finishes the proof. ∎

15.9 Appendix: some estimates in $G^3(\mathbb{R}^d)$

Proposition 15.61 *Let $g \in G^3(\mathbb{R}^d)$. Then, for some constant c,*
(i) $|\pi_2(g)| \le c\max_{i,j \text{ distinct}}\{|g^{i,j}|\} + |\pi_1(g)|^2$;
(ii) $|\pi_3(g)| \le c\max_{i,j,k \text{ distinct}}\{|g^{i,i,j}|, |g^{i,j,k}|\} + |\pi_1(g)|^3 + |\pi_2(g)|^{3/2}$.

Proof. Pick a path $x \in C_0^{1\text{-var}}([0,1], \mathbb{R}^d)$ such that $g = S_3(x)_{0,1} =: \mathbf{x}_{0,1}$. Then, statement (i) follows from the calculus identity

$$g^{i,j} = \mathbf{x}_{0,1}^{i,i} \equiv \int_0^1 x^i dx^i = \frac{1}{2}(x_1^i)^2 = \frac{1}{2}|\pi_1(g)|^2.$$

For (ii) we use the basic inequality $ab \le \frac{1}{3}a^3 + \frac{2}{3}b^{3/2}$ plus the identities

$$
\begin{aligned}
g^{j,i,i} &= g^j g^{i,i} - g^i g^{i,j} + g^{i,i,j}, & (15.40) \\
g^{i,j,i} &= g^j g^{i,i} - g^{i,i,j} - g^{j,i,i} & (15.41) \\
&= g^i g^{i,j} - 2g^{i,i,j},
\end{aligned}
$$

which we now establish by calculus. Indeed, (15.40) follows from

$$
\begin{aligned}
\mathbf{x}_{0,1}^{j,i,i} &= \int_{0<u_1<u_2<u_3<1} dx_{u_1}^j \, dx_{u_2}^i \, dx_{u_3}^i \\
&= \frac{1}{2} \int_{0<u<1} \left| x_{u,1}^i \right|^2 dx_u^j = \frac{1}{2} \int_{0<u<1} \left| x_{0,1}^i - x_{0,u}^i \right|^2 dx_u^j \\
&= \mathbf{x}_{0,1}^{i,i} x_{0,1}^j - x_{0,1}^i \mathbf{x}_{0,1}^{i,j} + \mathbf{x}_{0,1}^{i,i,j}
\end{aligned}
$$

whereas (15.41) follows from the fact that $\mathbf{x}_{0,1}^{i,j,i} + \mathbf{x}_{0,1}^{j,i,i}$ equals

$$
\int_{0<u<1} x_u^i x_u^j \, dx_u^i = \frac{1}{2} \left| x_{0,1}^i \right|^2 x_{0,1}^j - \frac{1}{2} \int_{0<u<1} \left| x_u^i \right|^2 dx_{0,u}^j = \mathbf{x}_{0,1}^{i,i} x_{0,1}^j - \mathbf{x}_{0,1}^{i,i,j}.
$$

The proof is finished. ∎

Proposition 15.62 *Let* $g, h \in G^3\left(\mathbb{R}^d\right)$ *with* $\|g\|, \|h\| \le M$ *for some positive constant* M. *Assume that for all distinct indices* $i, j, k \in \{1, \dots, d\}$

$$
\begin{aligned}
\left| g^i - h^i \right| &\le \varepsilon M, \\
\left| g^{i,j} - h^{i,j} \right| &\le \varepsilon M^2, \\
\left| g^{i,i,j} - h^{i,i,j} \right| &\le \varepsilon M^3, \\
\left| g^{i,j,k} - h^{i,j,k} \right| &\le \varepsilon M^3.
\end{aligned}
$$

Then $\left| \delta_{1/M} \left(g - h \right) \right| \le c\varepsilon$ *for some constant* c.

Proof. We may replace g, h by $\delta_{1/M} g, \delta_{1/M} h$ and hence there is no loss of generality assuming $M = 1$. The proof is now similar to the previous one. ∎

15.10 Comments

Our exposition here follows in essence Friz and Victoir [61]. The lift of certain Gaussian processes, including fractional Brownian motion with Hurst parameter $H > 1/4$, is due to Coutin and Qian [32] and based on piecewise linear approximations. The key role of (enough) decorrelation of increments for the existence of stochastic area was also pointed out by Lyons and Qian [120]. Karhunen–Loève approximations for fractional Brownian motion are studied by Millet and Sanz-Solé [130] and also Feyel and de la Pradelle [52], implicitly in Friz and Victoir [63]. We remark that equations (15.29), (15.30) explain why martingale arguments (see also Coutin and Victoir [33], Friz and Victoir [62] and Friz [69]) are enough to discuss the step-two case ($H > 1/3$), whereas equation (15.32) shows that the step-three case requires additional care. A large deviation principle for the lift of fractional Brownian motion was obtained by Millet and Sanz-Solé [129], for the Coutin–Qian class in Friz and Victoir [65]. Support statements for lifted fractional Brownian motion, for $H > 1/3$, appeared in Feyel and de la Pradelle [52] and Friz and Victoir [63]. Our Theorem 15.60 may also be obtained by applying the abstract support theorem of Aida–Kusuoka–Stroock [2, Corollary 1.13]. We conjecture that complementary Young regularity (Condition 15.56) is not needed for Theorem 15.60 to hold true.

16

Markov processes

We have seen in a previous chapter that Brownian motion B can be enhanced to a stochastic process $\mathbf{B} = \mathbf{B}(\omega)$ for which almost every realization is a geometric $1/p$-Hölder rough path, $p \in (2,3)$. As is well known,[1] d-dimensional Brownian motion B is a diffusion, i.e. a Markov process with continuous sample paths, with generator

$$\frac{1}{2}\Delta = \frac{1}{2}\sum_{i=1}^{d}\partial_i^2.$$

In the present chapter, our aim is to replace Brownian motion by a diffusion $X = X^a$ with uniformly elliptic generator in divergence form,

$$\frac{1}{2}\sum_{i,j=1}^{d}\partial_i\left(a^{ij}\partial_j\cdot\right),$$

followed by the construction of a suitable lifted process $\mathbf{X}^a$ with geometric rough (sample) paths. If $a = \left(a^{ij}\right)$ had enough regularity, one could effectively realize X^a as a semi-martingale, and then construct $\mathbf{X}^a$ as an enhanced semi-martingale. However, assuming no regularity (beyond measurability), this route fails. The generator for X^a itself is only defined in the "weak" sense and the focus must be on the bilinear ("Dirichlet") form $(f,g), (f,g) \mapsto \int_{\mathbb{R}^d} \sum_{i,j} a^{i,j}\partial_i f\ \partial_j g\,dx.$

The main tool in this chapter will be a suitable Dirichlet form that allows for a direct construction and analysis of $\mathbf{X}^a$. From Section 16.2 on, we shall rely heavily on the well-developed theory of Dirichlet forms. The essentials (for our purposes) are collected in Appendix E.

16.1   Motivation

As is common in the theory of partial differential equations, we shall assume that $a = a(x)$ is a symmetric matrix such that for some $\Lambda > 0$,

$$\forall \xi \in \mathbb{R}^d : \frac{1}{\Lambda}\left|\xi\right|^2 \leq \xi \cdot a(.)\xi \leq \Lambda\left|\xi\right|^2;$$

---

[1]See, for example, [143] Chapter VII, Proposition (1.11).

no regularity of $a = a\left(x\right)$ is assumed (besides measurability in $x \in \mathbb{R}^d$). Let us say again[2] that the study of such a diffusion process X^a, which in many ways behaves like a Brownian motion, relies heavily on analytic Dirichlet form techniques. In general, X^a cannot be constructed as a solution to a stochastic differential equation and need not be a semimartingale.

The main idea is roughly the following. Assume at first that $a\left(.\right)$ is smooth, with bounded derivatives of all orders. In this case, X^a *can be* constructed as a solution to a stochastic differential equation: it suffices to write the generator of X^a in non-divergent form as

$$\frac{1}{2} \sum_{i,j=1}^{d} \left[a^{ij} \partial_i \partial_j + \left(\partial_i a^{ij} \right) \partial_j \right] ;$$

then, knowing that a admits a Lipschitz square root,[3] say σ, so that $\sigma \sigma^T = a$, it is a standard exercise in Itô calculus to see that the diffusion constructed as a solution to the (Itô) stochastic differential equation

$$dX = \sigma\left(X\right) dB + b\left(X\right) dt, \quad \text{with } b = \left(b^j\right) = \left(\sum_{i=1}^{d} \partial_i a^{ij} \right)$$

has indeed generator $\frac{1}{2} \sum \partial_i \left(a^{ij} \partial_j \cdot \right)$. Moreover, $X = X^a$, the so-constructed process,[4] is plainly a semimartingale X and hence, following Section 14.1, there is a well-defined stochastic area process

$$t \mapsto A_{0,t}^{a;i,j} = \frac{1}{2} \left(\int_0^t X_{0,s}^{a;i} dX_s^{a;j} - \int_0^t X_{0,s}^{a;j} dX_s^{a;i} \right), \quad \text{with } 1 \leq i < j \leq d.$$

It is not hard to see (also using standard Itô calculus, cf. Exercise 16.15) that $t \mapsto \left(X_{0,t}^a, A_{0,t}^a \right)$ is a diffusion process on $\mathbb{R}^d \oplus so\left(d\right)$, started at 0, with generator given by

$$\mathcal{L}^a := \frac{1}{2} \sum_{i,j=1}^{d} \mathsf{u}_i \left(a^{ij} \mathsf{u}_j \cdot \right)$$

where, for $i = 1, \ldots, d$,

$$\mathsf{u}_i |_x = \partial_i + \frac{1}{2} \left(\sum_{1 \leq j < i \leq d} x^{1;j} \partial_{j,i} - \sum_{1 \leq i < j \leq d} x^{1;j} \partial_{i,j} \right). \tag{16.1}$$

---

[2] See, for example, [70, 158].

[3] See, for example, [162, Theorem 5.2.2].

[4] Strictly speaking, this construction depends on the choice of square root and one may prefer to write X^σ. However, we shall construct the lifted process in such a way that its generator (and hence its law) only depends on $\sigma \sigma^T = a$; thereby justifying our notation.

Here, ∂_i denotes the ith coordinate vector field on $\mathbb{R}^d$ and $\partial_{i,j}$ with $i < j$ the respective coordinate vector field on $so\,(d)$, identified with its upper diagonal elements. Following Sections 13.2 and 14.1, the enhancement to X^a is of the form

$$\mathbf{X}^a_{0,t} \equiv \left(X^a_{0,t}, A^a_{0,t}\right) \overset{\exp}{\underset{\log}{\rightleftarrows}} \left(1, X^a_{0,t}, \int_0^t X^a_{0,s} \otimes \circ dX^a_s\right)$$

where we can switch between the "path area view" and the "iterated Stratonovich integral view" using $\exp : \mathbb{R}^d \oplus so\,(d) \equiv \mathfrak{g}^2\left(\mathbb{R}^d\right) \to G^2\left(\mathbb{R}^d\right)$ and its inverse log, respectively.

This suggests constructing $\mathbf{X}^a$ *directly* as a $\mathfrak{g}^2\left(\mathbb{R}^d\right)$-valued Markov process with generator $\mathcal{L}^a$. In fact, for $f, g \in C_c^\infty\left(\mathfrak{g}^2\left(\mathbb{R}^d\right)\right)$ integration by parts shows that

$$\langle \mathcal{L}^a f, g \rangle = \int_{\mathfrak{g}^2\left(\mathbb{R}^d\right)} \sum_{i,j} a^{i,j} \mathfrak{u}_i f \; \mathfrak{u}_j g \, dm.$$

(Observe that $a\,(.)$ may indeed be a function on $\mathfrak{g}^2\left(\mathbb{R}^d\right)$ rather than only $\mathbb{R}^d$. In fact, this construction is carried out naturally on $\mathfrak{g}^N\left(\mathbb{R}^d\right)$, which allows for a direct "Markovian" modelling of the "higher-order" areas of $\mathbf{X}^a$.) The right-hand side above, which involves no derivatives of $\left(a^{ij}\right)$, is another instance of a Dirichlet form, and allows us to deal with measurable $\left(a^{ij}\right)$.

Remark 16.1 We shall find it more convenient in the present chapter to adopt the path area view and define the enhanced Markov process $\mathbf{X}^a$ as the $\mathfrak{g}^2\left(\mathbb{R}^d\right)$-valued process (X^a, A^a). Upon setting

$$
\begin{aligned}
(x, A) * (x', A') &\equiv \log\left(\exp\,(x, A) \otimes \left(\exp\,(x', A')\right)\right) \\
&= \left(x + x', A + A' + \frac{1}{2}\left(x \otimes x' - x' \otimes x\right)\right), \\
(x, A)^{-1} &= (-x, -A),
\end{aligned}
$$

we see that $\exp : \left(\mathfrak{g}^2\left(\mathbb{R}^d\right), *\right) \to \left(G^2\left(\mathbb{R}^d\right), \otimes\right)$ is a Lie group isomorphism. We then can and will work in $\mathfrak{g}^2\left(\mathbb{R}^d\right)$ identified with $G^2\left(\mathbb{R}^d\right)$, using identical notation. For instance, the Carnot–Caratheodory norm and distance are given by

$$
\begin{aligned}
\|(x, A)\| &= \|\exp\,(x, A)\| \sim |x| + |A|^{1/2}, \\
d\left((x, A), (x', A')\right) &= \left\|(x, A)^{-1} * (x', A')\right\|;
\end{aligned}
$$

elements in $C^{\alpha\text{-Höl}}\left([0, 1], \mathfrak{g}^2\left(\mathbb{R}^d\right)\right), \alpha \in (1/3, 1/2)$, are α-Hölder geometric rough paths and so forth. The same remarks apply when $\mathfrak{g}^2\left(\mathbb{R}^d\right)$ is replaced by $\mathfrak{g}^N\left(\mathbb{R}^d\right)$, noting that we can always use the exponential map to identify $\mathfrak{g}^N\left(\mathbb{R}^d\right)$ with $G^N\left(\mathbb{R}^d\right)$. For instance, the Lyons lift becomes

$$\mathfrak{s}_N : C^{\alpha\text{-Höl}}\left([0, 1], \mathfrak{g}^2\left(\mathbb{R}^d\right)\right) \to C^{\alpha\text{-Höl}}\left([0, 1], \mathfrak{g}^N\left(\mathbb{R}^d\right)\right)$$

where, writing $\exp_k$ for the exponential map from $\mathfrak{g}^k\left(\mathbb{R}^d\right)$ to $G^k\left(\mathbb{R}^d\right)$,

$$\mathfrak{s}_N := \exp_N^{-1} \circ S_N \circ \exp_2 .$$ $\qquad\square$

16.2 Uniformly subelliptic Dirichlet forms

The Lie algebra $\mathfrak{g} = \mathfrak{g}^N\left(\mathbb{R}^d\right)$ is naturally graded in the sense that it has a vector space decomposition

$$\mathfrak{g}^N\left(\mathbb{R}^d\right) = \mathfrak{V}_1 \oplus \cdots \oplus \mathfrak{V}_N$$

where $\mathfrak{V}_1 \cong \mathbb{R}^d$, and $\mathfrak{V}_2 \cong so\left(d\right)$, and $\mathfrak{V}_n \subset \left(\mathbb{R}^d\right)^{\otimes n}$ is given by[5]

$$\mathfrak{V}_n \cong \left[\mathbb{R}^d, \ldots, \left[\mathbb{R}^d, \mathbb{R}^d\right] \ldots\right] = \mathrm{span}\left\{\left[v_1, \ldots, \left[v_{n-1}, v_n\right]\right] : v_1, \ldots, v_n \in \mathbb{R}^d\right\}.$$

The Campbell–Baker–Hausdorff formula makes $(\mathfrak{g}, *)$ into a Lie group, isomorphic to $\left(G^N\left(\mathbb{R}^d\right), \otimes\right)$). There are left-invariant vector fields $\mathfrak{u}_1, \ldots, \mathfrak{u}_d$ on $\mathfrak{g}$ determined by

$$\mathfrak{u}_i|_0 = \partial_i|_0, \quad i = 1, \ldots, d,$$

where $\partial_i|_0$ are the coordinate vector fields associated with the canonical basis of $\mathfrak{V}_1 = \mathbb{R}^d$ and

$$\nabla^{\mathrm{hyp}} = \left(\mathfrak{u}_1, \ldots, \mathfrak{u}_d\right)^T$$

is the *hypoelliptic gradient* on $\mathfrak{g}$.

Example 16.2 When $N = 1$, $\mathfrak{g}^1\left(\mathbb{R}^d\right) \cong \mathbb{R}^d$ and the $\mathfrak{u}_i$ are precisely the standard coordinate vector fields ∂_i. When $N = 2$, we can identify $\mathfrak{g} = \mathfrak{g}^2\left(\mathbb{R}^d\right)$ with $\mathbb{R}^d \oplus so\left(d\right)$ and in this case $\mathfrak{u}_i$ takes the form given in (16.1). $\qquad\square$

Definition 16.3 $[\Xi\left(\Lambda\right)]$ For $\Lambda \geq 1$ we call $\Xi\left(\Lambda\right) = \Xi^{N,d}\left(\Lambda\right)$ the set of all measurable maps $a\left(.\right)$ from $\mathfrak{g} = \mathfrak{g}^N\left(\mathbb{R}^d\right)$ into the space of symmetric matrices such that

$$\forall \xi \in \mathbb{R}^d : \frac{1}{\Lambda}\left|\xi\right|^2 \leq \xi \cdot a\left(.\right)\xi \leq \Lambda\left|\xi\right|^2 .$$ $\qquad\square$

Theorem 16.4 *Fix* $\Lambda \geq 1$. *For* $f, g \in C_c^\infty\left(\mathfrak{g}, \mathbb{R}\right)$ *and* $a \in \Xi\left(\Lambda\right)$ *we define the carré-du-champ operator*

$$\Gamma^a\left(f, g\right) := \nabla^{\mathrm{hyp}} f \cdot a \nabla^{\mathrm{hyp}} g = \sum_{i,j=1}^d a^{i,j} \mathfrak{u}_i f\, \mathfrak{u}_j g$$

---

[5] Recall that $[a, b] = a \otimes b = b \otimes a$.

and $dm(x) = dx$ denotes the Lebesgue measure on $\mathfrak{g}$,

$$\mathcal{E}^a(f,g) := \left(\int_{\mathfrak{g}} \Gamma^a(f,g)\, dm \right).$$

When $a = I$, the identity matrix, we simply write Γ, $\mathcal{E}$ rather than $\Gamma^I, \mathcal{E}^I$. Then $\mathcal{E}^a$ extends to a regular Dirichlet form, as defined in Section E.2, Appendix E, which possesses $C_c^\infty(\mathfrak{g},\mathbb{R})$ as a core. The domain of $\mathcal{E}^a$, denoted by $W^{1,2}(\mathfrak{g},dm) := \mathcal{D}(\mathcal{E}^a)$, does not depend on the particular choice of $a \in \Xi(\Lambda)$ and is given as the closure of C_c^∞-functions with respect to

$$|f|^2_{W^{1,2}(\mathfrak{g},dm)} := \mathcal{E}(f,f) + \langle f, f \rangle_{L^2(\mathfrak{g},dm)}.$$

At last, $\mathcal{E}^a$ is strongly local (in the sense of Definition E.3).

Proof. We first discuss the case $a = I$. By invariance of the Lebesgue measure m under (left and right) multiplication on $(\mathfrak{g}, *)$, established in Proposition 16.40 in the appendix to this chapter, one sees that the vector fields $u_1, \ldots, u_d$ are formally skew-symmetric so that, for any $f, g \in C_c^\infty(\mathfrak{g},\mathbb{R})$,

$$\mathcal{E}(f,g) = \sum_{i=1}^d \langle u_i f, u_i g \rangle_{L^2} = -\sum_{i=1}^d \langle u_i u_i f, g \rangle_{L^2} = \langle \mathcal{L}f, g \rangle$$

where $\mathcal{L}$ is given in Hörmander form $\sum_{i=1}^2 u_i^2$. Consider now a sequence of C_c^∞-functions $f_n \to 0$ in L^2 so that $\mathcal{E}(f_n - f_m, f_n - f_m) \to 0$ as $n, m \to \infty$. To see that $\mathcal{E}$ is closeable with core $C_c^\infty(\mathfrak{g},\mathbb{R})$, and hence extends to a regular Dirichlet form, we need to check that $\mathcal{E}(f_n, f_n) \to 0$ as $n \to \infty$. To this end, fix $\varepsilon > 0$ and pick k large enough so that

$$\mathcal{E}(f_n - f_k, f_n - f_k) \leq 1$$

for all $n > k$. Using bilinearity and $\mathcal{E}(f_n, f_k) \leq \mathcal{E}(f_n, f_n)^{1/2}\,\mathcal{E}(f_k, f_k)^{1/2}$ it easily follows that

$$\sup_{n \in \{k,k+1,\ldots\}} \mathcal{E}(f_n, f_n) \leq C < \infty$$

where C depends on $\mathcal{E}(f_k, f_k)$ only. Moreover, for all $n > m > k$,

$$\begin{aligned}
\mathcal{E}(f_n, f_n) &= \langle \mathcal{L}f_m, f_n \rangle_{L^2} + \mathcal{E}(f_n - f_m, f_n) \\
&\leq |f_n|_{L^2}\,|\mathcal{L}f_m|_{L^2} + C\mathcal{E}(f_n - f_m, f_n - f_m)^{1/2}
\end{aligned}$$

so that we can first choose m large enough such that for all $n > m$,

$$C\mathcal{E}(f_n - f_m, f_n - f_m)^{1/2} < \varepsilon/2,$$

followed by taking n large enough so that $|f_n|_{L^2}\,|\mathcal{L}f_m|_{L^2} < \varepsilon/2$. But this shows that for n large enough $\mathcal{E}(f_n, f_n) < \varepsilon$, as required. For the discussion of $a \neq I$, let us note that $\mathcal{E}(f_n, f_n) \to 0$ can be equivalently expressed by saying

$$\forall i \in \{1, \ldots, d\} : |\mathfrak{u}_i f_n|^2_{L^2} \to 0 \text{ as } m \to \infty$$

and by passing to a subsequence we may assume that $\mathfrak{u}_i f_n \to 0$ a.e. for all $i = 1, \ldots, d$. Hence

$$
\begin{aligned}
\mathcal{E}^a(f_m, f_m) &= \int \lim_{n \to \infty} \left(\sum_{i,j=1}^{d} a^{i,j} \mathfrak{u}_i (f_n - f_m) \, \mathfrak{u}_j (f_n - f_m) \right) dm \\
&\leq \lim_{m \to \infty} \inf \mathcal{E}^a(f_n - f_m, f_n - f_m)
\end{aligned}
$$

by Fatou's lemma, which shows that $\mathcal{E}^a$ is also closeable. Strong locality of the resulting Dirichlet form, also denoted by $\mathcal{E}^a$, is a simple consequence of the fact that the $\mathfrak{u}_i$ are (pure) first-order differential operators. ∎

We now establish three important properties related to this setup.

Proposition 16.5 *Let* $a \in \Xi^{N,d}(\Lambda)$. *Then the following hold.*
(i) The intrinsic distance associated with $\mathcal{E}^a$,

$$d^a(x,y) = \sup \{f(x) - f(y) : f \in \mathcal{D}(\mathcal{E}^a) \cap C_c(\mathfrak{g}) \text{ and } \Gamma^a(f,f) \leq 1\},$$

defines a genuine metric on $\mathfrak{g}$. *When* $a = I$, *it coincides with the Carnot–Caratheodory metric* d *on* $\mathfrak{g}$. *Otherwise we have, for all* $x, y \in \mathfrak{g}$,

$$\frac{1}{\Lambda^{1/2}} d(x,y) \leq d^a(x,y) \leq \Lambda^{1/2} d(x,y).$$

In particular, the topology induced by d^a *coincides with the canonical topology on* $\mathfrak{g}$.
(ii) Set $B^a(x,r) = \{y \in \mathfrak{g} : d^a(x,y) < r\}$. *Then,*

$$\forall r \geq 0 \text{ and } x \in \mathfrak{g} : m(B^a(x,2r)) \leq 2^Q m(B^a(x,r))$$

with doubling constant Q *given by*

$$Q = (\dim_H \mathfrak{g})(1 + 2\ln \Lambda / \ln 2)$$

where $\dim_H(\mathfrak{g})$ *is the "homogenous" dimension[6] of* $\mathfrak{g}$ *defined by*

$$\dim_H(\mathfrak{g}) = \sum_{n=1}^{N} n \dim \mathfrak{V}_n.$$

(iii) The weak Poincaré inequality; that is, for all $r \geq 0$, $x \in \mathfrak{g}$ *and* $f \in \mathcal{D}(\mathcal{E})$,

$$\int_{B^a(x,r)} |f - \bar{f}|^2 \, dm \leq Cr^2 \int_{B^a(x,2r)} \Gamma(f,f) \, dm$$

---

[6] As a matter of fact, it is also the Haussdorff dimension of $\mathfrak{g}$ when equipped with Carnot–Caratheodory metric.

where $C = C(\Lambda, \dots)$ and $\bar{f}$ is the average of f over $B^a(x,r)$, i.e.

$$\bar{f} = m\left(B^a(x,r)\right)^{-1} \int_{B^a(x,r)} f\, dm.$$

Proof. Case 1: $a = I$, the identity matrix. In this case (i)–(iii) are well-known facts from *analysis on free nilpotent groups*. For the sake of completeness, statement (i) is shown in Section 16.9.3; (ii) follows by left-invariance and scaling; noting that on $\mathfrak{g} = \mathfrak{g}^N(\mathbb{R}^d)$, the dilation map $\delta_2 : \left(x^{(1)}, \dots, x^{(N)}\right) \mapsto \left(2x^{(1)}, \dots, 2^N x^{(N)}\right)$ has a Jacobian with value $2^{\dim \mathfrak{g}}$. A few more details are given in Section 16.9.1. At last, (iii) is a Poincaré inequality and the reader can find a self-contained proof in Section 16.9.2.

Case 2: $a \in \Xi(\Lambda)$ and no regularity assumptions beyond measurability. The key observation is that $\mathcal{E}^I$ and $\mathcal{E}^a$ are (obviously) quasi-isometric in the sense that

$$\frac{1}{\Lambda}\mathcal{E}(f,f) \le \mathcal{E}^a(f,f) \le \Lambda\mathcal{E}(f,f) \quad \text{or} \quad \frac{1}{\Lambda}\Gamma(f,f) \le \Gamma^a(f,f) \le \Lambda\Gamma(f,f)$$

and we conclude with invariance of properties (i)–(iii) under quasi-isometry (cf. Theorem E.8 in Appendix E). ∎

As it turns out (cf. Section E.4 in Appendix E for precise statements), the just-established properties (i)–(iii) allow us to use a highly developed, essentially analytic machinery. In particular, $\mathcal{E}^a$ determines a (non-positive) self-adjoint operator $\mathcal{L}^a$ on $L^2(\mathfrak{g}, dm)$ and weak (local) solutions to

$$\partial_t u = \mathcal{L}^a u$$

satisfy a parabolic Harnack inequality as well as Hölder regularity in space-time. More precisely, we have

Proposition 16.6 *(parabolic Harnack inequality) Let $a \in \Xi^{N,d}(\Lambda)$. There exists a constant[7] $C_H = C_H(\Lambda)$ such that*

$$\sup_{(s,y)\in Q^-} u(s,y) \le C_H \inf_{(s,y)\in Q^+} u(s,y),$$

whenever u is a non-negative weak solution of the parabolic partial differential equation $\partial_t u = \mathcal{L}^a u$ on some cylinder $Q = \left(t - 4r^2, t\right) \times B^a(x, 2r)$ for some reals $t, r > 0$. Here, $Q^- = \left(t - 3r^2, t - 2r^2\right) \times B(x,r)$ and $Q^+ = \left(t - r^2, t\right) \times B^a(x,r)$ are lower and upper subcylinders of Q separated by a lapse of time.

---

[7] As usual, dependence on N, d is not explicitly written.

Proposition 16.7 *(de Giorgi–Moser–Nash regularity) Let $a \in \Xi(\Lambda)$. Then there exist constants $\eta \in (0,1)$ and C_R, only depending on Λ, such that*

$$\sup_{(s,y),(s',y')\in Q_1} |u(s,y) - u(s',y')| \leq C_R \sup_{u\in Q_2} |u| \cdot \left(\frac{|s-s'|^{1/2} + d^a(y,y')}{r} \right)^{\eta}$$

whenever u is a non-negative weak solution of the parabolic partial differential equation $\partial_s u = \mathcal{L}^a u$ on some cylinder $Q_2 \equiv (t-4r^2,t) \times B(x,2r)$ for some reals $t, r > 0$. Here $Q_1 \equiv (t-r^2, t-2r^2) \times B(x,r)$ is a subcylinder of Q_2.

We also note that the L^2-semi-group $(\mathcal{P}_t^a : t \geq 0)$ associated with $\mathcal{L}^a$ resp. $\mathcal{E}^a$ admits a kernel representation[8] of the form

$$(\mathcal{P}_t^a f) = \int f(y) p^a(t,\cdot,y) \, dm(y) \qquad (16.2)$$

where the so-called heat-kernel p is a non-negative weak solution of the parabolic partial differential equation $\partial_s u = \mathcal{L}^a u$ with (distributional) initial data $u(0,\cdot) = \delta_x$. Thanks to self-adjointness of $\mathcal{L}^a$, the heat-kernel $p = p^a(t,x,y)$ is symmetric in x and y. As discussed in the generality of Section E.5, Appendix E, the heat-kernel allows for the construction of a continuous, symmetric diffusion process $\mathbf{X} = \mathbf{X}^{a,x}$ associated with $\mathcal{L}^a$ resp. $\mathcal{E}^a$ so that the finite-dimensional distributions of $\mathbf{X}$ are given by

$$\mathbb{P}^{a;x}[(\mathbf{X}_{t_1}, \ldots, \mathbf{X}_{t_n}) \in B] = \int_B p^a(t_1,x,y_1) \ldots p^a(t_n$$
$$- t_{n-1}, y_{n-1}, y_n) \, dy_1 \ldots dy_n.$$

We remark that $\mathbf{P}^{a,x} = \mathbf{X}_* \mathbb{P}^{a;x}$, the law of $\mathbf{X}$, can be viewed as a Borel measure on $C_x([0,\infty), \mathfrak{g})$. Although it is not always necessary to be specific about the underlying probability space, this allows us to realize $\mathbf{X}$ as a coordinate process on the path-space, i.e. $\mathbf{X}_t(\omega) = \omega_t$ for $\omega \in C_x([0,\infty), \mathfrak{g})$, equipped with $\mathbf{P}^{a,x}$.

Proposition 16.8 *(weak scaling) For any $a \in \Xi(\Lambda), r \neq 0$ set $a^r(x) := a(\delta_{1/r} x) \in \Xi(\Lambda)$, where we recall that δ denotes dilation on $\mathfrak{g}$,*

$$\left(\mathbf{X}_t^{a^r,x} : t \geq 0 \right) \overset{\mathcal{D}}{=} \left(\delta_r \mathbf{X}_{t/r^2}^{a,\delta_{1/r}(x)} : t \geq 0 \right).$$

Proof. It is easy to see, cf. Remark E.16, that $(\mathbf{X}_{\lambda t}^a : t \geq 0)$ is the symmetric diffusion associated with $\lambda^2 \mathcal{E}^a$. On the other hand, our state space has a structure that allows spatial scaling via dilation. Then the generator of

---

[8]See Exercise 16.10 below for a direct proof.

$(\delta_r \mathbf{X}_t^a : t \geq 0)$ is given by $r^2 \mathcal{L}^a\left(\delta_{1/r} \cdot\right) \equiv r^2 \mathcal{L}^{a^r}$ or, equivalently, the Dirichlet form $r^2 \mathcal{E}^{a^r}$. Combining these two transformations (take $\lambda = 1/r^2$) shows that $\left(\delta_r \mathbf{X}_{t/r^2}^a : t \geq 0\right)$ has associated Dirichlet form given by $\mathcal{E}^{a^r}$. It is also clear that starting $\delta_r \mathbf{X}_{t/r^2}^a$ at x is tantamount to starting $\mathbf{X}_{t/r^2}^{a^r}$ at $\delta_{1/r}$ (x).   ∎

Exercise 16.9 Let $\mathbf{B}$ be an enhanced Brownian motion.
(i) Show that
$$\mathbf{X}^{I/2;x} \equiv x * \log S_N\left((\mathbf{B})\right)$$

is a symmetric diffusion, started at $x \in \mathfrak{g}^N\left(\mathbb{R}^d\right)$, with generator $\sum_{i=1}^d \mathfrak{u}_i \circ \mathfrak{u}_i$.
(ii) Use scaling for enhanced Brownian motion to deduce the on-diagonal heat-kernel estimate

$$p\left(t, x, x\right) \leq t^{-\dim_H \mathfrak{g}/2} p\left(1, \delta_{t^{-1/2}} x, \delta_{t^{-1/2}} x\right) \leq ct^{-\dim_H \mathfrak{g}/2}.$$

(This is equivalent to $|\mathcal{P}_t|_{L^1 \to L^\infty} \leq ct^{-\dim_H \mathfrak{g}/2}$ where $\mathcal{P}_t$ is the associated semi-group.)   □

Exercise 16.10 *(i) Assume $\mathcal{E}$ is an abstract (symmetric) Dirichlet form and write $(\mathcal{P}_t)$ for the associated Markovian semi-group. Let $\nu \in (0, \infty)$. Show that the following two statements are equivalent:*
- there exists C_1 such that for all $t > 0$,

$$|\mathcal{P}_t|_{L^1 \to L^\infty} \leq C_1 t^{-\nu/2}; \tag{16.3}$$

*- **Nash's inequality** holds, i.e. there exists C_2 such that for all $f \in D(\mathcal{E}) \cap L^1$,*

$$|f|_{L^2}^{2+4/\nu} \leq C_2 \mathcal{E}(f, f) |f|_{L^1}^{2/\nu}. \tag{16.4}$$

(When switching between the two estimates, the constant C_j depends only on C_i and ν.)
(ii) Consider now $a \in \Xi^{N,d}(\Lambda)$ and the Dirichlet form $\mathcal{E}^a$ on $L^2(\mathfrak{g}, dm)$ where $\mathfrak{g} = \mathfrak{g}^N\left(\mathbb{R}^d\right)$ as usual. Use Exercise 16.9 and invariance of (16.4) under quasi-isometry to establish

$$|\mathcal{P}_t^a|_{L^1 \to L^\infty} \leq C_3 t^{-\dim_H \mathfrak{g}/2}.$$

(iii) Deduce the existence of a heat-kernel p^a, with the on-diagonal estimate

$$\forall t > 0, x \in \mathfrak{g} : p^a\left(t, x, x\right) \leq C_3 t^{-\dim_H \mathfrak{g}/2},$$

so that (16.2) holds for any $f \in L^2$.   □

16.3 Heat-kernel estimates

As in the previous section, write $\mathfrak{g} = \mathfrak{g}^N(\mathbb{R}^d)$. We now turn to "Gaussian" estimates of the heat-kernel

$$p^a : (0, \infty) \times \mathfrak{g} \times \mathfrak{g} \to [0, \infty).$$

Sharp estimates involve the intrinsic metric d^a on $\mathfrak{g}$, introduced in Proposition 16.5; although, for most (rough path) purposes one can use the Carnot–Caratheodory metric d. Once more, following Section E.4, Appendix E, all results of this Section are an automatic consequence of properties (i)–(iii) in Proposition 16.5. Nonetheless, it is instructive to note that an application of Harnack's inequality immediately leads to

$$p^a(t, x, x) \leq c_1 \frac{1}{m\left(B^a\left(x, t^{1/2}\right)\right)} \leq c_2 \frac{1}{m\left(B\left(x, t^{1/2}\right)\right)} \leq c_3 t^{-\dim_H \mathfrak{g}/2},$$

where c_1, c_2, c_3 depend only on Λ, in agreement with the conclusion of Exercise 16.10. We now state the full heat-kernel estimates.

Theorem 16.11 *Let $a \in \Xi(\Lambda)$. Then, for all $t > 0$ and $x, y \in \mathfrak{g}$ we have:*
(i) (upper heat-kernel bound) for any $\varepsilon > 0$ fixed there exists $C_u = C_u(\varepsilon, \Lambda)$ such that

$$p^a(t, x, y) \leq \frac{C_u}{\sqrt{t^{\dim_H \mathfrak{g}}}} \exp\left(-\frac{d^a(x, y)^2}{(4 + \varepsilon)t}\right);$$

(ii) (lower heat-kernel bound) there exists $C_l = C_l(\Lambda)$ such that

$$p^a(t, x, y) \geq \frac{1}{C_l} \frac{1}{\sqrt{t^{\dim_H \mathfrak{g}}}} \exp\left(-\frac{C_l d^a(x, y)^2}{t}\right).$$

Proof. An immediate corollary of the abstract heat-kernel estimates in Section E.4, Appendix E. ∎

Corollary 16.12 *For any $a \in \Xi(\Lambda)$, write $\mathbf{X} = \mathbf{X}^{a,x}$ for the (continuous) $\mathfrak{g}$-valued diffusion process associated with $\mathcal{E}^a$, started at $x \in \mathfrak{g}$. Then,*
(i) for all $\eta < 1/(4\Lambda)$ we have

$$M_\eta := \sup_{a \in \Xi(\Lambda)} \sup_{x \in \mathfrak{g}^2(\mathbb{R}^d)} \sup_{0 \leq s < t \leq 1} \mathbb{E}^{a,x}\left(\exp\left(\eta \frac{d(\mathbf{X}_t, \mathbf{X}_s)^2}{t - s}\right)\right) < \infty; \quad (16.5)$$

moreover, there exists $C(\Lambda)$ such that $M_\eta \leq 1 + C\eta \leq \exp(C\eta)$ for all $\eta \in \left[0, \frac{1}{16\Lambda}\right]$;
(ii) for any $\alpha \in (0, 1/2)$, there exists $c = c(\alpha, \Lambda)$ such that[9]

$$\sup_{a \in \Xi(\Lambda)} \sup_{x \in \mathfrak{g}^2(\mathbb{R}^d)} \mathbb{E}^{a,x} \exp\left(c\eta \|\mathbf{X}\|_{\alpha\text{-Höl};[0,1]}^2\right) < \infty;$$

---

[9] By convention, $\|\mathbf{X}\|_{\alpha\text{-Höl};[0,1]}$ is defined with respect to the Carnot–Caratheodory d on $\mathfrak{g}$.

(iii) there exists $c = c(\Lambda)$ such that

$$\sup_{a \in \Xi(\Lambda)} \sup_{x \in \mathfrak{g}^2(\mathbb{R}^d)} \mathbb{E}^{a,x} \exp\left(c\eta \|\mathbf{X}\|^2_{\psi_{2,1}\text{-var};[0,1]}\right) < \infty.$$

Proof. A straightforward computation shows that the upper heat-kernel estimate implies (16.5); this is not specific to the present setting and hence is carried out in a general context in Section E.6.1, Appendix E. The estimate on M_η for small $\lambda < 1/(16\Lambda)$ follows readily from the inequality $\exp(x) \le 1 + x \exp(x)$, for $x > 0$, and we obtain

$$M_\eta \le 1 + \eta \sup_{x \in \mathbb{R}^d} \sup_{s < t \in [0,1]} \mathbb{E}^{a,x}\left(\frac{d(\mathbf{X}_t, \mathbf{X}_s)^2}{t-s} \exp\left(\eta \frac{d(\mathbf{X}_t, \mathbf{X}_s)^2}{t-s}\right)\right).$$

Now it suffices to apply Cauchy–Schwarz, noting that $2\eta \le 1/(8\Lambda) < 1/(4\Lambda)$. The Fernique estimates for $\|\mathbf{X}\|_{\alpha\text{-Höl};[0,1]}$ are then a consequence of general principles, namely Theorem A.19 in the Appendix A. ∎

Recall that $p^a(t, x, y)\, dy$ is precisely the law of $\mathbf{X}_t^{a;x}$, i.e. the marginal law of a $\mathfrak{g}$-valued diffusion process. We now state a "localized" lower bound by considering the process $\mathbf{X}^{a;x}$ *killed* at its first exit from a fixed ball in $\mathfrak{g}$.

Theorem 16.13 *(localized lower heat-kernel bound) Let $a \in \Xi(\Lambda)$, $x, x_0 \in \mathfrak{g}, r > 0$ and write $\mathbf{X} = \mathbf{X}^{a,x}$ for the diffusion process associated with $\mathcal{E}^a$, started at x. Also set*

$$\xi_{B(x_0,r)} = \inf\{t \ge 0 : \mathbf{X}_t^{a;x} \notin B^a(x_0,r)\}.$$

Then the measure $\mathbb{P}^{a;x}\left(\mathbf{X}_t \in \cdot \,, \xi_{B(x_0,r)} > t\right)$ admits a density $p^a_{B(x_0,r)}$ $(t, x, y)\, dy$ with respect to the Lebesgue measure on $\mathfrak{g}$. Moreover, if x, y are two elements of $B^a(x_0, r)$ joined by a curve γ which is at a d^a-distance $R > 0$ of $\mathfrak{g}/B^a(x_0, r)$, there exists a constant $C_{ll} = C_{ll}(\Lambda)$ such that

$$p^a_{B(x_0,r)}(t, x, y) \ge \frac{1}{C_{ll}\delta^{d^2/2}} \exp\left(-C_{ll}\frac{d^a(x,y)^2}{t}\right) \exp\left(-\frac{C_{ll}t}{R^2}\right)$$

where $\delta = \min\{t, R^2\}$.

16.4 Markovian rough paths

The considerations of the previous section, with $\mathfrak{g} = \mathfrak{g}^N(\mathbb{R}^d)$ and uniformly elliptic matrix $a \in \Xi(\Lambda) = \Xi^{N,d}(\Lambda)$, apply to every fixed $N \in \{1, 2, \dots\}$. Corollary 16.12 tells us that the $\mathfrak{g}$-valued process $\mathbf{X}^a$ has a.s. sample paths of finite α-Hölder regularity (with respect to the Carnot–Caratheodory metric on $\mathfrak{g}$), for any $\alpha \in [0, 1/2)$.

For $N = 1$ and $a \in \Xi^{1,d}(\Lambda)$ we prefer to write X^a (instead of $\mathbf{X}^a \dots$) and note that X^a is an $\mathbb{R}^d$-valued Markov process. Similar to Brownian motion, its sample paths are not geometric rough paths. If a is smooth, X^a is a semi-martingale but for general $a \in \Xi^{1,d}(\Lambda)$, X^a need not be a semi-martingale. For $N \geq 2$ we can pick any $\alpha \in \left(\frac{1}{N+1}, \frac{1}{N}\right)$ and thus obtain a Markov process $\mathbf{X}^a$ whose sample paths are a.s. α-Hölder geometric rough paths. Of course, by means of the Lyons lift we have a *deterministic* one-to-one correspondence (cf. Theorem 9.12), applicable to almost every realization of $\mathbf{X}^a$,

$$\tilde{\mathbf{X}}^a \equiv (1, \pi_1(\mathbf{X}^a), \pi_2(\mathbf{X}^a)) \leftrightarrow \mathbf{X}^a,$$

and we can recover $\{\mathbf{X}^a_s : 0 \leq s \leq t\}$ from its level-two projection $\left\{\tilde{\mathbf{X}}^a_s : 0 \leq s \leq t\right\}$. On the other hand, the projected process $\tilde{\mathbf{X}}^a$ need not be Markovian: for instance, when $N = 3$, the future evolution of $\mathbf{X}^a$ (and thus of $\tilde{\mathbf{X}}^a$) will depend on the current state of $\mathbf{X}^a \in \mathfrak{g}^3(\mathbb{R}^d)$ and thus, in general, on $\pi_3(\mathbf{X}^a)$ which is not part of the state space for $\tilde{\mathbf{X}}^a$.

Definition 16.14 (i) Let $N \geq 2$ and $a \in \Xi^{N,d}(\Lambda)$. Almost every sample path of a Markov process $\mathbf{X}^a$, constructed from the Dirichlet form $\mathcal{E}^a$ on $L^2\left(\mathfrak{g}^N(\mathbb{R}^d)\right)$, $a \in \Xi^{N,d}(\Lambda)$, is an α-Hölder geometric rough path,[10] for some $\alpha < 1/2$, and is called a **Markovian rough path**.
(ii) Let $a \in \Xi^{1,d}(\Lambda)$, $a \circ \pi_1 \in \Xi^{2,d}(\Lambda)$ and fix $x \in \mathfrak{g}^2(\mathbb{R}^d)$. The $\mathfrak{g}^2(\mathbb{R}^d)$-valued Markov process $\mathbf{X}^{a \circ \pi_1, x}$, constructed from the Dirichlet form $\mathcal{E}^{a \circ \pi_1}$ on $L^2(\mathfrak{g}^2(\mathbb{R}^d))$, is called a **natural lift** of $X^{a; \pi_1(x)}$ or an **enhanced Markov process**. $\qquad \square$

The "naturality" of our definition of an enhanced Markov process comes from various points of view.

    (i) If $a \in \Xi^{1,d}(\Lambda)$ is smooth then X^a is a semi-martingale; following Section 14.1, we can then enhance X^a with its stochastic area, say A^a, given by iterated Stratonovich integration, and so obtain a $\mathfrak{g}^2(\mathbb{R}^d)$-valued lift of X^a which is seen to have the same law as $\mathbf{X}^{a \circ \pi_1}$, defined via the Dirichlet form $\mathcal{E}^{a \circ \pi_1}$. (See Exercise 16.15 below.)

    (ii) If a general $a \in \Xi^{1,d}(\Lambda)$ is the limit of smooth $a^n \subset \Xi^{1,d}(\Lambda)$, then $\mathbf{X}^{a_n \circ \pi_1}$ converges weakly and the limiting law coincides with that implied by $\mathcal{E}^{a \circ \pi_1}$. (See Exercise 16.15 below.)

(ii bis) This implies that, if for general $a \in \Xi^{1,d}(\Lambda)$ we construct X^a via $\mathcal{E}^a$ (for instance, on the canonical path space $C\left([0,1], \mathbb{R}^d\right)$ with appropriate measure P^a) and also $\mathbf{X}^{a \circ \pi_1}$ via $\mathcal{E}^{a \circ \pi_1}$ (for instance,

---

[10] ... on any interval $[0, T]$ ...

on the canonical path space $C\left([0,1],\mathfrak{g}^2\left(\mathbb{R}^d\right)\right)$ with appropriate measure $\mathbf{P}^{a\circ\pi_1}$), then

$$\pi_1\left(\mathbf{X}^{a\circ\pi_1}\right) \overset{\mathcal{D}}{=} X^a \quad \text{or, equivalently,} \quad \left(\pi_1\right)_* \mathbf{P}^{a\circ\pi_1} = P^a. \quad (16.6)$$

(See Exercise 16.15 below.)

(iii) For any $a \in \Xi^{2,d}\left(\Lambda\right)$ we can construct $\mathbf{X} = \mathbf{X}^a$ via $\mathcal{E}^a$ (for instance, on the canonical path space $C\left([0,1],\mathfrak{g}^2\left(\mathbb{R}^d\right)\right)$ with appropriate measure $\mathbf{P}^a$). We will see that, *on that same probabilty space,* for $\alpha < 1/2$,

$$d_{\alpha\text{-H\"ol};[0,1]}\left(S_2\left(\pi_1\left(\mathbf{X}\right)^{D_m}\right),\mathbf{X}\right) \to 0 \text{ in } \mathbf{P}^a\text{-probability.}$$

Here x^{D_m} denotes the piecewise linear approximations, and (D_m) $\subset \mathcal{D}\left[0,1\right]$ is a sequence of dissection with mesh $|D_m| \to 0$. (See Theorem 16.25 in Section 16.5.2.)

(iv) If $a \in \Xi^{1,d}\left(\Lambda\right)$ we can construct $X = X^a$ via $\mathcal{E}^a$ (for instance, on the canonical path space $C\left([0,1],\mathbb{R}^d\right)$ with appropriate measure P^a). We can then ask if there exists a process $\tilde{\mathbf{X}}$ *defined on the same probability space as* X such that

$$d_{\alpha\text{-H\"ol}}\left(S_2\left(X^{D_m}\right),\tilde{\mathbf{X}}\right) \to 0 \text{ in } P^a\text{-probability for } \alpha < 1/2.$$

The answer is affirmative. Indeed, from (iii) $\left(\pi_1\left(\mathbf{X}\right)^{D_m},m \in \mathbb{N}\right)$ is Cauchy with respect to $d_{\alpha\text{-H\"ol}}$ in $\mathbf{P}^a$-probability and hence, using (16.6), also Cauchy in P^a-probability, with limit $\tilde{\mathbf{X}}$ say, constructed on the same probability space as X. On the other hand, since $\pi_1\left(\mathbf{X}\right)^{D_m} \overset{\mathcal{D}}{=} X^{D_m}$ for all m we must have $\tilde{\mathbf{X}} \overset{\mathcal{D}}{=} \mathbf{X}$. (This also shows that the limit $\tilde{\mathbf{X}}$ *does not depend on the particular sequence* (D_m) underlying the construction of $\tilde{\mathbf{X}}$.)

Exercise 16.15 *Fix $x \in \mathfrak{g}^2\left(\mathbb{R}^d\right)$ and take, for simplicity of notation only, $x = 0$. Let $a \in \Xi^{1,d}\left(\Lambda\right)$ be smooth.*
(i) Construct $X^a = X^{a,0}$ as a solution to an Itô stochastic differential equation and verify that X^a is a semi-martingale.
(ii) From Section 14.1, we can enhance X^a with its stochastic area, say A^a, given by iterated Stratonovich integration, and so obtain a $\mathfrak{g}^2\left(\mathbb{R}^d\right)$-valued lift of X^a. Verify that this is consistent with the construction of $\mathbf{X}^{a\circ\pi_1} = \mathbf{X}^{a\circ\pi_1,0}$ via the Dirichlet form $\mathcal{E}^{a\circ\pi_1}$ on $L^2\left(\mathfrak{g}^2\left(\mathbb{R}^d\right)\right)$ in the sense that

$$\left(X^a,A^a\right) \overset{\mathcal{D}}{=} \mathbf{X}^{a\circ\pi_1}.$$

Deduce that, if $\mathbf{P}^{a\circ\pi_1}$ *denotes the law of* $\mathbf{X}^{a\circ\pi_1}$ *viewed as a Borel measure on* $C\left([0,1],\mathfrak{g}^2\left(\mathbb{R}^d\right)\right)$, *and similarly* P^a *denotes the law of* X^a *viewed as a Borel measure on* $C_0\left([0,1],\mathbb{R}^d\right)$, *then*

$$(\pi_1)_*\left(\mathbf{P}^{a\circ\pi_1}\right) = P^a.$$

(iii) Let $(\tilde{a}_n) \subset \Xi^{2,d}(\Lambda)$ *be smooth so that* $\tilde{a}_n \to \tilde{a} \in \Xi^{2,d}(\Lambda)$ *a.s. (which is always possible by using mollifier approximations for a given* $\tilde{a} \in \Xi^{2,d}(\Lambda)$). *As we shall see later in Section 16.6, this entails weak convergence (with respect to uniform topology, say)*

$$\mathbf{X}^{\tilde{a}_n} \implies \mathbf{X}^{\tilde{a}}.$$

Apply this convergence result to $\tilde{a}_n = a_n \circ \pi_1$, *where* $a_n \in \Xi^{1,d}$ *is a smooth mollifier approximation to* $a \in \Xi^{1,d}$. *Conclude that (16.6) remains valid for all* $a \in \Xi^{1,d}(\Lambda)$, *where* $\mathbf{P}^{a\circ\pi_1}$ *is fully determined by the (uniformly subelliptic) Dirichlet form* $\mathcal{E}^{a\circ\pi_1}$ *on* $L^2\left(\mathfrak{g}^2\left(\mathbb{R}^d\right)\right)$ *and* P^a *fully determined by the (uniformly elliptic) Dirichlet form* $\mathcal{E}^a$ *on* $L^2\left(\mathbb{R}^d\right)$.

16.5 Strong approximations

16.5.1 Geodesic approximations

Recall that $\mathfrak{g} = \mathfrak{g}^N\left(\mathbb{R}^d\right)$ equipped with Carnot–Caratheodory distance d is a geodesic space. Given a dissection D of $[0,1]$ and a deterministic path $\mathbf{x} \in C^{\alpha\text{-Höl}}\left([0,1],\mathfrak{g}^N\left(\mathbb{R}^d\right)\right)$ we can approximate $\mathbf{x}$ by its piecewise geodesic approximation, denoted by $\mathbf{x}^D$, obtained by connecting the points $(\mathbf{x}_{t_i} : t_i \in D)$ with geodesics run at unit speed. This was already discussed in Section 5.2, where we saw that

$$\sup_{D \in \mathcal{D}[0,1]} \left\|\mathbf{x}^D\right\|_{\alpha\text{-Höl};[0,1]} \le 3^{1-\alpha} \left\|\mathbf{x}\right\|_{\alpha\text{-Höl};[0,1]} \tag{16.7}$$

and also $\mathbf{x}^D \to \mathbf{x}$ uniformly on $[0,1]$ as $|D| \to 0$. Of course, our state space $\mathfrak{g}$, which may be identified with $G^N\left(\mathbb{R}^d\right)$, has additional structure. The geodesic approximation $\mathbf{x}^D$ has finite length and so has its projection $\pi_1\left(\mathbf{x}^D\right)$ to an $\mathbb{R}^d$-valued path. We can then recover $\mathbf{x}^D$ by computing area-integral(s) of $\pi_1\left(\mathbf{x}^D\right)$, formally

$$\mathbf{x}^D = \log S_N\left(\pi_1\left(\mathbf{x}^D\right)\right) \equiv \mathfrak{s}_N\left(\pi_1\left(\mathbf{x}^D\right)\right).$$

By interpolation (cf. Proposition 8.17) it is then clear that

$$d_{\alpha\text{-Höl}}\left(\log S_N\left(\pi_1\left(\mathbf{x}^D\right)\right), \mathbf{x}\right) \to 0 \text{ as } |D| \to 0.$$

Observe that $\pi_1\left(\mathbf{x}^D\right)$ is constructed based on knowledge of the entire $\mathfrak{g}$-valued path $\mathbf{x}$. In our present application, it would be enough to know only $\mathfrak{s}_2\left(\mathbf{x}\right)$, the projection of $\mathbf{x}$ to its first two levels. We have

Proposition 16.16 *Let* $N \geq 2$ *and* $\mathbf{x} \in C^{\alpha\text{-Höl}}\left([0,1], \mathfrak{g}^N\left(\mathbb{R}^d\right)\right)$ *for any* $\alpha < 1/2$. *Then for any* $k \in \{2, \ldots, N\}$,

$$d_{\alpha\text{-Höl}}\left(\mathfrak{s}_N\left(\pi_1\left(\mathfrak{s}_k\left(\mathbf{x}\right)^D\right)\right), \mathbf{x}\right) \to 0 \text{ as } |D| \to 0.$$

Remark 16.17 This proposition is purely deterministic and does not hold, in general, for $k = 1$. However, when $\mathbf{x} = \mathbf{x}(\omega)$ is a suitable sample path (see discussion below) then, for $k = 1$, convergence may hold almost surely.   □

Proof. It is enough to consider $k = 2$. Take $\alpha \in (1/3, 1/2)$ so that the projection $\mathfrak{s}_2(\mathbf{x}) \equiv (\pi_1(\mathbf{x}), \pi_2(\mathbf{x}))$ is a geometric α-Hölder rough path

which allows us to reconstruct the original path as the Lyons lift

$$\mathfrak{s}_N \circ \mathfrak{s}_2(\mathbf{x}) = \mathbf{x}.$$

Obviously, the geodesic approximations to $\mathfrak{s}_2(\mathbf{x})$, given by

$$[\mathfrak{s}_2(\mathbf{x})]^D = \mathfrak{s}_2 \circ \pi_1\left(\mathfrak{s}_2(\mathbf{x})^D\right),$$

converge uniformly with uniform α'-Hölder bounds, where $\alpha' \in (\alpha, 1/2)$ and then, by interpolation, in α-Hölder distance. By continuity of the Lyons lift, this implies

$$\mathfrak{s}_N \circ \pi_1\left(\mathfrak{s}_2(\mathbf{x})^D\right) \to \mathfrak{s}_N \circ \mathfrak{s}_2(\mathbf{x}) = \mathbf{x} \text{ as } |D| \to 0.$$

To see that this cannot be true for $k = 1$, it suffices to take a pure area rough path, say

$$t \mapsto (0, 0; t) \in \mathfrak{g}^2\left(\mathbb{R}^2\right)$$

which is $(1/2)$-Hölder. Obviously, no (lifted) geodesics approximation to $(0,0)$ can possibly recover the original path in $\mathfrak{g}^2\left(\mathbb{R}^2\right)$.   ∎

We emphasize that this approximation applies to Markovian rough paths $\mathbf{X}^{a;x}$ in a purely deterministic fashion, path-by-path, and requires (at least) a priori knowledge of *path and area*,

$$\mathfrak{s}_2\left(\mathbf{X}^{a;x}\right) \equiv \left(\pi_1\left(\mathbf{X}^{a;x}\right), \pi_2\left(\mathbf{X}^{a;x}\right)\right).$$

In contrast, we shall establish in the next section the *probabilistic statement* that (lifted) piecewise linear approximations to the $\mathbb{R}^d$-valued path $\pi_1\left(\mathbf{X}^{a;x}\right)$ also converge to $\mathbf{X}^{a;x}$, i.e.

$$d_{\alpha\text{-Höl}}\left(\mathfrak{s}_N\left(\pi_1(\mathbf{x})^D\right), \mathbf{x}\right) \to 0 \text{ as } |D| \to 0$$

in probability (and, in fact, in L^q for all $q < \infty$).

16.5.2 Piecewise linear approximations

In contrast to the just-discussed geodesic approximation, convergence of piecewise linear approximations, based on the $\mathbb{R}^d$-valued path $\pi_1\left(\mathbf{X}^{a,x}\right)$ alone and without a priori knowledge of the area $\pi_2\left(\mathbf{X}^{a,x}\right)$, is a genuine probabilistic statement and relies on subtle cancellations.

We maintain our standing notation, $\mathfrak{g} = \mathfrak{g}^N\left(\mathbb{R}^d\right)$, and $\Xi^{N,d}\left(\Lambda\right)$ denotes uniformly elliptic matrices defined on $\mathfrak{g}$. Recall that for any $a \in \Xi^{N,d}\left(\Lambda\right)$, we have constructed a $\mathfrak{g}$-valued diffusion process $\mathbf{X} = \mathbf{X}^a$, associated with the Dirichlet form $\mathcal{E}^a$. This process can be projected to an $\mathbb{R}^d$-valued process,

$$X = \pi_1\left(\mathbf{X}\right),$$

which will not be Markov in general.

Theorem 16.18 *Let $\alpha < 1/2$, $N \geq 2$ and $a \in \Xi^{N,d}\left(\Lambda\right)$. Then, for every $x \in \mathfrak{g}$ we have*

$$d_{\alpha\text{-Höl};[0,1]}\left(\mathfrak{s}_N\left(X^D\right),\mathbf{X}\right) \to 0 \text{ in } L^q\left(\mathbb{P}^{a,x}\right) \text{ as } |D| \to 0.$$

The proof stretches over the remainder of this section, and we shall just argue here how to reduce the proof to the seemingly simpler statements that

$$d_{\alpha\text{-Höl};[0,1]}\left(\mathfrak{s}_2\left(X^D\right),\tilde{\mathbf{X}}\right) \to 0 \text{ in probability}, \tag{16.8}$$

where $\tilde{\mathbf{X}} := \mathfrak{s}_2\left(\mathbf{X}\right) = \left(\pi_1\left(\mathbf{X}\right),\pi_2\left(\mathbf{X}\right)\right)$ is non-Markov in general, and

$$\sup_D \left\|\left\|\mathfrak{s}_2\left(X^D\right)\right\|_{\alpha\text{-Höl};[0,1]}\right\|_{L^q\left(\mathbb{P}^{a,x}\right)} < \infty. \tag{16.9}$$

(We will obtain (16.8), (16.9) in the forthcoming Theorems 16.25 and 16.19 below). Indeed, taking $\alpha \in \left(1/3,1/2\right)$ we can use continuity and basic estimates of the Lyons lift, to see that (16.8), (16.9) imply

$$d_{\alpha\text{-Höl};[0,1]}\left(\mathfrak{s}_N\left(X^D\right),\mathbf{X}\right) \to 0 \text{ in probability}, \tag{16.10}$$

$$\sup_D \left\|\left\|\mathfrak{s}_N\left(X^D\right)\right\|_{\alpha\text{-Höl};[0,1]}\right\|_{L^q\left(\mathbb{P}^{a,x}\right)} < \infty. \tag{16.11}$$

The convergence statement in Theorem 16.18 then follows a fortiori from general principles (based on interpolation), see Proposition A.15 in Appendix A. We now discuss the ideas that will lead us to the proof of (16.8) and (16.9).

The ideas

Fix a dissection $D = \{t_i : i\}$ of $[0,1]$ and $a \in \Xi\left(\Lambda\right)$. Let us project $\mathbf{X} = \mathbf{X}^a$ to the $\mathbb{R}^d$-valued process $X = X^a$ and consider piecewise linear approximations to X based on D, denoted by X^D. Of course, X^D has a canonically

defined area given by the usual iterated integrals and thus gives rise to a
$\mathfrak{g}$-valued path which we denote by $\mathfrak{s}_2\left(X^D\right)$. For $0 \leq \alpha < 1/2$ as usual, the
convergence

$$d_{\alpha\text{-Höl}}\left(\mathfrak{s}_2\left(X^D\right), \tilde{\mathbf{X}}\right) \to 0 \text{ in probability} \qquad (16.12)$$

as $|D| \to 0$ is a subtle problem and the difficulty is already present in the
pointwise convergence statement

$$\mathfrak{s}_2\left(X^D\right)_{0,t} \to \tilde{\mathbf{X}}_{0,t} \text{ as } |D| \to 0.$$

Our idea is simple. Noting that straight-line segments do not produce
area, it is an elementary application of the Campbell–Baker–Hausdorff formula to see that for $t \in D = \{t_i\}$,

$$\left(\mathfrak{s}_2\left(X^D\right)_{0,t}\right)^{-1} * \tilde{\mathbf{X}}_{0,t} = \sum_i A_{t_i,t_{i+1}}, \qquad (16.13)$$

where A is the area of $\tilde{\mathbf{X}}$ and $\cup_i\left[t_i, t_{i+1}\right] = [0, t]$. On the other hand, it is
relatively straightforward to show that the L^p norm of $\left\|\mathfrak{s}_2\left(X^D\right)\right\|_{\alpha\text{-Höl};[0,1]}$
is finite uniformly over all D. In essence, this reduces (16.12) to the point-
wise convergence statement, which we can rephrase as $\sum_i A_{t_i,t_{i+1}} \to 0$. It
is natural to show this in L^2, since this allows us to write[11]

$$\mathbb{E}\left[\left|\sum_i A_{t_i,t_{i+1}}\right|^2\right] = \sum_i \mathbb{E}\left(\left|A_{t_i,t_{i+1}}\right|^2\right) + 2\sum_{i<j} \mathbb{E}\left(A_{t_i,t_{i+1}} \cdot A_{t_j,t_{j+1}}\right).$$

For simplicity only, assume $t_{i+1} - t_i \equiv \delta$ for all i. As a sanity check, if X
were a Brownian motion and A the usual Lévy area, all off-diagonal terms
are zero and

$$\sum_i \mathbb{E}\left(\left|A_{t_i,t_{i+1}}\right|^2\right) \sim \sum_i \delta^2 \sim \frac{1}{\delta}\delta^2 \to 0 \text{ with } |D| = \delta \to 0,$$

which is what we want. Back to the general case of $\mathbf{X} = \mathbf{X}^a$, the plan must
be to cope with the off-diagonal sum. Since there are $\sim \delta^2/2$ terms, what
we need is

$$\mathbb{E}\left(A_{t_i,t_{i+1}} \cdot A_{t_j,t_{j+1}}\right) = o\left(\delta^2\right).$$

To this end, let us momentarily assume that

$$\sup_{x \in \mathfrak{g}} \mathbb{E}^{a,x}\left(A_{0,\delta}\right) = o\left(\delta\right) \qquad (16.14)$$

---

[11]Recall that $so\left(d\right) \subset \mathbb{R}^d \otimes \mathbb{R}^d$ has Euclidean structure, i.e. $A \cdot \tilde{A} = \sum_{k,l=1}^d A^{k,l}\tilde{A}^{k,l}$
and $|A|^2 = A \cdot A$. It may be instructive to consider $d = 2$, in which case A can be viewed
as scalar.

holds. Then, using the Markov property,[12]

$$\left| \mathbb{E}\left(A_{t_i,t_{i+1}} \cdot A_{t_j,t_{j+1}} \right) \right| \leq \mathbb{E}\left(\left| A_{t_i,t_{i+1}} \right| \times \left| \mathbb{E}^{\mathbf{X}_{t_j}} A_{0,\delta} \right| \right) = \mathbb{E}\left(\left| A_{t_i,t_{i+1}} \right| \right) \times o\left(\delta \right)$$

and since $\mathbb{E}\left(\left| A_{t_i,t_{i+1}} \right| \right) \sim \delta$, by a soft scaling argument, we are done. Unfortunately, (16.14) seems to be too strong to be true, but we are able to establish a weak version of (16.14) which is good enough to successfully implement what we just outlined. The key to all this (cf. the proof of the forthcoming Proposition 16.20) is a semi-group argument which leads to the desired cancellations.

Uniform Hölder bound

Let X^D denote the piecewise linear approximation to $X = X\left(\omega \right)$. We will need L^q-bounds, uniformly over all dissections D, of the homogenous α-Hölder norm of the path X^D and its area. That is, we want

$$\sup_D \left| \left\| \mathfrak{s}_2 \left(X^D \right) \right\|_{\alpha\text{-Höl};[0,1]} \right|_{L^q\left(\mathbb{P}^{a;x} \right)} < \infty.$$

This will follow a fortiori from the following uniform Fernique estimates.

Theorem 16.19 *There exists* $\eta = \eta\left(\Lambda \right) > 0$ *such that*

$$\sup_{a \in \Xi(\Lambda), x \in \mathfrak{g}} \sup_D \sup_{0 \leq s < t \leq 1} \mathbb{E}^{a,x}\left(\exp\left(\eta \frac{\left\| \mathfrak{s}_2 \left(X^D \right)_{s,t} \right\|^2}{t - s} \right) \right) < \infty. \qquad (16.15)$$

As a consequence, for any $\alpha \in [0, 1/2)$ *there exists* $C = C\left(\alpha, \Lambda \right) > 0$ *so that*

$$\sup_{a \in \Xi(\Lambda), x \in \mathfrak{g}} \sup_D \mathbb{E}^{a,x}\left(\exp\left(C \left\| \mathfrak{s}_2 \left(X^D \right) \right\|^2_{\alpha\text{-Höl};[0,1]} \right) \right) < \infty.$$

Proof. Estimate (16.15) shows that the process $\mathfrak{s}_2 \left(X^D \right)$ satisfies the Gaussian integrability condition put forward in Section A.4, Appendix A, uniformly over a, x, D as indicated. The consequence then follows from general principles, found in the same appendix. (We could also obtain uniform $\psi_{2,1}$-variation estimates.) In other words, we only have to establish (16.15). To this end, we recall from Corollary 16.12 that for $\eta \in [0, \frac{1}{4\Lambda})$,

$$M_\eta \equiv \sup_{a \in \Xi(\Lambda), x \in \mathfrak{g}} \sup_{0 \leq s < t \leq 1} \mathbb{E}^{a,x}\left(\exp\left(\eta \frac{\left\| \mathbf{X}_{s,t} \right\|^2}{t - s} \right) \right) < \infty.$$

---

[12]It is important that we condition with respect to $\mathbf{X}_{t_j} \in \mathfrak{g}^N\left(\mathbb{R}^d \right)$ and not $\hat{\mathbf{X}}_{t_j} \in \mathfrak{g}^2\left(\mathbb{R}^d \right)$, since $\mathbf{X}$ is Markov whereas, in general, $\hat{\mathbf{X}}$ is not.

Then, by the triangle inequality,[13]

$$\frac{\left\| \mathfrak{s}_2\left(X^D\right)_{s,t} \right\|}{\sqrt{t-s}} \leq \frac{\left\| \mathfrak{s}_2\left(X^D\right)_{s,s^D} \right\|}{\sqrt{s^D-s}} + \frac{\left\| \mathfrak{s}_2\left(X^D\right)_{s^D,t_D} \right\|}{\sqrt{t_D-s^D}} + \frac{\left\| \mathfrak{s}_2\left(X^D\right)_{t_D,t} \right\|}{\sqrt{t-t_D}}$$

$$\leq \frac{\left| X^D_{s,s^D} \right|}{\sqrt{s^D-s}} + \frac{\left\| \mathfrak{s}_2\left(X^D\right)_{s^D,t_D} \right\|}{\sqrt{t_D-s^D}} + \frac{\left| X^D_{t_D,t} \right|}{\sqrt{t-t_D}}$$

$$\leq \frac{\left\| \mathbf{X}_{s,s^D} \right\|}{\sqrt{s^D-s}} + \frac{\left\| \mathfrak{s}_2\left(X^D\right)_{s^D,t_D} \right\|}{\sqrt{t_D-s^D}} + \frac{\left\| \mathbf{X}_{t_D,t} \right\|}{\sqrt{t-t_D}}$$

$$\leq \left(\frac{3\left\| \mathbf{X}_{s,s^D} \right\|^2}{s^D-s} + \frac{3\left\| \mathfrak{s}_2\left(X^D\right)_{s^D,t_D} \right\|^2}{t_D-s^D} + \frac{3\left\| \mathbf{X}^x_{t_D,t} \right\|^2}{t-t_D} \right)^{1/2}.$$

Hence,

$$\mathbb{E}^{a,x}\left(\exp\left(\eta \frac{\left\| \mathfrak{s}_2\left(X^D\right)_{s,t} \right\|^2}{t-s} \right) \right)$$

$$\leq \mathbb{E}^{a,x}\left\{ \exp\left[\eta \left(\frac{3\left\| \mathbf{X}_{s,s^D} \right\|^2}{s^D-s} + \frac{3\left\| \mathfrak{s}_2\left(X^D\right)_{s^D,t_D} \right\|^2}{t_D-s^D} + \frac{3\left\| \mathbf{X}_{t_D,t} \right\|^2}{t-t_D} \right) \right] \right\}$$

$$\leq M_{6\eta}^2 \, \mathbb{E}^{a,x}\left(\exp\left(6\eta \frac{\left\| \mathfrak{s}_2\left(X^D\right)_{s^D,t_D} \right\|^2}{t_D-s^D} \right) \right)$$

and the proof is reduced to show that for some $\eta > 0$ small enough,

$$\sup_{a\in\Xi(\Lambda),x\in\mathfrak{g}} \sup_D \sup_{s<t\in D} \mathbb{E}^{a,x}\left(\exp\left(6\eta \frac{\left\| \mathfrak{s}_2\left(X^D\right)_{s,t} \right\|^2}{t-s} \right) \right) < \infty.$$

By the triangle inequality for the Carnot–Caratheodory distance, for $t_i, t_j \in D$,

$$\left\| \mathfrak{s}_2\left(X^D\right)_{t_i,t_j} \right\| \leq \left\| \tilde{\mathbf{X}}_{t_i,t_j} \right\| + d\left(\tilde{\mathbf{X}}_{t_i,t_j}, \mathfrak{s}_2\left(X^D\right)_{t_i,t_j} \right).$$

To proceed we note that, similar to equation (16.13),

$$\left(\mathfrak{s}_2\left(X^D\right)_{t_i,t_j} \right)^{-1} * \tilde{\mathbf{X}}_{t_i,t_j} = \sum_{k=i}^{j-1} A_{t_k,t_{k+1}}.$$

---

[13] Note that $\left\| \hat{\mathbf{X}}_{s,t} \right\| \leq \left\| \mathbf{X}_{s,t} \right\|$ for all $s < t$.

By left-invariance of the Carnot–Caratheodory distance d and equivalence of continuous homogenous norms (so that, in particular, $\|(x, A)\| \sim |x| + |A|^{1/2}$ where $|\cdot|$ denotes Euclidean norm on $\mathbb{R}^d$ resp. $\mathbb{R}^d \otimes \mathbb{R}^d$), there exists C such that

$$
\begin{aligned}
d\left(\tilde{\mathbf{X}}_{t_i, t_j}, \mathbf{S}_2\left(X^D\right)_{t_i, t_j}\right) &= \left\|\left(0, \sum_{k=i}^{j-1} A_{t_k, t_{k+1}}\right)\right\| \\
&\leq C\left|\sum_{k=i}^{j-1} A_{t_k, t_{k+1}}\right|^{1/2} \leq C\sqrt{\sum_{k=i}^{j-1}\left|A_{t_k, t_{k+1}}\right|} \\
&\leq C\sqrt{\sum_{k=i}^{j-1}\left\|\mathbf{X}_{t_k, t_{k+1}}\right\|^2}.
\end{aligned}
$$

By Cauchy–Schwarz,

$$
\begin{aligned}
&\mathbb{E}^{a, x}\left(\exp\left(6\eta\frac{\left\|\mathbf{S}_2\left(X^D\right)_{t_i, t_j}\right\|^2}{t_j - t_i}\right)\right) \\
&\leq \mathbb{E}^{a, x}\left(\exp\left(12\eta\frac{\left\|\mathbf{X}_{t_i, t_j}\right\|^2}{t_j - t_i}\right)\exp\left(12C\eta\frac{\sum_{k=i}^{j-1}\left\|\mathbf{X}_{t_k, t_{k+1}}\right\|^2}{t_j - t_i}\right)\right) \\
&\leq M_{24\eta}\mathbb{E}^{a, x}\left(\prod_{k=i}^{j-1}\exp\left(24C\eta\frac{\left\|\mathbf{X}_{t_k, t_{k+1}}\right\|^2}{t_j - t_i}\right)\right)
\end{aligned}
$$

and the $\mathbb{E}^{a, x}(\dots)$ term in the last line is estimated using the Markov property as follows:

$$
\begin{aligned}
&\mathbb{E}^{a, x}\left(\prod_{k=i}^{j-1}\exp\left(24C\eta\frac{\left\|\mathbf{X}_{t_k, t_{k+1}}\right\|^2}{t_j - t_i}\right)\right) \\
&\leq \prod_{k=i}^{j-1}\sup_{x\in\mathfrak{g}}\mathbb{E}^{a, x}\left(\exp\left(24C\eta\frac{t_{k+1}-t_k}{t_j - t_i}\frac{\left\|\mathbf{X}_{0, t_{k+1}-t_k}\right\|^2}{t_{k+1}-t_k}\right)\right) \\
&\leq \prod_{k=i}^{j-1}M_{24C\eta\frac{t_{k+1}-t_k}{t_j - t_i}} \\
&\leq \prod_{k=i}^{j-1}\exp\left(C' \times 24C\eta\frac{t_{k+1}-t_k}{t_j - t_i}\right) \quad \text{for } \eta \text{ small enough} \\
&= \exp\left(24C'C\eta\right) < \infty,
\end{aligned}
$$

where we used the "estimate on M_η" given in Corollary 16.12, valid for η small enough. The proof is then finished. ∎

The subtle cancellation

Let us define

$$
r_\delta(t, x) := \frac{1}{\delta}\mathbb{E}^{a, x}\left(A_{t, t+\delta}\right) \in so(d) \quad \text{and} \quad r_\delta(x) := r_\delta(0, x).
$$

For instance, (16.14) is now expressed as $\lim_{\delta \to 0} r_\delta(x) \to 0$ uniformly in x. Our goal here is to establish a weak version of this. We also recall that

$$A_{t,t+\delta} = \pi_2 \left(\mathbf{X}_{t,t+\delta} \right) = \pi_2 \left(\mathbf{X}_t^{-1} * \mathbf{X}_{t+\delta} \right).$$

Proposition 16.20 *(i) We have uniform boundedness of* $r_{\delta;t}(x)$,

$$\sup_{x \in \mathfrak{g}^2(R^d)} \sup_{\delta \in [0,1]} \sup_{t \in [0,1-\delta]} r_\delta(t,x) < \infty.$$

(ii) For all $h \in L^1(\mathfrak{g}, dm)$,

$$\lim_{\delta \to 0} \int_{\mathfrak{g}} dx\, h(x)\, r_\delta(x) \equiv 0.$$

Proof. (i) follows from Theorem 16.19. For (ii) we first note that it suffices to consider h smooth and compactly supported. Now the problem is local and we can assume that smooth locally bounded functions such as the coordinate projections $\pi_{1;j}$ and $\pi_{2;k,l}$ are in $D(\mathcal{E}^a)$. (More formally, we could smoothly truncate outside the support of h and work on a big torus.) Clearly, it is enough to show the componentwise statement

$$\lim_{\delta \to 0} \int_{\mathfrak{g}} dx\, h(x)\, \pi_{2;k,l}(r_\delta(x)) \equiv 0$$

for $k < l$ fixed in $\{1, \ldots, d\}$. To keep the notation short we set $f \equiv \pi_{2;k,l}(\cdot)$ and abuse it by writing A instead of $A^{k,l}$. We can then write

$$\mathbb{E}^{a,\cdot}(A_t) \equiv \mathbb{E}^{a,\cdot}(f(\mathbf{X}_t)) =: \mathcal{P}_t^a f(.)$$

and note that $\mathcal{P}_0^a f(x) = A$ when $x = (x^1, A) \in \mathfrak{g}$. Writing $\langle \cdot, \cdot \rangle$ for the usual inner product on $L^2(\mathfrak{g}, dx)$, we have

$$
\begin{aligned}
\langle h, \mathbb{E}^{a,\cdot} A_{0,t} \rangle &= \left\langle h, \mathbb{E}^{a,\cdot} f(\mathbf{X}_t) - A - \frac{1}{2} \mathbb{E}^{a,\cdot} \left([\pi_1(\cdot), \mathbf{X}_t^1] \right) \right\rangle \\
&= \langle h, \mathcal{P}_t^a f - \mathcal{P}_0^a f \rangle - \left\langle h, \frac{1}{2} \mathbb{E}^{a,\cdot} \left([\pi_1(\cdot), \mathbf{X}_t^1] \right) \right\rangle \\
&= \int_0^t \mathcal{E}^a(h, \mathcal{P}_s^a f) - \left\langle h, \frac{1}{2} \mathbb{E}^{a,\cdot} \left([\pi_1(\cdot), \mathbf{X}_t^1] \right) \right\rangle \\
&= \mathcal{E}^a(h, f) \times t - \left\langle h, \frac{1}{2} \mathbb{E}^{a,\cdot} \left([\pi_1(\cdot), \mathbf{X}_t^1] \right) \right\rangle + o(t).
\end{aligned}
$$

Here, again, we abused the notation by writing $[\cdot, \cdot]$ instead of picking out the (k,l) component and using the cumbersome notation $[\cdot, \cdot]^{k,l}$. Note that in general $\mathcal{E}^a(h, f) \times t \neq o(t)$ and our only hope is cancellation of $2\mathcal{E}^a(h, f)$ with the bracket term

$$\left\langle h, \mathbb{E}^{a,\cdot} \left([\pi_1(\cdot), \mathbf{X}_t^1] \right) \right\rangle \equiv \left\langle h, \mathbb{E}^{a,\cdot} \left([\pi_1(\cdot), \mathbf{X}_t^1]^{k,l} \right) \right\rangle.$$

To see this cancellation, we compute the bracket term

$$\langle h, \mathbb{E}^{a,\cdot}\left([\pi_1\left(\cdot\right), \mathbf{X}_t^1]^{k,l}\right)\rangle = \int dx\, h\left(x\right) \mathbb{E}^{a,x}\left(x^{1;k}\mathbf{X}_t^{1;l} - x^{1;l}\mathbf{X}_t^{1;k}\right)$$

$$= \int dx\, h\left(x\right)\left(\left(x^{1;k}\left[\mathcal{P}_t^a \pi_{1;l}\right]\left(x\right)\right. \right.$$
$$\left.\left. - x^{1;l}\left[\mathcal{P}_t^a \pi_{1;k}\right]\left(x\right)\right)\right),$$

and by adding and subtracting $x^{1;k}x^{1;l}$ inside the integral this rewrites as

$$\int dx\, h\left(x\right) x^{1;k}\left\{\left[\mathcal{P}_t^a \pi_{1;l}\right]\left(x\right) - \pi_{1;l}\left(x\right)\right\}$$
$$- \int dx\, h\left(x\right) x^{1;l}\left\{\left[\mathcal{P}_t^a \pi_{1;k}\right]\left(x\right) - \pi_{1;k}\left(x\right)\right\}.$$

It now follows, as earlier, that

$$\langle h, \mathbb{E}^{a,\cdot}\left([\pi_1\left(\cdot\right), \mathbf{X}_t^1]^{k,l}\right)\rangle = \left[\mathcal{E}^a\left(h\pi_{1;k}, \pi_{1;l}\right) - \mathcal{E}^a\left(h\pi_{1;l}, \pi_{1;k}\right)\right] \times t + o\left(t\right)$$

and we see that the required cancellation takes place if, for all h smooth and compactly supported,

$$\left[\mathcal{E}^a\left(h\pi_{1;k}, \pi_{1;l}\right) - \mathcal{E}^a\left(h\pi_{1;l}, \pi_{1;k}\right)\right] \underset{\text{(to be checked)}}{=} 2\mathcal{E}^a\left(h, \pi_{2;k,l}\right).$$

We will check this with a direct computation. First note that

$$\mathcal{E}^a\left(h\pi_{1;k}, \pi_{1;l}\right) - \mathcal{E}^a\left(h\pi_{1;l}, \pi_{1;k}\right) = \int \pi_{1,k}d\Gamma^a\left(h, \pi_{1,l}\right) - \int \pi_{1,l}d\Gamma^a\left(h, \pi_{1,k}\right)$$

which is immediately seen via symmetry of $d\Gamma^a\left(\cdot, \cdot\right)$, inherited from the symmetry of $\left(a^{ij}\right)$, and the Leibnitz formula

$$\mathcal{E}^a\left(gg', h\right) = \int g d\Gamma^a\left(g', h\right) + \int g' d\Gamma^a\left(g, h\right).$$

It is immediately checked from the definition of the vector fields $\mathfrak{u}_i$, see equation (16.1), that

$$\mathfrak{u}_i f \equiv \mathfrak{u}_i \pi_{2;k,l} = \begin{cases} -\left(1/2\right)\pi_{1;l} & \text{if } i = k \\ \left(1/2\right)\pi_{1;k} & \text{if } i = l \\ 0 & \text{otherwise} \end{cases}$$

so that (noting $\pi_{1,k} = 2\mathfrak{u}_i f$ and also using $\mathfrak{u}_j\pi_{1;l} = \delta_{jl}$, i.e. 1 if $j = l$ and 0 otherwise)

$$\int \pi_{1,k}d\Gamma^a\left(h, \pi_{1,l}\right) = \sum_{i,j}\int \pi_{1,k}a^{ij}\mathfrak{u}_i h \mathfrak{u}_j \pi_{1,l} = 2\sum_i \int \left(\mathfrak{u}_l f\right) a^{il}\left(\mathfrak{u}_i h\right)$$

and similarly

$$-\int \pi_{1,l} d\Gamma^a\left(h, \pi_{1,k}\right) = \sum_{i,j}\int\left(-\pi_{1,l}\right)a^{ij}\mathfrak{u}_i h\mathfrak{u}_j\pi_{1,k} = 2\sum_i\int\left(\mathfrak{u}_k f\right)a^{ik}\left(\mathfrak{u}_i h\right).$$

Therefore, using $\mathfrak{u}_j f = 0$ for $j \neq \{k, l\}$ in the second equality,

$$\begin{aligned}
\mathcal{E}^a\left(h\pi_{1;k}, \pi_{1;l}\right) - \mathcal{E}^a\left(h\pi_{1;l}, \pi_{1;k}\right) &= 2\sum_{j=k,l}\sum_i\int\left(\mathfrak{u}_j f\right)a^{ij}\left(\mathfrak{u}_i h\right) \\
&= 2\sum_{i,j}\int\left(\mathfrak{u}_j f\right)a^{ij}\left(\mathfrak{u}_i h\right)
\end{aligned}$$

and this equals precisely $2\mathcal{E}^a\left(h, f\right)$ as required.   ∎

Corollary 16.21 *For all $t \in [0,1)$ and all $h \in L^1\left(\mathfrak{g}, dx\right)$,*

$$\lim_{\delta\to 0}\int_{\mathfrak{g}} dx h\left(x\right)\mathbb{E}^{a,x}\left(\frac{A_{t,t+\delta}}{\delta}\right) \equiv 0.$$

Proof. We first write

$$\begin{aligned}
\int dx h\left(x\right)\mathbb{E}^{a,x}\left(\frac{A_{t,t+\delta}}{\delta}\right) &= \int\int h\left(x\right)p^a\left(t,x,y\right)r_\delta\left(y\right)dxdy \\
&= \int\left(\int h\left(x\right)p^a\left(t,x,y\right)dx\right)r_\delta\left(y\right)dy.
\end{aligned}$$

Then, noting that $y \mapsto \int h\left(x\right)p_t\left(x,y\right)dx$ is in $L^1\left(\mathfrak{g}, dx\right)$, the proof is finished by applying the previous proposition.   ∎

Theorem 16.22 *For all bounded sets $K \subset \mathfrak{g}$ and all $\sigma \in (0,1]$,*

$$\lim_{\delta\to 0}\sup_{t\in[\sigma,1]}\sup_{y\in K}\left|\mathbb{E}^{a,y}\left(\frac{A_{t,t+\delta}}{\delta}\right)\right| = 0.$$

Proof. It suffices to prove this for a compact ball $K = \bar{B}\left(0,R\right) \subset \mathfrak{g}$ of arbitrary radius $R > 0$. We fix $\sigma \in (0,1]$ and think of $r_\delta = r_\delta\left(t,y\right)$ as a family of maps, indexed by $\delta > 0$, defined on the cylinder $[\sigma,1] \times K$; that is,

$$\left(t,y\right) \in [\sigma,1] \times K \mapsto r_\delta\left(t,y\right) \in so\left(d\right).$$

By Proposition 16.20(i) we know that $\sup_{\delta>0}\left|r_\delta\right|_\infty < \infty$. We now show equicontinuity of $\{r_\delta : \delta > 0\}$. By the Markov property,

$$\begin{aligned}
r_\delta\left(t,y\right) &= \mathbb{E}^{a,y}\left(\frac{A_{t,t+\delta}}{\delta}\right) \\
&= \left\langle p^a\left(t,y,\cdot\right), \frac{\mathbb{E}^{a,\cdot}\left(A_{0,\delta}\right)}{\delta}\right\rangle \\
&= \left\langle p^a\left(t,y,\cdot\right), r_\delta\left(0,\cdot\right)\right\rangle,
\end{aligned}$$

so that, for all $(s, x), (t, y) \in [\sigma, 1] \times K$,

$$
\begin{aligned}
|r_\delta (s, x) - r_\delta (t, y)| &= |\langle p^a (s, x, \cdot) - p^a (t, y, \cdot), r_\delta (.) \rangle| \\
&\leq \left(\sup_{\delta \in (0,1]} |r_\delta|_\infty \right) |p^a (s, x, \cdot) - p^a (t, y, \cdot)|_{L^1} .
\end{aligned}
$$

From de Giorgi–Moser–Nash regularity (Proposition 16.7),

$$
(t, y) \in [\sigma, 1] \times K \mapsto p^a (t, y, z)
$$

is continuous for all z; the dominated convergence theorem then easily gives continuity of $(t, y) \mapsto p^a (t, y, \cdot) \in L^1$. In fact, this map is uniformly continuous when restricted to the compact $[\sigma, 1] \times K$, and it follows that $\{r_\delta : \delta > 0\}$ is equicontinuous as claimed. By Arzela–Ascoli, there exists a subsequence (δ^n) such that r_{δ^n} converges uniformly on $[\sigma, 1] \times K$ to some (continuous) function r. On the other hand Proposition 16.20 (ii), applied to $h = p^a (t, y, \cdot)$, shows that $r_\delta (t, y) \to 0$ as $\delta \to 0$ for all fixed $y, t > 0$. This shows that $r \equiv 0$ is the only limit point and hence

$$
\lim_{\delta \to 0} \sup_{t \in [\sigma, 1]} \sup_{y \in K} \left| \mathbb{E}^{a,y} \left(\frac{A_{t,t+\delta}}{\delta} \right) \right| = 0.
$$

∎

Convergence of the sum of the small areas

For fixed $a \in \Xi (\Lambda)$ and $x \in \mathfrak{g}$ let us define the real-valued quantity

$$
K_{\sigma, \delta} := \sup_{\substack{0 \leq u_1 < u_2 < v_1 < v_2 \leq 1: \\ v_1 - u_2 \geq \sigma, \\ |u_2 - u_1|, |v_2 - v_1| \leq \delta}} \frac{|\mathbb{E}^{a,x} (A_{u_1, u_2} \cdot A_{v_1, v_2})|}{(u_2 - u_1)(v_2 - v_1)}
$$

where $\delta, \sigma \in (0, 1)$. As above, $\cdot$ denotes the scalar product in $so (d)$.

Proposition 16.23 *For fixed $\sigma \in (0, 1)$, $k, l \in \{1, \dots, d\}$ we have $\lim_{\delta \to 0} K_{\sigma, \delta} = 0$.*

Proof. By the Markov property,[14]

$$\frac{\left|\mathbb{E}^{a,x}\left(A_{u_1,u_2}\cdot A_{v_1,v_2}\right)\right|}{(u_2-u_1)(v_2-v_1)}=$$

$$\frac{\left|\mathbb{E}^{a,x}\left(A_{u_1,u_2}\cdot\mathbb{E}^{a,\mathbf{X}_{u_2}}\left(A_{v_1-u_2,v_2-u_2}\right)\right)\right|}{(u_2-u_1)(v_2-v_1)}$$

$$\leq\frac{\left|\mathbb{E}^{a,x}\left(A_{u_1,u_2}\cdot\mathbb{E}^{a,\mathbf{X}_{u_2}}\left(A_{v_1-u_2,v_2-u_2};\|\mathbf{X}_{u_2}\|\leq R\right)\right)\right|}{(u_2-u_1)(v_2-v_1)}$$

$$+\frac{\left|\mathbb{E}^{a,x}\left(A_{u_1,u_2}\cdot\mathbb{E}^{a,\mathbf{X}_{u_2}}\left(A_{v_1-u_2,v_2-u_2};\|\mathbf{X}_{u_2}\|> R\right)\right)\right|}{(u_2-u_1)(v_2-v_1)}$$

$$\leq\frac{\mathbb{E}^{a,x}\left(|A_{u_1,u_2}|;\|\mathbf{X}_{u_2}\|\leq R\right)}{(u_2-u_1)}\sup_{\substack{\delta'\leq\delta\ \|y\|\leq R\\ u\in[\sigma,1]}}\frac{\left|\mathbb{E}^{a,y}\left(A_{u,u+\delta'}\right)\right|}{\delta'}$$

$$+\mathbb{E}^{a,x}\left(\frac{|A_{u_1,u_2}|}{u_2-u_1};\|\mathbf{X}_{u_2}\|> R\right)\sup_{\substack{\delta'\leq\delta\ \|y\|\leq R\\ u\in[\sigma,1]}}\frac{\mathbb{E}^{a,x}\left(|A_{u,u+\delta'}|\right)}{\delta'}$$

$$\leq\frac{\mathbb{E}^{a,x}\left(|A_{u_1,u_2}|\right)}{(u_2-u_1)}\sup_{\substack{\delta'\leq\delta\ \|y\|\leq R\\ u\in[\sigma,1]}}\frac{\left|\mathbb{E}^{a,y}\left(A_{u,u+\delta'}\right)\right|}{\delta'}$$

$$+\sqrt{\mathbb{P}^{a,x}\left(\|\mathbf{X}_{u_2}\|> R\right)}\sqrt{\mathbb{E}^{a,x}\left(\left|\frac{A_{u_1,u_2}}{u_2-u_1}\right|^2\right)}\sup_{\delta',u,x}\frac{\mathbb{E}^{a,x}\left(|A_{u,u+\delta'}|\right)}{\delta'}$$

$$\leq C\sup_{\substack{\delta'\leq\delta\ |y|\leq R\\ u\in[\sigma,1]}}\frac{\left|\mathbb{E}^{a,y}\left(A_{u,u+\delta'}\right)\right|}{\delta'}+C\sqrt{\mathbb{P}^{a,x}\left(\|\mathbf{X}_{u_2}\|> R\right)}$$

for some constant $C=C\left(\|x\|,\sigma,\Lambda\right)$ using Corollary 16.12 and Proposition 16.20(i). We then fix $\varepsilon>0$ and choose $R=R\left(\epsilon\right)$ large enough so that

$$C\sup_{u_2\in[0,1]}\sqrt{\mathbb{P}^{a,x}\left(\left|\mathbf{X}_{u_2}^x\right|> R\right)}\leq\varepsilon/2.$$

On the other hand, Theorem 16.22 shows that

$$C\sup_{\substack{\delta'\leq\delta\ |y|\leq R\\ u\in[\sigma,1]}}\frac{\left|\mathbb{E}^{a,y}\left(A_{u,u+\delta'}\right)\right|}{\delta'}\leq\frac{\varepsilon}{2}$$

for all δ small enough and the proof is finished. ∎

---

[14] Again, it is important to condition with respect to $\mathbf{X}.\in\mathfrak{g}^N\left(\mathbb{R}^d\right)$ and not $\hat{\mathbf{X}}.\in\mathfrak{g}^2\left(\mathbb{R}^d\right)$.

Corollary 16.24 *There exists* $C = C(\Lambda)$ *such that for all subdivisions* D *of* $[0,1]$, $s, t \in D$, *for any* $\sigma \in (0,1)$,

$$\mathbb{E}^{a,x}\left(\left|d\left(\mathfrak{s}_2\left(X^D\right)_{s,t}, \mathbf{X}_{s,t}\right)\right|^4\right) \leq C\left[(t-s)^2 K_{\sigma,|D|} + (t-s)\sigma\right].$$

Proof. Recalling the discussion around (16.13), equivalence of homogenous norms leads to

$$\mathbb{E}^{a,x}\left(\left|d\left(\mathfrak{s}_2\left(X^D\right)_{s,t}, \mathbf{X}_{s,t}\right)\right|^4\right) \leq c_1 \mathbb{E}^{a,x}\left(\left|\sum_{i:t_i \in D \cap [s,t)} A_{t_i,t_{i+1}}\right|^2\right).$$

Let us abbreviate $\sum_{i:t_i \in D \cap [s,t)}$ to $\sum_i$ in what follows. Clearly, $\mathbb{E}^{a,x}$ $\left(\left|\sum_i A_{t_i,t_{i+1}}\right|^2\right)$ is estimated by 2 times

$$\sum_{i \leq j} \mathbb{E}^{a,x}\left(A_{t_i,t_{i+1}} \cdot A_{t_j,t_{j+1}}\right)$$

$$\leq \sum_{\substack{i \leq j \\ t_j - t_{i+1} \geq \sigma}} \mathbb{E}^{a,x}\left(A_{t_i,t_{i+1}} \cdot A_{t_j,,tt_{j+1}}\right) + \sum_{\substack{i \leq j \\ t_j - t_{i+1} < \sigma}} \mathbb{E}^{a,x}\left(A_{t_i,t_{i+1}} \cdot A_{t_j,t_{j+1}}\right)$$

$$\leq K_{\sigma,|D|} \sum_{\substack{i \leq j \\ t_j - t_{i+1} \geq \sigma}} (t_{i+1} - t_i)(t_{j+1} - t_j)$$

$$+ \sum_{\substack{i \leq j \\ t_j - t_{i+1} < \sigma}} \sqrt{\mathbb{E}^{a,x}\left(\left|A_{t_i,t_{i+1}}\right|^2\right) \mathbb{E}^{a,x}\left(\left|A_{t_j,t_{j+1}}\right|^2\right)}$$

$$\leq K_{\sigma,|D|}(t-s)^2 + c_2 \sum_{\substack{i,j \\ t_j - t_{i+1} < \sigma}} (t_{i+1} - t_i)(t_{j+1} - t_j)$$

and the very last sum is estimated as follows:

$$\left|\sum_i (t_{i+1} - t_i) \sum_{\substack{j \\ t_j - t_{i+1} < \sigma}} (t_{j+1} - t_j)\right| \leq \sigma \sum_i (t_{i+1} - t_i) = \sigma(t-s).$$

The proof is finished.   ∎

Putting things together

Theorem 16.25 *Let* D *be a dissection of* $[0,1]$ *with mesh* $|D|$. *Then, for all* $1 \leq q < \infty$ *and* $0 \leq \alpha < 1/2$,

$$d_{\alpha\text{-Höl};[0,1]}\left(\mathfrak{s}_2\left(X^D\right), \mathbf{X}\right) \to 0 \text{ in } L^q\left(\mathbb{P}^{a,x}\right) \text{ as } |D| \to 0.$$

Proof. We first show pointwise convergence. We fix $\varepsilon > 0$ and apply Corollary 16.24 with $\sigma = \varepsilon/2C$. Then,

$$\sup_{\substack{s,t \in D \\ s < t}} \mathbb{E}^{a,x}\left(\left|d\left(\mathfrak{s}_2\left(X^D\right)_{s,t}, \mathbf{X}_{s,t}\right)\right|^4\right) \leq CK_{\sigma,|D|} + \frac{\varepsilon}{2}.$$

By Proposition 16.23 it then follows that, for $|D|$ small enough,

$$\sup_{\substack{s,t \in D \\ s < t}} \left\| d\left(\mathfrak{s}_2\left(X^D\right)_{s,t}, \mathbf{X}_{s,t}\right) \right\|_{L^4(\mathbb{P}^{a,x})}^4 \leq \varepsilon.$$

By Theorem 16.19 we have, for all $q \in [1, \infty)$,

$$\sup_{D} \left\| \left\| \mathfrak{s}_2\left(X^D\right) \right\|_{\alpha\text{-Höl};[0,1]} \right\|_{L^q(\mathbb{P}^{a,x})} + \left\| \left\| \mathbf{X} \right\|_{\alpha\text{-Höl};[0,1]} \right\|_{L^q(\mathbb{P}^{a,x})} < \infty \quad (16.16)$$

and both results combined yield

$$\lim_{|D| \to 0} \sup_{0 \leq s < t \leq 1} \left\| d\left(\mathfrak{s}_2\left(X^D\right)_{s,t}, \mathbf{X}_{s,t}\right) \right\|_{L^4(\mathbb{P}^{a,x})} = 0.$$

By Hölder's inequality the last statement remains valid even when we replace L^4 by L^q for any $q \in [1, \infty)$. We can then conclude by using Proposition A.15. ∎

16.6 Weak approximations

We maintain our standing notation, in particular $\mathfrak{g} = \mathfrak{g}^N\left(\mathbb{R}^d\right)$, and recall that for any $a \in \Xi^{N,d}(\Lambda)$, we have constructed a $\mathfrak{g}$-valued diffusion process $\mathbf{X}^a$, associated with the Dirichlet form $\mathcal{E}^a$.

16.6.1 *Tightness*

Proposition 16.26 *Let* $(a_n) \subset \Xi^{N,d}(\Lambda)$. *Then, for any starting point* $x \in \mathfrak{g}$ *and any* $\alpha \in [0, 1/2)$, *the family of processes* $(\mathbf{X}^{a_n,x} : n \in \mathbb{N})$ *is tight in the Polish space* $C_o^{0,\alpha\text{-Höl}}([0,1], \mathfrak{g})$.

Proof. Let us fix $\alpha' \in (\alpha, 1/2)$. From Proposition 8.17, $K_R = \{\mathbf{x} : \|\mathbf{x}\|_{\alpha'\text{-Höl}} \leq R\}$ is relatively compact in $C_o^{0,\alpha\text{-Höl}}([0,1], \mathfrak{g})$. The proof is then finished with the Fernique estimate from Corollary 16.12,

$$\sup_n \mathbb{P}\left(\mathbf{X}^{a_n} \in K_R\right) \leq c e^{-R^2/c}. \qquad \blacksquare$$

16.6.2 *Convergence*

In order to discuss weak convergence, let us first specialize some properties of non-negative quadratic forms to the present setting. From (E.1), using also quasi-isometry ("$\mathcal{E}^a \sim \mathcal{E}$"), we have that for all $f, g \in W^{1,2}(\mathfrak{g}, dm)$, the common domain of all $\mathcal{E}^a$ with $a \in \Xi^{N,d}(\Lambda)$, and $s > 0$,

$$\int |\nabla \mathcal{P}_s^a f|^2 \, dm \leq \Lambda \mathcal{E}^a (\mathcal{P}_s^a f, \mathcal{P}_s^a f)$$

$$\leq \Lambda \left(\frac{1}{2s} |f|_{L^2}^2 \wedge \mathcal{E}^a (f, f) \right)$$

$$\leq \frac{\Lambda}{2s} |f|_{L^2}^2 \wedge \left(\Lambda^2 \int |\nabla f|^2 \, dm \right). \qquad (16.17)$$

Lemma 16.27 *Let $(a_n : 1 \leq n \leq \infty) \subset \Xi^{N,d}(\Lambda)$ and assume $a_n \to a_\infty$ almost everywhere (with respect to Lebesgue measure on $\mathfrak{g}$). Set $\mathcal{E}^n = \mathcal{E}^{a_n}$ with associated semi-group $\mathcal{P}^n$. Assume $g, f \in W^{1,2}(\mathfrak{g}, dm)$, the common domain of all $\mathcal{E}^n$. Then*
(i) for every fixed $s \in [0, 1]$, assuming L^2-convergence $\mathcal{P}_s^n f$ to some limit, say $Q_s f$, and boundedness of $\{(\mathcal{P}_s^n f) : n\}$ in $W^{1,2}$, we have

$$\mathcal{E}^n (\mathcal{P}_s^n f, g) \to \mathcal{E}^\infty (Q_s f, g) \text{ as } n \to \infty;$$

(ii) we have

$$\sup_n \sup_{s \in [0,1]} |\mathcal{E}^n (\mathcal{P}_s^n f, g)| < \infty.$$

Proof. (i) Recall that $\mathcal{D}(\mathcal{E}^\infty)$ is a Hilbert space with inner product given by

$$\langle \cdot, \cdot \rangle_{\mathcal{E}^\infty} = \langle \cdot, \cdot \rangle_{L^2} + \mathcal{E}^\infty (\cdot, \cdot)$$

and (by quasi-isometry) $\Lambda^{-1} |f|_{W^{1,2}}^2 \leq \langle f, f \rangle_{\mathcal{E}^\infty} \leq \Lambda |f|_{W^{1,2}}^2$. By assumption, $\{(\mathcal{P}_s^n f) : n\} \subset W^{1,2}$ is bounded and hence

$$\|\mathcal{P}_s^n f\|_{\mathcal{E}^\infty}^2 = |f|_{L^2}^2 + \mathcal{E}^\infty (\mathcal{P}_s^n f, \mathcal{P}_s^n f)$$

is also uniformly bounded in n. Together with

$$\mathcal{P}_s^n f \to Q_s f \text{ as } n \to \infty$$

in L^2, an application of Lemma E.1 shows that this convergence holds weakly in $(\mathcal{D}(\mathcal{E}^\infty), \langle \cdot, \cdot \rangle_{\mathcal{E}^\infty})$. In particular, since $g \in W^{1,2}(\mathfrak{g}, dm) = \mathcal{D}(\mathcal{E}^\infty)$,

$$\mathcal{E}^\infty (\mathcal{P}_s^n f, g) \to \mathcal{E}^\infty (Q_s f, g) \text{ as } n \to \infty.$$

Thus, it only remains to see that

$$\delta(n) := \mathcal{E}^n (\mathcal{P}_s^n f, g) - \mathcal{E}^\infty (\mathcal{P}_s^n f, g) = \int \langle \nabla \mathcal{P}_s^n f, (a^n - a) \nabla g \rangle \, dm$$

converges to zero. From Cauchy–Schwarz we obtain

$$|\delta(n)| \leq \sqrt{\int |\nabla \mathcal{P}_s^n f|^2 \, dm} \sqrt{\int |a^n - a^\infty|^2 |\nabla g|^2 \, dm}.$$

It now suffices to note, from (16.17), and for fixed $s > 0$,

$$\sup_n \int |\nabla \mathcal{P}_s^n f|^2 \, dm < \infty;$$

$\delta(n) \to 0$ is then a consequence of $a^n \to a^\infty$ almost everywhere and bounded convergence.

(ii) Using quasi-isometry of $\mathcal{E}^n \sim \mathcal{E}$, Cauchy–Schwarz and (16.17),

$$
\begin{aligned}
|\mathcal{E}^n(\mathcal{P}_s^n f, g)| &\leq c_2 \sqrt{\int |\nabla \mathcal{P}_s^n f|^2 \, dm} \sqrt{\int^2 |\nabla g|^2 \, dm} \\
&\leq c_3 \sqrt{\int |\nabla f|^2 \, dm} \sqrt{\int^2 |\nabla g|^2 \, dm}
\end{aligned}
$$

and this bound is uniform in $s \in [0,1]$ and n. $\blacksquare$

Theorem 16.28 *Let* $x \in \mathfrak{g}$, $(a_n : 1 \leq n \leq \infty) \subset \Xi^{N,d}(\Lambda)$ *and assume* $a_n \to a_\infty$ *almost everywhere. Then* $\mathbf{X}^{a_n ; x}$ *converges weakly in* α-*Hölder topology to* $\mathbf{X}^{a_\infty ; x}$.

Proof. From Proposition 16.26 the family $(\mathbf{X}^{a_n, x})$ is tight. It then suffices to establish convergence of the finite-dimensional distributions. To this end, let us set $a := a_\infty$ and consider $p^{a_\infty} = p^{a_\infty}(t, x, y)$, the heat-kernel (or transition density) of $\mathbf{X}^{a_\infty, x}$. It will suffice to check that $p^{a_n} \to p^{a_\infty}$, uniformly on compacts in $(0, \infty) \times \mathfrak{g} \times \mathfrak{g}$. Since each heat-kernel p^a is a (weak) solution to the respective parabolic PDE $\partial_t p = \mathcal{L}_y^a p$, it follows from Hölder regularity of weak solutions (Proposition 16.7), uniformly over all $a \in \Xi^{N,d}(\Lambda)$, that

$$(t, y) \mapsto p^a(t, x, y)$$

is equicontinuous over sets of the form $(s, t) \times K \subset\subset (0, \infty) \times \mathfrak{g}$. More precisely, we cover $(s, t) \times K$ by finitely many "parabolic" cylinders Q_1^k so that $Q_1^k \subset Q_2^k \subset (0, \infty) \times \mathfrak{g}$ and note that $\max_k |p^a(\cdot, x, \cdot)|_{\infty; Q_2^k}$ is bounded by some constant (depending on Λ and the distance of $\cup^k Q_2^k$ to $\{0\} \times \mathfrak{g}$, which can be made close to $s > 0$ by taking k large enough), uniformly over $x \in K$. By symmetry, the same holds for $(t, x) \mapsto p^a(t, x, y)$ and from the triangle inequality,

$$
\begin{aligned}
|p^a(t, x, y) - p^a(t', x', y')| &\leq |p^a(t, x, y) - p^a(t', x, y')| \\
&\quad + |p^a(t', x, y') - p^a(t, x', y')|;
\end{aligned}
$$

we see that $(p^a : a \in \Xi^{N,d}(\Lambda))$ is equicontinuous on any compact set of form $(s, t) \times K \times K$. In conjunction with the heat-kernel bounds, it is clear from Arzela–Ascoli that there exists some $q \in C((0, \infty) \times \mathfrak{g} \times \mathfrak{g})$ so that, after switching to a subsequence if necessary, $p^{a_n} \to q$ uniformly on compacts.

Validity of the Chapman–Kolmogorov equation is preseved in this limit, and so

$$Q_t f := \int f(x) q(t, \cdot, y) \, dy, \quad f \in C_c^\infty$$

extends (uniquely) to strongly continuous semi-groups $(Q_t : t \geq 0)$ on $L^2(\mathfrak{g})$. Quite obviously then, at least for fixed t,

$$\mathcal{P}_t^{a_n} f \to Q_t f \quad \text{in } L^2.$$

Moreover, $(\mathcal{P}_t^{a_n} f : n \in \mathbb{N})$ is bounded in $W^{1,2}$, as is clear from (16.17) and so (Lemma E.1) $\mathcal{P}_t^{a_n} f \to Q_t f$ weakly in $W^{1,2}$. Thanks to Lemma 16.27 we can now pass to the limit in

$$\langle \mathcal{P}_t^{a_n} f, g \rangle_{L^2} = \langle f, g \rangle_{L^2} + \int_0^t \mathcal{E}^{a_n}(\mathcal{P}_s^{a_n} f, g) \, ds$$

and learn that

$$\langle Q_t f, g \rangle_{L^2} = \langle f, g \rangle_{L^2} + \int_0^t \mathcal{E}^{a \infty}(Q_s f, g) \, ds.$$

But this identifies (Q_t) as the semi-group associated with $\mathcal{E}^{a \infty}$. In particular, q must coincide with $p^{a \infty}$, which implies convergence of the finite-dimensional distributions. The proof is then finished. ∎

16.7 Large deviations

We maintain our standing notation, $\mathfrak{g} = \mathfrak{g}^N(\mathbb{R}^d)$, and $\Xi^{N,d}(\Lambda)$ denotes uniformly elliptic matrices defined on $\mathfrak{g}$.

Theorem 16.29 *Let $a \in \Xi^{N,d}(\Lambda)$ and $\mathbf{X}^{a;x}$ be the symmetric $\mathfrak{g}$-valued diffusion associated with the Dirichlet form[15]*

$$\frac{1}{2} \mathcal{E}^a,$$

started at some fixed point $x \in \mathfrak{g}$. Then the family $(\mathbf{X}^{a;x}(\varepsilon \cdot) : \varepsilon > 0)$ satisfies a large deviation principle with good rate function

$$I^a(\mathbf{h}) = \frac{1}{2} \sup_{D \subset \mathcal{D}[0,T]} \sum_{i:t_i \in D} \frac{\left| d^a\left(\mathbf{h}_{t_i}, \mathbf{h}_{t_{i+1}}\right) \right|^2}{|t_{i+1} - t_i|}$$

---

[15] The factor $1/2$ deviates from our previous convention but leads to a more familiar-looking rate function.

in α-Hölder topology, $\alpha \in [0, 1/2)$. More precisely, viewing $\mathbf{P}_\varepsilon^{a;x} = (\mathbb{P})_* \mathbf{X}_\varepsilon^{a;x}$, the law of $\mathbf{X}^{a;x}(\varepsilon \cdot)$, as a Borel measure on $C_x^{\alpha\text{-Höl}}([0,1], \mathfrak{g})$, the family $(\mathbf{P}_\varepsilon^{a;x} : \varepsilon > 0)$ satisfies a large deviation principle in $C_x^{\alpha\text{-Höl}}([0,1], \mathfrak{g})$ with good rate function I^a.

Proof. A large deviation principle in *uniform topology*, with rate function

$$I^a(\omega) = \frac{1}{2} |\omega|_{W^{1,2}([0,1], \mathfrak{g}^N(\mathbb{R}^d))}^2,$$

follows from an abstract Schilder theorem, Theorem E.20, applied to the specific context of the $\mathfrak{g}$-valued diffusion $\mathbf{X}^{a;x}$ associated with the Dirichlet form $\frac{1}{2}\mathcal{E}^a$. ∎

Remark 16.30 The rate function is precisely

$$\frac{1}{2} |\mathbf{h}|_{W^{1,2}([0,1], \mathfrak{g}^N(\mathbb{R}^d))}^2$$

relative to the metric space $\left(\mathfrak{g}^N(\mathbb{R}^d), d^a\right)$. When $a = I$, d^a is the Carnot–Caratheodory metric on $\mathfrak{g}^N(\mathbb{R}^d)$ and in this case, from Exercise 7.60 and basic facts of $W^{1,2}([0,1], \mathbb{R}^d)$,

$$|\mathbf{h}|_{W^{1,2}([0,1], \mathfrak{g}^N(\mathbb{R}^d))}^2 = |h|_{W^{1,2}([0,1], \mathbb{R}^d)}^2 = \int_0^1 \left|\dot{h}_u\right|^2 du$$

where we wrote $h = \pi_1(\mathbf{h})$. □

16.8 Support theorem

We maintain our standing notation, $\mathfrak{g} = \mathfrak{g}^N(\mathbb{R}^d)$, and $\Xi^{N,d}(\Lambda)$ denotes uniformly elliptic matrices defined on $\mathfrak{g}$. Recall that for any $a \in \Xi^{N,d}(\Lambda)$, we have constructed a $\mathfrak{g}$-valued diffusion process $\mathbf{X}^{a;x}$, started at some $x \in \mathfrak{g}$ associated with the Dirichlet form $\mathcal{E}^a$.

16.8.1 Uniform topology

Theorem 16.31 *There exists a constant $C = C(\Lambda, N)$ so that for any $h \in W_0^{1,2}([0,1], \mathbb{R}^d)$ and any $\varepsilon \in (0,1)$, we have*

$$\mathbb{P}^{a,0}\left(d_{\infty;[0,1]}(\mathbf{X}, S_N(h)) \le \varepsilon\right) \ge \exp\left(-\frac{C\left(1 + |h|_{W^{1,2};[0,1]}\right)^2}{\varepsilon^2}\right).$$

As a consequence, we have full support of $\mathbf{X}_t = \mathbf{X}_t^x$ in uniform topology. In other words,

$$supp(\mathbb{P}^{a,x}) = C_x([0,1], \mathfrak{g}).$$

Proof. This follows from general principles, Theorem E.20, applied to the specific $\mathfrak{g}$-valued process $\mathbf{X}^{a;x}$ associated with the Dirichlet form $\mathcal{E}^a$. ∎

16.8.2   Hölder topology

Recall that, modulo starting points, $W^{1,2}\left([0,1],\mathbb{R}^d\right)$ is in one-to-one correspondence to $W^{1,2}\left([0,1],\mathfrak{g}\right)$ where we write $\mathfrak{g} = \mathfrak{g}^N\left(\mathbb{R}^d\right)$ as usual. Indeed, any $\mathbb{R}^d$-valued $W^{1,2}$-path h lifts to

$$\mathfrak{s}_N\left(h\right) \equiv \log \circ S_N\left(h\right) \in W^{1,2}\left([0,1],\mathfrak{g}\right),$$

cf. Exercise 7.60; conversely, it suffices to project $\mathbf{h} \in W^{1,2}\left([0,1],\mathfrak{g}\right)$ to its first level $\pi_1\left(\mathbf{h}\right)$. Observe also that, for any $\alpha \in [0,1/2]$,

$$W^{1,2}_x\left([0,1],\mathfrak{g}\right) \subset C^{\alpha\text{-Höl}}_x\left([0,1],\mathfrak{g}\right).$$

Lemma 16.32 *Assume $\alpha \in [0,1/4)$. Fix $\mathbf{h} \in W^{1,2}_x\left([0,1],\mathfrak{g}\right)$, $\varepsilon > 0$ and define $B^{\mathbf{h}}_\varepsilon = \{\mathbf{x} : |\mathbf{x}|_{\alpha\text{-Höl}} \leq 2\,|\mathbf{h}|_{\alpha\text{-Höl}},\ d_\infty\left(\mathbf{x},\mathbf{h}\right) \leq \varepsilon\}$. Then*

$$\mathbb{P}^{a,x}\left(\mathbf{X} \in B^{\mathbf{h}}_\varepsilon\right) > 0.$$

Proof. Step 1: Taking $0 \leq \alpha < \beta \leq 1$ we claim that

$$\|\mathbf{x}\|_{\alpha\text{-Höl}} \lesssim \|\mathbf{x}\|^{\alpha/\beta}_{\beta\text{-Höl}}\, d_\infty\left(\mathbf{x},\mathbf{h}\right)^{1-\alpha/\beta} + \|\mathbf{h}\|_{\alpha\text{-Höl}}.$$

To see this, note that $d\left(\mathbf{x}_s,\mathbf{x}_t\right)$ can be estimated in two ways:

$$d\left(\mathbf{x}_s,\mathbf{x}_t\right) \leq 2d_\infty\left(\mathbf{x},\mathbf{h}\right) + \|\mathbf{h}\|_{\alpha\text{-Höl}}\,|t-s|^\alpha,$$

$$d\left(\mathbf{x}_s,\mathbf{x}_t\right) \leq \|\mathbf{x}\|_{\beta\text{-Höl}}\,|t-s|^\beta.$$

Given r, we can use the first estimate when $|t-s| \leq r$ and the second when $|t-s| > r$, so that

$$\sup_{0 \leq s \leq t \leq 1} \frac{d\left(\mathbf{x}_s,\mathbf{x}_t\right)}{|t-s|^\alpha} \leq \min\left(\|\mathbf{x}\|_{\beta\text{-Höl}}\,r^{\beta-\alpha}, \frac{2d_\infty\left(\mathbf{x},\mathbf{h}\right)}{r^\alpha} + \|\mathbf{h}\|_{\alpha\text{-Höl}}\right)$$

$$\leq \min\left(\frac{\|\mathbf{x}\|_{\beta\text{-Höl}}\,r^\beta}{r^\alpha}, \frac{2d_\infty\left(\mathbf{x},\mathbf{h}\right)}{r^\alpha}\right) + \|\mathbf{h}\|_{\alpha\text{-Höl}}.$$

Choosing r optimally, namely such that $r^\beta = 2d_\infty\left(\mathbf{x},\mathbf{h}\right)/\|\mathbf{x}\|_{\beta\text{-Höl}}$, then gives

$$\|\mathbf{x}\|_{\alpha\text{-Höl}} \leq 2^{1-\alpha/\beta}\left(d_\infty\left(\mathbf{x},\mathbf{h}\right)^{1-\alpha/\beta}\|\mathbf{x}\|^{\alpha/\beta}_{\beta\text{-Höl}}\right) + \|\mathbf{h}\|_{\alpha\text{-Höl}}.$$

Step 2: Now let $\alpha < \beta < 1/2$ so that we have Fernique estimates for $\overline{\|\mathbf{X}\|_{\beta\text{-Höl}}} = \|\mathbf{X}\left(\omega\right)\|_{\beta\text{-Höl};[0,1]}$. Then

$$\mathbb{P}^{a,x}\left(\mathbf{X} \in B^{\mathbf{h}}_\varepsilon\right) \geq \mathbb{P}^{a,x}\left(\left(2d_\infty\left(\mathbf{X},\mathbf{h}\right)\right)^{1-\alpha/\beta}|\mathbf{x}|^{\alpha/\beta}_{\beta\text{-Höl}} \leq |\mathbf{h}|_{\alpha\text{-Höl}}, d_\infty\left(\mathbf{X},\mathbf{h}\right) \leq \varepsilon\right)$$

$$= \mathbb{P}^{a,x}\left(\|\mathbf{x}\|^{\alpha/\beta}_{\beta\text{-Höl}} \leq \frac{\|\mathbf{h}\|_{\alpha\text{-Höl}}}{(2\varepsilon)^{1-\alpha/\beta}},\ d_\infty\left(X,h\right) \leq \varepsilon\right)$$

$$\geq \mathbb{P}^{a,x}\left(d_\infty\left(\mathbf{X},\mathbf{h}\right) \leq \varepsilon\right) - \mathbb{P}^{a,x}\left(\|\mathbf{x}\|_{\beta\text{-Höl}} > \frac{\|\mathbf{h}\|^{\beta/\alpha}_{\alpha\text{-Höl}}}{(2\varepsilon)^{\beta/\alpha-1}}\right)$$

$$= \Delta_1 - \Delta_2.$$

Obviously, both Δ_1 and Δ_2 tend to zero as $\varepsilon \to 0$. Positivity of $\mathbb{P}^{a,x}$ $\left(\mathbf{X} \in B_\varepsilon^{\mathbf{h}} \right)$ will follow from checking that $\Delta_2 / \Delta_1 \to 0$ as $\varepsilon \to 0$. Keeping h fixed, Theorem E.21 gives

$$\log \Delta_1 \geq -c_1 \left(\frac{1}{\varepsilon^2} \right)$$

while the Fernique estimates imply

$$\log \Delta_2 \leq -c_2 \left(\frac{1}{\varepsilon^{2\beta/\alpha - 2}} \right)$$

for some irrelevant positive constants c_1 and c_2. We see that $\Delta_2 / \Delta_1 \to 0$ if

$$2 < 2\beta/\alpha - 2$$

or equivalently $2\alpha < \beta$. Since $\beta < 1/2$ was needed to apply the Fernique estimates, we see that the argument works for any $\alpha \in [0, 1/4)$. ∎

Theorem 16.33 *Let $a \in \Xi^{N,d} (\Lambda)$ and $\mathbf{X} = \mathbf{X}^{a;x}$ be the $\mathfrak{g}$-valued symmetric diffusion process, started at some $x \in \mathfrak{g}$ and associated with the Dirichlet form $\mathcal{E}^a$. Fix $\mathbf{h} \in W_x^{1,2} ([0,1], \mathfrak{g})$. Then, for every $\alpha \in [0, 1/4)$ and every $\delta > 0$,*

$$\mathbb{P}^{a;x} \left(d_{\alpha\text{-H\"ol};[0,1]} (\mathbf{X}, \ \mathbf{h}) < \delta \right) > 0.$$

In particular, for every $\alpha \in [0, 1/4)$ the support of $\mathbf{X}^{a;x}$ in α-Hölder topology is precisely

$$C_x^{0,\alpha\text{-H\"ol}} \left([0,1], \mathfrak{g}^N \left(\mathbb{R}^d \right) \right).$$

Proof. Without loss of generality, $x = \mathbf{X}^a (0) = \mathbf{h}(0) = 0 \in \mathfrak{g}$. Pick $\alpha \in (\alpha', 1/4)$. By interpolation and the d_0/d_∞-estimate,

$$d_\infty (\mathbf{x}, \mathbf{y}) \leq d_0 (\mathbf{x}, \mathbf{y}) \lesssim \left(\|\mathbf{x}\|_\infty + \|\mathbf{y}\|_\infty \right)^{1 - 1/N} d_\infty (\mathbf{x}, \mathbf{y})^{1/N}$$

so that

$$d_{\alpha'\text{-H\"ol}} (\mathbf{X}^a, \mathbf{h}) \lesssim \left(|\mathbf{X}^a|_{\alpha\text{-H\"ol}} \right.$$
$$\left. + |\mathbf{h}|_{\alpha\text{-H\"ol}} \right)^{\alpha'/\alpha} \left(|\mathbf{X}^a|_\infty + |\mathbf{h}|_\infty \right)^{(1 - 1/N)(1 - \alpha'/\alpha)} d_\infty (\mathbf{X}^a, \mathbf{h})^{\frac{1 - \alpha'/\alpha}{N}}.$$

In particular, for $\mathbf{X}^a \in B_\varepsilon^{\mathbf{h}} = \{ \mathbf{x} : |\mathbf{x}|_{\alpha\text{-H\"ol}} \leq 2 |\mathbf{h}|_{\alpha\text{-H\"ol}}, \ d_\infty (\mathbf{x}, \mathbf{h}) \leq \varepsilon \}$ there exists c_1 (which may in particular depend on h, α', α) so that

$$d_{\alpha'\text{-H\"ol}} (\mathbf{X}^a, \mathbf{h}) \leq c_1 \varepsilon^{\frac{1 - \alpha'/\alpha}{N}}.$$

Fix $\delta > 0$ and take ε small enough such that $c_1 \varepsilon^{\frac{1 - \alpha'/\alpha}{N}} < \delta$. Clearly then

$$\mathbb{P}^x \left(d_{\alpha'\text{-H\"ol};[0,1]} (\mathbf{X}^a, \mathbf{h}) < \delta \right) \geq \mathbb{P}^{a,x} \left(\mathbf{X} \in B_\varepsilon^{\mathbf{h}} \right) > 0$$

where the final, strict inequality is due to Lemma 16.32. ∎

Remark 16.34 By taking $\alpha \in (1/5, 1/4)$ and $N = 4$ this yields a support characterization of $\mathbf{X}^{a;x}$ in α-Hölder rough path topology. Since $\mathbf{X}^{a;x}$ has sample paths which enjoy Hölder regularity for any exponent less than $1/2$, one suspects that the above support description holds true for any $a < 1/2$. Although we are able to show this when $\mathbf{h} \equiv 0$, see the following section, the extension to $\mathbf{h} \neq 0$ remains an open problem. □

16.8.3 *Hölder rough path topology: a conditional result*

We first study the probability that $\mathbf{X}^{a;x}$ stays in the bounded open domain $D \subset \mathfrak{g} = \mathfrak{g}^N \left(\mathbb{R}^d \right)$ for long times.

Proposition 16.35 *Let D be an open domain in $\mathfrak{g}$ with finite volume, no regularity assumptions are made about ∂D. Let $a \in \Xi(\Lambda)$ and $\mathbf{X}^a$ be the process associated with $\mathcal{E}^a$ started at $x \in \mathfrak{g}$ and assume $x \in D$. Then there exist positive constants $K_1 = K_1 (x, D, \Lambda)$ and $K_2 = K_2 (D, \Lambda)$ so that for all $t \geq 0$*

$$K_1 e^{-\lambda t} \leq \mathbb{P}\left[\mathbf{X}_s^{a,x} \in D \, \forall s : 0 \leq s \leq t\right] \leq K_2 e^{-\lambda t}$$

where $\lambda \equiv \lambda_1^a > 0$ is the simple and first Dirichlet eigenvalue of $-L^a$ on the domain D. Moreover,

$$\forall a \in \Xi(\Lambda) : 0 < \lambda_{\min} \leq \lambda_1^a \leq \lambda_{\max} < \infty$$

where $\lambda_{\min}, \lambda_{\max}$ depend only on Λ and D.

Remark 16.36 The proof will show that $K_1 \sim \psi_1^a (x)$. Noting that $\psi_1^a (x) e^{-\lambda_1^a t}$ solves the same PDE as $u^a (t, x)$, the above can be regarded as a "partial" parabolic boundary Harnack statement. □

Proof. If p_D^a denotes the Dirichlet heat-kernel for D, we can write

$$u^a (t, x) := \mathbb{P}^x \left[\mathbf{X}_s^a \in D \, \forall s : 0 \leq s \leq t\right] = \int_D p_D^a (t, x, y) \, dy.$$

As is well known,[16] p_D^a is the kernel for a semi-group $\mathcal{P}_D^a : [0, \infty) \times L^2 (D) \to L^2 (D)$ which corresponds to the Dirichlet form $(\mathcal{E}^a, \mathcal{F}_D)$ whose domain $\mathcal{F}_D$ consists of all $f \in \mathcal{F} \equiv D(\mathcal{E}^a)$ with quasicontinuous modifications equal to 0 q.e. on D^c. The infinitesimal generator of $\mathcal{P}_D^a$, denoted by L_D^a, is a self-adjoint, densely defined operator with spectrum $\sigma(-L_D^a) \subset [0, \infty)$. We now use an ultracontractivity argument to show that $\sigma(-L_D^a)$ is discrete. To this end, we note that the upper bound on p^a plainly implies $|p_D^a (t, \cdot, \cdot)|_\infty = O(t^{-d^2/2})$. Since $|D| < \infty$ it follows that $\|\mathcal{P}_D^a (t)\|_{L^1 \to L^\infty} < \infty$ which is, by definition, ultracontractivity of the semi-group $\mathcal{P}_D^a$. It is now

---

[16] See, for example, [70].

a standard consequence[17] that $\sigma\left(-L_D^a\right) = \{\lambda_1^a, \lambda_2^a, \dots\} \subset [0, \infty)$, listed in non-decreasing order. Moreover, it is clear that $\lambda_1^a \neq 0$; indeed, the heat kernel estimates are more than sufficient to guarantee that $\|\mathcal{P}_D^a(t)\|_{L^2 \to L^2} \to 0$ as $t \to \infty$ which contradicts the existence of non-zero $f \in L^2(D)$ so that $\mathcal{P}_D^a(t) f = f$ for all $t \geq 0$. Let us note that

$$
\begin{aligned}
\lambda_1^a &= \inf \sigma(H) \\
&= \inf \left\{ \mathcal{E}^a(f,f) : f \in \mathcal{F}_D \text{ with } |f|_{L^2(D)} = 1 \right\} \quad \text{(by Rayleigh–Ritz)} \\
&= \inf \left\{ \int_D \Gamma^a(f,f)\, dm : f \in \mathcal{F}_D \text{ with } |f|_{L^2(D)} = 1 \right\}
\end{aligned}
$$

and since $\Gamma^a(f,f)/\Gamma^I(f,f) \in [\Lambda^{-1}, \Lambda]$ for $f \neq 0$ it follows that $\lambda_1^a \in [\lambda_{\min}, \lambda_{\max}]$ for all $a \in \Xi(\Lambda)$ where we set

$$
\lambda_{\min} = \Lambda^{-1} \lambda_1^I, \quad \lambda_{\max} = \Lambda \lambda_1^I. \tag{16.18}
$$

The lower heat-kernel estimates for the killed process imply[18] irreducibility of the semi-group $\mathcal{P}_D^a$, hence simplicity of the first eigenvalue λ, and there is an a.s. strictly positive eigenfunction to $\lambda \equiv \lambda_1^a$, say $\psi \equiv \psi_1^a$, and by de Giorgi–Moser–Nash regularity we may assume that ψ is Hölder continuous and strictly positive away from the boundary (this follows also from Harnack's inequality). We also can (and will) assume that $\|\psi\|_{L^2(D)} = 1$.

Lower bound: Noting that $v(t,x) = e^{-\lambda t}\psi(x)$ is a weak solution of $\partial_t v = L_D^a v$ with $v(0, \cdot) = \psi$, we have

$$
v(t,x) = \int_D p_D^a(t,x,y)\,\psi(y)\,dy,
$$

at first for a.e. x but by using a Hölder-regular version of p_D^a the above holds for all $x \in D$. It follows that

$$
\begin{aligned}
0 &< \psi(x) \\
&= e^{\lambda t} \int_D p_D^a(t,x,y)\,\psi(y)\,dy \\
&\leq e^{\lambda(t+1)} \int_D p_D^a(t,x,y) \int_D p_D^a(1,y,z)\,\psi(z)\,dz\,dy \\
&\leq e^{\lambda(t+1)} \int_D \left(p_D^a(t,x,y) \sqrt{\int_D [p_D^a(1,y,z)]^2\, dz} \sqrt{\int \psi^2(z)\,dz} \right) dy \\
&\leq C(\Lambda,D)\, e^{\lambda(t+1)} u^a(t,x) \\
&= \left[C(\Lambda,D)\, e^{\lambda_{\max}} \right] \times e^{\lambda t} u^a(t,x)
\end{aligned}
$$

---

[17]See, for example, [38, Theorem 1.4.3].
[18]See, for example, [38, Theorem 1.4.3].

and this gives the lower bound with $K_1 = \psi(x) / \left[C(\Lambda, D) e^{\lambda_{\max}} \right]$. Clearly $\psi = \psi_1^a$ depends on a and so does K_1. Thus, what we need to show is that $\psi(x)$ can be bounded from below by a quantity which depends on a only through its ellipticity constant Λ. To this end, from

$$p_D^a(t, y, y) = \sum_{i=1}^{\infty} e^{-\lambda_i^a t} \left| \psi_i^a(y) \right|^2$$

evaluated at $t = 1$, say, we see that

$$\left| \psi(y) \right|^2 \leq e^{\lambda} p_D^a(1, y, y) \leq e^{\lambda_{\max}} p_D^a(1, y, y) \leq e^{\lambda_{\max}} p^a(1, y, y)$$

and by using our upper heat-kernel estimates for p^a we see that there is a constant $M = M(\Lambda, D)$ such that $|\psi|_\infty \leq M$. Given x and M we can find a compact set $\mathfrak{K} \subset D$ so that $m(D \backslash \mathfrak{K}) \leq 1/(4M^2)$ and $x \in \mathfrak{K}$ (recall that m is a Haar measure on $\mathfrak{g}$). By Harnack's inequality,

$$\sup_{\mathfrak{K}} \psi \leq C \psi(x)$$

for $C = C(\mathfrak{K}, \Lambda) = C(x, D, \Lambda)$. We then have

$$1 = |\psi|_{L^2} \leq M \sqrt{m(D \backslash \mathfrak{K})} + C\psi(x) \sqrt{m(\mathfrak{K})} \leq 1/2 + C\psi(x) \sqrt{m(D)}$$

which gives the required lower bound on $\psi(x) \equiv \psi_1^a(x)$, which only depends on x, D and Λ but not on a.

Upper bound: Recall that $-\lambda \equiv -\lambda_1^a$ denotes the first eigenvalue of L_D^a with associated semi-group $\mathcal{P}_D^a$. It follows that

$$\left| \mathcal{P}_D^a(t) f \right|_{L^2} \leq e^{-\lambda t} |f|_{L^2},$$

which may be rewritten as

$$\left| \int_D p_D^a(t, \cdot, z) f(z) \, dz \right|_{L^2} \leq e^{-\lambda t} |f|_{L^2}.$$

Let $t > 1$. Using Chapman–Kolmogorov and symmetry of the kernel,

$$
\begin{aligned}
u(t, x) &= \int_D p_D^a(t, x, z) \, dz = \int_D \int_D p_D^a(1, x, y) p_D^a(t-1, z, y) \, dy \, dz \\
&= \sqrt{m(D)} \left(\int_D \left(\int_D p_D^a(t-1, z, y) p_D^a(1, x, y) \, dy \right)^2 dz \right)^{1/2} \\
&= \sqrt{m(D)} \left| \left(\int_D p_D^a(t-1, \cdot, y) p_D^a(1, x, y) \, dy \right) \right|_{L^2(D)} \\
&= \sqrt{m(D)} \left| \mathcal{P}_D^a(t-1) p_D^a(1, x, \cdot) \right|_{L^2(D)} \\
&\leq \sqrt{m(D)} e^{-\lambda(t-1)} \left| p_D^a(1, x, \cdot) \right|_{L^2(D)} \\
&\leq \sqrt{m(D)} e^{\lambda_{\max}} e^{-\lambda t} \sqrt{p_D^a(2, x, x)} \\
&\leq K_2 e^{-\lambda t},
\end{aligned}
$$

where we used upper heat-kernel estimates in the last step to obtain $K_2 = K_2(D, \Lambda)$. ∎

Corollary 16.37 *Fix* $a \in \Xi(\Lambda)$. *There exists* $K = K(\Lambda)$ *and for all* $\varepsilon > 0$ *there exist* $\lambda = \lambda^{(\varepsilon)}$ *such that*

$$K^{-1} e^{-\lambda t \varepsilon^{-2}} \leq \mathbb{P}^{a,0} \left[\|\mathbf{X}\|_{0;[0,t]} < \varepsilon \right] \tag{16.19}$$

$$\forall x : \mathbb{P}^{a,x} \left[\|\mathbf{X}\|_{0;[0,t]} < \varepsilon \right] \leq K e^{-\lambda t \varepsilon^{-2}}. \tag{16.20}$$

Proof. A straightforward consequence of scaling and Proposition 16.35 applied to

$$D = B(0,1) = \{y : \|y\| < 1\}$$

where $\|\cdot\|$ is the standard Carnot–Caratheodory norm on $\mathfrak{g}$. Then λ is the first eigenvalue corresponding to a scaled by factor ε. ∎

Proposition 16.38 *Let* $\alpha \in [0, 1/2)$. *There exists a constant* $C_{16.38}$ *such that for all* $\varepsilon \in (0, 1]$ *and* $R > 0$,

$$\mathbb{P}^{a,0} \left(\sup_{|t-s| < \varepsilon^2} \frac{\|\mathbf{X}_{s,t}\|}{|t-s|^\alpha} > R \,\middle|\, \|\mathbf{X}\|_{0;[0,1]} < \varepsilon \right)$$
$$\leq C_{16.38} \exp \left(-\frac{1}{C_{16.38}} \frac{R^2}{\varepsilon^{2(1-2\alpha)}} \right).$$

Proof. There will be no confusion in writing $\mathbb{P}^\varepsilon$ for $\mathbb{P} \left(\cdot \,\middle|\, \|\mathbf{X}\|_{0;[0,1]} < \varepsilon \right)$. Suppose there exists a pair of times $s, t \in [0,1]$ such that

$$s < t, \ |t-s| < \varepsilon^2 \ \text{and} \ \frac{\|\mathbf{X}_{s,t}\|}{|t-s|^\alpha} > R.$$

Then there exists a $k \in \{1, \ldots, \lceil 1/\varepsilon^2 \rceil\}$ so that $[s,t] \subset [(k-1)\varepsilon^2, (k+1)\varepsilon^2]$. In particular, the probability that such a pair of times exists is at most

$$\sum_{k=1}^{\lceil 1/\varepsilon^2 \rceil} \mathbb{P}^\varepsilon \left(\|\mathbf{X}\|_{\alpha;[(k-1)\varepsilon^2, (k+1)\varepsilon^2]} > R \right).$$

Set $[(k-1)\varepsilon^2, (k+1)\varepsilon^2] =: [T_1, T_2]$. The rest of the proof is concerned with the existence of C such that

$$\mathbb{P}^\varepsilon \left(\|\mathbf{X}\|_{\alpha;[T_1, T_2]} > R \right) \leq C \exp \left(-C^{-1} \frac{R^2}{\varepsilon^{2(1-2\alpha)}} \right)$$

since the factor $\lceil 1/\varepsilon^2 \rceil$ can be absorbed in the exponential factor by making C bigger. We estimate

$$\mathbb{P}^0 \left(||\mathbf{X}||_{\alpha;[T_1,T_2]} > R \,\Big|\, ||\mathbf{X}||_{0;[0,1]} < \varepsilon \right)$$

$$\leq \frac{\mathbb{P}^0 \left(||\mathbf{X}||_{\alpha;[T_1,T_2]} > R; ||\mathbf{X}||_{0;[0,T_1]} < \varepsilon; ||\mathbf{X}||_{0;[T_2,1]} < \varepsilon \right)}{\mathbb{P}^0 \left[||\mathbf{X}||_{0;[0,1]} < \varepsilon \right]}.$$

By using the Markov property and the above lemma, writing $\lambda^{(\varepsilon)} = \lambda^{a;\varepsilon}$, this equals

$$\frac{\mathbb{E}^0 \left[\mathbb{P}^{\mathbf{X}_{T_2}} \left(||\mathbf{X}||_{0;[0,1-T_2]} < \varepsilon \right); ||\mathbf{X}||_{\alpha;[T_1,T_2]} > R; ||\mathbf{X}||_{0;[0,T_1]} < \varepsilon \right]}{\mathbb{P}^0 \left[||\mathbf{X}||_{0;[0,1]} < \varepsilon \right]}$$

$$\leq Ce^{\lambda^{(\varepsilon)}\varepsilon^{-2}} \mathbb{E}^0 \left[e^{-\lambda^{(\varepsilon)}(1-T_2)\varepsilon^{-2}}; ||\mathbf{X}||_{\alpha;[T_1,T_2]} > R; ||\mathbf{X}||_{0;[0,T_1]} < \varepsilon \right]$$

$$= Ce^{\lambda^{(\varepsilon)}T_2\varepsilon^{-2}} \mathbb{P}^0 \left[||\mathbf{X}||_{\alpha;[T_1,T_2]} > R; ||\mathbf{X}||_{0;[0,T_1]} < \varepsilon \right]$$

where constants were allowed to change in insignificant ways. If $\mathbf{X}$ had independent increments in the group (such as is the case for enhanced Brownian motion $\mathbf{B}$) $\mathbb{P}^0 [\dots]$ would split up immediately. This is not the case here, but the Markov property serves as a substitute; using the Dirichlet heat-kernel $p^a_{B(0,\varepsilon)}$ we can write

$$\mathbb{P}^0 \left[||\mathbf{X}||_{\alpha;[T_1,T_2]} > R; ||\mathbf{X}||_{0;[0,T_1]} < \varepsilon \right]$$

$$= \int_{B(0,\varepsilon)} dx \, p^a_{B(0,\varepsilon)} (T_1, 0, x) \, \mathbb{P}_x \left[||\mathbf{X}||_{\alpha;[0,T_2-T_1]} > R \right].$$

Then, scaling and the usual Fernique-type estimates for the Hölder norm of $\mathbf{X}$ give

$$\sup_x \mathbb{P}_x \left[||\mathbf{X}||_{\alpha;[0,T_2-T_1]} > R \right] \leq C \exp \left(-\frac{1}{C} \left(\frac{R}{\varepsilon^{1-2\alpha}} \right)^2 \right),$$

where we used $T_2 - T_1 = 2\varepsilon^2$, and we obtain

$$\mathbb{P}^0 \left[||\mathbf{X}||_{\alpha;[T_1,T_2]} > R; ||\mathbf{X}||_{0;[0,T_1]} < \varepsilon \right]$$

$$\leq C \exp \left(-\frac{1}{C} \left(\frac{R}{\varepsilon^{1-2\alpha}} \right)^2 \right) \mathbb{P}^0 \left[||\mathbf{X}||_{0;[0,T_1]} < \varepsilon \right]$$

$$\leq C \exp \left(-\frac{1}{C} \left(\frac{R}{\varepsilon^{1-2\alpha}} \right)^2 \right) e^{-\lambda^{(\varepsilon)}T_1\varepsilon^{-2}}.$$

Putting things together we have

$$\mathbb{P}^0 \left(\|\mathbf{X}\|_{\alpha;[T_1,T_2]} > R \,\Big|\, \|\mathbf{X}\|_{0;[0,1]} < \varepsilon \right)$$

$$\leq \ C e^{\lambda^{(\varepsilon)}(T_2 - T_1)\varepsilon^{-2}} \exp\left(-\frac{1}{C} \left(\frac{R}{\varepsilon^{1-2\alpha}} \right)^2 \right)$$

$$\leq \ C e^{2\lambda_{\max}} \exp\left(-\frac{1}{C} \left(\frac{R}{\varepsilon^{1-2\alpha}} \right)^2 \right)$$

and the proof is finished.   ∎

Theorem 16.39 *Let* $\alpha \in [0, 1/2)$. *For all* $R > 0$ *the ball* $\left\{ \mathbf{x} : \|\mathbf{x}\|_{\alpha\text{-Höl};[0,1]} < R \right\}$ *has positive* $\mathbb{P}^{a,0}$-*measure and*

$$\lim_{\epsilon \to 0} \mathbb{P}^{a,0} \left(\|\mathbf{X}\|_{\alpha\text{-Höl};[0,1]} < R \,\Big|\, \|\mathbf{X}\|_{0;[0,1]} < \varepsilon \right) \to 1. \qquad (16.21)$$

In particular, for any $\delta > 0$

$$\mathbb{P}^{a,0} \left(\|\mathbf{X}\|_{\alpha\text{-Höl};[0,1]} < \delta \Big| \right) > 0.$$

Proof. We first observe that the uniform conditioning allows us to localize the Hölder norm. More precisely, take $s < t$ in $[0,1]$ with $t - s \geq \epsilon^2$ and note that from $\|\mathbf{X}\|_{0;[0,1]} < \varepsilon$ we get

$$\frac{\|\mathbf{X}_{s,t}\|}{|t-s|^\alpha} \leq \epsilon^{1-2\alpha}.$$

It follows that for fixed R and ϵ small enough,

$$\mathbb{P}^{a,0} \left(\|\mathbf{X}\|_{\alpha\text{-Höl};[0,1]} \geq R \,\Big|\, \|\mathbf{X}\|_{0;[0,1]} < \varepsilon \right)$$

$$= \mathbb{P}^{a,0} \left(\sup_{|t-s| < \varepsilon^2} \frac{\|\mathbf{X}_{s,t}\|}{|t-s|^\alpha} \geq R \,\Big|\, \|\mathbf{X}\|_{0;[0,1]} < \varepsilon \right)$$

and the preceding proposition shows convergence to zero with ϵ and (16.21) follows. Finally,

$$\mathbb{P}^{a,0} \left(\|\mathbf{X}\|_{\alpha\text{-Höl};[0,1]} < R \right) \geq \mathbb{P}^{a,0} \left(\|\mathbf{X}\|_{\alpha\text{-Höl};[0,1]} < R \,\Big|\, \|\mathbf{X}\|_{0;[0,1]} < \varepsilon \right)$$

$$\times \ \mathbb{P}^{a,0} \left(\|\mathbf{X}\|_{0;[0,1]} < \varepsilon \right)$$

$$\geq \ \mathbb{P}^{a,0} \left(\|\mathbf{X}\|_{0;[0,1]} < \varepsilon \right) / 2 \quad \text{(for } \epsilon \text{ small enough)}$$

and this is positive by Proposition 16.37.   ∎

16.9 Appendix: analysis on free nilpotent groups

16.9.1 Haar measure

In Section 7.5 we introduced free nilpotent groups. More precisely, what we called $\left(G^N\left(\mathbb{R}^d\right),\otimes\right)$ was a particular representation, namely within the tensor algebra $\left(T^N\left(\mathbb{R}^d\right),+,\otimes\right)$, of an abstract (connected and simply connected Lie group) G associated with the Lie algebra $\mathfrak{g} = \mathfrak{g}^N\left(\mathbb{R}^d\right) \subset T^N\left(\mathbb{R}^d\right)$ with bracket given by

$$[u,v] = u \otimes v - v \otimes u.$$

The *abstract* exponential map from a Lie algebra to its associated (connected and simply connected) Lie group was then given explicitly by the exponential map on $T^N\left(\mathbb{R}^d\right)$ based on power series with respect to $\otimes$. Another representation of G is given by $(\mathfrak{g},*)$, where

$$* : \mathfrak{g} \times \mathfrak{g} \to \mathfrak{g}$$

is given by the Campbell–Baker–Hausdorff formula, as derived in Section 7.3. Thanks to nilpotency, $*$ is a polynomial map and $(\mathfrak{g},*)$ is indeed a realization of G and the abstract exponential map is merely the identity. In any case, G is uniquely determined by $\mathfrak{g}$ up to isomorphism and whatever concepts such as Carnot–Caratheodory norm/distance we have developed on $\left(G^N\left(\mathbb{R}^d\right),\otimes\right)$ are immediately transfered to $(\mathfrak{g},*)$, or indeed other representations of G, cf. Remark 16.1.

Let us recall a few facts about $\mathfrak{g}^N\left(\mathbb{R}^d\right)$. Using the terminology of Folland and Stein [54], we can say that $\mathfrak{g} = \mathfrak{g}^N\left(\mathbb{R}^d\right)$ is *graded* in the sense that

$$\mathfrak{g} = \mathfrak{V}_1 \oplus \cdots \oplus \mathfrak{V}_N,$$
$$[\mathfrak{V}_i,\mathfrak{V}_j] \subset \mathfrak{V}_{i+j},$$

with "jth level" given by $\mathfrak{V}_j = \mathfrak{g} \cap \left(\mathbb{R}^d\right)^{\otimes j}$. It is also *stratified* in the sense that $\mathfrak{V}_1$ generates $\mathfrak{g}$ as an algebra and so

$$[\mathfrak{V}_1,\mathfrak{V}_j] = \mathfrak{V}_{1+j}.$$

Moreover, a natural *family of dilations* on $\mathfrak{g}$ is given by $\{\delta_r : r > 0\}$, where

$$\delta_r\left(g_1,\ldots,g_N\right) = \left(rg_1, r^2 g_2,\ldots,r^N g_N\right), \quad \text{with } g_j \in \mathfrak{V}_j.$$

As already discussed (cf. Exercise 7.55), each dilation induces a group homomorphism of the form $\exp \circ \delta_r \circ \exp^{-1}$.

Proposition 16.40 *The Lebesgue measure λ on $\mathfrak{g} = \mathfrak{g}^N\left(\mathbb{R}^d\right)$ is the (unique up to a constant factor) left- and right-invariant Haar measure m on $(\mathfrak{g},*)$. Moreover, if G is the abstract Lie group associated with $\mathfrak{g}$, the left- and right-invariant Haar measure on G is given by*

$$m = (\exp)_* \lambda.$$

Remark 16.41 Where no confusion is possible we shall write $|A|$ instead of $m(A)$ for a measurable set $A \subset G$. $\qquad \square$

Proof. Set $n = \dim \mathfrak{g}^N(\mathbb{R}^d)$ and also

$$n_j = \dim(\mathfrak{V}_j \oplus \cdots \oplus \mathfrak{V}_N).$$

We can choose a basis $\{e_{n-n_N+1}, \ldots, e_n\}$ for $\mathfrak{V}_N$, extend it to a basis $\{e_{n-n_{N-1}+1}, \ldots, e_n\}$ for $\mathfrak{V}_{N-1} \oplus \mathfrak{V}_N$, and so forth, obtaining eventually a basis $\{e_1, \ldots, e_n\}$ for $\mathfrak{g}$. The dual basis $\{\xi_1, \ldots, \xi_n\}$ provides global coordinates for $\mathfrak{g}^N(\mathbb{R}^d)$ and, by the Campbell–Baker–Hausdorff formula,

$$\eta_k(x * y) = \eta_k(x) + \eta_k(y) + P_k(x, y)$$

where $P_k(x, y)$ is a polynomial which depends only on the coordinates $\eta_i(x), \eta_i(y)$ with $i < k$. Therefore, the differentials of the maps $x \mapsto x * y$ (with y fixed) and $y \mapsto x * y$ (with x fixed) are given with respect to the coordinates (η_k) by lower-triangular matrices with ones on the diagonal, and their determinants are therefore identically one. It follows that the volume form $d\eta_1 \ldots d\eta_n$, which corresponds to Lebesgue measure on $\mathfrak{g}$, is left- and right-invariant. $\qquad \blacksquare$

Remark 16.42 As the proof immediately reveals, this result holds true for an arbitrary (connected and simply connected) nilpotent Lie group G with Lie algebra $\mathfrak{g}$. $\qquad \square$

Definition 16.43 The homogenous dimension of $\mathfrak{g} = \mathfrak{g}^N(\mathbb{R}^d)$ is defined as

$$\dim_H \mathfrak{g} := \sum_{j=1}^N j(\dim \mathfrak{V}_j).$$

If G is the abstract Lie group associated with $\mathfrak{g}$, we equivalently write

$$\dim_H G. \qquad \square$$

Lemma 16.44 *Let G be the abstract Lie group associated with $\mathfrak{g} = \mathfrak{g}^N(\mathbb{R}^d)$. Then, for all $r > 0$ and measurable sets $E \subset G$,*

$$m(\delta_r E) = r^{\dim_H G} m(E).$$

In particular, for $B(x, r) = \{y \in G : d(x, y) < r\}$ we have

$$m(B(x, r)) = c r^{\dim_H \mathfrak{g}} \text{ with } c = m(B(0, 1)). \qquad (16.22)$$

Proof. By construction of the Haar measure and dilation on G, it suffices to compute everything in the Lie algebra with exponential coordinates

$$(x_{j,i})_{\substack{j=1,\ldots,N \\ i=1,\ldots,\dim \mathfrak{V}_j}}$$

and with respect to the Lebesgue measure. The image under δ_λ is precisely

$$\left(\lambda^j x_{j,i}\right)_{\substack{j=1,\ldots,N \\ i=1,\ldots,\dim \mathfrak{V}_j}}$$

and the determinant of the Jacobian of $(x_{j,i}) \mapsto \left(\lambda^j x_{j,i}\right)$ is obviously λ^Q. ∎

16.9.2 Jerison's Poincaré inequality

The Lie algebra of $\mathfrak{g} = \mathfrak{g}^N\left(\mathbb{R}^d\right)$ has a decomposition of the form $\mathfrak{g} = \mathfrak{V}_1 \oplus \cdots \oplus \mathfrak{V}_N$ and we shall represent the associated Lie group on the same space, e.g. $G = (\mathfrak{g}, *)$. There are left-invariant vector fields $\mathfrak{u}_1, \ldots, \mathfrak{u}_d$ on the group determined by

$$\mathfrak{u}_i|_0 = \partial_i|_0$$

where $\partial_i|_0$ are the coordinate vector fields associated with the canonical basis of $\mathfrak{V}_1 = \mathbb{R}^d$.

Example 16.45 When $N = 2$, we can identify $\mathfrak{g}$ with $\mathbb{R}^d \oplus so\,(d)$ and for $i = 1, \ldots, d$ we have

$$\mathfrak{u}_i|_x = \partial_i + \frac{1}{2}\left(\sum_{1 \le j < i \le d} x^{1;j}\partial_{j,i} - \sum_{1 \le i < j \le d} x^{1;j}\partial_{i,j}\right)$$

where ∂_i denotes the coordinate vector field on $\mathbb{R}^d$ and $\partial_{i,j}$ with $i < j$ the coordinate vector field on $so\,(d)$, identified with its upper-diagonal elements. □

Definition 16.46 We call $\nabla^{\mathrm{hyp}} = (\mathfrak{u}_1, \ldots, \mathfrak{u}_d)^T$ the **hypoelliptic gradient** on $G = \left(\mathfrak{g}^N\left(\mathbb{R}^d\right), *\right)$. When no confusion arises, we also write ∇. □

The following lemma is sometimes summarized by saying that the hypoelliptic gradient forms an "upper gradient" on G, equipped with Carnot–Caratheodory metric. Since $\mathfrak{g}$ and G have been identified, we state and prove it in the following form:

Lemma 16.47 (upper gradient lemma) *Let* $x, z \in \mathfrak{g}$. *For all compactly supported, smooth* $u : \mathfrak{g} \to \mathbb{R}$, *and* __admissible path__ Υ, *in the sense that*

$$\Upsilon \in C^{1\text{-Höl}}\left([0, \|z\|], \mathfrak{g}\right), \|\Upsilon\|_{1\text{-Höl}} \le 1),$$

which also has the property that $\Upsilon(0) = x, \Upsilon(\|z\|) = xz$, *we have*

$$|u(x) - u(xz)| \le \int_0^{\|z\|}\left|\nabla^{\mathrm{hyp}} u(\Upsilon_s)\right| ds.$$

Remark 16.48 The result extends to $u \in W^{1,2}\left(\mathfrak{g}, dm\right) \cap C\left(\mathfrak{g}\right)$, using the notation of Theorem 16.4. □

Proof. Let du denote the 1-form $\sum_j \partial_j u\left(.\right) dx^j$ with $j = 1, \ldots, \dim \mathfrak{g}$. By the fundamental theorem of calculus,

$$\left|u\left(x\right) - u\left(y\right)\right| \leq \int_0^{\|z\|} \left|\left\langle du\left(\Upsilon_s\right), \dot{\Upsilon}_s \right\rangle\right| ds.$$

But any $\Upsilon \in C^{1\text{-Höl}}$ is, viewed through the global "log"-chart, the step-N lift of an $\mathbb{R}^d$-valued 1-Hölder path γ and so

$$\dot{\Upsilon}_t = \sum_{i=1}^d \mathfrak{u}_i\left(\Upsilon_t\right) \dot{\gamma}_t^i.$$

(Namely, $\gamma = \pi_1\left(\Upsilon\right)$ is a 1-Hölder path $[0, \|z\|] \to \mathbb{R}^d$, at unit speed.) Then

$$\begin{aligned} \left\langle du\left(\Upsilon_s\right), \dot{\Upsilon}_s \right\rangle &= \sum_{i=1}^d \left\langle du\left(\Upsilon_s\right), \mathfrak{u}_i\left(\Upsilon_s\right)\right\rangle \dot{\gamma}_s^i \\ &= \left(\nabla u|_{\Upsilon_s}\right) \cdot \dot{\gamma}_s, \end{aligned}$$

where $\cdot$ is the inner product on $\mathbb{R}^d$. For a.e. s we have $|\dot{\gamma}_s| \leq 1$ and Cauchy–Schwarz on $\mathbb{R}^d$ shows that

$$\left|\left(\nabla u|_{\Upsilon_s}\right) \cdot \dot{\gamma}_s\right| \leq \left|\left(\nabla u|_{\Upsilon_s}\right)\right|.$$

■

Proposition 16.49 *(weak Poincaré inequality) There exists a constant C such that for all smooth $u : \mathfrak{g} \to \mathbb{R}$ and all $y \in \mathfrak{g}$ and $r > 0$,*

$$\int_{B(y,r)} \left(u\left(x\right) - \bar{u}_r\right)^2 dx \leq C r^2 \int_{B(y,2r)} \left|\nabla^{\text{hyp}} u\left(x\right)\right|^2 dx$$

where $\bar{u}_r = \int_{B(y,r)} u\left(x\right) dx$.

Proof. We may assume that $y = 0 \in \mathfrak{g}$ so that $B = B\left(y, r\right)$ is centred at the unit element in the group. We shall also write

$$\Upsilon^z : [0, \|z\|] \to \mathfrak{g}$$

for a geodesic which connects the unit element with z, parametrized to run at unit speed. It follows that

$$s \mapsto x \Upsilon_s^z$$

is a geodesic from x to xz, run at unit speed. From the "upper-gradient lemma"

$$|u(x) - u(xz)| \le \int_0^{\|z\|} \nabla u(x\Upsilon_s^z)\,ds$$

and by Cauchy–Schwarz,

$$|u(x) - u(xz)|^2 \le \|z\| \int_0^{\|z\|} |\nabla u|^2 (x\Upsilon_s^z)\,ds.$$

This and the left-invariance of the Lebesgue measure on $\mathfrak{g}$ yields

$$\int_B (u(x) - \bar{u})^2\,dx = \int_B \left(\frac{1}{|B|} \int_B (u(x) - u(y))\,dy \right)^2 dx$$

$$\le \frac{1}{|B|} \int_B \int_B ((u(x) - u(y)))^2\,dy dx$$

$$= \frac{1}{|B|} \int_{\mathfrak{g}} \int_{\mathfrak{g}} 1_B(x)\,1_B(xz)\,((u(x) - u(xz)))^2\,dz dx$$

$$\le \frac{1}{|B|} \int_{\mathfrak{g}} \int_{\mathfrak{g}} \int_0^{\|z\|} 1_B(x)\,1_B(xz)\,\|z\|\,|\nabla u|^2 (x\Upsilon_s^z)\,ds dz dx.$$

By right-invariance of the Lebesgue measure we obtain

$$\int_{\mathfrak{g}} 1_B(x)\,1_B(xz)\,\nabla u(x\Upsilon_s^z)^2\,dx = \int_{\mathfrak{g}} 1_{B\Upsilon_s^z}(\xi)\,1_{Bz^{-1}\Upsilon_s^z}(\xi)\,|\nabla u|^2(\xi)\,d\xi$$

$$\le 1_{2B}(z) \int_{2B} |\nabla u|^2(\xi)\,d\xi. \qquad (16.23)$$

Here we denote by Bh the right translation of B by h. The above inequality requires some explanation. If the expression under the sign of the middle interval has a non-zero value, then $\xi = x\Upsilon_s^z = yz^{-1}\Upsilon_s^z$ for some $x, y \in B$. Hence $z = x^{-1}y \in 2B$. Thus $\xi = x\Upsilon_s^{x^{-1}y}$ lies on a geodesic that joins x with y and so $d(x, \xi) + d(\xi, y) = d(x, y)$, which together with the triangle inequality implies $\xi \in 2B$. This leads to the estimate (16.23), as claimed. Then

$$\int_B (u(x) - \bar{u})^2\,dx \le \frac{1}{|B|} \int_{\mathfrak{g}} \int_0^{\|z\|} 1_{2B}(z)\,\|z\| \int_{2B} |\nabla u|^2(\xi)\,d\xi ds dz$$

$$= \frac{1}{|B|} \int_{2B} \int_{2B} \|z\|^2\,|\nabla u|^2(\xi)\,d\xi dz$$

and we conclude with $|B| = |B(0,1)|\,r^Q$, where Q is the homogenous dimension of $\mathfrak{g}$, cf (16.22), and

$$\int_{2B} \|z\|^2\,dz = |B(0,1)| \int_0^{2r} \rho^2 d(\rho^N) = (\text{const}) \times r^{Q+2}.$$

■

16.9.3   Carnot–Caratheodory metric as intrinstic metric

Following the notation of Theorem 16.4 we set

$$\Gamma(f,g) := \nabla^{\mathrm{hyp}} f \cdot \nabla^{\mathrm{hyp}} g = \sum_{i=1}^{d} \mathfrak{u}_i f \, \mathfrak{u}_i g, \ \mathcal{E}(f,g) = \int \Gamma(f,g) \, dm$$

for $f, g \in C_c^\infty(\mathfrak{g}, \mathbb{R})$ where $\mathfrak{g} \equiv \mathfrak{g}^N(\mathbb{R}^d)$ is equipped with Lebesgue measure dm. The domain of Γ resp. $\mathcal{E}$ naturally extends to $W^{1,2}(\mathfrak{g}, dm)$, the closure of C_c^∞ with respect to

$$\|f\|_{\mathcal{F}}^2 = \mathcal{E}(f,f) + \langle f, f \rangle_{L^2} \quad \text{with } L^2 = L^2(\mathfrak{g}, dm).$$

Let us define the *intrinsic metric* on $\mathfrak{g}$ by

$$
\begin{aligned}
\varrho(x,y) : \ &= \ \sup\{f(y) - f(x) : f \in C_c^\infty(\mathfrak{g},\mathbb{R}) \text{ and } |\Gamma(f,f)|_\infty \le 1\} \\
&= \ \sup\{f(y) - f(x) : f \in W^{1,2}(\mathfrak{g}, dm) \cap C \\
&\quad (\mathfrak{g},\mathbb{R}) \text{ and } |\Gamma(f,f)| \le 1 \text{ a.s.}\}
\end{aligned}
$$

and the *Carnot–Caratheodory distance* on $\mathfrak{g}$ by

$$
\begin{aligned}
d(x,y) \ &= \ \inf\left\{\int_0^1 |dh| : \ h \text{ Lipschitz, } \gamma(0) \right. \\
&= \ \left. x, \gamma(1) = y : d\gamma = \sum_{i=1}^{d} \mathfrak{u}_i(\gamma) \, dh^i \right\}. \quad (16.24)
\end{aligned}
$$

It is easy to see (cf. Remark 7.43) that this is precisely the Carnot–Caratheodory distance on $G = G^N(\mathbb{R}^d)$, cf. Definition 7.41, as seen through the log-chart.[19]

Theorem 16.50 *The Carnot–Caratheodory distance on $\mathfrak{g}$ coincides with the intrinsic metric.*

Proof. $\varrho(x,y) \le d(x,y)$: Fix $f \in C_c^\infty(\mathfrak{g}^N(\mathbb{R}^d), \mathbb{R})$ with $|\Gamma(f,f)|_\infty \le 1$ and h Lipschitz such that the solution to the ODE $d\gamma = \sum_{i=1}^{d} \mathfrak{u}_i(\gamma) \, dh^i$, $\gamma(0) = x$ satisfies $\gamma(1) = y$. Clearly,

$$
\begin{aligned}
f(y) - f(x) \ &= \ \int_0^1 Df(\gamma_t) \, d\gamma_t = \sum_{i=1}^{d} \int_0^1 (\mathfrak{u}_i f)_{\gamma_t} \, dh_t^i \\
&\le \ |\Gamma(f,f)|_\infty \int_0^1 |dh| \le \int_0^1 |dh|.
\end{aligned}
$$

---

[19]Strictly speaking, we should take 16.24 as the definition for $\tilde{d}(x,y)$, so that $\tilde{d}(\log g, \log h) = d(g,h)$ for all $g, h \in G^N(\mathbb{R}^d) = \log^{-1}(\mathfrak{g}^N(\mathbb{R}^d))$.

Passing the sup (resp. inf) over all such f (resp. h) we see that $\varrho(x,y) \leq d(x,y)$.

$\underline{\varrho(x,y) \geq d(x,y)}$: Assume momentarily that $d(x, \cdot)$ is an admissible function in the definition of ϱ. It would then follow that $\varrho(x,y) \geq d(x,y) - d(x,x) = d(x,y)$, which is what we seek. To make this rigorous we proceed in two steps. First we extend

$$\{\mathfrak{u}_i : i = 1, \ldots, d\}$$

to a full basis of $\mathfrak{g}^N(\mathbb{R}^d) \cong \mathbb{R}^m$ with $m = \dim \mathfrak{g}^N(\mathbb{R}^d)$,

$$\{\mathfrak{v}_j^\varepsilon : i = 1, \ldots, m\} := \{\mathfrak{u}_i : i = 1, \ldots, d\} \cup \left\{\varepsilon^{1/2} \partial_j\right\} \dim \mathfrak{g}^N(\mathbb{R}^d)$$

where $\{\partial_j\}$ denotes the coordinate vector fields corresponding to the last $(m - d)$ coordinates of $\mathbb{R}^m$. Replacing $\{\mathfrak{u}_i\}$ by $\{\mathfrak{v}_j^\varepsilon\}$ we can define an intrinsic distance $\varrho^\varepsilon(x,y)$ associated with $\Gamma^\varepsilon(f,g) := \sum_{i=j}^m \mathfrak{v}_j f \, \mathfrak{v}_j g$ and similarly a control distance $d^\varepsilon(x,y)$. Leaving straightforward mollification arguments to the reader (or [22, p. 285]), the class of admissible functions can be extended to include $d^\varepsilon(x, \cdot)$ and it follows that

$$d^\varepsilon(x,y) = \varrho^\varepsilon(x,y).$$

On the other hand, it is not hard to see that

$$d(x, y_\varepsilon) \leq d^\varepsilon(x,y) = \varrho^\varepsilon(x,y) \leq \varrho(x,y)$$

for some y_ε which converges to y as $\varepsilon \to 0$. Using continuity of the Carnot–Caratheodory distance it suffices to send $\varepsilon \to 0$ and the proof is finished. ∎

16.10 Comments

The Dirichlet form approach to Markovian rough paths was first adopted by Friz and Victoir [68] and already contains the bulk of the results in this chapter. (One could in fact bypass the use of abstract Dirichlet form theory and give a direct analytical treatment along the lines of Saloff-Coste and Stroock [151]). For some background on (symmetric) Dirichlet forms, the reader can consult Appendix E and the references cited therein. The theory of non-symmetric Dirichlet forms (e.g. in particular the heat-kernel estimates from Sturm [164]) would allow us to construct enhanced non-symmetric Markov process in a similar way. Exercise 16.10, concerning Nash's inequality and (on-diagonal) estimates for heat-kernels, is taken from Carlen *et al.* [22].

The construction of a stochastic area associated with a Markov process of the present type goes back to Lyons and Stoica [121]: they use forward–backward martingale decomposition to see that, in particular, an $\mathbb{R}^d$-valued (Markov) process X^a (with uniformly elliptic generator in divergence form) can be rendered accessible to Stratonovich integration; see also Rozkosz [146, 147] for related considerations. As a consequence of the Wong–Zakai theorem, (lifted) piecewise linear approximation will converge (to the Stratonovich lift) and so yield an enhanced Markov process $\mathbf{X}^a$. Lejay [105, 106] then establishes p-variation rough regularity of this enhancement $\mathbf{X}^a$. The link with our approach is made in comment (iv) after Definition 16.14 in Section 16.4. A version of Theorem 16.28 appears in Lejay [106], the author then also discusses applications to homogenization. Sample path large deviations for Markovian rough paths were established by Friz and Victoir [68]; we observed that the same arguments apply in the abstract setting of local Dirichlet spaces and therefore give the proof in the appropriate Section E.6 of Appendix E. (Although we are unaware of any reference to such a result in the context of Dirichlet space theory, J. Ramírez has shown us an unpublished preprint which covers our results.) Similarly, a support theorem was established by Friz and Victoir [68]; we have here outsourced the key estimates in the abstract setting of Section E.6 of Appendix E and obtained a slight sharpening of the result. The restriction to Hölder exponent $< 1/4$ in the (support) Theorem 16.33 is almost certainly a technical one though. Indeed, Theorem 16.39 shows that every ball around the trivial zero-path is charged in (rough-path) Hölder metric, for any exponent $< 1/2$, which is what one suspects. The problem, in contrast to the similar support discussion of Section 13.8, is the lack of the Cameron–Martin theorem in the present context. It is conjectured that the use of (time-dependent, non-symmetric) Dirichlet forms will allow us to generalize our discussion of Section 16.8.3 so as to obtain a support theorem for exponent $< 1/2$.

In Appendix 16.9 we collect some classical analytic results for free nilpotent groups. See Folland and Stein [54] for a general discussion. The Poincaré inequality (Proposition 16.49) appears explicitly in Jerison [93]; our simplified proof is a variation of a proof attributed to Varopoulos by Hajtasz and Koskela [83]. Theorem 16.50 on the consistency between intrinsic and Carnot–Caratheodory distance is taken from Carlen *et al.* [22].

Part IV

Applications to stochastic analysis

17

Stochastic differential equations and stochastic flows

We saw in Part III that large classes of multidimensional stochastic processes, including semi-martingales, Gaussian and Markov processes, are naturally enhanced to random rough paths, i.e. processes whose sample paths are almost surely geometric rough paths.

Clearly then, there is a *pathwise* notion of stochastic differential equation driven by such processes, simply by considering the rough differential equation driven by (geometric rough path) realizations of the enhanced process. As will be discussed in this chapter, there is a close link to (Stratonovich) stochastic differential equations, based on stochastic integration theory. But first, we start with a working summary of "rough path" continuity results (with precise references to the relevant statements in Part II).

17.1 Working summary on rough paths

17.1.1 Iterated integration

Let $p \geq 1$ and $\mathbf{x} \in C^{p\text{-var}}\left([0,T], G^{[p]}\left(\mathbb{R}^d\right)\right)$ be a weak geometric p-rough path. From Corollary 9.11, the Lyons-lifting map

$$S_N \quad : \quad C_o^{p\text{-var}}\left([0,T], G^{[p]}\left(\mathbb{R}^d\right)\right) \to C_o^{p\text{-var}}\left([0,T], G^N\left(\mathbb{R}^d\right)\right),$$

$$S_N \quad : \quad C_o^{1/p\text{-Höl}}\left([0,T], G^{[p]}\left(\mathbb{R}^d\right)\right) \to C_o^{1/p\text{-Höl}}\left([0,T], G^N\left(\mathbb{R}^d\right)\right)$$

is continuous in p-variation and $1/p$-Hölder (rough path) topology respectively. We think of $S_N(\mathbf{x})$ as attaching higher iterated integrals to $\mathbf{x}$. Indeed, when $\mathbf{x}$ consists of all iterated integrals (in the Riemann–Stieltjes sense) up to order $[p]$ of some $x \in C^{1\text{-var}}\left([0,T], \mathbb{R}^d\right)$, that is $\mathbf{x} = S_{[p]}(x)$, then $S_N(\mathbf{x}) = S_N(x)$.

17.1.2 Integration

Let $\gamma > p \geq 1$ and $\varphi = (\varphi_1, \ldots, \varphi_d) \subset \text{Lip}^{\gamma-1}\left(\mathbb{R}^d, \mathbb{R}^e\right)$. From Theorem 10.47, there is a unique rough integral of φ against $\mathbf{x}$ and[1]

---

[1] By convention, rough integrals are $G^{[p]}\left(\mathbb{R}^e\right)$-valued; π_1 is the projection to $\mathbb{R}^e$.

$$C_o^{p\text{-var}}\left([0,T],G^{[p]}\left(\mathbb{R}^d\right)\right) \quad \to \quad C_o^{p\text{-var}}\left([0,T],\mathbb{R}^e\right)$$

$$\mathbf{x} \quad \mapsto \quad \pi_1\left(\int_0^{\cdot} \varphi\left(x\right) d\mathbf{x}\right)$$

is continuous in p-variation topology. The same statement holds in $1/p$-Hölder topology. This integral generalizes the Riemann–Stieltjes integral in the sense that for $\mathbf{x} = S_{[p]}\left(x\right)$ with $x \in C^{1\text{-var}}\left([0,T],\mathbb{R}^d\right)$, we have

$$\pi_1\left(\int_0^{\cdot} \varphi\left(x\right) d\mathbf{x}\right) = \int_0^{\cdot} \varphi\left(x\right) dx$$

where the right-hand side is a well-defined Riemann–Stieltjes integral.

17.1.3 Differential equations

Let $\gamma > p \geq 1$ and $V = (V_1,\dots,V_d) \subset \text{Lip}^\gamma\left(\mathbb{R}^e\right)$, a collection of vector fields on $\mathbb{R}^e$. From Theorem 10.26 and its corollaries there is a unique solution $y = \pi_{(V)}\left(0,y_0;\mathbf{x}\right)$ to the RDE

$$dy = V\left(y\right) d\mathbf{x},$$

started at $y_0 \in \mathbb{R}^e$ and

$$C_o^{p\text{-var}}\left([0,T],G^{[p]}\left(\mathbb{R}^d\right)\right) \quad \to \quad C_o^{p\text{-var}}\left([0,T],\mathbb{R}^e\right)$$

$$\mathbf{x} \quad \mapsto \quad \pi_{(V)}\left(0,y_0;\mathbf{x}\right)$$

is continuous in p-variation topology. The same statement holds in $1/p$-Hölder topology. RDEs generalize ODEs in the sense that for $\mathbf{x} = S_{[p]}\left(x\right)$ with $x \in C^{1\text{-var}}\left([0,T],\mathbb{R}^d\right)$, we are dealing with a solution to the classical ODE

$$dy = V\left(y\right) dx,$$

understood as a Riemann–Stieltjes integral equation.

One has also the refined result that uniqueness/continuity holds when $V \subset \text{Lip}^p\left(\mathbb{R}^e\right)$ and $\mathbf{x}$ has finite p-variation. In fact, motivated by sample path properties of (enhanced) Brownian motion, namely

$$\mathbf{B}\left(\omega\right) \notin C^{2\text{-var}} \text{ but } \mathbf{B}\left(\omega\right) \subset C^{\psi_{2,1}\text{-var}} \cap C^{\alpha\text{-Höl}} \text{ a.s.}$$

for any $\alpha \in [0,1/2)$, the regularity assumption on $\mathbf{x}$ can be further relaxed to finite $\psi_{p,1}$-variation,

$$C^{p\text{-var}} \subset C^{\psi_{p,1}\text{-var}} \subset C^{(p+\varepsilon)\text{-var}},$$

cf. Definition 5.45. The continuity statements become somewhat more involved, but this will not be restrictive in applications. For $V \subset \text{Lip}^p\left(\mathbb{R}^e\right)$

and $1 \le p < p' < [p] + 1$ we have continuity

$$\mathbf{x} \in \left(C^{\psi_{p,1}\text{-var}} \left([0,T], G^{[p]} \left(\mathbb{R}^d \right) \right), d_{p'\text{-var}} \right)$$
$$\mapsto \pi_{(V)} (0, y_0; \mathbf{x}) \in \left(C^{p'\text{-var}} ([0,T], \mathbb{R}^e), |\cdot|_{p'\text{-var}} \right)$$

and, given $1 \le p < p' < p'' < [p] + 1$,

$$\mathbf{x} \in \left(C^{\psi_{p,1}\text{-var}} \cap C^{1/p'\text{-Höl}} \left([0,T], G^{[p]} \left(\mathbb{R}^d \right) \right), d_{1/p'\text{-Höl}} \right)$$
$$\mapsto \pi_{(V)} (0, y_0; \mathbf{x}) \in \left(C^{1/p'\text{-Höl}} ([0,T], \mathbb{R}^e), |\cdot|_{1/p''\text{-Höl}} \right).$$

Elements in the (non-complete, non-separable) metric space

$$\left(C^{\psi_{p,1}\text{-var}} \cap C^{1/p'\text{-Höl}} \left([0,T], G^{[p]} \left(\mathbb{R}^d \right) \right), d_{1/p'\text{-Höl}} \right)$$

are simply weak geometric $1/p'$-Hölder rough paths with additonal $\psi_{p,1}$-var regularity. To avoid measurability issues arising from non-separability, we can restrict attention to the (non-complete, separable) metric space

$$\left(C^{\psi_{p,1}\text{-var}} \cap C^{0,1/p'\text{-Höl}} \left([0,T], G^{[p]} \left(\mathbb{R}^d \right) \right), d_{1/p'\text{-Höl}} \right),$$

elements of which are geometric $1/p'$-Hölder rough paths with additional $\psi_{p,1}$-var regularity.

17.1.4 *Differential equations with drift*

Let us now consider two collections of vector fields,

$$V = (V_1, \ldots, V_d) \subset \text{Lip}^\gamma (\mathbb{R}^e),$$
$$W = (V_1, \ldots, V_{d'}) \subset \text{Lip}^\beta (\mathbb{R}^e),$$

driven by $\mathbf{x} \in C^{p\text{-var}} ([0,T], G^{[p]} (\mathbb{R}^d))$ and $h \in C^{1\text{-var}} \left([0,T], \mathbb{R}^{d'} \right)$, respectively. When $\gamma > p \ge 1$ and $\beta > 1$ it follows from Theorem 12.10 that there is a unique solution $y = \pi_{(V,W)} (0, y_0; (\mathbf{x}, h))$ to the RDE with drift

$$dy = V(y)\, d\mathbf{x} + W(y)\, dh, \quad y(0) = y_0 \in \mathbb{R}^e,$$

started at $y_0 \in \mathbb{R}^e$ and

$$\mathbf{x} \in \left(C^{p\text{-var}} \left([0,T], G^{[p]} \left(\mathbb{R}^d \right) \right), d_{p\text{-var}} \right)$$
$$\mapsto \pi_{(V,W)} (0, y_0; (\mathbf{x}, h)) \in C^{p\text{-var}} ([0,T], \mathbb{R}^e)$$

is continuous in p-variation topology. Again, the same statement holds in $1/p$-Hölder topology.

One has also the refined result that uniqueness/continuity holds when $\gamma = p, \beta = 1$ and the regularity assumption on $\mathbf{x}$ is relaxed to finite $\psi_{p,1}$-variation: from Theorem 12.11 (and the remarks afterwards) with $1 \leq p < p' < p'' < [p] + 1$ we have the following continuity statements:

$$\mathbf{x} \in \left(C^{\psi_{p,1}\text{-var}} \left([0,T], G^{[p]} \left(\mathbb{R}^d \right) \right), d_{p'\text{-var}} \right)$$

$$\mapsto \pi_{(V,W)} \left(0, y_0; (\mathbf{x}, h) \right) \in \left(C^{p'\text{-var}} \left([0,T], \mathbb{R}^e \right), |\cdot|_{p'\text{-var}} \right)$$

and

$$\mathbf{x} \in \left(C^{\psi_{p,1}\text{-var}} \cap C^{1/p'\text{-Höl}} \left([0,T], G^{[p]} \left(\mathbb{R}^d \right) \right), d_{1/p'\text{-Höl}} \right)$$

$$\mapsto \pi_{(V,W)} \left(0, y_0; (\mathbf{x}, h) \right) \in \left(C^{1/p'\text{-Höl}} \left([0,T], \mathbb{R}^e \right), |\cdot|_{1/p''\text{-Höl}} \right).$$

17.1.5 *Some further remarks*

For simplicity, we have only stated continuity of the RDE solutions as a function of $\mathbf{x}$. In fact, looking at the relevant statements in Part II shows continuity of

$$(y_0, \mathbf{x}; V) \mapsto \pi_{(V)} \left(0, y_0; \mathbf{x} \right)$$

and similar for RDEs with drift (see e.g. Corollary 10.40 and Theorem 12.11). It is also worth remarking that the assumption $V \subset \text{Lip}^\gamma \left(\mathbb{R}^e \right), \gamma > p$, allows for continuity results on the level of flows of diffeomorphisms, cf. Section 11.2 and Section 17.5 below.

17.2 Rough paths vs Stratonovich theory

In the present section we make the link to the theory of stochastic integration (resp. differential equations) with respect to (continuous) semi-martingales.[2]

17.2.1 *Stratonovich integration as rough integration*

We show in this subsection that rough integration along enhanced semi-martingales coincides with classical Stratonovich integration. The case of integration against (enhanced) Brownian motion is, of course, a special case.

---

[2]As is customary in this context, it is understood that there is an underlying prob-ability space, say $(\Omega, \mathfrak{F}, \mathbb{P})$, where $\mathfrak{F}$ is $\mathbb{P}$-complete and there will be a right-continuous filtration $(\mathfrak{F}_t)_{t \geq 0}$ with $\mathfrak{F}_0$ containing all $\mathbb{P}$-null sets (the "usual conditions").

Let N be a real-valued continuous semi-martingale, $M = \left(M^1, \ldots, M^d \right)$ an $\mathbb{R}^d$-valued continuous semi-martingale and $f \in C^1 \left(\mathbb{R}^n, \mathbb{R} \right)$. We fix a time-horizon $[0, T]$ and a sequence of dissections (D_n) with mesh $|D_n| \to 0$. As usual, N^{D_n} denotes the path obtained by piecewise linear approximation to the (sample) path $N = N \left(\cdot; \omega \right)$ and the same notation applies to $f(M) = f \left(M \left(\cdot; \omega \right) \right)$. It is a routine exercise in stochastic integration[3] to show that[4]

$$\lim_{n \to \infty} \int_0^t \left[f(M) \right]^{D_n} dN^{D_n} = \int_0^t f(M) \, dN + \frac{1}{2} \sum_{i=1}^d \int_0^t \partial_i f(M) \, d \left\langle M^i, N \right\rangle,$$

(17.1)

in probability and uniformly in $t \in [0, T]$. One can then use either side as a definition of the *Stratonovich integral* of $f(M)$ against N, denoted by

$$\int_0^t f(M) \circ dN.$$

This allows us in particular to define

$$\int_0^t \varphi(M) \circ dM \equiv \sum_{i=1}^d \int_0^t \varphi_i(M) \circ dM^i$$

where $\varphi = (\varphi_1, \ldots, \varphi_d)$ is a collection of $C^1 \left(\mathbb{R}^d, \mathbb{R}^e \right)$-functions. On the other hand, given $\gamma > 2$ we can pick $p \in (2, \min(3, \gamma))$ and know from Proposition 14.9 that M can be enhanced to a geometric p-rough path $\mathbf{M}$, i.e. $\mathbf{M}(\omega) \in C^{0, p\text{-var}} \left([0, T], G^2 \left(\mathbb{R}^d \right) \right)$. In particular, there is a well-defined rough integral in the sense of Section 10.6:

$$\int_0^t \varphi(M) \, d\mathbf{M}$$

provided $\varphi \in \mathrm{Lip}^{\gamma - 1}$ with $\gamma > p > 2$.

Proposition 17.1 *Let $\gamma > 2$ and $\varphi_i \in \mathrm{Lip}^{\gamma - 1} \left(\mathbb{R}^d, \mathbb{R}^e \right)$ for $i = 1, \ldots, d$, and M an $\mathbb{R}^d$-valued semi-martingale. Then the rough integral of the enhanced semi-martingale $\mathbf{M}$ against φ exists and, with probability one,*

$$\forall t \in [0, T] : \int_0^t \varphi(M) \circ dM = \pi_1 \left(\int_0^t \varphi(M) \, d\mathbf{M} \right).$$

Proof. The fact that $\int_0^t \varphi(M) \, d\mathbf{M}$ is well-defined follows from the fact that $\mathbf{M}$ is almost surely a geometric p-rough path, for any $p \in (2, \min(3, \gamma))$, see

---

[3] See, for example, Stroock [160, p. 229]; and recall that $[f(M)]^D$ denotes the piecewise linear approximation to the path $f(M \cdot)$, based on the dissection D.

[4] $\int f(M) \, dN$ denotes the Itô stochastic integral.

Theorem 14.12. Take a sequence of dissections (D_n) with mesh $|D_n| \to 0$. From Theorem 14.16, and the remarks at the beginning of that chapter, we know that $d_{p\text{-var}}\left(S_2\left(M^{D_n}\right), \mathbf{M}\right) \to_{n\to\infty} 0$ in probability and so, by basic continuity properties of rough integration (cf. Section 10.6),

$$d_{p\text{-var}}\left(\pi_1\left(\int_0^t \varphi(M)\, d\mathbf{M}\right), \int_0^t \varphi\left(M^{D_n}\right) dM^{D_n}\right) \to_{n\to\infty} 0$$

in probability. On the other hand, from (17.1) we have

$$d_\infty\left(\int_0^t \varphi(M) \circ dM, \int_0^t [\varphi(M)]^{D_n}\, dM^{D_n}\right) \to_{n\to\infty} 0$$

in probability. So, to conclude the proof, we only need to prove that

$$\sup_{t\in[0,T]}\left|\int_0^t \left(\varphi\left(M_s^{D_n}\right) - [\varphi(M)]_s^{D_n}\right) dM_s^{D_n}\right| \to_{n\to\infty} 0$$

in probability. Given a dissection $D = (t_i)$ and $s \in [0, t]$ we have

$$\varphi\left(M_s^D\right) = \varphi\left(M_{s_D} + \frac{s - s_D}{s^D - s_D} M_{s_D, s^D}\right)$$

$$= \varphi(M_{s_D}) + \varphi'(M_{s_D})\left(\frac{s - s_D}{s^D - s_D} M_{s_D, s^D}\right) + O\left(|M|_{0;[s_D, s^D]}^{\gamma-1}\right),$$

$$[\varphi(M)]_s^D = \varphi(M_{s_D}) + \frac{s - s_D}{s^D - s_D}\left[\varphi(M_{s^D}) - \varphi(M_{s_D})\right]$$

$$= \varphi(M_{s_D}) + \frac{s - s_D}{s^D - s_D}\left[\varphi'(M_{s_D}) M_{s_D, s^D}\right] + O\left(|M|_{0;[s_D, s^D]}^{\gamma-1}\right).$$

It follows that for $D_n = (t_i)$ and $s \in [t_i, t_{i+1}]$,

$$\varphi\left(M_s^{D_n}\right) - [\varphi(M)]_s^{D_n} = O\left(|M|_{0;[t_i, t_{i+1}]}^{\gamma-1}\right),$$

and hence that

$$\sup_{t\in[0,T]}\left|\int_0^t \left(\varphi\left(M_s^{D_n}\right) - [\varphi(M)]_s^{D_n}\right) dM_s^{D_n}\right| \le c_1 \sum_i |M|_{0;[t_i, t_{i+1}]}^{\gamma}$$

which converges to zero, even almost surely, with $|D_n| \to 0$, as is easily seen from Theorem 8.22, or directly from

$$\sum_i |M|_{0;[t_i, t_{i+1}]}^{\gamma} \le |M|_{p\text{-var};[0,T]}^p \left(\sup_{|t-s|\le|D_n|} |M|^{\gamma-p}\right)$$

and a.s. uniform continuity of $t \in [0, T] \mapsto M_t$. This concludes the proof. ∎

Exercise 17.2 *Let M be a continuous, $\mathbb{R}^d$-valued local martingale with lift* **M**.

(i) Show that the collection of iterated Stratonovich integrals up to level N, i.e.

$$S_N^{\text{Strato}}(M)_t := \left(1, M_t, \dots, \int_{\{0 < s_1 < \cdots s_N < t\}} \circ dM_{s_1} \otimes \cdots \otimes \circ dM_{s_N} \right)$$

viewed as a continuous path in $G^N\left(\mathbb{R}^d\right)$ coincides with the Lyons lift $S_N(\mathbf{M})$.

(ii) Let F be a moderate function (cf. Definition 14.4). Show that there exists $C = C(N, F, d)$ such that

$$\mathbb{E}\left(F\left(\sup_{0 \le t < \infty} \left| \int_{\{0 < s_1 < \cdots s_N < t\}} \circ dM_{s_1} \otimes \cdots \otimes \circ dM_{s_N} \right|^{1/N} \right) \right)$$

$$\le C\mathbb{E}\left(F\left(\left| \langle M \rangle_\infty \right|^{1/2} \right) \right).$$

Solution. *(i) is an easy consequence of Proposition 17.1 and we have in particular*

$$\int_{\{0 < s_1 < \cdots s_N < t\}} \circ dM_{s_1} \otimes \cdots \otimes \circ dM_{s_N} = \pi_N\left(S_N\left(\mathbf{M}\right) \right).$$

Clearly,

$$\sup_{0 \le t < \infty} \left| \int_{\{0 < s_1 < \cdots s_N < t\}} \circ dM_{s_1} \otimes \cdots \otimes \circ dM_{s_N} \right|^{1/N} \le \sup_{0 \le t < \infty} \left\| S_N\left(\mathbf{M}\right)_{0,t} \right\|$$

$$\le \left\| S_N\left(\mathbf{M}\right) \right\|_{p\text{-var};[0,\infty)}.$$

From Theorem 9.5, we have for some constant $c_1 = c_1(N, p)$,

$$\left\| S_N\left(\mathbf{M}\right) \right\|_{p\text{-var};[0,\infty)} \le c_1 \left\| \mathbf{M} \right\|_{p\text{-var};[0,\infty)}.$$

Hence, using Theorem 14.12 we see that

$$\mathbb{E}\left(F\left(\sup_{0 \le t < \infty} \left| \int_{\{0 < s_1 < \cdots s_N < t\}} \circ dM_{s_1} \otimes \cdots \otimes \circ dM_{s_N} \right|^{1/N} \right) \right)$$

$$\le \mathbb{E}\left(F\left(c_1 \left\| \mathbf{M} \right\|_{p\text{-var};[0,\infty)} \right) \right)$$

$$\le c_2 \mathbb{E}\left(F\left(\left| \langle M \rangle_\infty \right|^{1/2} \right) \right).$$

17.2.2 Stratonovich SDEs as RDEs

We extend the result of the previous section to differential equations. The main point is that solutions to RDEs driven by (enhanced) semi-martingales solve the corresponding Stratonovich stochastic differential

equation. Again, the case of (enhanced) Brownian motion is a special case
of this. We start by recalling that a solution to the *Stratonovich stochastic
differential equation*

$$dY = \sum_{i=1}^{d} V_i(Y) \circ dM^i, \qquad (17.2)$$

driven by a general (continuous) semi-martingale $M = (M^1, \dots, M^d)$ is,
by definition, a solution to the integral equation

$$Y_{0,t} = \sum_{i=1}^{d} \int_0^t V_i(Y) dM^i + \frac{1}{2} \sum_{i,j=1}^{d} \int_0^t V_i V_j(Y) d\langle M^i, M^j \rangle, \qquad (17.3)$$

assuming $V \in C^1$ so that $V_i V_j \equiv V_i^k \partial_k V_j$ is well-defined. Obviously then,
Y is a semi-martingale itself and from basic stochastic calculus Y is indeed
a solution to the Stratonovich integral equation

$$Y_{0,t} = \sum_{i=1}^{d} \int_0^t V_i(Y) \circ dM^i$$

where the Stratonovich integral on the right-hand side was defined in
the previous section. The extension to SDEs with drift-vector fields $W = (W_1, \dots, W_{d'})$, driven by an $\mathbb{R}^{d'}$-valued adapted process $H = (H_1, \dots, H_{d'})$
with sample paths in $C^{1\text{-var}}\left([0,T], \mathbb{R}^{d'}\right)$, is easy since H itself is a semi-martingale (with vanishing quadratic variation): a solution to the
Stratonovich SDE with drift

$$dY = \sum_{i=1}^{d} V_i(Y) \circ dM^i + \sum_{i=1}^{d'} W_j(Y) dH^j \qquad (17.4)$$

is then a solution to the equation

$$Y_{0,t} = \sum_{i=1}^{d} \int_0^t V_i(Y) dM^i + \frac{1}{2} \sum_{i,j=1}^{d} \int_0^t V_i V_j(Y) d\langle M^i, M^j \rangle$$

$$+ \sum_{j=1}^{d'} \int_0^t W_j(Y) dH^j. \qquad (17.5)$$

Theorem 17.3 *Let p, γ be such that $2 < p < \gamma$. Assume*
(i) $V = (V_i)_{1 \le i \le d}$ is a collection of vector fields in $\mathrm{Lip}^\gamma(\mathbb{R}^e)$;
(i bis) $W = (W_i)_{1 \le i \le d'}$ is a collection of vector fields in $\mathrm{Lip}^1(\mathbb{R}^e)$;
(ii) M is an $\mathbb{R}^d$-valued semi-martingale, enhanced to $\mathbf{M} = \mathbf{M}(\omega) \in C^{0,p\text{-var}}\left([0,T], G^2(\mathbb{R}^d)\right)$ almost surely;

(ii bis) H is an $\mathbb{R}^{d'}$-valued continuous, adapted process, so that $H(\omega) \in C^{1\text{-var}}\left([0,T], \mathbb{R}^{d'}\right)$ almost surely;

(iii) $y_0 \in \mathbb{R}^e$.

Then the (for a.e. ω well-defined) RDE solution

$$Y(\omega) = \pi_{(V,W)}(0, y_0; (\mathbf{M}(\omega), H(\omega)))$$

solves the Stratonovitch SDE

$$dY = \sum_{i=1}^{d} V_i(Y) \circ dM^i + \sum_{i=1}^{d'} W_j(Y) \, dH^j, \quad Y(0) = y_0. \qquad (17.6)$$

Remark 17.4 In the case of driving Brownian motion, $V \in \text{Lip}^2$ and $W \in \text{Lip}^1$ suffices to have an ω-wise uniquely defined RDE solution (which then solves the Stratonovich SDE driven by Brownian motion). Indeed, in the drift-free case ($W \equiv 0$) this follows from $\psi_{2,1}$-variation of Brownian motion and Theorem 10.41. In the drift case, we rely on Theorem 12.11. $\qquad \square$

Remark 17.5 Our proof of Theorem 17.3 does <u>not</u> rely on any existence results for Stratonovich SDEs (and in fact yields such a result). En passant, we obtain the classical **Wong–Zakai theorem** (e.g. [88], [160] or [97]), which asserts

$$\pi_{(V)}\left(0, y_0; M^{D_n}\right) \to Y \text{ in probability, uniformly on } [0, T],$$

as an immediate corollary of our Theorem 17.3.

Conversely, if one accepts the Wong–Zakai theorem then continuity of the Itô–Lyons map combined with $d_{p\text{-var}}\left(S_2\left(M^{D_n}\right), \mathbf{M}\right) \to_{n\to\infty} 0$ in probability (Theorem 14.16) immediately tells us that $\pi_{(V,W)}(0, y_0; (\mathbf{M}(\omega), H(\omega)))$ is a Stratonovich solution. $\qquad \square$

Proof. We may assume that M is a continuous local martingale (since its bounded variation part can always be added to the "drift"-signal H). By a localization argument we may assume that[5]

$$\langle M \rangle \equiv \left(\langle M^i, M^j \rangle : i, j \in \{1, \ldots, d\}\right)$$

and the p-variation of the enhanced martingale $\mathbf{M}$ remains bounded. We fix a sequence of dissections $D_n = (t_k^n)$ with $|D_n| \to 0$, and write, as

---

[5] The notation here is in slight contrast to Section 14.1, where we preferred to set $\langle M \rangle \equiv \left(\langle M^i, M^i \rangle : i = 1, \ldots, d\right)$. Let us remark, however, that the two quantities are comparable, as seen from the Kunita–Watanabe (or, in essence, the Cauchy–Schwarz) inequality; see, for example, [143, Chapter IV, Corollary (1.16)].

usual, M^{D_n}, H^{D_n} for the respective piecewise linear approximations of M, H based on D_n. Define

$$\tilde{Y}^n = \pi_{(V,W)}\left(0, y_0; (M^{D_n}, H^{D_n})\right),$$

and also the Euler approximation with "backbone $\tilde{Y}^n$" to (17.5), that is[6]

$$\begin{aligned}
Y^n_{t^n_{k+1}} &= Y^n_{t^n_k} + V\left(\tilde{Y}^n_{t^n_k}\right) M_{t^n_k, t^n_{k+1}} + \frac{1}{2} V^2\left(\tilde{Y}^n_{t^n_k}\right) \langle M \rangle_{t^n_k, t^n_{k+1}} \\
&\quad + W\left(\tilde{Y}^n_{t^n_k}\right) H_{t^n_k, t^n_{k+1}}
\end{aligned}$$

with $Y^n_{\cdot}$ defined within the intervals $[t^n_k, t^n_{k+1}]$ by linear interpolation. We see that for a fixed k,

$$\begin{aligned}
\left| Y^n_{t^n_k} - \tilde{Y}^n_{t^n_k} \right| &= \left| \sum_{j=0}^{k-1} Y^n_{t^n_j, t^n_{j+1}} - \tilde{Y}^n_{t^n_j, t^n_{j+1}} \right| \\
&\leq \sum_{j=0}^{k-1} \delta^n_j + \left| \sum_{j=0}^{k-1} X^n_j \right|
\end{aligned}$$

where

$$\begin{aligned}
\delta^n_j &= \left| \pi_{(V,W)}\left(t^n_j, \tilde{Y}^n_{t^n_j}; (M^{D_n}, H^{D_n})\right)_{t^n_j, t^n_{j+1}} - \mathcal{E}_{(V)}\left(\tilde{Y}^n_{t^n_j}, S_2\left(M^{D_n}\right)_{t^n_j, t^n_{j+1}}\right) \right. \\
&\quad \left. - \mathcal{E}_{(W)}\left(\tilde{Y}^n_{t^n_j}, H_{t^n_j, t^n_{j+1}}\right) \right|
\end{aligned}$$

and

$$X^n_j = \frac{1}{2} V^2\left(\tilde{Y}^n_{t^n_j}\right)\left(\langle M \rangle_{t^n_j, t^n_{j+1}} - M^{\otimes 2}_{t^n_j, t^n_{j+1}}\right).$$

We now apply Corollary 12.8 (or Davie's estimate, Lemma 10.7, in the case of no drift). Using only $\mathrm{Lip}^{\gamma-1}$ regularity for V and $\mathrm{Lip}^{\gamma-2}$ regularity for W we have that, for some $\theta > 1$,

$$\begin{aligned}
\sum_{j=0}^{k-1} \delta^n_j &\leq c_1 \sum_{j=0}^{k}\left(\left\|S_2\left(M^{D_n}\right)\right\|^p_{p\text{-var};[t^n_j, t^n_{j+1}]} + |H^{D_n}|_{1\text{-var};[t^n_j, t^n_{j+1}]}\right)^\theta \\
&\leq c_2 \sum_{j=0}^{k} |M|^{p\theta}_{p\text{-var};[t^n_j, t^n_{j+1}]} + |H|^\theta_{1\text{-var};[t^n_j, t^n_{j+1}]} \to 0 \text{ a.s.}
\end{aligned}$$

---

[6] In what follows $V(\cdot) M_{s,t}$ stands for $\sum_{i=1}^d V_i(\cdot) M^i_{s,t}$ and $V^2(\cdot) \langle M \rangle_{s,t}$ for $\sum_{i,j=1}^d V_i V_j(\cdot) \langle M^i, M^j \rangle_{s,t}$.

as $|D_n| \to 0$ where we used piecewise linearity of M^{D_n} on $[t_j^n, t_{j+1}^n]$ so that

$$\left\|S_2\left(M^{D_n}\right)\right\|_{p\text{-var};[t_j^n,t_{j+1}^n]} = \left|M^{D_n}\right|_{p\text{-var};[t_j^n,t_{j+1}^n]}$$
$$\leq 3^{1-1/p}\left|M\right|_{p\text{-var};[t_j^n,t_{j+1}^n]}.$$

On the other hand, $t \mapsto \langle M \rangle_t - M_t^{\otimes 2}$ is an ($\mathbb{R}^{d\times d}$-valued) martingale and since $V^2\left(\tilde{Y}_{t_j^n}^n\right)$ is $\mathfrak{F}_{t_j^n}$-measurable it follows that $\left(\sum_{j=0}^{k-1} X_j^n : k = 1, 2, \dots\right)$ is a martingale. Hence, using in particular Doob's L^2-inequality and orthogonality of martingale increments, we have

$$\mathbb{E}\left[\max_{k=1}^{\#D_n}\left|\sum_{j=0}^{k-1}X_j^n\right|^2\right]^{\frac{1}{2}} \leq 2\mathbb{E}\left[\left|\sum_{j=0}^{\#D_n-1}X_j^n\right|^2\right]^{\frac{1}{2}}$$

$$= 2\mathbb{E}\left[\sum_{j=0}^{\#D_n-1}\left|X_j^n\right|^2\right]^{\frac{1}{2}}$$

$$\leq 2\left|V\right|_{\text{Lip}^1}\mathbb{E}\left[\sum_{j=0}^{\#D_n-1}\left|\langle M\rangle_{t_j^n,t_{j+1}^n} - M_{t_j^n,t_{j+1}^n}^{\otimes 2}\right|^2\right]^{\frac{1}{2}}$$

$$\leq c_1\left|V\right|_{\text{Lip}^1}\mathbb{E}\left(\sum_{j=0}^{\#D_n-1}\left|\langle M\rangle_{t_j^n,t_{j+1}^n}\right|^2\right)^{\frac{1}{2}}$$

$$\to 0 \quad \text{as} \quad |D_n| \to 0.$$

The last estimate comes from the (classical) Burkholder–Davies–Gundy inequality (Theorem 14.6), and the final convergence is justified by $\sum_j |\langle M\rangle_{t_j^n,t_{j+1}^n}|^2 \to 0$ and bounded convergence. Switching to a subsequence, if necessary, we see that

$$\left|Y_{t_k^n}^n - \tilde{Y}_{t_k^n}^n\right| \to 0 \text{ a.s.}$$

and it is a small step to see that this implies

$$\left|Y^n - \tilde{Y}^n\right|_{\infty;[0,T]} \to 0 \text{ a.s.}$$

Now, if $V \in \text{Lip}^\gamma$ and $W \in \text{Lip}^1$, then $\tilde{Y}^n$ converges in probability (and uniformly on $[0,T]$) to the (pathwise unique) RDE solution

$$\tilde{Y} = \pi_{(V,W)}\left(0, y_0; (\mathbf{M}, H)\right)$$

and so we see that $\left| Y^n - \tilde{Y} \right|_{\infty;[0,T]} \to 0$ in probability.[7] On the other hand, from the definition of (Y^n) we have that for all $t \in (0,T]$,

$$Y^n_{0,t^{D_n}} = \sum_{k:t^n_k < t} \left(V\left(\tilde{Y}^n_{t^n_k}\right) M_{t^n_k, t^n_{k+1}} + \frac{1}{2} V^2\left(\tilde{Y}^n_{t^n_k}\right) \langle M \rangle_{t^n_k, t^n_{k+1}} \right.$$
$$\left. + \; W\left(\tilde{Y}^n_{t^n_k}\right) H_{t^n_k, t^n_{k+1}} \right), \tag{17.7}$$

where as usual t^{D_n} denotes the right-hand neighbour to t in D_n. We observe that $J^n_t := V\left(\tilde{Y}^n_t\right)$ is uniformly bounded by $|V|_\infty$ (and hence by any Lip^γ-norm $\ldots$), $\mathfrak{F}_t$-measurable for $t \in D_n$ (and $\mathfrak{F}_{t^{D_n}}$-measurable for a general $t \in [0,T]$). We note that $J_t := \lim_{n\to\infty} J^n_t = V\left(\tilde{Y}_t\right)$ exists in probability and uniformly in $t \in [0,T]$, and is adapted, thanks to right-continuity of $(\mathfrak{F}_t)$. Write $\left(J_{t_{D_n}} : t \in [0,T]\right)$ for the piecewise constant, left-point approximation; that is, equal to $J_{t^n_k}$ whenever $t \in (t^n_k, t^n_{k+1}]$. Similarly for $\left(J^n_{t_{D_n}} : t \in [0,T]\right)$. Then

$$\sum_{k:t^n_k < t} V\left(\tilde{Y}^n_{t^n_k}\right) M_{t^n_k, t^n_{k+1}}$$

$$= \int_0^{t^{D_n}} J^n_{s_{D_n}} \, dM_s$$

$$= \int_0^{t^{D_n}} J_{s_{D_n}} \, dM_s + \int_0^{t^{D_n}} \left(J^n_{s_{D_n}} - J_{s_{D_n}}\right) dM_s$$

$$\to \int_0^t V\left(\tilde{Y}_s\right) dM_s \text{ in probability as } n \to \infty;$$

where we used convergence of left-point Riemann–Stieltjes approximations to the Itô-integral, as well as

$$\int_0^{t^{D_n}} \left(J^n_{s_{D_n}} - J_{s_{D_n}}\right) dM_s \to 0 \text{ in probability as } n \to \infty,$$

as is easily seen from the dominated convergence theorem for stochastic integrals.[8] Similarly, but easier, the other two terms of the right-hand side of (17.7) are seen to converge to the Riemann-Stieltjes integrals

$$\frac{1}{2} \int_0^t V^2\left(\tilde{Y}_s\right) d\langle M \rangle_s + \int W\left(\tilde{Y}^n_s\right) dH_s.$$

---

[7] Whatever subsequence we have so far extracted we can also extract a further subsequence along which we have a.s. convergence, and this in fact implies that the original sequence converges in probability.

[8] See, for example, [143, Chapter IV, (2.12)].

At last, using $Y_{0,t^{D_n}}^n \to \tilde{Y}_{0,t}$ as $n \to \infty$, we see that

$$\tilde{Y}_{0,t} = \int_0^t V\left(\tilde{Y}_s\right) dM_s + \frac{1}{2}\int_0^t V^2\left(\tilde{Y}_s\right) d\langle M\rangle_s + \int_0^t W\left(\tilde{Y}_s\right) dH_s$$

and so the proof is finished. ∎

17.3 Stochastic differential equations driven by non-semi-martingales

As was seen in Part III, there are many multidimensional stochastic processes which allow for a natural enhancement to (random) geometric p-rough paths. These include Brownian motion and semi-martingales, but also non-semi-martingales such as certain Gaussian and Markov processes. RDE theory leads immediately to a pathwise notion of stochastic differential equations driven by such processes. It is reassuring that such solutions have a firm probabilistic justification. More precisely, if $\mathbf{X} = \mathbf{X}(\omega)$ denotes either an enhanced Brownian motion, semi-martingale, Gaussian or Markov process, then the abstract (random) RDE solution

$$\pi_{(V)}\left(0, y_0; \mathbf{X}(\omega)\right)$$

can be identified as a (strong or weak) limit of various natural approximations. In particular, in all cases (cf. Sections 13.3.3, 14.5, 15.5.1, 16.5.2) we have seen the validity of a **Wong–Zakai-type** result in the sense that, for any (deterministic) sequence of dissections $(D_n) \subset \mathcal{D}[0, T]$ with mesh $|D_n| \to 0$

$$\mathbf{X}^n \equiv S_{[p]}\left(X^{D_n}\right) \to \mathbf{X}, \quad \text{with } X \equiv \pi_1(\mathbf{X})$$

(in rough path topology and, at least, in probability) so that the abstract (random) RDE solution

$$\pi_{(V)}\left(0, y_0; \mathbf{X}\right)$$

is identified as a "Wong–Zakai" limit of solutions to the approximating ODEs

$$dy^n = V(y^n) dX^{D_n}, \quad y^n(0) = y_0,$$

and not dependent on the particular sequence (D_n). We can thus characterize RDE solutions driven by Gaussian and Markov processes in a completely elementary way.

Theorem 17.6 (differential equations driven by Gaussian signals)
Assume that
(i) $X = \left(X^1, \ldots, X^d\right)$ *is a centred continuous Gaussian process on* $[0, T]$ *with independent components;*

(ii) $\mathcal{H}$ denotes the Cameron–Martin space associated with X;
(iii) the covariance of X is of finite ρ-variation in 2D sense for some $\rho \in [1, 2)$;
(iv) $V = (V_1, \ldots, V_d)$ is a collection of Lip^γ-vector fields on $\mathbb{R}^e$, with $\gamma > 2\rho$;
(v) let $(D_n) \subset \mathcal{D}[0, T]$ with mesh $|D_n| \to 0$.
Then the (random) sequence of ODE solutions

$$\left(\pi_{(V)} \left(0, y_0; X^{D_n} \right) \right) \subset C \left([0, T], \mathbb{R}^e \right)$$

is Cauchy-in-probability with respect to uniform topology. The unique limit point, a $C \left([0, T], \mathbb{R}^e \right)$-valued random variable, does not depend on the particular sequence (D_n) and is identified as the (random) RDE solution

$$\pi_{(V)} \left(0, y_0; \mathbf{X} \right),$$

where $\mathbf{X}$ is the natural enhancement of X to a geometric p-rough path, $p \in (2\rho, \min(\gamma, 4))$.

Theorem 17.7 (differential equations driven by Markovian signals)
Assume that
(i) $X = \left(X^1, \ldots, X^d \right)$ is a (symmetric) Markov process with uniformly elliptic generator in divergence form[9]

$$\frac{1}{2} \sum_{i,j=1}^{d} \partial_i \left(a^{ij} \left(\cdot \right) \partial_j \cdot \right)$$

where $a \in \Xi^{1,d} \left(\Lambda \right)$, that is, measurable, symmetric and $\Lambda^{-1} I \leq a \leq \Lambda I$, for some $\Lambda > 0$, in the sense of positive definite matrices;
(ii) $V = (V_1, \ldots, V_d)$ is a collection of Lip^2-vector fields on $\mathbb{R}^e$;
(iii) let $(D_n) \subset \mathcal{D}[0, T]$ with mesh $|D_n| \to 0$.
Then the (random) sequence of ODE solutions

$$\left(\pi_{(V)} \left(0, y_0; X^{D_n} \right) \right) \subset C \left([0, T], \mathbb{R}^e \right)$$

is Cauchy-in-probability with respect to uniform topology. The unique limit point, a $C \left([0, T], \mathbb{R}^e \right)$-valued random variable, does not depend on the particular sequence (D_n) and is identified as the (random) RDE solution

$$\pi_{(V)} \left(0, y_0; \mathbf{X} \right),$$

where $\mathbf{X}$ is the enhancement of X to a geometric α-Hölder rough path, $\alpha \in (1/3, 1/2)$.

The proof of these theorems involves little more than combining the convergence results of Sections 15.5.1, 16.5.2 with continuity properties of

---

[9]Understood in the weak sense, i.e. via Dirichlet forms.

the Itô–Lyons map. Let us perhaps remark that enhanced Markov processes have finite $\psi_{2,1}$-variation (exactly as enhanced Brownian motion), so that we can use the "refined" continuity result with minimal Lip^2-regularity. (In fact, this would also work for enhanced Gaussian processes provided $\rho = 1$.)

17.4 Limit theorems

17.4.1 Strong limit theorems

Since almost sure convergence and convergence in probability are preserved under continuous maps, continuity results for RDE solutions (such as those recalled in Section 17.1) imply *immediately* corresponding probabilistic limit theorems. For the reader's convenience we spell out the following two cases; the (immediate) formulation for RDEs with drift is left to the reader.

Theorem 17.8 *Assume that*
(i) $(\mathbf{X}_n)$ *is a sequence of random geometric p-rough paths (resp. 1/p-Hölder rough paths) such that*

$$\mathbf{X}_n \to \mathbf{X}_\infty \quad a.s. \ [or: in \ probability, \ or: in \ L^q\,(\mathbb{P})\,\forall q < \infty]$$

in p-variation (resp. 1/p-Hölder) rough path topology;
(ii) $V = (V_1, \ldots, V_d) \in \mathrm{Lip}^\gamma\,(\mathbb{R}^e)\,, \gamma > p$ *and* $y_0 \in \mathbb{R}^e$.
Then

$$\boldsymbol{\pi}_{(V)}\,(0, y_0; \mathbf{X}_n) \to \boldsymbol{\pi}_{(V)}\,(0, y_0; \mathbf{X}_\infty) \quad a.s. \ [or: in \ probability, \ or: in \ L^q\,(\mathbb{P})$$
$$\forall q < \infty]$$

in p-variation (resp. 1/p-Hölder) rough path topology.

Proof. The case of almost sure convergence and convergence in probability is obvious from the above remarks. Stability of the Itô–Lyons map under $L^q\,(\mathbb{P})$-convergence, for all $q < \infty$, follows from the (purely) deterministic estimate (10.15) of Theorem 10.14. ∎

Theorem 17.9 *Assume that* $1 \le p < p' < p'' < [p] + 1$,
(i) $(\mathbf{X}_n) \subset C^{\psi_{p,1}\text{-var}}\left([0,T], G^{[p]}\left(\mathbb{R}^d\right)\right)$ *(resp.* $C^{\psi_{p,1}\text{-var}} \cap C^{1/p'\text{-Höl}}$*) a.s. such that*

$$\mathbf{X}_n \to \mathbf{X}_\infty \quad a.s. \ [or: in \ probability \,/\, L^q\,(\mathbb{P}) \ for \ all \ q \in [1, \infty)]$$

in p'-variation (resp. 1/p'-Hölder) rough path topology;
(ii) $V = (V_1, \ldots, V_d) \in \mathrm{Lip}^p\,(\mathbb{R}^e)$ *and* $y_0 \in \mathbb{R}^e$.

Then

$$\boldsymbol{\pi}_{(V)}\left(0, y_0; \mathbf{X}_n\right) \to \boldsymbol{\pi}_{(V)}\left(0, y_0; \mathbf{X}_\infty\right) \ \textit{a.s. [or: in probability} \ / \ L^q\left(\mathbb{P}\right)$$
$$\textit{for all } q \in [1, \infty)]$$

in p'-variation (resp. $1/p''$-Hölder) rough path topology.

Exercise 17.10 *In the context of either Theorem 17.8 or 17.9, assume $\mathbf{X}_n \to \mathbf{X}_\infty$ in $L^q\left(\mathbb{P}\right)$ for some fixed $q < \infty$. Use estimate (10.15) to discuss $L^{\tilde{q}}$ convergence of $\boldsymbol{\pi}_{(V)}\left(0, y_0; \mathbf{X}_n\right) \to \boldsymbol{\pi}_{(V)}\left(0, y_0; \mathbf{X}_\infty\right)$ for suitable $\tilde{q} = \tilde{q}\left(q\right) < \infty$.*

Theorem 17.8 applies in particular if $\mathbf{X}$ is a semi-martingale or an enhanced Brownian motion. In the latter case, as detailed in Theorem 17.9, the assumptions on V and $\mathbf{x}$ can be slightly weakened ($V \in \mathrm{Lip}^p\left(\mathbb{R}^e\right), \mathbf{x} \in C^{\psi_{p,1}\text{-var}} \ldots$); in all these cases the limiting RDE solutions are indeed classical Stratonovich solutions. But of course, these theorems can equally well be applied to (rough) differential equations driven by Gaussian or Markovian signals. There is no reason to list all possible approximation results: the reader may simply consult the catalogue of strong convergence results established in Sections 13.3, 14.5, noting that the **mollifier** and **Karhunen–Loéve** approximations of Section 15.5 are also applicable to enhanced Brownian motion.

Let us also draw attention to the existence of "non-standard" approximations X^n, which may be based upon knowing only finitely many points $\left(X_t : t \in D^n\right) \subset \mathbb{R}^d$, with the property that

$$X^n \to X,$$

say uniformly and in probability, but such that

$$\boldsymbol{\pi}_{(V)}\left(0, y_0; X^n\right) \nrightarrow \boldsymbol{\pi}_{(V)}\left(0, y_0; \mathbf{X}\right).$$

Indeed, in Theorem 13.24 we established a set of conditions under which $S_N\left(X^n\right)$ converges in probability/rough path sense[10] to a limit with possibly "modified area". The following corollaries are then an immediate consequence from Theorem 17.8.

Corollary 17.11 (McShane [126]) *Let B_n^{MSh} denote the McShane interpolation to 2-dimensional Brownian motion, as defined in Example 13.28, based on a fixed interpolation function $\phi = \left(\phi^1, \phi^2\right) \in C^1\left([0, 1], \mathbb{R}^2\right)$ with $\phi\left(0\right) = (0, 0)$ and $\phi\left(1\right) = (1, 1)$. Then, given $V = (V_1, V_2)$ of Lip^2-regularity, the solutions to*

$$dy^n = V\left(y^n\right) dB_n^{\mathrm{MSh}}, \quad y^n\left(0\right) = y_0$$

---

[10]Taking $\alpha \in (0, 1], N \geq [1/\alpha]$ and $\beta = 1/N$ in Theorem 13.24 will lead to γ-Hölder convergence for any $\gamma < \min\left(\alpha, 1/N\right)$. One can then pick γ large enough such that $[1/\gamma] = N$.

converge (in α-Hölder, $\alpha < 1/2/in$ probability) to the solution of the Stratonovich SDE

$$dy = V(y) \circ dB + c[V_1, V_2](y)\,dt, \quad y(0) = y_0$$

with $c = \frac{2}{\pi}\left(1 - 2\int_0^1 \dot{\phi}^1(s)\phi^2(s)\,ds\right)$.

Proof. It was verified in Example 13.28 that the assumptions of Theorem 13.24 are met with $\alpha \in (1/3, 1/2)$, $\beta = 1/2$ and $N = 2$. More precisely, $\mathbf{P}_t$ was identified as $\exp(t\boldsymbol{\Gamma})$ with

$$\boldsymbol{\Gamma} = \begin{pmatrix} 0 & \frac{2}{\pi}A_{0,1}^\phi \\ -\frac{2}{\pi}A_{0,1}^\phi & 0 \end{pmatrix}$$

and so the "correction" drift-vector field is of the form

$$[V_1, V_2]\,d\left(\frac{2}{\pi}A_{0,1}^\phi t\right) + [V_2, V_1]\,d\left(-\frac{2}{\pi}A_{0,1}^\phi t\right)$$

$$= \frac{4}{\pi}A_{0,1}^\phi[V_1, V_2]\,dt$$

$$= \frac{2}{\pi}\left(1 - 2\int_0^1 \dot{\phi}^1(s)\phi^2(s)\,ds\right)[V_1, V_2]\,dt.$$

∎

Corollary 17.12 (Sussmann [167]) *Let B_n^{Sm} denote Sussmann's approximation to d-dimensional Brownian motion, constructed in detail in Example 13.27 for some fixed $\mathbf{v} \in \mathfrak{g}^N(\mathbb{R}^d) \cap (\mathbb{R}^d)^{\otimes N}$, $N \in \{2, 3, \dots\}$, by (repeated) concatenation of linear chords and "geodesic loops" associated with $\mathbf{v}$. Then, given $V = (V_1, \dots, V_d)$ of Lip^N-regularity, the solutions to*

$$dy^n = V(y^n)\,dB_n^{\mathrm{Sm}}, \quad y^n(0) = y_0$$

converge (in α-Hölder, $\alpha < 1/N/in$ probability) to the solution of the Stratonovich SDE

$$dy = V(y) \circ dB + \left(\sum_{i_1, \dots, i_N} [V_{i_1}, [\dots, [V_{i_{N-1}}, V_{i_N}]]\dots]\big|_z \mathbf{v}^{i_1, \dots, i_N}\right)dt.$$

In particular, by suitable choice of N and $\mathbf{v}$ every possible Lie bracket of $\{V_1, \dots, V_d\}$ can be made to appear as a drift-vector field to the limiting SDE.

Proof. It was verified in Example 13.27 that the assumptions of Theorem 13.24 are met with $\alpha \in (1/3, 1/2)$, $\beta = 1/N$. ∎

17.4.2 Weak limit theorems

Similar to the previous section, a "weak" probabilistic formulation of Lyons' limit theorem is an immediate consequence of the fact that weak convergence is preserved under continuous maps. Again, the immediate extension to RDEs with drift is left to the reader.

Theorem 17.13 *Assume that*
(i) $(\mathbf{X}_n)$ *is a sequence of random geometric p-rough paths (resp. 1/p-Hölder rough paths), possibly defined on different probability spaces, such that*

$$\mathbf{X}_n \to \mathbf{X}_\infty \quad \text{weakly in p-variation (resp. 1/p-Hölder) topology;}$$

(ii) $V \in \mathrm{Lip}^\gamma (\mathbb{R}^e), \gamma > p,$ *and* $y_0 \in \mathbb{R}^e$.
Then
$$\pi_{(V)}(0, y_0; \mathbf{X}_n) \to \pi_{(V)}(0, y_0; \mathbf{X}_\infty)$$
weakly in p-variation (resp. 1/p-Hölder) rough path topology.

Theorem 17.14 *Assume that* $1 \le p < p' < p'',$
(i) $(\mathbf{X}_n) \subset C^{\psi_{p,1}\text{-var}}([0,T], G^{[p]}(\mathbb{R}^d))$ *(resp.* $C^{\psi_{p,1}\text{-var}} \cap C^{1/p'\text{-Höl}})$ *a.s. such that*

$$\mathbf{X}_n \to \mathbf{X}_\infty \quad \text{weakly in p'-variation (resp. 1/p'-Hölder) topology;}$$

(ii) $V = (V_1, \ldots, V_d) \in \mathrm{Lip}^p (\mathbb{R}^e),$ *and* $y_0 \in \mathbb{R}^e$.
Then
$$\pi_{(V)}(0, y_0; \mathbf{X}_n) \to \pi_{(V)}(0, y_0; \mathbf{X}_\infty)$$
weakly in p'-variation (resp. 1/p''-Hölder) rough path topology.

Again, there is no reason to list all of the possible weak approximation results: the reader may simply consult the catalogue of weak convergence results established for enhanced Brownian motion, Gaussian and Markov processes and apply Theorem 17.13. Nonetheless, let us list a few.

Example 17.15 (Donsker–Wong–Zakai) Let $(\xi_i : i = 1, 2, 3, \ldots)$ be a sequence of independent random variables, identically distributed, $\xi_i \overset{\mathcal{D}}{=} \xi$ with zero mean, unit covariance, and moments of all orders, i.e. $\mathbb{E}|\xi|^q < \infty$ for all $q < \infty$. Write $W_t^{(n)}$ for a rescaled, piecewise linearly connected, random walk

$$W_t^{(n)} = \frac{1}{n^{1/2}} \left(\xi_1 + \cdots + \xi_{[tn]} + (nt - [nt]) \, \xi_{[nt]+1} \right).$$

Also, let $V \in \mathrm{Lip}^2 (\mathbb{R}^e)$. Then $\pi_{(V)}(0, y_0; W^{(n)})$ converges weakly, in α-Hölder topology for any $\alpha \in [0, 1/2)$, to the Stratonovich solution of

$$dY = V(Y) \circ dB, \quad Y_0 = y_0 \in \mathbb{R}^e. \qquad \square$$

Example 17.16 (differential equations driven by Gaussian signals)
Consider enhanced fractional Brownian motion $\mathbf{B}^H$ and the resulting rough differential equations of the form

$$dY^H = V\left(Y^H\right) d\mathbf{B}^H,\ Y = y_0.$$

As an application of a general Gaussian approximation result, established in Section 15.6, we have $\mathbf{B}^H \to \mathbf{B}^{1/2}$ as $H \to 1/2$ weakly in α-Hölder topology for any $\alpha \in [0, 1/2)$ and hence weak convergence in α-Hölder topology for any $\alpha \in [0, 1/2)$, to the Stratonovich solution of

$$dY = V\left(Y\right) \circ dB,\ Y_0 = y_0 \in \mathbb{R}^e,$$

provided $V \in \mathrm{Lip}^\gamma\left(\mathbb{R}^e\right),\ \gamma > 2.$ □

Example 17.17 (differential equations driven by Markovian signals)
Let $a_n \in \Xi^{1,d}\left(\Lambda\right)$, smooth, so that

$$a_n \to a \in \Xi^{1,d}\left(\Lambda\right),$$

almost surely with respect to the Lebesgue measure on $\mathbb{R}^d$. Let X^n denote the diffusion with generator $\frac{1}{2} \sum_{i,j=1}^d \partial_i \left(a_n^{ij}\left(\cdot\right) \partial_j \cdot\right)$. Itô theory allows us to construct X^n as a semi-martingale (e.g. on d-dimensional Wiener space) and granted $V \in \mathrm{Lip}^2\left(\mathbb{R}^e\right)$, solutions to the Stratonovich SDEs

$$dY^n = V\left(Y^n\right) dX^n,\ Y_0 = y_0 \in \mathbb{R}^e$$

are given by $\pi_{(V)}\left(0, y_0; X^n\right)$ and converge weakly, in α-Hölder topology for any $\alpha \in [0, 1/2)$, to the (random) RDE solution

$$\pi_{(V)}\left(0, y_0; \mathbf{X}\right)$$

where $\mathbf{X}$ is the enhancement to the symmetric diffusion X with generator $\frac{1}{2} \sum_{i,j=1}^d \partial_i \left(a^{ij}\left(\cdot\right) \partial_j \cdot\right)$, understood via Dirichlet forms. □

17.5 Stochastic flows of diffeomorphisms

Recall from Section 11.2, Corollary 11.14, that for $V \in \mathrm{Lip}^{\gamma+k-1}$ with $\gamma > p \geq 1$ and $\mathbf{x}$ a geometric p-rough path $\pi_{(V)}\left(0, \cdot; \mathbf{x}\right)_t$ is a C^k-diffeomorphism and[11]

$$\mathbf{x} \mapsto \pi_{(V)}\left(0, \cdot; \mathbf{x}\right) \in \mathcal{D}_k\left(\mathbb{R}^e\right)$$

is continuous in the sense of flows of C^k-diffeomorphisms. Once more, this can be applied immediately in a purely pathwise fashion to almost every sample path $\mathbf{X} = \mathbf{X}\left(\omega\right)$ of an enhanced Brownian motion (or semi-martingale, Gaussian or Markov process) and every strong or weak

---

[11] The Polish space $\mathcal{D}_k\left(\mathbb{R}^e\right)$ was defined in Section 11.5.

approximation result for $\mathbf{X}$ leads to the corresponding *limit theorem for the stochastic flows*. (This kind of reasoning is exactly as in Section 17.4.)

To illustrate all this, consider an enhanced continuous semi-martingale $\mathbf{M}$ with sample path in $C^{0,p\text{-var}}\left([0,T], G^2\left(\mathbb{R}^d\right)\right)$ almost surely with $2 < p < \gamma$. We then learn, still assuming $V \in \text{Lip}^{\gamma+k-1}$, that

$$\pi_{(V)}\left(0, \cdot; \mathbf{M}\right) \in \mathcal{D}_k\left(\mathbb{R}^e\right)$$

where $\pi_{(V)}\left(0, \cdot; \mathbf{M}\left(\omega\right)\right)$ is not only a Stratonovich solution to the SDE $dY = V(Y) \circ dM$, with $M = \pi_1\left(\mathbf{M}\right)$, but now the solution flow to this equation (see [140, Section V.9] for a "classical" discussion of this). If we now assume that

$$\mathbf{M}^n \to \mathbf{M} \text{ (a.s., or: in probability)}$$

in p-variation rough path topology then we have[12] (also a.s., or: in probability)

$$\sup_{y_0 \in \mathbb{R}^e} \left|\partial_\alpha \pi_{(V)}\left(0, y_0; \mathbf{M}^n\right) - \partial_\alpha \pi_{(V)}\left(0, y_0; \mathbf{M}\right)\right|_{p\text{-var};[0,T]} \to 0 \text{ as } n \to \infty.$$

The classical case is when $\mathbf{M}^n = S_2\left(B^n\right)$, lifted dyadic piecewise linear approximations to Brownian motion. In this case we recover a classical (Wong–Zakai-type) limit theorem for stochastic flows, see [88] or [124]. The case $\mathbf{M}^n = S_2\left(M^n\right)$, lifted piecewise linear approximations to a generic semi-martingale, has been discussed in a classical context in [97].

We can also apply this to weak approximations. More precisely, if

$$\mathbf{M}^n \to \mathbf{M} \text{ weakly in } C^{0,p\text{-var}}\left([0,T], G^2\left(\mathbb{R}^d\right)\right)$$

in p-variation rough path topology, then still assuming $V \in \text{Lip}^{\gamma+k-1}$,

$$\pi_{(V)}\left(0, \cdot; \mathbf{M}^n\right) \to \pi_{(V)}\left(0, \cdot; \mathbf{M}\right) \text{ weakly in } \mathcal{D}_k\left(\mathbb{R}^e\right).$$

For instance, the weak (Donsker–Wong–Zakai-type) convergence of $\pi_{(V)}$ $\left(0, \cdot; W^{(n)}\right)$, where $W^{(n)}$ is a rescaled d-dimensional random walk (cf. Example 17.15), also holds in the sense of flows of C^k-diffeomorphisms as long as $V \in \text{Lip}^{\gamma+k-1}$ with $\gamma > p \geq 1$.

---

[12] The uniformity in $y_0 \in \mathbb{R}^e$ is a consequence of the invariance of the Lip^γ-norm under translation,

$$\forall y_0 \in \mathbb{R}^e, \gamma > 0 : |V|_{\text{Lip}^\gamma} = |V\left(y_0 + \cdot\right)|_{\text{Lip}^\gamma}.$$

17.6   Anticipating stochastic differential equations

17.6.1   Anticipating vector fields and initial condition

Because RDE solutions are constructed pathwise, it is clear that we can allow the vector fields V to be random as long as the appropriate Lipschitz-regularity holds with probability one. In particular, there is no problem if this randomness anticipates the randomness of the driving signal. The same remark applies for the initial condition. With the focus on enhanced Brownian motion, we have the following result:

Proposition 17.18 *Assume that*
(i) $\mathbf{B}$ *denotes a* $G^2\left(\mathbb{R}^d\right)$*-valued enhanced Brownian motion, lifting* $B = \pi_1\left(\mathbf{B}\right)$;
(ii) $V = (V_1, \ldots, V_d)$ *is a collection of random vector fields on* $\mathbb{R}^e$, *almost surely in* Lip^2;
(iii) y_0 *is an* $\mathbb{R}^e$*-valued random variable;*
(iv) the stochastic process Y *is defined as the RDE solution of*

$$dY = V\left(Y\right)d\mathbf{B}, \ Y_0 = y_0 \in \mathbb{R}^e;$$

(v) $(D_n) \subset \mathcal{D}\left[0, T\right]$ *with mesh* $|D_n| \to 0$.
Then, for any $\alpha \in \left[0, 1/2\right)$,

$$d_{\alpha\text{-Höl};\left[0,T\right]}\left(\pi_{(V)}\left(0, y_0; B^{D_n}\right), \pi_{(V)}\left(0, y_0; \mathbf{B}\right)\right) \to 0 \ as \ n \to \infty,$$

in probability and in L^q *for all* $q < \infty$. *If* (D_n) *is nested, e.g.* $D_n = (kT2^{-n} : k \in \{0, \ldots, 2^n\})$, *then convergence also holds almost surely.*

Inclusion of drift-vector fields is straightforward, as is the similar statement for full RDEs. It is also clear that any other strong convergence result for *enhanced* Brownian motion will yield a similar limit theorem for such "anticipating" stochastic differential equations. Under further regularity assumptions, $V \in \mathrm{Lip}^{\gamma+k-1}$ with $\gamma > 2$, the convergence holds at the level of C^k-flows. At last, the usual remark applies that $\mathbf{B}$ can be replaced by a variety of other rough paths, including enhanced semi-martingales, Gaussian and Markovian processes.

Remark 17.19 Following Nualart–Pardoux (cf. [135] and the references cited therein) one can say that Y is a solution to the (anticipating) Stratonovich equation

$$dY = V\left(Y\right) \circ dB, \ Y_0 = y_0 \in \mathbb{R}^e$$

if, by definition, for every sequence (D_n) with mesh tending to zero,

$$y_0 + \int_0^t V\left(Y\right) dB^{D_n} \to Y_t \ as \ t \to \infty,$$

in probability and uniformly in $t \in [0, T]$. It was verified in [30] that $\pi_{(V)}(0, y_0; \mathbf{B})$ is also a solution in the Nualart–Pardoux sense. $\qquad\square$

17.6.2 Stochastic delay differential equations

Let $\varepsilon \in (0, 1)$. A real-valued Brownian motion β, started at time -1 say, gives rise to the $\mathbb{R}^2$-valued process by setting

$$t \in [0, T] \mapsto (\beta_t^\varepsilon, \beta_t) := (\beta_{t-\varepsilon}, \beta_t).$$

On a sufficiently small time interval (of length $\leq \varepsilon$), it is clear that β^ε and β have independent Brownian increments so that

$$\left(\beta_{s,t}^\varepsilon, \beta_{s,t} : t \in [s, s+\varepsilon]\right)$$

has the distribution of a 2-dimensional standard Brownian motion $(B_t : t \in [0, \varepsilon])$ and so there is a unique solution to the Stratonovich SDE

$$dY_t^\varepsilon = V_1(Y_t^\varepsilon) \circ d\beta_t^\varepsilon + V_2(Y_t^\varepsilon) \circ d\beta_t, \quad Y^\varepsilon(0) = y_0 \in \mathbb{R}^e \qquad (17.8)$$

where $V_1, V_2 \in \mathrm{Lip}^2(\mathbb{R}^e)$. For $t > \varepsilon$, we are effectively dealing with an anticipating SDE but one can (classically) get around this by solving (17.8) as a stochastic flow: first on $[0, \varepsilon]$ then on $[\varepsilon, 2\varepsilon]$ and so on. By composition, we can then define a solution to (17.8) over $[0, T]$. It is easy to see that this construction is in precise agreement with solving the RDE

$$dY_t^\varepsilon = \begin{pmatrix} V_1 \\ V_2 \end{pmatrix} (Y_t^\varepsilon)\, d\mathbf{B}^\varepsilon, \quad Y^\varepsilon(0) = y_0 \in \mathbb{R}^e,$$

where $\mathbf{B}^\varepsilon$ is the lift of $(\beta^\varepsilon, \beta)$ constructed in Section 13.3.5. Indeed, from Theorem 17.3 (resp. Proposition 17.18) both constructions are consistent on $[0, \varepsilon]$, then $[\varepsilon, 2\varepsilon]$, etc. and hence on $[0, T]$.

Theorem 17.20 *Let* $\alpha \in (1/3, 1/2)$ *and* $V_1, V_2 \in \mathrm{Lip}^2$. *Define for* $\varepsilon > 0, Y^\varepsilon$ *to be the solution of the anticipating Stratonovich SDE*

$$dY_t^\varepsilon = V_1(Y_t^\varepsilon) \circ d\beta_t + V_2(Y_t^\varepsilon) \circ d\beta_{t-\varepsilon}, \quad Y^\varepsilon(0) = y_0,$$

and Z *to be the solution of the (standard) Stratonovich SDE*

$$dZ = (V_1(Z) + V_2(Z)) \circ d\beta_t - [V_1, V_2](Z)\, dt.$$

Then, for any $\alpha \in [0, 1/2)$,

$$|Y^\varepsilon - Z|_{\alpha\text{-Höl};[0,T]} \to 0 \text{ as } \varepsilon \to 0$$

in probability and in L^q *for all* $q < \infty$.

Proof. Set $V = (V_1, V_2)$. It is clear from the remarks preceding this theorem that $Y^\varepsilon = \pi_{(V)}(0, y_0; \mathbf{B}^\varepsilon)$ solves the given Stratonovich SDE for Y^ε. Similarly, setting $W := [V_1, V_2] \in \mathrm{Lip}^1$ and

$$\boldsymbol{\beta}_t = \exp\left((\beta_t, \beta_t); 0\right) \in G^2\left(\mathbb{R}^2\right)$$

we see (from Theorem 17.3 and the first remark thereafter) that $Z := \pi_{(V,W)}(0, y_0; (\boldsymbol{\beta}, t))$ solves the Stratonovich SDE for Z. But then Theorem 12.14 tells us that almost surely,

$$Z = \pi_{(V)}\left(0, y_0; \tilde{\mathbf{X}}\right),$$

where $\tilde{\mathbf{X}}_t = \exp((\beta_t, \beta_t); -t/2)$. For $\gamma > 2$, we conclude using Theorem 13.31 and continuity of the Itô–Lyons map. For $\gamma = 2$ we need Proposition 13.30 to ensure that every $\mathbf{B}^\varepsilon$ has finite $\psi_{2,1}$-variation. ∎

17.7 A class of stochastic partial differential equations

We now return to the setting of Section 11.3 where we studied non-linear evolution equations with "rough noise", such as a typical realization of d-dimensional Brownian motion and Lévy's area, $\mathbf{B}(\omega) = \exp(B, A)$. The equation then reads

$$
\begin{aligned}
du &= F\left(t, x, Du, D^2 u\right) dt + Du(t, x) \cdot V(x) \, d\mathbf{B}, \qquad (17.9) \\
u(0, \cdot) &= u_0 \in \mathrm{BUC}\left(\mathbb{R}^n\right),
\end{aligned}
$$

where $F = F(t, x, p, X) \in C\left([0, T], \mathbb{R}^n, \mathbb{R}^n, S^n\right)$ is assumed to be degenerate elliptic and $u = u(t, x) \in \mathrm{BUC}\left([0, T] \times \mathbb{R}^n\right)$ is a real-valued function of time and space. Under the assumption that $V = (V_1, \ldots, V_d) \subset \mathrm{Lip}^\gamma, \gamma > 4$, and that F satisfies $\Phi^{(3)}$-invariant comparison (as discussed in detail and with examples in Section 11.3), we then have a robust notion of a solution to the above stochastic partial differential equation. Indeed, combining Theorem 11.16 with convergence of (lifted) piecewise linear approximations of B to $\mathbf{B}$ suggests calling the so-obtained solutions to (17.9) *Stratonovich solutions*, writing also[13]

$$du = F\left(t, x, Du, D^2 u\right) dt + Du(t, x) \cdot V(x) \circ dB. \qquad (17.10)$$

Let us leave aside the first benefit of this approach, which is to deal with SPDEs with non-Brownian and even non-semi-martingale noise.[14] The

---

[13] Further justification for the "Stratonovich" notation (17.10) is possible, cf. the references at the end of this section.

[14] It suffices to replace $\mathbf{B}$ by some other (e.g. Gaussian or Markovian) rough path.

continuous dependence on the driving signal **B** in rough path topology implies various stability results (i.e. weak and strong limit theorems) for such SPDEs: it suffices that an approximation to B converges in rough path topology; examples beyond "piecewise linear" are mollifier and Karhunen–Lóeve approximations, as well as (weak) Donsker-type random walk approximations. A slightly more interesting example is left to the reader in the following

Exercise 17.21 *Let $V = (V_1, \ldots, V_d)$ be a collection of C^∞-bounded vector fields on $\mathbb{R}^n$ and B a d-dimensional standard Brownian motion. Show that, for every $\alpha = (\alpha_1, \ldots, \alpha_N) \in \{1, \ldots, d\}^N$, $N \geq 2$, there exist (piecewise) smooth approximations (z^k) to B, with each z^k only dependent on $\{B(t) : t \in D^k\}$ where (D^k) is a sequence of dissections of $[0, T]$ with mesh tending to zero, such that almost surely*

$$z^k \to B \text{ uniformly on } [0, T].$$

But u^k, the solutions to

$$\begin{aligned} du^k &= F\left(t, x, Du^k, D^2 u^k\right) dt + Du^k(t, x) \cdot V(x) dz^k, \quad u^k(0, \cdot) \\ &= u_0 \in \mathrm{BUC}(\mathbb{R}^n), \end{aligned}$$

converge almost surely locally uniformly to the solution of the "wrong" differential equation

$$du = \left[F\left(t, x, Du, D^2 u\right) + Du(t, x) \cdot V_\alpha(x)\right] dt + Du(t, x) \cdot V(x) \circ dB$$

*where V_α is the bracket-vector field given by $V_\alpha = \left[V_{\alpha_1}, \left[V_{\alpha_2}, \ldots \left[V_{\alpha_{N-1}}, V_{\alpha_N}\right]\right]\right]$. (Hint: Combine Sussmann's twisted approximations to Brownian motion (Exercise 13.27) with continuity of SPDE with respect to **B** in rough path topology.)*

17.8 Comments

Most of the material of **Section 17.2** belongs to the folklore of rough path theory. We note that the Itô stochastic differential equation

$$dY = V(Y) dB, \quad V = (V_1, \ldots, V_d) \subset \mathrm{Lip}^2$$

can be written in Stratonovich form and then solved pathwise as a (unique) RDE solution

$$dY = V(Y) d\mathbf{B} - \frac{1}{2} V^2(Y) d\langle M \rangle,$$

with "Lip^1-drift" given by $V^2(Y) d\langle M \rangle = \sum_{i,j=1}^d V_i V_j(Y) d\langle M^i, M^j \rangle$. Note that the existence of RDE solutions is ensured for $V \subset \mathrm{Lip}^\gamma, \gamma > 1$.

For a discussion of pathwise uniqueness under this assumption, we refer to Davie [37].

The classical Wong–Zakai theorem can be found, for instance, in Stroock [160], Kurtz *et al.* [97] or the classical monograph of Ikeda and Watanabe [88]. In the latter, the reader can also find a criterion for convergence of the Itô map with "modified" limit which covers McShane's example [126], but not Sussmann's example [167]. The material of **Section 17.3**, on SDEs driven by non-semi-martingales, consists of essentially trivial corollaries of the relevant results of Parts II and III; but we have tried to make the statements accessible to readers with no background in rough path theory. **Section 17.4.2** collects a number of weak limit theorems, including a "Donsker–Wong–Zakai" theorem which, perhaps, is known but for which we have failed to find a reference. The discussion of stochastic flows, **Section 17.5**, is rather immediate from the deterministic results of Chapter 11, nonetheless in striking contrast to the work required to obtain similar results previous to rough path theory (see Ikeda and Watanabe [88], Kunita [96] and Malliavin [124]). In the context of anticipating SDEs, **Section 17.6**, rough path theory was first exploited by Coutin *et al.* [30]. Theorem 17.20 concerning a simple delay equation appears (without proof) in Lyons [117] and is attributed to Ben Hoff [85]. See Neuenkirch *et al.* [133] for a recent study of "rough" delay equations. **Section 17.7** is a straightforward application of the deterministic results of Section 11.3 to SPDEs of the form

$$du = F\left(t, x, Du, D^2 u\right) dt + \sum_{i=1}^{d} H\left(x, Du\right) \circ dB^i,$$

with F fully non-linear but $H = (H_1, \ldots, H_d)$ linear (with respect to the derivatives of u); see Caruana *et al.* [24] (also for Exercise 17.21) and also Buckdahn and Ma [17]. In the case when both F and H are linear, (Wong–Zakai-type) approximations have been studied in great detail by Gyöngy [78–82] and in Caruana and Friz [23] with rough path methods. The above class of (fully non-linear) SPDEs, possibly generalized to $H = H(x, u, Du)$, is considered to be an important one (see Lions and Souganidis [109–112]) and the reader can find a variety of examples (drawing from fields as diverse as filtering and stochastic control theory, pathwise stochastic control, interest rate theory, front propagation and phase transition in random media, etc.) in the articles [110, 112]. See also the works of Buckdahn and Ma [17, 18, 19, 20].

Other classes of SPDEs (including a stochastic heat equation) can be studied using rough path methods; see Gubinelli and coworkers [76, 77], Teichmann [169] and also the relevant comments in Section 11.4.

18

Stochastic Taylor expansions

Our very approach to rough differential equations was based on good estimates of higher-order approximations, such as obtained in Davie's lemma. In particular, these can be written in the form of error estimates of higher-order Euler approximations (cf. Corollary 10.15). We shall now put these estimates in a stochastic context.

18.1 Azencott-type estimates

We now consider RDEs driven by a (random) geometric rough path. To this end, fix $p \geq 1$ and let us first consider a continuous $G^{[p]}$-valued process $\mathbf{X} = \mathbf{X}_t(\omega)$ which satisfies

$$\sup_{0 \leq s < t \leq T} \mathbb{E}\left(\exp\left(\eta \left[\frac{d(\mathbf{X}_s, \mathbf{X}_t)}{|t-s|^{1/p}}\right]^2\right)\right) = M < \infty. \qquad (18.1)$$

Recall that this assumption applies to enhanced Brownian motion and Markov processes with $p = 2$. It also applies to our class of enhanced Gaussian processes (although, in general, a deterministic time-change may be needed, cf. Exercise 15.36). From the results of Section A.4, Appendix A, this implies that $\mathbf{X}$ has a.s. finite $\psi_{p,p/2}$-variation and also finite $1/p'$-Hölder regularity for any $p' > p$. In particular, by choosing p' small enough (such that $[p] = [p']$) it is clear that $\mathbf{X}$ is a geometric $1/p'$-Hölder rough path. As a consequence, for any integer $N \geq [p]$, there is a well-defined Lyons lift of $\mathbf{X}$ to a G^N-valued path, denoted by $S_N(\mathbf{X})$.

Theorem 18.1 (Azencott-type estimates) *Let $\gamma > p \geq 1$ and let $\mathbf{X}$ be a continuous $G^{[p]}$-valued process which satisfies (18.1). Let $V = (V_i)_{1 \leq i \leq d}$ be a collection of $\mathrm{Lip}^{\gamma-1}$ vector fields in $\mathbb{R}^e$. Then, for any fixed interval $[s,t] \subset [0,T]$ and time-s initial condition $y_s \in \mathbb{R}^e$,*

$$\mathbb{P}\left(\sup_{r \in [s,t]} \left|\pi_{(V)}(s, y_s; \mathbf{X})_{s,r} - \mathcal{E}_{(V)}\left(y_s, S_{\lfloor \gamma \rfloor}(\mathbf{X})_{s,r}\right)\right| > R|t-s|^{\gamma/p}\right)$$

$$\leq C \exp\left\{-\frac{1}{|V|^2_{\mathrm{Lip}^{\gamma-1}(\mathbb{R}^e)}}\left(\frac{R}{C}\right)^{2/\gamma}\right\}$$

where $C = C(M, \eta, \gamma, p)$. Under the weaker assumption $V \in \mathrm{Lip}^{\gamma-1}_{\mathrm{loc}}$ one *has*[1]

$$\varlimsup_{t \to 0} \mathbb{P} \left(\sup_{r \in [0,t]} \left| \pi_{(V)} \left(0, y_0; \mathbf{X}\right)_{0,r} - \mathcal{E}_{(V)} \left(y_0, S_{\lfloor \gamma \rfloor} \left(\mathbf{X}\right)_{0,r}\right) \right| > R t^{\gamma/p} \right)$$

$$\leq \ C \exp \left\{ -\frac{1}{|V|^2_{\text{Lip}^{\gamma-1}(B(y_0,1))}} \left(\frac{R}{C}\right)^{2/\gamma} \right\}. \tag{18.2}$$

In particular, we see that

$$\frac{\sup_{r \in [0,t]} \left| \pi_{(V)} \left(0, y_0; \mathbf{X}\right)_{0,r} - \mathcal{E}_{(V)} \left(y_0, S_{\lfloor \gamma \rfloor} \left(\mathbf{X}\right)_{0,r}\right) \right|}{t^{\gamma/p}}$$

is bounded in probability as $t \to 0$ *and for all* $\theta \in [0, \gamma/p)$ *we have*

$$\frac{\sup_{r \in [0,t]} \left| \pi_{(V)} \left(0, y_0; \mathbf{X}\right)_{0,r} - \mathcal{E}_{(V)} \left(y_0, S_{\lfloor \gamma \rfloor} \left(\mathbf{X}\right)_{0,r}\right) \right|}{t^{\theta}} \to 0 \tag{18.3}$$

in probability as $t \to 0$.

Proof. Let us fix $p' \in (p, \gamma)$, e.g. (for the sake of tracking the constants) $p' = (p + \gamma)/2$. Then a.e. $\mathbf{X}(\omega)$ is a geometric p'-rough path and there exist $\tilde{\eta}, \tilde{M} > 0$, depending on η, M only, such that

$$\mathbb{E} \left(\exp \left(\tilde{\eta} \left(\frac{\|\mathbf{X}\|_{p'\text{-var};[s,t]}}{|t-s|^{1/p}} \right)^2 \right) \right) = \tilde{M} < \infty; \tag{18.4}$$

see equation (A.20) in Appendix A. Thanks to $\text{Lip}^{\gamma-1}$-regularity of the vector fields, $\gamma > p'$, we have existence of an RDE solution, i.e. $\pi_{(V)}(s, y_s; \mathbf{X}) \neq \varnothing$. As usual, we abuse the notation and write $\pi_{(V)}(s, y_s; \mathbf{X})$ for any such (not necessarily unique) RDE solution. From our *Euler RDE estimates*, Corollary 10.15, there exists $c_1 = c_1(p', \gamma)$ such that

$$\mathbb{P} \left(\sup_{r \in [s,t]} \left| \pi_{(V)} \left(s, y_s; \mathbf{X}\right)_{s,r} - \mathcal{E}_{(V)} \left(y_s, S_{\lfloor \gamma \rfloor} \left(\mathbf{X}\right)_{s,r}\right) \right| > R |t-s|^{\gamma/p} \right)$$

$$\leq \ \mathbb{P} \left(c_1 |V|^{\gamma}_{\text{Lip}^{\gamma-1}} \|\mathbf{X}\|^{\gamma}_{p'\text{-var};[s,t]} > R |t-s|^{\gamma/p} \right)$$

$$= \ \mathbb{P} \left(\frac{\|\mathbf{X}\|_{p'\text{-var};[s,t]}}{|t-s|^{1/p}} > \frac{1}{|V|_{\text{Lip}^{\gamma-1}}} \left(\frac{R}{c_1}\right)^{1/\gamma} \right)$$

$$\leq \ \tilde{M} \exp \left(-\eta \frac{1}{|V|^2_{\text{Lip}^{\gamma-1}}} \left(\frac{R}{c_1}\right)^{2/\gamma} \right).$$

---

[1] If explosion happens, we agree that $\left| \pi_{(V)} \left(0, y_0; \mathbf{X}\right)_{0,s} - \mathcal{E}_{(V)} \left(y_0, S_N \left(\mathbf{X}\right)_{0,s}\right) \right| = +\infty$.

At last, consider the case of $\mathrm{Lip}_{loc}^{\gamma-1}$-vector fields V. For fixed y_0, we can then find $\tilde{V} \in \mathrm{Lip}^{\gamma-1}$ so that $\tilde{V} \equiv V$ on a unit ball around y_0. Setting $Y_t \equiv \pi_{(V)}(0, y_0; \mathbf{X})_t$ and $\tilde{Y}_t \equiv \pi_{(\tilde{V})}(0, y_0; \mathbf{X})_t$ we see that

$$\mathbb{P}\left(\sup_{r \in [0,t]} \left| Y_{0,r} - \mathcal{E}_{(V)}\left(y_0, S_{\lfloor \gamma \rfloor}(\mathbf{X})_{0,r}\right) \right| > Rt^{\gamma/p} \right)$$

$$\leq \mathbb{P}\left(\sup_{r \in [0,t]} \left| Y_{0,r} - \mathcal{E}_{(V)}\left(y_0, S_{\lfloor \gamma \rfloor}(\mathbf{X})_{0,r}\right) \right| > Rt^{\gamma/p}; \ \sup_{r \in [0,t]} |Y_r - y_0| < 1 \right)$$

$$+ \mathbb{P}\left(\sup_{r \in [0,t]} |Y_r - y_0| \geq 1 \right)$$

$$\leq c_2 \exp\left\{ -\left(\frac{R}{c_2}\right)^{2/\gamma} \right\} + \mathbb{P}\left(\sup_{r \in [0,t]} |Y_r - y_0| \geq 1 \right)$$

where c_2 depends on M, η, γ, p and $|V|_{\mathrm{Lip}^{\gamma-1}; B(y_0,1)}$. Noting that $\mathbb{P}\left(|Y - y_0|_{\infty;[0,t]} \geq 1 \right) \to 0$ as $t \to 0$, the claimed estimate now follows. At last, observe that (thanks to $\theta < \gamma/p$) for every fixed $\varepsilon, R > 0$ there exists $Rt^{\gamma/p} \leq \varepsilon t^\theta$ for t small enough. It follows that, if $z_t = \sup_{r \in [0,t]} \left| \pi_{(V)} (0, y_0; \mathbf{X})_{s,r} - \mathcal{E}_{(V)}\left(y_0, S_{\lfloor \gamma \rfloor}(\mathbf{X})_{0,r}\right) \right|$,

$$\varlimsup_{t \to 0} \mathbb{P}\left(z_t > \varepsilon t^\theta \right) \leq \varlimsup_{t \to 0} \mathbb{P}\left(z_t > Rt^{\gamma/p} \right) \leq c_2 \exp\left\{ -\left(\frac{R}{c_2}\right)^{2/\gamma} \right\}$$

and since we take R arbitrarily large, we see that the lim sup is zero (and therefore a genuine lim). ∎

Example 18.2 Consider an enhanced Gaussian process $\mathbf{X}$, which satifies

$$\sup_{0 \leq s < t \leq T} \mathbb{E}\left(\exp\left(\eta \left[\frac{d(\mathbf{X}_s, \mathbf{X}_t)}{|t-s|^H} \right]^2 \right) \right) < \infty$$

for some $H \in (1/4, 1/2]$. (After setting $\rho = 1/(2H)$, it was pointed out in Exercise 15.36 that this holds for all enhanced Gaussian processes run at the correct time-scale.) Let $V = (V_1, \ldots, V_d)$ be a collection of smooth (possibly unbounded) vector fields. Then, for all $N \in \{2, 3, \ldots\}$, we may apply Theorem 18.1 with $\gamma = N + 1$ and $p = 1/H$ to see that, for every fixed $\varepsilon > 0$,

$$\mathbb{P}\left(\left| \pi_{(V)}(0, y_0; \mathbf{X})_{0,t} - \mathcal{E}_{(V)}\left(y_0, S_N(\mathbf{X})_{0,t}\right) \right| > \varepsilon t^{HN} \right) \to 0$$

as $t \to 0+$. This applies in particular to enhanced fractional Brownian motion with Hurst parameter H. □

Let us now give a variation of Theorem 18.1 applicable to enhanced martingales.

Theorem 18.3 *Let V, γ, p be as above and let $y = y(\omega) = \pi(0, y_0; \mathbf{M})$ be the (pathwise unique) random RDE solution to $dy = V(y) \, d\mathbf{M}$ where $\mathbf{M} = \mathbf{M}(\omega)$ is an enhanced L^q-martingale, $q \in [1, \infty)$. Then for any fixed $t \in (0, 1]$ and*

$$\phi(t) := \mathbb{E}\left(|\langle M \rangle_t|^{q/2} \right)^{2/q} = |\langle M \rangle_t|_{L^{q/2}}$$

we have for $C = C(q, \gamma, p)$,

$$\mathbb{P}\left(\sup_{0 \le s \le t} \left| \pi(0, y_0; \mathbf{M})_{0,s} - \mathcal{E}_{(V)}\left(y_0, S_{\lfloor \gamma \rfloor}(\mathbf{M})_{s,t} \right) \right| > R\phi(t)^{\frac{\gamma}{2}} \right) \le C\left(\frac{1}{R} \right)^{\frac{q}{\gamma}}.$$

Proof. Similar to the proof of Theorem 18.1 and left to the reader. ∎

18.2 Weak remainder estimates

Recall that the Euler approximation $\mathcal{E}_{(V)}(\dots)$ came from setting $f = I$, the identity function, in the Taylor expansion

$$f\left(\pi_{(V)}(0, y_0; x)_t \right) = f(y_0) + \sum_{k=1}^{N} \sum_{\substack{i_1, \dots, i_k \\ \in \{1, \dots, d\}}} V_{i_1} \cdots V_{i_k} f(y) \, \mathbf{x}_{0,t}^{k, i_1, \dots, i_k} + R_N(t, f; x);$$

valid, at least, for sufficiently smooth f and $x \in C^{1\text{-var}}([0, T], \mathbb{R}^d)$ with canonically defined kth iterated integrals $\mathbf{x}^k$. This obviously makes sense for RDEs and we can ask for an estimate of the remainder term

$$R_N(t, f; \mathbf{X}) := f\left(\pi_{(V)}(0, y_0; \mathbf{X})_t \right)$$

$$- \left(f(y_0) + \sum_{k=1}^{N} \sum_{\substack{i_1, \dots, i_k \\ \in \{1, \dots, d\}}} V_{i_1} \cdots V_{i_k} f(y) \, \mathbf{X}_{0,t}^{k, i_1, \dots, i_k} \right),$$

where we have abused the notation by writing $\mathbf{X}$ instead of $S_N(\mathbf{X})$.

Theorem 18.4 *Let $\gamma > p \ge 1$ and let $\mathbf{X}$ be a continuous $G^{[p]}$-valued process which satisfies (18.1). Let $V = (V_i)_{1 \le i \le d}$ be a collection of $\mathrm{Lip}^{\gamma-1}$ vector fields in $\mathbb{R}^e$. Then, for any function $f \in \mathrm{Lip}^{\gamma}(\mathbb{R}^e, \mathbb{R})$ we have*

$$\forall q \in [1, \infty) : \left\| R_{\lfloor \gamma \rfloor}(t, f; \mathbf{X}) \right\|_{L^q} = O\left(t^{\gamma/p} \right) \quad \text{as } t \to 0.$$

In the case of $\mathrm{Lip}_{\mathrm{loc}}^{\gamma-1}$ vector fields V and $f \in \mathrm{Lip}_{\mathrm{loc}}^{\gamma}$, we still have that for any $\theta \in [0, \gamma/p)$,

$$\frac{R_{\lfloor \gamma \rfloor}(t, f; \mathbf{X})}{t^{\theta}} \to 0 \quad \text{in probability as } t \to 0.$$

Proof. It it clear that $Y \equiv \pi_{(V)} (0, y_0; \mathbf{X})_t$ and $f(Y)$ can be written jointly as an RDE solution; say

$$dz = \tilde{V}(z) \, d\mathbf{X}$$

with $\mathrm{Lip}^{\gamma-1}$ vector fields $\tilde{V}$ obtained from writing the ODE system

$$
\begin{aligned}
dz^{(1)} &= V\left(z^{(1)}\right) dx, \\
dz^{(2)} &= f'\left(z^{(1)}\right) dz^{(1)} = f'\left(V\left(z^{(1)}\right) dx\right)
\end{aligned}
$$

in the form $dz = \tilde{V}(z) \, dx$, with $z = \left(z^{(1)}, z^{(2)}\right) \in \mathbb{R}^{e+1}$. It follows that $R_{\lfloor \gamma \rfloor}(t, f; \mathbf{X})$ is precisely the $(e+1)$th component of

$$\left| \pi_{(\tilde{V})} (0, y_0; \mathbf{X})_{0,t} - \mathcal{E}_{(\tilde{V})} \left(y_0, S_{\lfloor \gamma \rfloor} (\mathbf{X})_{0,t} \right) \right|$$

and the estimate of Theorem 18.1 is more than enough to ensure that the random variable $\hat{R}_t := \left| R_{\lfloor \gamma \rfloor}(t, f; \mathbf{X}) \right| / t^{\gamma/p}$ has moments of all orders, uniformly in $t \in (0, 1]$. But then, for all $t \in (0, 1]$,

$$\left\| R_{\lfloor \gamma \rfloor}(t, f; \mathbf{X}) \right\|_{L^q(\mathbb{P})} \leq \left\| \sup_{t \in (0,1]} \hat{R} \right\|_{L^q(\mathbb{P})} \times t^{\gamma/p}$$

and the proof is finished. In the case of $\mathrm{Lip}_{\mathrm{loc}}^{\gamma-1}$ vector fields V and $f \in \mathrm{Lip}_{\mathrm{loc}}^{\gamma}$ the same construction yields $\tilde{V} \in \mathrm{Lip}_{\mathrm{loc}}^{\gamma}$ and we conclude again with Theorem 18.1. ∎

18.3   Comments

In a Brownian – and semi-martingale – context, the estimates of **Section 18.1** go back to Azencott [4], Ben Arous [10] and Platen [138]. Estimate (18.2) plays an important role in subsequent developments, such as Castell [27]. For related works in a fractional Brownian rough path context, we mention Baudoin and Coutin [8]. Our presentations of Sections 18.1 and 18.2 improve on earlier results by the authors obtained in [67].

Let us also mention Aida [3] and then Inahama and Kawabi [89, 92], where the authors are led to somewhat different (stochastic) Taylor expansions for rough differential equations (in essence, asymptotic development in ε of solutions to $dy^{\varepsilon} = V(y^{\varepsilon}, \varepsilon) \, d\mathbf{x}$).

19

Support theorem and large deviations

We now discuss some classical results of diffusion theory: the Stroock–Varadhan support theorem and Freidlin–Wentzell large deviation estimates. Everything relies on the fact that the Stratonovich SDE

$$dY = \sum_{i=1}^{d} V_i(Y) \circ dB^i + V_0(Y)\, dt, \quad Y_0 = y_0 \in \mathbb{R}^e,$$

can be solved as an RDE solution which depends continuously on enhanced Brownian motion in rough path topology, subject to the suitable Lip-regularity assumptions on the vector fields. (A summary of the relevant continuity results was given in Section 17.1.)

19.1   Support theorem for SDEs driven by Brownian motion

Theorem 19.1 (Stroock–Varadhan support theorem) *Assume that* $V = (V_1, \ldots, V_d)$ *is a collection of* Lip^2*-vector fields on* $\mathbb{R}^e$, *and* V_0 *is a* Lip^1*-vector field on* $\mathbb{R}^e$. *Let* B *be a* d*-dimensional Brownian motion and consider the unique (up to indistinguishability) Stratonovich SDE solution* Y *on* $[0, T]$ *to*

$$dY = \sum_{i=1}^{d} V_i(Y) \circ dB^i + V_0(Y)\, dt, \quad Y_0 = y_0 \in \mathbb{R}^e. \qquad (19.1)$$

Let us write $y^h = \pi_{(V, V_0)}(0, y_0; (h, t))$ *for the ODE solution to*

$$dy = \sum_{i=1}^{d} V_i(Y)\, dh^i + V_0(Y)\, dt$$

started at $y_0 \in \mathbb{R}^e$ *where* h *is a Cameron–Martin path, i.e.* $h \in W_0^{1,2}([0, T], \mathbb{R}^d)$. *Then, for any* $\alpha \in [0, 1/2)$ *and any* $\delta > 0$,

$$\lim_{\varepsilon \to 0} \mathbb{P}\left(\left. \left| Y - y^h \right|_{\alpha\text{-H\"ol};[0,T]} < \delta \right| \left| B - h \right|_{\infty;[0,T]} < \varepsilon \right) \to 1$$

(where the Euclidean norm is used for conditioning $\left| B - h \right|_{\infty;[0,T]} < \varepsilon$*).*

Proof. Without loss of generality, $\alpha \in (1/3, 1/2)$. Let us write, for a fixed Cameron–Martin path h,

$$\mathbb{P}^\varepsilon (\cdot) \equiv \mathbb{P} \left(\cdot \mid \; |B - h|_{\infty;[0,T]} < \varepsilon \right).$$

From Theorem 17.3, there is a unique solution $\pi_{(V,V_0)}(0, y_0; (\mathbf{B}, t))$ to the RDE with drift

$$dY = V(Y) \, d\mathbf{B} + V_0(Y) \, dt, \;\; Y_0 = y_0$$

which then solves the Stratonovich equation (19.1). Set $\mathbf{h} \equiv S_2(h)$ and observe that $\mathbf{h}$ is of finite 1-variation and 1/2-Hölder, hence

$$\mathbf{h} \in C^{\psi_{2,1}\text{-var}}\left([0,T], G^2\left(\mathbb{R}^d\right)\right) \cap C^{0,\alpha\text{-Höl}}\left([0,T], G^2\left(\mathbb{R}^d\right)\right).$$

Take $\alpha' \in (1/3, \alpha)$. We now use continuity of the Itô–Lyons map from

$$C^{\psi_{2,1}\text{-var}}\left([0,T], G^2\left(\mathbb{R}^d\right)\right) \cap C^{0,\alpha\text{-Höl}}\left([0,T], G^2\left(\mathbb{R}^d\right)\right) \to C^{0,\alpha'\text{-Höl}}\left([0,T], \mathbb{R}^e\right),$$

in (rough path) α-Hölder to (classical) α'-Hölder topology, respectively, at the point $\mathbf{h}$. Given $\delta > 0$ fixed, there exists $\eta = \eta(h, \delta)$ small enough such that for

$$\forall \mathbf{B} \in C^{\psi_{2,1}\text{-var}} \cap C^{0,\alpha\text{-Höl}} : d_{\alpha\text{-Höl}}(\mathbf{B}, \mathbf{h}) < \eta \implies \left|Y - y^h\right|_{\alpha\text{-Höl}} < \delta.$$

In particular, using the fact that $\mathbf{B} \in C^{\psi_{2,1}\text{-var}} \cap C^{0,\alpha\text{-Höl}}$ almost surely,

$$\mathbb{P}^\varepsilon \left(\left|Y - y^h\right|_{\alpha\text{-Höl}} < \delta \right) \geq \mathbb{P}^\varepsilon \left(d_{\alpha\text{-Höl}}(\mathbf{B}, \mathbf{h}) < \eta \right)$$
$$\to 1 \text{ as } \varepsilon \to 0$$

thanks to Theorem 13.66. ∎

Remark 19.2 The regularity assumptions of Theorem 19.9 are optimal in the sense that $V \subset \mathrm{Lip}^2$ and $V_0 \in \mathrm{Lip}^1$ are needed for a unique Stratonovich solution. □

As an immediate corollary, we obtain the characterization of the support of the law of the solution of a Stratonovich SDE.

Corollary 19.3 *Assume that $V = (V_1, \ldots, V_d)$ is a collection of Lip^2-vector fields on $\mathbb{R}^e$, V_0 a Lip^1-vector field on $\mathbb{R}^e$, and B a d-dimensional Brownian motion. Consider the unique (up to indistinguishability) Stratonovich SDE solution on $[0,T]$ to*

$$dY = \sum_{i=1}^{d} V_i(Y) \circ dB^i + V_0(Y) \, dt$$

started at some $y_0 \in \mathbb{R}^e$. Then, for any $\alpha \in [0, 1/2)$, the topological support of Y, viewed as a $C^{0,\alpha}\left([0,T];\mathbb{R}^e\right)$-valued random variable, is precisely the α-Hölder closure of

$$\mathcal{S} = \left\{ \pi_{(V,V_0)}\left(0, y_0; (h,t)\right), \ h \in W_0^{1,2} \right\}.$$

Proof. The first inclusion, supp (law of $\pi(0, y_0; \mathbf{B})) \subset \overline{\mathcal{S}}$, is straight-forward from the Wong–Zakai theorem (equivalently: use Theorem 17.3 with Remarks 17.4 and 17.5).

For the other inclusion (usually considered the difficult one), it suffices to show that for every Cameron–Martin path h and every $\delta > 0$, the event

$$A^{h,\delta} = \left\{ \left| Y - y^h \right|_{\alpha\text{-Höl}} < \delta \right\} = \left\{ \left| \pi(0, y_0; \mathbf{B}) - \pi(0, y_0; h) \right|_{\alpha\text{-Höl}} < \delta \right\}$$

has positive probability. But this is an obvious consequence of Theorem 19.1. ∎

Remark 19.4 If one is only interested in the conclusion of Corollary 19.3, one bypasses the "conditional" consideration of Theorem 13.66 on which our proof of Theorem 19.1 relied. Indeed, in Section 13.7 we obtained *with much less work* (Theorem 13.54) the qualitative statement

$$\text{supp (law of } \mathbf{B}) = \overline{\left\{ S_2(h) : h \in W_0^{1,2} \right\}} \tag{19.2}$$

(support and closure with respect to α-Hölder rough path topology) so that for any $\eta > 0$ and $h \in W_0^{1,2}$,

$$\mathbb{P}\left(d_{\alpha\text{-Höl}}\left(\mathbf{B}, S_2(h)\right) < \eta\right) > 0.$$

Given $\delta > 0$ fixed, there exists $\eta = \eta(h, \delta)$ small enough such that $d_{\alpha\text{-Höl}}(\mathbf{B}, \mathbf{h}) < \eta \implies \left|Y - y^h\right|_{\alpha\text{-Höl}} < \delta$. Hence

$$\mathbb{P}\left(\left|Y - y^h\right|_{\alpha\text{-Höl}} < \delta\right) \geq \mathbb{P}\left(d_{\alpha\text{-Höl}}(\mathbf{B}, \mathbf{h}) < \eta\right) > 0,$$

which yields the (difficult) inclusion in the Stroock–Varadhan support theorem. □

We can also deal with the support at the level of stochastic flows (as discussed in Section 17.5).

Theorem 19.5 *(support for stochastic flows)* *Assume that $V = (V_1, \ldots, V_d)$ is a collection of $\text{Lip}^{\gamma+k-1}$-vector fields on $\mathbb{R}^e$, $\gamma > p$, so that $\pi_{(V)}(0, \cdot; \mathbf{B})$, the solution flow to the Stratonovich SDE*

$$Y = V(Y) \circ dB,$$

induces a C^k-flow of diffeomorphisms and we can view $\pi_{(V)}(0,\cdot;\mathbf{B})$ as a $\mathcal{D}_k(\mathbb{R}^e)$-valued random variable.[1] Then, for any $h \in W_0^{1,2}([0,T],\mathbb{R}^d)$,

$$\lim_{\varepsilon \to 0} \mathbb{P}\left(d_{\mathcal{D}_k(\mathbb{R}^e)}\left(\pi_{(V)}(0,\cdot;\mathbf{B}),\pi_{(V)}(0,\cdot;h)\right) < \delta \bigg| \; |B-h|_{\infty;[0,T]} < \varepsilon \right) \to 1$$

and

$$\text{supp } (\text{law of } \pi(0,\cdot;\mathbf{B})) = \overline{\mathcal{S}}$$

where $\mathcal{S} = \left\{ \pi(0,\cdot;h), \; h \in W_0^{1,2}([0,T],\mathbb{R}^d) \right\} \subset \mathcal{D}_k(\mathbb{R}^e)$.

Proof. The argument is then similar to the proof of Theorem 19.1. Let $\alpha \in (1/3,1/2)$. Let us write, for a fixed Cameron–Martin path h,

$$\mathbb{P}^\varepsilon(\cdot) \equiv \mathbb{P}\left(\cdot \mid \; |B-h|_{\infty;[0,T]} < \varepsilon \right).$$

Thanks to Corollary 11.14, $\text{Lip}^{\gamma+k-1}$-vector fields imply continuity of

$$C^{0,\alpha\text{-Höl}}\left([0,T],G^2(\mathbb{R}^d)\right) \;\to\; \mathcal{D}_k(\mathbb{R}^e)$$
$$\mathbf{x} \;\mapsto\; \pi_{(V)}(0,\cdot,\mathbf{x})$$

and we simply use it at the point $\mathbf{h} \equiv S_2(h)$. Given $\delta > 0$ fixed, there exists $\eta = \eta(h,\delta)$ small enough such that for

$$\forall \mathbf{B} \in C^{0,\alpha\text{-Höl}} : d_{\alpha\text{-Höl}}(\mathbf{B},\mathbf{h}) < \eta \implies |Y - y^h|_{\alpha\text{-Höl}} < \delta.$$

It follows, thanks to Theorem 13.66, that

$$\mathbb{P}^\varepsilon\left(|Y-y^h|_{\alpha\text{-Höl}} < \delta \right) \geq \mathbb{P}^\varepsilon\left(d_{\alpha\text{-Höl}}(\mathbf{B},\mathbf{h}) < \eta \right) \to 1$$

as $\varepsilon \to 0$. The proof is finished.   ∎

19.2   Support theorem for SDEs driven by other stochastic processes

The reader will have noticed the proofs of the previous section are essentially trivial corollaries of a suitable support description of enhanced Brownian motion in rough path topology, followed by appealing to continuity of the Itô–Lyons map. We have seen in Part II (more precisely, Theorems 15.60 and 16.33) that similar support descriptions hold for enhanced Gaussian and Markovian processes. As a consequence, the very same arguments lead us to support theorems for stochastic differential equations driven by Gaussian and Markovian signals. We have

---

[1] The Polish space $\mathcal{D}_k(\mathbb{R}^e)$ was defined in Section 11.5.

Proposition 19.6 *Assume that*
(i) $X = \left(X^1, \ldots, X^d \right)$ *is a centred continuous Gaussian process on* $[0,1]$ *with independent components;*
(ii) $\mathcal{H}$ *denotes the Cameron–Martin space associated with* X;
(iii) the covariance of X *is of finite* ρ-*variation dominated by some 2D control* ω, *for some* $\rho \in [1,2)$;
(iv) $\mathbf{X}$ *denotes the natural lift of* X *to a* $G^{[2\rho]}\left(\mathbb{R}^d \right)$-*valued process (with geometric rough sample paths);*
(v) $V = (V_1, \ldots, V_d)$ *is a collection of* Lip^γ-*vector fields on* $\mathbb{R}^e$, *with* $\gamma > 2\rho$.
Then, for any $p > 2\rho$, *the topological support of the solution to*

$$dY = V(Y)\, d\mathbf{X}, \ Y(0) = y_0 \in \mathbb{R}^e,$$

viewed as a $C^{0,p\text{-var}}\left([0,T] ; \mathbb{R}^e \right)$-*valued random variable, is precisely the* p-*variation closure of*

$$\mathcal{S} = \left\{ \pi_{(V)}(0, y_0; h), \ h \in \mathcal{H} \right\}.$$

If ω *is Hölder-dominated, the topological support of* Y, *viewed as a* $C^{0,1/p\text{-Höl}}$ $\left([0,T] ; \mathbb{R}^e \right)$-*valued random variable, is precisely the* $1/p$-*Hölder closure of*

$$\mathcal{S} = \left\{ \pi_{(V)}(0, y_0; h), \ h \in \mathcal{H} \right\}.$$

Proof. Left to the reader.   ∎

Proposition 19.7 *Assume that*
(i) $X = \left(X^1, \ldots, X^d \right)$ *is a Markov process with uniformly elliptic generator in divergence form,* $\frac{1}{2} \sum_{i,j=1}^d \partial_i \left(a^{ij}(\cdot)\, \partial_j \cdot \right)$ *understood in a weak sense (i.e. via Dirichlet forms) where* $a \in \Xi^{1,d}(\Lambda)$, *that is, measurable, symmetric and* $\Lambda^{-1} I \leq a \leq \Lambda I$ *in the sense of positive definite matrices;*
(ii) $\mathbf{X}$ *denotes the natural lift of* X *to a* $G^2\left(\mathbb{R}^d \right)$-*valued process (with geometric rough sample paths);*
(iii) $V = (V_1, \ldots, V_d)$ *is a collection of* Lip^4-*vector fields on* $\mathbb{R}^e$.
Then, for any $\alpha \in [0, 1/4)$, *the topological support of the solution to*

$$dY = V(Y)\, d\mathbf{X}, \ Y(0) = y_0 \in \mathbb{R}^e,$$

viewed as a $C^{0,\alpha}\left([0,T] ; \mathbb{R}^e \right)$-*valued random variable, is precisely the* α-*Hölder closure of*

$$\mathcal{S} = \left\{ \pi_{(V)}(0, y_0; h), \ h \in W_0^{1,2}\left([0,T], \mathbb{R}^d \right) \right\}.$$

Proof. Set $\tilde{\mathbf{X}} = S_4(\mathbf{X})$, the step-4 Lyons lift of $\mathbf{X}$. Clearly,

$$Y = \pi_{(V)}(0, y_0; \mathbf{X}) = \pi_{(V)}\left(0, y_0; \tilde{\mathbf{X}} \right)$$

and $\tilde{\mathbf{X}} \in C^{\psi_{4,1}\text{-var}} \cap C^{\alpha\text{-Höl}}$. It then suffices to use continuity of the solution map $\tilde{\mathbf{X}} \mapsto Y$ in α-Hölder topology for any $\alpha \in (1/5, 1/4)$, following the precise argument given in Remark 19.4.   ∎

It should be noted that the restriction to Hölder exponent $< 1/4$ (and Lip^4- rather than Lip^2-regularity) in Proposition 19.7 is a consequence of Theorem 16.33 where a support characterization of the enhanced Markov process $\mathbf{X}$ in α-Hölder rough path topology was only established for $\alpha < 1/4$. As noted in the comments of Section 16.10, the result is conjectured to hold for $\alpha < 1/2$ in which case we would have, for all $h \in W_0^{1,2}\left([0,T],\mathbb{R}^d\right)$ and $\delta > 0$,

$$\mathbb{P}\left(\left|\pi_{(V)}(0,y_0;\mathbf{X}) - \pi_{(V)}(0,y_0;h)\right|_{\alpha\text{-Höl};[0,T]} < \delta\right) > 0.$$

In fact, we can show this for $h = 0$.

Proposition 19.8 *Under the assumptions of Proposition 19.7 but with the weakened regularity assumption $V = (V_1,\ldots,V_d) \subset \text{Lip}^2(\mathbb{R}^e)$ we have, for any $\alpha \in [0,1/2)$,*

$$\lim_{\varepsilon \to 0} \mathbb{P}\left(\left.\left|\pi_{(V)}(0,y_0;\mathbf{X})\right|_{\alpha\text{-Höl};[0,T]} < \delta\right| \; \|\mathbf{X}\|_{\infty;[0,T]} < \varepsilon\right) \to 1.$$

As a consequence, for all $\delta > 0$,

$$\mathbb{P}\left(\left.\left|\pi_{(V)}(0,y_0;\mathbf{X})\right|_{\alpha\text{-Höl};[0,T]} < \delta\right.\right) > 0.$$

Proof. This follows readily from Theorem 16.39. ∎

19.3 Large deviations for SDEs driven by Brownian motion

In Theorem 13.42, we saw that Schilder's theorem holds for enhanced Brownian motion. That is, if $\mathbf{B}$ denotes a $G^2\left(\mathbb{R}^d\right)$-valued enhanced Brownian motion on $[0,T]$, then for any $\alpha \in [0,1/2)$, the family $(\delta_\varepsilon \mathbf{B} : \varepsilon > 0)$ satisfies a large deviation $C^{0,\alpha\text{-Höl}}\left([0,T],G^2\left(\mathbb{R}^d\right)\right)$ with rate function given by $I(\pi_1(\cdot))$, where $\pi_1(\cdot)$ denotes the projection of a $G^2\left(\mathbb{R}^d\right)$-valued path to an $\mathbb{R}^d$-valued path and

$$I(h) = \begin{cases} \int_0^T \left|\dot{h}_t\right|^2 dt & \text{for } h \in W_0^{1,2}\left([0,T],\mathbb{R}^d\right) \\ +\infty & \text{otherwise} \end{cases}. \tag{19.3}$$

Since $\pi_{(V)}(0,y_0;\mathbf{B})$, a Stratonovich solution to $dY = V(Y)\,d\mathbf{B}$, depends continuously on $\mathbf{B}$ in this α-Hölder rough path topology, we can apply the contraction principle to deduce (without any further work) a large deviation principle for solution of stochastic differential equations; better known as Freidlin–Wentzell estimates. More precisely, also including a drift term, we have

Theorem 19.9 (Freidlin–Wentzell large deviations) *Assume that* $V = (V_1, \ldots, V_d)$ *is a collection of* Lip^2*-vector fields on* $\mathbb{R}^e$, *and* V_0 *is a* Lip^1*-vector field on* $\mathbb{R}^e$. *Let B be a d-dimensional Brownian motion and consider the unique (up to indistinguishability) Stratonovich SDE solution on $[0, T]$ to*

$$dY^\varepsilon = \sum_{i=1}^{d} V_i(Y) \circ \varepsilon dB^i + V_0(Y) \, dt$$

started at y_0. Let $\alpha \in [0, 1/2)$. Then Y^ε satisfies a large deviation principle (in α-Hölder topology) with good rate function given by

$$J(y) = \inf \left\{ I(h) : \ \pi_{(V,V_0)}(0, y_0; (h, t)) = y \right\}$$

where I is given in (19.3).

Proof. Let $\alpha \in (1/3, 1/2)$ without loss of generality. The Stratonovich solution is given by the random RDE solution $\pi_{(V,V_0)}(0, y_0; (\delta_\varepsilon \mathbf{B}, t))$ and depends continuously (see Theorem 12.10, or Theorem 10.26 in the absence of a drift term) on[2]

$$\delta_\varepsilon \mathbf{B} \in C^{\psi_{2,1}\text{-var}}\left([0, T], G^2\left(\mathbb{R}^d\right)\right) \cap C^{0, \alpha\text{-Höl}}\left([0, T], G^2\left(\mathbb{R}^d\right)\right)$$
$$\equiv C^{\psi_{2,1}\text{-var}} \cap C^{0, \alpha\text{-Höl}}$$

with respect to α-Hölder rough path topology. Since $(\delta_\varepsilon \mathbf{B} : \varepsilon > 0)$ satisfies a large deviation principle in $C^{0, \alpha\text{-Höl}}\left([0, T], G^2\left(\mathbb{R}^d\right)\right)$ with good rate function I and

$$\mathbb{P}\left[\delta_\varepsilon \mathbf{B} \in C^{\psi_{2,1}\text{-var}}\left([0, T], G^2\left(\mathbb{R}^d\right)\right)\right] = 1,$$

it follows from Proposition C.5 that $(\delta_\varepsilon \mathbf{B} : \varepsilon > 0)$ satisfies a large deviation principle in the (non-complete, separable) metric space

$$\left(C^{\psi_{2,1}\text{-var}} \cap C^{0, \alpha\text{-Höl}}, d_{\alpha\text{-Höl}}\right)$$

with identical rate function. We conclude with the contraction principle, Theorem C.6. ∎

Remark 19.10 The regularity assumptions of Theorem 19.9 are optimal in the sense that Lip^2-regularity is needed for a unique Stratonovich solution. If one deals with Itô stochastic differential equations,

$$dY^\varepsilon = \sum_{i=1}^{d} V_i(Y) \varepsilon dB^i + V_0(Y) \, dt,$$

---

[2]Let us remark that under slightly stronger "$\mathrm{Lip}^\gamma, \gamma > 2$" regularity assumptions one can ignore $\psi_{2,1}$-variation.

it is well known that Lip^1-regularity suffices for existence/uniqueness. In this case, the large deviation estimates of Theorem 19.9 are known (e.g. [41, Lemma 4.1.6]) to be valid with identical rate function.   □

Exercise 19.11 *Assume $V_0 = 0$, $d = e$ such that $V_1, \ldots, V_d$ span the tangent space at every point, a Riemannian metric $\langle ., . \rangle_{(V)}$ is defined by declaring $V_1, \ldots, V_d$ orthonormal. Express $J(y)$ as the energy of the path y, i.e.*

$$J(y) = \frac{1}{2} \int_0^1 \langle \dot{y}_t, \dot{y}_t \rangle_{(V)} \, dt.$$

Let us discuss various extensions of this. As noted in Section 17.6, the rough path approach has no reliance whatsoever on adaptedness, and hence anticipating SDEs does not require a separate analysis. We can state the following large deviation principle for a class of such anticipating stochastic differential equations. For simplicity of notation only, we take $V_0 = 0$ here.

Theorem 19.12 *Let $\mathbf{B}$ be a $G^2\left(\mathbb{R}^d\right)$-valued enhanced Brownian motion. Let also $(Y_0^\varepsilon(\omega) : \varepsilon \geq 0)$ be a family of random elements of $\mathbb{R}^d$,*

$$V^\varepsilon(\omega) \equiv (V_1^\varepsilon(\omega), \ldots, V_d^\varepsilon(\omega) : \varepsilon \geq 0)$$

be a random collection of Lip^2-vector fields, both deterministic for $\varepsilon = 0$, such that for all $\delta > 0$

$$\lim_{\varepsilon \to 0} \varepsilon^2 \log \mathbb{P} \left(\max_{1 \leq i \leq d} \left| V_i^\varepsilon - V_i^0 \right|_{\text{Lip}^2} > \delta \right) = -\infty,$$

$$\lim_{\varepsilon \to 0} \varepsilon \log \mathbb{P} \left(\left| y_0^\varepsilon - y_0^0 \right| > \delta \right) = -\infty.$$

Let

$$Y^\varepsilon(\omega) = \pi_{(V^\varepsilon(\omega))}(0, Y_0^\varepsilon(\omega); \delta_\varepsilon \mathbf{B}(\omega))$$

denote the unique ω-wise defined RDE solution to

$$dy^\varepsilon = \varepsilon V^\varepsilon(y^\varepsilon) \, d\mathbf{B}_t \tag{19.4}$$

started from $Y_0^\varepsilon(\omega)$. Then $(Y^\varepsilon : \varepsilon > 0)$ satisfies a large deviation principle in α-Hölder topology, for any $\alpha \in [0, 1/2)$, with good rate function

$$J(y) = \inf \left\{ I(h) : \pi_{(V^0)}\left(0, y_0^0; h\right) = y \right\}.$$

Proof. Without loss of generality, assume $\alpha \in (1/3, 1/2)$ and take $\alpha' \in (\alpha, 1/2)$. We know that $\{\delta_\varepsilon \mathbf{B} : \varepsilon \geq 0\}$ satisfies a large deviation principle in α'-Hölder rough path topology. The assumptions on the vector fields and the initial conditions give that $\left(y_0^\varepsilon, (V_i^\varepsilon)_{1 \leq i \leq d}, \delta_\varepsilon \mathbf{B}\right)$ satisfies a large deviation principle in

$$\mathbb{R}^d \times \text{Lip}^2 \times C^{0, \alpha'\text{-Höl}}\left([0, T], G^2\left(\mathbb{R}^d\right)\right).$$

From continuity of $(y_0, V, \mathbf{x}) \mapsto \pi_{(V)} (0, y_0; \mathbf{x})$, see Theorem 10.41 (or Corollary 10.27 when $V \subset \text{Lip}^\gamma$ with $\gamma > 2$), we conclude with the contraction principle. ∎

We can also deal with large deviations at the level of stochastic flows. To this end, recall from Corollary 11.14 that, granted sufficient regularity, we have continuity of $\mathbf{x} \mapsto \pi_V (0, \cdot; \mathbf{x}) \in \mathcal{D}_k (\mathbb{R}^e)$ in the sense of flows of C^k-diffeomorphisms.[3] Using the large deviation principle for $(\delta_\varepsilon \mathbf{B})$, it is an immediate application of the contraction principle to obtain

Theorem 19.13 *(large deviations for stochastic flows)* *Assume* $V \in \text{Lip}^{\gamma+k-1}$ *with* $\gamma > 2$ *and* $k \in \{1, 2, \dots\}$. *Then, the* $\mathcal{D}_k (\mathbb{R}^e)$*-valued random variable given by* $\pi_V (0, \cdot; \delta_\varepsilon \mathbf{B})$, *i.e. the stochastic flow of the Stratonovich equation* $dY^\varepsilon = \sum_{i=1}^d V_i (Y) \circ \varepsilon dB^i$, *satisfies a large deviation principle with good rate function*

$$ J(\varphi) = \inf \left\{ I(h) : \pi_{(V)} (0, \cdot; h) = \varphi \in \mathcal{D}_k (\mathbb{R}^e) \right\}. $$

19.4 Large deviations for SDEs driven by other stochastic processes

Using the large deviation results for enhanced Gaussian and Markov processes established in Section 15.7 resp. Section 16.7, we can generalize the previous section to RDEs driven by Gaussian and Markovian signals. The proofs are the same:

Proposition 19.14 *Assume that*
(i) $X = (X^1, \dots, X^d)$ *is a centred continuous Gaussian process on* $[0, 1]$ *with independent components;*
(ii) $\mathcal{H}$ *denotes the Cameron–Martin space associated with* X;
(iii) the covariance of X *is of finite* ρ*-variation dominated by some 2D control* ω, *for some* $\rho \in [1, 2)$;
(iv) $\mathbf{X}$ *denotes the natural lift of* X *to a* $G^{[2\rho]} (\mathbb{R}^d)$*-valued process (with geometric rough sample paths);*
(v) $V = (V_1, \dots, V_d)$ *is a collection of* Lip^γ*-vector fields on* $\mathbb{R}^e$, *with* $\gamma > 2\rho$;
(vi) $Y^\varepsilon = \pi_{(V)} (0, y_0; \delta_\varepsilon \mathbf{X})$ *is the RDE solution to*

$$ dY^\varepsilon = \varepsilon V (Y^\varepsilon) \, d\mathbf{X}, \, Y(0) = y_0 \in \mathbb{R}^e. $$

Then, for any $p > 2\rho$, $(Y^\varepsilon : \varepsilon > 0)$ *satisfies a large deviation principle in* p*-variation topology, with good rate given by*

$$ J(y) = \inf \left\{ \frac{1}{2} |h|_{\mathcal{H}}^2 : \pi_{(V)} (0, y_0; h) = y \right\} $$

where we agree that $|h|_{\mathcal{H}}^2 = +\infty$ *when* $h \notin \mathcal{H}$.

---

[3] The Polish space $\mathcal{D}_k (\mathbb{R}^e)$ was defined in Section 11.5.

If ω is Hölder-dominated, then the above large deviation principle also holds in $1/p$-Hölder topology.

Proposition 19.15 *Assume that*
(i) $X = \left(X^1, \ldots, X^d \right)$ is a Markov process with uniformly elliptic generator in divergence form, $\frac{1}{2} \sum_{i,j=1}^{d} \partial_i \left(a^{ij} \left(\cdot \right) \partial_j \cdot \right)$ understood in a weak sense (i.e. via Dirichlet forms) where $a \in \Xi^{1,d} \left(\Lambda \right)$, that is, measurable, symmetric and $\Lambda^{-1} I \leq a \leq \Lambda I$ in the sense of positive definite matrices;
(ii) $\mathbf{X}$ denotes the natural lift of X to a $G^2 \left(\mathbb{R}^d \right)$-valued process (with geometric rough sample paths);
(iii) $V = \left(V_1, \ldots, V_d \right)$ is a collection of Lip^2-vector fields on $\mathbb{R}^e$;
(iv) $Y^\varepsilon = \pi_{(V)} \left(0, y_0; \mathbf{X}^\varepsilon \right)$ is the RDE solution to

$$ dY^\varepsilon = V \left(Y^\varepsilon \right) d\mathbf{X}^\varepsilon, \, Y \left(0 \right) = y_0 \in \mathbb{R}^e $$

where $\mathbf{X}^\varepsilon \left(\cdot \right) \equiv \mathbf{X} \left(\varepsilon \cdot \right)$.
Then, $\left(Y^\varepsilon : \varepsilon > 0 \right)$ satisfies a large deviation principle in α-Hölder topology, for any $\alpha \in [0, 1/2)$, with good rate function

$$ J(y) = \inf \left\{ I^a \left(h \right) : \pi_{(V)} \left(0, y_0; h \right) = y \right\} $$

where

$$ I^a \left(h \right) = \frac{1}{2} \sup_{D \subset \mathcal{D}[0,T]} \sum_{i : t_i \in D} \frac{\left| d^a \left(h_{t_i}, h_{t_{i+1}} \right) \right|^2}{\left| t_{i+1} - t_i \right|} $$

and d^a is the intrinsic distance on $\mathbb{R}^d$ associated with a.

It is clear that Propositions 19.14 and 19.15 can also be formulated for stochastic flows or with random vector fields, with initial conditions along the lines of Theorems 19.13 and 19.13. Inclusion of drift-vector fields is a similarly straightforward matter.

19.5 Support theorem and large deviations for a class of SPDEs

Let us return to the study of SPDEs of the form

$$ \begin{aligned} du &= F \left(t, x, Du, D^2 u \right) dt + Du \left(t, x \right) \cdot V \left(x \right) \circ dB, \quad (19.5) \\ u \left(0, \cdot \right) &= u_0 \in \mathrm{BUC} \left(\mathbb{R}^n \right), \end{aligned} $$

understood as RPDE (cf. Sections 11.3 and 17.7),

$$ du = F \left(t, x, Du, D^2 u \right) dt + Du \left(t, x \right) \cdot V \left(x \right) d\mathbf{B} \left(\omega \right) $$

with (enhanced) Brownian noise signal $\mathbf{B} \left(\omega \right) = \exp \left(B, A \right)$. To this end we assume that $F = F \left(t, x, p, X \right) \in C \left([0, T], \mathbb{R}^n, \mathbb{R}^n, S^n \right)$ is a degenerate elliptic and $\Phi^{(3)}$-invariant comparison (as discussed in detail and with examples in Section 11.3). We also assume that $V = (V_1, \ldots, V_d) \subset \mathrm{Lip}^\gamma, \gamma > 4$. Under these assumptions we saw that $(u_0, \mathbf{B}) \mapsto u \in \mathrm{BUC}([0, T] \times \mathbb{R}^n)$ is continuous.[4] The following two theorems concerning large deviations and support descriptions for such SPDEs are then proved, mutatis mutandis, with the arguments that we have already used for SDEs. We have

Theorem 19.16 *(large deviations for stochastic partial differential equations) Let $U_\varepsilon = U_\varepsilon(t, x; \mathbf{B}(\omega))$ denote the (ω-wise unique BUC) solution to*

$$dU_\varepsilon = F\left(t, x, DU_\varepsilon, D^2 U_\varepsilon\right) dt + DU_\varepsilon(t, x) \cdot V(x) \circ \varepsilon dB, \quad (19.6)$$
$$u(0, \cdot) = u_0 \in \mathrm{BUC}(\mathbb{R}^n).$$

Then the family $(U_\varepsilon : \varepsilon > 0)$ of $\mathrm{BUC}([0, T] \times \mathbb{R}^n)$-valued random variables satisfies a large deviation principle with good rate function

$$J(v) = \inf_{h \in \mathcal{H}} \left\{ I(h) : v = u^h \right\}$$

where $\mathcal{H} = W_0^{1,2}([0, T], \mathbb{R}^d)$ and u^h is the unique $\mathrm{BUC}([0, T] \times \mathbb{R}^n)$-solution[5] to

$$du^h = F\left(t, x, Du^h, D^2 u^h\right) dt + Du^h(t, x) \cdot V(x) dh,$$
$$u^h(0, \cdot) = u_0 \in \mathrm{BUC}(\mathbb{R}^n).$$

Theorem 19.17 *(support theorem for stochastic partial differential equations) Let $U = u(t, x; \mathbf{B}(\omega))$ be the (ω-wise unique BUC) solution to*

$$du = F\left(t, x, Du, D^2 u\right) dt + Du(t, x) \cdot V(x) \circ dB, \quad (19.7)$$
$$u(0, \cdot) = u_0 \in \mathrm{BUC}(\mathbb{R}^n).$$

Then, for any $\delta > 0$,

$$\lim_{\varepsilon \to 0} \mathbb{P}\left(\left. |U - u^h|_{\infty;[0,T]} < \delta \right| \ |B - h|_{\infty;[0,T]} < \varepsilon \right) \to 1.$$

(The Euclidean norm is used for conditioning $|B - h|_{\infty;[0,T]} < \varepsilon$.) In particular, the topological support of the law of U, viewed as a Borel measure on $\mathrm{BUC}([0, T] \times \mathbb{R}^n)$, is precisely $\{u^h : h \in \mathcal{H}\}$, where the closure is with respect to locally uniform convergence.

---

[4] Unless otherwise stated, BUC spaces are equipped with the topology of locally uniform convergence.

[5] In viscosity sense, cf. Section 11.3.

19.6   Comments

The rough path approach to the Stroock–Varadhan support description
[161] (see Aida *et al.* [2], Ben Arous *et al.* [12], Millet and Sanz-Solé
[128]) and the Freidlin–Wentzell large deviation estimates (e.g. Dembo and
Zeitouni [41], Deuschel and Stroock [42] and the references therein) was
first carried out by Ledoux *et al.* [101], by establishing the relevant support
and large deviation properties for EBM (in p-variation rough path topology, $p > 2$). Our Theorem 19.1 is based on the (conditional) support result
for EBM in Hölder rough path topology, as obtained in Section 13.7 (which
itself follows Friz *et al.* [57], see comments at the end of Chapter 13). Theorem 19.5 on the support of stochastic flows is known (e.g. Kunita [96]),
although our (rough path) proof appears to be new. As seen in this chapter, the "rough path" pattern of proof for these support theorems extends
without changes to other (Gaussian, Markovian) driving signals, for which
support descriptions in rough path topology are available. Such results were
obtained in Sections 15.8 and 16.8, and we refer to the comments sections
in these chapters for pointers to the literature. Support theorems for (simple) differential equations driven by Gaussian processes have been used in
a financial context to construct markets without arbitrage under transaction costs (Guasoni [73]). It will also play a role when discussing RDEs
driven by Gaussian signals under Hörmander's condition, to be discussed
in Section 20.4.

Theorem 19.9 is a classical result in the theory of large deviations (e.g.
Baldi *et al.* [5], Dembo and Zeitouni [41], Deuschel and Stroock [42]); so
are large deviation results for anticipating SDEs (Millet *et al.* [127]) (with
a rough path proof, cf. Theorem 19.12, taken from Coutin *et al.* [30]) and
stochastic flows (Ben Arous and Castell [11]) (the rough path proof of
Theorem 19.13 is new).

Support theorems for classes of (linear) stochastic differential equations
appear in Gyöngy [79] and also in Kunita [96]. We are unaware of large
deviation results for SPDEs of the type in Theorem 19.16.

Malliavin calculus for RDEs

We consider stochastic differential equations driven by a d-dimensional Gaussian process in the rough path sense, cf. Section 17.3. Examples to have in mind include Brownian motion, the Ornstein–Uhlenbeck process, fractional Brownian motion with Hurst parameter $H > 1/4$ and various (Brownian or other Gaussian) bridge processes. Let us note that if the driving signal is also a semi-martingale (e.g. in the case of Brownian motion or the Ornstein–Uhlenbeck process), it follows from Theorem 17.3 that we actually work with classical stochastic differential equations in the Stratonovich sense.

The (driving) Gaussian process induces a Gaussian measure on $C\left([0,1],\mathbb{R}^d\right)$ and can be viewed as an abstract Wiener space, which serves as the underlying probability space on which the enhanced Gaussian process was constructed, cf. Section 15.3.3. Solving a rough differential equation thus yields an (abstract) Wiener functional and is, a priori, accessible to methods of Malliavin calculus. In particular, we shall see in this chapter that, subject to certain non-degeneracy conditions, solutions to stochastic differential equations driven by Gaussian processes in the rough path sense admit a density with respect to the Lebesgue measure.

20.1 $\mathcal{H}$-regularity of RDE solutions

We assume $X = \left(X^1, \ldots, X^d\right)$ is a centred continuous Gaussian process on $[0,T]$ with independent components. The associated Cameron–Martin space is denoted by

$$\mathcal{H} \subset C\left([0,T],\mathbb{R}^d\right).$$

Recall from Proposition 15.7 that $\mathcal{H} \hookrightarrow C^{\rho\text{-var}}\left([0,T],\mathbb{R}^d\right)$ if we assume that covariance of X is of finite ρ-variation in 2D sense. Let us also recall that $\rho < 2$ is a sufficient (and essentially sharp) condition for X to admit a natural enhancement $\mathbf{X}$ to a geometric p-rough path, $p \in (2\rho, 4)$. It is important to understand perturbations of $\mathbf{X}$ in Cameron–Martin directions. More specifically, having realized X as the coordinate process on the path space, $X_t(\omega) = \omega_t$, we want to understand $\mathbf{X}(\omega + h)$. It is clear from the Cameron–Martin theorem that, for every fixed $h \in \mathcal{H}$, $\omega \mapsto \mathbf{X}(\omega + h)$ is a well-defined Wiener functional. On the other hand, the formal computation

$$\int (X+h) \otimes d(X+h) = \int X \otimes dX + \int h \otimes dX + \int X \otimes dh + \int h \otimes dh$$

suggests that $\mathbf{X}(\omega + h)$ can be expressed in terms of $\mathbf{X}(\omega)$ and cross-integrals of X and h. Under the standing assumption that $\rho < 2$, the last integral $\int h \otimes dh$ is obviously a well-defined Young integral. On the other hand, the integral

$$\int h \otimes dX \tag{20.1}$$

may not be a (pathwise defined) Young integral since $1/\rho + 1/p \not> 1$, in general when $\rho < 2$ and $p > 2\rho$. The following condition, already encountered in Section 15.8, is designed to ensure that (20.1) does make sense as a Young integral.

Condition 20.1 Let $X = (X^1, \ldots, X^d)$ be a centred continuous Gaussian process on $[0, T]$ with independent components which admits a natural lift in the sense of Section 15.3.3 to a (random) geometric p-rough path $\mathbf{X}$. We assume that $\mathcal{H}$ **has complementary Young regularity to** X, by which we mean that

$$\mathcal{H} \hookrightarrow C^{q\text{-var}}\left([0, T], \mathbb{R}^d\right)$$

for some $q \geq 1$ with $1/p + 1/q > 1$. □

For instance, Condition 20.1 is satisfied if the covariance of X has finite ρ-variation for some $\rho < 3/2$. This covers in particular Brownian motion (where we can take $p = 2 + \varepsilon$ and $q = 1$) and fractional Brownian motion with $H > 1/3$. In fact, thanks to a certain Besov regularity of $\mathcal{H}^H$, the Cameron–Martin space associated with fractional Brownian motion, we can also cover the regime $H \in (1/4, 1/3]$, despite the fact (cf. Proposition 15.5) that $H \in (1/3, 1/4)$ corresponds to $\rho \in (3/2, 2)$. Part (i) of Exercise 20.2 below gives a hint of what happens in the case $\rho \in [3/2, 2)$.

Exercise 20.2 *Throughout this exercise, assume X is a centred continuous Gaussian process on $[0, T]$ with independent components, with covariance of finite ρ-variation in 2D sense.*
(i) Assuming $\rho \in [1, 2)$, show that (20.1) makes sense as a Young–Wiener integral in the sense of Proposition 15.39.
(ii) Assuming $\rho \in [1, 3/2)$, show that Condition 20.1 is satisfied (in particular, (20.1) makes sense as a classical Young integral).
(iii) Consider d-dimensional fractional Brownian motion with Hurst parameter H. Using the fact (cf. Remark 15.10) that

$$\mathcal{H}^H \hookrightarrow C^{q\text{-var}} \text{ for any } q > (H + 1/2)^{-1}$$

show that, for any $H > 1/4$, Condition 20.1 is satisfied.

Exercise 20.3 *Let $h \in C^{\rho\text{-var}}([0, T], \mathbb{R})$ and X be a real-valued, centred continuous Gaussian process with covariance $R = R(s, t)$ of finite ρ-variation in 2D sense. Assume $\rho \in [1, 2)$ and show that the Young–Wiener* *integral*

$$\int_0^T h\,dX$$

is a Gaussian random variable with zero-mean and variance given by the 2D Young integral

$$\int_0^T \int_0^T h\left(s\right) h\left(t\right) dR\left(s,t\right).$$

A convenient consequence of Condition 20.1 is the possibility of considering $\mathbf{X}\left(\omega + h\right)$ simultaneously for all $h \in \mathcal{H}$; in contrast to the general case, where $\omega \mapsto \mathbf{X}\left(\omega + h\right)$ is only definent up to h-dependent null-sets. For the reader's convenience, we now recall Lemma 15.58.

Lemma 20.4 *Assume X is a Gaussian process which satisfies Condition 20.1. Then, writing $\mathbf{X}$ for the natural lift of X, the event*

$$\{\omega : \mathbf{X}\left(\omega + h\right) \equiv T_h \mathbf{X}\left(\omega\right) \text{ for all } h \in \mathcal{H}\} \tag{20.2}$$

has probability one.

Proposition 20.5 *Assume $X = \left(X^1, \ldots, X^d\right)$ is a Gaussian process which satisfies Condition 20.1 and write $\mathbf{X}$ for its natural lift, a (random) geometric p-rough path $\mathbf{X}$. Assume $V = \left(V_1, \ldots, V_d\right) \subset \operatorname{Lip}^\gamma\left(\mathbb{R}^e\right)$ with $\gamma > p$. Then, the (unique) RDE solution to*

$$dY = V\left(Y\right) d\mathbf{X}, \ Y_0 = y_0 \in \mathbb{R}^e$$

is almost surely continuously $\mathcal{H}$-differentiable; i.e. for almost every ω, the map

$$h \in \mathcal{H} \mapsto \pi_{(V)}\left(0, \cdot; \mathbf{X}\left(\omega + h\right)\right) \in C^{p\text{-var}}\left(\left[0,T\right], \mathbb{R}^e\right)$$

is continuously differentiable in Fréchet sense. In particular, the $\mathbb{R}^e$-valued Wiener functional

$$Y_t = Y_t\left(\omega\right) = \pi_{(V)}\left(0, y_0; \mathbf{X}\left(\omega\right)\right)_t$$

admits an $\mathcal{H}$-valued derivative $DY_t = DY_t\left(\omega\right)$ with the property that, with probability one,

$$\forall h \in \mathcal{H} : D_h Y_t := \langle DY_t, h \rangle_{\mathcal{H}} = \int_0^t \sum_{j=1}^d J^{\mathbf{X}}_{t \leftarrow s}\left(V_j\left(Y_s\right)\right) dh_s^j \tag{20.3}$$

where $J^{\mathbf{X}}_{t \leftarrow s} = J^{\mathbf{X}}_{t \leftarrow s}\left(\omega\right)$ denotes the Jacobian of $\pi_{(V)}\left(s, \cdot; \mathbf{X}\left(\omega\right)\right)_t$. (The integral above makes sense as a Young integral since the integrand has finite p-variation regularity.)

Proof. By assumption, $\mathcal{H} \hookrightarrow C^{q\text{-var}}\left([0,T],\mathbb{R}^d\right)$, with $1/q + 1/p > 1$, the embedding is (trivially) Fréchet smooth. On the other hand, for any ω in a set of full measure on which (20.2) holds we have

$$\pi_{(V)}\left(0, y_0; T_h\mathbf{X}\left(\omega\right)\right) = \pi_{(V)}\left(0, y_0; \mathbf{X}\left(\omega + h\right)\right).$$

Using Fréchet regularity of the Itô map, as detailed in Section 11.1.2, we see that

$$h \in C^{q\text{-var}} \mapsto \pi_{(V)}\left(0, y_0; T_h\mathbf{X}\left(\omega\right)\right) \in C^{p\text{-var}}\left([0,T],\mathbb{R}^e\right),$$

and hence also

$$h \in \mathcal{H} \mapsto \pi_{(V)}\left(0, y_0; \mathbf{X}\left(\omega + h\right)\right) \in C^{p\text{-var}}\left([0,T],\mathbb{R}^e\right),$$

must be continuously $\mathcal{H}$-differentiable. At last, time-t evaluation on $C^{p\text{-var}}$ $\left([0,T],\mathbb{R}^e\right)$ is trivially Fréchet smooth, so that

$$h \in \mathcal{H} \mapsto \pi_{(V)}\left(0, y_0; \mathbf{X}\left(\omega + h\right)\right)_t \in \mathbb{R}^e$$

is also continuously $\mathcal{H}$-differentiable. The representation (20.3) then follows from the fact that, with probability one,

$$\forall h \in \mathcal{H} : \pi_{(V)}\left(s, \cdot; \mathbf{X}\left(\omega + h\right)\right)_t = \pi_{(V)}\left(s, \cdot; T_h\mathbf{X}\left(\omega\right)\right)_t$$

and Duhamel's principle holds, as discussed in Exercise 11.9. The proof is now finished.   ∎

Exercise 20.6 *What will be needed in the sequel is the conclusion from Proposition 20.5 that Y_t is a.s. continuously $\mathcal{H}$-differentiable with explicit representation given in (20.3). One can arrive at this conclusion without relying on the Fréchet smoothness results of Section 11.1.2 but only relying on some knowledge about directional derivatives. To this end, recall from Section 11.1.1 that, given a geometric p-rough path $\mathbf{x}$, $q \geq 1 : 1/p + 1/q > 1$ and $V \subset \text{Lip}^\gamma\left(\mathbb{R}^e\right), \gamma > p$,*

$$h \in C^{q\text{-var}}\left([0,T],\mathbb{R}^d\right) \mapsto \pi_{(V)}\left(0, y_0; T_h\mathbf{x}\right)_t \in \mathbb{R}^e$$

admits directional derivatives $D_hY_t := \left\{\frac{d}{d\varepsilon}\pi_{(V)}\left(0, y_0; T_{\varepsilon h}\mathbf{x}\right)_t\right\}_{\varepsilon=0}$ with the representation formula

$$D_hY_t = \int_0^t \sum_{i=1}^d J_{t \leftarrow s}\left(V_i\left(Y_s\right)\right) dh_s^i,$$

where $J_{t \leftarrow s}$ is the Jacobian of $\pi_{(V)}\left(s, \cdot; \mathbf{x}\right)_t$.
(i) Use this representation formula to deduce the existence of an $\mathcal{H}$-valued derivative DY so that $D_hY_t = \langle DY, h\rangle_{\mathcal{H}}$.
(ii) Use continuity of the Itô–Lyons map to show that Y_t is a.s. continuously $\mathcal{H}$-differentiable.

20.2 Non-degenerate Gaussian driving signals

We remain in the framework of the previous section. In particular, $X = \left(X^1, \ldots, X^d\right)$ is again a centred continuous Gaussian process on $[0,T]$ with independent components which admits a lift $\mathbf{X}$ to a (random) geometric p-rough path. The overall aim of this chapter is to find sufficient criteria on the process X *and* vector fields $V = \left(V_1, \ldots, V_d\right)$ so that, for every $t \in (0,T]$, the $\mathbb{R}^e$-valued random variable

$$\pi_{(V)}\left(0, y_0; \mathbf{X}\left(\omega\right)\right)_t$$

admits a density with respect to the Lebesgue measure on $\mathbb{R}^e$. To this end, Condition 20.1 on the regularity of the Cameron–Martin space, namely

$$\mathcal{H} \hookrightarrow C^{q\text{-var}}\left([0,T], \mathbb{R}^d\right), \ 1/p + 1/q > 1,$$

will be in force throughout. As a simple consequence, thanks to Young's inequality, we have

$$C^{p\text{-var}}\left([0,T], \mathbb{R}^d\right) \hookrightarrow \mathcal{H}^* \cong \mathcal{H} \hookrightarrow C^{q\text{-var}}\left([0,T], \mathbb{R}^d\right),$$

where every $f \in C^{p\text{-var}}\left([0,T], \mathbb{R}^d\right)$ is viewed as an element in $\mathcal{H}^*$ via

$$h \in \mathcal{H} \mapsto \int_0^T f dh \equiv \sum_{k=1}^d \int_0^T f_k dh^k.$$

Condition 20.7 We assume **non-degeneracy of the Gaussian process** X **on** $[0,T]$ in the sense that for any $f \in C^{p\text{-var}}\left([0,T], \mathbb{R}^d\right)$,

$$\left(\int_0^T f dh = 0 \forall h \in \mathcal{H}\right) \implies f \equiv 0.$$

(Note that non-degeneracy on $[0,T]$ implies non-degeneracy on $[0,t]$ for any $t \in (0,T]$.) □

It is instructive to see how this condition rules out the Brownian bridge returning to the origin at time T or earlier; a Brownian bridge which returns to zero after time T is allowed. The Ornstein–Uhlenbeck process is another example for which Condition 20.7 is satisfied; and so is fractional Brownian motion for any value of its Hurst parameter H, simply because

$$C_0^\infty\left([0,T], \mathbb{R}^d\right) \subset \mathcal{H}^H,$$

as was pointed out in (15.5), Remark 15.10, and thanks to

Exercise 20.8 *Show that Condition 20.7 is satisfied if every smooth path $(h_t : t \in [0,T])$, possibly pinned at zero, is contained in $\mathcal{H}$.*

Solution. *We see that f is orthogonal to any $\dot{h} \in C^\infty\left([0,T],\mathbb{R}^d\right)$, hence must be zero in $L^2\left([0,T],\mathbb{R}^d\right)$ and thus is identically equal to zero by continuity).* □

Lemma 20.9 *The requirement that $\int_0^T f dh = 0 \forall h \in \mathcal{H}$ in Condition 20.7 can be relaxed to the quantifier "for all h in some orthonormal basis of $\mathcal{H}$".*

Proof. It suffices to check that

$$\left\{ \int_0^T f dh = 0 \forall h \in \mathcal{H} \right\} \Leftrightarrow \left\{ \int_0^T f dh_k = 0 \forall k \in \mathbb{N} \right\}$$

where (h_k) is an orthonormal basis for $\mathcal{H}$. Only the "$\Leftarrow$" direction is not-trivial. Assuming $\int_0^T f dh_k = 0$ for all k implies

$$\int_0^T f dh^{[n]} = 0 \text{ for all } n,$$

where $h^{[n]} \equiv \sum_{k=1}^n \langle h_k, h \rangle h_k$ is the Fourier expansion of h. It obviously converges in $\mathcal{H}$ (and hence also in $C^{q\text{-var}}$) to h and we conclude by continuity of the Young integral. ∎

20.3 Densities for RDEs under ellipticity conditions

We have the following density result for RDEs driven by a Gaussian rough path **X**.

Theorem 20.10 *Let $X = \left(X^1, \ldots, X^d\right)$ be a centred continuous Gaussian process on $[0,T]$ with independent components which admits a natural lift in the sense of Section 15.3.3 to a (random) geometric p-rough path **X**. Assume that*
(i) $\mathcal{H}$ has complementary Young regularity to X, i.e. $\mathcal{H} \hookrightarrow C^{q\text{-var}}([0,T], \mathbb{R}^d)$ with $1/q + 1/p > 1$;
(ii) X is non-degenerate in the sense of Condition 20.7;
(iii) $y_0 \in \mathbb{R}^e$ is a fixed "starting" point;
(iv) there are vector fields $V = (V_1, \ldots, V_d) \subset \text{Lip}^\gamma(\mathbb{R}^e)$, $\gamma > p$, which satisfy the following ellipticity condition at the starting point,

$$(E) : \text{span}\{V_1, \ldots, V_d\}\big|_{y_0} = \mathcal{T}_{y_0}\mathbb{R}^e \cong \mathbb{R}^e.$$

Then, for every $t \in (0, T]$, the $\mathbb{R}^e$-valued RDE solution

$$\pi_{(V)}\left(0, y_0; \mathbf{X}\left(\omega\right)\right)_t$$

admits a density with respect to the Lebesgue measure on $\mathbb{R}^e$.

Proof. Fix $t \in (0, T]$. From Proposition 20.5 we know that the $\mathbb{R}^e$-valued Wiener functional

$$Y_t\left(\omega\right) := \pi_{(V)}\left(0, y_0; \mathbf{X}\left(\omega\right)\right)_t$$

is a.s. continuously $\mathcal{H}$-differentiable. By a well-known criterion due to Bouleau–Hirsch, cf. Section D.5 of Appendix D, all we have to do is show that the so-called Malliavin covariance matrix

$$\sigma_t\left(\omega\right) := \left(\left\langle DY_t^i, DY_t^j \right\rangle_{\mathcal{H}}\right)_{i,j=1,\ldots,e} \in \mathbb{R}^{e \times e}$$

is invertible with probability one. To see this, assume there exists a (random) vector $v \in \mathbb{R}^e$ which annihilates the quadratic form σ_t. Then[1]

$$0 = v^T \sigma_t v = \left|\sum_{i=1}^e v_i DY_t^i\right|_{\mathcal{H}}^2 \quad \text{and so } v^T DY_t \equiv \sum_{i=1}^e v_i DY_t^i \in 0 \in \mathcal{H}.$$

Using the representation formula (20.3) this says precisely that

$$\forall h \in \mathcal{H} : v^T D_h Y_t = \int_0^t \sum_{j=1}^d v^T J_{t \leftarrow s}^{\mathbf{X}}\left(V_j\left(Y_s\right)\right) dh_s^j = 0, \tag{20.4}$$

where the last integral makes sense as a Young integral since the (continuous) integrand has finite p-variation regularity. Noting that the non-degeneracy condition on $[0, T]$ implies the same non-degeneracy condition on $[0, t]$, we see that the integrand in (20.4) must be zero on $[0, t]$ and evaluation at time 0 shows that for all $j = 1, \ldots, d$,

$$v^T J_{t \leftarrow 0}^{\mathbf{X}}\left(V_j\left(y_0\right)\right) = 0.$$

It follows that the vector $v^T J_{t \leftarrow 0}^{\mathbf{X}}$ is orthogonal to $V_j\left(y_0\right)$, $j = 1, \ldots, d$ and hence zero. Since $J_{t \leftarrow 0}^{\mathbf{X}}$ is invertible, this follows immediately from the chain rule and

$$\pi_{(V)}\left(0, \pi_{(V)}\left(0, \cdot; \mathbf{x}\right)_t; \overleftarrow{\mathbf{x}}\right)_t = \mathrm{Id}|_{\mathbb{R}^e} \text{ where } \overleftarrow{\mathbf{x}}\left(\cdot\right) = \mathbf{x}\left(t - \cdot\right),$$

we see that $v = 0$. The proof is finished.   ∎

---

[1] Upper T denotes the transpose of a vector or matrix.

Exercise 20.11 *Let $\sigma_t(\omega) := \left(\left\langle DY_t^i, DY_t^j \right\rangle_{\mathcal{H}} \right)_{i,j=1,\ldots,e} \in \mathbb{R}^{e \times e}$ denote the Malliavin covariance matrix of the RDE solution $Y_t \equiv \pi_{(V)}(0, y_0; \mathbf{X}(\omega))$, where $\mathbf{X}(\omega)$ is the lift of some Gaussian process $\left(X^1, \ldots, X^d \right)$ with covariance of finite ρ-variation for $\rho \in [1, 3/2)$. Show that*

$$\sigma_t = \sum_{k=1}^d \int_0^t \int_0^t J_{t \leftarrow s}^{\mathbf{X}}(V_k(Y_s)) \otimes J_{t \leftarrow s'}^{\mathbf{X}}(V_k(Y_{s'})) \, dR_{X^k}(s, s')$$

where the right-hand side is a well-defined 2D Young integral. Let $p \in (2\rho, 3)$ and show that $\sigma_t(\omega) = \hat{\sigma}(\mathbf{X}(\omega))$, where $\hat{\sigma}$ is a continuous map from $C^{p\text{-var}}\left([0, T], G^2\left(\mathbb{R}^d\right)\right)$ to $\mathbb{R}^{e \times e}$.

Solution. Let $\left(h_n^{(k)} : n \right)$ be an orthonormal basis of $\mathcal{H}^{(k)}$, the Cameron–Martin space associated with X^k. It follows that $\left(h_n^{(k)} : n=1, 2, \ldots; k = 1, \ldots, d \right)$ is an orthonormal basis of $\mathcal{H} = \oplus_{i=1}^d \mathcal{H}^{(k)}$ where we identify

$$h_n^{(1)} \in \mathcal{H}^{(1)} \equiv \begin{pmatrix} h_n^{(1)} \\ 0 \\ \ldots \\ 0 \end{pmatrix} \in \mathcal{H}$$

and similarly for $k = 2, \ldots, d$. From Parseval's identity,

$$
\begin{aligned}
\sigma_t &= \left(\left\langle DY_t^i, DY_t^j \right\rangle_{\mathcal{H}} \right)_{i,j=1,\ldots,e} \\
&= \sum_{n,k} \left\langle DY_t, h_n^{(k)} \right\rangle_{\mathcal{H}} \otimes \left\langle DY_t, h_n^{(k)} \right\rangle_{\mathcal{H}} \\
&= \sum_k \sum_n \int_0^t J_{t \leftarrow s}^{\mathbf{X}}(V_k(Y_s)) \, dh_{n,s}^{(k)} \otimes \int_0^t J_{t \leftarrow s}^{\mathbf{X}}(V_k(Y_s)) \, dh_{n,s}^{(k)} \\
&= \sum_k \int_0^t \int_0^t J_{t \leftarrow s}^{\mathbf{X}}(V_k(Y_s)) \otimes J_{t \leftarrow s'}^{\mathbf{X}}(V_k(Y_{s'})) \, dR^{(k)}(s, s').
\end{aligned}
$$

For the last step we used the fact that

$$\sum_n \int_0^T f \, dh_n \int_0^T g \, dh_n = \int_0^T \int_0^T f(s) g(t) \, dR(s, t)$$

whenever $f = f(t, \omega)$ and g are such that the integrals are a.s. well-defined Young integrals. We then conclude with $R(s, t) = \mathbb{E}(X_s X_t)$ and the L^2-expansion of the Gaussian process X,

$$X(t) = \sum_n \xi(h_n) h_n(t)$$

where the $\xi(h_n)$ form an IID family of standard Gaussians. $\qquad \square$

20.4 Densities for RDEs under Hörmander's condition

In the case of driving Brownian motion, it is well known that solutions to SDEs of the form

$$dY = \sum_{i=1}^{d} V_i\left(Y\right) dB^i,$$

started at $y_0 \in \mathbb{R}^e$, admit a (smooth) density provided the vector fields, now assusmed to be C^∞-bounded, satisfy *Hörmander's condition* at the starting point. By this we mean that the linear span of $\{V_1, \ldots, V_d\}$ *and all iterated Lie brackets* at y_0 have full span. (There is a well-known extension to SDEs with drift-vector field V_0 in which case (H) is replaced by the condition of full span by $\{V_1, \ldots, V_d\}$ *and all iterated Lie brackets* of $\{V_0, V_1, \ldots, V_d\}$ at y_0.) The aim of this section is to establish a similar density result for RDEs driven by a Gaussian rough path. Our focus will be the drift-free case, i.e. when $V_0 \equiv 0$. Also, the conditions on the underlying Gaussian driving signals are somewhat more involved than those required in Theorem 20.10, but still remain checkable for many familiar Gaussian processes. We have

Theorem 20.12 Let $\left(X_t^1, \ldots, X_t^d\right) = \left(X_t : t \in [0,T]\right)$ be a continuous, centred Gaussian process with independent components $X^1, \ldots, X^d$. Assume X satisfies the conditions listed in Section 20.4.1 below. (In particular, X is assumed to admit a natural lift $\mathbf{X}$ to a random geometric rough path.) Let $V = (V_1, \ldots, V_d)$ be a collection of C^∞-bounded vector fields on $\mathbb{R}^e$ which satisfies Hörmander's condition

$$(H): \mathrm{Lie}\left[V_1, \ldots, V_d\right]\big|_{y_0} = \mathcal{T}_{y_0}\mathbb{R}^e \cong \mathbb{R}^e$$

at some point $y_0 \in \mathbb{R}^e$. Then the random RDE solution

$$Y_t\left(\omega\right) = \pi_{(V)}\left(0, y_0; \mathbf{X}\left(\omega\right)\right)_t$$

admits a density with respect to the Lebesgue measure on $\mathbb{R}^e$ for all times $t \in (0,T]$.

Note that when $(X_t : t \in [0,T])$ happens to be a semi-martingale (such as a Brownian motion, an Ornstein–Uhlenbeck process, a Brownian bridge returning to the origin after time T, etc.), Theorem 20.12 really yields information about classical solutions to the Stratonovich SDE

$$dY = \sum_{i=1}^{d} V_i\left(Y\right) \circ dB^i, \quad Y_0 = y_0 \in \mathbb{R}^e.$$

20.4.1 Conditions on the Gaussian process

We assume that $X = (X^1, \ldots, X^d)$ is a centred continuous Gaussian process on $[0, T]$ with independent components which admits a natural lift in the sense of Section 15.3.3. Recall that this requires

$$\exists \rho \in [1, 2) : |R|_{\rho\text{-var};[0,1]^2} < \infty$$

where R denotes the covariance function of X. Equivalently, setting $H = 1/(2\rho)$, this condition may be stated as

$$\exists H \in (1/4, 1/2] : |R|_{(1/2H)\text{-var};[0,1]^2} < \infty.$$

Given such a process X one can, cf. Exercise 15.36, find a determinstic time-change $\tau : [0, T] \to [0, T]$ such that $\tilde{R}$, the covariance function of $\tilde{X} = X \circ \tau$, satisfies

$$\forall s < t \text{ in } [0, T] : \left|\tilde{R}\right|_{(1/2H)\text{-var};[s,t]^2} < (\text{const}) \times |t - s|^{2H}. \qquad (20.5)$$

Since the conclusion of Theorem 20.12 is invariant under such a time-change we can in fact assume, without loss of generality, that the covariance of X itself has Hölder-dominated $1/(2H)$-variation in the sense of (20.5). It now follows from Theorem 15.33 that the natural lift $\mathbf{X}$ has $1/p$-Hölder sample paths for any $p > 1/H$ and also that

$$\sup_{0 \le s < t \le T} \mathbb{E}\left(\exp\left(\eta \left[\frac{d(\mathbf{X}_s, \mathbf{X}_t)}{|t - s|^H}\right]^2\right)\right) < \infty.$$

Although the parameter H is reminiscent of fractional Brownian motion with Hurst parameter H we insist that, up to this point, our assumption covers every enhanced Gaussian process (up to irrelevant deterministic time-change).

Condition 20.13 $\mathcal{H}$ has complementary Young regularity to X, i.e. $\mathcal{H} \hookrightarrow C^{q\text{-var}}([0, T], \mathbb{R}^d)$ with $1/q + 1/p > 1$. $\qquad \square$

Condition 20.14 X is non-degenerate on $[0, T]$; recall that this means that for any $f \in C^{p\text{-var}}([0, T], \mathbb{R}^d)$,

$$\left(\int_0^T f \, dh \equiv \sum_{k=1}^d \int_0^T f_k \, dh^k = 0 \forall h \in \mathcal{H}\right) \implies f \equiv 0.$$

$\qquad \square$

Condition 20.15 X obeys a **Blumenthal zero–one law** in the sense that the germ σ-algebra $\cap_{t>0}\sigma(X_s : s \in [0, t])$ contains only events of probability zero or one. $\qquad \square$

Condition 20.16 Let $\mathbf{X}$ denote the natural lift of X and assume that for all $N \geq [p]$, the step-N Lyons lift of $\mathbf{X}$ has H-**rescaled full support in the small time limit**, by which we mean that for all $g \in G^N\left(\mathbb{R}^d\right)$ and for all $\varepsilon > 0$,

$$\liminf_{t \to 0} \mathbb{P}\left(d\left(\delta_{t^{-H}} S_N\left(\mathbf{X}\right)_{0,t}, g\right) < \varepsilon\right) > 0.$$ □

Some remarks are in order.

- Conditions 20.13 and 20.14 were already in force in our "elliptic" discussion and are just repeated for the sake of completeness.

- Condition 20.15 holds whenever X can be written as an adapted functional of Brownian motion. This includes fractional Brownian motion and more generally all examples in which X has a so-called Volterra presentation[2] of the form

$$X_t = \int_0^t K\left(t, s\right) dB_s \quad \text{(Itô integral)}.$$

It also includes (non-Volterra) examples in which $(X_t : t \in [0, T])$ is given as a strong solution of an SDE driven by Brownian motion, such as a Brownian bridge returning to the origin after time T, say. An example where the 0–1 law fails is given by the *random-ray* $X : t \mapsto tB_T\left(\omega\right)$, in which case the germ-event $\{\omega : dX_t\left(\omega\right)/dt|_{t=0+} \geq 0\}$ has probability $1/2$. (In fact, sample path differentiability at $0+$ implies non-triviality of the germ σ-algebra; see [46] and the references cited therein.) We observe that the random-ray example (a) is already ruled out by Condition 20.14 and (b) should be ruled out anyway since it does not trigger the bracket phenomenon needed for a Hörmander statement.

- Condition 20.16 says – in essence – that the driving signal must have, at least approximately and for small times, a fractional scaling behaviour similar to fractional Brownian motion with Hurst parameter H. As will be seen in the following proposition (see also Exercise 20.18), this condition can be elegantly verifed via the *support theorem* obtained in Section 15.8.

Proposition 20.17 *Let* B^H *denote d-dimensional fractional Brownian motion with fixed Hurst parameter* $H \in (1/4, 1/2]$ *and consider the lift to a (random) geometric p-rough path, denoted by* $\mathbf{X} = \mathbf{B}^H$, *with* $p \in (2, 4)$. *Then it satisfies Condition 20.16.*

---

[2] See [40].

Proof. Let us observe that B^H (and then $\mathbf{B}^H$) has finite $1/p$-Hölder sample paths for any $p > 1/H$. To keep the notation simple, we shall write $B, \mathbf{B}$ rather than $B^H, \mathbf{B}^H$ and also set $\tilde{\mathbf{B}} = S_N(\mathbf{B})$. From the support Theorem for Gaussian rough paths, Theorem 15.60, we know that the support of the law of $\mathbf{B}$ in p-variation topology is precisely

$$\overline{S_{[p]}(\mathcal{H})} = C_0^{0,p\text{-var}}\left([0,T], G^{[p]}(\mathbb{R}^d)\right),$$

using, for instance, the fact that $C_0^\infty([0,T],\mathbb{R}^d) \subset \mathcal{H}$. By continuity of the Lyons lift $S_N : \mathbf{B} \mapsto \tilde{\mathbf{B}}$, followed by evaluation of the path at time 1, it is clear that $\tilde{\mathbf{B}}_1$ has full support; that is,

$$\forall g \in G^N(\mathbb{R}^d), \varepsilon > 0 : \mathbb{P}\left(d\left(\tilde{\mathbf{B}}_1, g\right) < \varepsilon\right) > 0.$$

On the other hand, fractional scaling $\left(n^H B_{t/n} : t \geq 0\right) \overset{\mathcal{D}}{=} (B_t : t \geq 0)$ implies $\delta_{n^H}\tilde{\mathbf{B}}_{1/n} \overset{\mathcal{D}}{=} \tilde{\mathbf{B}}_1$ and so, thanks to full support of $\tilde{\mathbf{B}}_1$,

$$\liminf_{n \to \infty}\mathbb{P}\left(d\left(\delta_{n^H}\tilde{\mathbf{B}}_{1/n}, g\right) < \varepsilon\right) = \mathbb{P}\left(d\left(\tilde{\mathbf{B}}_1, g\right) < \varepsilon\right) > 0.$$

∎

Exercise 20.18 *Show that Condition 20.16 holds for any centred, continuous Gaussian process X which is assumed to be **asymptotically comparable to B^H in the small time limit** in the following sense:*
(i) there exists a probability space,[3] such that X and fractional Brownian motion can be realized jointly as a $(2d)$-dimensional Gaussian process

$$\left(X, B^H\right) = \left(X^1, B^{H;1}, \ldots, X^d, B^{H;d}\right),$$

with $\left(X^i, B^{H;i}\right)$ independent for $i = 1, \ldots, d$;
(ii) the $(2d)$-dimensional Gaussian process $\left(X, B^H\right)$ has covariance of finite $(1/2H)$-variation in the 2D sense;
(iii) we have

$$t^{-2H}\left|R_{X-B^H}\right|_{\infty;[0,t]^2} \to 0 \ \text{as } t \to 0 \tag{20.6}$$

where R_{X-B^H} is the covariance of the $\mathbb{R}^d$-valued Gaussian process $\left(X - B^H\right)$.

Proof. To keep the notation simple, we shall write again $B, \mathbf{B}$ rather than $B^H, \mathbf{B}^H$. Let us also set $1/n = t$. By independence of the pairs $\left(X^i, B^i\right)$, the covariance matrix R_{X-B^H} is diagonal and we focus on one entry. With mild abuse of notation (writing X, B instead of $X^i, B^i \ldots$) we have

$$n^{-2H}\left|R_{X-B}\right|_{\infty;[0,1/n]^2}$$
$$= \sup_{s,t\in[0,1]} \mathbb{E}[n^H\left(X_{s/n} - B_{s/n}\right)n^H\left(X_{t/n} - B_{t/n}\right)]$$

---

[3]Effectively, a Gaussian measure on $C\left([0,T],\mathbb{R}^{2d}\right)$.

which can be rewritten in terms of the rescaled process $X^{(n)} = n^H X_{./n}$, and similarly for B, as

$$\sup_{s,t \in [0,1]} \mathbb{E}\left[\left(X_s^{(n)} - B_s^{(n)}\right)\left(X_t^{(n)} - B_t^{(n)}\right)\right] = \left|R_{X^{(n)} - B^{(n)}}\right|_{\infty;[0,1]^2}.$$

By assumption, in particular (20.6) and continuity estimates for Gaussian rough paths obtained in Theorem 15.37, we see that

$$d_{p\text{-var}}\left(\mathbf{X}^{(n)}, \mathbf{B}^{(n)}\right) \to 0 \text{ as } n \to \infty \text{ in probability.}$$

By continuity of S_N, still writing $\tilde{\mathbf{X}}^{(n)} = S_N\left(\mathbf{X}^{(n)}\right)$ for fixed N, and similarly for $\mathbf{B}^{(n)}$, we have

$$d\left(\tilde{\mathbf{X}}_1^{(n)}, \tilde{\mathbf{B}}_1^{(n)}\right) \leq d_{p\text{-var};[0,1]}\left(\tilde{\mathbf{X}}^{(n)}, \tilde{\mathbf{B}}^{(n)}\right) \to 0 \text{ in probability.}$$

But then

$$\mathbb{P}\left(d\left(\delta_{n^H}\tilde{\mathbf{X}}_{1/n}, g\right) < \varepsilon\right)$$
$$= \mathbb{P}\left(d\left(\tilde{\mathbf{X}}_1^{(n)}, g\right) < \varepsilon\right)$$
$$\geq \mathbb{P}\left(d\left(\tilde{\mathbf{X}}_1^{(n)}, \tilde{\mathbf{B}}_1^{(n)}\right) + d\left(\tilde{\mathbf{B}}_1^{(n)}, g\right) < \varepsilon\right)$$
$$\geq \mathbb{P}\left(d\left(\tilde{\mathbf{B}}_1^{(n)}, g\right) < \varepsilon/2\right)$$
$$- \mathbb{P}\left(d\left(\tilde{\mathbf{X}}_1^{(n)}, \tilde{\mathbf{B}}_1^{(n)}\right) > \varepsilon/2\right)$$

and so

$$\liminf_{n \to \infty} \mathbb{P}\left(d\left(\delta_{n^H}\tilde{\mathbf{X}}_{1/n}, g\right) < \varepsilon\right) \geq \liminf_{n \to \infty} \mathbb{P}\left(d\left(\tilde{\mathbf{B}}_1^{(n)}, g\right) < \varepsilon/2\right)$$

and this is positive thanks to Proposition 20.17. ∎

Exercise 20.19 *Show that Theorem 20.12 applies to the following multidimensional Gaussian driving processes:*
(i) Brownian motion B;
(ii) fractional Brownian motion B^H with Hurst parameter, $H > 1/4$;
(iii) the Ornstein–Uhlenbeck process, realized (for instance) by Wiener–Itô integration,

$$X_t^i = \int_0^t e^{-(t-r)} dB_r^i \text{ with } i = 1, \ldots, d;$$

(iv) a Brownian bridge returning to zero after time T, e.g. $X^{T+\varepsilon}$ with $\varepsilon > 0$ where

$$X_t^{T+\varepsilon} := B_t - \frac{t}{T+\varepsilon}B_{T+\varepsilon} \text{ for } t \in [0,T].$$

Solution. (i), (ii) are immediate from the comments made above and Proposition 20.17.

(iii) We leave it to the reader to check that $\mathcal{H} \hookrightarrow C^{1\text{-var}}\left([0,T],\mathbb{R}^d\right)$ and the process is non-degenerate on $[0,T]$. To see validity of the zero–one law it suffices to note that X has Volterra structure. At last, Condition 20.16 is satisfied since X is asymptotically comparable to Brownian motion in the small time limit in the sense of Exercise 20.18. Indeed, take $r,s \in [0,t]$ and compute, with focus on one non-diagonal entry,

$$
\begin{aligned}
R_{X-B}\left(r,s\right) &\equiv \mathbb{E}[\left(X_r - B_r\right)\left(X_s - B_s\right)] \\
&= \int_0^t \left(e^{-(r-s)}-1\right)\left(e^{-(r-s)}-1\right)dr = O\left(t^3\right).
\end{aligned}
$$

(iv) Again, $\mathcal{H} \hookrightarrow C^{1\text{-var}}\left([0,T],\mathbb{R}^d\right)$ and non-degeneracy on $[0,T]$ are easy to see. Validity of the zero–one law follows by writing $X = X^{T+\varepsilon}$ as a strong solution to an SDE driven by Brownian motion with (well-behaved) drift (on $[0,T]$). At last, take $r,s \in [0,t] \subset [0,1]$ and compute

$$
\begin{aligned}
R_{X-B}\left(r,s\right) &\equiv \mathbb{E}[\left(X_r - B_r\right)\left(X_s - B_s\right)] \\
&= \int_0^t r\frac{B_{T+\varepsilon}}{T+\varepsilon}s\frac{B_{T+\varepsilon}}{T+\varepsilon}dr \leq \frac{t^3}{T+\varepsilon}
\end{aligned}
$$

so that, as one expects, the Brownian bridge is asymptotically comparable to Brownian motion in the small time limit.

20.4.2 Taylor expansions for rough differential equations

Given a smooth vector field W and smooth driving signal $x\left(\cdot\right)$ for the ODE $dy = V\left(y\right)dx$, it follows from basic calculus that

$$
J_{0\leftarrow t}^x \left(W\left(y_t^x\right)\right) = W\left(y_0\right) + \int_0^t J_{0\leftarrow s}^x \left([V_i,W]\left(y_s^x\right)\right)dx_s^i,
$$

where Einstein's summation convention is used throughout. Iterated use of this leads to the Taylor expansion

$$
\begin{aligned}
J_{0\leftarrow t}^x \left(W\left(y_t^x\right)\right) &= W|_{y_0} + [V_i,W]|_{y_0}\mathbf{x}_{0,t}^{1;i} \\
&\quad + [V_i,[V_j,W]]|_{y_0}\mathbf{x}_{0,t}^{2;i,j} \\
&\quad + \ldots \\
&\quad + [V_{i_1},\ldots[V_{i_k},W]]|_{y_0}\mathbf{x}_{0,t}^{k;i_1,\ldots,i_k} \\
&\quad + \cdots
\end{aligned}
\tag{20.7}
$$

where we write

$$
\mathbf{x}_{0,t}^{k;i_1,\ldots,i_k} = \int_0^t \int_0^{u_k} \cdots \int_0^{u_2} dx_{u_1}^{i_1} \ldots dx_{u_k}^{i_k}.
$$

Note that such an expansion makes immediate sense when x is replaced by a weak geometric p-rough path $\mathbf{x}$. In this case, $\mathbf{x}_{0,t}^k = \pi_k \left(S_k \left(\mathbf{x}_{0,t} \right) \right)$ where S_k denotes the Lyons lift. Remainder estimates can be obtained via *Euler* estimates, at least when expressing $J_{0 \leftarrow t}^x \left(W \left(y_t^x \right) \right)$ as a solution to a differential equation (ODE resp. RDE) of the form

$$dz = \hat{V}(z) \, dx.$$

(Cf. Proposition 10.3 and Corollary 10.15 and the resulting stochastic corollaries of Section 18.1.) This is accomplished by setting

$$z := \left(z^1, z^2, z^3 \right) := \left(y^x, J_{0 \leftarrow t}^x, J_{0 \leftarrow t}^x \left(W \left(y_t^x \right) \right) \right) \in \mathbb{R}^e \oplus \mathbb{R}^{e \times e} \oplus \mathbb{R}^e.$$

Noting that $J_{0 \leftarrow t}^x \left(W \left(y_t^x \right) \right)$ is given by $z^2 \cdot W \left(z^1 \right)$ in terms of matrix multiplication we have

$$
\begin{aligned}
dz^1 &= V_i \left(z^1 \right) dx^i \\
dz^2 &= -z^2 \cdot DV_i \left(z^1 \right) dx^i \\
dz^3 &= \left(dz^2 \right) \cdot W \left(z^1 \right) + z^2 \cdot d \left(W \left(z^1 \right) \right) \\
&= z^2 \cdot \left(-DV_i \left(z^1 \right) \cdot W \left(z^1 \right) + DW \left(z^1 \right) \cdot V_i \left(z^1 \right) \right) dx^i \\
&= z^2 \cdot \left[V_i, W \right] \big|_{z^1} dx^i
\end{aligned}
$$

started from $z_0 = \left(y_0, I, W \left(y_0 \right) \right)$ where I denotes the identity matrix in $\mathbb{R}^{e \times e}$ and we see that $\hat{V}$ is given by

$$
\hat{V}_i \left(z^1, z^2, z^3 \right) = \begin{pmatrix} V_i \left(z^1 \right) \\ -z^2 \cdot DV_i \left(z^1 \right) \\ z^2 \cdot \left[V_i, W \right] \left(z_1 \right) \end{pmatrix}, \quad i = 1, \ldots, d. \tag{20.8}
$$

We now consider the corresponding rough differential equation, $dz = \hat{V}(z) \, d\mathbf{x}$, where x is a weak geometric p-rough path. From the very construction of $\hat{V}$ it is clear that an expansion of the form

$$z_t = z_0 + \hat{V}(z_0) \mathbf{x}_{0,t}^1 + \hat{V}\hat{V}(z_0) \mathbf{x}_{0,t}^2 + \cdots,$$

after projection to the third component of $z_t = \left(z_t^1, z_t^2, z_t^3 \right)$, yields precisely the expansion (20.7). To be more precise, let us recall that, given smooth vector fields $V = \left(V_1, \ldots, V_d \right)$ on $\mathbb{R}^e$, an element $\mathbf{g} \in \oplus_{k=0}^m \left(\mathbb{R}^d \right)^{\otimes k}$ and $y \in \mathbb{R}^e$ we write

$$\mathcal{E}_{(V)}(y, \mathbf{g}) := \sum_{k=1}^m \sum_{\substack{i_1, \ldots, i_k \\ \in \{1, \ldots, d\}}} \mathbf{g}^{k; i_1, \ldots, i_k} V_{i_1} \cdots V_{i_k} I(y).$$

(Here I denotes the identity function on $\mathbb{R}^e$ and vector fields are identified with first-order differential operators.) In a similar spirit, given another sufficiently smooth vector field W, we first set

$$\left[V_{i_1}, V_{i_2}, \ldots, V_{i_k}, W \right] := \left[V_{i_1}, \left[V_{i_2}, \ldots \left[V_{i_k}, W \right] \ldots \right] \right]$$

and then

$$\mathbf{g}^k \cdot [V, \ldots, V, W]\big|_{y_0} := \sum_{\substack{i_1,\ldots,i_k \\ \in \{1,\ldots,d\}}} \mathbf{g}^{k,i_1,\ldots,i_k} [V_{i_1}, V_{i_2}, \ldots, V_{i_k}, W]\, I\,(y_0) \quad (20.9)$$

with the convention that $\mathbf{g}^0 \cdot V_k = V_k$. We can then state the following lemma.

Lemma 20.20 *Write* $p_3 : \mathbb{R}^e \oplus \mathbb{R}^{e \times e} \oplus \mathbb{R}^e \to \mathbb{R}^e$ *for the projection given by* $(z^1, z^2, z^3) \mapsto z^3$. *Let* f *be a smooth function on* $\mathbb{R}^e$ *lifted to* $\hat{f} = f \circ p_3$, *a smooth function on* $\mathbb{R}^e \oplus \mathbb{R}^{e \times e} \oplus \mathbb{R}^e$. *With the vector fields* $\hat{V}$ *as defined in (20.8) and* $z_0 = (y_0, I, W(y_0))$, *we have*

$$\hat{V}_{i_1} \cdots \hat{V}_{i_N}\big|_{z_0}\, \hat{f} = [V_{i_1}, \ldots, V_{i_N}, W]\big|_{y_0}\, f.$$

As a consequence, for any $\mathbf{g} \in G^m\left(\mathbb{R}^d\right)$,

$$p_3\left[\mathcal{E}_{(\hat{V})}(z_0, \mathbf{g})\right] = W\big|_{y_0} - \sum_{k=1}^m \boldsymbol{\pi}_k(\mathbf{g}) \cdot [V, \ldots, V, W]\big|_{y_0}.$$

Proof. Taylor expansion of the evolution equation of $z^3(t)$ shows that $\hat{V}_{i_1} \cdots \hat{V}_{i_N}\big|_{z_0} f = [V_{i_1}, \ldots, V_{i_N}, W]\big|_{y_0} f$, as required. ∎

Corollary 20.21 *Fix* $a \in T_{y_0}\mathbb{R}^e \cong \mathbb{R}^e$ *with* $|a| = 1$. *Let* $H \in (1/4, 1/2]$ *and* $\mathbf{X}$ *be a Gaussian rough path with the covariance of the underlying Gaussian process satisfying (20.5). Then, writing* y *for the solution to the random RDE* $dy = V(y)\, d\mathbf{X}$ *started at* y_0, *and* J *for the Jacobian of its flow, we have that for all* $\varepsilon > 0$

$$\lim_{n \to \infty} \mathbb{P}\left[\left|a^T J_{0 \leftarrow t}(W(y_t))\right.\right.$$

$$\left.\left. - \sum_{k=0}^m a^T\left(\mathbf{X}_{0,t}^k \cdot [V, \ldots, V, W]\big|_{y_0}\right)\right|_{t=1/n} > \frac{\varepsilon}{2} n^{-mH}\right] = 0.$$

Proof. As discussed at the beginning of Section 20.4.1, assumption (20.5) implies

$$\sup_{0 \leq s < t \leq T} \mathbb{E}\left(\exp\left(\eta\left[\frac{d(\mathbf{X}_s, \mathbf{X}_t)}{|t-s|^H}\right]^2\right)\right) < \infty$$

which was the standing assumption for remainder estimates of Azencott-type established in Theorem 18.1. Thanks to $|a| = 1$ and the previous lemma, applied with $\mathbf{g} = S_m(\mathbf{X}_{0,t})$, we can write

$$\mathbb{P}\left[\left|a^T J_{0 \leftarrow t}(W(y_t)) - \sum_{k=0}^m a^T\left(\mathbf{X}_{0,t}^k \cdot [V, \ldots, V, W]\big|_{y_0}\right)\right|_{t=1/n} > \frac{\varepsilon}{2} n^{-mH}\right]$$

$$\leq \mathbb{P}\left[\left|\pi_{(\hat{V})}(0, z_0, \mathbf{X})_{0,1/n} - \mathcal{E}_{(\hat{V})}\left(z_0, S_m(\mathbf{X})_{0,1/n}\right)\right| > \frac{\varepsilon}{2} n^{-mH}\right].$$

The vector fields $\hat{V}$ as defined in (20.8) are smooth but, in general, unbounded. Using the remainder estimates given in Theorem 18.1 in the small time limit (valid for $\mathrm{Lip}_{\mathrm{loc}}$-vector fields) we then obtain (as pointed out explicitly in Example 18.2) the required convergence. ∎

20.4.3 Hörmander's condition revisited

Let $V = (V_1, \ldots, V_d)$ denote a collection of smooth vector fields defined in a neighbourhood of $y_0 \in \mathbb{R}^e$. Given a multi-index $I = (i_1, \ldots, i_k) \in \{1, \ldots, d\}^k$, with length $|I| = k$, the vector field V_I is defined by iterated Lie brackets

$$V_I := [V_{i_1}, V_{i_2}, \ldots, V_{i_k}] \equiv [V_{i_1}, [V_{i_2}, \ldots, [V_{i_{k-1}}, V_{i_k}] \ldots]. \qquad (20.10)$$

If W is another smooth vector field defined in a neighbourhood of $y_0 \in \mathbb{R}^e$ we write[4]

$$\underbrace{a}_{\in (\mathbb{R}^d)^{\otimes(k-1)}} \cdot \underbrace{[V, \ldots, V, W]}_{\text{length } k} := \sum_{\substack{i_1, \ldots, i_{k-1} \\ \in \{1, \ldots, d\}}} a^{i_1, \ldots, i_{k-1}} [V_{i_1}, V_{i_2}, \ldots, V_{i_{k-1}}, W].$$

Recall that the step-r free nilpotent group with d generators, $G^r(\mathbb{R}^d)$, was realized as a submanifold of the tensor algebra

$$T^{(r)}(\mathbb{R}^d) \equiv \oplus_{k=0}^{r} (\mathbb{R}^d)^{\otimes k}.$$

Definition 20.22 Given $r \in \mathbb{N}$ we say that condition $(\mathbf{H})_r$ holds at $y_0 \in \mathbb{R}^e$ if

$$\mathrm{span}\{V_I|_{y_0} : |I| \le r\} = \mathcal{T}_{y_0}\mathbb{R}^e \cong \mathbb{R}^e; \qquad (20.11)$$

we say that **Hörmander's condition (H)** is satisfied at y_0 if $(\mathrm{H})_r$ holds for some $r \in \mathbb{N}$. □

An element $\mathbf{g} \in \oplus_{k=0}^{\infty} (\mathbb{R}^d)^{\otimes k}$ is called *group-like* iff for any $N \in \mathbb{N}$,

$$(\pi_0(\mathbf{g}), \ldots, \pi_N(\mathbf{g})) \in G^N(\mathbb{R}^d) \subset \oplus_{k=0}^{N} (\mathbb{R}^d)^{\otimes k}.$$

The following result tells us that Hörmander's condition is equivalent to a (seemingly stronger, "Hörmander-type") condition that involves only Lie brackets of V contracted against group-like elements. It will be important in carrying out the crucial induction step in the proof of Theorem 20.12.

---

[4]We introduced this notation in the previous section, cf. (20.9).

Definition 20.23 Given $r \in \mathbb{N}$ we say that condition $(\mathbf{HT})_r$ holds at $y_0 \in \mathbb{R}^e$ if the linear span of

$$\left\{ \pi_{k-1}(\mathbf{g}) \cdot \underbrace{[V, \ldots, V, V_i]}_{\text{length } k} \Big|_{y_0} : k = 1, \ldots, r; i = 1, \ldots, d, \ \mathbf{g} \in G^{r-1}(\mathbb{R}^d) \right\} \tag{20.12}$$

is full; that is, equal to $\mathcal{T}_{y_0} \mathbb{R}^e \cong \mathbb{R}^e$. $\qquad\square$

Proposition 20.24 *Let $r \in \mathbb{N}$ and $V = (V_1, \ldots, V_d)$ be a collection of smooth vector fields defined in a neighbourhood of $y_0 \in \mathbb{R}^e$. Then the $(H)_r$-span, by which we mean the linear span of (20.11), equals the $(HT)_r$-span, that is, the linear span of (20.12). In particular, Hörmander's condition (H) is satisfied at y_0 if and only if the span of (20.12) is full for some r large enough.*

Proof. We first make the trivial observation that $(\mathrm{HT})_r$ implies $(\mathrm{H})_r$ for any $r \in \mathbb{N}$. For the converse, fixing a multi-index $I = (i_1, \ldots, i_{k-1}, i_k)$ of length $k \leq r$ and writing $e_1, \ldots, e_d$ for the canonical basis of $\mathbb{R}^d$, we define

$$\begin{aligned} \mathbf{g} &= \mathbf{g}(t_1, \ldots, t_{k-1}) \\ &= \exp(t_1 e_{i_1}) \otimes \cdots \otimes \exp(t_{k-1} e_{i_{k-1}}) \\ &\in G^{r-1}(\mathbb{R}^d) \subset T^{r-1}(\mathbb{R}^d). \end{aligned}$$

It follows that any

$$\pi_{k-1}(\mathbf{g}) \cdot \underbrace{[V, \ldots, V, V_{i_k}]}_{\text{length } k} \Big|_{y_0}$$

lies in the $(\mathrm{HT})_r$-span. Now, the $(\mathrm{HT})_r$-span is a closed linear subspace of $\mathcal{T}_{y_0} \mathbb{R}^e \cong \mathbb{R}^e$ and so it is clear that any element of the form

$$\pi_{k-1}(\partial_\alpha \mathbf{g}) \cdot \underbrace{[V, \ldots, V, V_{i_k}]}_{\text{length } k} \Big|_{y_0},$$

where ∂_α stands for any higher-order partial derivative with respect to $t_1, \ldots, t_{k-1}$, i.e.

$$\partial_\alpha = \left(\frac{\partial}{\partial t_1} \right)^{\alpha_1} \cdots \left(\frac{\partial}{\partial t_{k-1}} \right)^{\alpha_{k-1}} \qquad \text{with } \alpha \in (\mathbb{N} \cup \{0\})^{k-1},$$

is also in the $(HT)_r$-span for any $t_1, \ldots, t_{k-1}$ and, in particular, when evaluated at $t_1 = \cdots = t_{k-1} = 0$. For the particular choice $\alpha = (1, \ldots, 1)$ we have

$$\frac{\partial^{k-1}}{\partial t_1 \ldots \partial t_{k-1}} \mathbf{g} \big|_{t_1 = 0, \ldots, t_{k-1} = 0} = e_{i_1} \otimes \cdots \otimes e_{i_{k-1}} =: \mathbf{h}$$

where $\mathbf{h}$ is an element of $T^{r-1}\left(\mathbb{R}^d\right)$ with the only non-zero entry arising on the $(k-1)$th tensor level, i.e.

$$\pi_{k-1}\left(\mathbf{h}\right) = e_{i_1} \otimes \cdots \otimes e_{i_{k-1}}.$$

Thus,

$$\pi_{k-1}\left(\mathbf{h}\right) \cdot \underbrace{\left[V, \ldots, V, V_{i_k}\right]}_{\text{length } k} \big|_{y_0} = \left[V_{i_1}, \ldots, V_{i_{k-1}}, V_{i_k}\right] \big|_{y_0}$$

is in our $(HT)_r$-span. But this says precisely that, for any multi-index I of length $k \leq r$, the bracket vector field evaluated at y_0, i.e. $V_I|_{y_0}$, is an element of our $(HT)_r$-span. ∎

20.4.4 Proof of Theorem 20.12

We are now in a position to give the proof of Theorem 20.12.
Proof. We fix $t \in (0, T]$. As usual, it suffices to show a.s. invertibility of

$$\sigma_t = \left(\left\langle DY_t^i, DY_t^j \right\rangle_{\mathcal{H}}\right)_{i,j=1,\ldots,e} \in \mathbb{R}^{e \times e}.$$

In terms of an orthonormal basis (h_n) of the Cameron–Martin space we can write

$$\sigma_t = \sum_n \langle DY_t, h_n \rangle_{\mathcal{H}} \otimes \langle DY_t, h_n \rangle_{\mathcal{H}} \qquad (20.13)$$

$$= \sum_n \int_0^t J_{t \leftarrow s}^{\mathbf{X}}\left(V_k\left(Y_s\right)\right) dh_{n,s}^k \otimes \int_0^t J_{t \leftarrow s}^{\mathbf{X}}\left(V_l\left(Y_s\right)\right) dh_{n,s}^l.$$

(Summation over up–down indices is from here on tacitly assumed.) Invertibility of σ is equivalent to invertibility of the reduced covariance matrix

$$C_t := \sum_n \int_0^t J_{0 \leftarrow s}^{\mathbf{X}}\left(V_k\left(Y_s\right)\right) dh_{n,s}^k \otimes \int_0^t J_{0 \leftarrow s}^{\mathbf{X}}\left(V_l\left(Y_s\right)\right) dh_{n,s}^l,$$

which has the advantage of being adapted, i.e. being $\sigma\left(X_s : s \in [0, t]\right)$-measurable. We now assume that

$$\mathbb{P}\left(\det C_t = 0\right) > 0$$

and will see that this leads to a contradiction with Hörmander's condition.
Step 1: Let K_s be the random subspace of $T_{y_0}\mathbb{R}^e \cong \mathbb{R}^e$, spanned by

$$\left\{ J^{\mathbf{X}}_{0 \leftarrow r}\left(V_k\left(Y_r\right)\right); r \in [0,s], k = 1,\ldots,d \right\}.$$

The subspace $K_{0^+} = \cap_{s>0} K_s$ is measurable with respect to the germ σ-algebra and by our "0–1 law" assumption, deterministic with probability one. A random time is defined by

$$\Theta = \inf\left\{ s \in (0,t] : \dim K_s > \dim K_{0^+} \right\} \wedge t, \qquad (20.14)$$

and we note that $\Theta > 0$ a.s. For any vector $v \in \mathbb{R}^e$ we have

$$v^T C_t v = \sum_n \left| \int_0^t v^T J^{\mathbf{X}}_{0 \leftarrow s}\left(V_k\left(Y_s\right)\right) dh^k_{n,s} \right|^2.$$

Assuming $v^T C_t v = 0$ implies

$$\forall n : \int_0^t v^T J^{\mathbf{X}}_{0 \leftarrow s}\left(V_k\left(Y_s\right)\right) dh^k_{n,s} = 0$$

and hence, by our non-degeneracy condition on the Gaussian process and Lemma 20.9,

$$v^T J^{\mathbf{X}}_{0 \leftarrow s}\left(V_k\left(Y_s\right)\right) = 0$$

for any $s \in [0,t]$ and any $k = 1,\ldots,d$, which implies that v is orthogonal to K_t. Therefore, $K_{0^+} \neq \mathbb{R}^e$, otherwise $K_s = \mathbb{R}^e$ for every $s > 0$ so that v must be zero, which implies C_t is invertible a.s., in contradiction of our hypothesis.
Step 2: We saw that K_{0^+} is a deterministic and linear subspace of $\mathbb{R}^e$ with strict inclusion $K_{0^+} \subsetneq \mathbb{R}^e$ In particular, there exists a deterministic vector $z \in \mathbb{R}^e \setminus \{0\}$ which is orthogonal to K_{0^+}. We will show that z is orthogonal to all vector fields and (suitable) brackets evaluated at y_0, thereby contradicting the fact that our vector fields satisfy Hörmander's condition. By definition (20.14) of Θ, $K_{0^+} \equiv K_t$ for $0 \leq t < \Theta$ and so for every $k = 1,\ldots,d$,

$$z^T J^{\mathbf{X}}_{0 \leftarrow t}\left(V_k\left(Y_t\right)\right) = 0 \text{ for } t \leq \Theta. \qquad (20.15)$$

Observe that, by evaluation at $t = 0$, this implies $z \perp \mathrm{span}\{V_1,\ldots,V_d\}\big|_{y_0}$.
Step 3: In view of Proposition 20.24, it suffices to show that z is orthogonal to all iterated Lie brackets of $V = (V_1,\ldots,V_d)$ contracted against group-like elements. To this end, we keep $k \in \{1,\ldots,d\}$ fixed and make the **induction hypothesis** $I(m-1)$:

$$\forall \mathbf{g} \text{ group-like}, j \leq m-1 : z^T \pi_j(\mathbf{g})\left[V,\ldots,V;V_k\right]\big|_{y_0} = 0.$$

We can now take the shortest path $\gamma^n : [0, 1/n] \to \mathbb{R}^d$ such that $S_m(\gamma^n)$ equals $\pi_{1,\ldots,m}(\mathbf{g})$, the projection of $\mathbf{g}$ to the free step-m nilpotent group with d generators, denoted $G^m(\mathbb{R}^d)$. Then

$$|\gamma^n|_{1\text{-var};[0,1/n]} = \|\pi_{1,\ldots,m}(\mathbf{g})\|_{G^m(\mathbb{R}^d)} < \infty$$

and the scaled path

$$h^n(t) = n^{-H}\gamma^n(t), \quad H \in (0,1)$$

has length (over the interval $[0, 1/n]$) proportional to n^{-H}, which tends to 0 as $n \to \infty$. Our plan is to show that

$$\forall \varepsilon > 0 : \liminf_{n \to \infty} \mathbb{P}\left(\left|z^T J^{h^n}_{0 \leftarrow 1/n}\left(V_k\left(y^{h^n}_{1/n}\right)\right)\right| < \varepsilon/n^{mH}\right) > 0 \qquad (20.16)$$

which, since the event involved is deterministic, really says that

$$\left|n^{mH} z^T J^{h^n}_{0 \leftarrow 1/n}\left(V_k\left(y^{h^n}_{1/n}\right)\right)\right| < \varepsilon$$

holds true for all $n \geq n_0(\varepsilon)$ large enough. Then, sending $n \to \infty$, a Taylor expansion and $I(m-1)$ shows that the left-hand side converges to

$$\left|z^T \underbrace{n^{mH}\pi_m\left(S_m\left(h^n\right)\right)}_{=\pi_m(\mathbf{g})} \cdot [V, \ldots, V; V_k]\big|_{y_0}\right| < \varepsilon$$

and since $\varepsilon > 0$ is arbitrary we showed $I(m)$, which completes the induction step.

Step 4: The only thing left to show is (20.16); that is, positivity of $\liminf$ of

$$\mathbb{P}\left(\left|z^T J^{h^n}_{0 \leftarrow 1/n}\left(V_k\left(y^{h^n}_{1/n}\right)\right)\right| < \varepsilon/n^{mH}\right)$$

$$\geq \mathbb{P}\left(\left|z^T J^{\mathbf{X}}_{0 \leftarrow \cdot}\left(V_k(y_\cdot)\right) - z^T J^{h^n}_{0 \leftarrow \cdot}\left(V_k\left(y^{h^n}_\cdot\right)\right)\right|_{\cdot = 1/n} < \varepsilon/n^{mH}\right)$$

$$- \mathbb{P}(\Theta \leq 1/n)$$

and since $\Theta > 0$ a.s. it is enough to show that

$$\liminf_{n \to \infty} \mathbb{P}\left(\left|z^T J^{\mathbf{X}}_{0 \leftarrow \cdot}\left(V_k(y_\cdot)\right) - z^T J^{h^n}_{0 \leftarrow \cdot}\left(V_k\left(y^{h^n}_\cdot\right)\right)\right|_{\cdot = 1/n} < \varepsilon/n^{mH}\right) > 0.$$

Using $I(m-1)$ + stochastic Taylor expansion (more precisely, Corollary 20.21) this is equivalent to showing positivity of $\liminf$ of

$$\mathbb{P}\left(\left|z^T \mathbf{X}^m_{0,\cdot}[V, \ldots, V; V_k] - z^T J^{h^n}_{0 \leftarrow \cdot}\left(V_k\left(y^{h^n}_\cdot\right)\right)\right|_{\cdot = 1/n} < \frac{\varepsilon}{2}/n^{mH}\right).$$

(Let us remark that the assumption $Hp < 1 + 1/m$ needed to apply Corollary 20.21 is satisfied thanks to Condition 20.16, part (ii), and the remark that our induction stops when m has reached r, the number of brackets needed in Hörmander's condition.) Rewriting things, we need to show positivity of $\liminf$ of

$$\mathbb{P}\bigg(|n^{mH}z^T\,[V,\dots,V;V_k]\,\mathbf{X}_{0,1/n}^m \underbrace{- z^T n^{mH} J_{0\leftarrow 1/n}^{h^n}\left(V_k\left(y_{1/n}^{h^n}\right)\right)}_{\to z^T\,[V,\dots,V;V_k]\pi_m(\mathbf{g})}| < \frac{\varepsilon}{2}\bigg)$$

or, equivalently, that

$$\lim_{n\to\infty}\inf\,\mathbb{P}\left(\left|z^T\,[V,\dots,V;V_k]\right|_{y_0}\left(n^{mH}\mathbf{X}_{0,1/n}^m - \pi_m(\mathbf{g})\right)\right| < \frac{\varepsilon}{2}\right) > 0.$$

But this is implied by Condition 20.16 and so the proof is finished. ∎

20.5 Comments

The bulk of the material of Sections 20.1, 20.2 and 20.3 is taken from Cass *et al.* [26]. Let us note that $\mathcal{H}$-differentiability of a Wiener functional implies $\mathbb{D}_{loc}^{1,2}$-regularity where $\mathbb{D}^{1,2}$ is defined as the subspace of $L^2(\mathbb{P})$ obtained as the closure of "nice" Wiener functionals with respect to $\|F\|_{\mathbb{D}^{1,2}} = |F|_{L^2(\mathbb{P})} + |DF|_{L^2(\mathbb{P},\mathcal{H})}$. In particular, the $\mathcal{H}$-derivative is then precisely the Malliavin derivative; some details and references on this are given in Section D.5, Appendix D. Exercise 20.11 on the representation of the Malliavian covariance matrix has well-known special cases: in the case of Brownian motion, $dR(s,s')$ is a Dirac measure on the diagonal $\{s = s'\}$ and the double integral reduces to a (well-known) single-integral expression; in the case of fractional Brownian motion with $H > 1/2$, $dR_H(s,t) \sim |t - s|^{2H-2}\,dsdt$, which is integrable at zero iff $H > 1/2$. (The resulting double-integral representation of the Malliavin covariance is also well known and appears, for instance, in Baudoin and Hairer [9], Saussereau and Nualart [152], Hu and Nualart [87].)

Our discussion of RDEs under Hörmander's condition follows closely Cass and Friz [25]. In the case of driving Brownian motion all this is, of course, classical and closely related to Hörmander's work on hypoellipticity. Previous works in this direction were focused on driving fractional Brownian motion B^H, with Hurst parameter $H > 1/2$, so that $dY = V(Y)\,dB^H$ makes sense as a Young differential equation. A density result under the ellipticity condition on V appeared in Saussereau and Nualart [152]. The deterministic estimate (cf. Exercise 11.10),

$$\left|J_{\cdot\leftarrow 0}^{y_0,\mathbf{X}}\right|_{p\text{-var};[0,T]} \leq C\exp\left(C\,\|\mathbf{X}\|_{p\text{-var};[0,T]}^p\right) \qquad (20.17)$$

can be applied in a step-1 setting[5] to see that $\left| J_{t \leftarrow 0}^{y_0, \mathbf{X}} \right| \in L^q(\mathbb{P})$ for all $q < \infty$ of all orders, thanks to Gaussian integrability of $\left| B^H(\omega) \right|_{p\text{-var};[0,T]}$. This allows us to obtain L^q-estimates on the inverse of the covariance matrix from which one obtains existence of a *smooth* density. This was carried out, again under the ellipticity condition on V, by Hu and Nualart [87]. Existence of a smooth density under Hörmander's condition was then obtained by Baudoin and Hairer [9], relying on some specific properties of fractional Brownian motion. At present, the question of how to obtain L^q-estimates in the regime $\rho \in [1, 2)$ is open. It is worthwhile noting (Friz and Oberhauser [59]) that the deterministic estimate (20.17) is optimal, so that L^q-estimates in the regime for $\rho \in [1, 2)$ will require further probabilistic input, presumably in the form of Gaussian chaos integrability.

---

[5] That is, with $p \in (1/H, 2)$ and $\|\mathbf{X}\|_{p\text{-var};[0,T]} = \left| B^H(\omega) \right|_{p\text{-var};[0,T]}$.

Part V

Appendices

Appendix A: Sample path regularity and related topics

A.1 Continuous processes as random variables

A.1.1 Generalities

A *stochastic process* with values in some measure space $(E, \mathcal{E})$ is a collection of random variables, i.e. measurable maps $X_t : (\Omega, \mathcal{A}, \mathbb{P}) \to (E, \mathcal{E})$, indexed by t in some set $\mathbf{T}$. Equivalently, X is a measurable map $X : (\Omega, \mathcal{A}, \mathbb{P}) \to (E^{\mathbf{T}}, \mathcal{E}^{\mathbf{T}})$ where $E^{\mathbf{T}}$ is the space of E-valued functions on $\mathbf{T}$ and $\mathcal{E}^{\mathbf{T}}$ is the smallest σ-algebra such that all projections $\pi_t : E^{\mathbf{T}} \to E$, defined by $f \mapsto f_t$, are measurable maps. The *law* or *distribution* of X is the image measure $\mathbb{P}^X := X_* \mathbb{P}$ defined on $(E^{\mathbf{T}}, \mathcal{E}^{\mathbf{T}})$. Since the underlying probability model $(\Omega, \mathcal{A}, \mathbb{P})$ is usually irrelevant it can be replaced by the *canonical* model $(E^{\mathbf{T}}, \mathcal{E}^{\mathbf{T}}, \mathbb{P}^X)$ and $Y_t : (E^{\mathbf{T}}, \mathcal{E}^{\mathbf{T}}, \mathbb{P}^X) \to (E, \mathcal{E})$ given by $f \mapsto \pi_t(f) = f_t$ is the *canonical version* of the stochastic process X. More precisely, X and Y are versions of each other in the following sense. One says that two processes X and X' defined respectively on the probability spaces $(\Omega, \mathcal{A}, \mathbb{P})$ and $(\Omega', \mathcal{A}', \mathbb{P}')$, having the same state space $(E, \mathcal{E})$, are *versions* of each other – or that they are "versions of the same process" – if for any finite sequence $t_1, \ldots, t_n$ and sets $A_i \in \mathcal{E}$,

$$ \mathbb{P}\left[X_{t_1} \in A_1, \ldots, X_{t_n} \in A_n \right] = \mathbb{P}'\left[X'_{t_1} \in A_1, \ldots, X'_{t_n} \in A_n \right]. $$

Two processes X and X' defined on the same probability space are said to be *modifications* of each other if $\mathbb{P}\left[X_t = X'_t \right] = 1$ for all t. At last, they are said to be *indistinguishable* if $\mathbb{P}\left[X_t = X'_t \text{ for all } t \right] = 1$.

Let us now assume that $\mathbf{T} = [0, T]$ and E is Polish (with $\mathcal{E} = \mathcal{B}_E$, the Borel σ-algebra). It is natural to ask if

$$ C = \left\{ f \in E^{\mathbf{T}} : f : \mathbf{T} \to E \text{ is continuous} \right\} $$

has $\mathbb{P}^X$-measure one. Unfortunately, the above set need not be $\mathcal{E}^{\mathbf{T}}$-measurable but we can still ask if C has full outer measure. If this is the case then, still writing $Y_t(f) = \pi_t(f) = f_t$, the measure $\mathbb{P}^X$ on $E^{\mathbf{T}}$ induces a probability measure $\mathbb{Q}$ on C, defined on the σ-algebra $\mathcal{C} \equiv \sigma(Y_t : t \in \mathbf{T}) = \mathcal{E}^{\mathbf{T}} \cap C$ by setting $\mathbb{Q}(\Gamma) := \mathbb{P}^X\left(\tilde{\Gamma} \right)$ where $\tilde{\Gamma}$ is any set in $\mathcal{E}^{\mathbf{T}}$ such that $\Gamma = \tilde{\Gamma} \cap C$. Obviously the process Y defined on $(C, \mathcal{C}, \mathbb{Q})$, better denoted by

$$\tilde{Y}_t : \left\{ \begin{array}{l} (C, \mathcal{C}, \mathbb{Q}) \rightarrow (E, \mathcal{E}) \\ f \mapsto \pi_t (f) = f_t \end{array} \right.$$

is another version of X and, moreover, a genuine $(C, \mathcal{C})$-valued random variable. This version again is defined on a space of functions and is made up of coordinate mappings and will also be referred to as *canonical*; we will also write $\mathbb{P}^X$ or $X_*\mathbb{P}$ instead of $\mathbb{Q}$; this causes no confusion so long as we know the space we work in. We say that a stochastic process $X :$ $(\Omega, \mathcal{A}, \mathbb{P}) \rightarrow \left(E^{\mathbf{T}}, \mathcal{E}^{\mathbf{T}} \right)$ is *continuous* if the set of continuous functions from $\mathbf{T} \rightarrow E$ has (outer) measure one. In this case, there is a version of X which is a genuine $(C, \mathcal{C})$-valued random variable. Let X denote this continuous version. Then

$$X : (\Omega, \mathcal{A}, \mathbb{P}) \rightarrow (C([0, T], E), \mathcal{C}),$$

where $\mathcal{C}$ is the σ-algebra generated by the coordinate maps in $C([0, T], E)$. On the other hand, $C([0, T], E)$ is a Polish space under the topology induced by uniform distance[1] and there is a natural Borel σ-algebra $\mathcal{B}$ generated by the open sets. The law of X, i.e. $X_*\mathbb{P}$, defines in fact a Borel measure on $\mathcal{B}$. This follows from $\mathcal{B} = \mathcal{C}$, which is easy to see. (All coordinate maps π_t are continuous, which shows that $\mathcal{C} \subset \mathcal{B}$. Conversely, $C([0, T], E)$ has a countable basis for its topology of the form $\cap_{t \in [0,T] \cap \mathbb{Q}} \left\{ f : d \left(f_t, \tilde{f}_t \right) < \varepsilon \right\} \in$ $\mathcal{C}$, $\tilde{f} \in C([0, T], E)$.

If E has a compatible[2] group structure, we can define increments

$$f_{s,t} = f_s^{-1} f_t$$

and then Hölder metrics of the form $d(f_0, g_0) + d_{\alpha\text{-Höl};[0,T]}(f, g)$ where

$$d_{\alpha\text{-Höl};[0,T]}(f, g) = \sup_{0 \le s < t \le T} \frac{d(f_{s,t}, g_{s,t})}{|t - s|^\alpha}.$$

The resulting path space $C^{\alpha\text{-Höl}}([0, T], E)$ is not separable in general but, at least when $E = \mathbb{R}^d$ or $G^N(\mathbb{R}^d)$, there are Polish subspaces (cf. Sections 5.3 and 8.6), denoted by $C^{0,\alpha\text{-Höl}}([0, T], E)$, defined as the set of all $f \in$ $C([0, T], E)$ such that

$$|f|_{\alpha\text{-Höl};[0,T]} \equiv \sup_{0 \le s < t \le T} \frac{d(f_s, f_t)}{|t - s|^\alpha} < \infty \quad \text{and} \quad \overline{\lim}_{\varepsilon \to 0} \sup_{0 < t - s < \varepsilon} \frac{d(f_s, f_t)}{|t - s|^\alpha} = 0.$$

By restricting s, t, ε to rationals one easily sees that $C^{0,\alpha\text{-Höl}}([0, T], E)$ and the σ-algebra $\mathcal{C}^{0,\alpha}$, generated by the coordinate maps in $C^{0,\alpha\text{-Höl}}([0, T], E)$, coincides with $\left\{ A \cap C^{0,\alpha} : A \in \mathcal{C} \right\}$. On the other hand, there is a natural

---

[1] See Stroock [159], for instance.
[2] That is, all group operations are continuous.

Borel σ-algebra denoted by $\mathcal{B}^{0,\alpha}$ on $C^{0,\alpha\text{-H\"ol}}\left([0,T],E\right)$. Using, in particular, separability of $C^{0,\alpha\text{-H\"ol}}\left([0,T],E\right)$ and continuity of the group operations, it is easy to see that the Borel σ-algebra $\mathcal{B}^{0,\alpha}$ equals $\mathcal{C}^{0,\alpha}$. In summary, a continuous process

$$X : (\Omega, \mathcal{A}, \mathbb{P}) \to (C\left([0,T],E\right), \mathcal{C})$$

which assigns full measure to $C^{0,\alpha\text{-H\"ol}}$ can be regarded as a $C^{0,\alpha\text{-H\"ol}}\left([0,T],\right.$ $\left. E \right)$-valued random variable whose law is a well-defined Borel measure on $\mathcal{B}^{0,\alpha}$. Identical remarks apply to p-variation spaces.

A.2   The Garsia–Rodemich–Rumsey estimate

A.2.1   Garsia–Rodemich–Rumsey on metric spaces

We discuss the Garsia–Rodemich–Rumsey result and several consequences, including a frequently used Besov–Hölder embedding and a simple proof of Kolmogorov's tightness criterion. Unless otherwise stated, (E,d) denotes a complete metric space.

Theorem A.1 (Garsia–Rodemich–Rumsey) *Consider* $f \in C\left([0,T],\right.$ $\left. E\right)$ *where* (E,d) *is a metric space. Let* Ψ *and* p *be continuous, strictly increasing functions on* $[0,\infty)$ *with* $\mathrm{p}(0) = \Psi(0) = 0$ *and* $\Psi(x) \to \infty$ *as* $x \to \infty$. *Then*

$$\int_0^T \int_0^T \Psi\left(\frac{d\left(f_s, f_t\right)}{\mathrm{p}(|t-s|)}\right) ds\, dt \le F \tag{A.1}$$

implies, for $0 \le s < t \le T$,

$$d\left(f_s, f_t\right) \le 8 \int_0^{t-s} \Psi^{-1}\left(\frac{4F}{u^2}\right) d\mathrm{p}(u). \tag{A.2}$$

In particular, if $\mathrm{osc}\left(f, \delta\right) \equiv \sup\left\{d\left(f_s, f_t\right) : s,t \in [0,T], |t-s| \le \delta\right\}$ *denotes the modulus of continuity of* f, *we have*

$$\mathrm{osc}\left(f, \delta\right) \le 8 \int_0^\delta \Psi^{-1}\left(\frac{4F}{u^2}\right) d\mathrm{p}(u).$$

Proof. Given $f \in C\left([0,T],E\right)$ and $\mathrm{p}(\cdot)$ we can set $\tilde{f}(\cdot) = f(T\cdot)$ and $\tilde{\mathrm{p}}(\cdot) = \mathrm{p}(T\cdot)$ and a simple change of variable shows that the "$T = 1$"-estimates obtained for $\tilde{f}, \tilde{\mathrm{p}}$ imply the required estimates for f,p. Thus, we can and will take $T = 1$ in the remainder of the proof. Define $I(t) = \int_0^1 \Psi\left(d\left(f_s, f_t\right)/\mathrm{p}(|t-s|)\right) ds$. Since $\int_0^1 I(t) = F$, there exists $t_0 \in (0,1)$ such that $I(t_0) \le F$. We shall prove that

$$d\left(f_{t_0}, f_0\right) \le 4 \int_0^1 \Psi^{-1}\left(\frac{4F}{u^2}\right) d\mathrm{p}(u). \tag{A.3}$$

By a similar argument

$$d\left(f_1, f_{t_0}\right) \leq 4 \int_0^1 \Psi^{-1}\left(\frac{4F}{u^2}\right) dp(u),$$

and combining the two we have (A.2); first for $s = 0, t = 0$ and then for arbitrary $0 \leq s < t \leq 1$ by reparametrization. To prove (A.3) we shall pick recursively two sequences $\{u_n\}$ and $\{t_n\}$ satisfying

$$t_0 > u_1 > t_1 > u_2 > \cdots > t_{n-1} > u_n > t_n > u_{n+1} > \ldots,$$

so that $t_n, u_n \searrow 0$ as $n \to \infty$, in the following manner. By induction, if t_{n-1} has already been chosen, pick

$$u_n \in (0, t_n) : \mathrm{p}\left(u_n\right) = \frac{1}{2}\mathrm{p}\left(t_{n-1}\right).$$

Trivially then, $\int_0^{u_n} I\left(t\right) dt \leq F$ and also $\int_0^{u_n} J\left(s\right) ds \leq \int_0^1 J\left(s\right) ds = I\left(t_{n-1}\right)$ where we set

$$J\left(s\right) := \Psi\left(d\left(f_s, f_{t_{n-1}}\right) / \mathrm{p}(|t_{n-1} - s|)\right).$$

Now, $t_n \in (0, u_n)$ is chosen so that $I\left(t_n\right) \leq 2F/u_n$ and also so that $J\left(t_n\right) \leq 2I\left(t_{n-1}\right)/u_n$. (To see that this is possible, assume the contrary so that $(0, u_n) = T_1 \cup T_2$ where

$$
\begin{aligned}
T_1 &= \left\{t \in (0, u_n) : I\left(t\right) > 2F/u_n\right\}, \\
T_2 &= \left\{t \in (0, u_n) : J\left(t\right) > 2I\left(t_{n-1}\right)/u_n\right\}.
\end{aligned}
$$

Then $|T_1| 2F/u_n \leq \int_{T_1} I\left(t\right) dt \leq F$ and since the inequality is strict if $|T_1| > 0$, we have $|T_1| < u_n/2$. The same argument gives $|T_2| < u_n/2$ and we have the desired contraction $|T_1 \cup T_2| < u_n$.) Having completed the construction of $\{u_n\}$ and $\{t_n\}$, we note that, by the defining properties of $\{t_n\}$,

$$\Psi\left(\frac{d\left(f_{t_n}, f_{t_{n-1}}\right)}{\mathrm{p}\left(t_{n-1} - t_n\right)}\right) \leq 2\frac{I\left(t_{n-1}\right)}{u_n} \leq \frac{4F}{u_{n-1} u_n} \leq \frac{4F}{u_n^2}$$

and this implies, using $\mathrm{p}\left(t_{n-1} - t_n\right) \leq \mathrm{p}\left(t_{n-1}\right) \leq 2\mathrm{p}\left(u_n\right) \leq 4\left(\mathrm{p}\left(u_n\right) - \mathrm{p}\left(u_{n+1}\right)\right)$,

$$
\begin{aligned}
d\left(f_{t_n}, f_{t_{n-1}}\right) &\leq \psi^{-1}\left(\frac{4F}{u_n^2}\right) \mathrm{p}\left(|t_{n-1} - t_n|\right) \\
&\leq 4\psi^{-1}\left(\frac{4F}{u_n^2}\right)\left(\mathrm{p}\left(u_n\right) - \mathrm{p}\left(u_{n+1}\right)\right) \\
&\leq 4 \int_{u_{n+1}}^{u_n} \psi^{-1}\left(\frac{4F}{u^2}\right) dp\left(u\right).
\end{aligned}
$$

Using continuity of f, summation over $n = 1, 2, \ldots$ we get

$$d\left(f_0, f_{t_0}\right) \leq 4 \int_0^{u_1} \psi^{-1}\left(\frac{4F}{u^2}\right) d\mathrm{p}\left(u\right)$$

and we are done. ∎

Corollary A.2 (Besov–Hölder embedding) *Let* $q > 1, \alpha \in (1/q, 1)$ *and* $x \in C\left([0, T], E\right)$ *and set*

$$F_{s,t} := |x|_{W^{\alpha,q};[s,t]}^q := \int\int_{[s,t]^2} \frac{d\left(x_v, x_u\right)^q}{|v - u|^{1 + q\alpha}} du\,dv.$$

Then there exists $C = C\left(\alpha, q\right)$ *such that for all* $0 \leq s < t \leq T$,

$$d\left(x_s, x_t\right)^q \quad \leq \quad C\left|t - s\right|^{q\alpha - 1} F_{s,t}, \qquad \text{(A.4)}$$

$$\text{or } |x|_{(\alpha-1/q)\text{-Höl};[s,t]} \quad \leq \quad C|x|_{W^{\alpha,q};[s,t]}, \qquad \text{(A.5)}$$

and a possible choice of the constant is $C = 32\left(\alpha + 1/q\right) / \left(\alpha - 1/q\right)$.

Proof. We take $\Psi\left(u\right) = u^q$ and $\mathrm{p}\left(u\right) = u^{\alpha + 1/q}$ in (A.2) and a simple computation yields the claimed estimate with constant $8.4^{1/q}\left(\alpha + 1q\right) / \left(\alpha - 1/q\right)$ $\leq 32\left(\alpha + 1q\right) / \left(\alpha - 1/q\right)$. ∎

Corollary A.3 (Besov-variation embedding) *Under the assumptions of the previous statement we have, for all* $0 \leq s < t \leq T$,

$$|x|_{(1/\alpha)\text{-var};[s,t]} \leq C\left|t - s\right|^{\alpha - 1/q} |x|_{W^{\alpha,q};[s,t]}.$$

Proof. From (A.4),

$$d\left(x_s, x_t\right)^{1/\alpha} \leq C\left|t - s\right|^{1 - \frac{1}{q\alpha}} F_{s,t}^{\frac{1}{q\alpha}}.$$

Obviously, $\omega_1\left(s, t\right) = t - s$ and $\omega_2\left(s, t\right) = F_{s,t}$ are controls. But then

$$\omega_1^{1-\varepsilon} \omega_2^{\varepsilon}$$

with $\varepsilon = 1/\left(q\alpha\right) \in (0, 1)$ is also a control and so we can replace $d\left(x_s, x_t\right)^{1/\alpha}$ by $\left(|x|_{(1/\alpha)\text{-var};[s,t]}\right)^{1/\alpha}$ in the above estimate. ∎

Exercise A.4 *Assume* $|x|_{(\alpha-1/q)\text{-Höl}} \equiv K < \infty$. *Give a direct proof of (A.4) with* $C = C\left(\alpha, q\right)$, *but not dependent on* K.

Solution. For brevity, let us write $|x_{s,t}|$ instead of $d\left(x_s, x_t\right)$. By the triangle inequality,

$$|x_{s,t}| \leq |x_{s,v}| + |x_{v,u}| + |x_{u,t}|$$

and we average this over $v \in [s, s+\varepsilon]$ and $u \in [t-\varepsilon, t]$, where ε will be chosen later (in fact, equal to $\varepsilon = (1/4)^{\frac{1}{\alpha-1/q}} |t-s|$). This yields

$$|x_{s,t}| \leq \frac{1}{\varepsilon} \int_s^{s+\varepsilon} |x_{s,v}| \, dv + \frac{1}{\varepsilon} \int_{t-\varepsilon}^t |x_{u,t}| \, du + \frac{1}{\varepsilon^2} \int_s^t \int_s^t |x_{v,u}| \, du \, dv.$$

Using the $(1/q - \alpha)$-Hölder modulus of x, the first two integrals on the right-hand side are estimated by $2K\varepsilon^{\alpha-1/q}$. For the last term, we write the double integral as

$$\int_s^t \int_s^t \left(\frac{|x_{v,u}|^q}{|u-v|^{1+q\alpha}} \right)^{\frac{1}{q}} \left(|u-v|^{(1/q+\alpha)q'} \right)^{\frac{1}{q'}} du \, dv$$

$$\leq \quad |x|_{W^{\alpha,q};[s,t]} \left(\int_s^t \int_s^t \left(|u-v|^{(1/q+\alpha)q'} \right) du \, dv \right)^{\frac{1}{q'}}$$

$$\leq \quad |x|_{W^{\alpha,q};[s,t]} |t-s|^{\left(\frac{1}{q}+\alpha \right) + \frac{2}{q'}}$$

where we used Hölder's inequality with $q' = q/(q-1)$. Putting things together, we have

$$|x_{s,t}| \leq 2K\varepsilon^{\alpha-1/q} + \frac{1}{\varepsilon^2} |t-s|^{2+\alpha-1/q} |x|_{W^{\alpha,q};[s,t]}.$$

Choosing $\varepsilon = \delta |t-s|$ makes the right-hand side a multiple of $|t-s|^{\alpha-\frac{1}{q}}$, from which we learn that $K \leq 2\delta^{\alpha-1/q} K + \delta^{-2} |x|_{W^{\alpha,q};[s,t]}$. Choosing δ such that $2\delta^{\alpha-1/q} = 1/2$ turns this into an estimate on K, namely $K \leq 2\delta^{-2} |x|_{W^{\alpha,q};[s,t]}$ and the proof is finished. □

Corollary A.5 (Besov–Lévy modulus embedding) *Let* $x \in C([0,T], E)$, $p > 1$ *and assume*

$$\exists \eta > 0 : \int \int_{[s,t]^2} \exp \left[\eta \left(\frac{d(x_u, x_v)}{|v-u|^{1/p}} \right)^2 \right] du \, dv = F_{s,t} < \infty.$$

Then there exists a constant C *depending on* p *only such that for all* $0 \leq s < t \leq T$,

$$d(x_s, x_t) \leq C \frac{1}{\eta^{1/2}} \zeta(t-s) \times \sqrt{\log(F_{s,t} \vee 4)}$$

where $\zeta(h) = \int_0^h u^{1/p-1} \sqrt{\log(1+1/u^2)} du$. *As a consequence,*

$$\exp \left[\frac{\eta}{C^2} \left(\sup_{s \leq u < v \leq t} \frac{d(x_u, x_v)}{\zeta(v-u)} \right)^2 \right] \leq F_{s,t} \vee 4.$$

Remark A.6 $\zeta(h) \sim h^{1/p}\sqrt{\log 1/h}$ as $h \to 0+$ in the sense that their ratio converges to a constant $c \in (0, \infty)$. $\square$

Remark A.7 *From monotonicity of ζ and $(s,t) \mapsto F_{s,t}$ we see that for any $[u,v] \subset [s,t]$,*

$$|x|_{0;[u,v]} \le C\frac{1}{\eta^{1/2}}\zeta(u-v) \times \sqrt{\log(F_{s,t} \vee 4)}$$

so that

$$\exp\left[\frac{\eta}{C^2}\left(\sup_{s \le u < v \le t} \frac{|x|_{0;[u,v]}}{\zeta(v-u)}\right)^2\right] \le F_{s,t} \vee 4.$$
$\square$

Proof. Using $\Psi(u) = e^{\eta u^2} - 1$ and $p(u) = u^{1/p}$ in (A.2) leads to an estimate of the form

$$d(x_s, x_t) \le 8/p \int_0^{t-s} u^{1/p-1}\sqrt{\frac{1}{\eta}\log(1+4F_{s,t}/u^2)}du.$$

Obviously, we may replace $F_{s,t}$ by $\tilde{F}_{s,t} = F_{s,t} \vee 4$ in the above estimate. Then, by a change of variable,

$$d(x_s, x_t) \le \frac{8}{p\eta^{1/2}}\tilde{F}_{s,t}^{1/(2p)}\zeta\left(\frac{t-s}{\sqrt{\tilde{F}_{s,t}}}\right).$$

It is easy to check that for suitable constants $c_1, c_2 > 0$

$$\zeta(uv) \le c_1\zeta(u)\zeta(v) \quad \text{for } u,v \in (0,1),$$
$$\zeta(u) \le c_2 u^{1/p}\sqrt{\log 1/u} \quad \text{for } u \in (0,1/2).$$

We see that $\zeta\left((t-s)/\sqrt{\tilde{F}_{s,t}}\right) \le (\text{const}) \times \zeta(t-s) \times \tilde{F}_{s,t}^{-1/2p}\sqrt{\log \tilde{F}_{s,t}}$ and so

$$d(x_s, x_t) \le c_3\eta^{1/2}\zeta(t-s)\sqrt{\log\left(\tilde{F}_{s,t}\right)},$$

as claimed. $\blacksquare$

A.2.2 *Garsia–Rodemich–Rumsey on $G^N(\mathbb{R}^d)$-valued paths*

We now specialize from the general metric setting to the free step-N nilpotent group $(G^N(\mathbb{R}^d), \otimes)$, a metric space under $(g,h) \mapsto d(g,h) = \|g^{-1} \otimes h\|$. The resulting path spaces were discussed in Section 8.1.

Proposition A.8 *Let $q > 2r \ge 2$ and $\mathbf{x} \in C([0,T], G^N(\mathbb{R}^d))$ such that*

$$\int_0^T \int_0^T \frac{d(x_s, x_t)^q}{|t-s|^{q/r}}dsdt \le M^q.$$

Set $p = \left(\frac{1}{r} - \frac{2}{q}\right)^{-1} > 1$. Then, there exists $C = C(q,r)$, which can be chosen non-increasing in $q \in (2r, \infty)$,

$$\|x\|_{1/p\text{-Höl};[0,T]} \leq CM.$$

Proof. An immediate consequence of Corollary A.2, which also shows that a possible choice of C is given by $C(q,r) = \frac{32}{r} / \left(\frac{1}{r} - \frac{2}{q}\right)$. ∎

We observe that the constant in the previous proposition does not depend on T. One can reconfirm this with the following scaling argument: rescale by defining $\tilde{x}(\cdot) := x(T\cdot)$ so that

$$\|\tilde{x}\|_{1/p\text{-Höl};[0,1]} = T^{1/p} \|x\|_{1/p\text{-Höl};[0,T]}$$

$$\left(\int_0^1 \int_0^1 \frac{d(\tilde{x}_s, \tilde{x}_t)^q}{|t-s|^{q/r}} ds\, dt\right)^{\frac{1}{q}} = T^{1/r - 2/q} \left(\int_0^T \int_0^T \frac{d(x_s, x_t)^q}{|t-s|^{q/r}} ds\, dt\right)^{\frac{1}{q}},$$

with identical scaling in T since $1/p = 1/r - 2/q$. Similarly, we have

$$\left(\int_0^1 \int_0^1 \frac{|\pi_k(\mathbf{x}_{s,t} - \mathbf{y}_{s,t})|^{q/k}}{|t-s|^{q/r}} ds\, dt\right)^{k/q}$$

$$= T^{k/r - 2k/q} \left(\int_0^T \int_0^T \frac{|\pi_k(\mathbf{x}_{s,t} - \mathbf{y}_{s,t})|^q}{|t-s|^{qk/r}} ds\, dt\right)^{1/q}$$

$$= T^{k/p} \left(\int_0^T \int_0^T \frac{|\pi_k(\mathbf{x}_{s,t} - \mathbf{y}_{s,t})|^q}{|t-s|^{qk/r}} ds\, dt\right)^{1/q},$$

$$\sup_{0 \leq s < t \leq 1} \frac{|\pi_k(\mathbf{x}_{s,t} - \mathbf{y}_{s,t})|}{|t-s|^{k/p}} = T^{k/p} \sup_{0 \leq s < t \leq T} \frac{|\pi_k(\mathbf{x}_{s,t} - \mathbf{y}_{s,t})|}{|t-s|^{k/p}},$$

which reduces the proof of the following result to $T = 1$.

Proposition A.9 *Let $q > 2r \geq 2$ and $\mathbf{x}, \mathbf{y} \in C\left([0,T], G^N(\mathbb{R}^d)\right)$ such that, for non-negative constants M and ε,*

$$\left(\int_0^T \int_0^T \frac{\|\mathbf{x}_{s,t}\|^q}{|t-s|^{q/r}} ds\, dt\right)^{1/q} \leq M \text{ and } \left(\int_0^T \int_0^T \frac{\|\mathbf{y}_{s,t}\|^q}{|t-s|^{q/r}} ds\, dt\right)^{1/q} \leq M,$$
$$\text{(A.6)}$$

$$\left(\int_0^T \int_0^T \frac{|\pi_k(\mathbf{x}_{s,t} - \mathbf{y}_{s,t})|^{q/k}}{|t-s|^{q/r}} ds\, dt\right)^{k/q} \leq \varepsilon M, \text{ for } k = 1, \ldots, N. \quad \text{(A.7)}$$

Then, setting $p = \left(\frac{1}{r} - \frac{2}{q}\right)^{-1} > 1$, there exists $C = C(q,r)$, non-increasing in q, such that

$$\|\mathbf{x}\|_{1/p\text{-Höl};[0,T]} \leq CM \text{ and } \|\mathbf{y}\|_{1/p\text{-Höl};[0,T]} \leq CM \quad \text{(A.8)}$$

and, for all $k = 1, \ldots, N,$

$$\sup_{0 \leq s < t \leq T} \frac{|\pi_k\left(\mathbf{x}_{s,t} - \mathbf{y}_{s,t}\right)|}{|t - s|^{k/p}} \leq \varepsilon C M^k \qquad (A.9)$$

where C *can be chosen non-increasing in* $q \in (2r, \infty)$.

Proof. The above scaling argument, as already pointed out, allows us to assume that $T = 1$. Moreover, at the cost of replacing $\mathbf{x}$ and $\mathbf{y}$ by $\delta_{1/M}\mathbf{x}$ and $\delta_{1/M}\mathbf{y}$, we can and will assume that $M = 1$.

In this proof, all the constants c_i, when dependent on q, but will be non-decreasing in q.

That inequality (A.8) holds true follows from Proposition A.8. We now prove (A.9) by induction over the level $k \in \{1, \ldots, N\}$. The case $k = 1$ is, again, a consequence of Proposition A.8. Let us now assume that it is true for all levels $1, \ldots, k - 1$ and establish the estimate for level k. Fix $s < t \in [0, 1]$, and define

$$z_r^s = \pi_k\left(\mathbf{x}_{s,s+r} - \mathbf{y}_{s,s+r}\right).$$

Fix $u < v$ in $[0, t - s]$. Using $\mathbf{x}_{s,s+v} - \mathbf{x}_{s,s+u} = \mathbf{x}_{s,s+u} \otimes (\mathbf{x}_{s+u,s+v} - 1)$ and $\pi_j\left(\mathbf{x}_{s+u,s+v} - 1\right) = 0$ (or $\pi_j\left(\mathbf{x}_{s+u,s+v}\right)$) for $j = 0$ (or $j > 0$), we have

$$
\begin{aligned}
z_v^s - z_u^s &= \pi_k\left(\mathbf{x}_{s,s+v} - \mathbf{x}_{s,s+u}\right) - \pi_k\left(\mathbf{y}_{s,s+v} - \mathbf{y}_{s,s+u}\right) \\
&= \sum_{j=1}^{k} \pi_{k-j}\left(\mathbf{x}_{s,s+u}\right) \otimes \pi_j\left(\mathbf{x}_{s+u,s+v}\right) - \sum_{j=1}^{k} \pi_{k-j}\left(\mathbf{y}_{s,s+u}\right) \\
&\qquad \otimes \pi_j\left(\mathbf{y}_{s+u,s+v}\right) \\
&= \sum_{j=1}^{k} \pi_{k-j}\left(\mathbf{x}_{s,s+u}\right) \otimes \pi_j\left(\mathbf{x}_{s+u,s+v} - \mathbf{y}_{s+u,s+v}\right) \\
&\qquad + \sum_{j=1}^{k} \pi_{k-j}\left(\mathbf{x}_{s,s+u} - \mathbf{y}_{s,s+u}\right) \otimes \pi_j\left(\mathbf{y}_{s+u,s+v}\right).
\end{aligned}
$$

Furthermore, using $\pi_0\left(\mathbf{x}_{s,s+u} - \mathbf{y}_{s,s+u}\right) = 0$, we obtain

$$
\begin{aligned}
z_v^s - z_u^s &= \sum_{j=1}^{k-1} \pi_{k-j}\left(\mathbf{x}_{s,s+u}\right) \otimes \pi_j\left(\mathbf{x}_{s+u,s+v} - \mathbf{y}_{s+u,s+v}\right) \\
&\quad + \sum_{j=1}^{k-1} \pi_{k-j}\left(\mathbf{x}_{s,s+u} - \mathbf{y}_{s,s+u}\right) \otimes \pi_j\left(\mathbf{y}_{s+u,s+v}\right) \\
&\quad + \pi_k\left(\mathbf{x}_{s+u,s+v} - \mathbf{y}_{s+u,s+v}\right).
\end{aligned}
$$

Hence,

$$\left(\int_0^{t-s} \int_0^{t-s} \frac{|z_v^s - z_u^s|^q}{|v - u|^{q/r}} dv du\right)^{1/q} \leq \Delta_1 + \Delta_2 + \Delta_3$$

where

$$\Delta_1 =$$

$$c_1 \sum_{j=1}^{k-1} \left(\int_0^{t-s} \int_0^{t-s} \|\mathbf{x}_{s,s+u}\|^{q(k-j)} \frac{|\pi_j \left(\mathbf{x}_{s+u,s+v} - \mathbf{y}_{s+u,s+v} \right)|^q}{|v-u|^{q/r}} du\, dv \right)^{1/q},$$

$$\Delta_2 =$$

$$c_2 \sum_{j=1}^{k-1} \left(\int_0^{t-s} \int_0^{t-s} |\pi_{k-j} \left(\mathbf{x}_{s,s+u} - \mathbf{y}_{s,s+u} \right)|^q \frac{\|\mathbf{y}_{s+u,s+v}\|^{qj}}{|v-u|^{q/r}} du\, dv \right)^{1/q},$$

$$\Delta_3 =$$

$$c_3 \left(\int_0^{t-s} \int_0^{t-s} \frac{|\pi_k \left(\mathbf{x}_{s+u,s+v} - \mathbf{y}_{s+u,s+v} \right)|^q}{|v-u|^{q/r}} du\, dv \right)^{1/q}.$$

Now, $\|\mathbf{x}\|_{1/p\text{-H\"ol};[0,1]}$, $\|\mathbf{y}\|_{1/p\text{-H\"ol};[0,1]} \leq C_1\left(q,r\right)$ by Proposition A.8 for a constant C_1, non-increasing in q. Hence,

$$\Delta_1 \leq c_4 \sum_{j=1}^{k-1} |t-s|^{\frac{(k-j)}{p}} \left(\int_0^{t-s} \int_0^{t-s} \frac{|\pi_j \left(\mathbf{x}_{s+u,s+v} - \mathbf{y}_{s+u,s+v} \right)|^q}{|v-u|^{q/r}} du\, dv \right)^{1/q}.$$

From the induction hypothesis, we have

$$|\pi_j \left(\mathbf{x}_{s+u,s+v} - \mathbf{y}_{s+u,s+v} \right)|^{q\left(1-\frac{1}{j}\right)} \leq c_5 \varepsilon^{q\left(1-\frac{1}{j}\right)} |t-s|^{\frac{q}{p}(j-1)}. \qquad (A.10)$$

Hence, we obtain

$$
\begin{aligned}
\Delta_1 \;\leq\; & c_6 \sum_{j=1}^{k-1} |t-s|^{\frac{(k-1)}{p}} \varepsilon^{\left(1-\frac{1}{j}\right)} \\
& \left(\int_0^{t-s} \int_0^{t-s} \frac{|\pi_j \left(\mathbf{x}_{s+u,s+v} - \mathbf{y}_{s+u,s+v} \right)|^{q/j}}{|v-u|^{qj/r}} du\, dv \right)^{1/q} \\
\leq\; & c_6 \varepsilon \sum_{j=1}^{k-1} |t-s|^{\frac{(k-1)}{p}} \quad \text{by assumption (A.7)}.
\end{aligned}
$$

For $u < v < t-s$ we also have $\|\mathbf{y}_{s+u,s+v}\|^{qj} \leq C_1\left(q,r\right)^q |t-s|^{q(j-1)/p} \|\mathbf{y}_{s+u,s+v}\|^q$ so that

$$
\begin{aligned}
\Delta_2 \;\leq\; & c_7 \sum_{j=1}^{k-1} |t-s|^{\frac{(j-1)}{p}} \\
& \left(\int_0^{t-s} \int_0^{t-s} |\pi_{k-j} \left(\mathbf{x}_{s,s+u} - \mathbf{y}_{s,s+u} \right)|^q \frac{\|\mathbf{y}_{s+u,s+v}\|^q}{|v-u|^{q/r}} du\, dv \right)^{1/q}.
\end{aligned}
$$

From the induction hypothesis, we have

$$\left|\pi_{k-j}\left(\mathbf{x}_{s,s+u}-\mathbf{y}_{s,s+u}\right)\right| \leq c_8 \varepsilon u^{\frac{k-j}{p}} \leq c_8 \varepsilon \left|t-s\right|^{\frac{k-j}{p}};$$

in particular, we see that

$$\Delta_2 \leq c_9 \varepsilon \left|t-s\right|^{\frac{(k-1)}{p}} \sum_{j=1}^{k-1}\left(\int_0^{t-s}\int_0^{t-s}\frac{\left\|\mathbf{y}_{s+u,s+v}\right\|^q}{\left|v-u\right|^{q/r}}dudv\right)^{1/q}$$

$$\leq c_{10} \varepsilon \left|t-s\right|^{\frac{(k-1)}{p}} \text{ by assumption on } \mathbf{y}.$$

Finally, using assumption (A.7), we have, defining $\Upsilon_{s,t} = \sup_{u,v\in[s,t]}$
$\left(\frac{\left|\pi_k\left(\mathbf{x}_{u,v}-\mathbf{y}_{u,v}\right)\right|}{\left|v-u\right|^{k/p}}\right)$,

$$\Delta_3 = c_3\left(\int_0^{t-s}\int_0^{t-s}\frac{\left|\pi_k\left(\mathbf{x}_{s+u,s+v}-\mathbf{y}_{s+u,s+v}\right)\right|^{q/k}}{\left|v-u\right|^{q/r}}dudv\right)^{1/q}\Upsilon_{s,t}^{1-1/k}$$

$$\leq c_3\varepsilon^{1/k}\sup_{u,v\in[s,t]}\left|\pi_k\left(\mathbf{x}_{u,v}-\mathbf{y}_{u,v}\right)\right|^{\left(1-\frac{1}{k}\right)}\left|t-s\right|^{\frac{k-1}{p}}$$

(with $c_9 = \max\left(c_3,1\right)$, not dependent on q). Hence, we see that

$$\frac{1}{\left|t-s\right|^{\frac{k-1}{p}}}\left(\int_0^{t-s}\int_0^{t-s}\frac{\left|z_v^s-z_u^s\right|^q}{\left|v-u\right|^{q/r}}dvdu\right)^{1/q} \leq c_{11}\left(\varepsilon+\varepsilon^{1/k}\Upsilon_{s,t}^{1-1/k}\right).$$

Another application of Proposition A.8 gives

$$\frac{\left|z_t^s-z_s^s\right|}{\left|t-s\right|^{k/p}} \leq c_{12}\left(\varepsilon+\varepsilon^{1/k}\Upsilon_{s,t}^{1-1/k}\right),$$

i.e. we prove that for all $s,t\in[0,T]$,

$$\frac{\left|\pi_k\left(\mathbf{x}_{s,t}-\mathbf{y}_{s,t}\right)\right|}{\left|t-s\right|^{k/p}} \leq c_{12}\left(\varepsilon+\varepsilon^{1/k}\Upsilon_{s,t}^{1-1/k}\right),$$

which readily implies that

$$\Upsilon_{s,t} \leq c_{12}\left(\varepsilon+\varepsilon^{1/k}\Upsilon_{s,t}^{1-1/k}\right),$$

or

$$\frac{\Upsilon_{s,t}}{\varepsilon} \leq c_{12}\left[1+\left(\frac{\Upsilon_{s,t}}{\varepsilon}\right)^{1-\frac{1}{k}}\right].$$

This last inequality implies that $\frac{\Upsilon_{s,t}}{\varepsilon} \leq c_{13}$, which concludes the induction. ∎

A.3 Kolmogorov-type corollaries

A.3.1 Hölder regularity and tightness

Let $(X_t : t \in [0,T])$ be a *stochastic process* with values in some Polish space (E,d) and assume there exist positive constants a, b, c such that for all $s, t \in [0,T]$,

$$\mathbb{E}\left(d\left(X_s, X_t\right)^a\right) \le c\,|t-s|^{1+b}.$$

Kolmogorov's criterion asserts that X then has a continuous version, see [144] for instance, which we shall also denote by X without further notice. In fact, as is well known (and will be seen below), X can be chosen with γ-Hölder continuous sample paths for any $\gamma < b/a$. The above condition is equivalent to the existence of $M > 0$ and $q > r \ge 1$ such that

$$|d\left(X_s, X_t\right)|_{L^q(\mathbb{P})} \le M\,|t-s|^{1/r} \tag{A.11}$$

and we find it convenient in applications to formulate the following criteria in this form.

Theorem A.10 (Kolmogorov) *Let* $M > 0, q > r \ge 1$ *and assume (A.11) holds for all* $s, t \in [0,T]$. *Then, for any* $\gamma \in [0, 1/r - 1/q)$ *there exists* $C = C\left(r, q, \gamma, T\right)$ *non-increasing in* q, *such that*

$$\mathbb{E}\left(|X|^q_{\gamma\text{-H\"ol};[0,T]}\right)^{\frac{1}{q}} \le CM.$$

Proof. Fix $\gamma \in [0, 1/r - 1/q)$. Since $q > 1$ and $\alpha := 1/q + \gamma < 1/r \le 1$, the Besov–Hölder embedding, established in Corollary A.2 (equivalently: Proposition A.8) shows that there exists a constant c_1 such that

$$\left|\frac{d\left(X_s, X_t\right)}{|t-s|^\gamma}\right|^q \le c_1 \int\int_{[0,T]^2} \frac{d\left(X_s, X_t\right)^q}{|t-s|^{1+q\alpha}}\,ds\,dt = c_1 \int\int_{[0,T]^2} \frac{d\left(X_s, X_t\right)^q}{|t-s|^{2+q\gamma}}\,ds\,dt.$$

After taking $\sup_{s,t\in[0,T]}$ and expectations, we have

$$\mathbb{E}\left(|X|^q_{\gamma\text{-H\"ol};[0,T]}\right) \le (c_1 M)^q \int\int_{[0,T]^2} |t-s|^{-2+q(1/r-\gamma)}\,ds\,dt < \infty,$$

using $1/r - \gamma > 1/q$ to see that the last double integral is finite. ∎

Assume now that E has enough structure so that bounded sets in $C^{\gamma'\text{-H\"ol}}([0,T], E)$ are precompact in $C^{\gamma\text{-H\"ol}}([0,T], E)$ for $\gamma < \gamma'$. This requires interpolation and Arzela–Ascoli and holds true, for instance, when $E = \mathbb{R}$ or more generally $G^N\left(\mathbb{R}^d\right)$, the step-$N$ free nilpotent group over $\mathbb{R}^d$, equipped with Carnot–Caratheodory distance. It will be enough for us to focus on this case. We have

Corollary A.11 (Kolmogorov–Lamperti tightness criterion) *Let* $(\mathbf{X}_t^n : t \in [0,T])$ *be a sequence of continuous* $G^N(\mathbb{R}^d)$*-valued processes. Let* $M > 0, q > r \geq 1$ *and assume*

$$\sup_n |(\mathbf{X}_s^n, \mathbf{X}_t^n)|_{L^q(\mathbb{P})} \leq M\,|t-s|^{1/r}$$

holds for all $s,t \in [0,T]$. *Then* $(\mathbf{X}^n)$ *is tight in* $C^{\gamma\text{-Höl}}([0,1], G^N(\mathbb{R}^d))$ *for any* $\gamma \in [0, 1/r - 1/q)$.

Proof. Take $\gamma < \gamma' < 1/r - 1/q$. By the previous theorem

$$\sup_n \mathbb{E}\left(\|\mathbf{X}^n\|_{\gamma'\text{-Höl};[0,T]}^q\right) \leq (CM)^q < \infty.$$

Writing $B_R = \left\{ \mathbf{x} \in C^{\gamma'\text{-Höl}}([0,T], G^N(\mathbb{R}^d)) : \|\mathbf{x}\|_{\gamma'\text{-Höl};[0,T]} < R \right\}$ it is clear from Chebyshev's inequality that

$$\sup_n \mathbb{P}[\mathbf{X}^n \in B_R] \to 0 \text{ as } R \to \infty.$$

The proof is then finished with the remark that B_R is precompact with respect to γ-Hölder topology in $C^{\gamma\text{-Höl}}([0,1], G^N(\mathbb{R}^d))$. ∎

Although not strictly necessary for the next result, we remain in the setting of $G^N(\mathbb{R}^d)$-valued processes for the remainder of this section. The following is a variation of Theorem A.10 with the feature that the constant C does not depend on q. This will be important since in a typical application M itself will be taken as a function of q (e.g. $M = O(q^{1/2})$ as $q \to \infty$ in a Gaussian setting). and it is important that C_2 does not depend on q.

Theorem A.12 *Let* $(\mathbf{X}_t : t \in [0,T])$ *be a continuous* $G^N(\mathbb{R}^d)$*-valued process. For any* $\gamma \in [0, 1/r)$ *there exist* $q_0(r, \gamma)$ *and* $C = C(r, \gamma, T)$ *such that if*

$$\forall s,t \in [0,T] : |d(\mathbf{X}_s, \mathbf{X}_t)|_{L^q(\mathbb{P})} \leq M\,|t-s|^{1/r}$$

holds for some $q \geq q_0$, *then we also have*

$$\mathbb{E}\left(\|\mathbf{X}\|_{\gamma\text{-Höl};[0,T]}^q\right)^{\frac{1}{q}} \leq CM$$

or, equivalently, for all $k = 1, \dots, N$,

$$\left| \sup_{0 \leq s < t \leq T} \frac{|\pi_k(\mathbf{X}_{s,t})|}{|t-s|^{k\gamma}} \right|_{L^{q/k}} \leq (CM)^k.$$

Proof. Pick $q_0 = q_0(\gamma, r)$ large enough so that for all $q \geq q_0 : \gamma < 1/r - 2/q$. It follows that

$$\|\mathbf{X}\|_{\gamma\text{-Höl};[0,T]}^q \leq c_1^q \|\mathbf{X}\|_{(1/r-2/q)\text{-Höl};[0,T]}^q$$

$$\leq c_2^q \int_0^T \int_0^T \frac{d(\mathbf{X}_s, \mathbf{X}_t)^q}{|t-s|^{q/r}} ds dt$$

with c_1, c_2 dependent on γ, r, T but not on q (we used the fact that the constant in Proposition A.8 can be chosen non-increasing in q). After taking $\sup_{s,t \in [0,T]}$ and expectations, we have

$$\mathbb{E}\left(\|\mathbf{X}\|_{\gamma\text{-Höl};[0,T]}^{q}\right) \le (c_2 M)^q \, T^2 < \infty. \qquad \blacksquare$$

If the previous result was about γ-Hölder sample path regularity of a $G^N\left(\mathbb{R}^d\right)$-valued process, the following is about the γ-Hölder distance of two such processes.

Theorem A.13 *Let* $\mathbf{X}, \mathbf{Y}$ *be continuous* $G^N\left(\mathbb{R}^d\right)$*-valued processes. Let* $\gamma \in [0, 1/r)$. *Assume that for some constant* M *and for* $q \ge q_0\left(r, \gamma\right)$, *we have for all* $s, t \in [0, T]$,

$$|d\left(\mathbf{X}_s, \mathbf{X}_t\right)|_{L^q(\mathbb{P})} \le M |t - s|^{1/r}, \qquad (A.12)$$

$$|d\left(\mathbf{Y}_s, \mathbf{Y}_t\right)|_{L^q(\mathbb{P})} \le M |t - s|^{1/r}, \qquad (A.13)$$

$$|\pi_k\left(\mathbf{X}_{s,t} - \mathbf{Y}_{s,t}\right)|_{L^{q/k}(\mathbb{P})} \le \varepsilon M^k |t - s|^{k/r} \text{ for } k \in \{1, \ldots, N\}. \qquad (A.14)$$

Then, there exists $C = C\left(r, \gamma, T, N\right)$ *such that (i) for all* $k = 1, \ldots, N$,

$$\left|\sup_{0 \le s < t \le T} \frac{|\pi_k\left(\mathbf{X}_{s,t}\right)|}{|t - s|^{k\gamma}}\right|_{L^{q/k}}, \left|\sup_{0 \le s < t \le T} \frac{|\pi_k\left(\mathbf{Y}_{s,t}\right)|}{|t - s|^{k\gamma}}\right|_{L^{q/k}} \le (CM)^k$$

$$\left|\sup_{0 \le s < t \le T} \frac{|\pi_k\left(\mathbf{X}_{s,t} - \mathbf{Y}_{s,t}\right)|}{|t - s|^{k\gamma}}\right|_{L^{q/k}} \le \varepsilon\,(CM)^k;$$

and (ii)

$$|d_{\gamma\text{-Höl}}\left(\mathbf{X}, \mathbf{Y}\right)|_{L^{q/N}} \le C \max\left(\varepsilon, \varepsilon^{1/N}\right) M.$$

Remark A.14 *In a typical (Gaussian) application, assumptions (A.12), (A.13), (A.14) hold for all* q *with* $M = Cq^{1/2}$. *In this case, the conclusions take the form*

$$\forall k \in \{1, \ldots, N\} : \left|\sup_{0 \le s < t \le T} \frac{|\pi_k\left(\mathbf{X}_{s,t} - \mathbf{Y}_{s,t}\right)|}{|t - s|^{k\gamma}}\right|_{L^q} \le \varepsilon \tilde{C} q^{k/2}$$

for $\tilde{C} = \tilde{C}\left(r, \gamma, T, N\right)$ *and similarly,*

$$|d_{1/p\text{-Höl}}\left(\mathbf{X}, \mathbf{Y}\right)|_{L^q} \le \tilde{C} \max\left(\varepsilon, \varepsilon^{1/N}\right) q^{1/2}. \qquad (A.15)$$

Proof. Pick $q_0 = q_0\left(\gamma, r\right)$ large enough so that for all $q \ge q_0 : \gamma < 1/r - 2/q =: 1/p$. It follows from Proposition A.9 that there exists c_1 which can be chosen independent of q,

$$\frac{1}{\varepsilon} \sup_{0 \le s < t \le T} \frac{|\pi_k (\mathbf{X}_{s,t} - \mathbf{Y}_{s,t})|}{|t-s|^{k\gamma}}$$

$$\le \sup_{0 \le s < t \le T} \frac{|\pi_k (\mathbf{X}_{s,t} - \mathbf{Y}_{s,t})|}{|t-s|^{k/p}}$$

$$\le c_1 \max_{k=1,\dots,N} \left| \int_0^T \int_0^T \frac{|\pi_k (\mathbf{X}_{s,t})|^{q/k}}{|t-s|^{q/r}} ds dt \right|^{k/q}$$

$$+ c_1 \max_{k=1,\dots,N} \left| \int_0^T \int_0^T \frac{|\pi_k (\mathbf{Y}_{s,t})|^{q/k}}{|t-s|^{q/r}} ds dt \right|^{k/q}$$

$$+ c_1 \frac{1}{\varepsilon} \max_{k=1,\dots,N} \left| \int_0^T \int_0^T \frac{|\pi_k (\mathbf{X}_{s,t} - \mathbf{Y}_{s,t})|^{q/k}}{|t-s|^{q/r}} ds dt \right|^{k/q}.$$

Hence, if $\Delta = \left(\frac{1}{\varepsilon}\right)^{\frac{q}{k}} \mathbb{E} \left(\left| \sup_{0 \le s < t \le T} \frac{|\pi_k (\mathbf{X}_{s,t} - \mathbf{Y}_{s,t})|}{|t-s|^{k\gamma}} \right|^{\frac{q}{k}} \right)$, we have

$$\Delta \le c_2^{q/k} \max_{k=1,\dots,N} \int_0^T \int_0^T \frac{\mathbb{E}\left(|\pi_k (\mathbf{X}_{s,t})|^{q/k}\right)}{|t-s|^{q/r}} ds dt$$

$$+ c_2^{q/k} \max_{k=1,\dots,N} \int_0^T \int_0^T \frac{\mathbb{E}\left(|\pi_k (\mathbf{Y}_{s,t})|^{q/k}\right)}{|t-s|^{q/r}} ds dt$$

$$+ c_2^{q/k} \frac{1}{\varepsilon^{q/k}} \max_{k=1,\dots,N} \mathbb{E} \left(\int_0^T \int_0^T \frac{|\pi_k (\mathbf{X}_{s,t} - \mathbf{Y}_{s,t})|^{q/k}}{|t-s|^{q/r}} ds dt \right)$$

$$\le c_3 c_2^{q/k} M^q + \left(\frac{c_2}{\varepsilon}\right)^{q/k} \varepsilon^{q/k} M^q = (c_4 M)^q,$$

which is equivalent to $\left| \sup_{0 \le s < t \le T} \frac{|\pi_k (\mathbf{X}_{s,t} - \mathbf{Y}_{s,t})|}{|t-s|^{k\gamma}} \right|_{L^{q/k}} \le c_5 \varepsilon M^k$, which is what we wanted to prove.

(ii) Take $g = \delta_{1/\lambda} \mathbf{X}_{s,t}$ with $\lambda = CM |t-s|^{1/\gamma}$ and similarly $h = \delta_{1/\lambda} \mathbf{Y}_{s,t}$. Note that by part (i),

$$\left| \sup_{s,t} \frac{|\pi_k (\mathbf{X}_{s,t})|}{\lambda^k} \right|_{L^{q/k}}, \left| \sup_{s,t} \frac{|\pi_k (\mathbf{Y}_{s,t})|}{\lambda^k} \right|_{L^{q/k}} \le 1$$

$$\left| \sup_{s,t} \frac{|\pi_k (\mathbf{X}_{s,t} - \mathbf{Y}_{s,t})|}{\lambda^k} \right|_{L^{q/k}} \le \varepsilon.$$

From Proposition 7.49,

$$d(g,h) \le c_6 \left(|g-h| + |g-h|^{1/N} \max\left(1, \|g\|\right)^{1-1/N} \right)$$

and so

$$\sup_{s,t} \frac{|d\left(\mathbf{X}_{s,t}, \mathbf{Y}_{s,t}\right)|}{\lambda}$$

$$\leq c_6 \max_{k=1,\ldots,N} \sup_{s,t} \frac{|\pi_k\left(\mathbf{X}_{s,t} - \mathbf{Y}_{s,t}\right)|}{\lambda^k}$$

$$+ c_6 \max_{k=1,\ldots,N} \sup_{s,t} \left| \frac{|\pi_k\left(\mathbf{X}_{s,t} - \mathbf{Y}_{s,t}\right)|}{\lambda^k} \right|^{1/N}$$

$$\max\left(1, \max_{k=1,\ldots,N} \sup_{s,t} \left| \frac{|\pi_k\left(\mathbf{X}_{s,t}\right)|^{1/k}}{\lambda} \right| \right)^{1-1/N}.$$

From Hölder's inequality, $\left|A^{1/N}B^{(1-1/N)}\right|_{L^p} \leq |A|_{L^p}^{1/N} |B|_{L^p}^{(1-1/N)}$, we can then bound $\left|\sup_{s,t} |d\left(\mathbf{X}_{s,t}, \mathbf{Y}_{s,t}\right)|/\lambda\right|_{L^{q/N}}$ by a constant (namely c_6) times

$$\max_k \left| \sup_{s,t} \frac{|\pi_k\left(\mathbf{X}_{s,t} - \mathbf{Y}_{s,t}\right)|}{\lambda^k} \right|_{L^{q/N}}$$

$$+ \max_k \left| \sup_{s,t} \left| \frac{|\pi_k\left(\mathbf{X}_{s,t} - \mathbf{Y}_{s,t}\right)|}{\lambda^k} \right|^{1/N} \right.$$

$$\left. \max\left(1, \max_k \sup_{s,t} \left| \frac{|\pi_k\left(\mathbf{X}_{s,t}\right)|^{1/k}}{\lambda} \right| \right)^{1-1/N} \right|_{L^{q/N}}.$$

$$\leq \max_{k=1} \left| \sup_{s,t} \frac{|\pi_k\left(\mathbf{X}_{s,t} - \mathbf{Y}_{s,t}\right)|}{\lambda^k} \right|_{L^{q/k}}$$

$$+ \max_{k=1} \left| \sup_{s,t} \left| \frac{|\pi_k\left(\mathbf{X}_{s,t} - \mathbf{Y}_{s,t}\right)|}{\lambda^k} \right| \right|^{1/N}.$$

$$\max\left(1, \max_k \left| \sup_{s,t} \left| \frac{|\pi_k\left(\mathbf{X}_{s,t}\right)|^{1/k}}{\lambda} \right| \right|_{L^{q/k}} \right)^{1-1/N}$$

$$= \varepsilon + \varepsilon^{1/N} \qquad \text{using } \left| \sup_{s,t} \left| \frac{|\pi_k\left(\mathbf{X}_{s,t}\right)|^{1/k}}{\lambda} \right| \right|_{L^{q/k}} \leq 1.$$

Since $\lambda = CM |t-s|^{1/\gamma}$, the proof is finished. ∎

A.3.2 L^q-convergence for rough paths

Theorem A.13 can be obviously used to establish L^q-convergence (with quantitative estimates!) of a sequence of continuous $G^N\left(\mathbb{R}^d\right)$-valued processes. It can be useful to have the following "soft" criterion for L^q-convergence which only relies on basic interpolation estimates.

Proposition A.15 *Let* $\mathbf{X}^n, \mathbf{X}^\infty$ *be continuous* $G^N\left(\mathbb{R}^d\right)$*-valued processes defined on* $[0,T]$*. Let* $q \in [1,\infty)$ *and assume that we have pointwise convergence in* $L^q\left(\mathbb{P}\right)$*, i.e. for all* $t \in [0,T]$*,*

$$d\left(\mathbf{X}_t^n, \mathbf{X}_t^\infty\right) \to 0 \ in \ L^q\left(\mathbb{P}\right) \ as \ n \to \infty; \tag{A.16}$$

and uniform Hölder bounds, i.e.

$$\sup_{1 \le n \le \infty} \mathbb{E}\left(\|\mathbf{X}^n\|_{\alpha\text{-Höl};[0,T]}^q\right) < \infty. \tag{A.17}$$

Then for $\alpha' < \alpha$*,*

$$d_{\alpha'\text{-Höl};[0,T]}\left(\mathbf{X}^n, \mathbf{X}^\infty\right) \to 0 \ in \ L^q\left(\mathbb{P}\right).$$

Remark A.16 To check condition (A.17) one typically uses Theorem A.12. Let us also note that (A.16) may be replaced by the assumption of pointwise convergence in probability; in this case, it is clear from (A.17) that $d\left(\mathbf{X}_t^n, \mathbf{X}_t^\infty\right)^{q'}$ is uniformly integrable for any $q' \in [1,q)$ and hence the conclusion becomes

$$d_{\alpha'\text{-Höl}}\left(\mathbf{X}^n, \mathbf{X}^\infty\right) \to 0 \ in \ L^{q'}\left(\mathbb{P}\right).$$

Obviously, there is no need to distinguish between q and q' if (A.17) holds for all $q < \infty$. (This is typical in our applications.) $\square$

Proof. It is enough to show d_∞ convergence in L^q. Indeed, once we have d_∞ convergence in L^q, we have d_0 convergence in L^q, and then by interpolation, we have $d_{\alpha'\text{-Höl}}$ convergence in L^q. For any integer m,

$$2^{1-q}\mathbb{E}\left[d_\infty\left(\mathbf{X}^n, \mathbf{X}^\infty\right)^q\right]$$

$$\le \mathbb{E}\left[\sup_{i=1,\dots,m} d\left(\mathbf{X}_{\frac{i}{m}T}^n, \mathbf{X}_{\frac{i}{m}T}^\infty\right)^q\right] + \mathbb{E}\left[\sup_{|t-s|<\frac{T}{m}} \left(\|\mathbf{X}_{s,t}^n\|^q + \|\mathbf{X}_{s,t}^\infty\|^q\right)\right]$$

$$\le \sum_{i=1}^m \mathbb{E}\left(d\left(\mathbf{X}_{\frac{i}{m}}^n, \mathbf{X}_{\frac{i}{m}}^\infty\right)^q\right) + \left(\frac{T}{m}\right)^{\alpha q} \times 2 \sup_{1 \le n \le \infty} \mathbb{E}\left[\|\mathbf{X}^n\|_{\alpha\text{-Höl};[0,T]}^q\right].$$

By first choosing m large enough, followed by choosing n large enough, we see that $d_\infty\left(\mathbf{X}^n, \mathbf{X}^\infty\right) \to 0$ in L^q as required.   ∎

A.4   Sample path regularity under Gaussian assumptions

We start with a simple characterization of Gaussian integrability.

Lemma A.17 *For a real-valued non-negative variable Z, the following three conditions are equivalent:*
(i) (Gauss tail) there exists $\eta_1 > 0$ such that

$$\mathbb{P}\left[Z > x\right] \le \frac{1}{\eta_1} e^{-\eta_1 x^2};$$

(ii) (Gaussian integrability) there exists $\eta_2 > 0$ such that

$$\mathbb{E}\left(e^{\eta_2 Z^2}\right) < \infty;$$

(iii) (square-root growth of moments) there exists $\eta_3 > 0$ such that for all $q \in [1, \infty)$,

$$\left.|Z|\right|_{L^q(\mathbb{P})} \le \frac{1}{\eta_3} \sqrt{q} < \infty.$$

When switching from the ith to the jth statement, the constant η_j only depends on η_i.

Proof. (i) implies (ii) by Chebyshev's inequality and the converse holds by using the formula

$$\mathbb{E}\left[X\right] = \int_0^\infty \mathbb{P}\left[X \ge x\right] dx.$$

Using the same formula, (i) implies (iii) since

$$
\begin{aligned}
\mathbb{E}\left[Z^{2p}\right] &= \int_0^\infty \mathbb{P}\left[Z^2 \ge x\right] d\left(x^p\right) \\
&\le \frac{1}{\eta} \int_0^\infty e^{-\eta x} d\left(x^p\right) = \frac{p}{\eta^{p+1}} \Gamma\left(p\right).
\end{aligned}
$$

Stirling's approximation for the Gamma function is given by

$$
\begin{aligned}
\Gamma\left(p\right) &= \sqrt{2\pi/p}\left(\frac{p}{e}\right)^p \left(1 + O\left(1/p\right)\right) \\
&\le \left(\frac{p}{e}\right)^p \text{ for } p \text{ large enough.}
\end{aligned}
$$

It is then clear that $\left.|Z|\right|_{L^{2p}} \le c\sqrt{p}$ and by making $c = c\left(\eta\right)$ large enough this holds for all p. To see that (iii) implies (ii) we simply expand the exponential

$$
\begin{aligned}
\mathbb{E}\left(e^{\eta Z^2}\right) &= \sum_{n=1}^\infty \frac{1}{n!} \eta^n \left(\left.|Z|\right|_{L^{2n}}\right)^{2n} \\
&\le \sum_{n=1}^\infty \frac{\left(2c^2\eta\right)^n}{n!} n^n
\end{aligned}
$$

and we see with Stirling, $\Gamma\left(n+1\right) = n!$, that this sum is finite for $\eta = \eta\left(c\right)$ small enough. ∎

We shall now see that (often sharp) generalized Hölder or variation regularity of a stochastic process can be shown from the following simple condition Gaussian integrability condition. It is not only satisfied by Brownian motion, a generic class of Gaussian processes and Markov processes with uniform elliptic generator in divergence form, but also by all respective enhancements to rough paths (provided one works with homogenous "norms" and distances on the rough path spaces).

Condition A.18 (Gaussian integrability (p)**)** *Given a (continuous[3]) process X on $[0,T]$ with values in a Polish space (E,d), there exist $p \geq 1$, $\eta > 0$ such that*

$$\sup_{0 \leq s < t \leq T} \mathbb{E} \exp\left(\eta \left[\frac{d(X_s, X_t)}{|t-s|^{1/p}}\right]^2\right) < \infty. \qquad (A.18)$$

Let us note that, from Lemma A.17, condition (A.18) is equivalent to

$$\sup_{0 \leq s < t \leq T} \left|\frac{d(X_s, X_t)}{|t-s|^{1/p}}\right|_{L^q(\mathbb{P})} = O\left(\sqrt{q}\right) \text{ as } q \to \infty.$$

It turns out that this rather generic condition implies a number of sample path properties, many of which are well known in the setting of Gaussian processes (see [48] and the references cited therein). Given a "modulus" function $\zeta : [0, \infty) \to [0, \infty)$, 0 at 0 and strictly increasing, we set

$$|X|_{\zeta\text{-Höl};[0,T]} := \sup_{0 \leq s < t \leq T} \frac{d(X_s, X_t)}{\zeta(t-s)} = \sup_{0 \leq s < t \leq T} \frac{|X|_{0;[s,t]}}{\zeta(t-s)}. \qquad (A.19)$$

(The second equality, with $|X|_{0;[s,t]} = \sup_{u,v \in [s,t]} d(X_s, X_t)$, follows from monotonicity of ζ.)

Theorem A.19 *Assume $(X_t : t \in [0,T])$ is a continuous process with values in a Polish space (E,d) which satisfies the Gaussian integrability condition (A.18) with parameters p, η. Assume*

$$\limsup_{h \to 0} \frac{h^{1/p}\sqrt{\log 1/h}}{\zeta(h)} < \infty.$$

Then there exists $c = c(p, T) > 0$ such that

$$\mathbb{E} \exp\left(c\eta |X|_{\zeta\text{-Höl};[0,T]}^2\right) < \infty.$$

---

[3] Condition (A.18) is more than enough, with Kolmogorov's criterion, to guarantee the existence of a continuous version of X. We shall simply assume that X is continuous.

Proof. Without loss of generality, we may assume $\eta = 1$. (Otherwise, replace the distance d by $\eta^{-1/2} d$.) Condition (A.18) implies that

$$F(\omega) := \int \int_{[0,T]^2} \exp \left[\left(\frac{d(X_u, X_v)}{|v-u|^{1/p}} \right)^2 \right] du \, dv$$

has finite expectation. By Corollary A.5, there exists $c > 0$ such that

$$\mathbb{E} \exp \left(c \sup_{0 \leq s < t \leq T} \left(\frac{d(X_s, X_t)}{\hat{\zeta}(t-s)} \right)^2 \right) \leq \mathbb{E}(F(\omega) \vee 4) < \infty,$$

where $\hat{\zeta}(h) := \int_0^h u^{1/p-1} \sqrt{\log(1+1/u^2)} du \sim h^{1/p} \sqrt{\log 1/h}$. By assumption, there exist positive constants c_1, c_2 so that

$$\hat{\zeta}(h) \leq c_1 \zeta(h) \text{ for } h \in [0, c_2).$$

Moreover, by making c smaller if necessary, $\mathbb{E} \exp \left(c |X|_{0;[0,T]}^2 \right) < \infty$ and the general case follows from the split

$$|X|_{\zeta\text{-Höl};[0,T]} \leq \sup_{s,t : |t-s| \leq c_2} \frac{|X|_{0;[s,t]}}{\hat{\zeta}(t-s)/c_1} + \frac{1}{\hat{\zeta}(c_2)} |X|_{0;[0,T]}.$$

■

Exercise A.20 *Let $p' > p$. Show that, under the assumptions of Theorem A.19,*

$$\sup_{0 \leq s < t \leq T} \mathbb{E} \exp \left(c\eta \left[\frac{|X|_{p'\text{-var};[s,t]}^2}{|t-s|^{1/p}} \right]^2 \right) < \infty. \tag{A.20}$$

Theorem A.19 implies that (X_t) has ζ-modulus regularity where $\zeta(h) = h^{1/p} \sqrt{\ln 1/h}$. We know from examples (e.g. Brownian motion with $p = 2$) that this is the exact modulus, and in this sense Theorem A.19 is optimal. This should be compared with the following *law of iterated logarithm* in which, essentially, $\sup_{0 \leq s < t \leq T}$ in (A.19) is restricted to fixed $s = 0$.

Theorem A.21 *Assume $(X_t : t \in [0,T])$ is a process with values in a Polish space (E, d) which satisfies the Gaussian integrability condition (A.18) with parameters p, η. Then, there exists a (deterministic) constant $C = C(p, \eta) < \infty$ s.t.*

$$\limsup_{h \downarrow 0} \frac{|X|_{0;[0,h]}}{h^{1/p} \sqrt{\ln \ln 1/h}} \leq C \text{ a.s.}$$

Remark A.22 In the notation of Definition 5.45, $h^{1/p}\sqrt{\ln\ln 1/h} = \psi_{1/p,-1/2}(h)$ as $h \downarrow 0$. We insist that, in general, (X_t) will not have $\psi_{1/p,-1/2}$-modulus regularity. However, we will see below that (X_t) enjoys $\psi_{p,p/2}$-variation regularity where $\psi_{p,p/2}$ is Lipschitz equivalent (in the sense of Lemma 5.48) to the inverse of $\psi_{1/p,-1/2}$.

Proof. We start with a tail estimate on $|X|_{0;[0,T]} = |\tilde{X}|_{0;[0,1]}$, introducing the reparametrization $\tilde{X}(\cdot) := X(T\cdot) : [0,1] \to E$ and noting that $\tilde{X}$ satisfies the Gaussian integrability condition with parameters $p, \eta/T^{2/p}$. It is then clear from Theorem A.19 that $|X|_{0;[0,T]}/T^{1/p}$ enjoys Gaussian integrability (uniformly in T) and hence, by Chebyshev, has a Gaussian tail; that is, for c_1 large enough, not dependent on T,

$$\mathbb{P}\left[|X|_{0;[0,T]} \geq x\right] \leq c_1 \exp\left(-\frac{1}{c_1}\left(\frac{x}{T^{1/p}}\right)^2\right).$$

The main idea is now to scale by a geometric sequence. Fix $\varepsilon > 0$, $q \in (0,1)$, set $c_2 = \sqrt{(1+\varepsilon)c_1}$ and also set $\varphi(h) = h^{1/p}\sqrt{\ln\ln 1/h}$. Define the event

$$A_n = \left\{|X|_{0;[0,q^n]} \geq c_2\varphi(q^n)\right\}.$$

It follows that, for n large enough,

$$
\begin{aligned}
\mathbb{P}(A_n) &= \mathbb{P}\left[|X|_{0;[0,q^n]} \geq c_2\varphi(q^n)\right] \\
&\leq c_1 \exp\left(-\frac{1}{c_1}\left(\frac{c_2\varphi(q^n)}{q^{n/p}}\right)^2\right) \\
&= c_1 \exp\left(-\frac{1}{c_1}c_2^2\ln\ln q^n\right) \\
&= c_1(-n\ln q)^{-c_2^2/c_1}.
\end{aligned}
$$

This is summable in n and hence, by the Borel–Cantelli lemma, we get that only finitely many of these events occur; i.e. $|X(\omega)|_{0;[0,q^n]} < c_2\varphi(q^n)$ for all $n \geq n_0(\varepsilon,\omega)$ large enough. For all h small enough, pick n such that

$$q^{n+1} \leq h < q^n.$$

We then have

$$
\begin{aligned}
\limsup_{h\downarrow 0}\frac{|X|_{0;[0,h]}}{\varphi_{p,2}(h)} &\leq \limsup_{n\to\infty}\frac{\varphi_{p,2}(q^n)}{\varphi_{p,2}(q^{n+1})}\frac{|X|_{0;[0,q^n]}}{\varphi_{p,2}(q^n)}\frac{\varphi_{p,2}(q^{n+1})}{\varphi_{p,2}(h)} \\
&\leq q^{-1/p}\sqrt{(1+\varepsilon)c_1}
\end{aligned}
$$

and the proof is finished. (Sending $q \uparrow 1$, followed by $\varepsilon \downarrow 0$, actually shows that one can take $C = \sqrt{c_1}$.) ∎

We now turn to variational regularity of (X_t). Theorem A.19 readily implies ψ-variation regularity provided ψ is taken as the inverse of the modulus $\zeta(h) = h^{1/p}\sqrt{\ln 1/h}$; equivalently (cf. Lemma 5.48), $\psi(h) = (h/\sqrt{\ln 1/h})^p$. We shall establish a sharper result below (with ln replaced by the iterated logarithm ln ln). There are examples (e.g. Brownian motion, Theorem 13.69) in which this is the exact variation, and in this sense Theorem A.24 below is optimal.

Lemma A.23 *Let* $x : [0,1] \to \mathbb{R}^d$ *be a continuous path, and* $\phi : \mathbb{R}^+ \times \mathbb{R}^+ \to \mathbb{R}^+$ *a function increasing in the first dimension, and decreasing in the second dimension. Then,*

$$\sup_{(t_i) \in \mathcal{D}([0,1])} \sum_i \phi\left(d\left(x_{t_i}, x_{t_{i+1}}\right), t_{i+1} - t_i\right) \leq 2 \sum_{n=2}^{\infty} \sum_{k=0}^{2^n-4} \phi\left(|x|_{0;\left[\frac{k}{2^n}, \frac{k+4}{2^n}\right]}, 2^{-n}\right).$$

Proof. For $(s,t) \subset [0,1]$, we first find the integer $n^{s,t} \in \{0,1,2,\dots\}$ such that

$$2^{-n^{s,t}-1} < |t-s| \leq 2^{-n^{s,t}}.$$

We then cover (s,t) with the interval $I^{s,t} = [\sigma_{s,t}, \sigma^{s,t}]$, where

$$\sigma_{s,t} = \max\left\{k2^{-n^{s,t}-1}, \, k \in \left\{0,\dots,2^{n^{s,t}+1}\right\}, k2^{-n^{s,t}-1} \leq s\right\},$$

$$\sigma^{s,t} = \min\left\{k2^{-n^{s,t}-1}, \, k \in \left\{0,\dots,2^{n^{s,t}+1}\right\}, k2^{-n^{s,t}-1} \geq t\right\}.$$

Observe first that $|I^{s,t}| 2^{n^{s,t}+1}$ is equal to 2 or 3. Indeed, by definition of $n^{s,t}$, we have

$$\left|I^{s,t}\right| 2^{n^{s,t}+1} \geq |t-s| 2^{n^{s,t}+1} > 1,$$

and as both $\sigma^{s,t} 2^{n^{s,t}+1}$ and $\sigma_{s,t} 2^{n^{s,t}+1}$ are integers, we must have $\left|I^{s,t}\right| 2^{n^{s,t}+1} \geq 2$. Also, if $\left|I^{s,t}\right| 2^{n^{s,t}+1} \geq 4$, it means that the interval $\left[\sigma_{s,t} + \frac{1}{2^{n^{s,t}+1}}, \sigma^{s,t} - \frac{1}{2^{n^{s,t}+1}}\right]$ is of length greater than or equal to $2^{-n^{s,t}}$, and hence, as $\left[\sigma_s + \frac{1}{2^{n^{s,t}+1}}, \sigma^{s,t} - \frac{1}{2^{n^{s,t}+1}}\right] \subset [s,t]$, so is $[s,t]$. This contradicts the definition of $n^{s,t}$.

Observe that if (s,t) and (u,v) are two disjoint intervals, then $I^{s,t}$ and $I^{u,v}$ are not identical. Assume this were not the case. Without loss of generality, we can assume $u \geq t$. Necessarily, $n^{s,t} = n^{u,v}$. As $u \geq t$, we obtain $\sigma_{u,v} \geq \sigma^{s,t} - \frac{1}{2^{n+1}} \geq \sigma_{s,t} + \frac{1}{2^{n+1}}$, which contradicts the assumption that $I^{s,t} = I^{u,v}$.

For a fixed dissection $(t_i) \in \mathcal{D}([0,1])$, we have from the monotonicity assumption on ϕ,

$$\sum_i \phi\left(d\left(x_{t_i}, x_{t_{i+1}}\right), t_{i+1} - t_i\right) \leq \sum_i \phi\left(|x|_{0,I^{t_i,t_{i+1}}}, 2^{-n^{t_i,t_{i+1}}-1}\right)$$

and then define $\alpha_{n,k}^{(2)} = \# \left\{ I^{t_i, t_{i+1}} = \left[\frac{k}{2^{n+1}}, \frac{k+2}{2^{n+1}} \right] \right\}$, $\alpha_{n,k}^{(3)} = \# \left\{ I^{t_i, t_{i+1}} = \left[\frac{k}{2^{n+1}}, \frac{k+3}{2^{n+1}} \right] \right\}$. We have just seen that $\alpha_{n,k}^{(2)}, \alpha_{n,k}^{(3)} \in \{0,1\}$ and in fact $\alpha_{n,k}^{(2)} + \alpha_{n,k}^{(3)} \leq 1$. We therefore obtain

$$
\sum_i \phi \left(d \left(x_{t_i}, x_{t_{i+1}} \right), t_{i+1} - t_i \right)
$$

$$
\leq \sum_{n=0}^{\infty} \sum_{k=0}^{2^{n+1}-2} \alpha_{n,k}^{(2)} \phi \left(|x|_{0; \left[\frac{k}{2^{n+1}}, \frac{k+2}{2^{n+1}} \right]}, 2^{-(n+1)} \right)
$$

$$
+ \sum_{n=1}^{\infty} \sum_{k=0}^{2^{n+1}-3} \alpha_{n,k}^{(3)} \phi \left(|x|_{0; \left[\frac{k}{2^{n+1}}, \frac{k+3}{2^{n+1}} \right]}, 2^{-(n+1)} \right)
$$

$$
\leq 2 \sum_{n=1}^{\infty} \sum_{k=0}^{2^{n+1}-4} \phi \left(|x|_{0; \left[\frac{k}{2^{n+1}}, \frac{k+4}{2^{n+1}} \right]}, 2^{-(n+1)} \right),
$$

using $\alpha_{n,k}^{(2)} + \alpha_{n,k}^{(3)} \leq 1$. It then suffices to take the supremum over all dissections. ∎

The next result deals with generalized variation and we recall (cf. Definition 5.45) that for $p \geq 1$,

$$
\psi_{p,p/2} (t) = \left| \frac{t}{\sqrt{\ln^* \ln^* 1/t}} \right|^p \quad \text{with } \ln^* (h) = \max (1, \ln h).
$$

Theorem A.24 *Assume* $(X_t : t \in [0,T])$ *is a process with values in a Polish space* (E,d) *which satisifies the Gaussian integrability condition (A.18) with parameters* p, η. *Then, there exists* $c = c_p > 0$ *such that*

$$
\mathbb{E} \left(\exp \left(c \frac{\eta}{T^{2/p}} |X|^2_{\psi_{p,p/2}\text{-var};[0,T]} \right) \right) < \infty.
$$

Proof. The reparametrization $\tilde{X} (\cdot) := X (T \cdot) : [0,1] \to E$ satisfies the Gaussian integrability condition with parameters $p, \eta/T^{2/p}$. At the same time,

$$
\left| \tilde{X} \right|_{\psi_{p,p/2}\text{-var};[0,1]} = |X|_{\psi_{p,p/2}\text{-var};[0,T]}
$$

and so we can assume without loss of generality that $T = 1$. Furthermore, at the price of replacing the distance d by $\eta^{-1/2}d$, we may assume $\eta = 1$. After these preliminary remarks, let us define ϕ_α by

$$
\phi_\alpha (r) = \psi_{p,p/2} (r) 1_{\psi_{p,p/2}(r) > \alpha}.
$$

For a fixed $M > 0$ and a dissection (t_i) of $[0,1]$, and a fixed continuous path x, we have

$$\sum_i \psi_{p,p/2}\left(\frac{d\left(x_{t_i}, x_{t_{i+1}}\right)}{M}\right)$$

$$= \sum_i \psi_{p,p/2}\left(\frac{d\left(x_{t_i}, x_{t_{i+1}}\right)}{M}\right) 1_{\psi_{p,p/2}\left(\frac{d\left(x_{t_i}, x_{t_{i+1}}\right)}{M}\right) \leq \frac{t_{i+1}-t_i}{2}}$$

$$+ \sum_i \phi_{\frac{t_{i+1}-t_i}{2}}\left(\frac{d\left(x_{t_i}, x_{t_{i+1}}\right)}{M}\right)$$

$$\leq \frac{1}{2} + \sum_i \phi_{\frac{t_{i+1}-t_i}{2}}\left(\frac{d\left(x_{t_i}, x_{t_{i+1}}\right)}{M}\right).$$

Taking the supremum over all dissections, we see that

$$|x|_{\psi_{p,p/2}\text{-var}} \leq \inf\left\{M > 0, \sup_{(t_i)\in\mathcal{D}([0,1])} \phi_{\frac{t_{i+1}-t_i}{2}}\left(\frac{d\left(x_{t_i}, x_{t_{i+1}}\right)}{M}\right) \leq \frac{1}{2}\right\}.$$

Using the previous lemma, we obtain

$$|x|_{\psi_{p,p/2}\text{-var}} \leq \inf\left\{M > 0, \sum_{n=2}^{\infty}\sum_{k=0}^{2^n-4} \phi_{2^{-n-1}}\left(\frac{|x|_{0;[k2^{-n},(k+4)2^{-n}]}}{M}\right) \leq \frac{1}{4}\right\},$$

and in particular, that

$$\mathbb{P}\left(|X|_{\psi_{p,p/2}\text{-var}} \geq M\right) \leq \mathbb{P}\left(\sum_{n=2}^{\infty}\sum_{k=0}^{2^n-4} \phi_{2^{-n-1}}\left(\frac{|X|_{0;[k2^{-n},(k+4)2^{-n}]}}{M}\right) > \frac{1}{4}\right).$$

From Theorem A.19 and (A.19) there exists $c_1 > 0$ such that $\mathbb{P}(\Omega_M^c) \leq c_1 \exp\left(-M^2/c_1\right)$, where

$$\Omega_M = \left\{\sup_{0\leq s<t\leq 1} \frac{|X|_{0;[s,t]}}{|t-s|^{1/p}\left(\ln^*\frac{1}{t-s}\right)^{1/2}} < M\right\}.$$

Now, on the set Ω_M,

$$\phi_{2^{-n-1}}\left(\frac{|X|_{0;[k2^{-n},(k+4)2^{-n}]}}{M}\right)$$

$$\leq \psi_{p,p/2}\left(\left(2^{-(n-2)}\right)^{1/p}\left(\ln^* 2^{n-2}\right)^{1/2}\right) 1_{\psi_{p,p/2}\left(\frac{|X|_{0;[k2^{-n},(k+4)2^{-n}]}}{M}\right) > 2^{-n-1}}.$$

For $n \geq 2$ we have

$$
\begin{aligned}
\psi_{p,p/2}\left(\left(2^{-(n-2)}\right)^{1/p}\left(\ln^* 2^{n-2}\right)^{1/2}\right) &\leq \psi_{p,0}\left(\left(2^{-(n-2)}\right)^{1/p}\left(\ln^* 2^{n-2}\right)^{1/2}\right) \\
&= 2^{-(n-2)}\left(\ln^* 2^{n-2}\right)^{p/2} \\
&\leq 4.2^{-n}\max\left(1,(\ln 2)^{p/2}(n-2)^{p/2}\right) \\
&\leq 4.2^{-n}(n-1)^{p/2},
\end{aligned}
$$

and so

$$
\begin{aligned}
\mathbb{P}&\left(\left.|X|\right._{\psi_{p,p/2}\text{-var}} \geq M\right) - \mathbb{P}\left(\Omega_M^C\right) \\
&\leq \mathbb{P}\left(\left.|X|\right._{\psi_{p,p/2}\text{-var}} \geq M \cap \Omega_M\right) \\
&\leq \mathbb{P}\left(\sum_{n=2}^{\infty} 2^{-(n-2)}(n-1)^{p/2} \right. \\
&\qquad \left. \times \sum_{k=0}^{2^n-4} 1_{\psi_{p,p/2}\left(\frac{|X|_{0;[k2^{-n},(k+4)2^{-n}]}}{M}\right) > 2^{-n-1}} 1_{\Omega_M} > \frac{1}{4}\right).
\end{aligned}
$$

Using $\mathbb{P}\left(\sum \alpha_i 1_{A_i} > \beta\right) \leq \sum \frac{1}{\beta}\mathbb{E}\left(\alpha_i 1_{A_i}\right) = \sum_i \frac{\alpha_i}{\beta}\mathbb{P}(A_i)$, we obtain

$$
\begin{aligned}
\mathbb{P}&\left(\left.|X|\right._{\psi_{p,p/2}\text{-var}} \geq M\right) - \mathbb{P}\left(\Omega_M^C\right) \\
&\leq \sum_{n=2}^{\infty} 2^{4-n}(n-1)^{p/2} \\
&\qquad \times \sum_{k=0}^{2^n-4}\mathbb{P}\left(\psi_{p,p/2}\left(\frac{|X|_{0;[k2^{-n},(k+4)2^{-n}]}}{M}\right) > 2^{-n-1} \cap \Omega_M\right) \\
&\leq \sum_{n=0}^{\infty} 2^{4-n}(n-1)^{p/2}\sum_{k=0}^{2^n-4}\mathbb{P}\left(\psi_{p,p/2}\left(\frac{|X|_{0;[k2^{-n},(k+4)2^{-n}]}}{M}\right) > 2^{-n-1}\right).
\end{aligned}
$$

So it only remains to bound

$$
\begin{aligned}
\mathbb{P}&\left(\psi_{p,p/2}\left(\frac{|X|_{0;[k2^{-n},(k+4)2^{-n}]}}{M}\right) > 2^{-n-1}\right) \\
&= \mathbb{P}\left(\frac{|X|_{0;[k2^{-n},(k+4)2^{-n}]}}{(4.2^{-n})^{1/p}} > \frac{M}{(4.2^{-n})^{1/p}}\psi_{p,p/2}^{-1}\left(2^{-n-1}\right)\right).
\end{aligned}
$$

Now, we have seen in Theorem A.19 that, for a positive constant c_2,

$$
\sup_{0\leq s<t\leq 1}\mathbb{E}\exp\left(c_2\left(\frac{|X|_{0;[s,t]}}{|t-s|^{1/p}}\right)^2\right) < \infty.
$$

Then, from Chebyshev's inequality and for large enough constant c_3,

$$\mathbb{P}\left(\psi_{p,p/2}\left(\frac{|X|_{0;[k2^{-n},(k+3)2^{-n}]}}{M}\right) > 2^{-n-1}\right)$$

$$\leq c_3 \exp\left(-\frac{M^2}{c_3}\left|2^{n/p}\psi_{p,p/2}^{-1}\left(2^{-n-2}\right)\right|^2\right),$$

and so

$$\mathbb{P}\left(|X|_{\psi_{p,p/2}\text{-var}} \geq M\right)$$

$$\leq c_3 \sum_{n=2}^{\infty} (n-1)^{p/2} \exp\left(-\frac{M^2}{c_3}\left|2^{n/p}\psi_{p,p/2}^{-1}\left(2^{-n-2}\right)\right|^2\right)$$

$$+ c_1 \exp\left(-M^2/c_1\right).$$

We have seen in Lemma 5.48 that $c_5\psi_{p,-1/2} \leq \psi_{p,p/2}^{-1}$, which implies that for $n \geq 2$,

$$\left|2^{n/p}\psi_{p,p/2}^{-1}\left(2^{-n-2}\right)\right|^2 \geq c_5 \ln^* \ln^* 2^{n-2} \geq c_6 \left(1 + \ln n\right).$$

Hence, for a positive constant c_7,

$$\exp\left(-\frac{M^2}{c_3}\left|2^{n/p}\psi_{p,p/2}^{-1}\left(2^{-n-2}\right)\right|^2\right) \leq \exp\left(-c_7 M^2\right) \cdot \left(\frac{1}{n}\right)^{c_7 M^2}.$$

For M large enough, we have $c_3 \sum_{n\geq 2} (n-1)^{p/2} \, n^{-c_7 M^2} = c_8 < \infty$ and so

$$\mathbb{P}\left(|X|_{\psi_{p,p/2}\text{-var}} \geq M\right) \leq c_8 \exp\left(-c_7 M^2\right) + c_1 \exp\left(-M^2/c_1\right).$$

The proof is now finished.   ∎

A.5 Comments

The Garsia–Rodemich–Rumsey result is well known, e.g. Stroock [159], and so are the resulting Besov–Hölder and Besov–Lévy modulus embeddings. The Besov-variation embedding is a more recent insight, Friz and Victoir [64]. Exercise A.4 is due to Krylov [95].

Everything up to and including Kolmogorov's criterion is standard, see Revuz and Yor [143] or Stroock [159], for instance. The ψ-variation sample path behaviour of a generic process under the Gaussian integrability assumption of Condition of A.18 is essentially taken from Friz and Oberhauser [59]; it implies $\psi_{2,1}$-variation regularity of Brownian motion, a classical result of Taylor [168].

Appendix B: Banach calculus

Throughout this appendix, $E, F, \ldots$ are Banach spaces with respective norms $|\cdot|_E$ and $|\cdot|_F$ and we simply write $|\cdot|$ when no confusion is possible.

B.1 Preliminaries

We say that a map f from $(a, b) \subset \mathbb{R}$ into a Banach space E is continuously differentiable, in symbols $f \in C^1((a, b), E)$, if and only if $\dot{f}(t) := \frac{df}{dt}(t) := \lim_{\varepsilon \to 0} (f(t + \varepsilon) - f(x))/\varepsilon$ exists (as a strong limit in E) for all $t \in (a, b)$ with $\dot{f} \in C((a, b), E)$. Similarly, the Riemann integral of a continuous function can be defined as a strong limit of Riemann sum approximations and the fundamental theorem of calculus is valid; see [45] for instance.

The following proposition is useful in showing that the directional derivatives of an ODE (resp. RDE) solution, as a function of starting point and driving signal, exist as strong limits in $C^{1\text{-var}}([0, 1], \mathbb{R})$ (resp. $C^{p\text{-var}}([0, 1], \mathbb{R})$), simply using the continuous embedding $C^{p\text{-var}}([0, 1], \mathbb{R}) \hookrightarrow C([0, 1], \mathbb{R})$.

Proposition B.1 *Assume $E \hookrightarrow F$, i.e. E is continuously embedded in F. Assume $f \in C^1((a, b), F)$ such that its derivative $\dot{f}$ (defined as a strong limit in F) actually takes values in E and extends to a continuous function from $[a, b]$ into E. Assume furthermore that $f(a) \in E$. Then $f \in C^1((a, b), E)$ with derivative given by $\dot{f}$.*

Proof. By assumption, $\dot{f}$ extends to a continuous function from $[a, b]$ into $E \hookrightarrow F$. By the fundamental theorem of calculus, for all $t \in (a, b)$,

$$f(t) - f(a) = \int_a^t \dot{f}(s)\, ds,$$

where the definite integral is the strong limit in F of approximating Riemann sums. On the other hand, $\dot{f} \in C([a, b], E)$ and so the definite integral also exists as a strong limit in E of approximating Riemann sums. Since convergence in E implies convergence in F, these integrals coincide and in particular

$$f(t) = f(a) + \int_a^t \dot{f}(s)\, ds \in E$$

for all $t \in (a, b)$. By the fundamental theorem of calculus, we now see that f is continuously differentiable in E with derivative given by $\dot{f}$. In other words, $\dot{f}$ (which was defined as a strong limit of difference quotients with convergence in F) is actually convergent in E.   ∎

B.2   Directional and Fréchet derivatives

The space of linear, continuous maps from E to F is denoted by $L(E, F)$ and is itself a Banach space under the operator norm

$$|f| := \sup_{y \in E : |y| \leq 1} |f(y)|.$$

Definition B.2 Let U be an open set in E, and $f : U \to F$ is Fréchet differentiable at $x \in U$ iff there exists $Df(x) \in L(E, F)$ s.t. for all $h \in E$

$$|f(x+h) - f(x) - Df(x)h| = o(|h|).$$

It is said to be Fréchet differentiable on U if $Df(x)$ exists for all $x \in U$. If $x \mapsto Df(x)$ is continuous[1] we say that f is C^1 in the Fréchet sense and write $f \in C^1(U, F)$.   □

Definition B.3 We say that $f : U \subset_o E \to F$ has directional derivative at $x \in U$ in direction $h \in E$ if the following limit exists:

$$\lim_{t \to 0} \frac{f(x+th) - f(x)}{t} = D_h f(x).$$ □

Directional derivatives are automatically homogenous in h in the sense that

$$D_{\lambda h} f(x) = \lambda \lim_{t \to 0} \frac{f(x + t\lambda h) - f(x)}{t\lambda} = \lambda D_h f(x), \quad \lambda \in \mathbb{R}.$$

However, existence of $D_h f(x)$ for all $h \in E$, need not imply linearity in h, as is seen in the example $f(0,0) = 0$; $f(x, y) \mapsto x^2 y / (x^2 + y^2)$. Obviously, Fréchet differentiability implies the existence of directional derivatives in all directions and $D_h f(x) = Df(x)h$. In applications one is often interested in the converse. The following two propositions are useful criteria for this purpose.

Proposition B.4 *Let U be an open set in E, $f : U \to F$ a function that has directional derivatives in all directions, and $A : U \to L(E, F)$ a continuous map such that*

$$D_h f(x) = A(x)h$$

for all $x \in U, h \in E$. Then $f \in C^1(U, F)$ and $Df(x)h = A(x)h$.

---

[1] With respect to the operator norm on $L(E, F)$.

Proof. By the fundamental theorem of calculus,

$$f(x+h) - f(x) = \int_0^1 \frac{df(x+th)}{dt} dt = \int_0^1 D_h f(x+th)\, dt = \int_0^1 A(x+th)\, dt.$$

It follows that

$$f(x+h) - f(x) - A(x)h = \varepsilon(h)h$$

with $\varepsilon(h) \equiv \int_0^1 (A(x+th) - A(x))\, dt$ and

$$\begin{aligned}
\|\varepsilon(h)\| &\leq \int_0^1 \|A(x+th) - A(x)\|\, dt \\
&\leq \max_{t\in[0,1]} \|A(x+th) - A(x)\| \to 0 \text{ as } h \to 0
\end{aligned}$$

by continuity of A. ∎

Proposition B.5 *Let U be an open set in E, and $f : U \to F$ be a continuous map that admits directional derivatives at all points and in all directions; more precisely, for all $x \in U$ and $h \in E$,*

$$D_h f(x) = \lim_{\varepsilon \to 0} \frac{f(x + \varepsilon h) - f(x)}{\varepsilon} = \frac{\partial}{\partial \varepsilon} \{f(x + \varepsilon h)\}_{\varepsilon=0}$$

exists (as a strong limit in F). Assume further that

$$(x, h) \in U \times E \mapsto D_h f(x) \in F$$

is uniformly continuous on bounded sets. Then f is C^1 in the Fréchet sense.

We prepare the proof with the following:

Lemma B.6 *Let $U \subset_o E$ and $\varphi : U \times E \to F$ be uniformly continuous on bounded sets such that for all $x \in U$, the map*

$$h \mapsto \varphi(x, h) =: \varphi(x)h$$

is linear. Then the map

$$\begin{aligned}
\tilde{\varphi} \;:\; & U \to L(E, F) \\
& x \mapsto (h \mapsto \varphi(x)h)
\end{aligned}$$

is well-defined and uniformly continuous on bounded sets.

Proof. Fixing x, by assumption $h \mapsto \varphi(x)h$ is linear; by the assumption on uniform continuity on bounded sets $h \mapsto \varphi(x)h$ is also continuous and hence a well-defined element of $L(E, F)$. By the assumption of uniform continuity on bounded sets, for every $R > 0$ and $\varepsilon > 0$ there exists δ such that for $x, x' \in U$ and $h, h' \in E$ with $|x|, |x'| \leq R$ and $|h|, |h'| \leq R$,

$$|x - x'| + |h - h'| < \delta \implies |\varphi(x, h) - \varphi(x', h')| < \varepsilon.$$

Restricting attention to $R > 1$, given $x, x' \in U$ with $|x|, |x'| \leq R$ and $|x - x'| < \delta$ implies

$$\left| \tilde{\varphi}(x) - \tilde{\varphi}(x') \right|_{\mathrm{op}} \equiv \sup_{h \in E : |h| = 1} \left| \varphi(x) h - \varphi(x') h \right| < \varepsilon.$$

This says precisely that $\tilde{\varphi}$ is uniformly continuous on bounded sets. ∎

Proof (Proposition B.5). We first show that $\varphi(x, h) := D_h f(x)$ is linear in h. As remarked after the definition of the directional derivative, homogeneity in h is clear. Given $g, h \in E$ we have

$$\underbrace{\frac{f(x+\varepsilon(g+h)) - f(x)}{\varepsilon}}_{\rightarrow\, D_{g+h} f(x) \text{ as } \varepsilon \rightarrow 0} = \frac{f(x+\varepsilon(g+h)) - f(x+\varepsilon g)}{\varepsilon} + \underbrace{\frac{f(x+\varepsilon g) - f(x)}{\varepsilon}}_{\rightarrow\, D_g f(x) \text{ as } \varepsilon \rightarrow 0}.$$

The first term on the right-hand side hence converges as $\varepsilon \rightarrow 0$. We claim it equals $D_h f(x)$. To this end, using the fundamental theorem of calculus and homogeneity,

$$
\begin{aligned}
f(x + \varepsilon(g + h)) - f(x + \varepsilon g) &= \int_0^1 \frac{d}{dt} f(x + \varepsilon g + t\varepsilon h)\, dt \\
&= \varepsilon \int_0^1 D_h f(x + \varepsilon g + th)\, dt.
\end{aligned}
$$

It follows that

$$
\left| \frac{f(x + \varepsilon(g + h)) - f(x + \varepsilon g)}{\varepsilon} - D_h f(x) \right|
$$

$$
\leq \int_0^1 \left| D_h f(x + \varepsilon(g + th)) - D_h f(x) \right| dt \rightarrow 0 \text{ as } \varepsilon \rightarrow 0,
$$

where we used in the last step uniform continuity of $D.f(\cdot)$ on bounded sets. This completes the proof that $D_h f(x)$ is linear in h.

By Lemma B.6 linearity of $D_h f(x)$ in h together with the (uniform) continuity (on bounded sets) assumption of $(x, h) \mapsto D_h f(x)$ then implies that $x \mapsto \{h \mapsto D_h f(x)\} \in L(E, F)$ is continuous and by Proposition B.4 we can conclude that $f \in C^1(U, E)$. ∎

The following result is sometimes referred to as "closedness of the differentiation operator".

Proposition B.7 *Assume $f_n \in C^1(U, F)$, where U is an open set in E and $f_n \rightarrow f$ uniformly on bounded sets in U (which implies a priori $f \in C(U, F)$). Let $g \in C(U, L(E, F))$ and assume that*

$$Df_n \rightarrow g$$

also uniformly on bounded sets. Then $f \in C^1(U, F)$ and $Df = g$.

Proof. Fix $x \in U$ and $h \in E$. Then

$$
\begin{aligned}
f\left(x+\varepsilon h\right)-f\left(x\right) &= \lim_{n\to\infty} f_n\left(x+\varepsilon h\right)-f_n\left(x\right) \\
&= \lim_{n\to\infty} \int_0^\varepsilon Df_n\left(x+th\right)h\,dt \\
&= \int_0^\varepsilon g\left(x+th\right)h\,dt
\end{aligned}
$$

thanks to $Df_n \to g$ uniformly on bounded sets. By continuity of g,

$$
D_h f\left(x\right) = \frac{\partial}{\partial \varepsilon} \left\{ \frac{f\left(x+\varepsilon h\right)-f\left(x\right)}{\varepsilon} \right\}_{\varepsilon=0}
$$

exists and equals $D_h f\left(x\right) = g\left(x\right)h$ and so we can conclude with Proposition B.4. ∎

B.3   Higher-order differentiability

Definition B.8 We fix U, an open set of E. We say that $f : U \to F$ has a directional derivative at $x \in U$ in direction $(h_1,\ldots,h_k) \in E^k$ if

$$
D_{(h_1,\ldots,h_k)}f\left(x\right) := D_{h_1} \cdots D_{h_k} f\left(x\right) \tag{B.1}
$$

exists. □

The calculus example $f(0,0) = 0$ and $f(x,y) = xy\left(x^2 - y^2\right)/x^2 + y^2$ otherwise (in which $1 = \partial_x \partial_y f(0,0) \neq \partial_y \partial_x f(0,0) = -1$) shows that the order of $h_1,\ldots,h_k$ can matter. Nonetheless, under reasonable conditions (namely continuity of the kth directional derivatives), the order does not matter and $D_{(h_1,\ldots,h_l)}f(x)$ behaves multilinearly in $(h_1,\ldots,h_k)$.

Higher-order Fréchet differentiability is defined inductively as follows.

Definition B.9 Let $k \in \{1,2,\ldots\}$, and U an open set of E. A function $f : U \to F$ is $(k+1)$-times Fréchet differentiable on U if it is Fréchet differentiable on U and

$$
Df : U \to L\left(E, F\right)
$$

is k-times Fréchet differentiable on U. The kth-order differential is a map

$$
D^k f : U \to L\left(E,\ldots,L\left(E,L\left(E,F\right)\right)\right) \cong L\left(E^{\otimes k}, F\right)
$$

where $L\left(E^{\otimes k}, F\right)$ is the space of multilinear bounded maps from $E \times \cdots \times E$ (k times) into F. If $D^k f$ is continuous then we say that f is C^k in the Fréchet sense and write $f \in C^k\left(U, F\right)$.

A map which is C^k Fréchet for all $k \geq 1$ is said to be Fréchet smooth.

Given $A \in L\left(E^{\otimes k}, F\right)$ we shall indicate multilinearity by writing A $\langle h_1, \ldots, h_k \rangle$ instead of $A\left(h_1, \ldots, h_k\right)$. The criteria we have seen to establish that a function is C^1 in the Fréchet sense all extend to the case of C^k.

Proposition B.10 *Suppose* $k \in \{1, 2, \ldots\}$ *and* U *is an open set of* E. *Assume that* $f : U \to F$ *is a function such that* $D_{h_1} \cdots D_{h_l} f(x)$ *exists for all* $x \in U$ *and* $h_1, \ldots, h_l \in E$, *and* $l = 1, 2, \ldots, k$. *Further assume there exist continuous functions* $A_l : U \to L\left(E^{\otimes l}, F\right)$ *such that*

$$D_{h_1} \cdots D_{h_l} f(x) = A_l(x) \langle h_1, \ldots, h_l \rangle$$

for all $x \in U$, $h_1, \ldots, h_l \in E$, *and* $l = 1, 2, \ldots, k$. *Then* $f : U \to F$ *is* C^k *in the Fréchet sense.*

Proposition B.11 *Suppose* $k \in \{1, 2, \ldots\}$ *and* U *is an open set of* E. *Assume that* $f : U \to F$ *is a function such that* $D_{h_1} \cdots D_{h_l} f(x)$ *exists for all* $x \in U$ *and* $h_1, \ldots, h_l \in E$, *and* $l = 1, 2, \ldots, k$. *Assume further that*

$$(x; h_1, \ldots, k_k) \in U \times E^k \mapsto D_{h_1} \cdots D_{h_l} f(x) \in F$$

is uniformly continuous on bounded sets. Then f *is* C^k *in the Fréchet sense.*

Proof. Take $l \in \{1, \ldots, k\}$ and define $g\left(\varepsilon_1, \ldots, \varepsilon_l\right) = f\left(x + \sum_{j=1}^k \varepsilon_j h_j\right)$ and note that

$$D_{(h_1, \ldots, h_k)} f\left(x + \sum_{j=1}^k \varepsilon_j h_j\right) = \frac{\partial^k g}{\partial \varepsilon_1 \ldots \partial \varepsilon_k}.$$

Since the order of the partial derivatives does not matter here, we have $D_{h_1} \cdots D_{h_l} f(x) = D_{h_{\pi(1)}} \cdots D_{h_{\pi(l)}} f(x)$ for any permutation π of $\{1, \ldots, l\}$. In view of Proposition B.5, it is clear that $f \in C^1$ in the Fréchet sense and so

$$D_{h_{\pi(1)}} \cdots D_{h_{\pi(l-1)}} \left(D_{h_{\pi(l)}} f(x)\right) = D_{h_{\pi(1)}} \cdots D_{h_{\pi(l-1)}} \left(Df(x) h_{\pi(l)}\right).$$

This shows multilinearity in $h_1, \ldots, h_l$. By the assumption of uniform continuity on bounded sets,

$$A_l(x) \langle h_1, \ldots, h_l \rangle := D_{h_1} \cdots D_{h_l} f(x)$$

defines a continuous map from $U \to L\left(E^{\otimes l}, F\right)$, and we conclude with Proposition B.10. ∎

B.4   Comments

Fréchet regularity is a classical topic in non-linear functional analysis. Propositions B.5 and B.10 appear in Driver [45], for instance. We are unaware of any reference for Propositions B.5, B.11.

Appendix C: Large deviations

C.1   Definition and basic properties

Let $\mathcal{X}$ be a topological space. A *rate function* is a lower semicontinuous mapping $I : \mathcal{X} \to [0, \infty]$, i.e. a mapping so that all level sets $\{x \in \mathcal{X} : I(x) \leq \Lambda\}$ are closed. A *good rate function* is a rate function for which all level sets are compact subsets of $\mathcal{X}$. The set $\mathcal{D}_I := \{x \in E : I(x) < \infty\}$ is called the *domain of I*. Given $A \subset E$ we also set

$$I(A) = \inf_{x \in A} I(x).$$

Lemma C.1 *Let I be a good rate function. Then for each closed set F in E,*

$$I(F) = \lim_{\delta \downarrow 0} I(F^\delta)$$

where the open δ-neighbourhood of a set $A \subset E$ is defined as

$$A^\delta = \cup \{B(x, \delta) : x \in A\}, \; B(x, \delta) = \{y \in E : d(x, y) < \delta\}.$$

Proof. [41, Lemma 4.1.6]. ∎

Unless otherwise stated, we assume that probability measures on $\mathcal{X}$ are defined on the Borel sets, i.e. the smallest σ-algebra generated by the open sets in $\mathcal{X}$.

Definition C.2 A family $\{\mu_\varepsilon : \varepsilon > 0\}$ of probability measures on $\mathcal{X}$ satisfies the large deviation principle (LDP) with good rate function I if, for every Borel set A,

$$-I(A^\circ) \leq \liminf_{\varepsilon \to 0} \varepsilon \log \mu_\varepsilon(A) \leq \limsup_{\varepsilon \to 0} \varepsilon \log \mu_\varepsilon(A) \leq -I(\bar{A}).$$

□

Remark C.3 Sometimes it is practical to parametrize the family of probability measures so as to consider $\varepsilon^2 \log \mu_\varepsilon(A)$. □

Before turing to (various) contraction principles, we state two basic properties of LDPs and refer to [41, Lemma 4.1.6] for proofs.

Proposition C.4 *A family $\{\mu_\varepsilon : \varepsilon > 0\}$ of probability measures on a regular topological space can have at most one rate function associated with its LDP.*

Proposition C.5 *Let $\mathcal{E}$ be a measurable subset of $\mathcal{X}$ such that $\mu_\varepsilon(\mathcal{E}) = 1$ for all $\varepsilon > 0$. Suppose that $\mathcal{E}$ is equipped with the topology induced by $\mathcal{X}$. If $\{\mu_\varepsilon\}$ satisfies the LDP in $\mathcal{E}$ with (good) rate function I and $\mathcal{D}_I \subset \mathcal{E}$, then the same LDP holds in $\mathcal{E}$.*

C.2   Contraction principles

Theorem C.6 (contraction principle) *Let $\mathcal{X}$ and $\mathcal{Y}$ be Hausdorff topological spaces. Suppose $f : \mathcal{X} \to \mathcal{Y}$ is a continuous map. If $\{\mu_\varepsilon\}$ satisfies an LDP on $\mathcal{X}$ with good rate function I, then the image measures $\{f_*\mu_\varepsilon\}$, where $f_*\mu_\varepsilon \equiv \mu_\varepsilon \circ f^{-1}$, satisfy an LDP on $\mathcal{Y}$ with good rate function*

$$J(y) = \inf\{I(x) : x \in \mathcal{X} \text{ and } f(x) = y\}.$$

Proof. [41, Lemma 4.1.6]. ∎

Definition C.7 *A family $\{\mu_\varepsilon : \varepsilon > 0\}$ of probability measures on a topological space $\mathcal{X}$ is exponentially tight if for every $M < \infty$, there exists a compact[1] set K_M such that*

$$\limsup_{\varepsilon \to 0} \varepsilon \log \mu_\varepsilon(K_M^c) < -M.$$

Theorem C.8 (inverse contraction principle) *Let $\mathcal{X}$ and $\mathcal{Y}$ be Hausdorff topological spaces. Suppose $g : \mathcal{Y} \to \mathcal{X}$ is a continuous injection and that $\{\nu_\varepsilon\}$ is an exponentially tight family of probability measures on $\mathcal{Y}$. If $\{g_*\nu_\varepsilon\}$ satisfies an LDP in $\mathcal{X}$ with rate function $I : \mathcal{X} \to [0,\infty]$, then $\{\nu_\varepsilon\}$ satisfies an LDP in $\mathcal{Y}$ with good rate function $I' \equiv I \circ g$.*

Proof. [41, Theorem 4.2.2] combined with Proposition C.5 and the remark that $\mathcal{D}_I \subset g(\mathcal{Y})$. ∎

Theorem C.9 (extended contraction principle) *Let $\{\mu_\varepsilon\}$ be a family of probability measures that satisfies an LDP with good rate function I on a Hausdorff topological space $\mathcal{X}$. For $m = 1, 2, \ldots$, let $f^m : \mathcal{X} \to \mathcal{Y}$ be continuous maps, with $(\mathcal{Y}, d)$ a separable metric space. Assume there exists a measurable map $f : \mathcal{X} \to \mathcal{Y}$ such that for every $\Lambda < \infty$,*

$$\lim_{m \to \infty} \sup_{\{x:I(x)\le\Lambda\}} d(f^m(x), f(x)) = 0.$$

---

[1] Since $\overline{K_M}^c \subset K_M^c$ it is enough to require that K_M is precompact.

Assume that $\{f_*^m \mu_\varepsilon\}$ are exponentially good approximations of $\{f_* \mu_\varepsilon\}$ and in the sense that[2]

$$\lim_{m \to \infty} \limsup_{\varepsilon \to 0} \varepsilon^2 \log \mu_\varepsilon \left(\{x : d(f^m(x), f(x)) > \delta\} \right) = -\infty.$$

Then $\{f_* \mu_\varepsilon\}$ satisfies an LDP in $\mathcal{Y}$ with good rate function $I' \equiv \inf \{I(x) : y = f(x)\}$.

Proof. [41, Theorem 4.2.23]. ∎

---

[2] The separability on $\mathcal{Y}$ guarantees measurability of $\{x : d(f^m(x), f(x)) > \delta\}$.

Appendix D: Gaussian analysis

D.1   Preliminaries

We start with a description of the general set-up of Gaussian analysis on a Banach space, following closely Ledoux's Saint Flour notes [102]. Other references with a similar point of view are [103] and [42, Chapter 4].

A mean-zero Gaussian measure μ on a real separable Banach space E equipped with its Borel σ-algebra $\mathcal{B}$ and norm $|\cdot|$ is a Borel probability measure on $(E, \mathcal{B})$ such that the law of each continuous linear functional on E is a zero-mean Gaussian random variable. We first claim that

$$\sigma^2 = \sup_{\xi \in E^*, |\xi| \leq 1} \int \langle \xi, x \rangle^2 \, d\mu(x) < \infty.$$

Indeed, writing $i : E^* \to L^2(\mu) = L^2(E, \mathcal{B}, \mu; \mathbb{R})$ for the injection map, σ is the operator norm of i which is bounded by the closed graph theorem. Since E is separable, the Banach norm $|\cdot|$ may be described as a supremum over a countable set $(\xi_n)_{n \geq 1}$ of elements of the unit ball of the dual space E^*; that is, for every $x \in E$,

$$|x| = \sup_{n \geq 1} \langle \xi_n, x \rangle$$

and in particular, the norm is a measurable map on $(E, \mathcal{B})$. There is an abstract Wiener space factorization of the form

$$E^* \xrightarrow{\ i\ } L^2(\mu) \xrightarrow{\ j\ } E.$$

Here i denotes the embedding of E^* into $L^2(\mu)$ and the linear, continuous map j is constructed so that $i^* = j$, provided $L^2(\mu)$ is identified with its dual. By linearity, the construction of j is readily reduced to defining $j(\varphi)$, where $\varphi \in L^2(\mu)$ is non-negative, with total mass one, so that $\varphi(x) \, d\mu(x)$ yields a probability measure. The integrand $x \mapsto x$ being trivially continuous, one has existence of the *Bochner integral*[1]

$$j(\varphi) := \int x \varphi(x) \, d\mu(x)$$

---

[1] Following [150], one may prefer to construct the Bochner integral over any compact K. Taking a compact exhaustion (K_n) of E it is easy to see that $j(\varphi I_{K_n})$ is Cauchy and we write $j(\varphi)$ for the limit.

as the unique element $j(\varphi) \in E$, so that for all $\lambda \in E^*$,

$$\langle \lambda, j(\varphi) \rangle_{E^*,E} = \int \langle \lambda, x \rangle_{E^*,E} \, \varphi(x) \, d\mu(x).$$

One defines E_2^* to be the closure of E^*, or more precisely: the closure of $i(E^*)$, in $L^2(\mu)$. The reproducing kernel Hilbert space $\mathcal{H}$ of μ is then defined as

$$\mathcal{H} := j(E_2^*) \subset j(L^2(\mu)) \subset E.$$

The map j restricted to E_2^* is linear and bijective onto $\mathcal{H}$ and induces a Hilbert structure

$$\langle h, g \rangle_{\mathcal{H}} := \left\langle \tilde{h}, \tilde{g} \right\rangle_{L^2(\mu)} \qquad \forall h, g \in \mathcal{H}$$

where

$$h \in \mathcal{H} \mapsto \tilde{h} \equiv \left(j|_{E_2^*} \right)^{-1}(h) \in L^2(\mu)$$

is also known as a *Paley–Wiener map*.[2] To summarize, we have the picture

$$E^* \xrightarrow{i} i(E^*) \quad \subset \quad \overline{i(E^*)} =: E_2^* \subset L^2(\mu) \xrightarrow{j} E$$
$$\text{and } j|_{E_2^*} \quad : \quad E_2^* \longleftrightarrow \mathcal{H} \subset E$$

and the triplet $(E, \mathcal{H}, \mu)$ is known as an *abstract Wiener space*. Under μ, the map $x \mapsto \tilde{h}(x)$ is a Gaussian random variable with variance $|h|_{\mathcal{H}}^2 = \langle h, h \rangle_{\mathcal{H}}$. Note that σ is also given by $\sup_{x \in \mathcal{K}} |x|$, where $\mathcal{K}$ is the closed unit ball of H. In particular, for every $x \in \mathcal{H}$,

$$|x| \le \sigma \, |x|_{\mathcal{H}}.$$

Moreover, $\mathcal{K}$ is a compact subset of $\mathcal{H}$. (To this end, use weak compactness of E^* to show that j is compact and conclude that j^* is also compact.)

Definition D.1 The triplet $(E, \mathcal{H}, \mu)$ is called an abstract Wiener space. □

Theorem D.2 (Cameron–Martin) *For any $h \in \mathcal{H}$, the probability measure $\mu(h + \cdot)$ is absolutely continuous with respect to μ, with density given*

by the formula

$$\mu(h + A) = \exp\left(-\frac{|h|_{\mathcal{H}}^2}{2} \right) \int_A \exp\left(-\tilde{h} \right) d\mu.$$

Proof. [102] or [42, Chapter 4]. ∎

---

[2] We also use the notation $\xi(h) := \tilde{h}$.

Example D.3 Take $E = C_0\left([0,1],\mathbb{R}\right)$ and μ to be the Wiener measure, i.e. the distribution of a standard Brownian motion started at the origin. If m is a finitely supported measure on $[0,1]$, say $m = \sum c_i \delta_{t_i}$ with $\{c_i\} \subset \mathbb{R}$ and $\{t_i\} \subset [0,1]$, then clearly $h = j^*j(m)$ is the element of E given by

$$h(t) = \sum c_i \min(t_i, t);$$

it satisfies

$$\int_0^1 \dot{h}_t^2 dt = \int \langle m, x \rangle^2 d\mu(x) = |h|_{\mathcal{H}}^2.$$

By a standard extension, we can then identify $\mathcal{H}$ with the Sobolev space $W_0^{1,2}\left([0,1],\mathbb{R}\right)$. Observe that for $h \in \mathcal{H}$, we have

$$\tilde{h} = \left(j^*|_{E_2^*}\right)^{-1}(h) = \int_0^1 \dot{h}_t dW_t.$$

(While we equipped the Wiener space $C_0\left([0,1],R\right)$ with uniform topology, other choices are possible.) □

D.2 Isoperimetry and concentration of measure

Gaussian measures enjoy a remarkable isoperimetric property. Following [102], we state it in the form due to C. Borell.

Theorem D.4 (Borell's inequality) *Let $(E, \mathcal{H}, \mu)$ be an abstract Wiener space and $A \subset E$ a measurable Borel set with $\mu(A) > 0$. Take $a \in (-\infty, \infty]$ such that*

$$\mu(A) = \int_{-\infty}^a \frac{1}{\sqrt{2\pi}} e^{-x^2/2} dx =: \Phi(a).$$

Then, if $\mathcal{K}$ denotes the unit ball in $\mathcal{H}$ and μ_* stands for the inner measure[3] associated with μ, then for every $r \geq 0$,

$$\mu_*(A + r\mathcal{K}) = \mu_*\{x + rh : x \in A, h \in \mathcal{K}\} \geq \Phi(a + r). \tag{D.1}$$

The following corollary is applicable in a Gaussian rough path context.

Corollary D.5 (generalized Fernique estimate) *Let $(E, \mathcal{H}, \mu)$ be an abstract Wiener space and $A \subset E$ a measurable Borel set with $\mu(A) > 0$. Assume $f : E \to \mathbb{R} \cup \{-\infty, \infty\}$ is a measurable map and $N \subset E$ a μ-null-set such that for all $x \notin N$,*

$$|f(x)| < \infty \tag{D.2}$$

---

[3]Measurability of the so-called Minkowski sum $A + r\mathcal{K}$ is a delicate topic. Use of the inner measure bypasses this issue and is not restrictive in applications.

and for some positive constant c,

$$\forall h \in \mathcal{H}: \ |f(x)| \leq c \{|(f(x-h))| + \sigma |h|_{\mathcal{H}}\}. \tag{D.3}$$

Then

$$\int \exp\left(\eta |f(x)|^2\right) d\mu(x) < \infty \ \ \text{if } \eta < \frac{1}{2c^2\sigma^2}.$$

Proof. We have for all $x \notin N$ and all $h \in r\mathcal{K}$, where $\mathcal{K}$ denotes the unit ball of $\mathcal{H}$ and $r > 0$,

$$
\begin{aligned}
\{x : |f(x)| \leq M\} \ &\supset \ \{x : c(|f(x-h)| + \sigma |h|_{\mathcal{H}}) \leq M\} \\
&\supset \ \{x : c(|f(x-h)| + \sigma r) \leq M\} \\
&= \ \{x + h : |f(x)| \leq M/c - \sigma r\}.
\end{aligned}
$$

Since $h \in r\mathcal{K}$ was arbitrary,

$$
\begin{aligned}
\{x : |f(x)| \leq M\} \ &\supset \ \cup_{h \in r\mathcal{K}} \{x + h : |f(x)| \leq M/c - \sigma r\} \\
&= \ \{x : |f(x)| \leq M/c - \sigma r\} + r\mathcal{K}
\end{aligned}
$$

and we see that

$$
\begin{aligned}
\mu[|f(x)| \leq M] \ &= \ \mu_*[|f(x)| \leq M] \\
&\geq \ \mu_*(\{x : |f(x)| \leq M/c - \sigma r\} + r\mathcal{K}).
\end{aligned}
$$

We can take $M = (1+\varepsilon)c\sigma r$ and obtain

$$\mu[|f(x)| \leq (1+\varepsilon)c\sigma r] \geq \mu_*(\{x : |f(x)| \leq \varepsilon\sigma r\} + r\mathcal{K}).$$

Keeping ε fixed, take $r \geq r_0$ where r_0 is chosen large enough such that

$$\mu[\{x : |f(x)| \leq \varepsilon\sigma r_0\}] > 0.$$

Letting Φ denote the distribution function of a standard Gaussian, it follows from Borell's inequality that

$$\mu[|f(x)| \leq (1+\varepsilon)c\sigma r] \geq \Phi(a + r)$$

for some $a > -\infty$. Equivalently,

$$\mu[|f(x)| \geq x] \leq \bar{\Phi}\left(a + \frac{x}{(1+\varepsilon)c\sigma}\right)$$

with $\bar{\Phi} \equiv 1 - \Phi$ and using $\bar{\Phi}(z) \lesssim \exp\left(-z^2/2\right)$ we see that this implies

$$\int \exp\left(\eta |f(x)|^2\right) d\mu(x) < \infty$$

provided

$$\eta < \frac{1}{2}\left(\frac{1}{(1+\varepsilon)c\sigma}\right)^2.$$

Sending $\varepsilon \to 0$ finishes the proof. ∎

Corollary D.6 (large deviations) *The family* $\mu_\varepsilon (\cdot) = \mu \left(\varepsilon^{-1} (\cdot) \right)$ *satisfies an LDP with good rate function*

$$I (A) = \inf \left\{ \frac{1}{2} |h|_{\mathcal{H}}^2 ; h \in A \cap \mathcal{H} \right\}.$$

Proof. (Sketch) Borell's inequality quickly leads to the upper bound, Cameron–Martin to the lower bound. See [102] for details. ∎

D.3 L^2-expansions

Recall the picture

$$E^* \xrightarrow{\iota^*} \iota^* (E^*) \quad \subset \quad \overline{\iota^* (E^*)} =: E_2^* \subset L^2 (\mu) \xrightarrow{\iota} E$$
$$\text{and } \iota|_{E_2^*} \quad : \quad E_2^* \longleftrightarrow \mathcal{H} \subset E$$

Given $h \in \mathcal{H}$, the map $x \mapsto \tilde{h} (x)$ is a Gaussian random variable (with variance $|h|_{\mathcal{H}}^2$) under the measure μ. We can think of $x \mapsto X (x) = x$ as an E-valued random variable with law μ. Then, for any ONB $(h_k) \subset \mathcal{H}$ we have the L^2-*expansion*

$$X (x) = \lim_{m \to \infty} \sum_{k=1}^m \tilde{h}_k (x) h_k \text{ a.s.}$$

where the sum converges in E for μ-a.e. x and in all $L^p (\mu)$-spaces, $p < \infty$. For any $A \subset \mathbb{N}$ we define

$$X^A (x) = \sum_{k \in A} \tilde{h}_k (x) h_k$$
$$\overset{\text{a.s.}}{=} \mathbb{E}^\mu \left(X | \sigma \left(\tilde{h}_k : k \in A \right) \right).$$

Note that for $|A| < \infty$, this is a finite sum with values in $\mathcal{H}$; if $|A| = \infty$ this sum converges in E for μ-almost every x and in every $L^p (\mu)$. All this follows from X^A being the conditional expectation of X given $\{\tilde{h}_k : k \in A\}$ and suitable (vector-valued) martingale convergence results, cf. [102] and the references cited therein.

D.4   Wiener–Itô chaos

Let $(E, \mathcal{H}, \mu)$ be an abstract Wiener space. Let (h_k) be the sequence of Hermite polymomials defined via $e^{\lambda x - \lambda^2 / 2} = \sum \lambda^k h_k (x)$ so that $\left(\sqrt{k!} h_k \right)$ is an orthonormal basis of $L^2(\gamma_1)$ where γ_1 is the canonical Gaussian measure on $\mathbb{R}$. For any multi-index $\alpha = (\alpha_0, \alpha_1, \dots) \in \mathbb{N}^{\mathbb{N}}$ with $|\alpha| = \alpha_0 + \alpha_1 + \cdots < \infty$ we set

$$H_\alpha = \sqrt{\alpha!}\,\Pi_i h_{\alpha_i} \circ \xi_i$$

(where $\alpha! = \alpha_0!\alpha_1!\dots$). Then the family (H_α) constitutes an orthonormal basis of $L^2(\mu)$.

Definition D.7 The (real-valued) homogenous Wiener chaos $\mathcal{W}^{(n)}(\mu)$ of degree n is defined as[4]

$$
\begin{aligned}
\mathcal{W}^{(n)}(\mu, \mathbb{R}) &= \left\{ \phi \in L^2(\mu) : \langle \phi, H_\alpha \rangle = 0 \text{ for all } \alpha : |\alpha| \neq n \right\} \\
&= \overline{\text{span}\{H_\alpha : |\alpha| = n\}} \quad \text{with closure in } L^2(\mu).
\end{aligned}
$$

Any element $\psi \in \mathcal{W}^{(n)}(\mu)$ can be written as.[5]

$$\psi = \sum_{\alpha : |\alpha| = n} \langle \psi, H_\alpha \rangle\, H_\alpha.$$

The (real-valued, non-homogenous) Wiener chaos of degree n is defined as

$$\mathcal{C}^{(n)}(\mu) = \oplus_{i=0}^n \mathcal{W}^{(n)}(\mu). \qquad \square$$

Obviously, any $\psi \in \mathcal{C}^{(n)}(\mu)$ can be written as

$$\psi = \sum_{\alpha : |\alpha| \leq n} \langle \psi, H_\alpha \rangle\, H_\alpha = L^2\text{-}\lim_{m \to \infty} \sum_{\substack{\alpha : |\alpha| \leq n \\ \alpha_k = 0, k > m}} \langle \psi, H_\alpha \rangle\, H_\alpha.$$

Since $\sum_{\alpha : |\alpha| \leq n, \alpha_k = 0, k > m} \langle \psi, H_\alpha \rangle\, H_\alpha$ is a polynomial of degree $\leq n$ in the variables $\xi_1, \dots, \xi_m$, we see that $\mathcal{C}^{(n)}(\mu)$ is precisely the L^2-closure of all polynomials in ξ_i of degree less than or equal to n.

Theorem D.8 (Wiener–Itô chaos integrability) *(i) For $\psi \in \mathcal{W}^{(n)}(\mu)$ and $1 < p < q < \infty$ we have*

$$|\psi|_{L^p(\mathbb{P})} \leq |\psi|_{L^q(\mathbb{P})} \leq \left(\frac{q-1}{p-1} \right)^{\frac{n}{2}} |\psi|_{L^p(\mathbb{P})}. \tag{D.4}$$

(ii) For $\psi \in \mathcal{C}^{(n)}(\mu)$ and $1 < p < q < \infty$ we have[6]

$$|\psi|_{L^p(\mathbb{P})} \leq |\psi|_{L^q(\mathbb{P})} \leq \left\{ (n+1)(q-1)^{n/2} \max\left(1, (p-1)^{-n}\right) \right\} |\psi|_{L^p(\mathbb{P})}. \tag{D.5}$$

---

[4] If F denotes a real separable Banach space, we can define the F-valued homogenous chaos $\mathcal{W}^{(n)}(\mu, F)$ as $\{\phi \in L^2(\mu; F) : \langle \phi, H_\alpha \rangle = 0 \text{ for all } \alpha : |\alpha| \neq n\}$.

[5] This sum is convergent μ-a.s. and in $L^2(\mu)$.

[6] By a somewhat more involved argument [39, Theorem 3.2.5], the constant on the right-hand side of (D.5) can be taken in the form $C_n \left(\frac{q-1}{p-1} \right)^{\frac{n}{2}}$ but this will be of no advantage to us.

(iii) Let $\psi \in \mathcal{C}^{(n)}(\mu)$ and $0 < p < q < \infty$. Then there exists $C = C(n, p, q)$ such that

$$|\psi|_{L^p(\mathbb{P})} \leq |\psi|_{L^q(\mathbb{P})} \leq C |\psi|_{L^p(\mathbb{P})}.$$

Proof. The estimate in (i) for $1 < p < q < \infty$ is a well-known consequence of the hypercontractivity of the Ornstein–Uhlenbeck process on abstract Wiener spaces.

Ad (ii). Take $k \in \{0, \ldots, n\}$ and call $J_k : \mathcal{C}^{(n)}(\mu) \to \mathcal{W}^{(k)}(\mu)$ the L^2-projection on the kth homogenous chaos. Then

$$|J_k \psi|_{L^p} \leq \begin{cases} (p-1)^{k/2} |\psi|_{L^p} & \text{if } p \geq 2 \\ (p-1)^{-k/2} |\psi|_{L^p} & \text{if } p < 2 \end{cases},$$

which is easily seen from (D.4) when $p > 2$ and from a duality argument for $1 < p < 2$. In particular, $J_k : L^p \to L^p$ is a bounded linear operator for any $1 < p < \infty$. From $\psi = \sum_{k=0}^{n} J_k \psi$, we have $|\psi|_{L^q} \leq \sum_{k=0}^{n} |J_k \psi|_{L^q}$ and hence

$$
\begin{aligned}
|\psi|_{L^q} &\leq \sum_{k=0}^{n} \left(\frac{q-1}{p-1} \right)^{\frac{k}{2}} |J_k \psi|_{L^p} \\
&\leq |\psi|_{L^p} \sum_{k=0}^{n} \begin{cases} (q-1)^{k/2} & \text{if } p \geq 2 \\ (q-1)^{k/2} (p-1)^{-k} & \text{if } p < 2. \end{cases} \\
&\leq |\psi|_{L^p} (n+1) (q-1)^{n/2} \max \left(1, (p-1)^{-n} \right),
\end{aligned}
$$

as required.

(iii) For the extension to $0 < p < q < \infty$ it suffices to consider the case $0 < p \leq 1, q = 2$ and $\psi \in \mathcal{C}^{(n)}(\mu)$. Using Cauchy–Schwarz,

$$
\begin{aligned}
\mathbb{E}\left(|\psi|^2 \right) &= \mathbb{E}\left(|\psi|^{p/2} |\psi|^{2-p/2} \right) \\
&\leq \left(\mathbb{E}(|\psi|^p) \right)^{1/2} \mathbb{E}\left(|\psi|^{4-p} \right)^{1/2} \\
&\leq \left(\mathbb{E}(|\psi|^p) \right)^{1/2} \left\{ (n+1)(3-p)^{n/2} \right\}^{(2-p/2)} \mathbb{E}\left(|\psi|^2 \right)^{1-p/4},
\end{aligned}
$$

we obtain $|\psi|_{L^2(\mathbb{P})} \leq c |\psi|_{L^p(\mathbb{P})}$ for some constant $c = c(p, n)$. ∎

A practical corollary is that, for $\psi, \psi' \in \mathcal{C}^{(n)}(\mu)$, we have $|\psi \psi'|_{L^2(\mathbb{P})} \leq c |\psi|_{L^2(\mathbb{P})} |\psi'|_{L^2(\mathbb{P})}$ for $c = c(n)$. (A direct proof of this can be found in [124, Proposition 1.7.2] where it is used to establish equivalence of all moments.) If the previous theorem implies the qualitative statement

$$|\psi|_{L^p(\mathbb{P})} \sim |\psi|_{L^q(\mathbb{P})} \text{ for all } p, q > 0 \text{ and } \psi \in \mathcal{C}^{(n)}(\mu), \tag{D.6}$$

the following result can be viewed as an extension to $p = 0$, where L^0-convergence is understood as convergence in μ-probability.

Theorem D.9 *(i) For any $p \in [0, \infty)$ the (non-homogenous) Wiener chaos $\mathcal{C}^{(n)}(\mu)$ is the L^p-closure of polynomials in ξ_i of degree less than or equal to n.*
(ii) For any $p \in (0, \infty)$ and any sequence of random variables in $\mathcal{C}^{(n)}(\mu)$, convergence in μ-probability is equivalent to L^p-convergence.

Proof. It suffices to check that a Cauchy sequence in probability, say X_n, also converges in L^p for $p > 0$. We argue by contradiction and assume it does not converge in L^p, for some $p > 0$. Then there exists $\varepsilon > 0$, such that for arbitrarily large m, n one has $|X_n - X_m|_{L^p} > \varepsilon$. Let $\delta \in (0, 1)$. It follows that

$$\mathbb{P}\left[\frac{|X_n - X_m|}{|X_n - X_m|_{L^p}} > \delta\right] \leq \mathbb{P}\left[|X_n - X_m| > \delta\varepsilon\right]$$

which, by assumption, tends to zero as $m, n \to \infty$. On the other hand, using equivalence of $|\cdot|_{L^q(\mathbb{P})}$ and $|\cdot|_{L^p(\mathbb{P})}$ on $\mathcal{C}^{(n)}(\mu)$, for any $q > p$, the next lemma applied with $\xi = |X_n - X_m|$ implies that

$$\inf_{n,m} \mathbb{P}\left[|X_n - X_m| > \delta |X_n - X_m|_{L^p}\right] > 0$$

and so yields the desired contradiction. ∎

Lemma D.10 (Paley–Zygmund inequality) *Let $0 < p < q < \infty$ and $\xi \geq 0$ be a random variable in $L^q(\mathbb{P})$. Then, for any $\delta \in (0, 1)$,*

$$\mathbb{P}\left[\xi > \delta |\xi|_{L^p(\mathbb{P})}\right] \geq \left((1 - \delta^p)\left(\frac{|\xi|_{L^p(\mathbb{P})}}{|\xi|_{L^q(\mathbb{P})}}\right)^p\right)^{\frac{q}{q-p}}.$$

Proof. Set $t = \delta |\xi|_{L^p(\mathbb{P})}$. From $\mathbb{E}\xi^p \leq t^p + \mathbb{E}(\xi^p; \xi \geq t) \leq t^p + \mathbb{E}(\xi^q)^{p/q} \mathbb{P}[\xi > t]^{1-p/q}$ it follows that

$$\mathbb{P}[\xi > t]^{\frac{q-p}{p}} \geq \frac{|\xi|^p_{L^p(\mathbb{P})} - t^p}{|\xi|^p_{L^q(\mathbb{P})}} = \left(\frac{|\xi|_{L^p(\mathbb{P})}}{|\xi|_{L^q(\mathbb{P})}}\right)^p (1 - \delta^p).$$

∎

D.5   Malliavin calculus

Following [99], [136, Section 4.1.3] or [171, Section 3.3] we have the following notion of $\mathcal{H}$-regularity for a Wiener functional F.

Definition D.11 Given an abstract Wiener space $(E, \mathcal{H}, \mu)$, a random variable (i.e. measurable map) $F : E \to \mathbb{R}$ is a.s. continuously $\mathcal{H}$-differentiable, in symbols $F \in C^1_{\mathcal{H}}$ a.s., if for μ-almost every ω, the map

$$h \in \mathcal{H} \mapsto F(\omega + h) \tag{D.7}$$

is continuously Fréchet differentiable. A vector-valued r.v. $F = \left(F^1, \ldots, F^n\right)$: $E \to \mathbb{R}^n$ is a.s. continuously $\mathcal{H}$-differentiable iff each F^i is a.s. continuously $\mathcal{H}$-differentiable.

Similarly, if (D.7) is a.s. k-times Fréchet differentiable, we write $F \in C_{\mathcal{H}}^k$ a.s. and say that F is k-times a.s. continuously $\mathcal{H}$-differentiable. When $k = \infty$ we say that F is a.s. $\mathcal{H}$-smooth.   □

The notion of $\mathcal{H}$-differentiability was introduced in [99] and plays a fundamental role in the study of transformation of measure on Wiener space. Integrability properties of F and DF aside, $C_{\mathcal{H}}^1$-regularity is stronger than Malliavin differentiability in the usual sense. Indeed, by [136, Theorem 4.1.3] (see also [99], [171, Section 3.3]) $C_{\mathcal{H}}^1$ implies $\mathbb{D}_{\mathrm{loc}}^{1,2}$-regularity where the definition of $\mathbb{D}_{\mathrm{loc}}^{1,2}$ is based on the commonly used Shigekawa Sobolev space $\mathbb{D}^{1,p}$. (Our notation here follows [136, Sections 1.2, 1.3.4]). This remark will be important to us since it justifies the use of Bouleau–Hirsch's criterion (e.g. [136, Section 2.1.2]) for establishing absolute continuity of F.

Proposition D.12 (Bouleau–Hirsch) *Let $(E, \mathcal{H}, \mu)$ be an abstract Wiener space and $F = \left(F^1, \ldots, F^n\right) : E \to \mathbb{R}^n$ a measurable map. Assume $F \in C_{\mathcal{H}}^1$ is weakly non-degenerate, by which we mean that the Malliavin covariance matrix*

$$\sigma\left(\omega\right) := \left(\left\langle DF^i, DF^j \right\rangle_{\mathcal{H}}\right)_{i,j=1,\ldots,n} \in \mathbb{R}^{n \times n}$$

is μ-almost surely non-degenerate. Then F, viewed as an $\mathbb{R}^n$-valued random variable under μ, admits a density with respect to Lebesgue measure on $\mathbb{R}^n$.

Proof. From [136, Section 4.1.3], $F \in C_{\mathcal{H}}^1$ implies that $f \in \mathbb{D}_{loc}^{1,2}$ and the usual Bouleau–Hirsch criterion [136, Section 4.1.3] applies.   ∎

D.6   Comments

Section D.1 follows closely Ledoux's Saint Flour notes [102]. Other references with a similar point of view are Ledoux and Talagrand [103] and Deuschel and Stroock [42, Chapter 4]. **Sections D.2** and **D.3** follow Ledoux [102]. The generalized Fernique estimate is taken from Friz and Oberhauser [60]. The basic definitions of **Section D.4** also follow Ledoux [102, Section 5]. Integrability of the Wiener–Itô chaos via hypercontractivity of the Ornstein–Uhlenbeck semi-group is classical, see Ledoux [102, Section 8] as well as most references on Malliavin calculus, such as Nualast [135]. Theorem D.9 appears in Schreiber [153], our proof is taken from de la Peña and Giné [39].

Section D.5:. There are many good books on Malliavin calculus, including Malliavin [124], Nualart [135] and Shigekawa [155]. The concept of $\mathcal{H}$-differentiability is due to Kusuoka [99]; see Nualart [135] and in particular Süleyman Üstünel and Zakai [171, Section 3.3].

Appendix E: Analysis on local dirichlet spaces

E.1 Quadratic forms

Consider a Hilbert space $(H, \langle \cdot, \cdot \rangle)$ with a non-positive self-adjoint operator $\mathcal{L}$, defined on a dense linear subspace $\mathcal{D}(\mathcal{L})$. Spectral calculus[1] allows us to define the self-adjoint operator $\sqrt{-\mathcal{L}}$, with domain $\mathcal{D}(\sqrt{-\mathcal{L}})$. A quadratic form, defined on $\mathcal{D} := \mathcal{D}(Q) := \mathcal{D}(\sqrt{-\mathcal{L}})$, is then given by

$$Q(f, f) = \left\langle \sqrt{-\mathcal{L}}f, \sqrt{-\mathcal{L}}f \right\rangle$$

and this form is **non-negative** in the sense that $Q(f, f) \geq 0$ for all f. (By polarization, this induces a symmetric bilinear form Q, defined on $\mathcal{D} \times \mathcal{D}$, so that $Q(f, g) = \langle \sqrt{-\mathcal{L}}f, \sqrt{-\mathcal{L}}g \rangle$.) It is well known[2] that this yields a **closed** form in the sense that whenever $(f_n) \subset \mathcal{D}$ such that

$$f_n \to f \text{ in } H \text{ with } n \to \infty \text{ and } Q(f_n - f_m, f_n - f_m) \to 0 \text{ with } n, m \to \infty$$

then $f \in \mathcal{D}$ and $Q(f_n - f, f_n - f) \to 0$ with $n \to \infty$. Conversely, every such form arises in this way from a (non-positive) self-adjoint operator $\mathcal{L}$. In many applications one has forms which are not closed but *closable* in the sense that whenever $(f_n) \subset \mathcal{D}$ is such that

$$f_n \to 0 \text{ in } H \text{ with } n \to \infty \text{ and } Q(f_n - f_m, f_n - f_m) \to 0 \text{ with } n, m \to \infty$$

then $Q(f_n, f_n) \to 0$ with $n \to \infty$. In this case, Q admits a (minimal) extension to a closed form $\overline{Q}$; we shall not distinguish between Q and $\overline{Q}$. Let us further recall that the domain $\mathcal{D}$ of a (symmetric, closed, non-negative) form is a Hilbert space under the inner product

$$\langle f, g \rangle_Q = \langle f, g \rangle + Q(f, g).$$

By spectral calculus, one defines $\mathcal{P}_t = e^{t\mathcal{L}} : H \to H$, which yields a (strongly continuous, contraction) semi-group $(\mathcal{P}_t : t \geq 0)$ on H with infinitesimal generator given by $\mathcal{L}$. For $t > 0$, $\mathcal{P}_t$ maps H into $\mathcal{D}$ and

$$\langle \mathcal{P}_t f, g \rangle = \langle f, g \rangle - \int_0^t Q(\mathcal{P}_s f, g) \, ds \quad \text{for} \quad f, g \in \mathcal{D};$$

---

[1] See, for example, Yosida [176, Chapter XI].
[2] See, for example, [38].

as may be seen by integrating the equality

$$e^{-t\lambda} = 1 - \int_0^t \sqrt{\lambda} e^{-\lambda s} \sqrt{\lambda} ds, \quad \lambda \in [0, \infty)$$

against $d\langle E_\lambda f, g \rangle$, where $\{E_\lambda : \lambda \in [0, \infty)\}$ is the spectral resolution of the (non-negative, self-adjoint) operator $-\mathcal{L}$. The following lemma is based on similar ideas.

Lemma E.1 *(i) For all $f \in \mathcal{D} = \mathcal{D}\left(\sqrt{-\mathcal{L}}\right)$,*

$$Q\left(\mathcal{P}_t f, \mathcal{P}_t f\right) \leq \left(\frac{1}{2t} \langle f, f \rangle\right) \wedge Q\left(f, f\right). \tag{E.1}$$

(ii) Assume $f_n \to f$ in H and assume that (f_n) is bounded in $\left(\mathcal{D}, \langle \cdot, \cdot \rangle_Q\right)$;

$$\sup_n \langle f_n, f_n \rangle_Q < \infty.$$

Then $f \in \mathcal{D}$ and $f_n \to f$ weakly in $\left(\mathcal{D}, \langle \cdot, \cdot \rangle_Q\right)$.

Proof. (i) It suffices to integrate the elementary inequality

$$\lambda e^{-2t\lambda} \leq \frac{1}{2t} \wedge \lambda, \quad \lambda \in [0, \infty)$$

against $d\langle E_\lambda f, f \rangle$, where $\{E_\lambda : \lambda \in [0, \infty)\}$ is the spectral resolution of $-\mathcal{L}$. (ii) Step 1. Let us first assume that $f \in \mathcal{D}$. For every $h \in \mathcal{D}(\mathcal{L}) \subset \mathcal{D} = \mathcal{D}\left(\sqrt{-\mathcal{L}}\right)$ we have

$$-Q\left(f_n, h\right) = \langle f_n, \mathcal{L}h \rangle \to \langle f, \mathcal{L}h \rangle$$

and so, for all $h \in \mathcal{D}(\mathcal{L})$,

$$\langle f_n, h \rangle_Q \to \langle f, h \rangle_Q.$$

What we want is that this convergence holds for all $h \in \mathcal{D}$. If we can show density of $\mathcal{D}(\mathcal{L})$ in $\left(\mathcal{D}, \langle \cdot, \cdot \rangle_Q\right)$, the extension to all $h \in \mathcal{D}$ is straightforward. But this density statement is also easy to see: for instance, given $h \in \mathcal{D}$ one has $P_t h \in \mathcal{D}(\mathcal{L})$ and $P_t h \to h$ in $\mathcal{D}$ since

$$Q\langle P_t h - h, P_t h - h \rangle = \int_{[0,\infty)} \lambda \left(e^{-\lambda t} - 1\right)^2 d\langle E_\lambda h, h \rangle \to 0$$

by bounded convergence, using $\int \lambda d\langle E_\lambda h, h \rangle = Q\left(h, h\right) < \infty$ since $h \in \mathcal{D}$. Step 2. Let us now consider an arbitrary $f \in H$. We mollify using the semigroup. For $t > 0$ we set $f_t := \mathcal{P}_t f$ and similarly $f_t^n := \mathcal{P}_t f^n$. It is easy to see that $f_t^n \to f_t$ in H as $n \to \infty$ and also that (f_t^n) is uniformly bounded in the sense that

$$M := \sup_{n, t \in (0,1]} \langle f_t^n, f_t^n \rangle_Q < \infty.$$

Apply step 1 to see that $f_t^n \to f_t$ weakly in $\left(\mathcal{D}, \langle \cdot, \cdot \rangle_Q \right)$. In particular,

$$\langle f_t, f_t \rangle_Q \le \liminf_{n \to \infty} \langle f_t^n, f_t^n \rangle_Q \le M.$$

Hence $\sup_{t \in (0,1]} \langle f_t, f_t \rangle_{\mathcal{D}} \le M < \infty$ and this entails that $f \in \mathcal{D}$. Indeed, by monotone convergence

$$
\begin{aligned}
\int_{[0,\infty)} \lambda d \langle E_\lambda f, f \rangle &= \lim_{t \downarrow 0} \int_{[0,\infty)} \lambda e^{-2\lambda t} d \langle E_\lambda f, f \rangle \\
&\le \lim_{t \downarrow 0} \langle f_t, f_t \rangle_Q < \infty,
\end{aligned}
$$

which shows that $f \in \mathcal{D}$. We can now appeal to step 1 to conclude the proof. ∎

E.2 Symmetric Markovian semi-groups and Dirichlet forms

Let us now consider a quadratic (equivalently: symmetric bilinear) form $\mathcal{E}(\cdot, \cdot)$ on the Hilbert space $L^2(E, m)$ where E is assumed to be a locally compact Polish space, m is a Radon measure on E of full support. The classical example to have in mind is $E = \mathbb{R}^d$, equipped with Lebesgue measure, and $\mathcal{E}(f, f) = \int |\nabla f(x)|^2 \, dx$, defined on $\mathcal{D}(\mathcal{E}) = W^{1,2}(\mathbb{R}^d)$, the usual Sobolev space of L^2-functions on $\mathbb{R}^d$ with weak derivatives in L^2.

Following Fukushima *et al.* [70] we have the following abstract

Definition E.2 A non-negative definite symmetric bilinear form $\mathcal{E}$, densely defined on $L^2(E, m)$, is called a **Dirichlet form** if it is
(i) **closed** in the sense of quadratic forms; i.e. whenever $(f_n) \subset \mathcal{D}(\mathcal{E})$ is such that

$$f_n \to f \text{ in } L^2 \text{ with } n \to \infty \text{ and } \mathcal{E}(f_n - f_m, f_n - f_m) \to 0 \text{ with } n, m \to \infty$$

then $f \in \mathcal{D}(\mathcal{E})$ and $\mathcal{E}(f_n - f, f_n - f) \to 0$ with $n \to \infty$;
(ii) **Markovian** *in the sense that*

$$f \in \mathcal{D}(\mathcal{E}), g = (0 \vee f) \wedge 1 \implies g \in \mathcal{D}(\mathcal{E}), \mathcal{E}(g, g) \le \mathcal{E}(f, f).$$

The pair $(\mathcal{E}, \mathcal{D}(\mathcal{E}))$ is called a Dirichlet space relative to $L^2(E, m)$. □

Everything said in the previous section applies to Dirichlet forms: in particular, there exists a non-positive self-adjoint operator $\mathcal{L}$ on $L^2(E, m)$ so that

$$\mathcal{D}(\mathcal{E}) = \mathcal{D}\left(\sqrt{-\mathcal{L}}\right),$$
$$\mathcal{E}(f, g) = \left\langle \sqrt{-\mathcal{L}}f, \sqrt{-\mathcal{L}}g \right\rangle_{L^2}$$

and there is a (strongly continuous, contraction) semi-group $(\mathcal{P}_t : t \geq 0)$ on $L^2(E, m)$ with infinitesimal generator given by $\mathcal{L}$, so that

$$\langle \mathcal{P}_t f, g \rangle_{L^2} = \langle f, g \rangle_{L^2} - \int_0^t \mathcal{E}(\mathcal{P}_s f, g)\, ds \quad \text{for} \quad f, g \in \mathcal{D}(\mathcal{E})$$

and Lemma E.1 is also valid. The Markovian property of $\mathcal{E}$ is equivalent to *Markovianity* of the associated L^2-semi-group in the sense that

$$\forall t > 0 : f \in L^2(E, m), 0 \leq f \leq 1 \ m\text{-a.e.} \implies 0 \leq \mathcal{P}_t f \leq 1 \ m\text{-a.e.}$$

The Dirichlet forms interesting to us enjoy further properties of the following kind.

Definition E.3 A Dirichlet form $\mathcal{E}$ is called **regular** if there exists a core; that is, a subset $\mathcal{C} \subset \mathcal{D}(\mathcal{E}) \cap C_c(E)$ which is dense in $\mathcal{D}(\mathcal{E})$ with respect to $\mathcal{E}_1$ and dense in $C_c(E)$ with respect to uniform norm. It is called **strongly local** if $\mathcal{E}(f, g)$ is zero whenever $f \in \mathcal{D}(\mathcal{E})$ is constant on a neighbourhood of the support of $g \in \mathcal{D}(\mathcal{E})$.   □

Strong locality of $\mathcal{E}$ has the interpretation of no killing and no jumps (e.g. within its Beurling–Deny decomposition [70, Section 3.2]). Any such Dirichlet form can be written as

$$\mathcal{E}(f, g) = \int_E d\Gamma(f, g)$$

where Γ is the so-called *energy measure*, a positive semi-definite, symmetric bilinear form $\mathcal{D}(\mathcal{E})$ with values in the signed Radon measures on E. It can be defined by

$$\int \varphi d\Gamma(f, f) = \mathcal{E}(f, \varphi f) - \frac{1}{2}\mathcal{E}(f^2, \varphi)$$

for every $f \in \mathcal{D}(\mathcal{E}) \cap L^\infty$ and $\varphi \in \mathcal{D}(\mathcal{E}) \cap C_c$.

In all our applications, $\Gamma(f, g)$ will be absolutely continuous with respect to m and we shall simply write

$$d\Gamma(f, g) = \Gamma(f, g)\, dm$$

and call the map $f, g \mapsto \Gamma(f, g)$, $\mathcal{D}(\mathcal{E}) \times \mathcal{D}(\mathcal{E}) \to L^1(E, m)$, the *carré du champ operator*. The *intrinsic metric* of $\mathcal{E}$ is then defined as

$$d(x, y) = \sup\{f(x) - f(y) : f \in \mathcal{D}(\mathcal{E}) \text{ and continuous,}$$
$$d\Gamma(f, f)/dm \le 1 \ m\text{-a.e.}\}.$$

In general, d can be degenerate, i.e. $d(x, y) = \infty$ or $\rho(x, y) = 0$ for some $x \ne y$.

Definition E.4 A strongly local Dirichlet form $\mathcal{E}$ with domain $\mathcal{D}(\mathcal{E})$ is called **strongly regular** if it is regular and its intrisic metric is a genuine metric on E whose topology coincides with the original one.   □

Let us recall (cf. Definition 5.17) that a *geodesic* (or geodesic path) joining two points $x, y \in E$ is a continuous path $\gamma : [0, 1] \to E$ such that $\gamma(0) = x, \gamma(1) = y$ and

$$d(\gamma_s, \gamma_t) = |t - s| d(x, y) \quad \forall 0 \le s < t \le 1$$

and that E is called a *geodesic space* if all points x, y can be joined by a geodesic. Observe that $z := \gamma_{1/2}$ is a *midpoint* of x, y in the sense that

$$d(x, z) = d(z, y) = \frac{1}{2} d(x, y).$$

If all $x, y \in E$ have a midpoint we say that E has the *midpoint property*. In fact, any complete metric space with the midpoint property is geodesic: given x, y, iterated use of the midpoint property yields $\gamma_{1/2}$ then $\gamma_{1/4}, \gamma_{3/4}$ and so on, which yields (a candidate for) a geodesic, defined on all dyadic rationals. The extension of $\gamma = \gamma_t$ to all $t \in [0, 1]$ is possible by the completeness assumption; continuity of γ is then easy to check.

Proposition E.5 *Assume $\mathcal{E}$ is a strongly local, strongly regular Dirichlet form on E so that (E, d) is a complete metric space. Then (E, d) is a geodesic space.*

Proof. We follow [165]. Fix arbitrary elements $x, y \in E$ and set $R = d(x, y)$. It then suffices to show the midpoint property

$$\exists z \in E : d(x, z) = d(z, y) = R/2.$$

We argue by contradiction. Assuming that there is no midpoint z, we have $\bar{B}_{R/2}(x) \cap \bar{B}_{R/2}(y) = \varnothing$. By compactness, these sets have a positive distance, say $3\varepsilon > 0$, and it is also clear that

$$d(\bar{B}_{R/2+\varepsilon}(x), \bar{B}_{R/2+\varepsilon}(y)) > \varepsilon.$$

We now define the continuous function

$$f_0 = \begin{cases} d(x, \cdot) - (R/2 + \varepsilon) & \text{on } \bar{B}_{R/2+\varepsilon}(x) \\ (R/2 + \varepsilon) - d(y, \cdot) & \text{on } \bar{B}_{R/2+\varepsilon}(y) \\ 0 & \text{else} \end{cases}$$

and note that

$$d\Gamma\left(f_0, f_0\right) = 1_{\bar{B}_{R/2+\varepsilon}(x)} d\Gamma\left(d\left(x, \cdot\right), d\left(x, \cdot\right)\right) + 1_{\bar{B}_{R/2+\varepsilon}(y)} d\Gamma\left(d\left(y, \cdot\right), d\left(y, \cdot\right)\right)$$
$$\leq dm,$$

using the fact that $\Gamma\left(d\left(x, \cdot\right), d\left(x, \cdot\right)\right) \leq 1$ for all x. Moreover, $f_0\left(x\right) - f_0\left(y\right) = R + 2\varepsilon > R$. But this is a contradiction to

$$R = \sup\left\{f\left(x\right) - f\left(y\right) : f \in \mathcal{D}\left(\mathcal{E}\right) \text{ and } f \text{ continuous}, \Gamma\left(f, f\right) \leq 1\right\}.$$

∎

E.3 Doubling, Poincaré and quasi-isometry

We now make the standing assumption that $\mathcal{E}$ is a strongly local, strongly regular Dirichlet form on E. As we shall see, the following properties (I)–(III) have remarkably powerful implications.

Definition E.6 Let $(\mathcal{E}, \mathcal{D}\left(\mathcal{E}\right))$ be a Dirichlet space relative to $L^2\left(E, m\right)$. Assume that $\mathcal{E}$ is a strongly local and strongly regular Dirichlet form, with intrinsic metric d. We then say that $\mathcal{E}$ has (or satisfies) the
(I) **completeness property** if the metric space (E, d) is a complete metric (and hence by Proposition E.5 a geodesic) space;
(II) **doubling property** if there exists a doubling constant $N = N\left(\mathcal{E}\right)$ such that

$$\forall r \geq 0, x \in E : m\left(B\left(x, 2r\right)\right) \leq 2^N m\left(B\left(x, r\right)\right);$$

(III) **weak Poincaré inequality** if there exists $C_P = C_P\left(\mathcal{E}\right)$ such that for all $r \geq 0$ and $f \in D\left(\mathcal{E}\right)$,

$$\int_{B(x,r)} \left|f - \bar{f}_r\right|^2 dm \leq C_P r^2 \int_{B(x,2r)} \Gamma\left(f, f\right) dm$$

where

$$\bar{f}_r = m\left(B\left(x, r\right)\right)^{-1} \int_{B(x,r)} f \, dm. \qquad \square$$

Let us make two useful remarks. First, the doubling property (II) readily implies

$$\forall 0 < r < r' : m\left(B\left(x, r'\right)\right) \leq \left(2r'/r\right)^N m\left(B, r\right); \qquad (\text{E.2})$$

and second, the right-hand side in the weak Poincaré inequality can be written as

$$\int_{B(x,r)} \left|f - \bar{f}_r\right|^2 dm = \inf_{\alpha \in \mathbb{R}} \int_{B(x,r)} \left|f - \alpha\right|^2 dm. \qquad (\text{E.3})$$

Definition E.7 Two (strongly local and strongly regular) Dirichlet forms $\mathcal{E}$ and $\tilde{\mathcal{E}}$ are **quasi-isometric** if $\mathcal{D}(\mathcal{E}) = \mathcal{D}\left(\tilde{\mathcal{E}}\right)$ and there exists $\Lambda \geq 1$ such that for all f in the common domain,

$$\frac{1}{\Lambda}\mathcal{E}(f,f) \leq \tilde{\mathcal{E}}(f,f) \leq \Lambda\mathcal{E}(f,f).$$ □

If we write $d, \tilde{d}$ for the respective intrinsic metrics associated with $\mathcal{E}$ and $\tilde{\mathcal{E}}$, it is clear that the metrics are Lipschitz equivalent in the sense that

$$\frac{1}{\Lambda^{1/2}}d(x,y) \leq \tilde{d}(x,y) \leq \Lambda^{1/2}d(x,y)$$

for all $x, y \in E$.

Theorem E.8 *Let $\mathcal{E}$ satisfy properties (I)–(III). Then, assuming $\mathcal{E}$ and $\tilde{\mathcal{E}}$ are quasi-isometric, $\tilde{\mathcal{E}}$ also satisfies properties (I)–(III), with new (doubling and Poincaré) constants depending on Λ.*

Proof. Invariance of the completeness property is clear from Lipschitz equivalence of d and $\tilde{d}$.

Second, we assume doubling for balls with respect to d. Then, using (E.2),

$$
\begin{aligned}
m\left(\tilde{B}(x,2r)\right) &\leq m\left(B\left(x,2\Lambda^{1/2}r\right)\right) = m\left(B\left(x,2\Lambda r/\Lambda^{1/2}\right)\right) \\
&\leq (4\Lambda)^N m\left(B\left(x,r/\Lambda^{1/2}\right)\right) \\
&\leq (4\Lambda)^N m\left(\tilde{B}(x,r)\right)
\end{aligned}
$$

where $\tilde{B}(x,r) = \left\{y \in E : \tilde{d}(x.y) < r\right\}$. At last, let us write

$$\tilde{f}_r = m\left(\tilde{B}(x,r)\right)^{-1} \int_{\tilde{B}(x,r)} f \, dm$$

for the average of f over $\tilde{B}(x,r)$. Using (E.3) and the weak Poincaré inequality for $\mathcal{E}$, we see that

$$
\begin{aligned}
\int_{\tilde{B}(x,r)} \left|f - \tilde{f}_r\right|^2 dm &\leq \inf_{\alpha \in \mathbb{R}} \int_{B(x,r\Lambda^{1/2})} |f - \alpha|^2 dm \\
&= \int_{B(x,r\Lambda^{1/2})} \left|f - \bar{f}_{r\Lambda^{1/2}}\right|^2 dm \\
&\leq C_P r^2 \Lambda \int_{B(x,2r\Lambda^{1/2})} d\Gamma(f,f) \\
&\leq C_P r^2 \Lambda^2 \int_{\tilde{B}(x,2r\Lambda)} d\tilde{\Gamma}(f,f)
\end{aligned}
$$

where $\tilde{\Gamma}$ denotes the energy measure of $\tilde{\mathcal{E}}$. By a covering argument, derived from (I), (II), this implies the weak Poincaré property for $\tilde{\mathcal{E}}$ on $L^2(E, m)$. ∎

A special case of quasi-isometry arises from scaling.

Proposition E.9 (scaling) *Let $\mathcal{E}$ be a Dirichlet form on $L^2(E, m)$ with*

doubling and Poincaré constants N, C_P respectively. Then, for every $\varepsilon > 0$, the scaled Dirichlet form

$$\mathcal{E}^\varepsilon \equiv \varepsilon\mathcal{E}$$

satisfies (I)–(III) with the same doubling and Poincaré constants N, C_P and with intrinsic metric given by

$$\forall x, y \in E : d^\varepsilon(x, y) = \frac{1}{\varepsilon^{1/2}} d(x, y).$$

Proof. The relation $d^\varepsilon \equiv d/\varepsilon^{1/2}$ is an obvious consequence of the definition. We only need to check the behaviour of doubling and Poincaré constants and this quasi-isometry. Writing B^ε for balls with respect to d^ε, we obviously have

$$B^\varepsilon(x, r) = B\left(x, r\varepsilon^{1/2}\right)$$

for any $x \in E, r > 0$. Clearly then,

$$m\left(B^\varepsilon(x, 2r)\right) \leq 2^Q m\left(B\left(x, r\varepsilon^{1/2}\right)\right) = 2^Q m\left(B^\varepsilon(x, r)\right)$$

and so Q is also the doubling constant for $\mathcal{E}^\varepsilon$. Finally, note that

$$f_r^\varepsilon := m\left(B^\varepsilon(x, r)\right)^{-1} \int_{B^\varepsilon(x,r)} f\, dm = \bar{f}_{r\varepsilon^{1/2}}$$

and so

$$\begin{aligned}
\int_{B^\varepsilon(x,r)} |f - f_r^\varepsilon|^2\, dm &= \int_{B(x,r\varepsilon^{1/2})} |f - \bar{f}_{r\varepsilon^{1/2}}|^2\, dm \\
&\leq C_P r^2 \varepsilon \int_{B(x,2r\varepsilon^{1/2})} d\Gamma(f, f) \\
&= C_P r^2 \int_{B^\varepsilon(x,2r)} d\Gamma^\varepsilon(f, f)
\end{aligned}$$

where $\Gamma^\varepsilon = \varepsilon\Gamma$ is the energy measure of $\mathcal{E}^\varepsilon$. We see that C_P is the Poincaré constant for $\mathcal{E}^\varepsilon$ and the proof is finished. ∎

E.4   Parabolic equations and heat-kernels

Recall the standing assumption that $\mathcal{E}$ is a strongly local, strongly regular Dirichlet form on $L^2\left(E, m\right)$. There is an associated non-positive self-adjoint operator $\mathcal{L}$ on $L^2\left(E, m\right)$ and a strongly continuous semi-group $\left(\mathcal{P}_t : t \geq 0\right)$. We can now consider *weak solutions* to the parabolic partial differential equation

$$\partial_t u = \mathcal{L} u;$$

that is, a function $u : t \mapsto u\left(t, \cdot\right) \in \mathcal{D}\left(\mathcal{E}\right)$ so that

$$\forall g \in \mathcal{D}\left(\mathcal{E}\right) : \langle u\left(t, \cdot\right), g \rangle_{L^2} = \langle u\left(0, \cdot\right), g \rangle_{L^2} - \int_0^t \mathcal{E}\left(u\left(s, \cdot\right), g\right) ds.$$

What one has in mind is that actually $u = u\left(t, x\right)$ for $\left(t, x\right) \in [0, \infty) \times E$, regarded as a one-parameter family $\left(u\left(t, \cdot\right) : t \geq 0\right)$ depending only on space. (Obviously, the semi-group operator $\mathcal{P}_t u_0$ yields a solution to this PDE with intial date $u\left(0, \cdot\right) = u_0$.) The notion of solution can be localized. Indeed, by restricting ourselves to times in some interval $I \subset [0, \infty)$ and a test function g compactly supported on some open set $G \subset E$, we can speak of *(local) weak solutions* to

$$\partial_t u = \mathcal{L} u \ \text{ on } Q$$

where $Q = I \times G$ is a *(parabolic) cylinder*. All of the following four theorems are classical, proofs can be found in [163] and [164, p. 304].

Theorem E.10 (de Giorgio–Moser–Nash regularity) *Assume $\mathcal{E}$ is a strongly local, strongly regular Dirichlet form on $L^2\left(E, m\right)$ which satisfies (I)–(III). Then there exist constants $\eta_R \in \left(0, 1\right)$ and C_R (depending only on N, C_P, i.e. the doubling and Poincaré constants of $\mathcal{E}$) so that*[3]

$$\sup_{(s,y),(s',y')\in Q_1} |u\left(s, y\right) - u\left(s', y'\right)| \leq C_R \sup_{u \in Q_2} |u| \cdot \left(\frac{|s - s'|^{1/2} + d\left(y, y'\right)}{r}\right)^{\eta}$$

whenever u is a non-negative weak solution of the parabolic partial differential equation $\partial_s u = \mathcal{L} u$ on some cylinder $Q_2 \equiv \left(t - 4r^2, t\right) \times B\left(x, 2r\right)$ for some reals $t, r > 0$. Here, $Q_1 \equiv \left(t - r^2, t - 2r^2\right) \times B\left(x, r\right)$ is a subcylinder of Q_2.

Theorem E.11 (parabolic Harnack inequality) *Assume $\mathcal{E}$ is a strongly local, strongly regular Dirichlet form on $L^2\left(E, m\right)$ which satisfies*

---

[3] Strictly speaking, the statement is that any non-negative weak solution to $\partial_s u = \mathcal{L} u$ on Q_2 has an m almost-identical version that enjoys this regularity.

(I)–(III). Then there exists a constant C_H which depends only on N, C_P (the doubling and Poincaré constants of $\mathcal{E}$) such that

$$\sup_{(s,y)\in Q^-} u\,(s,y) \le C_H \inf_{(s,y)\in Q^+} u\,(s,y)$$

whenever u is a non-negative weak solution of the parabolic partial differential equation $\partial_t u = \mathcal{L}u$ on some cylinder $Q = \left(t - 4r^2, t\right) \times B\left(x, 2r\right)$ for some reals $t, r > 0$. Here, $Q^- = \left(t - 3r^2, t - 2r^2\right) \times B\left(x, r\right)$ and $Q^+ = \left(t - r^2, t\right) \times B\left(x, r\right)$ are lower and upper subcylinders of Q separated by some elapse of time.

Theorem E.12 (heat-kernel) *Assume $\mathcal{E}$ is a strongly local, strongly regular Dirichlet form on $L^2\left(E, m\right)$ which satisfies (I)–(III). Let $\mathcal{L}$ and $(\mathcal{P}_t)$ denote the associated self-adjoint operator resp. Markovian semi-group. Then*
*(i) there exists a continuous function, called the **heat-kernel**,*

$$p : (0, \infty) \times E \times E \to [0, \infty),$$

symmetric in the last two arguments, i.e. $p\,(t, x, y) = p\,(t, y, x)$, so that

$$\forall t > 0 : \mathcal{P}_t f = \int p\,(t, \cdot, y)\, f\,(y)\, dm\,(y)\,, \quad f \in L^2;$$

(ii) for every fixed $x \in E$, the map $(t, y) \mapsto p\,(t, x, y)$ is a (global, weak) solution to $\partial_t u = \mathcal{L}u$ on $(0, \infty) \times E$;
*(iii) it satisfies the **Chapman–Kolmogorov equations**: for all $s < t$ and $x, y \in E$,*

$$p\,(t, x, z) = \int p\,(s, x, y)\, p\,(t - s, y, z)\, dy.$$

The proof of the heat-kernel existence (cf. [164, p. 304]) follows immediately from an estimate on the operator norm $\|\mathcal{P}_t\|_{L^1 \to L^\infty}$, which in turn follows from suitable Sobolev inequalities. Let us note that $(t, y) \mapsto p\,(t, x, y)$ is a weak solution to $\partial_t u = \mathcal{L}_y u$ and so, by Harnack's inequality, we have

$$p\,(t, x, x) \le c_H \inf \left\{ p\,(2t, x, y) : y \in B\left(x, t^{1/2}\right) \right\}.$$

Integrating this estimate over the ball $B\left(x, t^{1/2}\right)$, we obtain

$$m\left(B\left(x, t^{1/2}\right)\right) p\,(t, x, x) \le c_H \int_{B^a\left(x, t^{1/2}\right)} p\,(2t, x, y)\, dy \le c_H,$$

which leads to an on-diagonal estimate for the heat-kernel. Let us now state the full estimates.

Theorem E.13 (Aronson heat-kernel estimates) *Assume $\mathcal{E}$ is a strongly local, strongly regular Dirichlet form on $L^2(E, m)$ which satisfies (I)–(III) and let p denotes its heat kernel. Then*
(i) for every $\varepsilon > 0$, there exists a constant C_U which depends only on ε, N, C_P (the doubling and Poincaré constants of $\mathcal{E}$) such that the following upper bound holds,

$$p(t, x, y) \leq \frac{C_U}{\sqrt{m\left(B\left(x, t^{1/2}\right)\right) m\left(B\left(y, t^{1/2}\right)\right)}} \exp\left(-\frac{d(x, y)^2}{(4 + \varepsilon) t}\right);$$

(ii) there exists $C_L = C_L(\mathcal{E})$ such that the following lower bound holds,

$$p(t, x, y) \geq \frac{1}{C_L} \frac{1}{m\left(B\left(x, t^{1/2}\right)\right)} \exp\left(-\frac{C_L d(x, y)^2}{t}\right),$$

always for all $t, x, y \in (0, \infty) \times E \times E$.

One should observe that the exponent in the upper heat-kernel bound does not involve any constant and implies

$$\limsup_{t \to 0} t \log p(t, x, y) \leq -\frac{d(x, y)^2}{4},$$

as is seen by sending $\varepsilon \to 0$ after taking log, multiplying by t and taking the lim sup. (It is known, however, that one cannot take $\varepsilon = 0$ in the actual upper heat-kernel bound.) A famous result due to Varadhan [172], in the setting of diffusions on Euclidean space with elliptic generator, states that the lim sup above can be replaced by a genuine limit and equality holds. An extension to free nilpotent groups was given by Varopoulos [173], the extension to the present abstract setting was obtained by Ramírez [141].

Theorem E.14 (Varadhan–Ramírez formula) *Assume $\mathcal{E}$ is a strongly local, strongly regular Dirichlet form on $L^2(E, m)$ which satisfies (I)–(III) and let p denotes its heat kernel. Then, for all $x, y \in E$,*

$$4t \log p(t, x, y) \to -d(x, y)^2 \quad \text{as } t \to 0.$$

E.5 Symmetric diffusions

As is well known, a Dirichlet form (always assumed to be symmetric) induces a symmetric Markov process, although the construction, e.g. [70, Chapter 7], involves some subtleties. In the present context it is easier to proceed directly, i.e. by using the heat-kernel associated with $\mathcal{E}$ as the transition density of a time-homogenous diffusion.

Proposition E.15 *(associated Markov process) Assume $\mathcal{E}$ is a strongly local, strongly regular Dirichlet form on $L^2\left(E, m\right)$ which satisfies (I)–(III) and let p denotes its heat kernel. For every $x \in E$, there exists a Markov process $X = X^x$, defined on some probability space $\left(\Omega, \mathfrak{F}, \mathbb{P}\right)$, $\mathbb{P} = \mathbb{P}^x$, with the property that, for any $0 \leq t_1 < \cdots < t_n \leq 1$ and any measurable subset B of the n-fold product of E,*

$$\mathbb{P}\left[\left(X_{t_1}, \ldots, X_{t_n}\right) \in B\right] = \int_B p\left(t_1, x, y_1\right) \ldots p\left(t_n - t_{n-1}, y_{n-1}, y_n\right) dy_1 \ldots dy_n.$$

In fact, we may take $\mathbb{P}$ as a Borel measure on $\Omega = C_x\left([0, \infty), E\right)$ so that X can be realized as the canonical coordinate process, $X_t\left(\omega\right) \equiv \omega_t$.

Proof. This is classical and we shall be brief. Thanks to the Chapman–Kolmogorov equations,

$$\mu_{t_1, \ldots, t_n}\left(B\right) := \int_B p\left(t_1, x, y_1\right) \ldots p\left(t_n - t_{n-1}, y_{n-1}, y_n\right) dy_1 \ldots dy_n$$

defines a consistent set of finite-dimensional distributions. By Kolmogorov's extension theorem, there exists a unique probability measure on $E^{[0, \infty)}$ which has the correct finite-dimensional distributions and $\omega : E^{[0, \infty)} \to E$, $\omega \mapsto \omega_t$ is a realization of X with $X_0 = x$. It is easy to see that Kolmogorov's criterion is satisfied (this follows a fortiori from the upper heat-kernel bounds, although softer arguments are possible) and so we can switch to a version of X with a.s. continuous sample path. The law of this process is indeed a Borel measure on $C_x\left([0, \infty), E\right)$, and the coordinate process on that space has the same law. ∎

Remark E.16 Let $\mathcal{E}$ be as in the previous proposition, with doubling and Poincaré constant given by N, C_P respectively. If X is the symmetric diffusion associated with $\mathcal{E}$, started at some fixed point $x \in E$, then the scaling process

$$X^\varepsilon\left(\cdot\right) = X\left(\varepsilon\cdot\right)$$

is the symmetric diffusion associated with the scaled Dirichlet form $\varepsilon\mathcal{E}$. In this context, recall from Proposition E.9 that the associated intrinsic metric was precisely

$$d^\varepsilon \equiv d/\varepsilon^{1/2}$$

and that doubling/Poincaré holds for $\varepsilon\mathcal{E}$ with identical constants N, C_P. □

Proposition E.17 (localized lower heat-kernel bounds) *[163]*
Assume $\mathcal{E}$ is a strongly local, strongly regular Dirichlet form on $L^2\left(E, m\right)$ which satisfies (I)–(III). Write X^x for the associated symmetric diffusion. For $x_0 \in E$ and $r > 0$, define

$$\xi^x_{B(x_0, r)} = \inf\left\{t \geq 0 : X_t^x \notin B\left(x_0, r\right)\right\}.$$

Then the measure $\mathbb{P}\left(X_t^x \in \cdot \; ; \xi_{B(x_0,r)}^x > t\right)$ *admits a density with respect to* m; *we call it* $p_{B(x_0,r)}(t,x,y)\,dy$. *Moreover, if* x,y *are two elements of* $B(x_0,r)$ *joined by a curve* γ *which is at a* d-*distance* $R > 0$ *of* $E/B(x_0,r)$, *then there exists a constant* C *which depends only on* N, C_P *(the doubling and Poincaré constants of* $\mathcal{E}$*) such that*

$$p_{B(x_0,r)}(t,x,y) \geq \frac{1}{C} \frac{1}{m\left(B\left(x,\delta^{1/2}\right)\right)} \exp\left(-C\frac{d(x,y)^2}{t}\right)\exp\left(-\frac{Ct}{R^2}\right),$$

where $\delta = \min\{t, R^2\}$.

E.6   Stochastic analysis

Let us assume, throughout, that $\mathcal{E}$ is a strongly local, strongly regular Dirichlet form on $L^2(E, m)$ which satisfies (I)–(III) and write X for the associated symmetric diffusion process. It should be no surprise that the strong Gaussian tail estimates for the heat-kernel imply sample paths regularity reminiscent of Brownian sample paths. Moreover, we will establish an abstract Schilder and support theorem.

E.6.1   *Fernique estimates*

Lemma E.18 *For every* $\eta < 1/4$ *there exists* M *only dependent on* η *and* N, C_P *(the doubling and Poincaré constants of* $\mathcal{E}$*) so that*

$$\sup_{x \in E} \sup_{0 \leq s < t \leq 1} \mathbb{E}^x\left(\exp\left(\eta\frac{d(X_s, X_t)^2}{t-s}\right)\right) \leq M < \infty.$$

In other words, X *satisfies the Gaussian integrability condition A.18, uniformly over all possible starting points.*

Proof. Since (X_t) is a (time-homogenous) Markov prosess, we clearly have

$$\mathbb{E}^x\left(\exp\left(\eta\frac{d(X_s, X_t)^2}{t-s}\right)\right) \leq \sup_{x \in E}\mathbb{E}^x\left(\exp\left(\eta\frac{d(x, X_{t-s})^2}{t-s}\right)\right).$$

We now fix $s < t$ in $[0,1]$ and consider the scaled process $\tilde{X}(\cdot) \equiv X((t-s)\cdot)$. Following Remark E.16, the corresponding scaled (intrinsic) metric is $\tilde{d} \equiv d/|t-s|^{1/2}$ and so

$$\mathbb{E}^x\left(\exp\left(\eta\frac{d(X_s, X_t)^2}{t-s}\right)\right) \leq \sup_{x \in E}\mathbb{E}^x\left(\exp\left(\eta\tilde{d}\left(\tilde{X}_0, \tilde{X}_1\right)^2\right)\right).$$

Now, the heat-kernel estimates for $\tilde{X}$ hold with constants only depending on N, C_P (i.e. independent of the scaling) and so we obtain

$$\mathbb{E}^x \left(\exp \left(\eta \tilde{d} \left(\tilde{X}_0, \tilde{X}_1 \right)^2 \right) \right) \leq c_1 \int_E \exp \left(- \left(\frac{1}{4(1+\varepsilon)} - \eta \right) d\left(x,y\right)^2 \right) dy$$

where $\eta < 1/4$, $\varepsilon > 0$ small enough so that $\eta < \frac{1}{4(1+\varepsilon)}$ and $c_1 = c_1\left(\varepsilon, \eta\right)$. The last integral is of the form

$$(*) = \int_E f\left(d\left(x,y\right)\right) dy \leq N \int_0^\infty f\left(r\right) r^{N-1} dr < \infty$$

where $f\left(r\right) = e^{-c_2 r^2}$, $c_2 = \left(\frac{1}{4(1+\varepsilon)} - \eta \right)$ and N denotes the doubling constant. To see this, let us first remark that the doubling property (II) implies

$$\forall r \geq 1, x \in E: m\left(B\left(x,r\right)\right) \leq \left(2r\right)^N m\left(B\left(x,1\right)\right); \qquad \text{(E.4)}$$

as is seen by taking N as the smallest integer such that $r \leq 2^N$, so that

$$m\left(B\left(x,r\right)\right) \leq \left(2.2^{N-1}\right)^N m\left(B\left(x,r/2^N\right)\right) \leq \left(2r\right)^N m\left(B\left(x,1\right)\right).$$

We then have

$$\int_E f\left(d\left(x,y\right)\right) dy$$

$$= \lim_{R\to\infty} \int_0^R f\left(\cdot\right) d\left(m\left(B\left(x,\cdot\right)\right)\right) \qquad \text{as a Riemann–Stieltjes integral}$$

$$= -\int_0^\infty f'\left(r\right) \left(m\left(B\left(x,r\right)\right)\right) dr \qquad \text{using integration by parts}$$

$$\leq c_3 \int_0^\infty f'\left(r\right) r^N dr \qquad \text{from (E.4) and } -f' \equiv |f'|$$

$$\leq c_3 N \int_0^\infty r^{N-1} f\left(r\right) dr \qquad \text{using integration by parts.}$$

∎

Proposition E.19 *For every $\alpha \in [0, 1/2)$, there exists $\eta > 0$, only depending on N, C_P (the doubling and Poincaré constants of $\mathcal{E}$), so that*

$$\sup_{x\in E} \mathbb{E}^x \left(\exp \left(\eta \left| X \right|^2_{\alpha\text{-Höl};[0,1]} \right) \right) < \infty.$$

Proof. Immediate from Section A.4 of Appendix A.   ∎

E.6.2 Schilder's theorem

We can now prove a sample path large deviation statement for the family $(X(\varepsilon\cdot) : \varepsilon > 0)$. To this end, let us recall our notation

$$|x|^2_{W^{1,2}} \equiv \sup_{D \subset \mathcal{D}[0,T]} \sum_{i:t_i \in D} \frac{\left|d\left(x_{t_i}, x_{t_{i+1}}\right)\right|^2}{|t_{i+1} - t_i|}, \quad x \in C\left([0,1], E\right),$$

and, writing x^D for the piecewise geodesic approximation based on some $D = (t_i)$,

$$\left|x^D\right|^2_{W^{1,2}} = \sum_{i:t_i \in D} \frac{\left|d\left(x_{t_i}, x_{t_{i+1}}\right)\right|^2}{|t_{i+1} - t_i|},$$

as was seen in Exercise 5.24. We then have

Theorem E.20 (Schilder's theorem) *Assume $\mathcal{E}$ is a strongly local, strongly regular Dirichlet form on $L^2(E, m)$ which satisfies (I)–(III). Write X for the symmetric diffusion associated with $\mathcal{E}$, started at some fixed point $o \in E$, and set $X^\varepsilon(t) = X(\varepsilon t)$. Then the family $(X^\varepsilon(t) : \varepsilon > 0)$ satisfies a large deviation principle. More precisely, if $P_\varepsilon = (X^\varepsilon)_* \mathbb{P}$ denotes the law of X^ε, viewed as a Borel measure on the Polish space $(C_o([0,1], E), d_\infty)$, then $(P_\varepsilon : \varepsilon > 0)$ satisfies a large deviation principle on this space with good rate function given by*[4]

$$I(x) = \frac{1}{4} |x|^2_{W^{1,2};[0,1]} \in [0, \infty], \tag{E.5}$$

defined for any $x \in C_o([0,T], E), d_\infty$.

Proof. (Upper bound[5]**)** Write x for a generic path in $C_o([0,1], E)$ and let x^m denote the piecewise geodesic approximation of x interpolated at points in $D_m = \{i/m : i = 0, \ldots, m\}$. For brevity, write $\mathcal{H}$ for $W^{1,2}_o([0,1], E)$ and

$$|x|^2_{\mathcal{H}} := |x|^2_{W^{1,2}} = \sup_{D \subset [0,1]} \sum \frac{d\left(x_{t_i}, x_{t_{i+1}}\right)^2}{|t_{i+1} - t_i|}.$$

<u>Step 1:</u> For G open and non-empty, $l := \inf\left\{|h|^2_{\mathcal{H}} : h \in G \cap \mathcal{H}\right\} < \infty$ and so

$$P_\varepsilon[x^m \in G] = P_\varepsilon[x^m \in G \cap \mathcal{H}] \le P_\varepsilon\left[|x^m|^2_{\mathcal{H}} \ge l\right].$$

---

[4] Recall that $|x|^2_{W^{1,2};[0,1]} \equiv \sup_{D \subset [0,1]} \sum \frac{d\left(x_{t_i}, x_{t_{i+1}}\right)^2}{|t_{i+1} - t_i|}$.

[5] The argument follows closely the argument of Schilder's theorem for Brownian motion, Theorem 13.38.

By Chebyshev's inequality, it follows that $P_\varepsilon\left[x^m \in G\right]$ is bounded by[6]

$$e^{-l\eta/\varepsilon}E_\varepsilon\left(\exp\left(\frac{\eta}{\varepsilon}\left|x^m\right|_{\mathcal{H}}^2\right)\right) = e^{-l\eta/\varepsilon}E_\varepsilon\exp\left(\frac{\eta}{\varepsilon}\sum_{i=1}^m d\left(x_{\frac{i-1}{m},\frac{i}{m}}\right)^2 \middle/ \frac{1}{m}\right)$$

$$= e^{-l\eta/\varepsilon}\mathbb{E}\exp\left(\eta\sum_{i=1}^m d\left(X_{\varepsilon\frac{i-1}{m},\varepsilon\frac{i}{m}}\right)^2 \middle/ \frac{\varepsilon}{m}\right)$$

$$\leq e^{-l\eta/\varepsilon}M_\eta^m,$$

where we used the Markov property in the last estimate; M_η is the constant of Lemma E.18, finite for any $\eta < 1/4$. It follows that

$$\limsup_{\varepsilon\to 0} \varepsilon \log P_\varepsilon\left[x^m \in G\right] \leq -l\eta,$$

and upon sending $\eta \uparrow 1/4$ shows that

$$\limsup_{\varepsilon\to 0} \varepsilon \log P_\varepsilon\left[x^m \in G\right] \leq -\frac{1}{4}I\left(G\right).$$

Step 2: We show that geodesic approximation to X^m is an exponentially good approximation to X in the sense that for every $\delta > 0$,

$$\limsup_{\varepsilon\to 0} \varepsilon \log P_\varepsilon\left[d_{\infty;[0,1]}\left(x^m,x\right) \geq \delta\right] \leq -\infty \text{ as } m \to \infty.$$

Indeed, fix $\alpha \in [0,1/2)$ and observe that $|X|_{\alpha\text{-Höl};[0,1]}$ has a Gaussian tail. Using

$$\sup_D \left|X^D\right|_{\alpha\text{-Höl};[0,1]} \leq 3\left|X\right|_{\alpha\text{-Höl};[0,1]}$$

(thanks to Proposition 5.20) and $d\left(X_t,X_t^D\right) \leq d\left(X_t,X_{t_D}\right) + d\left(X_{t_D},X_t^D\right)$ is readily follows that

$$P_\varepsilon\left[d_{\infty;[0,1]}\left(x^m,x\right) \geq \delta\right] \leq \mathbb{P}\left[d_{\infty;[0,1]}\left(X^m\left(\varepsilon\cdot\right),X\left(\varepsilon\cdot\right)\right) \geq \delta\right]$$

$$= \mathbb{P}\left[\sup_{t\in[0,1]} d^\varepsilon\left(X_{\varepsilon t}^m,X_{\varepsilon t}\right) \geq \frac{\delta}{\varepsilon^{1/2}}\right]$$

$$\leq \mathbb{P}\left[\left(\sup_{0\leq s<t\leq 1} \frac{d^\varepsilon\left(X_s^{(\varepsilon)},X_t^{(\varepsilon)}\right)}{|t-s|^\alpha}\right) \geq \frac{\delta}{\varepsilon^{1/2}}m^\alpha\right].$$

The proof is then easily finished noting that for $\alpha \in (0,1/2)$, the corresponding α-Hölder "norm" of $X^{(\varepsilon)}$, with respect to d^ε, has a Gaussian tail which only depends on the doubling and Poincare constants, both of which are independent of ε.

---

[6] E_ε denotes expectations with respect to P_ε.

Step 3: Exactly as in the Brownian motion case, Theorem 13.38.

Proof. (Lower bound) It is enough to consider an open ball of fixed radius, say 2δ, centred at some $h \in \mathcal{H}$. Write again $D_m = \{i/m : i = 0, \dots, m\}$ and set

$$\bar{B}^m(h, \delta) = \{x \in C_o([0, 1], E) : \forall t \in D_m : |x(t) - h(t)| \le \delta\}.$$

Writing $B(h, 2\delta) \subset C([0, 1], E)$ for the open ball of radius 2δ in the uniform distance, centred at h, we can estimate

$$P_\varepsilon[B(h, 2\delta)] \ge P_\varepsilon[\bar{B}^m(h, \delta)] - P_\varepsilon[\bar{B}^m(h, \delta) \setminus B(h, 2\delta)].$$

The second term can be handled with the upper bound already proven. Indeed, let us assume that m is large enough so that

$$\max_{i=1,\dots,m} |h|_{0;\left[\frac{i-1}{m}, \frac{i}{m}\right]} < \delta/2.$$

It then follows that for any $x(\cdot) \in \bar{B}^m(h, \delta) \setminus B(h, 2\delta)$, there exists some time $t \in \left[\frac{i-1}{m}, \frac{i}{m}\right]$ so that

$$d\left(x_t, x_{\frac{i-1}{m}}\right), d\left(x_t, x_{\frac{i}{m}}\right) \ge \delta/2.$$

Since either $\left|t - \frac{i-1}{m}\right| \ge \frac{1}{2m}$ or $\left|t - \frac{i}{m}\right| \ge \frac{1}{2m}$ we see that $|x|^2_{W^{1,2}} \ge \frac{m\delta^2}{2}$. Hence, using the upper bound with the closed set $\bar{B}^m(h, \delta) \setminus B(h, 2\delta)$, we see that

$$\limsup_{\varepsilon \to 0} \varepsilon \log P_\varepsilon[\bar{B}^m(h, \delta) \setminus B(h, 2\delta)] \le -\frac{1}{4} \frac{m\delta^2}{2} \to -\infty \text{ as } m \to \infty$$

so that the other term, $P_\varepsilon[\bar{B}^m(h, \delta)]$, gives the main contribution. Writing

$$P_\varepsilon[\bar{B}^m(h, \delta)] = \int_{A_1} \dots \int_{A_m} \prod_{i=1}^m p\left(\frac{\varepsilon}{m}, x_{i-1}, x_i\right) dx_1 \dots dx_m$$

we can normalize the measure on each ball $A_i = \bar{B}\left(h\left(\frac{i}{m}\right), \delta\right)$, by dividing through $|A_i|$, so that by Jensen's inequality $\log P_\varepsilon[\bar{B}^m(h, \delta)]$ is bounded from below by

$$\log\left(\prod_{i=1}^m |A_i|\right) + \frac{1}{A_1 \cdots A_m} \int_{A_1} \dots \int_{A_m} \prod_{i=1}^m \log p\left(\frac{\varepsilon}{m}, x_{i-1}, x_i\right) dx_1 \dots dx_m.$$

Then

$$\underline{\lim}_{\varepsilon \to 0} \varepsilon \log P_\varepsilon[B^m(h, \delta)]$$

$$\ge \lim_{\varepsilon \to 0} \frac{1}{A_1 \cdots A_m} \int_{A_1} \dots \int_{A_m} \prod_{i=1}^m \varepsilon \log p\left(\frac{\varepsilon}{m}, x_{i-1}, x_i\right) dx_1 \dots dx_m$$

$$\ge -\frac{m}{4} \frac{1}{A_1 \cdots A_m} \int_{A_1} \dots \int_{A_m} \sum_{i=1}^m d(x_{i-1}, x_i)^2 dx_1 \dots dx_m$$

and by continuity of d, we can now send $\delta \to 0$ and see that

$$
\begin{aligned}
\lim_{\varepsilon \to 0} \varepsilon \log P_\varepsilon \left[B^m \left(h, \delta \right) \right] &\geq -\frac{1}{4} \sum_{i=1}^{m} d \left(h_{i-1}, h_i \right)^2 \Big/ \frac{1}{m} \\
&\geq -\frac{1}{4} \left| h \right|_{W^{1,2};[0,1]}^2 = -\frac{1}{4} I \left(h \right).
\end{aligned}
$$

The proof is then finished.   ∎

E.6.3 Support theorem

Theorem E.21 (support) *Assume $\mathcal{E}$ is a strongly local, strongly regular Dirichlet form on $L^2 \left(E, m \right)$ which satisfies (I)–(III). Write X for the symmetric diffusion associated with $\mathcal{E}$, started at some fixed point $x \in E$. Then there exists a constant C only dependent on N, C_P (the doubling and Poincaré constants of $\mathcal{E}$) so that for any path $h \in W_x^{1,2} \left([0,1], E \right)$ and any $\varepsilon \in (0,1)$, we have*

$$
\mathbb{P}^x \left(\sup_{t \in [0,1]} d \left(X_t, h_t \right) \leq \varepsilon \right) \geq \exp \left(-\frac{C \left(1 + \left| h \right|_{W^{1,2};[0,1]} \right)^2}{\varepsilon^2} \right).
$$

In particular, if $P^x = X_* \mathbb{P}^x$ denotes the law of X^x, viewed as a Borel measure on $C_x \left([0,1], E \right)$, then

$$
\operatorname{supp} \left(\mathrm{P}^x \right) = \mathrm{C}_x \left([0,1], \mathrm{E} \right).
$$

Proof. As a preliminary remark, let us note that for any $M \geq 1$ we have

$$
\inf_{\substack{r > 0 \\ a, b \in E : d(a,b) \leq Mr}} \frac{m \left(B \left(a, r \right) \right)}{m \left(B \left(b, r \right) \right)} \geq \frac{1}{(4M)^N} > 0
$$

as is seen from (E.2) and

$$
\begin{aligned}
m \left(B \left(b, r \right) \right) &\leq m \left(B \left(a, r + d \left(a, b \right) \right) \right) \\
&\leq m \left(B \left(a, \left(M + 1 \right) r \right) \right) \\
&\leq \left(2 \left(M + 1 \right) \right)^N m \left(B \left(a, r \right) \right).
\end{aligned}
$$

We now turn to the actual proof, given in three steps.

<u>Step 1:</u> From the very definition of $W^{1,2}$-regularity, $(s,t) \mapsto \left| h \right|_{W^{1,2};[s,t]}^2$ is super-additive (in fact, additive) and

$$
d \left(h_s, h_t \right) \leq \left| t - s \right|^{1/2} \left| h \right|_{W^{1,2};[0,1]}. \tag{E.6}
$$

By the Markov property, and defining $y_0 = x$ and $t_i = i/n$ where δ will be fixed later (as a function of ε),

$$\mathbb{P}^x \left(\forall i = 1, \ldots, n : d\left(X_{t_i}, h_{t_i}\right) < \frac{1}{n^{\frac{1}{2}}} \text{ and } \sup_{t_{i-1} \le t \le t_i} d\left(X_t, h_{t_{i-1}}\right) < \varepsilon \right)$$

$$= \int_{B\left(h_{t_1}, n^{-1/2}\right)} \cdots \int_{B\left(h_{t_n}, n^{-1/2}\right)} p_{B\left(h_{t_0}, \varepsilon\right)} \left(\frac{1}{n}, y_0, y_1\right) \cdots$$

$$\cdots p_{B\left(h_{t_{n-1}}, \varepsilon\right)} \left(\frac{1}{n}, y_{n-1}, y_n\right) dy_1 \cdots dy_n.$$

$$=: q_{\varepsilon, n}.$$

We join the points y_i and y_{i+1} by the curve γ_i, which is the concatenation of three geodesic curves joining first y_i with h_{t_i}, then h_{t_i} with $h_{t_{i+1}}$ and finally $h_{t_{i+1}}$ with $y_{t_{i+1}}$. Using $d\left(y_i, h_{t_i}\right) \le n^{-1/2}$ for all i, we see that

$$d\left(y_i, y_{i+1}\right) \le \text{length}\left(\gamma_i\right) \le 2n^{-1/2} + d\left(h_{t_i}, h_{t_{i+1}}\right), \tag{E.7}$$

and also that γ_i remains in the ball

$$B\left(h_{t_i}, n^{-1/2} + d\left(h_{t_i}, h_{t_{i+1}}\right)\right)$$

$$\subset \quad B\left(h_{t_i}, n^{-1/2} + n^{-1/2} |h|_{W^{1,2};[0,1]}\right)$$

where the last inclusion is due to (E.6). Choose n as the smallest integer such that $\varepsilon \ge n^{-1/2} \left(2 + |h|_{W^{1,2};[0,1]}\right)$. The curve γ_i then stays inside $B\left(t_i, \varepsilon\right)$; more precisely,

$$R_i \quad \equiv \quad d\left(\gamma_i, B\left(h_{t_i}, \varepsilon\right)^c\right)$$

$$\ge \quad \varepsilon - n^{-1/2} \left(1 + |h|_{W^{1,2};[0,1]}\right) \ge n^{-1/2}.$$

In particular, $\delta := \min\left(\frac{1}{n}, R_i^2\right) = \frac{1}{n}$ which also implies $nR_i^2 \ge 1$. If c_1 denotes the constant whose existence is guaranteed by the localized lower heat-kernel bound, then

$$p_{B\left(h_{t_i}, \varepsilon\right)} \left(\frac{1}{n}, y_i, y_{i+1}\right)$$

$$\ge \quad \frac{1}{c_1} \frac{1}{m\left(B\left(y_i, \delta^{1/2}\right)\right)} e^{-c_1 n d(y_i, y_{i+1})^2} e^{-c_1 \frac{1}{nR_i^2}}$$

$$\ge \quad \underbrace{\frac{e^{-9c_1}}{c_1}}_{=:\exp(c_2)} \frac{1}{m\left(B\left(y_i, n^{-1/2}\right)\right)} \exp(-2c_1 |h|^2_{W^{1,2};[t_i, t_{i+1}]})$$

using (E.7) $\implies$ $nd\left(y_i, y_{i+1}\right)^2 \le 8 + 2\left|h\right|_{W^{1,2};[t_i,t_{i+1}]}^2$ in the last line. With this lower bound at hand, and the previous lemma, noting that

$$
\begin{aligned}
d\left(h_{t_{i+1}}, y_{t_i}\right) &\le n^{-1/2} + d\left(h_{t_i}, h_{t_{i+1}}\right) \\
&\le n^{-1/2}\left(1 + \left|h\right|_{W^{1,2};[t_i,t_{i+1}]}\right) =: Mn^{-1/2},
\end{aligned}
$$

it follows immediately from the definition of $q_{\varepsilon,n}$ that

$$
\begin{aligned}
q_{\varepsilon,n} &\ge \prod_{i=0}^{n-1} e^{-c_2}\frac{m\left(B\left(h_{t_{i+1}}, n^{-1/2}\right)\right)}{m\left(B\left(y_{t_i}, n^{-1/2}\right)\right)} e^{-2c_1\left|h\right|_{W^{1,2};[t_i,t_{i+1}]}^2} \\
&\ge \prod_{i=0}^{n-1} \frac{e^{-c_2}}{4^N}\left(1 + \left|h\right|_{W^{1,2};[t_i,t_{i+1}]}\right)^{-N} e^{-2c_1\left|h\right|_{W^{1,2};[t_i,t_{i+1}]}^2}.
\end{aligned}
$$

Of course, by making c_1 larger we can absorb the polynomial factor $(1 + \left|h\right|_{W^{1,2};[t_i,t_{i+1}]})^{-N}$ into the exponential factor. Thus, for c_3 large enough, also chosen such that $e^{-c_2}/4^N \ge e^{-c_3}$, we have

$$
\begin{aligned}
q_{\varepsilon,n} &\ge e^{-nc_3}\exp\left(-2c_3\sum_{i=0}^{n-1}\left|h\right|_{W^{1,2};[t_i,t_{i+1}]}^2\right) \\
&\ge e^{-nc_3}\exp\left(-2c_3\left|h\right|_{W^{1,2};[0,1]}^2\right).
\end{aligned}
$$

We chose n such that $(n-1)^{-1/2} > \varepsilon/\left(2 + \left|h\right|_{W^{1,2};[0,1]}\right) \ge n^{-1/2}$. Hence

$$
\begin{aligned}
q_{\varepsilon,n} &\ge e^{-c_3}\exp\left(-c_3\frac{\left(2 + \left|h\right|_{W^{1,2};[0,1]}\right)^2}{\varepsilon^2}\right)\exp\left(-2c_3\left|h\right|_{W^{1,2};[0,1]}^2\right) \\
&= \exp\left(-c_4\frac{\left(2 + \left|h\right|_{W^{1,2};[0,1]}\right)^2}{\varepsilon^2}\right)\exp\left(-2c_4\left|h\right|_{W^{1,2};[0,1]}^2\right).
\end{aligned}
$$

Step 2: We first note that from $t_i = i/n$ and our choice of n,

$$
d\left(h_{t_{i-1}}, h_t\right) \le n^{-1/2}\left|h\right|_{W^{1,2};[0,1]} \le \varepsilon.
$$

The probability of $\sup_{0 \le t \le 1} d\left(X_t, h_t\right) < 2\varepsilon$ equals

$$
\begin{aligned}
&\mathbb{P}^x\left(\max_{i=1,\dots,d}\sup_{t_{i-1}\le t\le t_i} d\left(X_t, h_t\right) < 2\varepsilon\right) \\
\ge\ &\mathbb{P}^x\left(\max_{i=1,\dots,d}\sup_{t_{i-1}\le t\le t_i} d\left(X_t, h_{t_{i-1}}\right) + d\left(h_{t_{i-1}}, h_{t_i}\right) < 2\varepsilon\right) \\
\ge\ &\mathbb{P}^x\left(\forall i = 1,\dots,n:\ \sup_{t_{i-1}\le t\le t_i} d\left(X_t, h_{t_{i-1}}\right) < \varepsilon\right) \ge q_{\varepsilon,n},
\end{aligned}
$$

which is plainly estimated from below by q^ε and the last probability is what we estimate in step 1. This finishes the proof of the estimate in the statement of the theorem.

Step 3: The quantitative estimate established in step 2 plainly implies that $\overline{W^{1,2}} \subset \mathrm{supp}(X_*\mathbb{P}^x)$ and hence $\overline{W^{1,2}} \subset \mathrm{supp}(X_*\mathbb{P}^x)$, by passing to the uniform closure. To see the converse, we use again the geodesic nature of the state space and recall from Lemma 5.19 that $X^D \to X$ uniformly on $[0,1]$ as $|D| \to 0$. Since $X^D(\omega) \in W^{1,2}$ for every ω this easily implies that $\mathrm{supp}(X_*\mathbb{P}^x) \subset \overline{W^{1,2}}$. ∎

E.7 Comments

The basic theory on quadratic forms appears in Davies [38], for instance. We then follow Fukushima *et al.* [70], Varopoulos *et al.* [174] and especially Strum [163]. Proposition E.5 is taken from Sturm [165]. We are unaware of any precise references to the material of Section E.6.2.

Frequently used notation

Finite-dimensional objects

$\mathbb{R}^d$ d-dimensional Euclidean space with basis $\{e_1, \ldots, e_d\}$

$\left(\mathbb{R}^d\right)^{\otimes k}$ k-tensors over $\mathbb{R}^d$, see page 129

$T^N\left(\mathbb{R}^d\right)$ step-N truncated tensor algebra, see page 129

π_k projection from $T^N\left(\mathbb{R}^d\right) \to \left(\mathbb{R}^d\right)^{\otimes k}$

δ_λ dilation map, see page 133

$G^N\left(\mathbb{R}^d\right)$ step-N free nilpotent group over page $\mathbb{R}^d$, see page 142

$\mathfrak{g}^N\left(\mathbb{R}^d\right)$ step-N free nilpotent Lie algebra over page $\mathbb{R}^d$, see page 139

$\mathfrak{t}^N\left(\mathbb{R}^d\right), 1 + \mathfrak{t}^N\left(\mathbb{R}^d\right)$ see page 134

$|\cdot|$ Euclidean norm, on $\mathbb{R}^d$ or $\left(\mathbb{R}^d\right)^{\otimes k}$ for some $k \in \{1, \ldots, N\}$

$\|\cdot\|$ Carnot–Caratheodory norm, on $G^N\left(\mathbb{R}^d\right)$, see page 144

$U_1, \ldots, U_d$ invariant vector fields $G^N\left(\mathbb{R}^d\right)$, see page 149

$\mathfrak{u}_1, \ldots, \mathfrak{u}_d$ invariant vector fields $G^N\left(\mathbb{R}^d\right)$, see page 457

Paths and path-spaces

$x = (x_t : t \in [0,T])$ a generic path with values in some metric space

D a dissection (t_j) of $[0,T]$

$|D|$ the mesh of D, i.e. $\max_j |t_{j+1} - t_j|$

$\mathcal{D}[0,T]$ the set of all dissections of $[0,T]$

x^D a piecewise linear or geodesic approximation, see page 32,

$C\left([0,T], E\right)$ continuous paths with values in a metric space, see page 19

$C^{\alpha\text{-Höl}}\left([0,T], E\right)$ Hölder continuous paths, with exponent α, see page 77

$C^{p\text{-var}}\left([0,T], E\right)$ continuous paths of finite p-variation, see page 77

$W^{1,p}\left([0,T], E\right)$ paths with $W^{1,p}$-Sobolev regularity, see page 42

$W^{\delta,p}\left([0,T], E\right)$ fractional Sobolev (or Besov) paths, see page 87

$\omega(s,t)$ a control function, see page 21

$\omega\left([s,t], [u,v]\right)$ a 2D control function, see page 105

$f^{D,\tilde{D}}, f^{\mu,\tilde{\mu}}$ approximations to $f = f(s,t)$, see pages 108, 110

Rough paths and rough path-spaces

$\mathbf{x} = (\mathbf{x}_t)$ a path with values in the group $\left(G^N\left(\mathbb{R}^d\right), \otimes\right)$

$\mathbf{x}_{s,t} = \mathbf{x}_s^{-1} \otimes \mathbf{x}_s$ (group) increment of $\mathbf{x}$

$C^{1/p\text{-Höl}}\left([0,T], G^{[p]}\left(\mathbb{R}^d\right)\right), C^{p\text{-var}}\left([0,T], G^{[p]}\left(\mathbb{R}^d\right)\right)$ see page 195

$C^{0,1/p\text{-Höl}}\left([0,T], G^{[p]}\left(\mathbb{R}^d\right)\right), C^{0,p\text{-var}}\left([0,T], G^{[p]}\left(\mathbb{R}^d\right)\right)$ see page 195

$\|\cdot\|_{p\text{-var}}, \|\cdot\|_{1/p\text{-Höl}}$ see page 165

$d_{p\text{-var}}, d_{1/p\text{-Höl}}$ homogenous distances, see page 166

$\rho_{p\text{-var}}, \rho_{1/p\text{-Höl};[0,T]}$ inhomogenous distances, see page 170

$C^{(p,q)\text{-var}}\left([0,T], \mathbb{R}^d \oplus \mathbb{R}^{d'}\right), \|\cdot\|_{p,q\text{-}\omega_1,\omega_2}, \rho_{p,q\text{-}\omega_1,\omega_2}$ see pages 197, 170

Operations on rough path-spaces

$S_N(\mathbf{x})$ Lyons lift, see page 186

$S_N(\mathbf{x}, h)$ Young pairing, see page 204

$T_h(\mathbf{x})$ translation operator, see page 209

Differential equations

$V = (V_1, \ldots, V_d)$ a collection of vector fields

$x, \mathbf{x}$ (smooth, rough) driving signal

$\pi_{(V)}(0, y_0; x), \pi_{(V)}(0, y_0; \mathbf{x})$ ODE, RDE solution, see pages 55, 224

$\boldsymbol{\pi}_{(V)}(0, y_0; \mathbf{x})$ full RDE solution, see page 241

$\int \varphi(x)\, d\mathbf{x}$ rough integral, see page 253

Stochastic processes

β real-valued standard Brownian motion, see page 327

B $\mathbb{R}^d$-valued standard Brownian motion, see page 327

$dB, \circ dB$ Itô, Stratonovich differential

$\mathbf{B}$ $G^2(\mathbb{R}^d)$-valued enhanced Brownian motion, see page 333

$\mathcal{M}^c_{0,\mathrm{loc}}(\mathbb{R}^d)$ class of $\mathbb{R}^d$-valued continuous local martingales, see page 386

M $\mathbb{R}^d$-valued continuous (semi-)martingale, see page 386

$dM, \circ dM$ Itô, Stratonovich differential

$\langle M \rangle$ quadratic variation process, componentwise defined, see page 386

$\mathbf{M}$ $G^2(\mathbb{R}^d)$-valued enhanced (semi-)martingale, see page 387

X $\mathbb{R}^d$-valued Gaussian process

$\mathcal{H}$ Cameron–Martin (reproducing kernel Hilbert) space to X

R covariance of a Gaussian process, typically of finite ρ-variation, $\rho \geq 1$

β^H real-valued fractional Brownian motion, see page 405

B^H $\mathbb{R}^d$-valued fractional Brownian motion, see page 431

$\mathbf{B}^H$ $G^{[1/H]}(\mathbb{R}^d)$-valued enhanced fractional Brownian motion, page 431

$\mathbf{X}$ $G^{[2\rho]}(\mathbb{R}^d)$-valued enhanced Gaussian process, see page 429

$X^a, X^{a,x}$ $\mathbb{R}^d$-valued Markov process, generator $\partial_i\left(a^{ij}\partial_j\right)$, see page 454

$\mathbf{X}^a, \mathbf{X}^{a,x}$ $G^2(\mathbb{R}^d)$-valued Markov process, see page 461

$p^a(t, x, y)$ heat kernel, see page 461

$\mathcal{E}^a$ Dirichlet form, see page 457

$\mathcal{L}^a$ generator (in divergence form), see page 460

References

[1] A. A. Agrachev. Introduction to optimal control theory. In *Mathematical Control Theory, Part 1, 2 (Trieste, 2001)*, ICTP Lecture Notes, VIII, pages 453–513 (electronic). Abdus Salam International Center for Theoretical Physics, Trieste, 2002.

[2] S. Aida, S. Kusuoka and D. Stroock. On the support of Wiener functionals. In *Asymptotic Problems in Probability Theory: Wiener Functionals and Asymptotics (Sanda/Kyoto, 1990)*, Volume 284 of Pitman Research Notes in Mathematics Series, pages 3–34. Longman Science and Technology, Harlow, 1993.

[3] S. Aida. Semi-classical limit of the bottom of spectrum of a Schrödinger operator on a path space over a compact Riemannian manifold. *J. Funct. Anal.*, 251(1):59–121, 2007.

[4] R. Azencott. Formule de Taylor stochastique et développement asymptotique d'intégrales de Feynman. In *Seminar on Probability, XVI, Supplement*, pages 237–285. Springer, Berlin, 1982.

[5] P. Baldi, G. Ben Arous and G. Kerkyacharian. Large deviations and the Strassen theorem in Hölder norm. *Stochastic Process. Appl.*, 42(1):171–180, 1992.

[6] R. F. Bass and T. Kumagai. Laws of the iterated logarithm for some symmetric diffusion processes. *Osaka J. Math.*, 37(3):625–650, 2000.

[7] F. Baudoin. *An Introduction to the Geometry of Stochastic Flows*. Imperial College Press, London, 2004.

[8] F. Baudoin and L. Coutin. Operators associated with a stochastic differential equation driven by fractional Brownian motions. *Stochastic Process. Appl.*, 117(5):550–574, 2007.

[9] F. Baudoin and M. Hairer. A version of Hörmander's theorem for the fractional Brownian motion. *Probab. Theory Related Fields*, 139(3–4):373–395, 2007.

[10] G. Ben Arous. Flots et séries de Taylor stochastiques. *Probab. Theory Related Fields*, 81(1):29–77, 1989.

[11] G. Ben Arous and F. Castell. Flow decomposition and large deviations. *J. Funct. Anal.*, 140(1):23–67, 1996.

[12] G. Ben Arous, M. Grădinaru and M. Ledoux. Hölder norms and the support theorem for diffusions. *Ann. Inst. H. Poincaré Probab. Statist.*, 30(3):415–436, 1994.

[13] P. Billingsley. *Convergence of Probability Measures*. John Wiley & Sons Inc., New York, 1968.

[14] R. L. Bishop and R. J. Crittenden. *Geometry of Manifolds*. AMS Chelsea Publishing, Providence, RI, 2001. Reprint of the 1964 original.

[15] J.-M. Bismut. *Large Deviations and the Malliavin Calculus*, Volume 45 of Progress in Mathematics. Birkhäuser Boston Inc., Boston, MA, 1984.

[16] E. Breuillard, P. Friz and M. Huesmann. From random walks to rough paths, *Proc. Amer. Math. Soc.*, 137:3487–3496, 2009.

[17] R. Buckdahn and J. Ma. Pathwise stochastic control problems and stochastic HJB equations. *SIAM J. Control Optim.*, 45(6):2224–2256 (electronic), 2007.

[18] R. Buckdahn and J. Ma. Stochastic viscosity solutions for fully non-linear SPDEs (I). *Stochastic Process. Appl.*, 93:181–204, 2001.

[19] R. Buckdahn and J. Ma. Stochastic viscosity solutions for fully non-linear SPDEs (II). *Stochastic Process. Appl.*, 93:205–228, 2001.

[20] R. Buckdahn and J. Ma. Pathwise stochastic Taylor expansion and stochastic viscosity solution for fully nonlinear SPDEs. *Ann. Prob.*, 30(3):1131–1171, 2002.

[21] D. Burago, Y. Burago and S. Ivanov. *A Course in Metric Geometry*, Volume 33 of Graduate Studies in Mathematics. American Mathematical Society, Providence, RI, 2001.

[22] E. A. Carlen, S. Kusuoka and D. W. Stroock. Upper bounds for symmetric Markov transition functions. *Ann. Inst. H. Poincaré Probab. Statist.*, 23(2, suppl.):245–287, 1987.

[23] M. Caruana and P. Friz. Partial differential equations driven by rough paths. *J. Different. Equations*, 247(1):140–173, 2009.

[24] M. Caruana, P. Friz and H. Oberhauser. A (rough) pathwise approach to a class of non-linear stochastic partial differential equations. arXiv:0902.3352v2, 2009.

[25] T. Cass and P. Friz. Densities for rough differential equations under Hoermander's condition. *Ann. Math.*, accepted, 2008. [Available for download at http://pjm.math.berkeley.edu/scripts/coming.php?.jpath=annals]

[26] T. Cass, P. Friz and N. Victoir. Non-degeneracy of Wiener functionals arising from rough differential equations. *Trans. Amer. Math. Soc.*, 361:3359–3371, 2009.

[27] F. Castell. Asymptotic expansion of stochastic flows. *Probab. Theory Related Fields*, 96(2):225–239, 1993.

[28] K. T. Chen. Integration of paths, geometric invariants and a generalized Baker–Hausdorff formula. *Ann. Math. (2)*, 65:163–178, 1957.

[29] K. T. Chen. Integration of paths—a faithful representation of paths by non-commutative formal power series. *Trans. Amer. Math. Soc.*, 89:395–407, 1958.

[30] L. Coutin, P. Friz and N. Victoir. Good rough path sequences and applications to anticipating stochastic calculus. *Ann. Probab.*, 35(3):1172–1193, 2007.

[31] L. Coutin and A. Lejay. Semi-martingales and rough paths theory. *Electron. J. Probab.*, 10(23):761–785 (electronic), 2005.

[32] L. Coutin and Z. Qian. Stochastic analysis, rough path analysis and fractional Brownian motions. *Probab. Theory Related Fields*, 122(1):108–140, 2002.

[33] L. Coutin and N. Victoir. Enhanced Gaussian processes and applications. Preprint, 2005.

[34] M. G. Crandall, H. Ishii and P.-L. Lions. User's guide to viscosity solutions of second order partial differential equations. *Bull. Amer. Math. Soc. (N.S.)*, 27(1):1–67, 1992.

[35] S. Das Gupta, M. L. Eaton, I. Olkin, M. Perlman, L. J. Savage and M. Sobel. Inequalities on the probability content of convex regions for elliptically contoured distributions. In *Proceedings of the Sixth Berkeley Symposium on Mathematical Statistics and Probability (University of California, Berkeley, CA, 1970/1971), Vol. II: Probability Theory*, pages 241–265. University of California Press, Berkeley, CA, 1972.

[36] P. Crépel and A. Raugi. Théorème central limite sur les groupes nilpotents. *Ann. Inst. H. Poincaré Sect. B (N.S.)*, 14(2):145–164, 1978.

[37] A. M. Davie. Differential equations driven by rough paths: an approach via discrete approximation. *Appl. Math. Res. Express. AMRX*, (2):Art. ID abm009, 40, 2007.

[38] E. B. Davies. *Heat Kernels and Spectral Theory*, Volume 92 of Cambridge Tracts in Mathematics. Cambridge University Press, Cambridge, 1989.

[39] V. H. de la Peña and E. Giné. *Decoupling*. Springer-Verlag, New York, 1999.

[40] L. Decreusefond. Stochastic integration with respect to Volterra processes. *Ann. Inst. H. Poincaré Probab. Statist.*, 41(2):123–149, 2005.

[41] A. Dembo and O. Zeitouni. *Large Deviations Techniques and Applications*, Volume 38 of Applications of Mathematics (New York). Springer-Verlag, New York, second edition, 1998.

[42] J.-D. Deuschel and D. W. Stroock. *Large Deviations*, Volume 137 of Pure and Applied Mathematics. Academic Press Inc., Boston, MA, 1989.

[43] J. Dieudonné. *Foundations of Modern Analysis*. Academic Press, New York, 1969. Enlarged and corrected printing, *Pure Appl. Math.*, 10-I.

[44] H. Doss. Liens entre équations différentielles stochastiques et ordinaires. *C. R. Acad. Sci. Paris Sér. A–B*, 283(13):Ai, A939–A942, 1976.

[45] B. K. Driver. Analysis tools with applications. Draft, 2003.

[46] R. M. Dudley. Sample functions of the Gaussian process. *Ann. Prob.*, 1(1):66–103, 1973.

[47] R. M. Dudley and R. Norvaiša. *Differentiability of Six Operators on Nonsmooth Functions and p-Variation*, Volume 1703 of Lecture Notes in Mathematics. Springer-Verlag, Berlin, 1999. With the collaboration of Jinghua Qian.

[48] R. M. Dudley and R. Norvaiša. An introduction to p-variation and Young integrals – with emphasis on sample functions of stochastic processes. Lecture given at the Centre for Mathematical Physics and Stochastics, Department of Mathematical Sciences, University of Aarhus, 1998.

[49] C. Feng and H. Zhao. Rough path integral of local time. *C. R. Math. Acad. Sci. Paris*, 346(7–8):431–434, 2008.

[50] D. Feyel and A. de la Pradelle. Curvilinear integrals along enriched paths. *Electron J. Probab.*, 11:860–892, 2006.

[51] D. Feyel and A. de la Pradelle. On fractional Brownian processes. *Potential Anal.*, 10(3):273–288, 1999.

[52] D. Feyel and A. de la Pradelle. Curvilinear integrals along enriched paths. *Electron. J. Probab.*, 11(34):860–892 (electronic), 2006.

[53] W. H. Fleming and H. Mete Soner. *Controlled Markov Processes and Viscosity Solutions*, Volume 25 of Stochastic Modelling and Applied Probability. Springer, New York, second edition, 2006.

[54] G. B. Folland and E. M. Stein. *Hardy Spaces on Homogeneous Groups*, Volume 28 of Mathematical Notes. Princeton University Press, Princeton, NJ, 1982.

[55] G. B. Folland. *Real Analysis*. John Wiley & Sons Inc., New York, second edition, 1999.

[56] D. Freedman. *Brownian Motion and Diffusion*. Springer-Verlag, New York, second edition, 1983.

[57] P. Friz, T. Lyons and D. Stroock. Lévy's area under conditioning. *Ann. Inst. H. Poincaré Probab. Statist.*, 42(1):89–101, 2006.

[58] P. Friz and H. Oberhauser. Rough path limits of the Wong–Zakai type with a modified drift term. *J. Funct. Anal.*, 256(10):3236–3256, 2009.

[59] P. Friz and H. Oberhauser. Isoperimetry and rough path regularity, 2007.

[60] P. Friz and H. Oberhauser. A generalized Fernique theorem and applications. Preprint, 2009.

[61] P. Friz and N. Victoir. Differential equations driven by Gaussian signals. *Ann. Inst. H. Poincaré Probab. Statist.*, 46 (DOI 10.1214/09-AIHP202).

[62] P. Friz and N. Victoir. Approximations of the Brownian rough path with applications to stochastic analysis. *Ann. Inst. H. Poincaré Probab. Statist.*, 41(4):703–724, 2005.

[63] P. Friz and N. Victoir. A note on the notion of geometric rough paths. *Probab. Theory Related Fields*, 136(3):395–416, 2006.

[64] P. Friz and N. Victoir. A variation embedding theorem and applications. *J. Funct. Anal.*, 239(2):631–637, 2006.

[65] P. Friz and N. Victoir. Large deviation principle for enhanced Gaussian processes. *Ann. Inst. H. Poincaré Probab. Statist.*, 43(6):775–785, 2007.

[66] P. Friz and N. Victoir. The Burkholder–Davis–Gundy inequality for enhanced martingales. In *Séminaire de Probabilités XLI*, Volume 1934 of Lecture Notes in Mathematics Springer, Berlin, 2008.

[67] P. Friz and N. Victoir. Euler estimates for rough differential equations. *J. Different. Equations*, 244(2):388–412, 2008.

[68] P. Friz and N. Victoir. On uniformly subelliptic operators and stochastic area. *Probab. Theory Related Fields*, 142(3–4):475–523, 2008.

[69] P. K. Friz. Continuity of the Itô-map for Hölder rough paths with applications to the support theorem in Hölder norm. In *Probability and Partial Differential Equations in Modern Applied Mathematics*, Volume 140 of IMA Volumes in Mathematical Applications, pages 117–135. Springer, New York, 2005.

[70] M. Fukushima, Y. Ōshima and M. Takeda. *Dirichlet Forms and Symmetric Markov Processes*, Volume 19 of de Gruyter Studies in Mathematics. Walter de Gruyter & Co., Berlin, 1994.

[71] D. Gilbarg and N. S. Trudinger. *Elliptic Partial Differential Equations of Second Order*. Classics in Mathematics. Springer-Verlag, Berlin, 2001. Reprint of the 1998 edition.

[72] M. Gromov. *Metric Structures for Riemannian and Non-Riemannian Spaces*. Modern Birkhäuser Classics. Birkhäuser Boston Inc., Boston, MA, English Edition, 2007. Based on the 1981 French original, with appendices by M. Katz, P. Pansu and S. Semmes, translated from the French by S. M. Bates.

[73] P. Guasoni. No arbitrage under transaction costs, with fractional Brownian motion and beyond. *Math. Finance*, 16(3):569–582, 2006.

[74] M. Gubinelli. Controlling rough paths. *J. Funct. Anal.*, 216(1):86–140, 2004.

[75] M. Gubinelli. Ramifications of rough paths, arXiv:math/060300v1, to appear in *J. Different. Equations*.

[76] M. Gubinelli and S. Tindel. Rough evolution equations, arXiv:0803.0552v1, to appear in *Annals of Probability*, 2008.

[77] M. Gubinelli, A. Lejay and S. Tindel. Young integrals and SPDEs. *Potential Anal.*, 25(4):307–326, 2006.

[78] I. Gyöngy. The stability of stochastic partial differential equations and applications. I. *Stochastics Stochastics Rep.*, 27(2):129–150, 1989.

[79] I. Gyöngy. The stability of stochastic partial differential equations and applications. Theorems on supports. In *Stochastic Partial Differential Equations and Applications, II (Trento, 1988)*, Volume 1390 of Lecture Notes in Mathematics, pages 91–118. Springer, Berlin, 1989.

[80] I. Gyöngy. The stability of stochastic partial differential equations. II. *Stochastics Stochastics Rep.*, 27(3):189–233, 1989.

[81] I. Gyöngy. The approximation of stochastic partial differential equations and applications in nonlinear filtering. *Comput. Math. Appl.*, 19(1):47–63, 1990.

[82] I. Gyöngy. On stochastic partial differential equations. Results on approximations. In *Topics in Stochastic Systems: Modelling, estimation and adaptive control*, Volume 161 of Lecture Notes in Control and Information Science, pages 116–136. Springer, Berlin, 1991.

[83] P. Hajłasz and P. Koskela. Sobolev met Poincaré. *Mem. Amer. Math. Soc.*, 145(688):x+101, 2000.

[84] P. Hartman. *Ordinary Differential Equations*, Volume 38 of Classics in Applied Mathematics. Society for Industrial and Applied Mathematics (SIAM), Philadelphia, PA, 2002. Corrected reprint of the second (1982) edition [Birkhäuser, Boston, MA; MR0658490 (83e:34002)], with a foreword by P. Bates.

[85] B. Hoff. The Brownian frame process as a rough path, arXiv:math/0602008v1, 2006.

[86] Y. Hu and D. Nualart. Rough path analysis via fractional calculus, *Trans. Amer. Math. Soc.* 361(5):2689–2718, 2009.

[87] Y. Hu and D. Nualart. Differential equations driven by Hölder continuous functions of order greater than 1/2, *Stochast. Anal. Appl.*, 2:399–413, 2007.

[88] N. Ikeda and S. Watanabe. *Stochastic Differential Equations and Diffusion Processes*. North-Holland Publishing Co., Amsterdam, second edition, 1989.

[89] Y. Inahama. A stochastic Taylor-like expansion in the rough path theory. Preprint, 2008.

[90] Y. Inahama. Laplace's method for the laws of heat processes on loop spaces. *J. Funct. Anal.*, 232(1):148–194, 2006.

[91] Y. Inahama and H. Kawabi. Large deviations for heat kernel measures on loop spaces via rough paths. *J. London Math. Soc. (2)*, 73(3):797–816, 2006.

[92] Y. Inahama and H. Kawabi. Asymptotic expansions for the Laplace approximations for Itô functionals of Brownian rough paths. *J. Funct. Anal.*, 243(1):270–322, 2007.

[93] D. Jerison. The Poincaré inequality for vector fields satisfying Hörmander's condition. *Duke Math. J.*, 53(2):503–523, 1986.

[94] I. Karatzas and S. E. Shreve. *Brownian Motion and Stochastic Calculus*, Volume 113 of Graduate Texts in Mathematics. Springer-Verlag, New York, second edition, 1991.

[95] N. V. Krylov. *Introduction to the Theory of Diffusion Processes*, Volume 142 of Translations of Mathematical Monographs. American Mathematical Society, Providence, RI, 1995. Translated from the Russian manuscript by V. Khidekel and G. Pasechnik.

[96] H. Kunita. *Stochastic Flows and Stochastic Differential Equations*, Volume 24 of Cambridge Studies in Advanced Mathematics. Cambridge University Press, Cambridge, 1997. Reprint of the 1990 original.

[97] T. G. Kurtz, É. Pardoux and P. Protter. Stratonovich stochastic differential equations driven by general semimartingales. *Ann. Inst. H. Poincaré Probab. Statist.*, 31(2):351–377, 1995.

[98] S. Kusuoka. On the regularity of solutions to SDE. In *Asymptotic Problems in Probability Theory: Wiener functionals and asymptotics (Sanda/Kyoto, 1990)*, Volume 284 of Pitman Research Notes in Mathematics Series, pages 90–103. Longman Science and Technology, Harlow, 1993.

[99] S. Kusuoka. The nonlinear transformation of Gaussian measure on Banach space and absolute continuity. I. *J. Fac. Sci. Univ. Tokyo Sect. IA Math.*, 29(3):567–597, 1982.

[100] J. Lamperti. On convergence of stochastic processes. *Trans. Amer. Math. Soc.*, 104:430–435, 1962.

[101] M. Ledoux, Z. Qian and T. Zhang. Large deviations and support theorem for diffusion processes via rough paths. *Stochastic Process. Appl.*, 102(2):265–283, 2002.

[102] M. Ledoux. Isoperimetry and Gaussian analysis. In *Lectures on Probability Theory and Statistics (Saint-Flour, 1994)*, Volume 1648 of Lecture Notes in Mathematics, pages 165–294. Springer, Berlin, 1996.

[103] M. Ledoux and M. Talagrand. *Probability in Banach Spaces*, Volume 23 of Ergebnisse der Mathematik und ihrer Grenzgebiete (3) [Results in Mathematics and Related Areas (3)]. Springer-Verlag, Berlin, 1991.

[104] A. Lejay. An introduction to rough paths. In *Séminaire de Probabilités XXXVII*, Volume 1832 of Lecture Notes in Mathematics, pages 1–59. Springer, Berlin, 2003.

[105] A. Lejay. Stochastic differential equations driven by processes generated by divergence form operators. I. A Wong–Zakai theorem. *ESAIM Probab. Stat.*, 10:356–379 (electronic), 2006.

[106] A. Lejay. Stochastic differential equations driven by processes generated by divergence form operators. II: Convergence results. *ESAIM Probab. Stat.*, 12:387–411 (electronic), 2008.

[107] A. Lejay and N. Victoir. On (p,q)-rough paths. *J. Different. Equations*, 225(1):103–133, 2006.

[108] D. Lépingle. La variation d'ordre p des semi-martingales. *Z. Wahrscheinlichkeitstheorie und Verw. Gebiete*, 36(4):295–316, 1976.

[109] X.-D. Li and T. J. Lyons. Smoothness of Itô maps and diffusion processes on path spaces. I. *Ann. Sci. École Norm. Sup. (4)*, 39(4):649–677, 2006.

[110] P.-L. Lions and P. E. Souganidis. Viscosity solutions of fully nonlinear stochastic partial differential equations. *Sūrikaisekikenkyūsho Kōkyūroku*, (1287):58–65, 2002. Viscosity solutions of differential equations and related topics (in Japanese) (Kyoto, 2001).

[111] P.-L. Lions and P. E. Souganidis. Fully nonlinear stochastic partial differential equations. *C. R. Acad. Sci. Paris Sér. I Math.*, 326(9):1085–1092, 1998.

[112] P.-L. Lions and P. E. Souganidis. Fully nonlinear stochastic partial differential equations: non-smooth equations and applications. *C. R. Acad. Sci. Paris Sér. I Math.*, 327(8):735–741, 1998.

[113] P.-L. Lions and P. E. Souganidis. Uniqueness of weak solutions of fully nonlinear stochastic partial differential equations. *C. R. Acad. Sci. Paris Sér. I Math.*, 331(10):783–790, 2000.

[114] E. R. Love. A generalization of absolute continuity. *J. London Math. Soc.*, 26:1–13, 1951.

[115] T. Lyons. The interpretation and solution of ordinary differential equations driven by rough signals. In *Stochastic Analysis (Ithaca, NY, 1993)*, pages 115–128. American Mathematical Society, Providence, RI, 1995.

[116] T. Lyons. Differential equations driven by rough signals. *Rev. Mat. Iberoamericana*, 14(2):215–310, 1998.

[117] T. Lyons. Systems controlled by rough paths. In *European Congress of Mathematics*, pages 269–281. European Mathematical Society, Zürich, 2005.

[118] T. Lyons and Z. Qian. Calculus of variation for multiplicative functionals. In *New Trends in Stochastic Analysis (Charingworth, 1994)*, pages 348–374. World Scientific Publishing, River Edge, NJ, 1997.

[119] T. Lyons and Z. Qian. Flow of diffeomorphisms induced by a geometric multiplicative functional. *Probab. Theory Related Fields*, 112(1):91–119, 1998.

[120] T. Lyons and Z. Qian. *System Control and Rough Paths*. Oxford University Press, 2002.

[121] T. Lyons and L. Stoica. The limits of stochastic integrals of differential forms. *Ann. Probab.*, 27(1):1–49, 1999.

[122] T. Lyons and N. Victoir. An extension theorem to rough paths. *Ann. Inst. H. Poincaré Anal. Non Linéaire*, 24(5):835–847, 2007.

[123] T. J. Lyons, M. Caruana and T. Lévy. *Differential Equations Driven by Rough Paths*, Volume 1908 of Lecture Notes in Mathematics. Springer, Berlin, 2007. Lectures from the 34th Summer School on Probability Theory held in Saint-Flour, July 6–24, 2004, with an introduction concerning the Summer School by Jean Picard.

[124] P. Malliavin. *Stochastic Analysis*, Volume 313 of Grundlehren der Mathematischen Wissenschaften [Fundamental Principles of Mathematical Sciences]. Springer-Verlag, Berlin, 1997.

[125] H. P. McKean. *Stochastic Integrals*. AMS Chelsea Publishing, Providence, RI, 2005. Reprint of the 1969 edition, with errata.

[126] E. J. McShane. Stochastic differential equations and models of random processes. In *Proceedings of the Sixth Berkeley Symposium on Mathematical Statistics and Probability (University of California, Berkeley, CA, 1970/1971), Vol. III: Probability Theory*, pages 263–294. University of California Press, Berkeley, CA, 1972.

[127] A. Millet, D. Nualart and M. Sanz. Large deviations for a class of anticipating stochastic differential equations. *Ann. Probab.*, 20(4):1902–1931, 1992.

[128] A. Millet and M. Sanz-Solé. A simple proof of the support theorem for diffusion processes. In *Séminaire de Probabilités, XXVIII*, Volume 1583 of Lecture Notes in Mathematics, pages 36–48. Springer, Berlin, 1994.

[129] A. Millet and M. Sanz-Solé. Large deviations for rough paths of the fractional Brownian motion. *Ann. Inst. H. Poincaré Probab. Statist.*, 42(2):245–271, 2006.

[130] A. Millet and M. Sanz-Solé. Approximation of rough paths of fractional Brownian motion. In *Seminar on Stochastic Analysis, Random Fields and Application*, Volume 59 of Progress in Probability, pages 275–303. Birkhuser, Basel, 2008.

[131] R. Montgomery. *A Tour of SubRiemannian Geometries, their Geodesics and Applications*, Volume 91 of Mathematical Surveys and Monographs. American Mathematical Society, Providence, RI, 2002.

[132] J. Musielak and Z. Semadeni. Some classes of Banach spaces depending on a parameter. *Studia Math.*, 20:271–284, 1961.

[133] A. Neuenkirch, I. Nourdin and S. Tindel. Delay equations driven by rough paths, 2007.

[134] D. Neuenschwander. *Probabilities on the Heisenberg Group*, Volume 1630 of Lecture Notes in Mathematics. Springer-Verlag, Berlin, 1996.

[135] D. Nualart. *The Malliavin Calculus and Related Topics*. Springer-Verlag, New York, 1995.

[136] D. Nualart. *The Malliavin Calculus and Related Topics*. Springer-Verlag, Berlin, second edition, 2006.

[137] G. Pisier and Q. H. Xu. The strong p-variation of martingales and orthogonal series. *Probab. Theory Related Fields*, 77(4):497–514, 1988.

[138] E. Platen. A Taylor formula for semimartingales solving a stochastic equation. In *Stochastic Differential Systems (Visegrád, 1980)*, pages 157–164. Springer, Berlin, 1981.

[139] M. H. Protter and C. B. Morrey, Jr. *A First Course in Real Analysis*. Undergraduate Texts in Mathematics. Springer-Verlag, New York, second edition, 1991.

[140] P. E. Protter. *Stochastic Integration and Differential Equations*, Volume 21 of Stochastic Modelling and Applied Probability. Springer-Verlag, Berlin, 2005. Second edition, version 2.1, corrected third printing.

[141] J. A. Ramírez. Short-time asymptotics in Dirichlet spaces. *Comm. Pure Appl. Math.*, 54(3):259–293, 2001.

[142] C. Reutenauer. *Free Lie Algebras*. Clarendon Press, New York, 1993.

[143] D. Revuz and M. Yor. *Continuous Martingales and Brownian Motion*, Volume 293 of Grundlehren der Mathematischen Wissenschaften [Fundamental Principles of Mathematical Sciences]. Springer-Verlag, Berlin, third edition, 1999.

[144] L. C. G. Rogers and D. Williams. *Diffusions, Markov Processes, and Martingales. Vol. 1*. Cambridge University Press, Cambridge, 2000. Reprint of the second (1994) edition.

[145] L. C. G. Rogers and D. Williams. *Diffusions, Markov Processes, and Martingales. Vol. 2*. Cambridge University Press, Cambridge, 2000. Reprint of the second (1994) edition.

[146] A. Rozkosz. Stochastic representation of diffusions corresponding to divergence form operators. *Stochastic Process. Appl.*, 63(1):11–33, 1996.

[147] A. Rozkosz. Weak convergence of diffusions corresponding to divergence form operators. *Stochastics Stochastics Rep.*, 57(1–2):129–157, 1996.

[148] W. Rudin. *Principles of Mathematical Analysis*. McGraw-Hill, New York, third edition, 1976.

[149] W. Rudin. *Real and Complex Analysis*. McGraw-Hill, New York, third edition, 1987.

[150] W. Rudin. *Functional Analysis*. McGraw-Hill, New York, second edition, 1991.

[151] L. Saloff-Coste and D. W. Stroock. Opérateurs uniformément sous-elliptiques sur les groupes de Lie. *J. Funct. Anal.*, 98(1):97–121, 1991.

[152] B. Saussereau and D. Nualart. Malliavin calculus for stochastic differential equations driven by a fractional Brownian motion. *Stochastic Process Appl.*, 119(2):391–409, 2009.

[153] M. Schreiber. Fermeture en probabilité de certains sous-espaces d'un espace L^2. Application aux chaos de Wiener. *Z. Wahrscheinlichkeitstheorie und Verw. Gebiete*, 14:36–48, 1969/70.

[154] L. A. Shepp and O. Zeitouni. A note on conditional exponential moments and Onsager–Machlup functionals. *Ann. Probab.*, 20(2):652–654, 1992.

[155] I. Shigekawa. *Stochastic Analysis*, Volume 224 of Translations of Mathematical Monographs. American Mathematical Society, Providence, RI, 2004. Translated from the 1998 Japanese original by the author.

[156] E.-M. Sipiläinen. A pathwise view of solutions of stochastic differential equations. PhD thesis, University of Edinburgh, 1993.

[157] R. S. Strichartz. The Campbell–Baker–Hausdorff–Dynkin formula and solutions of differential equations. *J. Funct. Anal.*, 72(2):320–345, 1987.

[158] D. W. Stroock. Diffusion semigroups corresponding to uniformly elliptic divergence form operators. In *Séminaire de Probabilités, XXII*, Volume 1321 of Lecture Notes in Mathematics, pages 316–347. Springer, Berlin, 1988.

[159] D. W. Stroock. *Probability Theory, An Analytic View*. Cambridge University Press, Cambridge, 1993.

[160] D. W. Stroock. *Markov Processes from K. Itô's Perspective*, Volume 155 of Annals of Mathematics Studies. Princeton University Press, Princeton, NJ, 2003.

[161] D. W. Stroock and S. R. S. Varadhan. On the support of diffusion processes with applications to the strong maximum principle. In *Proceedings of the Sixth Berkeley Symposium on Mathematical Statistics and Probability (University of California, Berkeley, CA, 1970/1971), Vol. III: Probability Theory*, pages 333–359. University of California Press, Berkeley, CA, 1972.

[162] D. W. Stroock and S. R. Srinivasa Varadhan. *Multidimensional Diffusion Processes*, Volume 233 of Grundlehren der Mathematischen Wissenschaften [Fundamental Principles of Mathematical Sciences]. Springer-Verlag, Berlin, 1979.

[163] K. T. Sturm. Analysis on local Dirichlet spaces. III. The parabolic Harnack inequality. *J. Math. Pure Appl. (9)*, 75(3):273–297, 1996.

[164] K.-T. Sturm. Analysis on local Dirichlet spaces. II. Upper Gaussian estimates for the fundamental solutions of parabolic equations. *Osaka J. Math.*, 32(2):275–312, 1995.

[165] K.-T. Sturm. On the geometry defined by Dirichlet forms. In *Seminar on Stochastic Analysis, Random Fields and Applications (Ascona, 1993)*, Volume 36 of Progress Probability, pages 231–242. Birkhäuser, Basel, 1995.

[166] H. J. Sussmann. On the gap between deterministic and stochastic ordinary differential equations. *Ann. Probability*, 6(1):19–41, 1978.

[167] H. J. Sussmann. Limits of the Wong–Zakai type with a modified drift term. In *Stochastic Analysis*, pages 475–493. Academic Press, Boston, MA, 1991.

[168] S. J. Taylor. Exact asymptotic estimates of Brownian path variation. *Duke Math. J.*, 39:219–241, 1972.

[169] J. Teichmann. Another approach to some rough and stochastic partial differential equations. arXiv:0908.2814v1, 2009.

[170] N. Towghi. Multidimensional extension of L. C. Young's inequality. *JIPAM J. Inequal. Pure Appl. Math.*, 3(2):Article 22, 13 pp. (electronic), 2002.

[171] A. Süleyman Üstünel and M. Zakai. *Transformation of measure on Wiener space*. Springer Monographs in Mathematics. Springer-Verlag, Berlin, 2000.

[172] S. R. S. Varadhan. On the behavior of the fundamental solution of the heat equation with variable coefficients. *Comm. Pure Appl. Math.*, 20:431–455, 1967.

[173] N. Th. Varopoulos. Small time Gaussian estimates of heat diffusion kernels. II. The theory of large deviations. *J. Funct. Anal.*, 93(1):1–33, 1990.

[174] N. Th. Varopoulos, L. Saloff-Coste and T. Coulhon. *Analysis and Geometry on Groups*, Volume 100 of Cambridge Tracts in Mathematics. Cambridge University Press, Cambridge, 1992.

[175] F. W. Warner. *Foundations of Differentiable Manifolds and Lie Groups*, Volume 94 of Graduate Texts in Mathematics. Springer-Verlag, New York, 1983. Corrected reprint of the 1971 edition.

[176] K. Yosida. *Functional Analysis*, second edition. Die Grundlehren der mathematischen Wissenschaften, Band 123. Springer-Verlag, New York, 1968.

[177] L. C. Young. An inequality of Hölder type connected with Stieltjes integration. *Acta Math.*, (67):251–282, 1936.

Index

support theorem (*cont.*)
    for SDE driven by Markovian
        signal, 537
    for SPDE, 543
    for stochastic flow, 535
    Stroock–Varadhan, 533
symmetric diffusion, 625

time reversal of paths, 61
translation operator, 209

upper gradient lemma, 495

vector fields
    C^k bounded, 68
    1-Lipschitz continuous, 53

Lipschitz regular in sense of
    Stein, 213
non-explosion condition on, 69

Wiener's characterization, 96
Wong–Zakai theorem, 511

Young integral, 116
Young pairing, 204
Young regularity, complementary,
    449, 546
Young–Lóeve estimate, 114
Young–Lóeve–Towghi estimate,
    122
Young–Wiener integral, 434

Printed in the United States
by Baker & Taylor Publisher Services